PHYSICS
IN
MEDICINE & BIOLOGY
ENCYCLOPEDIA

In Two Volumes

PHYSICS IN MEDICINE & BIOLOGY ENCYCLOPEDIA

ENCYCLOPEDIA

Medical Physics, Bioengineering and Biophysics

Editor

T. F. McAinsh

Department of Clinical Physics and Bio-Engineering, Glasgow, UK

VOLUME 1
A-M

PERGAMON PRESS

OXFORD · NEW YORK · TORONTO · SYDNEY · FRANKFURT

PHYSICS

7183-6846

U.K.	Pergamon Press Ltd., Headington Hill Hall, Oxford OX3 0BW, England
U.S.A.	Pergamon Press Inc., Maxwell House, Fairview Park, Elmsford, New York 10523, U.S.A.
CANADA	Pergamon Press Canada Ltd., Suite 104, 150 Consumers Road, Willowdale, Ontario M2J 1P9, Canada
AUSTRALIA	Pergamon Press (Aust.) Pty. Ltd., P.O. Box 544, Potts Point, N.S.W. 2011, Australia
FEDERAL REPUBLIC OF GERMANY	Pergamon Press GmbH, Hammerweg 6, D-6242 Kronberg, Federal Republic of Germany
JAPAN	Pergamon Press Ltd., 8th Floor, Matsuoka Central Building, 1-7-1 Nishishinjuku, Shinjuku-ku, Tokyo 160, Japan
BRAZIL	Pergamon Editora Ltda., Rua Eça de Queiros, 346, CEP 04011, São Paulo, Brazil
PEOPLE'S REPUBLIC OF CHINA	Pergamon Press, Qianmen Hotel, Beijing, People's Republic of China

First edition 1986

Library of Congress Cataloging in Publication Data

Main entry under title:
Physics in medicine & biology encyclopedia.
1. Medical physics — Dictionaries. 2. Biological physics — Dictionaries. 3. Biomedical engineering — Dictionaries. I. McAinsh, T. F. II. Title: Physics in medicine and biology encyclopedia. [DNLM: 1. Biophysics — encyclopedias. QT 13 P578]
R895.A3P47 1986   574.19′1′0321   85-28379

British Library Cataloguing in Publication Data

Physics in medicine & biology encyclopedia.
1. Medical physics 2. Biological physics
I. McAinsh, T. F.
610′.1′53   R895
ISBN 0-08-026497-2

Printed in Great Britain by A. Wheaton & Co. Ltd., Exeter

Contents

Honorary Editorial Advisory Board

Foreword

Medicine and biology advance by revolutions. The nineteenth century saw several, including (in medicine) hygiene, asepsis and anesthesia and (in biology) Darwinism and bacteriology. The first half of the twentieth century saw the rise of chemotherapy, which began when aspirin emerged from the Bayer factory in 1899, reached a climax with antibiotics, analgesics, tranquilizers and hypotensive drugs and now seems to have stabilized. During the same period genetics and biochemistry grew into major scientific disciplines. In parallel with these developments medicine and biology are being increasingly transformed by the incorporation of ideas and techniques derived from physics. The Encyclopedia is a handbook and guide to this new revolution which will, in time to come, be seen as one of the major achievements of twentieth century science and technology.

The origins of this revolution can be traced back to the seventeenth century proposition that science was not merely the theoretical discipline established by the Greeks but could actually be useful to society. Medicine was one of the first proving grounds for this new experimental approach. Enthusiasm was inspired, as always, by the latest scientific advances — mechanics in the seventeenth century and electricity in the eighteenth century. But these early aspirations, even when reinforced by the ingenious craftsmanship of the nineteenth century, did little for the health of the people. It was the union of the intellectual resources of the physical sciences with the abundant technology of the twentieth century that produced the latest, and perhaps the greatest revolution in health care. Since the discoveries of x rays and radioactivity, physicists have been prominent in this movement. Having been recruited since the 1920s to deal with the problem of radiation dosimetry, they were in the right place when the broader opportunities arose — particularly in the climate of increased technological activity generated by World War II. Indeed nuclear medicine grew by the exploitation of measurement techniques based on instruments and methods devised by physicists in the wartime atomic energy projects. The design of the linear accelerators which marked the effective start of supervoltage radiotherapy depended largely on microwave techniques originally used in radar.

The diagnostic and therapeutic uses of ionizing radiation still provide a significant part of the physicist's contribution to medicine. The first major success of computers in health care was in dosage estimation and treatment planning for radiotherapy. More recently the domain of the radiologists has been greatly extended by the incorporation of new imaging techniques, some involving radioactive isotopes and others derived from physics-based techniques including thermography, ultrasonics and nuclear magnetic resonance.

But the uses of radiation no longer dominate medical physics, as they did before the present revolution began. Physicists are now to be found as members of clinical teams in ophthalmology, urology, neurology, neurosurgery, respiratory medicine, medical and surgical cardiology, orthopedics, child health, psychiatry, otology and many other specialities. Clinical chemists have greatly extended their repertoire by the incorporation of techniques such as atomic absorption spectrophotometry, x-ray fluorescence and neutron activation analysis. In the use of these and related techniques, physicists, biochemists and clinicians have made important contributions to the study and control of industrial and environmental hazards in a wider context. Moreover, the physics-based techniques of x-ray diffraction, electron microscopy, radioactive tracers, electronics signal processing and computing have accelerated the progress of biochemistry and physiology and the emergence (out of these two sources) of molecular biology and genetic en-

ix

gineering. In some disciplines, including radiobiology, virology, immunology and pharmacology, medicine and biology have come closely together.

In assembling the text the Editor-in-Chief, Mr Tom McAinsh, has shown great skill and insight in identifying the subjects to be treated and in choosing contributors who have written in a style that is both readable and authoritative. The successful efforts of editor and contributors have produced a work which will be of great interest and value to scientists and engineers working in the realm of medicine, to clinicians in virtually every specialty and to the new generation of biologists. It will be useful also to teachers seeking to illustrate the ideas and methods of physics by reference to biological problems as well as to students of physics and engineering who may be attracted by the prospect of working in the clinical realm and to biologists wishing to secure fuller understanding of the scientific techniques used in their own and adjacent disciplines. An important feature is that contributors have reviewed not only the current situation in their specialized fields but also the scientific foundations on which future progress will be based. The Encyclopedia will therefore be of lasting value as a guide to the development and application of the advances in clinical physics and bioengineering which will have a distinctive influence on the progress of health care in the years ahead.

J. LENIHAN
University of Glasgow

Preface

Each of us has an identity that heredity, environment, and circumstance have somehow fashioned. Activity that gives expression to this identity brings particular satisfaction and all the more so if the activity is socially useful and therefore attracts financial support. I enjoy the gaining of understanding — of scientific concepts, processes, and procedures — and the passing on of that understanding to others, and this, I can only assume, is the reason why I have undertaken the sole editorship of an Encyclopedia containing over two hundred articles on diverse medical/scientific topics and written by almost as many authors. There was, of course, from the outset, the additional, obvious if implicit incentive, deeply appealing to a medical physicist, that such work would give much needed publicity to the substantial, present-day involvement of physical scientists in medicine. The public, even the broad scientific community, and certainly the influential politicians and administrators who control public health finances, little realise the importance or the nature of the role of such scientists in health care. They may have a deep and justified respect for the doctors who treat them when they are sick, but somehow they do not connect the technologically sophisticated x-ray equipment, the nuclear medicine apparatus, the lasers, the ultrasound scanners, the clever electronic aids for the handicapped, the fetal monitors, electrocardiographs, and all the other paraphernalia of modern medicine with the engineers and physicists who alone (if the matter is given but a moment's thought) could have designed, built and developed them, or who alone, by virtue of their training and expertise, can ensure their continuing, effective physical use and application.

Such has been the encroachment and impact of high technology in medicine that in many spheres diagnosis and therapy can only be effected by the united efforts of "medics" and scientists, working together in multidisciplinary teams. One of my peripheral hopes for this Encyclopedia is that it will give the reader who is a college undergraduate, or recent graduate, in Physics, or Electronic/Electrical or Mechanical Engineering an insight into the opportunities for interesting, useful and gainful activity that a career in medicine provides.

I should explain how the Encyclopedia came about and how it grew within the framework of the institution in which I have spent my own career as a medical physicist, the West of Scotland Health Boards' Department of Clinical Physics and Bio-Engineering. The Department is the largest organization of its kind in the world, having a present staff complement that includes 90 graduate physicists and engineers, and 130 fully qualified medical physics technicians. It is a regional department. Thus, while some of the staff work in the central laboratories in the heart of Glasgow, most perform their daily tasks in teams in a variety of clinical and diagnostic departments in hospitals throughout the West of Scotland, serving a total population of some three million. Further, although the Department is primarily a branch of the National Health Service it has staff based in many professorial departments of the Faculty of Medicine of Glasgow University. It was to Professor John Lenihan, who until his retirement in September 1983 was Director of the Department of Clinical Physics and Bio-Engineering and Professor of Clinical Physics at Glasgow University, that Pergamon Press initially turned for help and advice on a Physics and Medicine in Biology Encyclopedia. Professor Lenihan subsequently recommended me to Pergamon Press as Editor, having overcome my initial "swithering" (a good Scottish word, meaning "to hesitate over a decision") by referring to the substantial support I could expect from my numerous and knowledgeable colleagues.

Both Professor Lenihan and Professor Joe McKie who succeeded Professor Lenihan in office in 1983, and indeed the entire Department, have maintained a friendly and constructive interest in the Encyclopedia throughout the period of its preparation. As the reader may have observed, Professor Lenihan has kindly written the Foreword. Several of my eminent colleagues have contributed articles as have many friends and associates who work in various departments at Glasgow University, including the Beatson Institute for Cancer Research, and in the Bioengineering Unit of the University of Strathclyde, Glasgow's other University. If these contributions have added a certain Scottish flavour to the Encyclopedia, may its pages bring its readers that incomparable sense of illumination, stimulation and enlightenment that one associates (so I am informed) with a "taste of Scotland".

In compiling the Encyclopedia my intention has been to provide a comprehensive and convenient source of enlightenment for individual hospital physicists, medical technologists, and clinicians who wish to be informed on any of the numerous, important topics of which they themselves have little or even no knowledge but which nevertheless provide routine activity for their fellow professionals who work in other university and hospital departments. Each article is therefore written for the reader who has a basic grounding in physics, but no particular knowledge of the topic under discussion — whatever his knowledge or distinction in his own field, be it medical or scientific. Because the Encyclopedia is aimed essentially at the novice, it will make excellent reading for college students who are studying science or medicine, the bibliographies and reference lists associated with each article pointing the way to deeper study and involvement. I would emphasise, nevertheless, that the Encyclopedia is neither a training book nor a text book. For this reason, and having regard to the wide readership for which the Encyclopedia is intended, I have, in the main, omitted topics which by their very nature would necessitate a highly mathematical treatment. On the other hand because many of the articles included are predominantly clinical in nature (not a few have been written by eminent physicians and surgeons) a glossary has been added for the benefit of readers who are students or practitioners of physical science and who may have scant knowledge of anatomy, physiology, and pathology.

The topicality of the Encyclopedia is, of course, an important consideration, and I have therefore included techniques which have recently blossomed in importance, such as the clinical applications of nuclear magnetic resonance spectrometry, digital fluorography, Doppler techniques for measuring blood flow, and the use of heavy ions and mesons in radiotherapy. Medical imaging in its theoretical and its practical aspects is broadly discussed, and in all its modes — computerized axial tomography, nuclear magnetic resonance, ultrasound scanning, positron emission tomography and radionuclide imaging in general. Topics which continue to provide bread-and-butter activity for hospital physicists have not been forgotten: conventional radiotherapy in all its aspects, radiology, radioactivity measurement, and protection from the possible hazards of ionizing and other types of radiation. Also represented are topics which hospital physicists in small hospital departments are unlikely to encounter in their day-to-day work. These include activation analysis, spectroscopy and chromatography in forensic medicine, physics in dentistry, electromyography, communication aids for the physically handicapped, and radiopharmaceuticals preparation, to name but a few. Finally, research topics such as radiobiology, membrane physics, cell electrophoresis, x-ray diffraction in molecular biology, electron microscopy, bioelectricity and genetic engineering have been included, since it is essential to gain a deeper understanding of the processes which underlie disease if diseases are to be conquered rather than, at best, treated and contained.

My foremost duty as Editor as I conceive it, has been to represent the interests of the reader. Books, after all, are for readers, as hospitals are for patients. In retrospect I realise that as Editor I have been an "impedance matcher", optimizing the flow of information from writer to reader and minimizing the amount of heat and vexation generated at the interface! In this connection, I owe a specific debt of gratitude to several of the Encyclopedia's distinguished contributors. All of them are specialists in their own field. Many are authors of books; almost all are regular contributors to the professional literature. They are mainly accustomed, however, to writing for their peers. Writing for the reader whom I have been wont to call the "intelligent ignoramus" can pose particular problems. If a lady in the course of a conversation refers to her son's German mistress she will possibly have emphasised the word "German", or perhaps those to whom she speaks know, either from her previous remarks or from knowledge otherwise gained, that her son is a twelve-year old, not unusually precocious schoolboy who is studying the German language. I tell this little tale to illustrate how easy it is for the reader who does not have the benefit of background knowledge to misinterpret or even fail to understand what might appear, prima facie, a clear, simple and unambiguous statement. This is a condition which I am sure we all remember from our student days. Incidentally, it is somewhat alarming how easily a simple omission, say, of a comma, or a particular placement of a phrase or clause in a sentence, or a perhaps unfortunate choice of preposition or conjunction, can impede communication. On those occasions when I have sought confirmation of an interpretation — from one of my many local experts — it has been interesting to note how, in not a few instances, they fail to see what the problem is, at least at a first reading. Their minds, have been conditioned to "lock on" to a particular interpretation — in their case, the correct one. As I have indicated, I am extremely grateful to those contributors — for their understanding and forbearance — who have allowed me, with their approval, to rearrange their material. I hope that my interceding in this way has been to the benefit of the reader. It may also have conferred a certain uniformity of style to the Encyclopedia.

It remains for me to thank all who in one way or another have assisted in the evolution and production of the Encyclopedia. I am grateful to have been supported by the distinguished members of the Honorary Editorial Board and I thank them in particular for their constructive advice and recommendations a propos the final compilation of articles. I would acknowledge, too, in the same respect, the helpful and constructive criticisms and suggestions of Dr Dick Mould, of the Westminster Hospital, London. I remember, too, the help of my colleague, Dr R C Lawson and that of Professor P W Horton, Professor of Medical Physics, University of Surrey, England, in preparing the initial compilation. I would thank, also, J Stewart Orr, Professor and Director of Medical Physics, Hammersmith Hospital, London, Dr N C Spurway, Senior Lecturer in Physiology, University of Glasgow, and my colleague Dr T E Wheldon, who all were characteristically helpful in introducing me to many notable and appropriate authors. Many of my colleagues have put their expertise at my disposal — the names of Dr Aled Evans, Dr A T Elliott, Dr D J Mackinnon and Dr R G Bessent come particularly to mind. Of course, I owe a special debt of gratitude to Dr Bill Martin for his excellent Index. My thanks, also, to Mrs Jeanette Mackinnon who somehow deciphered and typed my editorial scribblings. And I would acknowledge once more my indebtedness to the Department of Clinical Physics and Bio-Engineering, to its former Director, Professor John Lenihan, and to its present Director, Professor Joe McKie particularly for his kind support and interest in the latter period of the Encyclopedia's preparation.

I have, of course, to thank the Publishers, Pergamon Press, who have made it all possible. One of the most pleasant aspects of my task has been my association with the hard working copy editors of Pergamon's Encyclopedias group. Salutations therefore to Mr Peter Strickland, Miss Debbie Puleston, and colleagues. It has been a particular privilege, too, to have been associated with Pergamon's Managing Editor, Encyclopedias, Dr Philip Maxwell. Dr Maxwell's conscientious professionalism and single-minded dedication to the job-in-hand is outmatched, I have observed, only by his remarkable and unfailing politeness to everyone with whom he deals.

And, finally, my thanks to my dear wife, Muriel, who surely must be the most patient listener in all the world.

T. F. McAinsh
Editor-in-Chief
*Department of Clinical Physics
and Bio-Engineering, Glasgow, UK*

Classified List of Articles

The Classified List of Articles groups the contents of the Encyclopedia by article title into a number of broad fields, alphabetically organized, from Audiology to Vision. The reader is thus presented with a general overview of the contents of the Encyclopedia. Some articles inevitably relate equally well to more than one heading. Rather than make an arbitrary decision as to which section they belong, each article has been listed wherever appropriate. For example, "Computers in Neurology" is listed under both "**Computers in Medicine**" and "**Neurological Sciences.**"

The main topics covered are:

Audiology
Biomaterials and Biomechanical
   Engineering
Biophysics
Blood
Cardiology
Computers in Medicine
Gastroenterology, Nephrology
   and Urology
Imaging

Laboratory Techniques
Mathematics in Medicine
   and Biology
Molecular and Cell Biology
Neurological Sciences
Nuclear Magnetic Resonance
Nuclear Medicine
Physics in Dentistry
Physiological Measurement
   and Monitoring

Radiobiology
Radiology
Radiotherapy
Respiratory Physics
Safety in Medicine
Therapeutic Aids and Techniques
Ultrasonics
Units
Vision

Audiology
Acoustic Impedance Audiometry; Artificial Ear; Artificial Mastoid; Audiometers; Brain-Stem Electric Response Audiometry; Ear Anatomy and Physiology; Electric Response Audiometry; Electrocochleography; Electrodermal Audiometry; Electroencephalic Audiometry; Hearing Aids; Objective Audiometry; Sound: Biological Effects; Speech Spectrography

Biomaterials and Biomechanical Engineering
Artificial Joints, Implanted; Artificial Limbs and Locomotor Aids: Evaluation; Artificial Membranes; Biomechanics; Bone: Mechanical Properties; Dental Force Analysis; Dental Materials; Gait Analysis; Heart Valves; Implanted Prostheses: Tissue Response; Mechanical Devices in Medicine and Rehabilitation; Prostheses, Myoelectrically Controlled; Soft Connective Tissues: Mechanical Behavior

Biophysics
Animal Calorimetry; Bioelectricity; Biological Control Theory; Biometry; Biophysics; Cryobiology; Cybernetics, Biological; Decision Theory; Hormesis and Homeostasis; Hypothermia; Mathematical Modelling in Biology; Muscle; Thermodynamics, Classical

Blood
Blood Cell Analysis: Automatic Counting and Sizing; Blood Cell Analysis: Morphological and Related Characteristics; Blood Flow: Invasive and Noninvasive Measurement; Blood Gas Analysis; Blood Gas Tensions: Continuous Measurement; Blood Pressure: Invasive and Noninvasive Measurement; Blood Viscosity Measurement; Doppler Blood Flow Measurement; Hemodynamics; Plethysmography; Thrombosis

Cardiology
Ambulatory Monitoring; Cardiac Catheterization; Cardiac Function: Noninvasive Assessment; Cardiac Output Measurement; Cardiac Pacemakers: Computerized Data Handling; Cardiac Pacemakers, Implantable; Cardiac Pacemakers, Temporary; Computers in Cardiology;

Defibrillators; Dynamic Cardiac Studies; Echocardiography; Electrocardiography; Heart Valves; Heart–Lung Machines; Intra-Aortic Balloon Pumps; Monitoring Equipment in Coronary and Intensive Care; Vectorcardiography

Computers in Medicine
Biological Kinetics: Computerized Data Analysis; Cardiac Pacemakers: Computerized Data Handling; Computer-Aided Diagnosis; Computerized Axial Tomography; Computerized Image Analysis in Radiology; Computers in Cardiology; Computers in Clinical Biochemistry; Computers in Neurology; Microcomputers; Microcomputers: An Application in the Clinical Laboratory

Gastroenterology, Nephrology and Urology
Endoscopes; Fiber Endoscopy; Renal Dialysis; Renal Function: Diagnostic Measurement; Urology: Fluid Flow and Pressure Measurement

Imaging
Cerebral Blood Flow: Regional Measurement; Computerized Axial Tomography; Computerized Image Analysis in Radiology; Digital Fluorography; Dynamic Cardiac Studies; Image Analysis: Receiver-Operating-Characteristic Curves; Image Analysis: Transfer Functions; Image Analysis: Extraction of Quantitative Diagnostic Information; Mammography; Neutron Radiography; Nuclear Magnetic Resonance Imaging; Positron Emission Tomography; Radiography and Fluoroscopy in Medicine; Radionuclide Imaging; Radionuclide Brain Imaging; Scanning Electron Microscopy; Single-Photon Emission Tomography; Thermography; Transmission Electron Microscopy; Ultrasonic Image Analysis; Ultrasound in Medicine; Ultrasound in Obstetrics

Laboratory Techniques
Blood Cell Analysis: Automatic Counting and Sizing; Blood Cell Analysis: Morphological and Related Characteristics; Centrifuges: Principles and Applications; Cervical Cytology: Automation; Chromatography; Chromosome Analysis, Automatic; Clinical Biochemistry: Automation; Clinical Chemistry: Physics and Instrumentation; Colorimetry; Computers in Clinical Biochemistry; Electron Microprobe Analysis; Electron Microscopy: Freeze-Fracture Replication; Fluorimetry; Forensic Applications of Chromatography; Forensic Applications of Spectroscopy; Microcomputers: An Application in the Clinical Laboratory; Microphotometry; Neutron Activation Analysis; Particle-Induced X-Ray Emission Analysis; Photon Activation Analysis; Radioimmunoassay; Scanning Electron Microscopy; Spectroscopy; Transmission Electron Microscopy

Mathematics in Medicine and Biology
Biometry; Biological Control Theory; Biostatistics; Cancer Statistics; Decision Theory; Image Analysis: Transfer Functions; Mathematical Modelling in Biology; Signal Analysis Techniques; Statistical Methods in Medicine

Molecular and Cell Biology
Cell Electrophoresis; Cervical Cytology: Automation; Chemical Carcinogenesis; Chromosome Analysis, Automatic; Genetic Code; Genetic Engineering; X-Ray Diffraction in Molecular Biology

Neurological Sciences
Cerebral Blood Flow: Regional Measurement; Computers in Neurology; Electroencephalography; Electromyography; Evoked Potentials; Intracerebral Electrodes; Intracranial Pressure Measurement; Neurosurgery: Physiological Monitoring; Radionuclide Brain Imaging; Radionuclide Cisternography; Visual Cortical Neurophysiology

Nuclear Magnetic Resonance
Nuclear Magnetic Resonance: General Principles; Nuclear Magnetic Resonance Imaging; Nuclear Magnetic Resonance Spectroscopy

Nuclear Medicine
Beta-Particle Detection; Cerebal Blood Flow: Regional Measurement; Cyclotrons; Dosimetry of Internally Administered Radioactive Substances; Dynamic Cardiac Studies; Gamma-Ray Detectors; Neutron Sources; Nuclear Medicine Department Design and Equipment; Occupancy Principle; Phantoms in Nuclear Medicine; Positron Emission Tomography; Quality Assurance in Nuclear Medicine; Radiation Quantities and Units; Radioactivity Measurement; Radioactivity Measurement: Counting Statistics; Radioiodine: Clinical Uses; Radionuclide Brain Imaging; Radionuclide Cisternography; Radionuclide Generators; Radionuclide Imaging; Radionuclides: Clinical Uses; Radionuclides: Whole-Body Monitors; Radiopharmaceuticals: Preparation and Quality Assurance; Renal Function: Diagnostic Measurements; Single-Photon Emission Tomography

Physics in Dentistry
Dental Diagnosis; Dental Enamel: Crystallography; Dental Fluoridation; Dental Force Analysis; Dental Materials; Preventive Dentistry; Teeth: Electron Microscopy Studies; Teeth: Physical Properties

Physiological Measurement and Monitoring
Ambulatory Monitoring; Biotelemetry; Blood Flow: Invasive and Noninvasive Measurement; Blood Gas Analysis; Blood Gas Tensions: Continuous Measurement; Blood Pressure: Invasive and Noninvasive Measurement; Blood Viscosity Measurement; Cardiac Catheterization; Cardiac Function: Noninvasive Assessment; Cardiac Output Measurement; Cerebral Blood Flow: Regional Measurement; Clinical Temperature Measurement; Doppler Blood Flow Measurement; Echocardiography; Electrocardiography; Electrocochleography; Electroencephalography; Electromyography; Electrooculography; Electroretinography; Evoked Potentials; Fetal Monitoring; Intracerebral Electrodes; Intracranial Pressure Measurement; Medical Gases: Measurement and Analysis; Medical Photography; Monitoring Equipment in Coronary and Intensive Care; Neonatal Intensive Care Equipment; Neurosurgery: Physiological Monitoring; Neutron Activation Analysis *In Vivo;* Photogrammetry; Physiological Measurement; Plethysmography; Recording and Display Devices; Renal Function: Diagnostic Measurement; Respiratory Function: Physiology; Respiratory Function: Methods of Assessment; Respiratory Function Measurement: Equipment; Signal Analysis Techniques; Space Biology and Physiology; Thermography; Urology: Fluid Flow and Pressure Measurement; Vectorcardiography

Radiobiology
Cell Population Kinetics; Microdosimetry; Radiation Carcinogenesis; Radiation Chemistry; Radiobiology: Charged and Uncharged Particles; Radiobiology: Kinetic Basis of Normal-Tissue Response to Radiation; Radiobiology: Prenatal and Perinatal Irradiation; Radiosensitizers; Target Theory and Repair Models in Cellular Radiobiology; X Rays: Biological Effects

Radiology
Computerized Axial Tomography; Computerized Image Analysis in Radiology; Digital Fluorography; Mammography; Neutron Radiography; Quality Assurance in Diagnostic Radiology; Radiography and Fluoroscopy in Medicine; Radiography and Fluoroscopy in Phonetics; Speech Spectrography; X-Ray Production; X Rays in Medicine: Early History

Radiotherapy

Brachytherapy; Cancer Statistics; Electron Linear Accelerators; Fast Neutron Therapy; Hyperthermia in Cancer Treatment; Ionizing Radiation: Absorption in Body Tissues; Neutron-Capture Therapy; Neutron Dosimetry; Neutron Kerma Values; Radiation Dosimetry; Radiation Quantities and Units; Radiation Quantities: Measurement; Radiosensitizers; Radiotherapy: Afterloading Techniques; Radiotherapy: Beta Particles; Radiotherapy: Cobalt Treatment; Radiotherapy: Computer-Aided Treatment Planning; Radiotherapy: Heavy Ions, Mesons, Neutrons and Protons; Radiotherapy: Isodose Charts; Radiotherapy: Linear Accelerators; Radiotherapy: Radiation Dose, Time and Fraction Number Formulae; Radiotherapy: Treatment Planning; Radium in Medicine: Early History

Respiratory Physics

Anesthesia Physics; Medical Gases: Measurement and Analysis; Respiratory Function: Physiology; Respiratory Function: Methods of Assessment; Respiratory Function Measurement: Equipment

Safety in Medicine

Medical Electrical Equipment: Safety Aspects; Nonionizing Electromagnetic Radiation: Potential Hazards; Radiation Protection and Personnel Monitoring; Radiation Protection: External Exposure; Radiation Protection: Internal Exposure; Radioactive Waste Disposal: Hospital Practice; Radio-Frequency and Microwave Radiation: Potential Hazards; Sound: Biological Effects; Static Electricity in Hospitals; Ultrasound: Potential Hazards; Ultraviolet Radiation and the Skin; Ultraviolet Radiation: Potential Hazards

Therapeutic Aids and Techniques

Biofeedback; Cardiac Pacemakers: Computerized Data Handling; Cardiac Pacemakers, Implantable; Cardiac Pacemakers, Temporary; Communication Aids for the Physically Handicapped; Cryobiology; Defibrillators; Electric and Magnetic Fields: Biological Effects; Electroconvulsive Therapy; Electrosurgery; Endoscopes; Fiber Endoscopy; Hearing Aids; Heart–Lung Machines; Hyperbaric Medicine; Intra-Aortic Balloon Pumps; Lasers in Medicine; Laser Physics; Mechanical Devices in Medicine and Rehabilitation; Nebulizer Therapy; Prostheses, Myoelectrically Controlled; Renal Dialysis; Ultraviolet Radiation and the Skin

Ultrasonics

Doppler Blood Flow Measurement; Echocardiography; Ultrasonic Image Analysis; Ultrasound in Medicine; Ultrasound in Obstetrics; Ultrasound: Potential Hazards; Ultrasound Therapy; Ultrasound: Tissue Characterization; Ultrasound: Transmission and Scattering in Human Tissue

Units

Radiation Quantities and Units; Radiation Quantities: Measurement; SI Units

Vision

Binocular Vision; Color Blindness; Color Vision; Electrooculography; Electroretinography; Intraocular Fluid Dynamics; Vision; Visual Cortical Neurophysiology; Visual Fields and Thresholds

Alphabetic List of Articles

VOLUME 2

A

Acoustic Impedance Audiometry

Acoustic impedance (admittance) audiometry describes a group of procedures used clinically in the diagnosis of hearing disorders. The procedures all make use of an impedance or admittance meter or bridge which measures the admittance at the tympanic membrane presented to incident low-frequency sound waves. This measure characterizes the mobility of the middle-ear structures, which is modified by certain disorders such as secretory otitis media.

The measured admittance is also modified by reflex contractions of the stapedius muscle in the middle ear. Examination of this reflex provides further valuable diagnostic information regarding the middle ear and other parts of the auditory system which contribute to the reflex arc.

1. Definitions

Acoustic impedance is the complex ratio of sound pressure to volume velocity and is measured in $Pa\,s\,m^{-3}$ or cgs acoustical ohms. Its reciprocal is termed acoustic admittance. The latter is sometimes resolved into the in-phase and quadrature components of conductance and susceptance. The unit of measurement of the admittance quantities is $m^3\,Pa^{-1}\,s^{-1}$ or cgs acoustical mhos.

Acoustic compliance is the scalar ratio of volume displacement to sound pressure and has units of $m^3\,Pa^{-1}$.

2. Instrumentation

The design of a typical impedance meter is illustrated in Fig. 1. The probe is sealed in the ear canal by means of a soft plastic cuff. A sound wave (probe tone) at a frequency of typically 220 Hz is delivered to the probe by a miniature receiver and the sound pressure it generates in the ear canal is measured via a miniature microphone. There is also a pneumatic line which allows the ambient pressure in the ear canal to be varied.

The probe tone originates from the oscillator and passes via the automatic volume-control (AVC) circuit before reaching the receiver. The AVC is also connected to the microphone and set to maintain a constant sound pressure level (SPL) of about 85 dB in the ear canal. Since the volume velocity of the probe tone is proportional to the voltage V driving the receiver and the sound pressure is constant, the acoustic admittance at the tip of the probe is simply proportional to V. Thus the voltmeter connected to the receiver driving signal gives a direct reading of admittance modulus.

More elaborate circuitry is necessary to resolve this signal into in-phase and quadrature components in

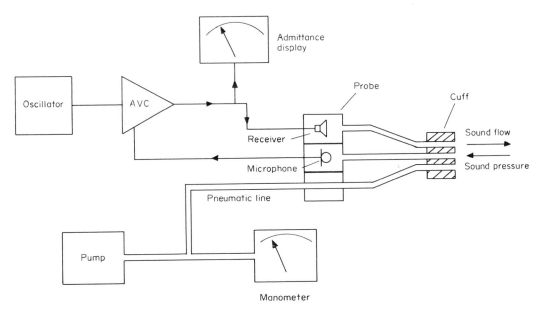

Figure 1
Typical design of impedance or admittance meter

instruments which measure conductance and susceptance.

3. Relevant Anatomy of the Middle Ear

Figure 2 illustrates the functional anatomy of the middle ear, whereby incident sound waves in the ear canal are transformed into a pressure wave in the cochlear fluids, thus stimulating the sensory receptors therein (see *Ear Anatomy and Physiology*).

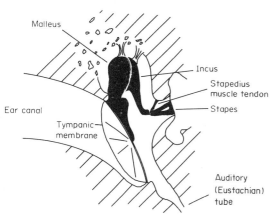

Figure 2
Functional anatomy of the middle ear

Among the supporting structures of the tiny bones of the middle ear (ossicles) is the stapedius muscle which is attached to the neck of the stapes. The function of this muscle is unknown. Its major property is that it contracts as a reflex to sounds above about 80 dB SPL. This acoustic reflex is bilateral in that monaural stimulation causes the muscles of both ears to contract. This reflex is utilized frequently in audiological assessment.

4. Clinical Applications of Acoustic Admittance Measurements

It is convenient to divide acoustic admittance measurements into two classes: active measurements which involve the acoustic reflex, and passive measurements which do not.

The major passive measurement procedure is tympanometry in which the measured admittance is plotted as a function of the ambient pressure in the ear canal for a range of pressures usually between 2 kPa above and below atmospheric pressure. The resultant graph is termed a tympanogram.

The effect of varying the ear-canal pressure is to alter the pressure differential between the ear canal and middle-ear cavity. This pressure differential tends to stiffen the tympanic membrane and thus reduce its acoustic admittance, as illustrated by the tympanogram

in Fig. 3. The pressure at which the admittance is maximal represents zero pressure differential across the tympanic membrane. Measurement of that pressure gives an indication of the middle-ear pressure, which tends to differ from atmospheric pressure in certain disease states which interfere with the ventilation of the middle ear via the auditory (Eustachian) tube (see Fig. 2). The normal range of middle-ear pressures is −1.0 to 0.5 kPa.

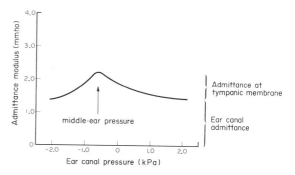

Figure 3
Typical tympanogram illustrating middle-ear pressure, ear canal admittance and admittance at tympanic membrane

At the extreme positive and negative pressures used in tympanometry, the admittance at the tympanic membrane approaches zero. At these pressures the measured admittance is approximately equal to the admittance of the volume of air between the probe and the tympanic membrane. This value can be subtracted from the admittance corresponding to the tympanogram peak to derive the admittance at the tympanic membrane under conditions of zero pressure differential. The normal range is 0.2–1.5 mmhos. It is reduced considerably when the middle ear contains fluid and increased when the ossicles are fractured or disarticulated or when the tympanic membrane is scarred. Otosclerosis, which is a bony growth occurring mainly around the stapes footplate, reduces the mobility of the ossicles but does not usually reduce the admittance at the tympanic membrane sufficiently to be distinguishable from normal variation.

If the tympanic membrane is perforated, variation of the ear-canal pressure does not produce a pressure differential across the tympanic membrane and therefore the tympanogram is completely flat.

Active acoustic admittance measurements involve examination of the acoustic reflex of the stapedius muscle (see Fig. 2). Contraction of the stapedius muscle stiffens the ossicular chain and therefore reduces the admittance at the tympanic membrane. This admittance change can be detected by the instrumentation described above. The anatomy of the reflex arc is illustrated in Fig. 4. Measurable effects on this acoustic reflex are caused by pathologies which affect either the sensory and afferent sections of the reflex arc (ossicles, cochlea and auditory nerve), the reflex center in the brain stem, or the efferent

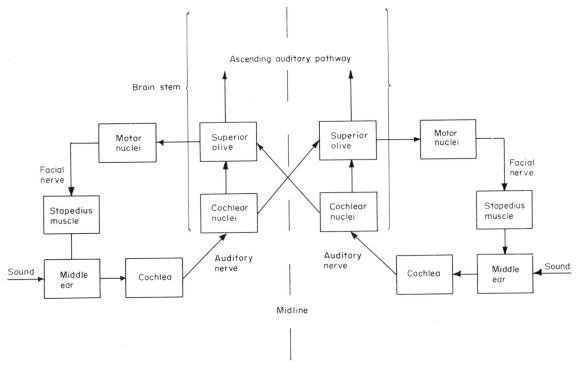

Figure 4
Organization of the acoustic reflex arc showing ipsilateral and contralateral pathways

and motor sections of the reflex arc (facial nerve, stapedius muscle and ossicles).

The reflex may be actuated by applying an additional stimulus, either ipsilaterally via the measurement probe, or contralaterally using an auxiliary earphone on the opposite ear. Thus, four measurement conditions exist: (a) probe in the left ear, ipsilateral stimulus; (b) probe left, contralateral; (c) probe right, ipsilateral; and (d) probe right, contralateral.

When there is any significant impediment to ossicular mobility in an ear, no acoustic reflex is recordable with the probe in that ear whether stimulated ipsilaterally or contralaterally. Therefore, the presence of a recordable acoustic reflex is a reliable indication that there is no significant conductive hearing loss. A conductive hearing loss of as little as 10 dB is sufficient to prevent the recording of an acoustic reflex.

A cochlear hearing loss may alter the acoustic reflex when the stimulus is presented to the affected side, although the sensitivity of the acoustic reflex is reduced far less than is the hearing sensitivity measured by conventional pure-tone audiometry. This difference illustrates a phenomenon termed recruitment, whereby sensitivity to high-intensity sounds is affected less than sensitivity to low-intensity sounds. Recruitment tends to be greater when the site of the hearing loss is the cochlea than when it is the auditory nerve. Therefore, comparison of acoustic reflex thresholds and auditory

thresholds provides information regarding the site of dysfunction (Priede and Coles 1974).

Adaptation of the acoustic reflex during continuous stimulation also provides an index of the site of dysfunction. The time taken for the acoustic reflex response to decay by 50% for a 1000 Hz tone is usually measured; times less than 5 s are an indication of an auditory nerve lesion (Anderson et al. 1970).

The acoustic reflex is more sensitive to random-noise stimuli than to pure tones (Niemeyer and Sesterhenn 1974). This difference in sensitivity diminishes with increasing cochlear hearing loss. Measurement of the difference can provide an approximate estimate of the severity of a hearing loss in patients who are unable or unwilling to make voluntary responses.

5. Application of Acoustic Impedance Measurements to School Screening

Acoustic impedance measurements provide a rapid objective method of screening school children for hearing disorders. A combination of tympanometry to detect middle-ear disease and an acoustic-reflex test to detect a conductive hearing loss or a severe cochlear or neural hearing loss is sometimes used.

Improved detection of less-severe cochlear and neural hearing losses may be achieved in future by including pure-tone and noise stimulation of the acoustic reflex to estimate hearing thresholds.

6. Standardization of Acoustic Impedance Measurements

Currently subcommittee 29C of the International Electrotechnical Commission (IEC) is developing a standard to cover impedance or admittance instruments used in audiology entitled *Aural Impedance/Admittance Instruments*. It was made available as a draft for public comment in January 1984.

See also: Audiometers; Objective Audiometry

Bibliography

Anderson H, Barr B, Wedenberg E 1970 The early detection of acoustic tumours by the stapedius reflex test. In: Wolstenholme G E W, Knight J (eds.) 1970 *Sensorineural Hearing Loss.* Churchill, London, pp. 275–89
Feldman A S, Wilber L A 1976 *Acoustic Impedance and Admittance: The Measurement of Middle Ear Function.* Williams and Wilkins, Baltimore, Maryland
International Electrotechnical Commission 1984 *Aural Impedance/Admittance Instruments* (draft). International Electrotechnical Commission, Geneva
Kinsler L E, Fey A R 1962 *Fundamentals of Acoustics.* Wiley, New York
Niemeyer W, Sesterhenn G 1974 Calculating the hearing threshold from the stapedius reflex threshold for different sound stimuli. *Audiology* 13: 421–27
Priede V M, Coles R R A 1974 Interpretation of loudness recruitment tests: Some new concepts and criteria. *J. Laryngol. Otol.* 88: 641–62

M. E. Lutman

Ambulatory Monitoring

The purpose of ambulatory monitoring is to collect physiological data while the patient is following a normal daily routine. Recordings are generally continuous for 24 h but may be intermittent during a longer period. Any parameter that can be converted into an electrical signal may be recorded but the electrical activity of the heart, brain and eye, as well as arterial blood pressure and respiration are the most frequently studied.

1. Recording Systems

All continuous recordings are made by miniature tape recorders operating at a speed such that data collected in 24 h can be stored on a C 120 cassette or miniature reel. Radiotelemetry and transmission over the telephone system have also been used, but these methods are not considered to constitute ambulatory monitoring.

Tape recorders actuated by the patient at the onset of symptoms are generally less useful than continuous records.

The parameters commonly monitored contain frequencies in the range ~ 0–100 Hz and obviously not all of these can be recorded directly on magnetic tape. Direct recording can be used to obtain a bandwidth of 0.15–100 Hz provided there is a replay/record speed ratio of 25:1 or greater. For lower frequencies, pulse-width modulation or frequency modulation is used. Digital recording on magnetic tape is still in the development stage and the instruments at present are too bulky to be acceptable, as are the "bubble memories" needed to accommodate the 10–20 Mbyte of data per channel of ECG every 24 h.

Recently, the availability of CMOS microprocessors with their low-voltage and low-power requirements has led to the development of the so-called "intelligent" ambulatory ECG recorders. In these, the incoming data are analyzed in real time and then stored, thus dispensing with conventional retrospective analysis or scanning.

2. ECG Replay and Analysis Systems

During 24 h the heart generates some 100 000 electrical signals or blood-pressure pulses, while even more electrical activity occurs in the brain. To attempt to analyze this amount of information unaided would obviously be so time consuming as to be impractical. If, however, the tape is replayed at high speed, not only is it possible to recover low-frequency signals but the data can be scanned in an hour or less.

The earliest scanners provided a triggered electrocardiogram (ECG) displayed on an oscilloscope with successive waveforms superimposed. This was later augmented by a series of vertical lines, the heights of which were inversely proportional to heart rate. The combined display facilitated the detection of arrhythmias but not necessarily their classification. For ten years or so, effort continued to be concentrated on developing aids for the analyst rather than on automatic systems which might replace him. In the meantime, however, analog and digital computer techniques were being increasingly used to classify ECG abnormalities found in static patients from coronary care units, wards and outpatient departments.

By the early 1970s these methods were beginning to be adapted for use in ambulatory monitoring where more limited objectives had to be accepted. Two factors were responsible for this. First, at replay/record speed ratios of 60:1 or 120:1, present day microcomputers do not have time for detailed processing, and secondly, ECGs from ambulant subjects are obtained from fewer sites and contain more artefacts than those from resting patients. A result of these problems is that automatic analyses for ambulatory ECGs concentrate on detecting changes in rhythm and in the morphology of the most easily detectable component of the ECG waveform, the QRS complex (see *Electrocardiography*).

Rhythm is comparatively easy to classify by comparing beat-to-beat intervals with a running average and displaying the results as a histogram supplemented by selected print-outs of the analog signal. Changes in the morphology of the QRS complex are currently detected by various means including the comparison of their duration, area or frequency spectrum with that of normal waves, or by more complex cross-correlation. Success rates as high as 98% have been reported, but until impartial test tapes of arrhythmias are more freely available, this figure may be considered somewhat optimistic. In addition to revealing transient arrhythmias, ambulatory monitoring can document episodes of cardiac ischemia associated with angina. Although clinically highly significant, they have received less attention than arrhythmias because some first-generation recorders could not reproduce the associated ECG changes reliably and the changes themselves could be mimicked by physiological variations.

3. Other Physiological Parameters

The signals displayed by an electroencephalogram (EEG) are one or two orders of magnitude smaller than ECG signals, and to minimize artefacts miniature encapsulated preamplifiers are placed near to the scalp electrodes (see *Electroencephalography*). Whereas two channels of ECG are adequate, four are virtually essential for EEG. Attempts at multiplexing to provide 16 channels have been made but efforts in this direction have been hampered by the limited high-frequency response of the recorder. Technical developments during the past five years have made it now possible to record simultaneously eight channels of EEG in a reasonably compact instrument without recourse to multiplexing. The EEG is difficult to analyze automatically and most recordings are replayed on to high-speed pen recorders or displayed in page mode on large screen oscilloscopes. Sections are then selected for visual analysis with possibly some assistance from bandpass filtering.

A recording of the electrical activity of the eye is made by placing electrodes at the external canthi of both eyes (see *Electrooculography*). Signal levels are of the same order as the EEG. In ambulatory monitoring, it is only used in conjunction with the ECG, EEG or blood-pressure measurement to give an indication of depth of sleep. Analysis is performed in the same fashion as for the EEG.

Although ambulatory blood pressure can be recorded from an automated form of sphygmomanometer, a catheter inserted into the brachial artery and attached to a combined miniature pressure recorder/infuser unit and four-channel tape recorder is more commonly used. Analysis is usually limited to the automatic measurement of systolic, diastolic and mean blood pressures, and heart rate. Histograms and trend plots are generated automatically in addition to some statistical manipulation.

Apparatus for monitoring respiration in ambulant subjects must be unobtrusive, thus precluding any direct connection with the nose and mouth (see *Respiratory Function Measurement: Equipment*). The respiratory inductive plethysmograph is currently the most promising device. It consists of two coils incorporated in an elastic vest, one coil being placed around the rib cage, the other around the abdomen. The coils are linked to oscillators whose frequencies change during the respiratory cycle. Frequency-to-voltage converters provide dc signals which are recorded on tape and subsequently replaced on to pen recorders.

See also: Physiological Measurement

Bibliography

Littler W A (ed.) 1980 *Clinical Ambulatory Monitoring.* Chapman and Hall, London
Marchesi C (ed.) 1984 Ambulatory monitoring—Real time analysis versus tape scanning systems. *Ambulatory Monitoring. Cardiovascular and Allied Applications.* Martinus Nijhoff, Boston
Stott F D, Raftery E B, Goulding L (eds.) 1980 *Proc. 3rd Int. Symp. Ambulatory Monitoring.* Academic Press, London

Further information on ambulatory monitoring can be found in papers presented at the *International Symposium on Ambulatory Monitoring*, held at Padua, Italy on 28–30 March, 1985

L. Goulding

Anesthesia Physics

To render patients insensitive to pain when undergoing surgery, it has become conventional to administer various anesthetic agents, most of them by way of the lungs. This inhalational method offers great flexibility in the choice of the mixtures and concentrations of agents administered and guarantees not only rapid absorption by the body but also fairly rapid excretion by the lungs, particularly in the early phase of recovery to consciousness. The administration of drugs in this way implies a need to store, pipe and mix gaseous anesthetics and ancillary agents (nitrous oxide, cyclopropane, oxygen, carbon dioxide), and also to provide various organic liquids which have an anesthetic effect when vaporized (diethyl ether, trichloroethylene, or trifluorochlorobromoethane, known commercially as halothane). Depending on the depth of unconsciousness or the degree of muscle relaxation required, the normally spontaneous process of breathing may need to be taken over by an anesthetic ventilator for the duration of the operation.

Trolley-mounted bulk storage gas cylinders (usually containing oxygen, nitrous oxide, cyclopropane and carbon dioxide) together with pressure regulators, flowmeters, and volatile liquid vaporizers make up an anesthetic machine. Gases from the storage cylinders (or from bulk supplies piped around the hospital) are each passed through a valve reducing the supply pressure to 414 kPa. At this reduced pressure the gases are piped through flowmeters into a mixing manifold and then through various pressure and oxygen failure alarm systems to the breathing circuit. The flowmeters are usually of the variable orifice type. Gas flow rate is controlled by a needle valve and its value indicated by the position of a cylindrical or spherical float suspended in the gas stream within a vertically mounted tapered tube.

Once the gases from the cylinders are mixed, various volatile liquids are vaporized into the gas stream using specially designed anesthetic vaporizers. The design of these vaporizers usually involves splitting off a controllable fraction of the carrier gas from the main stream, passing it over or bubbling it through the liquid to be vaporized, and then recombining the streams. This apparently simple arrangement has its complications. The extraction of latent heat of vaporization results in a fall of temperature and hence a fall in the vapor pressure of the anesthetic liquid depending on the volume of liquid present and the thermal conductivity of the container. The resulting tendency for the output of vaporizers to change with time has made necessary a variety of methods, either manual or automatic, which alter the by-pass fraction to compensate for temperature change. The widely varying saturated vapor pressures of the liquid anesthetics which may be used makes it necessary to design suitable vaporizers for each one. Vaporizers are usually mounted in series, downstream of the carrier gas mixing manifold. Although the various liquid anesthetics are usually used singly, the optimum order of mounting vaporizers involves calculations to establish the best compromise between vapor pressures, solubilities and chemical reactivity with breathing circuit components.

1. Anesthetic Breathing Circuits

In a spontaneously breathing adult, air flows in and out of the lungs in a "tidal" pattern. A graph of flow against time is almost sinusoidal in form. Flows build up from zero to around $1.5 \, \text{l s}^{-1}$ during inspiration, followed by a fall to zero flow and reverse flow during expiration. A cough can accelerate flow to $10 \, \text{l s}^{-1}$, but the average consumption of fresh gas mixture during breathing lies between 0.1 and $0.15 \, \text{l s}^{-1}$. It is the function of an anesthetic breathing circuit to duct a specific gas mixture from what is usually a constant flow supply, into the patient's intermittently filling lungs and to provide for the complete or partial elimination of the expired mixture.

Historically, patients inhaled their anesthetic through a gauze pad on to which liquid anesthetic was dropped. Such open breathing circuits have been superseded by T-shaped or circular arrangements of corrugated flexible

tubing through which the required anesthetics are ducted. The tubing is corrugated to prevent kinking and is usually made of neoprene rubber or of plastic. Lengths are joined by various nationally or internationally standardized push-fit methods. These circuits may be disposable or sterilizable for reuse. The diffusion of gases into and out of the wall materials has occupied a number of workers concerned with losses from the circuit, cross contamination of anesthetics, and the pollution of the atmosphere in operating theaters.

A great variety of anesthetic circuit configurations have been developed. In the case of a simple T-shaped arrangement, with the patient on one limb of the T, the fresh gas supplied to another and the third limb open to atmosphere (see Fig. 1), it is evident that the fresh-gas supply flow rate determines whether some entrainment of air from the open limb occurs during inspiration. Other factors of significance include the length of the open limb, which has a bearing on the availability of any residual mixture from a previous exhalation for rebreathing. Similarly, the relative diameters of the limbs of the T (which affect their flow resistance) have an influence not only on the movement of mixtures within the T but also at the lung–blood interface within the patient's lung.

The need to economize in gas mixture has led to a number of variations in the fundamental T-shaped circuit. The inclusion of a flexible reservoir bag in the inspiratory or in the expiratory limb of the T-shape provides a means of buffering to cope with momentarily

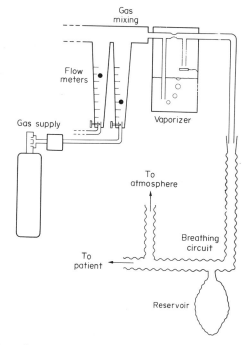

Figure 1
An anesthetic machine and breathing circuit

high inspirational flow whilst at the same time providing the anesthetist with a manual method of inflating the lungs if for any reason spontaneous breathing stops.

Ultimate economies in fresh-gas supply can be achieved with circular breathing circuits. The use of valving on the fundamental T circuit can guarantee separation of inspiratory and expiratory flows so that the expiratory limb can be ducted full circle and into the fresh-gas supply. With such a circular arrangement, a canister packed with soda lime and having as low as a resistance to flow as possible must be used to absorb expired carbon dioxide. For a completely closed circle, fresh-gas mixture need only be supplied at a rate equal to the patient's net rate of absorption of the various gaseous components. Calculation of the concentration of the mixture present in the circle is complicated because, in the lung, oxygen is exchanged for carbon dioxide in unequal proportions and different anesthetic components are absorbed at differing rates.

The functional analysis of breathing circuit behavior both by physical measurement and by computer simulation has received much attention. Variables such as circuit dimensions, fresh-gas flows, and the placement of components such as valves, reservoir bags and absorbers can be related to consequential rebreathing, anesthetic and oxygen concentrations in the lung, and fresh-gas consumption.

2. Anesthetic Ventilators

As previously stated, the patient may need to be very deeply anesthetized or may need muscle-relaxant drugs. In both instances the spontaneous breathing reflex is absent and automatic lung ventilators are necessary to take over the work of breathing. The anesthetic gas mixtures are produced as previously described but are supplied to a mechanical piston or bellows arrangement which produces alternate inflation and deflation of the patient's lungs. A positive pressure of around 1 Pa applied by way of a breathing circuit to an adult lung may induce an inspiratory movement of about 0.05 l of gas. The volume increase of a lung per unit increase of pressure is termed the compliance of the lung.

An anesthetic ventilator applies a positive pressure to inflate the patient's lungs. At some preset limit the anesthetic ventilator must switch from a lung filling phase to an expiratory phase. In some ventilators the elastic recoil of the chest is allowed to expel gas naturally at expiration, but in other designs an actual subatmospheric pressure is induced. In this latter circumstance the ventilator is said to have a negative phase.

The switching of a ventilator from inspiration to expiration phase is achieved by various means. Some designs make use of either simple mechanical or, more recently, solid-state timing devices. Preset inspiration and expiration times are not necessarily equal. In practice, a purely timing arrangement is not very convenient because gas flow rate into the lung, which depends on supply pressure and the patient's lung compliance, must

be taken into account. Other cycling mechanisms used are volume cycling, where the changeover to expiration occurs when some preset volume of mixture has entered the lungs, and pressure cycling, where expiration starts after a certain pressure at the mouth has been reached. Yet another system uses a flow detection method, changeover taking place when inspiratory flow has fallen to zero.

Ventilators can be crudely classified as low-output impedance or high-output impedance devices. Low-output impedance ventilators often utilize a bellows or piston arrangement, which after filling, discharges its contents into the patient, with the bellows or piston being driven solely by a weight acting under gravity. A typical generated pressure of a low-output impedance ventilator is 12 Pa; the internal resistance to the ventilator is some 2 Pa s l^{-1}. A high-output impedance ventilator uses either a positive drive on the bellows or piston, or even the direct ducting of the supply gas through a control valve into the patient's lungs. A high-output impedance ventilator has an output pressure above 50 Pa. Typically its output resistance is some 60 Pa s l^{-1}. A very high output impedance ventilator generates a pressure of some 4 kPa and has an output resistance of 8 kPa s l^{-1}.

Low-output impedance ventilators are sensitive to loading on being connected to the patient. An adult lung with compliance of about 0.05 l Pa^{-1} and input resistance of 6 Pa s l^{-1} when connected to the low-output impedance ventilator produces an exponentially falling input flow. In contrast, the high-output impedance systems are insensitive to loading but pressure relief and cycling mechanisms must be very reliable if a patient is to be protected from excessive pressures.

The theoretical analysis of ventilator function has occupied a number of authorities. Some of the terminology used is not always unambiguous. Compliance as referred to above is the total chest compliance but this can sometimes be subdivided into chest-wall compliance and pulmonary compliance. Pressures referred to above are machine-generated pressures but in practice the resistances throughout the ventilator–lung system produce differing pressures at different points of the system. The term tidal volume meaning the volume of gas delivered in one inspiration has been mentioned. The term ventilation rate or respiratory frequency represents the number of breaths per unit time. However, the unqualified term "ventilation" is often used synonymously with the term "minute volume" and means the volume of fresh gas delivered over some unit time. To a first approximation, the variables can be linked as follows:

$$\text{minute volume} = \text{ventilation rate} \times \text{tidal volume}$$
$$\text{tidal volume} = \text{inflation pressure} \times \text{compliance}$$

3. Monitoring

It is imperative that the anesthetist closely monitors the physiological state of the deeply anesthetized patient and supervises the state of the life-support apparatus. This often involves automatic detection of body temperature (which is sometimes artificially reduced) and heart activity monitoring by means of the ECG signal or by using a photoelectric pulse monitor. For some procedures automatic detection of blood pressure is desirable but for nonelaborate surgery simple cuff occlusion methods of pressure measurement are adequate. For some research purposes and occasionally for routine purposes, cardiac output and the state of muscle relaxation may be monitored.

The supervision of anesthetic apparatus includes not only the adjustment of gas composition and ventilator settings. The anesthetist who connects a patient to physiological monitoring apparatus must be aware of electrical safety considerations and of gas ignition possibilities from electrostatic discharges or surgical diathermy sparks.

In the context of apparatus assessment, various gas analysis devices are of use (see *Medical Gases: Measurement and Analysis*). To follow breath-by-breath changes in anesthetic composition, stable and fast-response gas analyzers have always been in demand. Paramagnetic oxygen analyzers were once popular because of their specificity, but polarographic detectors have been developed which have a reduced sensitivity to nitrous oxide and make possible a probe type of detector. For the measurement of anesthetics such as nitrous oxide or halothane, infrared and ultraviolet absorption analyzers have been developed with adequately fast response times. Over recent years, both magnetic sector and quadrupole mass spectrometers have been developed for respiratory and anesthetic gas measurements.

4. Anesthetic Uptake

Consider the uptake of an anesthetic, say halothane, by an individual organ such as the brain. For any vessel of volume V liters and perfused by a steady gas flow of F liters per minute of gas, consider the addition of M_0 liters of anesthetic as a bolus into the gas flow. If mixing in the vessel is perfect, the concentration in the vessel is given by

$$-\frac{dM}{dt} = \frac{M}{V} F$$

hence

$$M = M_0 \exp(-F/V)t$$

For an organ such as the brain, if it is perfused by blood in which halothane is dissolved, the tissue–blood partition coefficient λ must be taken into account. The effective volume of the organ for the vapor is then $V\lambda$. Assuming that the perfusing blood contains a steady concentration C_{in} milliliters of halothane per 100 ml of blood and that C_{out} is the output concentration in equilibrium with the organ, the wash-in equation is

$$C_{out} = C_{in}[1 - \exp(-F/V\lambda)t]$$

The rate of uptake of anesthetic is given by

$$\left(\frac{d}{dt}C_{out}\right)V\lambda = 10FC_{in}\exp(-F/V\lambda)t$$

where the factor of 10 converts C_{in} into milliliters per liter.

It can be deduced separately that the capacity of the organ for vapor is $10V\lambda C_{in}$ milliliters and that, since organ content = organ capacity × organ saturation (C_{out}/C_{in}), the organ content is given by:

$$10V\lambda C_{in}[1 - \exp(-F/V\lambda)t]$$

Given now that about 1 vol % of halothane administered to the lungs produces 2.38 ml of vapor per 100 ml of blood and given that, for the brain, $\lambda = 2.3$, it is possible to calculate the brain uptake rate. A typical brain volume is 2.1 l and it receives about 14% of cardiac output (i.e., 0.98 l min^{-1}). Substituting in the above formulae gives

$$\text{brain capacity for halothane} = 10 \times 2.1 \times 2.3 \times 2.38$$
$$= 115\ \text{ml}$$

$$\text{brain uptake rate (when } t = 1\ \text{min)} = 3.9\ \text{ml min}^{-1}$$

$$\text{time constant} = 4.9\ \text{min}$$

$$\text{brain content after 1 min} = 21.1\ \text{ml vapor}$$

Figures such as these can be related to empirically determined levels of anesthetics that produce insensibility to pain. It has been established that using halothane only, the brain must be 75.8% saturated, taking about 7 min breathing of a 1% mixture, before a patient has a 50% probability of not feeling a surgical incision. In practice, other agents such as sodium pentothal are used intravenously to speed the induction of anesthesia.

See also: Respiratory Function: Physiology; Respiratory Function: Methods of Assessment; Respiratory Function Measurement: Equipment; Physiological Measurement

Bibliography

Dorsch J A, Dorsch S E 1975 *Understanding Anaesthesia Equipment: Construction, Care and Complications.* Williams and Wilkins, Baltimore, Maryland
Hill D W 1980 *Physics Applied to Anaesthesia*, 4th edn. Butterworth, London
Mushin W W, Rendell-Baker L, Thompson P W, Mapleson W W 1980 *Automatic Ventilation of the Lungs*, 3rd edn. Blackwell, Oxford
Saidman L J, Smith N T 1984 *Monitoring in Anaesthesia.* Butterworth, New York
Sykos M K, Vickers M D, Hull C J 1981 *Principles of Clinical Measurement.* Blackwell, London

I. B. Monk

Animal Calorimetry

Animal calorimetry is the measurement of heat produced by, and the energy stored within, animals in the course of their metabolism. Distinction is made between direct calorimetry, in which heat emission by the animal is measured, and indirect calorimetry, in which the amounts of heat generated by the animal or laid down in the tissues as growth are inferred from quantitative measurements of metabolic reactants. Animal calorimetry is widely used in agricultural science to study the efficiency of food conversion by farm animals, and in clinical medicine.

Animal calorimetry originates from the work of Lavoisier who demonstrated that animal heat is a result of oxidative processes within the body. He enclosed a guinea pig in an ice chamber and found an equivalence between the amount of ice melted and the amount of carbon dioxide produced. A century later, Rubner's experiments on dogs gave practical proof of the close agreement between direct and indirect calorimetry. Such agreement is now seen as an inevitable consequence of the laws of thermodynamics, but uncertainty still remains as to the time period and other conditions over which comparisons are valid.

1. Indirect Calorimetry

Indirect calorimetry is founded on the principle of conservation of energy and the Hess law of constant heat summation. These laws justify the application of simple calorific values of foodstuffs (determined by bomb calorimetry on the laboratory bench) in calculating the level of an animal's metabolism from its respiratory gas exchange.

The word metabolism, meaning change, refers to the sum total of all the chemical reactions which take place in the body in the process of living. These include a host of anabolic (synthetic) and catabolic (degradative) reactions in which substrates are synthesized from food for later utilization at the same or at another body site. The net result of all these reactions is that food energy is converted into other forms of energy, especially heat.

1.1 Theory

Accurate indirect calorimetry is possible because: (a) practically all digestible food and all animal tissues and products consist mainly of either carbohydrate, fat, or protein, and (b) for each of these categories the amount of heat produced on oxidation is in fixed proportion to the amount of oxygen consumed. For example, combustion of cellulose in a bomb calorimeter yields the result:

$$\underset{(162\,g)}{C_6H_{10}O_5} + \underset{(134.4\,l)}{6O_2} \rightarrow \underset{(134.4\,l)}{6CO_2} + 5H_2O + \underset{(2850\,kJ)}{heat}$$

The amounts of heat and carbon dioxide produced per liter of oxygen consumed are virtually constant for all carbohydrate materials. The same is true for the complete oxidation of fats and for the combustion of proteins to carbon dioxide, urea and water. Values are summarized in Table 1. (The values quoted in Table 1 and throughout this article conform with those agreed at the 3rd Symposium on Energy Metabolism organized by the European Association for Animal Production in May 1964 (Blaxter 1965).) It is apparent from the last column of Table 1 that the amount of heat produced must be about 20 kJ per liter of oxygen consumed regardless of what mixture of carbohydrate, fat, and protein is being metabolized. This immediately provides a simple, approximate basis for indirect animal calorimetry. If the oxygen consumption in liters is represented by O_2 (at standard temperature and pressure), then the heat produced M (in kJ) is given by

$$M \simeq 20.5\,O_2$$

For more accurate measurements of heat production it is necessary to consider the balance between the metabolism of carbohydrate, fat, and protein. The nitrogen from metabolized protein appears almost exclusively as urea excreted in urine, hence the amount of protein metabolized, P, may be determined from the amount of nitrogen in the urine N using the agreed value formula:

$$P = 6.25\,N$$

Thus the amounts of oxygen and carbon dioxide associated with the metabolism of protein may be calculated. The remaining oxygen consumption and carbon dioxide production of an animal are associated with carbohydrate and fat metabolism, the relative proportions of which may be assessed from consideration of the respiratory quotient, RQ (equal to carbon dioxide produced/oxygen consumed). This calculation is tedious

Table 1
The amounts of heat and carbon dioxide produced and the amount of oxygen consumed in the oxidation of various foodstuffs

Substance oxidized	Heat produced (kJ)	O_2 consumed (liter)	CO_2 produced (liter)	Respiratory quotient	Heat produced/ O_2 consumed (kJ liter^{-1})
Carbohydrate (1 g)	17.6	0.829	0.829	1.000	21.2
Fat (1 g)	39.8	2.013	1.431	0.711	19.8
Protein (1 g)	18.4	0.957	0.774	0.810	19.2
CH_4 (1 liter)	39.6	2.000	1.000	0.500	19.8

and is further complicated in the case of ruminant animals that break down some foods by anaerobic fermentation in the rumen with consequent production of methane. The calorific constants for combustion of methane (CH_4) are given in Table 1.

Calculation was greatly simplified by Brouwer who pointed out that the information contained in Table 1 can be formulated into three equations. If the net amounts (in grams) of carbohydrate, fat and protein metabolized are K, F and P, respectively, and CH_4 represents the number of liters of methane synthesized by an animal, then the heat produced, M, must (from column 1 of Table 1) be:

$$M = 17.6\,K + 39.8\,F + 18.4\,P - 39.6\,CH_4$$

Similarly, the amounts of oxygen consumed and of carbon dioxide produced may be expressed by the analogous equations formed using the constants in columns 2 and 3 of Table 1. The three equations contain quantities which may be measured, namely O_2, CO_2, CH_4 and P (from urea nitrogen) and three unknown quantities K, F and M which can be eliminated by simple algebra leading to the solution:

$$M = 16.18\,O_2 + 5.02\,CO_2 - 2.17\,CH_4 - 5.99\,N$$

where N, urinary nitrogen, is expressed in grams.

This, the Brouwer equation, is much simpler to use than the traditional RQ method of calculating heat production, although the latter is still commonly employed. The problem of measuring an animal's heat production is thus reduced to the technical problem of determining the amounts of respiratory gaseous exchange.

1.2 Apparatus

Respiratory gaseous exchange is measured using an indirect calorimeter known as a respiration chamber. There are two types of chamber, using closed- and open-circuit principles. The first closed-circuit respiration apparatus was that of Regnault and Reiset in 1849 and its basic design differs little from that of modern instruments. Air is circulated by a pump in a closed loop through the animal chamber and through selective absorbers which remove carbon dioxide and water vapor (Fig. 1). Oxygen is admitted to the chamber from a calibrated reservoir (spirometer) so as to maintain a constant pressure. The amounts of carbon dioxide and water produced are obtained from the weight gains of the absorbers. Methane, if any, builds up inside the chamber and is determined from analysis of samples of chamber air taken at the start and end of the experimental period. These analyses also provide corrections for changes in oxygen and carbon dioxide concentrations within the chamber. The closed-circuit system is simple in concept and operation; its main disadvantage is that many hours are often needed to achieve accurately measurable weight changes. In some closed-circuit methods, the animal chamber is replaced by a face mask.

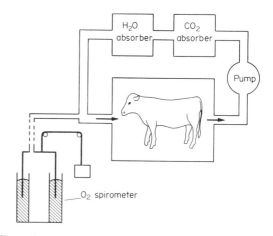

Figure 1
Closed-circuit indirect calorimeter

This imposes less restriction on the experimental subject, but avoidance of leaks becomes more difficult.

The open-circuit principle (Fig. 2) was first introduced practically by Pettenkofer in 1862. Many variations have since been used. The essential measurements are the quantity of air drawn through the system and the composition of inlet and outlet air. In some instruments gas samples are collected for later analysis: modern electronic gas analyzers which can be used "on line," in particular oxygen analyzers using the paramagnetic property of oxygen, have made the open-circuit calorimeter a very accurate and fast acting tool. Measurement of flow rate is the principal limitation on the accuracy of the method. The best type of flowmeter is one that measures volume flow rather than mass flow, as the effects of barometric pressure changes on the flowmeter and gas analyzers are then equal and opposite, and consequent corrections are reduced. The animal chamber may again be replaced by a face mask or hood over the head. Leaks are relatively unimportant provided that air flow and pressure are maintained at levels which ensure that all leaks are inward. For calculation of heat production it is convenient to use a rearranged form of the Brouwer equation, which expresses quantities in terms of differences in percentage concentration

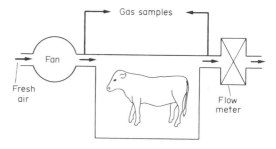

Figure 2
Open-circuit indirect calorimeter

between outlet and inlet air (ΔO_2, ΔCO_2, ΔCH_4) and outlet air flow rate F_o (in liters per second). The rate of heat production, $\dot{M}$ (in watts) is thus given by

$$\dot{M} = F_o(-204.7\,\Delta O_2 + 7.3\,\Delta CO_2 - 64.6\,\Delta CH_4) - 0.0693\,\dot{N}$$

where $\dot{N}$, the rate of nitrogen excretion, is expressed in g per 24 h. Note also that ΔO_2, the change in percentage oxygen concentration as air passes through the chamber, is negative, so that the first term is positive.

The ventilation rate of an open-circuit calorimeter must be kept low enough to produce accurately measurable gas concentration differences, say 0.5–1%. On the other hand, if it is very low, the depleted oxygen and raised carbon dioxide levels in the chamber can become harmful to the subject. Also, unless the chamber confines the animal very closely, the gas concentrations are slow to equilibrate and the response time is poor. The speed of response may be artificially increased by allowing for the rates at which gas concentrations are changing, by use of the equation:

$$\dot{M} = \left(F_o + V\frac{d}{dt}\right)(-204.7\,\Delta O_2 + 7.3\,\Delta CO_2 - 64.6\,\Delta CH_4) - 0.0693\,\dot{N}$$

where V is the volume of air in the system in liters. By this method, fluctuations in the level of metabolism of large animals can be followed with a time resolution of only a few minutes.

Another indirect method, confinement calorimetry, is a compromise between the closed- and open-circuit systems. The subject is confined in a completely closed chamber for a limited period, during which rates of change of gas concentrations in the chamber are noted. The chamber may be opened briefly, at intervals, for recharging with fresh air, thus making observations nearly continuous. This method avoids measurement of flow rate and does not use up gas absorbing materials. It has advantages for measurements on large animals.

1.3 Energy Balance

Indirect calorimetry is not confined to measurement of heat. It can also provide a complete analysis of the balance between the total energy supplied in the food, that laid down as body tissue in growth or as milk or eggs, that excreted as urine and feces, and that dissipated as heat. In deriving the Brouwer equation it was pointed out that there were two other unknowns, K and F, the net amounts of carbohydrate and fat metabolized. The word net is important in this context as it emphasizes that these quantities are not simply the amounts of dietary carbohydrate and fat metabolized; any change in body fat must also be considered. Similarly, P refers to the net amount of protein metabolized regardless of whether or not it is dietary in origin. Another approach to indirect calorimetry is the carbon–nitrogen balance technique. With this technique, all food, feces and urine are collected, weighed and analyzed for carbon and nitrogen content. To complete the carbon balance, it is still necessary to measure respiratory exchange of carbon dioxide. Excess carbon or nitrogen found to be entering but not leaving the body is assumed to have been retained in the body as growth (or, for example, milk and eggs, which can also be accounted for by analysis). Since all the retained nitrogen can be attributed to protein deposition, it is also possible to identify the amount of retained fat from that portion of the carbon not attributable to protein retention. Finally, the amount of energy E (in kJ) retained in the body is given by

$$E = 51.9\,C_0 - 19.4\,N_0$$

where C_0 and N_0 are carbon and nitrogen retained (in grams). During undernutrition or starvation, energy retention becomes negative. A complete analysis of the energy balance is thus possible.

2. Direct Calorimetry

Direct calorimeters are also classified into two types, adiabatic and isothermal. In adiabatic calorimeters (Fig. 3) the surfaces of the animal chamber are heavily insulated and heat from the animal is absorbed by a liquid-cooled heat exchanger situated inside the chamber. The flow of cooling liquid is regulated so that the ventilating air leaves the chamber at the same temperature as it enters. Sensible heat losses from the animal (i.e., radiative, convective and conductive losses) are found from the product of flow and temperature change of the cooling liquid, and insensible or evaporative heat loss from the product of humidity change, ventilating air flow, and latent heat of vaporization of water.

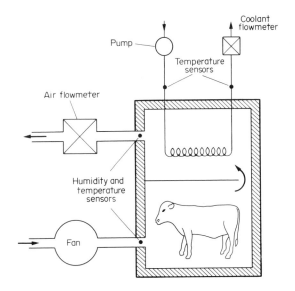

Figure 3
Adiabatic direct calorimeter

In isothermal calorimeters the sensible heat from the animal passes through the surfaces of the chamber into a surrounding water jacket. Heat flow is estimated from the temperature gradient which occurs across a surface barrier layer attached to the walls, floor and ceiling of the chamber. Early instruments of this type used an air gap as the surface barrier, but they were slow in response and the method fell out of use. It was revived by Benzinger in 1949 whose gradient-layer calorimeter used a surface barrier of thin plastic sheet, the mean temperature across which was measured by a network of thermocouples. He also introduced platemeters for recording evaporative heat output by measuring the heat from excess moisture condensing from the airstream on to gradient-layer-lined plates (Fig. 4). Air, first saturated with moisture vapor at T_d (the required dewpoint temperature), passes through the first platemeter (P1) which records the heat used to warm it to T_a (the required air temperature). Sensible heat added by the subject is removed either through the gradient layer surrounding the chamber or through that of platemeter P2, both of which are operated at T_a. The moisture added by the subject is recondensed in platemeter P3 where the air is returned to saturation at temperature T_d. The outputs of platemeters P1 and P3 added together thus represent the evaporative heat loss from the subject, and those of P2 and the chamber represent the sensible heat loss. The gradient-layer type of calorimeter, though complex to construct, is fast acting and provides direct output signals which are electrical analogs of the sensible, evaporative and total heat outputs from the subject.

A calorimeter suitable for large animals (e.g., cattle) has been constructed which combines the gradient-layer principle with open-circuit indirect calorimetry (McLean 1971). This permits simultaneous measurements of heat production (indirect) and heat loss (direct) with a time response for both of only a few minutes. This instrument has shown that in fixed temperature environments, heat production balances heat loss over periods as short as 10 min throughout the day, except for transient imbalances caused by feeding and drinking. Even the 20% increase in metabolism which occurs when an animal stands up after lying is mirrored by a 20% increase in heat loss. Such fine balance may not be maintained when the animal is subjected to a normal, fluctuating thermal environment.

Although it is theoretically possible to calculate the sensitivity of animal calorimeters, some form of calibration is nearly always necessary. For direct calorimetry, a known amount of heat may be generated electrically within the chamber, and if required part of this heat may be used to vaporize water so that both sensible and evaporative heat measurements may be calibrated. For indirect calorimetry, flowmeters and gas analyzers must be calibrated. Extreme accuracy is necessary for oxygen analysis as this measurement contributes by far the major proportion of the total estimate of heat production. Fortunately, fresh air provides an ideal calibration gas for oxygen analyzers. Since most gas analyzers are dependent on gas partial pressure, varying the total pressure of dried fresh air provides an easy method of calibration over the practical range. In the calibration of paramagnetic oxygen analyzers, allowance must be made for the diamagnetic effect of nitrogen when total pressure is altered. A combined calibration of both gas analyzers and flowmeters may be effected by injection and recovery of known amounts of the gases to be measured or of another gas (say nitrogen), so as to dilute the oxygen content of ventilating air. Finally, both direct and indirect calorimeters may be calibrated by burning a known quantity of alcohol inside the chamber; the main problem here is to ensure that complete combustion of the alcohol occurs.

Animal calorimeters, depending on their complexity, occasion the measurement of a great many parameters. Temperatures, heat flows, fluid flows, electrical power and gas concentrations may all have to be measured and processed at frequent intervals. This is an ideal application for automatic data logging. A small computer can be added to the system to scale, process and monitor the data, provide automatic control of calorimeter function, and issue alarms in the event of malfunction.

Bibliography

Blaxter K L 1962 *The Energy Metabolism of Ruminants.* Hutchinson, London

Blaxter K L (ed.) 1965 *Energy Metabolism (3rd Symp. on Energy Metabolism: European Association for Animal Production)*. Academic Press, London

Kleiber M 1961 *The Fire of Life: An Introduction to Animal Nutrition.* Wiley, New York

Lusk G 1928 *The Science of Nutrition*, 4th edn. Saunders, Philadelphia

McLean J A 1971 A gradient layer calorimeter for large animals. *J. Inst. Heat. Vent. Eng.* 39: 1–8

McLean J A 1972 On the calculation of heat production from open-circuit calorimetric measurements. *Br. J. Nutr.* 27: 597–600

McLean J A, Watts P R 1976 Analytical refinements in animal calorimetry. *J. Appl. Physiol.* 40: 827–39

Thorbeck G, Aersøe H (eds.) 1958 *1st Symposium on Energy Metabolism: Principles, Methods and General Aspects.* European Association for Animal Production, Rome

J. A. McLean

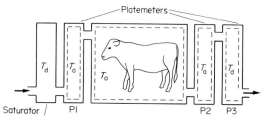

Figure 4
Gradient-layer calorimeter. T_a represents air temperature; T_d, dewpoint temperature; -----, heat-sensitive gradient layers

Artificial Ear

An artificial ear is a device designed to simulate the acoustical properties of an average human ear, and is used for the measurement and calibration of earphones. Practical devices approximate the desired characteristics to a greater or lesser degree and are usually termed acoustic couplers or ear simulators depending on the degree of simulation.

An acoustic coupler is a simple device consisting of a cavity of specified dimensions, terminated by a measuring microphone. The cavity has a volume approximating to that of the external ear enclosed by an earphone of the appropriate type. An ear simulator, however, incorporates acoustic networks in the form of resonant cavities and resistive elements which simulate the acoustic impedance presented by the eardrum coupled with the middle ear cavity and structure.

The demands in many fields for standard methods of measurement are such that provisional solutions in the form of acoustic couplers are often accepted while work proceeds towards the eventual goal of a universal artificial ear. Any practical device, however, cannot completely simulate an average human ear and only by direct measurement can the sound pressure level developed in an individual ear be determined with confidence.

1. General Applications

Artificial ears are required for measurements on earphones of three basic types: insert, supra-aural and circumaural, which are worn in, on and over the ear, respectively. Insert earphones are coupled to the ear canal, either directly or through an adaptor, usually in the form of an individually fitting earmold. Earphones of this type are used extensively with hearing aids. Supra-aural earphones are housed within an earcap which sits on the outer ear, an arrangement providing a more or less controlled degree of seal. Earphones of this type have enjoyed successful use for many years in the field of audiometry, for which well defined and reproducible methods of measurement are, of course, essential.

Circumaural earphones are widely used in communications but their popularity in the field of domestic hi-fi has strengthened the need for suitable methods of measurement, particularly to cover the frequency range above 5 kHz. An objective measurement of circumaural earphone performance requires the provision of a surface to mate with the earphone cushion to provide the required acoustic coupling. Measurements are sometimes made with a measuring microphone mounted flush with the surface of a rigid flat plate, the axis of the microphone being aligned with the axis of the transducer. Unfortunately in this basic configuration measurement repeatability is not good and the measured response does not correspond well with real-ear data. In theory these characteristics can be improved by the insertion of a small amount of acoustic damping

material into the cavity between plate and earcap in order to control any spurious resonances which may occur (Delany and Bazley 1979). It has, however, been difficult to find stable materials with the required acoustic properties. Circumaural earphones therefore pose the greatest measurement problems and it must be admitted that at present there is no satisfactory way to determine the effective real ear response other than by subjective means. If these measurement problems were indeed resolved, earphones of this type could find favor in audiometry because the lower acoustic source impedance would mean that ear canal sound pressure levels would be less dependent on the size of individual ears.

2. Artificial Ears in Audiometry

Pure-tone audiometry is based upon standard values for the normal threshold of hearing stated in terms of sound pressure levels developed in specified artificial ears or acoustic couplers. Generally these values have been obtained by transferring the earphone used to obtain the subjective thresholds to the chosen measuring device and reapplying for each frequency the averaged voltages required by individual thresholds. The major task is therefore one of maintaining these transferred calibration values for the audiometric frequencies in the range 125–8000 Hz. The nature of the acoustic load upon the earphone is of minor importance provided that an earphone of type exactly similar to that used to gather the original data is employed for subsequent threshold tests. The transferred coupler levels are known as reference equivalent threshold sound pressure levels and, in general, these will be different for each combination of earphone type and coupler design.

In the early days of electroacoustic audiometry there was significant disagreement over reference threshold levels, partly accounted for by differing methods of data collection and partly by the use of different measuring devices. For instance, in the UK, the standard BS 2497 (1954) gave pure-tone thresholds in terms of a contemporary artificial ear BS 2042 (1953). In this device a conical cavity of 3 cm^3 was terminated by a coiled tube containing graded wool fibers to provide an acoustic resistance of 120 Ω. The sound pressure level was measured by a probe tube microphone introduced into the cavity.

The situation, however, was normalized by the publication in 1964 of an international standard, ISO 389, which brought together the most reliable estimates of hearing threshold from a number of countries including the UK, USA and USSR. These estimates were expressed as sound pressure levels developed by certain standard earphones, in particular measuring devices maintained by the standards laboratories in those countries. ISO 389 was later augmented by an addendum, ISO 389A (1970), giving data for a number of earphone types in combination with the widely used 9-A coupler. The 9-A coupler was developed originally by the US National Bureau of Standards but the design was later standardized as the IEC provisional reference coupler (IEC 303, 1970). The enclosed coupler volume is approximately 6 cm³ and is terminated by a "one inch" condenser microphone. Although the earphone is attached without leakage, a small capillary is provided for static pressure equalization.

As the IEC provisional reference coupler does not provide the same acoustic load to an earphone as presented by the human ear, different calibration values are needed for each earphone type. A potential step forward came with the standardization of the so-called wide-band artificial ear (IEC 318, 1970), which was based on the work of Delany et al. (1967). All earphones of the common TDH 39 (+MX41/AR earcap) pattern may be calibrated to common threshold levels, appropriate values of which are presently in preparation. Figure 1 shows a sectional view of this artificial ear.

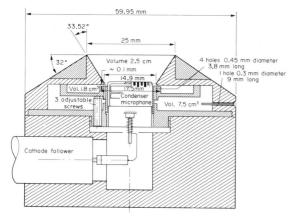

Figure 1
Sectional view of the International Electrotechnical Commission wide-band artificial ear

3. Artificial Ears for Hearing-Aid Measurements

The acoustic performance of hearing aids is measured in a suitable measuring device by placing the aid in a regulated sound field, either in anechoic space or in a purpose-built test enclosure, and recording its sound output for a specified range of frequencies (usually 100–5000 Hz) and input sound pressure levels (usually in the range 50–90 dB SPL). Generally measurements are made in accordance with some agreed procedure such as IEC 118-0 (1983).

IEC 118-0 stipulates the use of an occluded-ear simulator which is described in IEC 711 (1981) (see Fig. 2). This provides a two-branch acoustic network simulation of the acoustic impedance of the median human ear which gives a closer approximation to the real-ear characteristics than does the 2 cm³ acoustic coupler, specified in IEC 126 (1973) (see Fig. 3), and called up by IEC 118-7 (1983) for production inspection measurements. With the occluded-ear simulator an objective

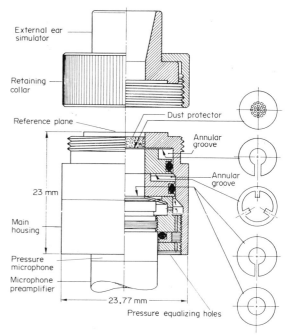

Figure 2
An example of one specific design of occluded-ear simulator

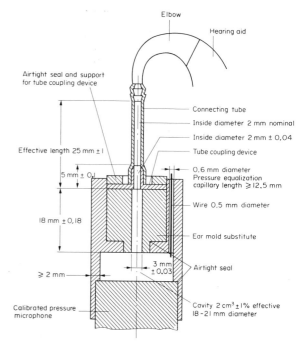

Figure 3
Arrangement of the 2 cm³ acoustic coupler for attachment to a behind-the-ear hearing aid

estimate can be made of the sound pressure level developed at the eardrum of a typical adult wearer for frequencies up to 10 kHz.

With either type of measuring device hearing aids not of the body-worn type require particular adaptations to the basic measuring device. For hearing aids worn behind the ear, a 25 mm length of 2 mm diameter tube is used to connect the hearing aid and coupler. All-in-the-ear models are mounted directly on the cavity with the earmold simulator removed.

4. Head-and-Torso Simulator

A more recent requirement has developed for an ear simulator to be used in free field, incorporated in a head and torso simulator (manikin), as a means of establishing the equivalent sound pressure level at the eardrum. Comparison of data obtained with a hearing aid being worn and the basic unaided manikin response can provide a measure of the insertion gain for the particular hearing aid. An IEC report is presently being prepared, describing such a manikin, based on anthropometric data presented by Burkhard and Sachs (1975). The manikin may be equipped with ear simulators conforming to IEC 711 (1981).

See also: Artificial Mastoid; Audiometers; Hearing Aids; Ear Anatomy and Physiology

Bibliography

British Standards Institution 1953 *An Artificial Ear for the Calibration of Earphones of the External Type*, BS 2042. British Standards Institution, London
British Standards Institution 1954 *The Normal Threshold of Hearing for Pure Tones by Earphone Listening*, BS 2497. British Standards Institution, London
Burkhard M D, Sachs R M 1975 Anthropometric manikin for acoustic research. *J. Acoust. Soc. Am.* 58: 214–22
Delany M E, Bazley E N 1979 *A Damped Flat Plate Coupler for Use in Calibrating Circumaural Earphones*, National Physical Laboratory Report No. AC 90. National Physical Laboratory, London
Delany M E, Whittle L S, Cook J P, Scott V 1967 Performance studies on a new artificial ear. *Acustica* 18: 231–37
International Electrotechnical Commission 1970 *An IEC Artificial Ear of the Wide Band Type for the Calibration of Earphones Used in Audiometry*, IEC 318. International Electrotechnical Commission, Geneva
International Electrotechnical Commission 1970 *IEC Provisional Reference Coupler for the Calibration of Earphones Used in Audiometry*, IEC 303. International Electrotechnical Commission, Geneva
International Electrotechnical Commission 1973 *IEC Reference Coupler for the Measurement of Hearing Aids Using Earphones Coupled to the Ear by Means of Ear Inserts*, IEC 126. International Electrotechnical Commission, Geneva
International Electrotechnical Commission 1981 *Occluded-Ear Simulator for the Measurement of Earphones Coupled to the Ear by Ear Inserts*, IEC 711. International Electrotechnical Commission, Geneva
International Electrotechnical Commission 1983 *Hearing Aids*, Pt. 0: *Methods for Measurement of Electro-Acoustical Characteristics*, IEC 118-0. International Electrotechnical Commission, Geneva
International Electrotechnical Commission 1983 *Hearing Aids*, Pt. 7: *Measurement of the Performance Characteristics of Hearing Aids for Quality Inspection for Delivery Purposes*, IEC 118-7. International Electrotechnical Commission, Geneva
International Organisation for Standardisation 1964 *Standard Reference Zero for the Calibration of Pure Tone Audiometers*, 389. International Organisation for Standardisation, Geneva
International Organisation for Standardisation 1970 *Additional Data in Conjunction with the 9-A Coupler*, 389, Addendum 1. International Organisation for Standardisation, Geneva

B. C. Grover

Artificial Joints, Implanted

The greatest advance in the treatment of arthritic disorders this century has been the development of an adequate hip prosthesis. Most other joints of the body have also had replacement implants developed. They are most commonly used in the treatment of patients with arthritis, with a joint damaged by injury, or occasionally with a tumor. Less commonly, they are employed for cosmetic purposes (e.g., in metacarpo-phalangeal joints).

Some of the major problems of endoprostheses are biological in character. For example, even in the highly successful hip arthroplasty, 1% of patients may die post-operatively, usually due to a pulmonary embolus from a deep-vein thrombosis. The second major hazard is infection, particularly delayed infection. The use of a clean-air tent within the operating theater has reduced this risk, but has not obviated it. This article deals with the physical aspects of implanted artificial joints.

1. Design

The specification is normally the responsibility of an orthopedic surgeon. Common requirements for artificial joints include strength of design, coupled with lightness and compactness; durable materials which are non-irritant; easy surgical application with a rock solid fixation; and easy extraction, leaving tissue available for salvage procedures. It is necessary to know the loads acting on the joints and their resultants, together with the articulating speeds. During the walking cycle the load on a normal hip joint rises to 5 times body weight, and in dropping 1 m the load at the knee rises to 25 times body weight. The range of angular movements (flexion, extension, abduction, adduction and rotation) and the stability of the joint have also to be considered (see *Gait Analysis*). The knee, for example, requires a range of flexion of 110° to facilitate rising from a chair. The elbow also has a rotatory component to its movement as well as flexion and extension. The extremely large range of motion at the shoulder poses problems in producing a stable shoulder joint. The most successful of these joints have reversed the natural geometry, using a captive head in which the ball of the joint is put on the glenoid component and the space within the bone of the tuberosity region of the humerus is used for the socket. Failure to appreciate the true modes of movement of the joint may result in a loosening of the prosthesis. Hinge replacements for the elbow have failed for this reason. Similar problems may occur with hinges at the knee. Although the knee has a hinge action, it also rolls and slides. These factors must be taken into account during the design.

2. Materials

The material used in the body must be nontoxic, as must any wear debris (i.e., it must be biocompatible). The material must have:

(a) adequate mechanical strength, both static and to loading in tension, compression and shear;

(b) an appropriate Young's modulus of elasticity;

(c) reasonable ductility;

(d) sufficient hardness;

(e) resistance to corrosion in the body's environment, whether used singly or in combination with other materials;

(f) long-term dimensional and chemical stability;

(g) wear resistance between the parts; and

(h) the capability of being sterilized.

The most commonly used combination is a metal such as stainless steel or chrome–cobalt and high-density, ultrahigh-molecular-weight polyethylene. Some prostheses use only one material, such as the Ring prosthesis and the McKee–Farrar prosthesis for the hip, which employ chrome–cobalt, or the Swanson prosthesis for the finger which is made of the ultrahigh-molecular-weight polyethylene. Polypropylene and silicone rubber have been used for finger joints, and ceramics are being increasingly used in Europe, with either metal or ceramic articulation.

Wear debris of Teflon, which was used because of its low coefficient of friction, induced granulomata, which often dissolved the neighboring bone. Concern has been expressed over the carcinogenic effect of metal debris in rats, but there is no evidence that this occurs in man.

Ductility is important in loaded implants because of the mechanical failure that can result from inevitable minor flaws in a stressed situation. Such failure is less likely to occur in a ductile material than a brittle one.

The ability of a ductile material to deform plastically enables local stress concentrations to be relieved by yielding, thereby avoiding the propagation of cracks. Ceramics have the disadvantage that they do not readily yield locally, so that concentration of load acting on a small area of the material produces extremely high local stresses, leading to crack propagation and failure.

Corrosion occurs mainly with the metallic components of a prosthesis. The reaction is electrochemical in nature. It is for this reason that only the same metals are used to articulate with each other. In consequence, the coefficient of friction is high and the likelihood of loosening is increased. Only those metals which have the ability to form a protective oxide layer are used in prostheses, and three such metals are titanium, chrome–cobalt alloy and, to a lesser extent, stainless steel (type 316). In a metal–plastic joint, wear occurs chiefly on the plastic by abrasive, adhesive and fatigue mechanisms. Abrasive wear is most important initially, but subsequently adhesive wear of the two components plays an increasingly important role. The type of loading, the counterface roughness and the manner in which the two surfaces move against each other all affect the wear process.

The advantages of high-density, ultrahigh-molecular-weight polyethylene are:

(a) it is the simplest polymer with no oxidizing or side groups;

(b) it has very long polymer chains which give it excellent mechanical properties and good wear resistance;

(c) it has a low coefficient of friction against steel;

(d) it does not appear to deteriorate chemically or mechanically, or swell appreciably in the body;

(e) it requires few additives;

(f) it is nontoxic and its wear debris is nontoxic; and

(g) it can be sterilized by γ-irradiation.

Problems may arise, however, from long-term environmental degeneration, wear and creep.

Attempts have been made to reinforce plastic materials with carbon fibers, e.g., the tibial plateau of the Sheehan knee prosthesis. It is hoped that by increasing the strength, the wear will be decreased. The problems of such reinforcement are that wear particles of carbon are sharp and can irritate the synovial tissue, as they do in cruciate ligament replacement. They also lodge in the draining lymph nodes and are not metabolized.

3. Wear

There are three main procedures which can be used in wear studies to predict the life of the prosthesis and to study basic wear mechanisms. The first is to monitor the performance, wear rate, and life of the complete assembly. This has been done in the human body for the rate of wear of polyethylene acetabular cups in the hip. The second is component testing in bearing or joint simulators in the laboratory. A number of these have been produced for the hip and the knee, but there are relatively few data using such simulators because of the difficulties and expense in designing, building and operating them. Thirdly, specimens and materials may be tested in the laboratory under well-controlled conditions. The most popular laboratory machines use test pins pressed against the periphery or flat plate of a rotating disk (pin-on-disk machine). If reciprocating motion is preferred the pin is simply held against a flat rectangular specimen which oscillates at a required amplitude and frequency. In certain circumstances a flat annular ring of a washer pressed against a rotating disk may be used.

The problem with the first method is that it calls for long-term investigation and, in isolation, fails to safeguard the patient against implants with excessive wear rates. The second method involves the design of complex and expensive equipment that can never achieve complete simulation of the working cycle and environment encountered in the body. The third method calls for great care in the laboratory testing of material specimens, and the data can sometimes be criticized on various grounds. The loads and speeds at which the tests are conducted are often constant, whereas physiological loads are impulsive and the speeds variable. Nevertheless, the pattern of wear achieved by wear-testing machines can be used to give useful predictions of bearing life in bioengineering applications. Using these methods it is estimated that the wear of a Charnley low-friction hip arthroplasty with stainless steel rubbing on high-density, ultrahigh-molecular-weight polyethylene would be 2 mm in 20 years. However, some modern knee prostheses, such as the polycentric design, may only have a life of two years.

4. Production

The basic forming processes may include forging, molding, casting and welding. Implant forgings are usually made by impact, or drop forging, in shaped dies. This is normally carried out on material which has been preheated to about 1150 °C, and produces forgings which have excellent mechanical properties. Hot, solid-phase compression molding, which is similar to forging, is used for the ultrahigh-molecular-weight polyethylene. The billet of preheated plastic is squeezed between shaped dies in a hydraulic press, and allowed to cool, whereupon it assumes the shape of the die.

The casting process which is nearly always used for endoprostheses is the lost wax investment process. A master (or pattern) is made from which further master molds are produced that are in effect female patterns. These are then used to make wax patterns which are identical to the master pattern. A number of wax patterns are then assembled to a central wax stem or "gate," and the whole assembly is coated with a smooth refractory substance by dipping or spraying. Successive coats of refractory material are applied until a strong shell is formed. The next stage involves heating the invested wax assembly so that the wax melts and runs out, hence leaving a hollow shell of refractory material. The shell is then subjected to a vacuum process to remove any residual wax. Molten metal is poured into the refractory shells, or molds, under very clean conditions in a vacuum, thus ensuring that no gas bubbles are present in the finished castings. The final stage is the removal of the refractory mold and the separation of individual castings from the gate.

The process most widely used for implant welding is argon-arc welding; electron-beam welding is also useful for this purpose. The machining and finishing processes may involve turning, milling, grinding, honing, lapping and polishing. The tolerances and surface finishes obtained for these processes are given in Table 1.

5. Fixation

In general, fixation to the bony skeleton may be achieved by mechanical fixation, by the use of cement and adhesives, or by the use of porous materials into which soft tissue or bone may grow.

Table 1

Tolerance and surface finish values for machining processes in artificial joint production

Machining process	Tolerance (mm)			Surface finish R_a [a]
	Nominal size 6 mm	Nominal size 18 mm	Nominal size 50 mm	
Forging	0.35–0.55	0.55–0.85	0.75–1.25	3.2–25
Investment casting	0.075–0.10	0.1–0.15	0.15–0.18	1.6–6.3
Drilling and rough turning	0.06–0.09	0.09–0.15	0.13–0.17	1.6–6.3
Milling	0.04–0.06	0.06–0.09	0.07–0.13	1.6–6.3
Fine turning	0.018–0.023	0.021–0.032	0.03–0.05	0.8–1.6
Grinding	0.003–0.006	0.004–0.009	0.005–0.013	0.1–0.8
Honing	0.0015–0.003	0.002–0.004	0.003–0.005	0.05–0.2
Lapping	0.0001–0.003	0.0001–0.003	0.0005–0.004	0.01–0.1

a R_a is a measure of the surface roughness in μm; the higher the number, the tougher the surface

Mechanical fixation is usually successful in fracture work, where, once the fracture has healed, the loads on the implant fixation are substantially reduced. The situation is entirely different in the fixation of an endoprosthesis: loads must be spread adequately so that local stresses do not exceed the capacity of the bone to withstand them. This type of fixation has been used at the hip (Austin Moore, Thompson and Ring designs), and to lesser extent at the knee (MacIntosh tibial plateau prosthesis and Walldius hinge) and in the fingers (Swanson prosthesis).

Adhesives have no real part to play in the fixation of joint endoprostheses, but cements have proved invaluable (Table 2). The most commonly employed is self-curing poly(methyl methacrylate), which acts as a grouting agent.

The feature which allows poly(methyl methacrylate) to be used as a prosthetic fixative is its ability to form an accurate cast of any surface with which it comes in contact while it is in the viscous state before polymerization. On polymerization the cast becomes locked into the methacrylate, irregularities on the outer surface of the prosthesis making the bond stronger. The optimal circumstances for the use of self-curing poly(methyl methacrylate) are that:

(a) the bone cavity is adequately prepared with a clean, rough, trabecular surface;

(b) the cement is well extruded into bone spaces;

(c) the cement is fully constrained;

(d) the prosthetic components are correctly positioned and centered;

(e) the loads imposed are mainly compressive; and

(f) there is low friction between the joint surfaces, minimizing torque.

Bone growth into a variety of porous surfaces has been stimulated. It occurs with sintered titanium fiber–metal composites and sintered Vitallium. For adequate bone ingrowth a porous material must have pore channels $>100\,\mu m$ in diameter with satisfactory pore continuity. Implantation requires intimate contact and lack of motion between the porous surface and host bone, which may prove difficult to achieve. Materials formed from Teflon and carbon have been used to coat the surfaces of implants to allow stabilization by tissue ingrowth. These differ from the porous metal surfaces in that they are of low modulus and, therefore, the conditions of intimate contact and lack of movement are less critical.

See also: Artificial Limbs and Locomotor Aids: Evaluation; Implanted Prostheses: Tissue Response; Bone: Mechanical Properties; Prostheses, Myoelectrically Controlled

Table 2
Compositions of various bone cements used as prosthetic fixatives

CMW bone cement	
(a) liquid	
methyl methacrylate	98.78%
ascorbic acid	0.01%
N,N-dimethyl-p-toluidine	1.20%
hydroquinone	5–10 ppm
(b) powder	
poly(methyl methacrylate)	97.00%
residual benzoyl peroxide	3.00%
(c) radiopaque agent	
barium sulfate BP	5 g
Surgical Simplex plain bone cement	
(a) liquid	
methyl methacrylate	97.4 vol%
N,N-dimethyl-p-toluidine	2.6 vol%
hydroquinone	75 ± 15 ppm
(b) powder[a]	
poly(methyl methacrylate)	16.7 wt%
methyl methacrylate–styrene copolymer	83.3 wt%
Surgical Simplex radiopaque bone cement	
(a) liquid	
methyl methacrylate	97.4 vol%
N,N-dimethyl-p-toluidine	2.6 vol%
hydroquinone	75 ± 15 ppm
(b) powder[a]	
poly(methyl methacrylate)	15.00 wt%
methyl methacrylate–styrene copolymer	75.00 wt%
barium sulfate USP	10.00 wt%

a The powder contains in chemical combination a peroxidic activity sufficient to initiate, in conjunction with the N,N-dimethyl-p-toluidine in the liquid, the polymerization reaction when the liquid is mixed with the powder

Bibliography

Atkinson J R, Dowling J M, Cicek R Z 1980 Materials for internal prostheses: The present position and possible future developments. *Biomaterials* 1: 89–96

Charnley J 1970 *Acrylic Cement in Orthopaedic Surgery.* Livingstone, London

Dowson D, Wright V (eds.) 1981 *An Introduction to the Biomechanics of Joints and Joint Replacement.* Mechanical Engineering Publications, London

Dumbleton J H 1981 *Tribology of Natural and Artificial Joints.* Elsevier, Amsterdam

Freeman M A R 1976 Current state of total joint replacement. *Br. Med. J.* 2: 1301–4

Rushton N, Dandy D J, Naylor C P E 1983 The clinical arthroscopic and histological findings after replacement of the anterior cruciate ligament with carbon fibre. *J. Bone Jt. Surg. Br.* Vol. 65: 308–9

Seedhom B B, Dowson D, Wright V 1973 Wear of solid phase formed high density polyethylene in relation to the use of artificial hips and knees. *Wear* 24: 33–51

Weightman B 1977 *The Scientific Basis of Joint Replacement.* Pitman Medical, Tunbridge Wells, UK

Williams D F (ed.) 1979 *Biocompatibility of Implant Materials.* Sector, London

Wright V 1982 Measurement of joint movement—An overview. *Clin. Rheum. Dis.* 8: 523–32

Wright V, Dowson D 1977 *The Evaluation of Artificial Joints with Particular Reference to Joint Simulators.* Sector, London

V. Wright

Artificial Limbs and Locomotor Aids: Evaluation

The evaluation of any artificial limb or locomotor aid is a comprehensive disciplined procedure aimed in general at assessing the value of the aid. Evaluation is an essential element in a process which starts with the initiation of an idea and finishes, hopefully, with the availability of a device or procedure in the clinic, useful to the patient and with the involved professionals trained in its application. The entire process involves research, development, evaluation, production, education and service. Unfortunately, the evaluation procedure is often omitted, or is carried out in only a partial or unstructured way. Although a great deal of money is spent each year in the UK on the provision of artificial limbs and locomotor aids, evaluation with the obvious essential questions it seeks to answer attracts only a very small proportion of this.

1. Evaluation

Evaluation seeks to establish whether the procedure or device satisfies the needs for which it is designed: (a) functionally, (b) clinically and (c) economically; it also seeks to determine how the device compares with other devices intended to satisfy the same need. This, in turn, implies an initial screening procedure to justify the funding and resources normally required for an effective evaluation program. There must be a clear statement of the function and other properties which are expected of the device, as this forms the basis of the evaluation.

Having defined the claims it becomes possible to examine the factors which must be considered in the evaluation, such as improved mobility, enhanced comfort, reduced energy cost, improved appearance, ease of manufacture, reduced cost and reduced weight. A decision may then be made as to those assessments and measurements which are required.

In general, the device must be measured against a control for the evaluation to be meaningful. Normally this would involve a comparison with other devices intended for the same purpose. However, if no other device exists, the evaluation may compare the function and characteristics of the device against an "ideal" as specified by the clinic team.

Evaluation is composed of a number of interrelated activities designed to determine the characteristics of the device and its clinical effectiveness. The determination of the device's characteristics for the most part involves objective measurement of factors such as mass properties, strength, corrosion resistance, electrical safety, toxicity, cost of material and cost of application. Conversely, clinical testing often depends heavily on subjective measurement by the patient, the physician, the prosthetist, the therapist and other members of the clinic team.

There are, of course, measuring techniques available for assessing certain aspects of patient performance, e.g., analysis of expired gas to measure energy consumption, heart-rate monitoring to measure energy cost, force plate/video studies to provide kinetic data, and electromyography to assess muscle function. It must be said that none of these measuring systems has yet provided more than background information against which clinical trials have been carried out. However, the importance of such measurement is that it provides a permanent record which may then be absolutely compared, within the limits of the system, to measurement at a different time.

The subjective assessment by the patient and the clinic team must be carefully structured. In general it utilizes questionnaires, structured conversations or a combination of both. The design is critical and requires careful preparation, with all the factors of importance being identified from the outset. Questions must not lead the subject towards any specific answer, but must be neutral. Scoring systems, where comparisons are sought, should generally be on a three- or five-point scale. This approach makes later comparisons easier than the use of phrases or words which do not have an identified relative meaning, or which do not employ a linear scale. A scoring system which employs a combination of questionnaires and structured conversations for the assessment of amputee performance is described by Day (1978). This type of systematic approach is a prerequisite to an effective evaluation.

2. Protocol

For a specific evaluation, the first step is the establishment of a formal protocol. This is a clear statement of the aims and means of the evaluation and normally includes the following:

(a) The detail of the device to be evaluated. This consists of a functional specification and, if appropriate, manufacturing and fitting instructions.

(b) Details of the devices or characteristics against which the evaluation is to be completed.

(c) The centers participating and the organizational procedures.

(d) The number of patients and devices to be involved and the timetable.

(e) The tests, assessments and measurements to be carried out, the manner of execution and the details of the records to be kept.

(f) The procedure for reporting.

3. Evaluation Centers

There is a general philosophy which may be applied to any evaluation in this field. It is unlikely that an evaluation carried out by the developers of a device will have universal acceptance. Naturally they will have made their own assessment, but if the device is to be made generally available, and if the evaluation is to be accepted by others as a guide to use and prescription, centers other than that where the device was developed must take part in the formal evaluation. They must have the acknowledged expertise, facilities, staff and willingness to be involved. The developers must instruct the participating centers in the techniques and procedures involved, and should then have a continuing involvement through the duration of the evaluation. An appropriate organization, or one of the participating centers, must adopt a coordinating role. Among the few documented examples of this approach are the evaluation of a sensory feedback device (Wannstedt and Craik 1978), and the evaluation of pneumatic orthoses (Quigley 1977). These examples serve as a model for the implementation of this philosophy.

4. Control of Variables

It cannot be emphasized too strongly that great care must be taken in identifying and, where possible, controlling the variables. The most variable element of all is the patient. As far as possible, patient selection must relate to the characteristics of the device under evaluation and take into account the claims of the developer. Female patients, for example, may in certain circumstances be much more concerned with appearance than male patients. Enfeebled patients may not be able to tolerate the weight of a device which substantially improves the function of a young active patient. There are two different aspects which influence patient selection, namely: is the device designed for a particular group of patients; or, alternatively, is one aim of the evaluation to establish the patient groups for which the device is suitable? In the first case the patients must conform to certain criteria, and in the second they must be representative of a wider, identified cross section.

Where the device to be tested is in intimate patient contact, as in a prosthesis, the influence of the attachment, or socket, must be eliminated or controlled. It is, for example, meaningless to evaluate different prosthetic knee mechanisms for above-knee amputees without first considering the effect of socket fit. It is only with a well designed and fitted socket that the amputee is able to control and use a knee mechanism effectively and make subjective judgements about its function. Objective measurements of gait characteristics would be similarly affected. In an evaluation of modular below-knee prostheses where each patient in the evaluation was fitted with six different types of prostheses consecutively, Solomonidis (1975) describes the use of a master mold to ensure that each socket is identical.

5. Measuring and Reporting the Result

In selecting the number of centers to be involved, the number of patients and the number of devices, the question of statistical validity must be examined. It must, however, be borne in mind that the evaluation should be completed within a timespan which renders the results useful.

The features of and claims made for the device under evaluation will dictate the factors to be measured in a particular case. As previously stated, these will involve physical measurement and subjective assessments, all of which must be carefully recorded. The analysis naturally focuses on those factors which in the first place indicated the need for an evaluation. Often, however, unexpected areas of interest emerge which should also be identified and reported.

The question of reporting deserves special mention. Much money is spent on evaluations, on clinical trials and on other related activities often with, it must be said, inadequate reporting procedures. For an evaluation to realize its full value, not only must the funding authority be satisfied with the results, but the material must be presented in such a way, and in appropriate places, that all the different groups involved may apply the results to the benefit of the patient. In this respect it should be recognized that the educators should be involved in the evaluation process from an early stage. This is the most effective way to ensure that a successful conclusion of an evaluation may be followed by an early introduction of the device or procedure into clinical service.

6. Concluding Remarks

Evaluation is a complex procedure which combines physical testing, functional testing, clinical trials, and cost assessments. It requires detailed planning, and is usually a lengthy, expensive exercise. However, it is only by promoting and encouraging this activity that some order and logic can emerge from the chaos which exists in many aspects of the practice in this field. The cost and quality benefits in improved design, prescription and clinical practice from an ongoing national evaluation program would be immense. An international evaluation program would multiply the benefits several fold. Evaluation can be the only sound basis for practice in this important area.

See also: Prostheses, Myoelectrically Controlled; Artificial Joints, Implanted

Bibliography

Day H J B 1978 Clinical factors influencing amputee activity. Its description and numeration. *Proc. Standards for Lower Limb Prostheses*. International Society for Prosthetics and Orthotics, Copenhagen, pp. 235–53
Hughes J 1972 Evaluation in Scotland. In: Wilson A B Jr 1972 *Report Int. Conf. Prosthetics and Orthotics*. International Society for Prosthetics and Orthotics, Copenhagen, pp. 338–42
Hughes J, Paul J P, Kenedi R M 1970 Control and movement of the lower limbs. *Mod. Trends Biomech.* 1: 147–79

Le Blanc M A 1973 Clinical evaluation of a comprehensive approach to below-knee orthotics. *Orthot. Prosthet.* 27(2): 1–16

Murdoch G, Hughes J 1973 Clinical and biomechanical aspects of current prosthetic practice. In: Kenedi R M (ed.) 1973 *Perspectives in Biomedical Engineering.* Macmillan, London, pp. 67–72

Quigley M J 1977 Evaluation of the Ortho-Walk Type B pneumatic orthosis on thirty-seven paraplegic patients. *Orthot. Prosthet.* 31(1): 29–50

Solomonidis S E (ed.) 1975 *Modular Artificial Limbs— Below-Knee Systems.* HMSO, Edinburgh

Solomonidis S E (ed.) 1980 *Modular Artificial Limbs— Above-Knee Systems.* HMSO, Edinburgh

Stallard J, Rose G K, Tait J H, Davies J B 1978 Assessment of orthoses by means of speed and heart rate. *J. Med. Eng. Technol.* 2: 22–24

Wannstedt G, Craik R L 1978 Clinical evaluation of a sensory feedback device: The limb load monitor. *Bull. Prosthet. Res.* 10-29: 8–49

J. Hughes

Artificial Mastoid

An artificial mastoid is a device used for the measurement and calibration of electromechanical bone vibrators worn either for bone-conduction audiometry or as output transducers for hearing aids. Designs are based on averaged data for the mechanical impedance presented to an appropriate transducer applied to the mastoid process. An alternative term, mechanical coupler, is commonly used in recognition of the difficulties involved in reducing human-mastoid impedance data to an easily simulated form.

1. Bone-Conduction Audiometry

Bone-conduction audiometry plays an important role in the clinical diagnosis of hearing disorder, particularly in the differentiation of conductive and sensorineural impairments. In the case of bone-conduction audiometry, tones are presented to the subject via an electromechanical vibrator applied to the mastoid bone. For some tests, forehead placement is used. Standardized values for the normal threshold of hearing by both air and bone conduction are therefore required if consistent judgements are to be made on different persons in different clinics. While an individual subjective calibration could be provided for each bone vibrator based on the average of thresholds obtained from a group of normally hearing persons, only an agreed objective calibration can assume an easily verified standard of performance without the need for constant selection of test crews. In practice, however, international agreement on suitable reference equivalent threshold values is only now being achieved, despite an appropriate mechanical coupler having been standardized since 1971.

2. Mechanical-Coupler Design

Early designs by Carlisle and Mundel (1944) and others employed simple viscoelastic pads to simulate the mechanical impedance presented to a bone vibrator. Despite certain refinements, these early designs suffered stability problems with the mechanical properties of the impedance matching elements and none found general favor. In 1960, Weiss described a new design in which damping was provided by means of a piston moving in an air-filled chamber. The device, which was claimed to overcome the problems of materials stability, was not widely used, partly owing to its relative fragility and partly because the mechanical load presented to a bone vibrator did not vary with applied static force in the same way as for a vibrator worn on the head.

More recently, a mechanical coupler has been standardized internationally (IEC 373, 1971). The specification follows a design by Whittle and Robinson (1967) which owed much to the measurements of human-mastoid impedance gathered by Dadson et al. (1954) and modified by later studies. The design, shown in Fig. 1, returns to the use of viscoelastic elements in the form of a butyl rubber–silicone rubber laminate, clamped to a 3.5 kg inertial mass which contains a piezoelectric transducer. Vibrators are applied to the rubber pad with a static force of 5.4 N and all vibrators having a nominal contact area of 175 mm^2 can be accommodated. Despite small adjustments now required by manufacturing experience gained over ten years, the specification will assume greater importance with the publication of appropriate reference equivalent threshold vibration levels (which will probably be stated in terms of force) for use in conjunction with this device.

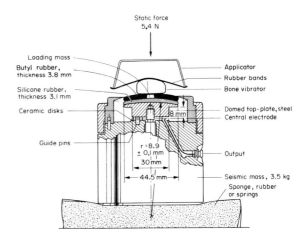

Figure 1
Constructional details of the IEC mechanical coupler. (Reproduced by permission of the International Electrotechnical Commission, Geneva, from whom complete documents can be obtained.)

3. Hearing-Aid Measurements

A further requirement exists for an artificial mastoid (mechanical coupler) for the evaluation of acousto-mechanical performance characteristics of hearing aids with bone conduction output. Measurement methods on hearing aids of this type have developed only slowly, perhaps because only a very small percentage of hearing aids employ coupling of this type. However, with the publication of IEC 118-9 (1985) it is now possible to use internationally agreed procedures to express the vibratory output from a hearing aid with reference to specified acoustic input conditions. The mechanical coupler described in IEC 373 is suitable for this particular application.

See also: Artificial Ear; Audiometers; Acoustic Impedance Audiometry; Hearing Aids

Bibliography

Carlisle R W, Mundel A B 1944 Practical hearing-aid measurements. *J. Acoust. Soc. Am.* 16: 45–51

Dadson R S, Robinson D W, Greig R G P 1954 The mechanical impedance of the human mastoid process. *J. Appl. Phys.* 5: 435–42

International Electrotechnical Commission 1971 *An IEC Mechanical Coupler for the Calibration of Bone Vibrators Having a Specified Contact Area and Being Applied with a Specified Static Force*, IEC 373. International Electrotechnical Commission, Geneva

International Electrotechnical Commission 1985 *Characteristics of Hearing Aids with Bone Vibrator Outputs*, IEC 118-9. International Electrotechnical Commission, Geneva

Weiss E 1960 An air damped artificial mastoid. *J. Acoust. Soc. Am.* 32: 1582–88

Whittle L S, Robinson D W 1967 An artificial mastoid for the calibration of bone vibrators. *Acustica* 19: 80–89

B. C. Grover

Artificial Membranes

Artificial membranes have made an important contribution to medicine. This article considers the utilization of the properties of artificial membranes for medical purposes. Relevant mass transfer aspects are discussed and the exploitation of membrane properties is illustrated by reference to particular medical applications. These applications cover solute and fluid removal in blood purification, gas transfer in blood oxygenation, combining membranes with adsorbents in artificial cells, controlled release, and skin substitutes.

1. Mass Transfer

This discussion of mass transfer concerns membranes that permit solute and fluid transfer in the artificial kidney and gaseous diffusion in the artificial lung.

Although such membranes operate in contact with complex solutions, permeation properties may be approximated by phenomenological mass transfer models, the parameters of which are obtained from *in vitro* testing with simple fluids.

In the phenomenological approach, the membrane is treated as a zone between two uniform fluid phases. The solute flux across the membrane is linearly related to driving forces causing the flux. One of the simplest models is that derived from Fick's first law:

$$J_A = -D_{AB} \nabla C_A \tag{1}$$

where J_A is the molar flux of species A relative to the molar average velocity of the system, ∇C_A is the gradient in concentration of A within the membrane and D_{AB} is the binary diffusion coefficient. In Eqn. (1), it is assumed that the membrane is a pseudo binary homogeneous phase, that the concentration of A is small, and the only driving force is the concentration difference. For a dialysis membrane, where the principal result is solute flow, the solute would be species A and the membrane and the solvent contained therein are taken together as the pseudo, second species B. If the membrane is thin such that the concentration profile is assumed linear and D_{AB} is independent of concentration, Eqn. (1) becomes:

$$J_A = D_{AB} (C_{A1} - C_{A2})/t_m \tag{2}$$

where C_{A1} and C_{A2} are, respectively, the high and low concentrations at the inner faces of a membrane of thickness t_m. In practice, the concentration C_A^0 in the uniform fluid phases bounding the membrane is measured. C_A^0 is related to C_A by the equilibrium partition coefficient K_A, as $K_A = C_A / C_A^0$.

Assuming Henry's law holds (i.e., K_A is independent of C_A^0) substitution for C_A gives:

$$J_A = D_{AB} K_A (C_{A1}^0 - C_{A2}^0)/t_m = P_r (C_{A1}^0 - C_{A2}^0)/t_m \tag{3}$$

where P_r is the permeability coefficient. Equation (3) has been widely used to describe the permeation of gases through hydrophobic homogeneous films such as polysiloxane and polyethylene that are above the glass transition temperature. For external gas phases, the ideal conditions assumed above are approached with the concentration C_A^0 replaced by the gas partial pressure and K_A replaced by S, the Bunsen solubility coefficient.

For liquid permeation systems such as dialysis, K_A and D_{AB} may be dependent on C_A^0. In addition, J_A may exhibit an approximate inverse proportionality to the membrane thickness. Under these circumstances, it is usual to replace P_r/t_m by P_m, the membrane permeability factor, and Eqn. (3) becomes:

$$J_A = P_m (C_{A1}^0 - C_{A2}^0) \tag{4}$$

Equation (4) cannot be applied when the solute flux is coupled with solvent flow, as in ultrafiltration, where the principal result is the transfer of solvent across the membrane under a pressure gradient driving force.

For a ternary membrane system, the net solute flux J_A and the net volume flux J_V across the membrane are given by:

$$J_A = P_m \Delta C_A^0 + \overline{C_A}(1 - \sigma) J_V \qquad (5)$$

$$J_V = L_p (\Delta P - \sigma \Delta \pi) \qquad (6)$$

where ΔC_A^0 and ΔP are the respective solute concentration and hydrostatic pressure differences across the membrane, $\overline{C_A}$ is the average solute concentration in the membrane, L_p is the hydraulic conductivity coefficient and σ is the Staverman reflection coefficient. $\Delta \pi$ is the osmotic pressure that would be exerted across the membrane if the membrane was completely impermeable to the solute. The derivation of Eqns. (5) and (6) and experimental methods for the determination of L_p and σ are given in Klein et al. (1977a).

Despite their apparently simple form, the phenomenological coefficients (P_m, P_r, L_p, σ) have proved useful in the theoretical analysis of artificial kidneys and artificial lungs. The coefficients have also served as a basis for the interpretation of membrane structure and in the characterization of membranes used in artificial cells, controlled release and as skin substitutes.

2. Extracorporeal Blood Purification

The importance of membranes to extracorporeal blood purification is derived from the fact that they permit separation and purification processes under conditions appropriate to the nature and properties of blood. The advantages of applying membrane separation processes to extracorporeal blood purification have found partial fulfilment in the development of the artificial kidney or hemodialysis procedure. The membrane separation processes relevant to the artificial kidney are dialysis, osmosis, and ultrafiltration.

Dialysis occurs when a permeable solute passes through a membrane separating solutions of different concentration, under the action of the concentration gradient driving force. In a multisolute system, each solute diffuses across the membrane under the influence of its particular concentration gradient.

Osmosis takes place when a solvent passes through a membrane separating a solvent from a solution containing an impermeable solute, under the action of the concentration gradient driving force. Osmotic pressure is defined as the excess pressure which must be applied to the solution to prevent solvent entry. Therefore, osmotic pressure is not a simple solution property since it is dependent on membrane and solute characteristics.

Ultrafiltration involves the use of a pressure gradient driving force to transfer solvent through a membrane. As fluid passes through the membrane, it may carry solute with it and this solute drag may contribute to an overall solute removal.

In the artificial kidney blood flows within a membrane envelope on the outside of which is passed a balanced isotonic solution, termed the dialysate. Solutes and water are removed from the blood by dialysis and ultrafiltration. Osmosis is generally avoided but can be used to accomplish water removal by the addition of solute to the dialysate. The membrane is polymeric and the envelope is produced from flat sheets, tubing or hollow fibers.

The relationship between solute flux and membrane permeability in the artificial kidney is described by Eqn. (5). An important feature in the artificial kidney is that membrane permeability P_m is not the only factor determining solute diffusion. The presence of liquid layers on either side of the membrane leads to the following relationship between P_m and the overall permeability P_o:

$$1/P_o = 1/P_m + 1/P_b + 1/P_d \qquad (7)$$

or in terms of mass transfer resistances

$$R_o = R_m + R_b + R_d \qquad (8)$$

where b and d denote blood and dialysate, respectively.

Volume flux in the artificial kidney is obtained from Eqn. (6). In the absence of osmotic effects:

$$L_p = J_V / \Delta P \qquad (9)$$

Equation (9) is an acceptable approximation for dilute solutions or when $\sigma \Delta \pi \ll \Delta P$.

The degree of ultrafiltration depends on the pressure gradient across the membrane and the ultrafiltration characteristics of the membrane. The applied pressure gradient is limited by the wet strength of the membrane and the design of any membrane-support structure. The ultrafiltration characteristics are determined by membrane structure, thickness, and swelling characteristics.

Dialysis and ultrafiltration are used in the artificial kidney as a treatment for chronic renal failure. Uncertainty exists regarding the identity of solutes which should be removed in this clinical situation. Suggestions have been made that, in addition to the removal of low-molecular-weight solutes, it is important to remove solutes with molecular weights in excess of 1500. With membranes in current clinical use, permeability is in inverse proportion to the size of the diffusing solute molecule and so a relationship between the membrane resistance R_m and solute molecular weight can be assumed. While R_m increases with solute molecular weight, there is no comparable increase in R_b and R_d, and for higher-molecular-weight solutes R_o and R_m can be taken as equal. Therefore, if treatment effectiveness is related to the removal of higher-molecular-weight solutes, it should be possible to reduce treatment time and compensate by increasing membrane area, for a given membrane. Indeed, this approach has been adopted clinically despite lack of verification on the role of higher-molecular-weight solutes in chronic renal failure.

If the treatment time is reduced and the membrane area increased, however, a limiting factor, clinically, is the inability to remove excess water in the shorter time, since rapid ultrafiltration often leads to hypotension. To overcome this difficulty, treatment involving an additional ultrafiltration stage has been proposed. Such treatment is termed sequential ultrafiltration–hemodialysis. The additional ultrafiltration stage may be carried out before or after the normal hemodialysis procedure.

A closer approximation to the natural operation of the human kidney is found in the hemofiltration process which involves ultrafiltration, convective solute transfer and dilution with physiological saline. By using membranes with different transport characteristics, hemofiltration performs a stronger ultrafiltration stage than hemodialysis, the loss of excess fluid being compensated for by the added saline. Dilution may take place before or after the ultrafiltration stage.

The application of extracorporeal blood purification with membranes to the removal of protein-bound solutes, as in hepatic support, or the removal of high molecular-weight species, such as immune complexes and macroglobulins, may prove successful with the technique of plasma separation. Using more permeable membranes than hemodialysis, plasma separation ensures the separation of plasma from whole blood by filtration. The separated plasma may be purified, or replaced by fresh plasma, and returned to the body.

3. Blood Oxygenation

Artificial lungs, commonly termed oxygenators, are basically oxygen and carbon dioxide gas exchangers. Oxygenators replace the respiratory function of the natural lung during open-heart surgery or provide long-term support for a failing lung. For adult requirements, approximately $5 \times 10^{-6} \, m^3 O_2 \, s^{-1}$ must be delivered and $4.2 \times 10^{-6} \, m^3 CO_2 \, s^{-1}$ removed, measured at STP. The oxygenators fall into two categories, the direct contact type and the membrane type. In the direct contact type, the ventilating gas is introduced directly into the blood, usually by bubbling, and in the membrane type, a membrane is interposed between blood and gas phases. Using pure oxygen at atmospheric pressure as the ventilating gas, the partial pressure gradients available for exchange are 93 and 6.7 kPa for oxygen and carbon dioxide, respectively.

Oxygenation membranes in current use or under development may be considered under three categories. These are: (a) homogeneous, (b) macroporous, and (c) composites of types (a) and (b).

Prior to the introduction of polysiloxane membranes, the use of homogeneous crystalline polymers such as polytetrafluoroethylene and polyethylene dictated that large membrane areas be used for adequate gas exchange. Polysiloxane possesses oxygen and carbon dioxide permeability coefficients several orders of magnitude greater than the other materials, primarily because of a lack of crystallinity. The superior gas permeability of polysiloxane elastomers is offset by poor mechanical strength, which necessitates incorporation of a reinforcing silica filler or casting the membrane onto a reinforcing fabric. Carbon dioxide is a more condensable gas than oxygen and thus exhibits a 5–6 fold greater solubility in polysiloxane. Since D_{AB} is similar for both gases, P_r for carbon dioxide is correspondingly higher ($P_r = D_{AB}S$). This selectivity compensates to some extent for the disparity in available partial pressure gradients. For laminar blood flow parallel to the membrane surface, the diffusional resistance of the blood phase accounts for 80–90% of the total transfer resistance in polysiloxane membrane oxygenators. Attempts have been directed towards reducing the blood film resistance by convective mixing of the blood adjacent to the membrane. Progress has encouraged the search for membranes more permeable than the polysiloxane elastomers and macroporous membranes have been investigated.

Meares (1976) defines macroporous membranes as those having permanent pores (> 5 nm diameter), which are deliberately introduced into the membrane during manufacture. Polytetrafluoroethylene and polypropylene macroporous membranes are used in commercial oxygenators. The polypropylene membrane is 25 µm thick and has elongated pores 200 nm long and 20 nm wide. Water flow through the membrane is achieved only at pressures greatly in excess of those encountered in the artificial lung. The pores, which extend from one surface of the membrane to the other, are derived from a precursor composed of microcrystalline lamellae. The precursor is uniaxially stretched to open up voids between the lamellae. For approximately isobaric conditions, the primary transport mode will be free diffusion of oxygen and carbon dioxide through the pores. Permeability coefficients obtained for the polypropylene membrane under isobaric conditions and with external gas phases are 1–2 orders of magnitude greater than those for polysiloxane. Unlike polysiloxane, the macroporous membrane shows little selectivity towards oxygen and carbon dioxide, since these gases are of similar molecular size. Keller and Shultis (1979) have measured oxygen transmission rates for both polymer membranes but with liquid in contact with one side. The polypropylene membrane gave transmission rates only 1–1.65 times that of polysiloxane (63 µm thick) suggesting that wetting of the pores occurs. This effect may be overcome with a composite membrane structure, which inhibits pore wetting.

Ketteringham et al. (1975) have described a composite membrane consisting of a 25 µm layer of homogeneous polysulfone cast onto the macroporous polypropylene membrane cited above. Tests made on blood in contact with the membrane indicated that the composite membrane would give at least a two-fold improvement in transmission rate in comparison to practical polysiloxane membranes.

In summary, composite membrane structures would appear to offer improved permeation rates by enabling advantage to be taken of the reduction in blood phase mass transfer resistance. Additionally, relatively impermeable materials possessing good biocompatibility may be cast onto the porous substrate to produce an acceptable oxygenation membrane.

4. Artificial Cells

Solute removal by adsorption or chemical reaction offers potential benefits to blood purification. Artificial cells utilize materials such as activated charcoal, ion-exchange resins and enzymes to give more rapid and extensive treatment than that possible by hemodialysis. However, most substances of interest cannot be used in direct contact with blood. Artificial cells overcome this problem by encasing the reactive core material in a semipermeable membrane, thus combining substances capable of rapid or selective solute removal with the blood compatibility and permeability properties of membranes.

Artificial cells range from microcapsules of 100 μm diameter, containing solutions or suspensions, to solid adsorbent granules > 1 mm diameter, coated with polymer. The solute removed from the blood may be completely trapped in the cell or, alternatively, after reaction in the cell may be allowed to pass back through the membrane. A much greater membrane surface area to volume ratio than can be obtained with the artificial kidney is possible with devices based on artificial cells.

Artificial cells have been considered for the treatment of acute poisoning, chronic renal failure, hepatic failure, diseases dependent on enzyme deficiency, and the removal of solutes from the gastrointestinal tract. The first clinical success was achieved in blood purification by hemoperfusion—extracorporeal passage of blood through a column containing activated charcoal granules coated with polymer.

5. Controlled Release

Controlled release involves the temporary restraint of a pharmacologically active agent by a polymer membrane, which may, if desired, be biodegradable. This contrasts with the operation of artificial cells, in which the reactive substances are shielded permanently from direct contact with the biological environment. In controlled release, the agent is released by diffusion through the membrane, by erosion of the membrane, or by a combination of diffusion and erosion. There are two basic controlled-release systems, which are termed reservoir and monolithic. In the reservoir system, the agent is encapsulated by a membrane, and in the monolithic system, the agent is dispersed throughout a membrane matrix.

The possibilities available for varying membrane structure and permeability characteristics permit controlled release to extend the application of membranes into the area of artificial glands.

Release agents which have been studied include drugs designed to achieve a particular therapeutic effect and prostaglandins for interaction with blood platelets and the consequent improvement in the compatibility of blood-contacting materials.

6. Skin Substitutes

It is not reasonable to expect an artificial membrane to perform all the vital functions of skin. However, the ability of membranes to permit selective permeability and controlled release make it possible to consider membranes as skin substitutes in certain situations. One area where membranes have been studied as temporary skin substitutes is that of burn wound coverings.

In the case of burning, the properties of the skin which ensure regulation of water and heat losses and the prevention of infection are lost. Fluid loss may be controlled by the selection of a membrane with a water vapor permeability rate similar to that of normal skin. Infection is avoided by the incorporation of an antibacterial agent into the membrane and its subsequent controlled release. The membrane is permeable to oxygen, thus maintaining a proper moisture balance above the wound, while allowing oxygen to contact the wound surface and promote tissue regrowth. The membrane wound covering is nonadherent as tissue ingrowth is undesirable in order to accomplish ready removal of the membrane for reepithelialization.

See also: Renal Dialysis; Heart–Lung Machines

Bibliography

Chang T M S 1972 *Artificial Cells*. Thomas, Springfield, Illinois

Courtney J M, Gaylor J D S, Gilchrist T 1977 Dialysis. In: McMullan J T (ed.) 1977 *Physical Techniques in Medicine*, Vol. 1. Wiley, London

Courtney J M, Gaylor J D S, Holtz M, Klinkmann H 1984 Polymer membranes. In: Hastings G W, Ducheyne P (eds.) 1984 *Macromolecular Biomaterials*. CRC Press, Boca Raton, Florida

Davies J W L 1983 Synthetic materials for covering burn wounds: Progress towards perfection, Pt. 1: Short term dressing materials. *Burns* 10: 94–103

Davies J W L 1983 Synthetic materials for covering burn wounds: Progress towards perfection, Pt. II: Longer term substitutes for skin. *Burns* 10: 104–8

Drukker W, Parsons F M, Maher J F (eds.) 1983 *Replacement of Renal Function by Dialysis*. Martinus Nijhoff, The Hague

Gardner C R 1976 Biomedical applications of membrane processes. In: Meares P (ed.) 1976 *Membrane Separation Processes*. Elsevier, Amsterdam

Keller K H, Shultis K L 1979 Oxygen permeability in ultrathin and microporous membranes during gas–liquid transfer. *Trans. Am. Soc. Artif. Intern. Organs* 25: 469–72

Ketteringham J, Zapol W, Birkett J, Nelsen L, Massucco A, Raith C 1975 A high permeability, nonporous, blood compatible membrane for membrane lungs: *in vivo* and *in vitro* performance. *Trans. Am. Soc. Artif. Intern. Organs* 21: 224–32

Klein E, Autian J, Bower J D, Buffaloe G, Centella L J, Colton C K, Darby T D, Farrell P C, Holland F F, Kennedy R S, Lipps B Jr, Mason R, Nolph K D, Villarroel F, Wathen R L 1977a Evaluation of hemodialyzers and dialysis membranes. Evaluation of hemodialysis membranes. *Artif. Organs* 1: 21–40

Klein E, Autian J, Bower J D, Buffaloe G, Centella L J, Colton C K, Darby T D, Farrel P C, Holland F F, Kennedy R S, Lipps B Jr, Mason R, Nolph K D,

Villarroel F, Wathen R L 1977b Evaluation of hemo-dialyzers and dialysis membranes. *In vitro* character-ization of hemodialyzers. *Artif. Organs* 1: 59–77

Lysaght M J, Gurland H J (eds.) 1983 *Plasma Separation and Plasma Fractionation. Current Status and Future Directions.* Karger, Basel

Meares P 1976 The physical chemistry of transport and separation by membranes. In: Meares P (ed.) 1976 *Membrane Separation Processes.* Elsevier, Amsterdam

Park G B 1978 Burn wound coverings—A review. *Biomater. Med. Devices. Artif. Organs* 6: 1–35

Sliwka W 1975 Microencapsulation. *Angew. Chem. Int. Ed. Engl.* 14: 539–50

J. M. Courtney and J. D. S. Gaylor

Audiometers

An audiometer is an instrument used for the measure-ment of hearing acuity. Only audiometers used for tests requiring a deliberate response from the subject are described in this article; the instrumentation for elec-trical response tests of hearing is quite different (see *Electric Response Audiometry*). Tests are usually made with tones or speech signals.

Although audiometry enjoys a history extending over 100 years, a lengthy gestation period delayed the accept-ance of audiometers for hearing assessment until the 1930s when vacuum tubes provided new possibilities in design. The earliest machines relied on the induction of electrical currents in movable coils to control the level of a simple "click" stimulus. Credit for the original inven-tion is attributed by different authorities to Hughes in the UK and Hartmann in Germany, both in 1879. The early development of audiometers has been reviewed by Bunch (1943) and later by Stephens (1979).

The performance of present-day audiometers is specified by an international standard, IEC 645 (1979 under revision), which groups instruments into func-tional categories numbered 1–5, based on the range of facilities required for different applications. Type 1 represents a very sophisticated audiometer, type 5 a very simple portable screening machine.

Manual audiometers require that an operator selects the appropriate test signal and records the subject's response. Automatic recording audiometers, however, are programmed to present a complete sequence of test signals to the subject without intervention of the oper-ator. A permanent record of the subject's response as the stimulus is tracked is plotted on a graphic recorder. The original instrument of this type was described by von Békésy (1947). The term speech audiometer may also be encountered, although in practice audiometry with speech signals is usually performed through a con-ventional audiometer equipped to accept speech signals from a source such as a prerecorded magnetic tape.

1. Essential Features of Pure-Tone Audiometers

The essential features of a pure-tone audiometer include an audio oscillator to generate tones over the required frequency range, an attenuator or hearing level control to adjust the intensity of the tone presented, an inter-ruptor switch enabling the operator to control the presentation of test signals and appropriate output transducer(s), headphones for hearing by air conduction, and a bone vibrator for hearing by bone conduction. Sometimes the signal can be fed to a loudspeaker to allow "free field" audiometry to be carried out.

1.1 Audio Oscillators

Most audiometers incorporate oscillators of a fixed frequency design which, depending on the type of audi-ometer, will deliver tones by earphone from the pre-ferred range of frequencies: 125, 250, 500, 750, 1000, 1500, 2000, 3000, 4000, 6000 and 8000 Hz. Facilities for bone-conduction measurements are provided only on type 1, 2 and 3 instruments with tones restricted to the range 250–4000 Hz. On some audiometers, continuously variable tones are produced by a beat frequency oscil-lator permitting, for example, pitch matching of tones presented binaurally.

1.2 Hearing-Level Control

Adjustment of the test signal strength is regulated by the hearing-level control which is normally graduated in steps of 5 dB over a range -10 dB to about $+120$ dB, relative to the normal threshold of hearing (depending on type of audiometer). In order to protect the trans-ducers, the maximum intensity may be curtailed at very high and low frequencies on air conduction and will be restricted to $\leqslant 70$ dB above threshold on bone conduc-tion. Continuously variable level control is provided on some audiometers. Other important features include masking signals (see Sect. 4), a patient response indi-cator, and switching facilities to control the passage of signals through the audiometer.

2. Specification of Audiometric Threshold

Pure-tone audiometry is based on the principle of agreed standard values for the normal threshold of hearing, stated usually in terms of sound pressure or vibration levels developed by the audiometer earphones or bone vibrator in appropriate measuring devices, known as acoustic or mechanical couplers.

These standard values derive, of course, from measurements on many individuals deemed to possess normal hearing. For this purpose normal hearing is considered to exist for persons in the age range 18–30 years, normal on otoscopic inspection, and with no history of hearing disorder or undue exposure to intense noise.

Data for air-conduction thresholds are specified by an international standard, ISO 389 (1964) and supple-ment ISO 389A (1970). Acoustical interaction between audiometric earphones and acoustic couplers has made it necessary to provide different sets of values for the various earphone–coupler combinations which may be used in practice. Even so, the publication of this standard marked a significant advance on an existing situation in which national standards showed clear differences. For example, the British Standard thresholds were some 10–15 dB lower than the figures adopted in the USA. Further improvement will be possible with the agreement of threshold values for use with the so-called IEC wide-band artificial ear (IEC 318, 1970) on which the outputs from all earphones of a common audiometric pattern can be compared to common reference levels.

In the case of bone conduction, general agreement on threshold values has been difficult to achieve and individual national standards have been prepared. This situation has arisen despite the fact that (a) a mechanical coupler for the measurement of signals from an audiometric bone vibrator has been standardized internationally since 1971 (IEC 373), and (b) the clinically measured difference between air and bone-conduction thresholds is commonly used as a diagnostic test of hearing disorder, which suggests that both air and bone conduction data should, ideally, be gathered from the same test persons. However, while rationalization of national standards will be forthcoming, it should be admitted that the restricted precision of most clinical procedures lessens the significance of small discrepancies in presently-used data.

Much emphasis is placed on the accuracy and stability of the signals developed by the headphones and bone vibrator of an audiometer. A regular program of maintenance is therefore required to ensure the accuracy of calibration not only in terms of output level but with regard to frequency and purity of tone.

3. Techniques of Audiometry

There is no universally agreed procedure for the measurement of audiometric threshold, although many schemes have been proposed (e.g., Hughson and Westlake 1944). All are based on the regular presentation of tone pulses in some prescribed order with the aim of approaching threshold in a repeatable fashion. The apparent dissimilarity between clinical tests of auditory threshold and real-life listening situations has been discussed by Robinson (1971), but it says much of this rather simple procedure that it remains predominant in the audiological repertoire. Results of air and bone conduction pure-tone audiometry are marked on a chart called an audiogram, giving a graph of hearing loss against frequency. The levels at which sounds become uncomfortably loud may also be recorded. Figure 1 shows a typical audiogram for a mixed hearing loss, that is, one in which both conductive and sensorineural elements are present.

In the case of automatic recording audiometry, the chart provides a continuous record of the tracking task performed as the subject attempts to regulate the level of the tone to his own auditory threshold. The signals may be presented either as a series of fixed frequencies or as

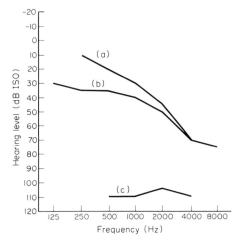

Figure 1
Typical audiogram for a mixed hearing loss showing results for: (a) bone conduction, (b) air conduction, and (c) loudness discomfort level

a gliding tone which covers the required frequency range in a few minutes.

A number of documented tests of hearing function can be performed with the facilities found on many clinical audiometers. Many are directed towards the abnormal growth and decay of loudness associated with dysfunction of the cochlea, and include the short increment sensitivity index (SISI), the sensorineural acuity level (SAL) and the tone decay test (Katz 1972).

4. Masking

An important feature built into all but simple screening audiometers is the provision of a masking stimulus. Masking provides the means to exclude from the assessment of one ear the influence of the other. Masking is required in the following situations:

(a) When a hearing-level difference by air conduction of greater than about 40 dB exists between the ears. Sound radiated from the earphone may then be received in the better ear or bone conducted sound may stimulate the cochlea of the better ear when the poorer ear is tested.

(b) When bone conduction is tested in a nonsymmetrical hearing loss. The bone conducted signal will be heard in the better ear irrespective of the placement of the vibrator. It is therefore essential to eliminate the influence of the better cochlea before reliable measurements can be made on the other ear.

Fletcher (1940) showed that a pure tone can be masked effectively by a small band of noise, the critical band, centered on the tonal frequency. IEC 645 therefore specifies noise bands falling between the limits of one-third and one-half octaves which provide the facility known as narrow-band masking.

Other signals may be provided for masking purposes. Broad-band noise can be used to mask pure tones but because so much of the energy falls outside the relevant critical band the presentation level must be higher than for narrow-band noise. Random noise, weighted to reflect the spectral characteristics of speech, may be provided as a masker for speech signals. Whatever masking signals are provided by the audiometer their levels should be calibrated to appropriate values.

The effectiveness of masking can, however, be tested by delivering a small level of masking to the nontest ear while making a first estimate of the hearing level. Repeat measurements of hearing level are then made while masking is increased in stages; when the measurement of hearing level has stabilized, masking is complete.

5. Speech Audiometry

Speech audiometry is undertaken mainly in the differential diagnosis of hearing disorder, and in the assessment of social handicap, providing a more direct test than is possible with pure-tone audiometry.

Although live-voice testing is feasible, prerecorded material is generally preferred because intertest variations are minimized. Many different speech materials are used including spondee words and sentences (real and nonsense), but most widely used are lists of monosyllabic words displaying phonemic balance. Among the measurements which can be made are:

(a) the level at which the presence of speech can just be detected, termed the speech reference level or speech detection threshold;

(b) the level at which 50% of the material is understood, termed the hearing level for speech or speech reception threshold; and

(c) the maximum discrimination score, obtained when speech is presented to the listener at the most favorable level.

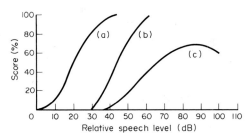

Figure 2
Speech audiogram, showing typical results for: (a) normal hearing, (b) conductive hearing loss, and (c) sensorineural hearing loss

A full speech discrimination curve, obtained for an appropriate range of intensities, will provide rather more information, especially in cases of sensorineural hearing loss for which the rate of discrimination increases with intensity and the maximum discrimination score diminishes with increased deafness. Figure 2 shows a speech audiogram with typical results for conductive and sensorineural hearing loss.

Comparative audiometric measurements can only be made with reference to normative values for the particular test. However, for speech audiometry, less agreement exists about the most appropriate reference levels to be used, a situation hindered by problems with terminology. A widely used standard, ANSI S 3.6 (1969), specifies a reference speech level of 19.5 dB (SPL) corresponding to a 0 dB (HL) setting of the hearing level control and representing an approximate threshold of detectability.

6. Computer Control in Audiometry

Most audiometric procedures require systematic and repeated presentations of test signals and are therefore amenable to computer control. A self-recording system based on the presentation of fixed frequencies has been described by Berry et al. (1979). Computerization is particularly suited to interactive procedures, those in which the presentation of a particular signal is dependent on the outcome of the preceding test.

Recently, systems have been demonstrated which allow programmed presentation of either closed sets of speech material or various forms of nonspeech signals and which provide automatic recording of subject responses together with their response times, thought to be an important factor. Immediate statistical analysis of the collected data provides an assessment of the subject's performance based on previously acquired normative data for the particular test. Tests such as these should show improved sensitivity over conventional procedures, particularly with regard to the consistency in application of auditory decision criteria.

Digital techniques are beginning to affect the design and operation of conventional pure-tone audiometers, and future advances in electronics and particularly the use of microcomputers will undoubtedly stimulate significant changes in the techniques of audiometry.

See also: Electrodermal Audiometry; Electroencephalic Audiometry; Objective Audiometry; Brain-Stem Electric Response Audiometry; Electric Response Audiometry; Electrocochleography

Bibliography

American National Standards Institute 1969 *American National Standard Specification for Audiometers*, ANSI S3.6—1969. American National Standards Institute, New York

Békésy G von 1947 A new audiometer. *Acta Oto-Laryngol.* 35: 411–22

Berry B F, John A J, Shipton M S 1979 A computer-controlled audiometry system. *Proc. Inst. Acoust.* (Spring Conf.): 20.3

Bunch C C 1943 *Clinical Audiometry*. Mosby, St Louis

Fletcher H 1940 Auditory patterns. *Rev. Mod. Phys.* 12: 47–65

Hughson W, Westlake H 1944 Manual for program outline for rehabilitation of aural casualties both military and civilian. *Trans. Am. Acad. Ophthal. Otolaryngol.* 48 (Suppl.): 1–15

International Electrotechnical Commission 1970 *An IEC Artificial Ear of the Wide Band Type for the Calibration of Earphones Used in Audiometry*, IEC 318. International Electrotechnical Commission, Geneva

International Electrotechnical Commission 1971 *An IEC Mechanical Coupler for the Calibration of Bone Vibrators Having a Specified Contact Area and Being Applied with a Specified Static Force*, IEC 373. International Electrotechnical Commission, Geneva

International Electrotechnical Commission 1979 *Audiometers*, IEC 645. International Electrotechnical Commission, Geneva

International Standards Organisation 1964 *Standard Reference Zero for the Calibration of Pure Tone Audiometers*, ISO 389. International Standards Organisation, Geneva

International Standards Organisation 1970 *Additional Data in Conjunction with the 9-A Coupler*, ISO 389, Addendum No. 1. International Standards Organisation, Geneva

Katz J (ed.) 1972 *Handbook of Clinical Audiology*. Williams and Wilkins, Baltimore, Maryland

Robinson D W 1971 A review of audiometry. *Phys. Med. Biol.* 16: 1–24

Stephens S D G 1979 Audiometers from Hughes to modern times. *Br. J. Audiol.* 13 (Suppl. 2): 17–23

B. C. Grover

B

Beta-Particle Detection

Beta-particle-emitting radionuclides are more difficult to detect than γ-ray emitters, because of the short range of electrons in matter. Thus purely β-emitting radionuclides are selected for tracer studies only when no suitable γ emitter exists. Table 1 lists the pure β-particle

Table 1
Some pure β-emitting radionuclides in biomedical use

Nuclide	Half-life	Particle energies (MeV)
^{3}H	12.35 years	0.0186
^{14}C	5730 years	0.156
^{32}P	14.29 days	1.710
^{35}S	87.39 days	0.167
^{90}Y	64.0 hours	2.284

emitters in common use. In radiotherapy, therapeutic advantage is taken of the short range of the radiation from the β-emitting isotopes ^{32}P, ^{90}Y, ^{131}I and ^{198}Au (see *Brachytherapy*). The only convenient radioactive isotopes of hydrogen, carbon and sulfur (^{3}H, ^{14}C and ^{35}S) are all low-energy β emitters, and are widely used for labelling biochemical compounds.

1. Liquid Scintillation Counting

The technique of liquid scintillation counting is the most widely used method of detecting low-energy β emitters. In this method, the radioactive sample is intimately mixed with a liquid scintillator. It is the method of choice for counting ^{3}H, ^{14}C and ^{35}S and can also be used

for high-energy β emitters such as ^{32}P, and for low-energy x- and γ-ray emitters such as ^{55}Fe and ^{125}I.

The liquid scintillator consists of a fluorescent solute contained in an organic solvent. The β particles lose their energy by ionization and excitation of the solvent, resulting in the emission of photons, normally in the blue or ultraviolet regions of the spectrum. These are detected by a photomultiplier-tube assembly.

Table 2 lists some of the scintillation solutes in common use, and their abbreviated chemical names. If necessary a secondary solute, or wavelength shifter, which absorbs the light and reemits it at a longer wavelength more suited to the spectral response of the photomultiplier tubes, is also added.

The β particles are often of very low energy and thus the light flashes from the scintillator are very weak. On average, one β particle from the decay of ^{3}H causes the emission of only a few electrons from the photomultiplier cathode. This is comparable with the number of electrons emitted thermally, thus the background count rate can be very high. The sample is contained in a transparent glass or plastic vial placed between two photomultiplier tubes. The light photons from the scintillator eject electrons from the photocathodes of both photomultipliers simultaneously, while thermal effects cause a random emission of electrons. Using suitable electronic circuitry, counts are registered only when pulses are detected in both photomultiplier tubes within a very short coincidence-revolving time (often < 10 ns). The background count rate for ^{3}H is thus reduced by orders of magnitude and may be reduced still further by cooling the whole assembly to about 4 °C. A schematic diagram of a modern liquid scintillation counter is shown in Fig. 1. In this case, two counting channels are provided to enable dual isotope studies to be undertaken. Automatic operation is normal using microcomputers or minicomputers to control the operations and to calculate the results.

Table 2
Scintillation solutes

Solute	Abbreviation	Type	Wavelength of maximum fluorescence (nm)
2,5-diphenyloxazole	PPO	primary	375
p-terphenyl	TP	primary	342
1,4-di-[2-(5-phenyloxazolyl)]benzene	POPOP	secondary	415
2-phenyl-5-(4-biphenylyl)- -1,3,4-oxadiazole	PBD	primary	375
2-(4′-*t*-butylphenyl)-5- -(4″-biphenylyl)-1,3,4-oxadiazole	butyl-PBD	primary	385

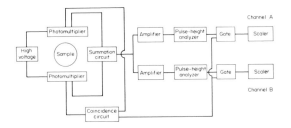

Figure 1
Schematic diagram of a dual-channel liquid
scintillation counter (after Parker 1978.
©Heinemann, London. Reproduced with permission)

2. Sample Preparation in Liquid Scintillation Counting

For routine liquid scintillation counting, sample preparation must be straightforward. Samples may consist of
solids, gases, or plant or animal tissues, as well as
liquids. Some materials, such as fatty acids, are readily
soluble in scintillator solvents such as toluene, and little
is required in the way of preparation. In other cases
chemical treatment is required.

A great number of labelled compounds containing ^{14}C
are available, and these are often assayed by converting
the carbon to carbon dioxide which is then absorbed in
the scintillator mix. Since the solubility of carbon dioxide in toluene is low, it is customary to absorb the gas
in a suitable base such as phenethylamine or methoxy-
ethylamine which is then dissolved in toluene or ethanol.

Tritium-labelled samples are more demanding because
they cannot be usefully converted into a simple gaseous
form. For biological materials such as tissues, amino
acids or RNA, solubilizers like hyamine are used to
facilitate incorporation in toluene. For water and for
aqueous samples it is now customary to use emulsion
counting. The emulsion system consists of an aromatic
solvent containing suitable scintillators mixed with a

detergent. The most suitable detergents are the
alkylphenylpolyethyleneglycol ethers known under the
trade name of Triton. These are available commercially
premixed with scintillators and they may incorporate up
to 50% water by volume while maintaining a counting
efficiency of 15–20% for tritium.

In preparing samples, care must be taken to ensure
that undue amounts of chemiluminescence are not created, giving anomalously high sample count rates.

3. Quench Corrections in Liquid Scintillation Counting

A number of substances may absorb radiation energy
before it is transferred to the scintillator. This phenomenon is known as chemical quenching and it is important
for aqueous samples. It is particularly noticeable if
oxygen is present. Oxygen can be removed by bubbling
an inert gas such as nitrogen through the scintillator
solution. Color quenching can also occur. Here, the
overall light transmission of the system is reduced.
Decoloration may be possible. If the count rate is
adequate it is often useful to dilute the sample.

Both chemical and color quenching reduce the voltage
pulse output of the counter (a function of the β energy
dissipated) causing an apparent downwards shift in
the pulse-height spectrum or, in the case of tritium, an
actual reduction in the number of β particles detected.
Figure 2 shows the pulse-height spectra for ^3H, ^{14}C and
^{36}Cl, together with those of ^3H and ^{14}C when quenched.
The effect of quenching on the pulse-height spectrum
and on the number of β particles detected is evident.
Since quenching is normally present, and may well vary
between individual samples, it is necessary to determine
the amount of quenching present and correct for its
effects. The two main methods of quench correction are
based on internal and external standards.

(a) The internal-standard method. The sample is first
counted on its own and then recounted following the

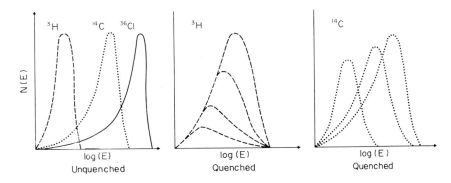

Figure 2
Quenched and unquenched β-ray spectra of different nuclides as obtained with a liquid scintillation counter (after
Parker 1978. © Heinemann, London. Reproduced with permission)

addition of a known activity of the radionuclide. The difference between the two counts enables the counting efficiency (and therefore the degree of quenching) to be determined. This is in turn used to correct the initial reading obtained with the sample alone. This method is capable of high accuracy, but is tedious.

(*b*) *The external-standard method.* The measured spectrum is divided into two counting channels, covering different ranges of pulse height. Quenching causes a shift in pulse height (see Fig. 2) and therefore alters the ratio of counts in the two channels. A calibration curve can thus be prepared of channel ratio versus counting efficiency provided a set of samples is available which are of known activity but exhibit different degrees of quenching.

The disadvantage of this technique is that high-activity samples are required to achieve adequate precision. This disadvantage is overcome if the sample is counted twice—once alone, and then with an external γ-ray emitter such as ^{137}Cs close to the counting vial. The Compton electrons resulting from the γ-ray interactions provide a high count rate. The degree of quenching can be estimated using the channel-ratio method applied to these "external" counts. The system is calibrated as before.

The external-standard, channel-ratio method can be automated, and if the calibration curve is stored in a microcomputer, corrected results can be calculated on-line.

4. Cerenkov Counting

High-energy β emitters (above 0.26 MeV, and preferably above 1 MeV in maximum energy), such as ^{32}P or ^{90}Y, can be detected using a liquid scintillation counter to detect the visible Cerenkov radiation which is emitted when an electron of sufficient energy passes through a medium such as water. Sample preparation is particularly easy as no scintillator is required and color quenching only is important as chemical quenching does not affect the Cerenkov emission.

5. Dual Isotope Studies

Since the pulse heights obtained using a liquid scintillation counter are a function of the β energy lost in the medium, different radionuclides will provide different pulse-height spectra. A technique similar to that employed in γ-ray spectroscopy, and employing two counting windows, is used in dual isotope studies. It is possible to assay mixtures containing ^{3}H and ^{14}C. Again, the calibration can be carried out using an on-line computer. Separate quench corrections are required for the ^{3}H and ^{14}C components.

6. Gas Counters

Although liquid scintillation counting is usually the method of choice in biomedical work, gas counters can be used for certain specialized applications.

High-energy β emitters can be detected with an end-window Geiger counter. If a thin mica window is used, ^{14}C can be measured, although sensitivity is poor. Gaseous sources may be introduced into the counting volume itself and the detector operated as a proportional counter or as a Geiger counter.

β-Particle ionization chambers are simple to construct and are sometimes used in the assay of solid sources. The corrections for self-absorption and backscatter can be considerable, so thin evaporated sources should be used. High-activity sources of the higher-energy β emitters can be estimated using an ionization chamber, such as the type 1383 A designed at the National Physical Laboratory in the UK, as a secondary instrument for both γ and β calibrations.

7. Low-Level Activity

Both liquid scintillation counters and gas counters can be used for low-activity measurements provided great care is used to optimize the counting efficiency and reduce the background count rate. The background count is in part due to internal radioactive contamination from environmentally occurring radionuclides such as ^{40}K and members of the radium and thorium series of radionuclides. External shielding and careful selection of materials can reduce the background count considerably. The background component due to cosmic radiation is best reduced by anticoincidence shielding, with the sample counter surrounded by a ring of other counters connected in parallel with one another but in anticoincidence with the sample counter. Cosmic-ray particles such as π mesons which trigger the sample counter will also be registered in the ring and will therefore be recognized as being due to cosmic radiation.

See also: Radiation Dosimetry; Gamma-Ray Detectors

Bibliography

Bransome E D (ed.) 1970 *The Current Status of Liquid Scintillation Counting.* Grune and Stratton, New York

Horrocks D L 1974 *Applications of Liquid Scintillation Counting.* Academic Press, New York

Horrocks D L, Peng C-T (eds.) 1971 *Organic Scintillators and Liquid Scintillation Counting.* Academic Press, New York

International Commission on Radiation Units and Measurements 1972 *Measurement of Low-Level Radioactivity*, ICRU Report No. 22. ICRU, Washington, DC

National Council on Radiation Protection and Measurements 1978 *A Handbook of Radioactivity Measurement Procedures: With Nuclear Data for Some Biologically Important Radionuclides*, NCRP Report No. 58. NCRP, Washington, DC

Parker R P 1978 The use and measurement of radio-isotopes. In: Williams D L, Nunn R F, Marks V (eds.) 1978 *Scientific Foundations of Clinical Biochemistry*. Heinemann, London, pp. 285–98

R. P. Parker

Binocular Vision

Each eye has a field of vision. Where the two fields overlap to form a single image, there is a field of binocular vision, a single mental impression of objects in three-dimensional space. This article covers various aspects of binocular vision, together with the associated neurophysiology.

1. Binocular Single Vision

Normal-sighted individuals see a limited region of the outside world as a single fused image rather than as two separate images. This region of binocularly fused single vision is called Panum's fusional area. Even in normal-sighted individuals, all objects within the binocular visual field that are not fused may be seen as double (Ogle 1972). Usually this doubling is not visually obtrusive, although it is easily demonstrated. (Look at the tip of one finger while moving a second finger away in depth. The second finger will appear double while the first finger continues to be seen in binocular single vision.)

In diplopic (double) vision the individual is conscious of a separate image from each eye. This condition is associated with misalignment of the direction of the eyes, as in strabismus (i.e., squint), and may also result from head injuries. In some patients double vision is avoided either by suppressing vision in one eye or by alternating vision between the two eyes.

2. Binocular Disparity

The retinal points in the two eyes that give rise to the same visual directions are defined as corresponding points. The loci of object locations, the images of which fall on corresponding points, is called the horopter. Images of a location in space that do not fall on corresponding points are said to be disparate. Disparity is usually specified in minutes of arc. In terms of disparity, Panum's area is rather greater in its horizontal extent than in its vertical extent. It is larger in peripheral than in central vision, and its size also depends on the spatial nature of the target and whether the target is stationary or moving (Ogle 1972). In normal conditions Panum's area seldom exceeds 20′ horizontally in central

vision. When the two eyes are accurately converged, the two images of the object being fixated fall on corresponding points at the center of the left and right foveae, and so have zero disparity. When some individuals attempt to fixate on an object the two retinal images do not fall on corresponding points. For such individuals the two images have a finite disparity, called a fixation disparity (Ogle et al. 1967). However, a region of binocular fused vision remains, provided that the fixation disparity is less than the horizontal extent of Panum's area.

3. Binocular Depth Perception

The left and right eyes are separated, and consequently receive slightly different images of three-dimensional objects. This geometrical difference between the left and right eye's images can be described by saying that the retinal images of objects at different depths have different binocular disparities (Foley 1978). Wheatstone (1838, 1852) first showed that binocular disparity is capable of eliciting a compelling impression of depth and three-dimensional solidity. This is the principle of the stereoscope. Two photographs are taken from slightly different locations, and viewed through mirrors or prisms so that the two photographs are seen in binocular fusion. Individuals who are not stereoblind see the photographed scene in vivid depth.

Binocular disparity alone does not indicate the absolute distance of an object from the observer. Since disparity is measured relative to the frontoparallel plane through the point of binocular convergence, perceived stereoscopic depth is relative to this plane, so that the angle of convergence of the eyes must be taken into account when estimating absolute depth. (A small correction due to fixation disparity is ignored.)

4. Cyclopean Vision

Photographs of real-world scenes viewed through a stereoscope commonly include monocularly available cues to depth in addition to the binocular cue of disparity. Julesz (1978) showed that depth perception can occur in the absence of monocular cues to either depth or form by devising a visual stimulus in which these monocular cues were absent. One type of so-called Julesz pattern consists of patterns of randomly located dots. The patterns viewed by the two eyes are identical except that in one eye's pattern a region of dots is bodily displaced to the left or right. When viewed monocularly, no depth or form can be seen, but when viewed in binocular fusion normal-sighted individuals see the displaced region of dots in vivid depth. This is an example of so-called cyclopean perception in which information as to form (i.e., the area seen in depth) is not available prior to binocular processing. A curious observation is that form and depth are not always seen immediately in Julesz patterns. When the binocularly perceived form is complex it may take several minutes to appear.

5. Stereoscopic Motion-in-Depth Perception

In addition to the classical stereoscopic system for relative position in depth, the human visual pathway seems to contain a second stereoscopic system for motion in depth (Regan et al. 1979). The stereoscopic motion system responds to the velocity ratio V_L/V_R, where V_L and V_R are the velocities of the left and right retinal images, respectively. (In contrast, as noted above, the classical stereoscopic system responds to a difference in position of retinal images, i.e., to disparity.) The stereoscopic motion system consists of four subsystems, each selectively sensitive to a different value of the ratio V_L/V_R, i.e., each is tuned to a different direction of motion in depth. In some individuals, areas of the visual field are selectively blind to disparity, while other areas are selectively blind to stereoscopic motion in depth (Regan et al. 1979).

6. Neurophysiology of Binocular and Stereoscopic Vision

The properties of individual neurons in the visual cortex of the monkey, cat and other animals have gone some way to provide a physiological basis for binocular fusion, for classical stereoscopic depth perception and for stereoscopic motion perception. Some cortical neurons respond most strongly to binocular stimulation when the target is at a particular distance in front of or behind the fixation point (i.e., when the left and right eye's images have a particular disparity). The critical value of disparity is different for different neurons (Hubel and Wiesel 1970). These neurons strongly respond to an edge (contour) viewed by left and right eyes, provided that the images on the two retinae are oriented parallel and have the same polarity of contrast (i.e., light/dark or dark/light). These binocular depth neurons provide a partial physiological basis for binocular fusion and classical stereopsis. Other neurons respond best to binocularly viewed motion when the direction of motion is along a line passing through the head, whereas yet other neurons respond best when motion is along a line passing wide of the head (Regan et al. 1979). The latter two types of neuron are relatively insensitive to disparity, and may go some way to providing a physiological basis for the stereoscopic perception of motion in depth.

See also: Vision; Visual Cortical Neurophysiology; Visual Fields and Thresholds

Bibliography

Bishop P O, Henry G H 1971 Spatial vision. *Annu Rev. Psychol.* 22: 119–60
Foley J M 1978 Primary distance perception. In: Held R, Leibowitz H W, Teuber H-L (eds.) 1978 *Handbook of Sensory Physiology*, Vol. 8. Springer, New York, pp. 181–213
Hubel D H, Wiesel T N 1970 Stereoscopic vision in macaque monkey. *Nature (London)* 225: 41–42
Julesz B 1978 Global stereopsis: cooperative phenomena in stereoscopic depth perception. In: Held R, Leibowitz H W, Teuber H-L (eds.) 1978 *Handbook of Sensory Physiology*, Vol. 8. Springer, New York, pp. 215–56
Ogle K N 1972 *Research in Binocular Vision*. Hafner, New York
Ogle K N, Martens T G, Dyer J A 1967 *Oculomotor Imbalance in Binocular Vision and Fixation Disparity*. Kimpton, London
Poggio G F, Motter B C, Squatrito S, Trotter Y 1985 Responses of neurons in visual cortex (V1 and V2) of the alert macaque to dynamic random-dot stereograms. *Vision Res.* 25: 397–406
Regan D, Beverley K I, Cynader M 1979 The visual perception of motion in depth. *Sci. Am.* 241: 136–51
Wheatstone C 1838/1852 Contributions to the physiology of vision—I, II. *Philos. Trans. R. Soc. London, Ser. B* 128: 371–94; 142: 1–18

D. Regan

Bioelectricity

"Animal electricity" is a concept with a long, if somewhat murky, history. Commentators have remarked upon the lifelike features of many electrostatic and discharge phenomena (see, for example, Gregory 1981). In turn, these electrical phenomena were often capable of potent biological effects; furthermore, some of these effects were identical to those produced by the attacks of certain fish. Thus the links were numerous, yet for many centuries they were too intricate for clear interpretation; even the seminal observations of Galvani (1737–98) upon the dancing of frogs' legs, touched by heterogeneous metals in salted water, were contemporarily explained by the flow through the metal of Vital Spirit, not mere electricity. The foundations of a modern understanding were only laid down—by du Bois-Reymond, Helmholtz and Bernstein—in mid-nineteenth century Germany.

1. Concentration Gradients Across Cell Membranes

The great majority of bioelectric phenomena have their origin in the potential differences which exist across cell membranes. These arise from inequalities of ionic concentration between the intra- and extracellular fluids, which are themselves caused as follows.

Metabolic energy, made available by the hydrolysis of adenosine triphosphate (ATP), fuels a mechanism not yet fully understood but known to reside in macro-molecular complexes embedded within the cell membrane. This drives out from the cell the most common cation of biological fluids, sodium (Na^+), and is termed the "sodium pump." The charge deficit which would arise if the outward pumping process went on unchecked—the inside of the cell being left massively negative with respect to the outside—is largely compensated by the inflow of the second most common cation, potassium (K^+). This ion permeates freely through cell membranes, and in the simplest instances is passively drawn in to make up the electrical deficit.

The chief anionic constituents of intracellular fluids are organic species, synthesized metabolically from small precursors, but themselves too large to diffuse back through the membrane. Extracellularly, the principal anion is chloride (Cl^-); this, though able to pass freely through many cell membranes, is repelled electrically from achieving a comparable concentration intracellularly. Thus, all cells contain predominantly K^+ and organic anions, in electrochemically comparable amounts, while the extracellular fluid contains chiefly Na^+ and Cl^-.

2. Membrane Potentials

A transmembrane potential difference (PD), known to physiologists as a "membrane potential," results from the cationic separation just described. To see that this must happen, suppose that the internal negativity produced by the pump were completely cancelled by K^+ uptake. This would eliminate the force tending to hold K^+ ions within the cell. Being freely permeating, they would diffuse out, leaving the inside negative again.

The thermodynamic equilibrium condition, for an ion under a concentration gradient, was derived by Nernst in the 1890s as:

$$E_j = \frac{RT}{z_j F} \ln \frac{[j]_o}{[j]_i}$$

where $[j]_i$ and $[j]_o$ represent concentrations of the ion j inside and outside the cell, respectively; z_j is the algebraic valency of j; E_j is the equilibrium membrane potential, measured inside relative to out (i.e., it is negative for a cation having higher concentration inside); F is the Faraday constant; R is the gas constant; and T is the absolute temperature. At 310 K (mammalian body temperature), RT/F equals 26.7. Thus, for a monovalent cation 30 times more concentrated inside the cell than out, the equilibrium internal potential would be about -90 mV. An approximately 30-fold concentration ratio characterizes K^+ distribution across the membranes of the principal electrically excitable cells, nerve and skeletal muscle; in these cells membrane potentials in the vicinity of 90 mV are in fact observed (note that as the membrane is only about 7 nm thick, this PD implies a transmembrane field $> 10^5$ V cm^{-1}).

Strictly, the membrane potential of a living cell in a normal environment is not a pure K^+ equilibrium potential, for at least two reasons. First, other ions have finite, though commonly small, permeabilities. In particular Na^+, with a steep inwardly directed concentration gradient and therefore an inside-positive equilibrium potential, has permeability one to two orders of magnitude less than that of K^+ in a resting (i.e., electrically unexcited) cell, but this suffices for it to reduce the inside-negative polarization by ~ 10 mV. Secondly, however, no quantitative allowance has been made for the continuing electrogenic (current-generating) effect of the Na^+ pump. Yet the kind of pump described above, steadily transporting positively charged ions out of the cell, contributes an emf which summates algebraically with the Nernstian influences, and acts in the hyperpolarizing direction (i.e., tends to drive the inside still more negative). Normally this effect contributes only a few millivolts but, in a cell recovering from Na^+ load, the figure may be tens of millivolts. A final complication is that a pure Na^+ pump is probably a rarity. Most pumps, in most conditions, appear to transport K^+ ions into the cell at the same time as transporting Na^+ out. (This is active transport, not passive diffusion; it apparently occurs through different membrane channels from those involved in thermodynamic equilibration, and the ions transported become free to participate in the Nernstian process only when released into the intracellular fluid.) Nevertheless, the coupling ratio, (Na^+ active efflux)/(K^+ active influx), is usually more than unity, so pumping is still electrogenic, though less markedly so than in the simple case.

Various equations have been proposed for the steady-state membrane potential of a cell through whose membrane K^+ and Na^+ can both permeate in parallel with an exchange pump. A simple and instructive (though incomplete) one, due to Mullins and Noda, is (see Junge 1981)

$$E_m = \frac{RT}{F} \ln \left(\frac{rP_{K+}[K^+]_o + P_{Na+}[Na^+]_o}{rP_{K+}[K^+]_i + P_{Na+}[Na^+]_i} \right)$$

where E_m is the membrane potential; r is the coupling ratio; and P_{K+} and P_{Na+} are the passive permeabilities of the membrane to K^+ and Na^+, respectively. When $r = 1$, this equation describes a neutral pump, which affects E_m only indirectly (through the distribution of ions). It was originally derived in this form by Goldman in the 1940s. For passive permeabilities deduced from rates of ion movement (transmembrane currents or fluxes) Goldman's expression applies only to the case of a constant intramembrane electric field; for values deduced directly from effects on potential, this limitation ceases. In two other situations—namely, when $P_{Na+} = 0$ and when $r = \infty$ (the pure Na^+ pump)—the Mullins–Noda equation reduces to the Nernst relation for K^+. The last consideration is the least intuitively evident: it indicates that, in the steady state, the depolarizing effect of Na^+ inward leak and the hyperpolarizing effect of its outward transport by an uncoupled pump would exactly balance.

In the whole of this section, it has been possible to ignore anions. Those too large to permeate cannot, of course, carry current. The chloride ion, by contrast, is a very small ion, and $P_{Cl^-} \approx P_{K^+}$ in some cells—notably skeletal muscle, which is the major bulk of the normal body. In just these instances, however, Cl^- is distributed passively, following the electrical forces set up by cations. Writing $z = -1$ in the Nernst relation, it is evident that an inside-negative potential implies $[Cl^-]_o > [Cl^-]_i$. Thus, the presence of Cl^- in a cell of high P_{Cl^-}, though it affects the rate at which steady-state E_m is attained after a perturbation, does not affect the final value.

3. Action Potentials

An action potential (AP) is a rapid swing of intracellular potential in the depolarizing direction, followed—typically within about 1 ms—by an almost equally rapid return towards resting condition. Although smaller, less reproducible "spikes" of depolarization do occur in some tissues, a true AP is of an amplitude that is dependent only on the ionic gradients across the cell membrane, and not on the stimulus strength. This property is represented by the classical physiological term, an "all-or-nothing" response.

Except where cell geometry changes significantly, APs propagate without decrement along the length of the cell. They are the means by which rapid and relatively interference-free conveyance of information is achieved down long, cablelike cells such as nerve and skeletal-muscle fibers. The all-or-nothing feature implies that signalling in such fibers is frequency coded: AP frequency commonly, but not always, increases with stimulus intensity. Also implicit in the all-or-nothing property, and crucial to the avoidance of interference, is the existence of a threshold stimulus strength, below which no response occurs.

Stimulation can be achieved by many forms of energy input (e.g., mechanical, chemical or thermal), but all have it in common that, if adequate to produce an AP, they achieve, as an immediate consequence of their application, a depolarization of not less than 15 mV or so. Nerve, muscle and sensory cells rapidly follow this stimulating depolarization with the much larger all-or-nothing response which constitutes the AP. (To distinguish between stimulus and response, note that an otherwise similar, but inexcitable, cell would show only the first of the two stages of depolarization just described.) Alternatively, the stimulating depolarization can be produced directly at the cathode of a pair of electrodes laid against the outside of the cell. Such direct, electrical excitation is the preferred tool for both experimental and clinical nerve excitation. It also simulates the natural excitation process over the main length of a fiber, for ahead of any point at which the peak of an AP is occurring, local current spread produces depolarization, and triggers the response in the new region. This is how propagation takes place.

Finally, the following three points are worth noting. First, a firm threshold can be stated only for square-wave stimuli; excitable cells accommodate to slowly rising ("ramp") depolarization, and almost all of them in the extreme do not respond at all. Second, very shortly after one AP, only a weak second response can be elicited, and this requires a stronger stimulus than usual: while thus seen to be incompletely recovered from the first response, the cell is said to be "refractory." Third—a merely terminological point—the membrane response to depolarization in an excitable cell is often described as "active." Context must distinguish between this electrical activity and the uphill transport of ions by metabolic activity.

3.1 Ionic Mechanism in the Cells First Studied

The key to understanding the AP mechanism was the intracellular electrode. This consisted first of a capillary, about 50 μm in diameter, filled with saline solution and inserted axially along an exceptionally large nerve fiber from a cut end. A little later, glass tubes drawn down to tip diameters <1 μm (microelectrodes) were found capable of penetrating radially through the membranes of a variety of cell types, and giving well-sealed electrical connection to the inside. The absolute values of trans-membrane potential difference, in a resting cell and at the AP peak, were revealed by these two forms of intracellular electrode. In giant, invertebrate nerve fibers (first studied by Hodgkin and Huxley in 1939) and subsequently in many other cells—notably mammalian nerve and skeletal muscle—the peak of the AP gave an inside-positive reading of some tens of millivolts (Fig. 1).

Earlier thinking, dating from the turn of the century, but based only on recordings from extracellular electrodes, had assumed that the AP consisted simply of a depolarization from resting potential towards zero; this was ascribed to a moment of general, ionic leakiness. The newly observed "overshoot" to positive potential required, instead, a switch of permselectivity: K^+ dominated the resting state and an ion for which the Nernst potential (E_j, above) was inside-positive must transiently dominate the active one. To carry the currents involved, it could only be an ion present in bulk, and the obvious candidate was Na^+. In the late 1940s, Hodgkin, Huxley and Katz found that reducing Na^+ concentration in the fluid bathing the nerve duly diminished both the rate and extent of the overshoot, and the essential mechanism of this class of AP was established. In terms of the general formula given above for E_m (membrane potential), P_{Na^+} swings briefly from a value of one to two orders of magnitude lower than P_{K^+} to a value about two orders greater.

Elaboration of the concept required that sophisticated feedback amplifiers be combined with two intracellular electrodes, to produce the "voltage clamp" (pioneered for giant nerve fibers by Cole and applied with conspicuous success by Hodgkin and co-workers at the start of the 1950s). One electrode passes current through a

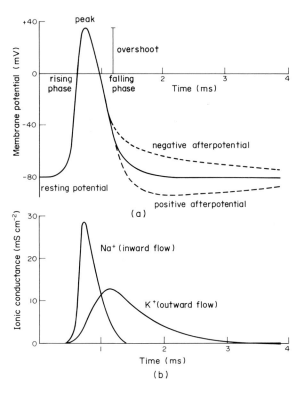

Figure 1
(a) Generalized AP for nerve or skeletal muscle. No one tissue is particularly intended, and calibrations are for guidance only. The continuous line is the idealized form; one or other "afterpotential" (dashed lines) is usually observed in practice. (b) Sketch curves indicate time-courses of conductance-variation in active channels of squid (unmyelinated) giant nerve fiber, computed by Hodgkin and Huxley (1952). Note that AP in mammalian, myelinated nerve fibers does not involve variations of K^+ conductance. Afterpotentials lasting matters of milliseconds are ascribable to nonresting ratios of the two conductances (e.g., conductances represented in (b) give positive afterpotential); similar phenomena lasting ~ 100 ms or more are commonly due to elevated rates of active transport

length of membrane, the other monitors E_m. A step-change of E_m (usually in the depolarizing direction) is imposed and then held clamped by the feedback system. Readout is of the current supplied to achieve this, which must be equal and opposite to membrane current at each moment. Voltage clamping thus provides quantitative, time-resolved information about the membrane currents flowing during the few milliseconds after a stimulus. Ions involved can be confirmed by repeating the experiments in different solutions. For the cells initially studied, the findings are that inward Na^+ current, peaking in perhaps $0.5-1$ ms and reducing to zero in $2-3$ ms, is superseded by an outward K^+ current. The effect of the latter, in an unclamped cell, is to hasten the falling phase of the AP.

Hodgkin and Huxley (1952) consummated their analysis of the AP by means of equations founded on voltage-clamp data (see Sect. 5). Already it may be seen nonquantitatively how the all-or-nothing and threshold features arise. Depolarization induces Na^+ entry, but Na^+ entry, being the inflow of positive charge, induces further depolarization. Any suprathreshold stimulus thus initiates a runaway, in which E_m swings from negative to positive, towards the value at which net cation entry ceases. The rate of reaching this positive peak increases with the strength of the stimulus, but the voltage attained does not. Subthreshold depolarizations also promote Na^+ inflow; however, the inward current (I_{Na+}) in these cases is less than the outward current of K^+ ions (I_{K+}) which flows through the resting-state P_{K+} as a consequence of the depolarization. An exact threshold stimulus is one following which inward I_{Na+} and outward I_{K+} are metastably equal for a period of perhaps several milliseconds, before randomly moving either way.

3.2 Other Action Potentials

The mechanism which was the subject of the last two sections operates where rapid propagation over long distances is required in uninsulated cells; high ionic current flows facilitate this, by accelerating the spread of depolarization ahead of the active site. All but the smallest of the nerve fibers found in vertebrates (fish to mammals) are, in addition, tightly wrapped in cylinders of fatty insulation (myelin). These reduce radial current leak, and also reduce the capacitance between intra-cellular and extracellular fluids, because the latter cannot penetrate to the nerve membrane under the myelin. Only at gaps termed "the nodes of Ranvier," which are a few micrometers wide and occur every 2 mm or so, is the membrane exposed to the Na^+-rich fluid and therefore actively able to repeat the AP. Propagation between nodes is a matter of passive, electrical conduction only. Because this is rapid, whereas there is detectable delay at each node, such propagation is likened to a series of jumps (saltatory). Recent work, particularly in Ritchie's laboratory, has demonstrated that there is no increase of P_{K+} during the falling phase of an AP in mammalian, myelinated nerve: it would confer negligible advantage there, because of the lower capacity (Chiu et al. 1979).

More radically different APs are found where propagation velocity is not limiting, as in the smaller cells of smooth muscle, glands and certain sense organs. Here, the role of the AP is usually to elicit some intracellular process, such as contraction, secretion, or chemical communication to another cell, for which calcium (Ca^{2+}) is the trigger. In many such instances, most or all of the inward current is itself carried by Ca^{2+} (an ion which is at two orders of magnitude lower concentration in extracellular fluids than Na^+, but with intracellular concentrations more than proportionally lower still). Often Na^+ can also enter, sometimes by the same membrane channel as Ca^{2+} and elsewhere by an independent channel with different voltage sensitivity and kinetic properties (see Junge 1981).

The most complex AP known is that of the mammalian cardiac ventricle. Here, fairly rapid propagation, Ca^{2+} entry, and exceptionally long duration are all advantageous. The long duration ensures that a new beat cannot be activated before the previous one has terminated—this is required in a pumping function. What occurs (Fig. 2) is a fast Na^+ overshoot, followed

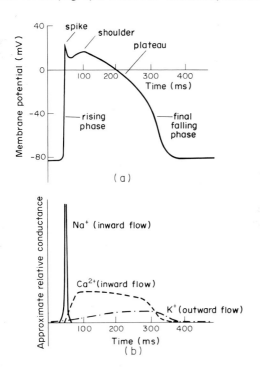

Figure 2
(a) Cardiac ventricular AP with differences from nerve AP somewhat emphasized. Note time scale.
(b) Sketch curves suggesting time courses of conductance-variation in active channels (derived from Beeler and Reuter 1977)

by a slow phase of Ca^{2+} entry, and only after about 250 ms the full activation of K^+ outflow. Other regions of the heart, such as the conducting bundles or the pacemaker, employ variations of this theme which impart greater propagation velocity, or spontaneous rhythmicity, and different responses to such chemical modifiers as adrenaline (Noble 1979).

4. Cables and Conduction

There is an evident analogy between a long nerve or muscle fiber and a submarine cable. This has been exploited to analyze the spread of activation in excitable

tissues, in terms of membrane capacitance and the resistances of core, membrane and external medium. The time constant, for charging a unit area of membrane, is proportional to the product of the capacitance and resistance of the area, while the length constant, for steady-state decrement of voltage with length, is given by

$$\left(\frac{\text{membrane resistance}}{\text{core resistance} + \text{medium resistance}} \right)^{1/2}$$

The dominant influence on propagation velocity, in cells of different size but the same type, is the rate at which activation spreads. Thus, the velocity can be increased by (a) myelination or (b) increasing the diameter of the fiber.

In myelinated nerves, velocity is proportional to external (myelin sheath) diameter; in nonmyelinated cells, it is proportional to the square root of fiber diameter. Mammalian, myelinated nerves range from about 1 to 20 μm in diameter and their conduction (propagation) velocities in m s^{-1} can be calculated by multiplying these diameters by a factor of 5–6. Myelination is so beneficial that even the giant (0.5–1 mm) invertebrate axons, studied by Hodgkin and co-workers, which have undoubtedly evolved to achieve the maximum velocity possible in the absence of myelination, only conduct at about 20 m s^{-1}. Jack et al. (1975) give a full account of cable theory and its applications.

5. Hodgkin–Huxley Equations for the Action Potential

The original equations of Hodgkin and Huxley (1952) refer explicitly to the giant nerve fiber of the squid. First, the voltage- and time-dependences of Na^+ and K^+ conductance, observed in voltage-clamp experiments, were approximated by empirically fitted expressions. Next, it was necessary to demonstrate that these expressions, obtained from a length of membrane responding as a unit to step-changes of voltage, could account for the so-called membrane AP—namely, the behavior of the same length of membrane left free to change its voltage in the milliseconds after a pulse stimulus. Finally, the same set of equations, describing current flow and voltage change in membrane units, were applied to infinitesimal units abutted axially to compose a cable. The form implied for the moving disturbance— the propagating AP—was thence computed.

The equations embody the basic tenets that Na^+ and K^+ are the only ions for which membrane conductance is not constant, and that individual ionic currents are simply additive (they do not interact) and ohmic. This implies that they are each expressible as the product of a conductance, specific for that ion, and a driving voltage—the latter being the difference between actual transmembrane potential E and the equilibrium potential for the ion concerned E_j. The kinetic responses of the above conductances to voltage change are described by equations of the form:

$$\mathrm{d}y/\mathrm{d}t = \alpha(1 - y) - \beta y$$

where α and β are rate constants varying with voltage in ways characterized empirically, and y (dimensionless) takes values between 0 and 1. Thus, y might be the fraction of a flux-controlling ("gating") particle in one state or position, $1 - y$ the fraction in the only alternative state, and α and β rates of entering and leaving the first state. Because Na^+ conductance first rises and then falls again on maintained depolarization, it is described as the product of two variables with the above kinetics, an "activation" variable m, and an "inactivation" variable h. Potassium ion conductance, which in squid nerve lacks the later fall, needs only an activation variable n. To embody the fact that both activations are delayed after a triggering depolarization (i.e., they follow sigmoid curves against time), the two activation variables are raised to powers >1 in the expressions for resulting conductance. Raising to the power 3, for instance, would have molecular meaning if three particles, each moving through the membrane with first-order kinetics after a voltage change, must cooperate to open the gate and let ions pass.

The final equation for membrane current density I_m (ignoring here a small leak of other ions) is

$$I_m = C_m \cdot \mathrm{d}E/\mathrm{d}t + \bar{g}_{Na+} m^3 h(E - E_{Na+})$$
$$+ \bar{g}_{K+} n^4(E - E_{K+})$$

where C_m is the membrane capacitance, and $\bar{g}_{Na+}$, $\bar{g}_{K+}$ represent the limiting attainable conductances to the indicated ions (gates fully open).

To derive the form of any AP, this equation must be solved simultaneously with the kinetic equations for the gating variables. These invoke a total of six subordinate expressions, representing the voltage dependences of the values of α and β. For the membrane AP, $I_m = 0$ once the triggering pulse is over as the ionic currents merely discharge and recharge C_m. For the propagating AP, the basic step is to reexpress I_m in terms of the second length derivative of E, using a fundamental relation from cable theory. Since the AP is a wave propagating with constant velocity θ, this may in turn be substituted by $(1/\theta^2)(\mathrm{d}^2 E/\mathrm{d}t^2)$, which makes the composite expression tractable. The equations are then solved numerically, for different values of θ, until a solution which gives a stable waveform is found. Values arrived at agree with experiment within relatively few percent, which, considering the inevitable approximations in the underlying curve fitting, is a satisfactory result.

Analyses other than Hodgkin and Huxley's have been applied to voltage-clamp results from giant nerve fibers, but it is doubtful if any fits the combination of original and subsequent biophysical data comparably well. The APs of many other tissues have in turn been subjected to analyses of the Hodgkin–Huxley type, that is, assuming independent ionic channels and considering the gating of each channel as a function separate from its ionic selectivity (its permeability to one ionic species rather than another). Conductance may be rectifying, not ohmic (e.g., at the nodes of Ranvier); activation variables may be raised to different powers (m^2 at nodes); I_{K+} may inactivate with time (e.g., in vertebrate skeletal muscles), or may not be present (mammalian nodes); inward current may be carried substantially or wholly by Ca^{2+} (e.g., Fig. 2 and many other preparations); yet the general approach seems remarkably fruitful.

6. Recent Developments in Membrane Biophysics

Apart from the voltage clamping of many other kinds of cell, the principal developments since 1952 have hinged around three new techniques. First came the replacement of the natural intracellular medium (cytoplasm) by artificial solutions. Initially achieved for squid nerve by Baker and Shaw in 1960 (see Junge 1981), this technique allows the ionic selectivity, pharmacology and other chemical properties of membrane channels to be much more fully investigated. Drawing largely upon such data, Hille (1978, 1984) has developed increasingly detailed molecular models of both the ion-filtering regions and the gates of the presumed transmembrane pores.

Secondly, the conductances of individual ionic channels have been investigated by two related techniques—noise analysis pioneered by Verween and by Stevens, and the "patch clamp," used notably by Neher (Neher and Stevens 1977). Electrical noise, due to fluctuations in the number of open ionic channels, was first studied with relatively conventional electrode techniques. More recently, areas of membrane $\sim 0.5\,\mu m$ in diameter have been isolated by suction onto microelectrode tips, and enable the modest number of ionic channels contained in such a patch (say 1–100, according to membrane and conditions) to be discriminated with substantially improved precision. Both approaches lead to values of $\sim 10\,pS$ for the conductance of a single cation channel. Encouragingly similar figures are obtained when conducting channels are introduced into artificial lipid films of thickness resembling natural membranes (black films) by the addition of pore-forming macromolecules.

Thirdly, the intramembrane displacement currents, flowing when the transmembrane PD is sharply altered, have been studied for indications of molecular events which might be associated with the opening and closing of ion gates. The summed effect of such events would show as asymmetries in the capacitative surge, when hyperpolarizing square waves (which do not open gates) are compared with depolarizing ones (which do open gates). To isolate intramembrane displacements from transmembrane ionic movements, experiments are performed in solutions lacking both Na^+ and K^+. "Asymmetry currents" have been studied in nerve by Armstrong (1981) and Keynes and Kimura (1983), and in muscle by Adrian (1978).

7. Synapses and Other Cell–Cell Junctions

Where one cell (excitable or otherwise) adjoins another there is a gap, measured in nanometers, between their membranes. The narrowest gap normally found, the "gap junction" (~ 1.5 nm) is bridged by characteristically arrayed macromolecular structures which appear to embody conducting channels. Such junctions offer low-resistance pathways for the electrical coupling of neighboring cells in most glands and in spontaneously active (cardiac and many smooth) muscles.

Synapses, the points of information transfer between nerve cells, are rarely of this direct, electrical type. Usually they employ chemical transmission across gaps of ~ 20 nm. Nerve–muscle and nerve–gland cell junctions are similar, though often have a wider gap. Many ideas about chemical transmission were, in fact, first worked out for the nerve–skeletal muscle junction, which is larger and more accessible than true synapses (Katz 1966).

Chemical "transmitter substance" is synthesized on one (the "presynaptic") side of the gap (e.g., acetylcholine at the terminal of a motor nerve to the skeletal muscle). From there it is released when an AP invades the terminal and diffuses across the gap. Transmitter release depends on Ca^{2+} entry. Experiments in low-Ca^{2+} media show that the release is quantized; however, with normal Ca^{2+} availability, many hundreds of quanta are released by every AP. Each quantum consists of thousands of molecules, and probably represents the content of one of numerous subcellular vesicles which characterize the ultrastructure of each presynaptic nerve terminal.

The effect of the transmitter when it reaches the other side of the gap (usually less than 1 ms after the AP invaded the presynaptic side) is to produce depolarization of the receiving ("subsynaptic") membrane. This is a region specialized to respond to its particular transmitter, and its response is not all-or-nothing but a depolarization graded in relation to the amount of chemical released. A graded response is possible because the chemically receptive region is not also excited by depolarization, so the runaway properly displayed by the AP is not present here. The chemically induced ionic response is a simultaneous increase of P_{Na^+} and P_{K^+}, much as had been assumed to occur in the AP itself till the intracellular electrode demonstrated the positive overshoot: accordingly, the maximum "postsynaptic" potential is a depolarization to an E_m which is still inside-negative by about 15 mV. From this locus of depolarization, however, current flows to the surrounding regions of membrane, which are once again electrically (and not chemically) excitable. If depolarization there exceeds threshold, a new AP arises and propagates onward along the postsynaptic cell. When the postsynaptic potential is sustained, there is in fact a train of APs, and because the rate of renewed triggering after any previous AP is proportional to the depolarizing voltage gradient, greater postsynaptic depolarization will lead to faster firing—the "frequency code."

It follows from their essential asymmetry that synapses act as "valves," which impose unidirectional information-flow on the nervous system: APs can propagate either way along a nerve fiber, but in normal biological function only ever arise at the postsynaptic end. It is also not surprising that most drugs and metabolic products influencing the nervous system do so principally at the synapses, which are chemically sensitive and are not all-or-nothing in their operation. In addition, though rare at nerve–skeletal muscle junctions, convergence of many synaptic inputs onto one cell is the rule in the central nervous system (CNS). This makes possible the summation of input effects, which are "spatial" when several synapses operate together and "temporal" when the same one operates more than once within the integration time of the subsynaptic membrane.

Another important process, which does not occur at vertebrate nerve–skeletal muscle junctions but is prominent elsewhere (particularly in the CNS), is inhibition. The commonest form of inhibition depends on certain transmitters which have direct, inhibitory effects on the postsynaptic cell. These effects are brought about by increases of K^+ and/or Cl^- permeability, which tend to hold the membrane either hyperpolarized or insufficiently depolarized for firing. Since central nerve cells often have thousands of synapses, a balance of excitatory and inhibitory influences must usually be reflected in the frequency at which each one fires. The lowest threshold for the initiation of APs exists where the main outgoing fiber (axon) leaves the nerve cell body. This, therefore, is the point at which the balance takes effect: the point at which what Sherrington, at the beginning of the century, called the "integrative action of the nervous system" must principally occur. The modern account of central excitatory and inhibitory mechanisms outlined above was pioneered by Eccles, a pupil of Sherrington, in the 1950s and 1960s.

Finally, in this section, it should be mentioned that many small nerve cells—arguably even the numerical majority in the CNS—might not possess the AP mechanism, but might rely entirely on synaptic interactions and the decremental spread of potential changes through their limited length. Such a possibility is technically hard to explore.

8. Electrical Behavior of Sensory Structures

Senses are classified as "general" or "special." General senses are those distributed throughout the body, such as touch, pain, and the sense of limb position. Special senses (e.g., vision, taste and balance) are each located in one or two specialized organs within the head. General senses employ the peripheral termination of the sensory nerve as their sensing element. Any accessory structure characteristically associated with a particular form of ending seems merely to modify the stimuli reaching that nerve termination. Special senses employ organized layers of detector cells which (except in the evolutionarily oldest case, smell) are not themselves neural; however, the nerves leading thence to the CNS perform not only "reporting" but analytic functions.

In the general senses, separate membrane specializations at the nerve terminal must be presumed to be part of each distinct form of sensitivity. The electrical consequences, however, seem similar in all cases, as demonstrated initially in the laboratories of Gray and Lowenstein in the 1950s and 1960s (see Schmidt 1978, Mountcastle 1980). Depolarization is produced in the sensitive ending, and its amplitude is graded to the stimulus. Further along the nerve, where the sensory specialization gives way to normal, electrical excitability, APs arise from any sufficient depolarization, and their frequency encodes the stimulus intensity. Note that the threshold of detection of a sensory stimulus by the CNS arises at this point. The analogy with the postsynaptic function of an excitatory synapse is strong; but the graded depolarization, which triggers the APs, is in the sensory case called a "generator potential."

The electrophysiology of special sense organs is a lot more complex. Complexity begins with the receptor cells: their response to a stimulus usually has an electrical component, but only in a minority of cases is this a depolarization leading simply (by an electrical or chemical synaptic mechanism) to nerve excitation. Many "receptor potentials" are indeed hyperpolarizations.

An example of how such responses arise is provided by studies of the vertebrate photoreceptors, rods and cones. On illumination they hyperpolarize to an extent which paradoxically increases with $\log[Na^+]_0$. Evidence from Hagins and Zuckerman suggests that, in the dark, Na^+ leaks rapidly into one end of the receptor and is pumped electrogenically out from the other (creating a "dark current") (see Mountcastle 1980). Light reduces P_{Na^+}, probably by releasing Ca^{2+} from the structures carrying the photopigments, and allows the pump to drive the inside of the receptor more negative. Communication to the earliest neural cells in the visual pathway (most of which in turn hyperpolarize) is by chemical transmission. Depolarizing transmitter is thought to be released continually from the photoreceptors onto the nerve-cell input terminals. When the receptors hyperpolarize they release less transmitter, so the nerve cells hyperpolarize. These early nerve cells, too, are small, and display only graded changes of E_m. Regenerative responses are not recorded earlier than the fourth cellular stage of the pathway.

One other class of bioelectric mechanism should be mentioned. It occurs in the inner ear. The primary sound-receptor cells ("hair cells") lie between two fluids which differ in potential by nearly 100 mV. Sound disturbance appears to modulate the resistance presented by the hair cell between the positive fluid and its own, extranegative interior. (The pitch of the sound signalled to the CNS is determined by *which* hair cells are active. This is a matter partly of acoustic tuning—as demonstrated by von Békésy and Davis in the 1950s

—and partly of electrical resonances, recently demonstrated in individual hair cells by several laboratories.) In any case, the resultant current variations, in phase with the incident sound ("cochlear microphonics"), contain more energy than the stimulus. This is one of the several steps by which the auditory pathway amplifies signals. Other amplifying steps, which all senses share, consist in the production of APs from original or synaptic stimuli. Most sensory pathways amplify threshold stimuli by a number of orders of magnitude. The ultimate energy source is of course metabolism, usually operating by means of the transmembrane ionic pumps.

9. Whole-Organ Alternating Current Signals

Gross, external electrical signals can be detected not only by the cochlea (as discussed above), but from most other active organs of special sense. The electroretinogram (ERG) first recorded by Helmholtz in the 1860s has been particularly employed as a research, and sometimes a diagnostic, tool. As might in part be deduced from the last section, its interpretation in relation to cellular events is problematic, despite major contributions, notably by Brown (see Mountcastle 1980). Phenomena of perhaps greater general interest are the signals from brain, muscles, and most especially, heart (see *Electrocardiography*).

The electroencephalogram (EEG), pioneered by Berger in the 1920s, consists of the massed, alternating signals of the whole brain, recorded from the scalp surface (though the electrocorticogram, recorded direct from an exposed area of brain surface during surgery, is often instructively similar). A sustained rhythm ($\sim 10 \, Hz$) of high amplitude, the "α rhythm," characterizes waking inactivity or very light sleep, and has been interpreted as a synchronized "hunt" for input. It is interrupted by attention or concentration, and may be replaced by pulses of electrical activity having higher frequency and usually lower amplitude. Though there is controversy as to the origins of EEG waveforms, they are used both clinically and in psychological research, as particular patterns are empirically indicative of particular states from epilepsy to dreaming (see *Electroencephalography*).

Electromyogram (EMG) records may be taken from the skin overlying the muscle concerned, or, with precision gained at the cost of comfort, using needle electrodes pushed into the tissue. They represent the summed APs of the muscle fibers, so that the degree of electrical activity, when it can be unambiguously recorded, exactly indicates the extent of muscle activation. Studies of biofeedback have recently made use of electromyography, though it is a technique which has been familiar to clinicians and exercise physiologists for some time (see *Electromyography*).

The most widely employed form of whole-organ recording is, however, electrocardiography. The first clear human electrocardiogram (ECG) was made with a string galvanometer by Einthoven in 1903. He recorded from three leads, one on each arm and one on a leg, usually the left leg. Modelling the body as a homogeneous, equilateral triangle with the heart as its center (the "Einthoven triangle"), one may calculate the vector orientation and magnitude of the single dipole to which the heart's summed emfs are equivalent at any instant. Indeed, in modern vectorcardiography, leads I and II (Fig. 3) may be taken to the *x* and *y* axes, respectively,

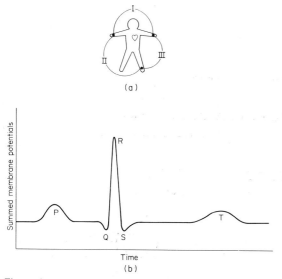

(a)

(b)

Figure 3
(a) Positions of classical leads for recording an ECG.
(b) Normal human ECG such as might be recorded from lead II

of an oscilloscope, and the cyclic excursions of the end of the vector traced out. The basic ECG (Fig. 3) is, however, easier for the nonexpert to interpret, and gives much useful information. The P wave represents depolarization of the first heart chambers to be active in each beat, the atria. The complex QRS wave represents depolarization of the main mass of muscle, the ventricles. Atrial repolarization is obscured by QRS, but ventricular repolarization produces the T wave; thus, the mean duration of the ventricular AP (Fig. 2) corresponds approximately to the QT interval. As heart rate increases (e.g., in exercise), this interval shortens; but the main reduction is in the interval between beats, TP. In early stages of the condition known as "heart block," where propagation of the AP from atria to ventricles is impaired, some P waves, particularly during exercise, are not followed by QRST; in total block, the two are entirely dissociated. This is one example of the diagnostic use of the basic ECG. Another notes vertical displacements of the ST trace, or even inversions of the T wave. These may represent abnormal sequences of repolarization across the population of ventricular cells,

and give warning that some probably have inadequate blood supply.

In modern clinical practice, the classical leads are supplemented by "precordial" leads, recording between an electrical mean of the Einthoven leads and one of six points on the front of the chest. These give general information about the "dorso-ventral" vector, which the classical leads cannot represent. In addition, simply by proximity, they may give localized information about the site of an abnormality (see *Vectorcardiography*).

10. External Electromagnetic Fields

Although a wide variety of species, from bacteria to birds, are sensitive to the earth's magnetic field, only in certain fish are electrical fields known to be used navigationally, or rather in local pilotage. Predators such as sharks can locate buried prey by the natural electric fields of the latter's body. Sensitivity is such that current densities in the sub-pA cm^{-2} range appear to be detectable. So-called "weak electric fish" themselves emit electrical signals pulsed at up to 1 kHz and steer in muddy water according to the sensed electrical characteristics of the materials near them. The detection sites, electroreceptors scattered across areas of the skin, have in both cases been identified.

A very different function is served by the strong shocks (up to 600 V) which can be delivered by the offensive electrical organs of a small number of predatory fish, such as electric rays. These organs, termed electroplaques, consist of thousands of modified nerve–muscle junctions, operating in series, and are activated by the release of acetylcholine from incoming nerves. The least understood feature of the mechanism is why the delivering fish itself is unaffected by the discharge.

Mammals and man are, of course, aware of the stimulation of pain receptors by electric shocks to the skin, but they have no conscious sensibility to weak electric fields. However, consequences of exposure to electromagnetic radiation, at frequencies below the ionizing level (e.g., microwave) and intensities below those likely to produce significant heating, have been reported for a generation in the Communist world and are now receiving some scientific credence in the West. Effects on circadian rhythms, bone and blood-cell growth, and behavioral indices have been reported; and sometimes an enzymic or ionic change capable of underlying the gross response has been identified. Investigation is complicated by apparent "windows" of effectiveness, in terms of modulation or carrier frequency and even of power. Threshold currents or fields are often only about 10^{-5} of those required to trigger, say, an AP in an isolated nerve cell. However, they are of the same order as the volume–conductor fields induced by an active cell at modest distances in its surrounding tissue—fields which, at least in the case of the brain, it has been independently suggested should not be ignored. Adey (1981) is among those who have speculated on cooperative phenomena, perhaps involving chemical movements along the surfaces of membranes rather than across them, which might allow such weak fields to influence cell masses. The importance of this topic for public health will be evident.

11. Epithelia and Glands

The discussion so far has been mostly concerned with so-called "excitable cells": nervous, sensory and muscular. Gland cells have been mentioned, but only in so far as they, too, may be excited (stimulated) by nerves to secrete. However, all cells have membrane potentials, because they all have inside–outside concentration differences maintained by metabolic pumps. In addition, there is a whole range of extracellular PDs—some having physiological importance in themselves and all being valuable indicators of tissue activity—produced by cells which are not, in any intended sense, excitable (i.e., do not give brief electrical responses to stimuli).

What is required, instead, is that the cells be organized in layers, with robust and relatively leak-resistant lateral adhesion, so that they separate one compartment of extracellular fluid from another. Such a layer is termed an epithelium, or if so convoluted as to enclose one of the compartments largely or completely and secrete fluid into it, it becomes a gland. Thus, glands are a subclass of epithelia, and usually employ comparable biophysical mechanisms.

Logically, there is one more requirement for an epithelium to produce a PD: the active transport processes at its two surfaces must be asymmetrical. However, since epithelia only form where fluids must be made or kept chemically different, this requirement is always met.

The first epithelium to be subject to rigorous biophysical analysis, by Ussing and co-workers in the 1950s, was frog skin (see Ussing and Thorn 1973). Living in an aqueous environment of many times lower salt concentration than its body fluids, the frog needs to take up salt and extrude water at rates sufficient to counter leaks in the opposite direction. Water is lost through the kidneys, but the salt uptake is through the skin. An early observation was that absorptive activity produced a positive potential on the inner surface, relative to the outer. This potential was greatest (many tens of millivolts in experimental situations) when the impermeant anion SO_4^{2-} was substituted in the outer fluid for Cl^-, but it was zero only when the outside solution contained no Na$^+$ or when known chemical inhibitors of Na$^+$ pumps had been applied. Ussing concluded that Na$^+$ was actively transported inwards, and Cl$^-$, when available, followed passively, drawn by the electrical gradient at the same steady-state rate as Na$^+$, but against sufficient resistance that it only diminished that gradient, rather than eliminating it.

A prime tool in Ussing's analysis was the short-circuit current. The skin PD was backed off, so that it was in effect voltage clamped to zero, and the current required could be taken as equal and opposite to that flowing through the skin. If, additionally, the solutions on the two sides of the skin had been made identical for the purposes of this experiment, no bulk driving force would be available to bring about net movements. Only ions directly subject to metabolically driven pumps would show movement (demonstrable radioisotopically) from one side to the other. In the frog skin, the one ion so moving was Na$^+$: its flux exactly balanced the short-circuit current.

The contents of many tubular organs, such as the gut, reproductive passages and urinary tract, are topologically exterior to the animal body, and the majority, though not quite all, show negative potentials relative to the body fluids, analogous to those of frog skin. Though other ions (e.g., Cl$^-$) are subject to direct active transport in specialized locations, it is much more common for a Na$^+$ pump to be the prime mover. Figure 4 illustrates the way the ordinary Na$^+$ pumps of the cell

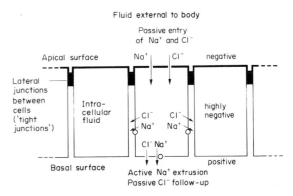

Figure 4
Representative actively transporting epithelium (detailed membrane morphology omitted). Components are labelled on cell at left, ionic movements indicated in center, and relative voltages shown by sign and size of symbols, right

membrane may be concentrated into one face only of an epithelial cell and interact with selective permeabilities, which are also asymmetrically located, to produce net transport. The diagram refers specifically to the active layer of frog skin, and to the epithelia lining parts of the intestines and kidneys of vertebrates generally, but only slightly modified mechanisms are widespread. In most locations, characteristic morphological developments increase the areas of membrane present, and bring the sites at which metabolic energy is made available very close to where it is used; these changes enhance transport capacity, but do not alter principles.

Three points should be noted. First, water transport in biological systems seems always to be an osmotic consequence of the primary active transport of solutes, usually NaCl. Secondly, certain solutes which are not charged, and therefore cannot be electrically associated with ionic pumping, are transported in the wake of Na^+ by virtue of a modified, apical, Na^+-entry system: in these instances at least, Na^+ appears to enter the cell not by simple pores but by carrier systems of such chemical constitution that they take the other solute across the membrane with the Na^+. Glucose absorption in the intestine is achieved by coupled transport of this type. Finally, although the ionic pumps of well-documented epithelia appear in all cases to be electrogenic, note that a neutral pump would in general produce a PD across an epithelium, just as it would across a cell membrane, because of the concentration differences it set up. Only where equal net permeabilities were presented by the epithelium to back-diffusion of the two pumped ions would no PD develop.

12. Injury and Growth Potentials in Animal Tissues

When any cell is cut, current flows from the exterior into the negatively charged cytoplasm, until either the membrane potential in the remaining, intact region runs down due to cessation of pumping, or the cut has been sealed by a high-resistance clot. On a grosser scale, if a wound is made in a pumping epithelium (or if a digit or limb, surrounded by such an epithelium, is cut through) current flows usually with the wound positive to the skin surface. "Injury currents" of both these types have long been known. In the last 20 years, however, there have been a number of reports of accelerated or enhanced wound healing by imposition of currents supplementary to the natural ones, and healing impairment by opposite currents. Clinical use of what appear to be similar phenomena has been made in the enhancement of bone repair. There is controversy, however, both as to whether or not nerve is involved in the natural process, and as to whether the electrical enhancements are achieved directly or by electrolytic products (Jaffe and Nuccitelli 1977, Becker 1978).

Bone in turn provides one of the best-known examples of the possibility that natural growth is electrically modulated. (Research into this was pioneered by Lund, in the 1930s and 1940s.) Bone thickens where it is stressed. Stress also induces piezoelectric potentials in bone: signals are generated in the protein (collagen) fibers and rectified at the fiber–mineral junctions. (This is one of the very few bioelectric potentials, above the molecular level, which does not have its origin in a transmembrane ionic pump). By mechanisms presumably related to those of the previous paragraph, the electrical effects might induce the growth. Among other possibilities is that the voltages existing across transporting epithelia might cause electrophoresis of membrane-dissolved molecules from base to apex, or vice versa. The few millivolts recorded across embryonic

cell layers would suffice, and the resulting intra-membrane gradients could arguably promote the bio-chemical and morphological differentiation of apical from basal surfaces, which characterizes mature transporting epithelia.

Finally, there are instances where changes in single-cell membrane potentials modulate development. There is no better example than the one with which development begins: fertilization of the ovum. When one sperm has entered, the rapid block to entry of another depends on a depolarization by $\sim 50\,mV$ taking place within a few seconds.

13. Bioelectric Mechanisms in Plants

In plants, as in animals, potential differences of tens of millivolts can often be recorded which appear to be secondary consequences of concentration differences, of pumping or (possibly important in plants) of streaming phenomena. Plant cells, like those of animals, have sodium pumps: they thus contain high concentrations of K^+ relative to extracellular fluids and display membrane potentials. Roots take up ions by vigorous active-transport mechanisms, analogous to those of epithelia (in particular, to the skins of freshwater animals). Plant biophysicists frequently use electrical methods as tools for studying these processes. Instances where the essential biological mechanism is electrical do not, however, appear to be nearly as numerous in plants as in animals.

Action potentials do occur in plants. Giant algal cells have been studied, notably by MacRobbie, almost as much as giant nerve fibers, and for similar reasons (see Hope 1971, Nobel 1974). In these, Cl^--based APs are seen, but their function is obscure. In higher plants, however, APs seem to be the triggers for some, though not all, rapid movements: the protective droop of mimosa is a well-documented instance, and certain insectivorous plants, such as the Venus fly-trap, probably signal from sensors to motile structures in similar ways.

Injury and growth potentials are widespread. Current flows inward at the growth-point of many single plant cells, and growth is stimulated by artificial imposition of such flow: the beneficial effect may be that the current takes Ca^{2+} into the cytoplasm. Phenomena which appear somewhat similar have been demonstrated in the growing roots of many higher plants. At first sight the growth potentials in aerial shoots such as corn coleoptiles also appear similar; when these turn against gravity or towards light, the faster-growing side (where the stimulating hormone or auxin accumulates) may become positive by as much as $80\,mV$ relative to the other side. However, in this instance (as Lund showed) the PD appears to have a negative-feedback effect, for imposed fields enhance growth on the negative side, perhaps by attracting auxin to it.

14. The Electrical Aspect of Adenosine Triphosphate Synthesis

An extremely important and general biological mechanism, which is most easily demonstrated in plants, is the build-up of PD (often in excess of 100 mV) across membranes which are actively synthesizing adenosine triphosphate (ATP), the universal intracellular energy conversion and storage molecule. In the plant, the synthesis is activated by light (photosynthesis). In the animal cell, the principal mode of synthesis is by fuel oxidation (oxidative phosphorylation). In either case, a gradient of protons (H^+) and a consequent PD are built up across the active, intracellular membrane concerned—that within the plant-cell organelle, the chloroplast, or the animal-cell organelle, the mitochondrion. According to the favored, modern view of Mitchell, it is the energy of this H^+ gradient which drives the formation of ATP (see Nicholls 1982).

Almost all the bioelectric phenomena discussed in earlier sections depend on metabolism supplying ATP to the ionic pumps located in cell membranes. If the H^+ gradient hypothesis is right, as seems highly likely, the overall relationship between bioelectricity and ATP is satisfyingly circular.

See also: Evoked Potentials; Muscle

Bibliography

Adey W R 1981 Tissue interactions with nonionizing electromagnetic fields. *Physiol. Rev.* 61: 435–514

Adrian R H, 1978 Charge movement in the membrane of striated muscle. *Annu. Rev. Biophys. Bioeng.* 7: 85–112

Adrian R H, Marshall M W 1976 Action potentials reconstructed in normal and myotonic muscle fibres. *J. Physiol.* 258: 125–43

Aidley D J 1978 *The Physiology of Excitable Cells*, 2nd edn. Cambridge University Press, Cambridge

Armstrong C M 1981 Sodium channels and gating currents. *Physiol. Rev.* 61: 644–83

Barlow H B, Mollon J D (eds.) 1982 *The Senses*. Cambridge University Press, Cambridge

Becker R O 1978 Electrical osteogenesis: Pro and con. *Calcif. Tissue Res.* 26: 93–97

Beeler G W, Reuter H 1977 Reconstruction of the action potential of ventricular myocardial fibres. *J. Physiol.* 268: 177–210

Chiu S Y, Ritchie J M, Rogart R B, Stagg D 1979 A quantitative description of membrane current in rabbit myelinated nerve. *J. Physiol.* 292: 149–66

Cole K S 1968 *Membranes, Ions and Impulses: A Chapter of Classical Biophysics*. University of California Press, Berkeley

Fain G L, Lisman J E 1981 Membrane conductances of photoreceptors. *Prog. Biophys. Mol. Biol.* 37: 91–147

Gregory R L 1981 *Mind in Science. A History of Explanations in Psychology and Physics*. Weidenfeld and Nicolson, London

Handbook of Physiology 1977 *et seq*. Sect. 1: *The Nervous System*; Sect. 2: *The Cardiovascular System*. American Physiological Society, Bethesda; Williams and Wilkins, Baltimore

Hille B 1978 Ionic channels in excitable membranes: Current problems and biophysical approaches. *Biophys. J.* 22: 283–94

Hille B 1984 *Ionic Channels of Excitable Membranes*. Sinauer, Sunderland, Massachusetts

Hodgkin A L, Huxley A F 1952 A quantitative description of membrane current and its application to conduction and excitation in nerve. *J. Physiol.* 117: 500–44

Hope A B 1971 *Ion Transport and Membranes: A Biophysical Outline*. Butterworth, London

Jack J J B, Noble D, Tsien R W 1975 *Electric Current Flow in Excitable Cells*. Clarendon, Oxford

Jaffe L F, Nuccitelli R 1977 Electrical controls of development. *Annu. Rev. Biophys. Bioeng.* 6: 445–76

Junge D 1981 *Nerve and Muscle Excitation*, 2nd edn. Sinauer, Sunderland, Massachusetts

Kandel E R, Schwartz J H 1981 *Principles of Neural Science*. Edward Arnold, London

Katz B 1966 *Nerve, Muscle, and Synapse*. McGraw-Hill, New York

Keynes R D, Kimura J E 1983 Kinetics of activation of the sodium conductance in the squid giant axon. *J. Physiol.* 336: 621–34

Mountcastle V B (ed.) 1980 *Medical Physiology*, 14th edn. Mosby, St. Louis

Neher E, Stevens C F 1977 Conductance fluctuations and ionic pores in membranes. *Annu. Rev. Biophys. Bioeng.* 6: 345–81

Nicholls D G 1982 *Bioenergetics: An Introduction to the Chemiosmotic Theory*. Academic Press, London

Nobel P S 1974 *Introduction of Biophysical Plant Physiology*. Freeman, San Francisco

Noble D 1979 *The Initiation of the Heartbeat*, 2nd edn. Clarendon, Oxford

Schmidt R F (ed.) 1978 *Fundamentals of Sensory Physiology*. Springer, Berlin

Ussing H H, Thorn N A (eds.) 1973 *Transport Mechanisms in Epithelia*, 5th Alfred Benzon Symposium. Munksgaard, Copenhagen

N. Spurway

Biofeedback

In biofeedback, the patient or subject, who is made aware of the ongoing measured value of some aspect of his bodily function, attempts to self-induce a beneficial change in that function, with or without some form of encouragement from the practitioner or experimenter. Typical examples of bodily processes said to be modified through biofeedback training are given in Table 1. In no instance is the patient given specific instructions on how to bring changes about. Indeed, it is not known what the patient does to change the functioning of a biological process or what response on his part is reinforced, should encouragement be given to him by the practitioner. Control may be based on greater conscious

Table 1
Some typical examples of biofeedback training

Biological process	Application
Alpha brain-wave activity	mental tranquility (Kamiya 1968)
Sensorimotor cortical rhythm	epilepsy (Lubar and Bahler 1976)
Heart rate	anxiety (Blanchard and Abel 1976)
Blood pressure	essential hypertension (Benson et al. 1971)
Galvanic skin response	emotional control (Klinge 1972)
Surface skin temperature	migraine (Turin and Johnson 1976)
Forehead muscle tension	tension headache (Budzinski et al. 1973)
Colon activity	functional diarrhoea (Furman 1973)
Respiration	asthma (Vachon and Rich 1976)
Muscle reeducation	foot-drop (Basmajian et al. 1975)
Visual acuity	myopia (Epstein et al. 1978)
Single motor units	none

awareness of the bodily process; on cognitive activity of varying sorts, for example, thinking of exciting or calming events, solving problems in mental arithmetic, thinking of one's limbs becoming tense or relaxed, or hot or cold; on unmediated and unconscious internal responding; or the action of a servocontrol system (Yates 1980).

Biofeedback has been systematically studied for less than 20 years. As an area of research, it came into being as a result of the convergence of several lines of investigation: the establishment of so-called voluntary control of electroencephalography (EEG) patterns, operant conditioning of the heart rate of rats whose skeletal muscles had been totally paralyzed, and the operant conditioning of the human heart and the galvanic skin response.

An area of biofeedback research and application that is fairly representative of work in the field is surface skin temperature of the hand (or finger) and its relation to migraine headaches. Basic research by Keefe (1975) indicated that subjects may be trained to raise or to lower finger surface skin temperature merely with the use of auditory and visual feedback. Changes were less than 1 °C. Control subjects who might have been given instructions to raise or lower skin temperature without the aid of biofeedback were not employed. In a review of the biofeedback process literature, King and Montgomery (1980) noted that such small changes had been typically found by a number of researchers, and that biofeedback-induced decreases in surface skin temperature had been more easily obtained than increases. Furthermore, King and Montgomery (1981) were able to find reliable changes in surface skin temperature only under conditions where subjects were permitted to engage in peripheral muscular movements, that is, as opposed to altering some more central mediating process to bring about changes in surface skin temperature of the

hand. Yet, Donald and Hovmand (1981) provided fairly convincing results leading them to the contrary conclusion that the autoregulation of skin temperature can be demonstrated in a virtually placid limb. Finally, King and Montgomery (1980) noted that although average surface skin temperature changes had been small, sizable individual differences in temperature control had been reported by a number of researchers; for example, some subjects have been able to increase their surface skin temperature by as much as 10 °C.

Applied research was inaugurated by the demonstration by Graham and Woolfe in 1938 that ergotamine tartrate, a vasoconstrictor which had been used in the medical treatment of migraine as early as 1884, reduced the amplitude of pulsations of the temporal artery with an associated decrease in head pain (Dalessio 1972). This was partly because ergot and various other medications, while proving effective for migraine in many cases, also produced undesirable side-effects.

On the basis of the hypothesis that migraine was a stress-related syndrome involving both dysfunctioning of vascular behavior in the head and periphery and of the sympathetic nervous system. Sargent et al. (1972, 1973) devised a biofeedback treatment utilizing the measurement of surface skin temperature of the hand by means of thermistors attached to the right index finger and forehead. Presumably this treatment approach was aimed at decreasing the likelihood of the painful dilation of the extracranial arteries commonly associated with migraine headache, somewhat indirectly, by training subjects to reduce peripheral vasoconstriction. Specifically, subjects were instructed to warm their hands making use of the visual display of the difference in temperature between their head and hands and with the help, supposedly, of autogenic phrases such as "I feel quite quiet; I am quite relaxed; my whole body is relaxed and my hands are warm, relaxed and warm." They were seen weekly and practised hand-warming at home. Some 74% of the subjects were judged to be improved at the end of a year—an impressive result. Unfortunately, a complete report of changes in surface skin temperature was not provided; essential control procedures were lacking. Subjects may have improved clinically as a result of the passage of time, nonspecific treatment effects including expectations on the part of the experimenter being transmitted to the subject, indirect or general relaxation training, specific biofeedback training, or any combination of these.

An improvement in experimental expertise was evidenced in studies by Wickramaskera (1973) and Turin and Johnson (1976). Wickramaskera treated two migraine subjects with a series of relaxation training sessions including daily home practice assisted by biofeedback training of the forehead muscles. No improvement in migraine activity was found. However, clinical improvement was subsequently attained with a series of hand-warming biofeedback training sessions. Turin and Johnson trained four migraine subjects to warm their hands, and three other subjects initially to cool their hands and later on to warm their hands in an attempt to control for nonspecific expectancy effects. Data were presented indicating that subjects cooled and warmed their hands. Clinical improvement only occurred with hand warming and not hand cooling. Subjects reported that they were able to use their training to prevent the onset of migraine or at least to reduce its intensity. Both studies provided some further support for the belief that migraine headache activity is a function of the surface temperature of the hands, in that nonspecific factors and relaxation appeared to have been nullified yet a decrease in migraine activity was found to be associated with an increase in surface skin temperature of the hand.

Further advance in experimental design was represented by the controlled group study of Blanchard et al. (1978) involving 37 migraine subjects being randomly assigned to each of three conditions: hand-warming, autogenic training, home practice; progressive relaxation training, home practice; and waiting-list control. Migraine headache activity decreased in all three groups but the decrease was greater for the treatment groups than for the control group. No difference was found between the two treatment groups. This study provided support for the efficacy of the treatment of Sargent et al. (1972, 1973) for migraine by controlling for the passage of time and some incidental nonspecific effects. These results led Blanchard and Epstein (1978) to conclude that thermal biofeedback training with regular home practice and relaxation training with regular home practice were equally worthwhile as treatments for migraine, and consequently they postulated that a final common pathway of action in migraine, for both relaxation training and biofeedback training, is a relaxation response which includes peripheral vasodilation and hence skin temperature increase. Unfortunately, Blanchard et al. (1978) did not present data pertaining to the actual level of performance attained in terms of increased surface skin temperature.

A double-blind controlled group study by Kewman and Roberts (1980) involved 34 migraine subjects being randomly assigned to three groups: biofeedback hand warming, handcooling, and self-monitoring only. All groups showed significant improvement with regard to migraine headache activity. However, there were no significant between-group differences. Furthermore, the data were regrouped in accordance with a learning criterion to form four new groups comprising subjects whose hand surface skin temperature was either increased, not affected by training, not involved in training, or decreased. Significant reductions in migraine headache activity were found for all but the temperature decrease group. Yet no differences were found among these three groups, that is, changes in migraine headache activity cannot be attributed to the specific treatment effect of raising the surface skin temperature of the hands. However, this conclusion should be tempered by the additional finding that migraine headache activity was significantly greater for the decrease temperature

group than for the other three groups combined. More recent studies (e.g., Blanchard et al. 1982, Daley et al. 1983) firmly point to the qualified conclusion that biofeedback training of surface skin temperature results in decreases in migraine headache activity for some subjects. However, missing in these reports are data specifically linking increases in hand temperature with these decreases in migraine headache activity.

To summarize, the above studies indicate that: (a) increases in surface skin temperature of the hands may be accomplished via some unspecified central process, and that these changes may indeed be large for some individuals; (b) reduction in migraine headache activity is associated wih surface skin temperature biofeedback training for some individuals; and (c) increases in surface skin temperature of the hands per se have not been shown to be linked with decreased migraine headache activity to a substantial degree.

These findings on surface-skin-temperature biofeedback training and migraine are typical for the field of biofeedback and self control. Specific biofeedback training to influence specific biological processes involving central control has been shown to be effective to varying degrees. The practical value of biofeedback training programs has been demonstrated in a number of areas, such as anxiety, asthma, hyperactivity, colitis, hypertension, insomnia, and tension headache for some subjects. However, the expectation that biofeedback training would be a kind of behavioral medicine that would improve the functioning of various bodily processes and physical health as a consequence of greater specific control of these processes is generally still unfulfilled. A major exception to this somewhat pessimistic conclusion has been the area of the rehabilitation of physical function where case studies and controlled experimental work have provided strong evidence for the specific effectiveness of biofeedback training (Yates 1980).

Nevertheless, there appears to be little doubt that biofeedback will continue as a field of psychological research. Perusal of recent issues of the journal *Biofeedback and Self Regulation* reveals a firm trend for rigorous work utilizing powerful, sophisticated experimental designs and novel experimental approaches. Whether such a development will lead to unequivocal demonstrations of substantial control of bodily processes utilizing biofeedback and whether the application of such control will prove to be efficacious is of course an empirical matter as in any other field of applied science.

See also: Biological Control Theory

Bibliography

Basmajian J V, Kukulka C G, Narayan M G, Takebe K 1975 Biofeedback treatment of foot-drop after stroke compared with standard rehabilitation techniques: Effects on voluntary control and strength. *Arch. Phys. Med. Rehabil.* 56: 231–36

Benson H, Shapiro D, Tursky B, Schwartz G E 1971 Decreased systolic blood pressure through operant conditioning techniques in patients with essential hypertension. *Science* 173: 740–42

Blanchard E B, Abel G G 1976 An experimental case study of the biofeedback treatment of a rape-induced psychophysiological cardiovascular disorder. *Beh. Ther.* 7: 113–19

Blanchard E G, Andrasik F, Neff D F, Arena J G, Ahles T A, Jurish S E, Pallmeyer T P, Saunders N L, Teders S J, Barron K D, Rodichok L D 1982 Biofeedback and relaxation training with three kinds of headache: Treatment effects and their prediction. *J. Consult. Clin. Psychol.* 50: 562–75

Blanchard E B, Epstein L H 1978 *A Biofeedback Primer.* Addison-Wesley, London

Blanchard E B, Theobald D E, Williamson D A, Silver B V, Brown D A 1978 Temperature biofeedback in the treatment of migraine headaches. *Arch. Gen. Psychiat.* 35: 581–88

Budzinski T H, Adler C S, Mullaney D J 1973 EMG biofeedback and tension headache: A controlled outcome study. *Psychosom. Med.* 6: 509–14

Dalessio D J 1972 *Wolff's Headache and other Head Pain.* Oxford University Press, New York

Daly E J, Donn P A, Galliher M J, Zimmerman J S 1983 Biofeedback applications to migraine and tension headaches: A double blinded outcome study. *Biofeedback Self Regul.* 8: 135–52

Donald M W, Hovmand J 1981 Autoregulation of skin temperature with feedback-assisted relaxation of the target limb, and controlled variation in local air temperature. *Percept. Mot. Skills* 53: 799–809

Epstein L H, Collins F L, Hannay H J, Looney R 1978 Fading and feedback in the modification of visual acuity. *J. Behav. Med.* 1: 273–87

Furman S 1973 Intestinal biofeedback in functional diarrhea: A preliminary report. *J. Beh. Ther. Exp. Psychiat.* 4: 317–21

Kamiya J 1968 Conscious control of brain waves. *Psychol. Today* 1: 57–60

Keefe F J 1975 Conditioning changes in differential skin temperature. *Percep. Mot. Skills* 40: 283–88

Kewman D, Roberts A H 1980 Skin temperature biofeedback and migraine headaches: A double-blind study. *Biofeedback Self Regul.* 5: 327–45

King N J, Montgomery R B 1980 Biofeedback-induced control of human peripheral temperature: A critical review of the literature. *Psychol. Bull.* 88: 738–52

King N J, Montgomery R B 1981 The self-control of human peripheral (finger) temperature: An exploration of somatic maneuvers as aids to biofeedback training. *Beh. Ther.* 12: 263–73

Klinge V 1972 Effects of exteroceptive feedback and instructions on control of spontaneous galvanic skin response. *Psychophysiol.* 9: 305–17

Lubar J F, Bahler W W 1976 Behavioral management of epileptic seizures following EEG biofeedback training of the sensorimotor rhythm. *Biofeedback Self Regul.* 1: 77–104

Sargent J D, Green E E, Walters E D 1972 The use of autogenic feedback training in a pilot study of migraine and tension headaches. *Headache* 35: 129–35

Sargent J D, Green E E, Walters E D 1973 Preliminary report on the use of autogenic feedback training in the treatment of migraine and tension headaches. *Psychosom. Med.* 35: 129–35

Turin A, Johnson W G 1976 Biofeedback therapy for migraine headaches. *Arch. Gen. Psychiat.* 33: 517–19

Vachon L, Rich E S 1976 Visceral learning in asthma. *Psychosom. Med.* 38: 122–30

Wickramaskera I 1973 Temperature feedback for the control of migraine. *J. Beh. Ther. Exp. Psychiat.* 4: 343–45

Yates A J 1980 *Biofeedback and the Modification of Behavior.* Plenum, New York

R. Nicki

Biological Control Theory

Claude Bernard in his lectures to the Royal College of France in the second half of the nineteenth century demonstrated one of the earliest examples of a systematic approach to the analysis of physiological systems. Bernard divided all life into three forms. The lowest form comprised animals that are totally dependent upon the nature of the external environment; for example, eels have been known to return to normal physiological function after being removed from ice. The next category contained animals partially dependent on the environment; cold-blooded animals, poikilotherms, are an example. The third and highest category referred to those animals possessing an internal environment which is largely independent of the external environment. Man lies in this third category. The maintenance of a constant internal environment requires the existence of a number of effective biological regulatory and/or control systems. The study of such systems advanced with the work of Cannon in the 1920s (Cannon 1929). Truly quantitative analysis of biological control systems probably dates from the work of Weiner (1948) when in his classic book *Cybernetics* he provided the conceptual and theoretical framework for much of the work which has followed.

The material presented in this outline of biological control theory is intended to provide a general view of the topic rather than a discussion of specific biological systems. For additional information on specific topics the reader is referred to the works of Linkens (1979), Schwan (1969) and Brown and Gann (1973).

A basic model of a biological control system is shown in Fig. 1. In the figure, X and Y are the system input and

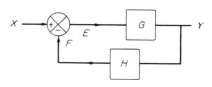

Figure 1
Canonical form of a biological control system

output. The output or controlled variable is maintained at a fixed level which is defined by the input. The variable Y is regulated in the following way. Assume that its value is forcibly changed by some external influence; then the new level of Y is relayed back to the comparator by the feedback or afferent pathway H. The feedback signal F is then subtracted from the input X to define the error signal E. The error signal in turn activates the forward or efferent pathway G which is capable of returning the output Y to its original value. These relationships can be defined algebraically by stepping around the loop: $F = HY$, $E = X - F$ and $Y = GE$. Hence the relationship of Y to X can be defined as:

$$Y = \frac{G}{1 + GH} X \tag{1}$$

The term $G/(1 + GH)$ is often called the closed loop transfer function. This control strategy is for the system operating as a regulator. The same configuration becomes a control system if X is varied. Under these circumstances the control strategy will cause the output Y to follow the input X. For those unfamiliar with systems theory, Eqn. (1) may appear either deceptively simple or obscure. The real question is how Eqn. (1) can be used to describe the behavior of a real biological system. To answer this question, the problem will be addressed first qualitatively and secondly quantitatively.

Consider the physiological example of blood pressure control. Assume that the variable Y is aortic blood pressure which is being observed immediately following the onset of heavy exercise. Under these conditions Y will be below its normal level. This information is fed back to the blood pressure center in the brain by the baroreceptors situated in the aortic arch and carotid sinuses. The feedback signal F is then compared with the set value for aortic blood pressure X, and an error signal E defined. The signal E activates the appropriate efferent regulatory mechanisms, in this case increased heart rate, stroke volume or peripheral vascular resistance (Hyndman 1973). Even from this simple description of a physiological response it is clear that the simple block diagram of Fig. 1 is inadequate since there are at least five parallel afferent pathways and three parallel efferent pathways. Hence the block diagram of Fig. 1 should be expanded in terms of series elements and parallel pathways as required.

Two other important factors which emerge from this simple analysis are the existence of a set point and the nature of the physiological signals.

The existence of set points in biological control systems has been an issue which has caused considerable discussion over an extended period. The core of the problem is that while from the control theory standpoint a set point is necessary, it is often difficult to envisage what this might mean physiologically. In some biological control systems the problem can be potentially resolved by the existence of two antagonistic afferent signals which sum to produce the error signal. An example is the human thermal system. Hammel (1968) has supported the existence of a set point for body temperature (core temperature).

The structure of a physiological system is a matter of considerable importance in the study of control strategy in biological systems. Frequently the afferent and efferent pathways of a system process signals of different types. For example, a chemoreceptor may receive an input which represents changes in the partial pressure of oxygen (p_{O_2}) and carbon dioxide (p_{CO_2}) while the output is encoded in the form of a pulse train. The analysis of biological control systems therefore involves the need for techniques which can study information coded in different ways.

The nature of nerve impulses is of particular importance. In any single nerve fiber, the magnitude of the signal that the nerve fiber is transmitting may be measured in terms of the time interval between successive impulses or, if this is constant, in terms of the frequency of the impulses. The number of nerve fibers involved may be no more than about ten, but in vertebrates particularly, it is much more likely to be several hundred, or perhaps several thousand. In control systems analysis it is therefore necessary to make such inferences as we can about the properties of populations of receptor and effector cells from the properties of single cells. These properties (in suitable circumstances) can be measured, and show that there must be limits beyond which populations of any of these kinds of cell cannot give responses which are linearly related to the excitations that induce them, and suggest what kind of nonlinearity may be present between these limits.

The earliest attempts to measure the response–excitation relations of receptor cells were based on sensations experienced by human subjects. The stimulation of a receptor cell by the appropriate physical or chemical change in the surroundings usually leads to the appearance of a receptor potential which itself acts as a generator potential and initiates nerve impulses, or which gives rise to a separate generator potential. Receptor potentials (and generator potentials if they are different) vary in magnitude continuously with the variations in the strength of the excitatory stimulus—they are not "all-or-nothing." Nerve impulses are initiated only when the receptor potential is large enough, and the larger it is the more frequently do nerve impulses appear; there is, therefore, a threshold in the excitation of a receptor cell below which the response is zero (Katz 1966).

When the excitatory stimuli are large the phenomenon of saturation occurs, that is, the response reaches a limiting value which is not exceeded however large the stimulus. Each nerve impulse lasts a discrete period of time: once a nerve cell has fired, it cannot fire again for a predefined time. In the "myelinated" nerves of warm-blooded animals, the maximum frequency at which impulses can be transmitted is about 1000 per second; in the nervous tissue of cold-blooded animals, the maximum frequency may be less than 1 per second.

These two kinds of nonlinearity (i.e., when the stimuli

are very small and very large, respectively) occur in groups or associations of receptor cells, as well as in single cells. In a group, however, different cells will reach threshold and saturation at somewhat different values of stimulus, so that the transitions between the graded response–excitation relation in the middle of the range of stimuli, with the zero response at one end and a constant response at the other end, will be rounded off; the complete response–excitation curve will be sigmoidal.

In trying to elucidate the response–excitation characteristics of effector pathways, there are three points it is important to consider: (a) the relation between the mechanical or chemical force developed and the magnitude of the excitatory stimulus applied; (b) the effect of the inherent damping, or internal velocity feedback, on the force developed which comes in as soon as there is any movement of the controlled quantity at the output stage; and (c) the nature of the time delay between the beginning of the excitation and the development of the maximum response.

Consider, for example, a muscle that is allowed to shorten when it is stimulated. The force produced is less than the isometric force corresponding to its length at any moment by an amount which depends on the speed at which it moves. If a muscle is working against a load, it has to generate a force equal to that exerted by the load before any movement occurs; if the force of this load is equal to the isometric force produced by a given stimulus, there is no movement; as the opposing force becomes smaller, the speed of movement increases up to a maximum value which is reached when there is no opposing load. For any given muscle, there will be a family of curves all of the same shape but with different intercepts on the axes depending on the strength of the stimulus. For each strength of stimulus, there will be another family of curves according to the length of the muscle when the measurements are made.

When a muscle is suddenly stimulated by the arrival of a train of nerve impulses, the force developed rises gradually; when the excitation is suddenly stopped, the force falls gradually. There is thus some kind of time delay between stimulus and response. This delay may be approximated by two simple exponential decays: the time constants may be nearly the same, or one may be much smaller than the other, depending on the kind of muscle involved. This approximation implies that the processes within the muscle which give rise to the delay have linear properties. These processes are known, however, to be nonlinear, and the time course of the development of force cannot be accurately represented by a combination of two, or any number of, exponential curves. The time relations of the fall in the force exerted after the stimulus ceases may be very nearly exponential, but as a rule, the force falls more rapidly than it rises.

Only a limited amount is known about the delays associated with the onset of secretion or active transport of transmitters. The rate of secretion rises gradually after the beginning of a sudden excitation, and falls gradually after the excitation stops; the delays may be considered, therefore, to be at least roughly exponential.

1. Biological Control and the Nervous System

Efferent signals may often represent the sum of several afferent signals or the difference between them; they may depend on the rate of change of afferent signals, or their sum over some interval of time. It is not difficult to establish this in a qualitative way, but it is usually extremely difficult to determine quantitatively whether, and in what conditions, the relationships between efferent and afferent signals are linear.

In the analysis of a particular biological control system, the system gain and stiffness, time constants of delays, and differentiating and integrating mechanisms must all be considered. However, in addition to these considerations, some biological systems are "self-adaptive" or "self-optimizing," that is, parameters change automatically in such a way as to restore the system equilibrium. Self-adaptive biological systems require the concept of a figure of merit and the ability to determine when optimal conditions have been met. From this point of view nervous systems undoubtedly possess the properties necessary for them to be called self-adaptive.

The word autonomic defines that part of the nervous system which acts as "controller" in the regulation of the activities of the organs involved in the circulation of the blood, the digestion of food, and so forth. This system is not really "autonomous," but is closely linked and integrated with the other parts of the nervous system.

Given the complex nature of living systems, it is clear that simple linear control theory techniques are an inadequate basis for analysis. It is frequently the case that the system under investigation may well contain both hard and soft nonlinearities, for example switch and sigmoidal elements, respectively. The system may also exhibit spontaneous oscillations, in which case phase-plane techniques or models based on the Van der Pol equation are often employed (Linkens 1979). Nonlinear analyses are frequently based on techniques like the describing function method (Gibson 1963).

See also: Biofeedback

Bibliography

Brown J H U, Gann D S 1973 *Engineering Principles in Physiology*. Academic Press, New York

Cannon W B 1929 Organisation for physiological homeostasis. *Physiol. Rev.* 9: 399–431

Gibson J E 1963 *Nonlinear Automatic Control*. McGraw-Hill, New York

Hammel H T 1968 Regulation of internal body temperature. *Annu. Rev. Physiol.* 30: 641–710

Hyndman B W 1973 An example of digital computer simulation, investigation of the human cardiovascular system. In: Haga E (ed.) 1973 *Computer Techniques in Biomedicine and Medicine.* Auerbach, Philadelphia, Pennsylvania, pp. 98–120

Katz B 1966 *Nerve, Muscle and Synapse.* McGraw-Hill, New York

Linkens D A 1979 Modelling of circadian and related biological rhythms. In: Linkens D A (ed). 1979 *Biological Systems, Modelling and Control.* Peter Peregrinus, London, pp. 275–305

Schwan H P 1969 *Biological Engineering.* McGraw-Hill, New York

Weiner N 1948 *Cybernetics.* Wiley, New York

R. I. Kitney

Biological Kinetics: Computerized Data Analysis

The collection of data from biological experiments may now be readily achieved using computers; the problems are usually only those of timing or of selecting storage for large quantities of raw data. The analysis of data by computer presents problems in that value judgements are implicit in the programs, a fact often unappreciated by the experimental worker when fitting data routinely.

Kinetic experiments that have been analyzed sufficiently often for some measure of agreement to exist include (a) those involving so-called compartmental models, which are used to provide a theoretical structure to describe the behavior of foreign substances introduced into a biological system; (b) measurements of steady-state enzyme kinetics; and more recently (c) measurement of the synthesis of a specific protein or DNA/RNA molecule. The behavior of a foreign substance covers the movements of substances (e.g., drugs) introduced into the biological system and labelled (usually with radioisotopes) so that they can be traced. A compartmental model leads to a system of first-order linear differential equations with constant coefficients:

$$q_i = -\sum_{i=1}^{m} k_{ij} q_i + \sum_{1}^{m} k_{ji} q_j \qquad (1)$$

where k_{ij} and k_{ji} are outward and inward rate constants, and q_i and q_j are quantities of tracer or drug. The equation may be written in terms of concentrations rather than quantities, in which case the coefficients become fractional transfer coefficients. Since the behavior of a normal component of the biological system is usually under investigation, there must also be a parallel set of equations for the unlabelled moiety of the chemical. These equations are usually but not always linear. The solution of differential equations such as Eqn. (1) is a set of exponential terms:

$$q(t_j) = \sum_{1}^{m} A_i e^{\lambda_i t_j} \quad j = 1, n \qquad (2)$$

where A_i and λ_i are secondary constants. This solution is usually known as the sum of exponentials. (For the theory of compartmental systems, see Atkins 1974, Shipley and Clark 1972.)

Similar equations can be applied to pre-steady-state (fast-reaction) enzyme kinetics. It is also possible to treat such equation systems by Laplace transforms, which offer the possibility of introducing discontinuous events (step functions). A discussion of these methods may be found in Finkelstein and Carson (1979).

Steady-state enzyme kinetics and the analysis of rates of synthesis on a nucleic acid template have special equation forms, which are considered in Sect. 2. Other examples of biological kinetics are estimates of protein turnover in whole animals, and microbiological reactor systems.

1. Nonlinear Regression Techniques for Parameter Estimation

1.1 Distribution of Error

Figures 1 and 2 portray types of exponential curve from which estimates of secondary constants A_i, λ_i may be extracted. (Note that these may only be related in a complicated way to the primary rate constants k_{ij}, or the fractional turnover coefficients.) The problem is that the crude data will have some kind of inherent error (Figs. 2 and 3), the distorting effects of which must be minimized if the parameter estimates are to be of value. Thus a measurement q_j at time t_j is described by Eqn. (3):

$$q_j = \sum_{1}^{m} A_i e^{-\lambda_i t_j} + \epsilon_j \qquad (3)$$

In the absence of precise information, which is usually

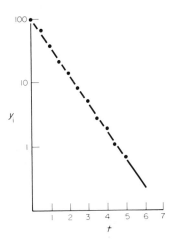

Figure 1
Semilog plot of $y_i = 100\, e^{-t} + \epsilon_i$. The relative error ϵ_i/y_i is constant with a value of 0.05 (after Jacquez 1972)

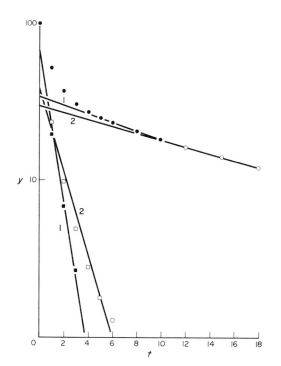

Figure 2
Semilog plot of $y = 50\,e^{-1.5t} + 20\,e^{-0.5t} + 30\,e^{-0.05t}$.
The two straight lines labelled "1" were derived by
curve-peeling from the full circles on the original
curve; the lines labelled "2" were derived from the
complete set of points

the case, it is assumed that ϵ_j is a random sample of error
from some distribution with zero mean and variance
σ_j^2. The variances may differ for the different samples,
but a variable may be defined such that $\eta_j = \epsilon_j/\sigma_j$, where

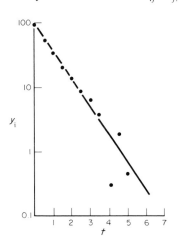

Figure 3
Semilog plot of $y_i = 100\,e^{-t} + \epsilon_i$. The error ϵ_i has zero
mean and standard deviation 2 (after Jacquez 1972)

η comes from a distribution with zero mean and
variance $= 1$. Then, η_j may be expressed by:

$$\eta_j = \frac{1}{\sigma_j}\left(q_j - \sum A_i\,e^{-\lambda_i t_j}\right) \qquad (4)$$

and having made an assumption about the distribution
of η, it can be asserted that an unbiased estimate of A_i
and λ_i will be arrived at if the sum of squares of these
deviations S is minimized, where

$$S = \sum_{j=1}^{n} \frac{1}{\sigma_j^2}\left(q_j - \sum A_i\,e^{-\lambda_i t_i}\right)^2 \qquad (5)$$

Very little is known about how σ_j varies with $q(t_j)$: in any
case, it probably behaves differently in each experiment,
and perhaps even for each measured variable. There are
three practical approaches:

(a) If there are one or more replicates at at least some
time points, the sample variance at these points may
be estimated, and interpolations for σ_j at the other
points may be used.

(b) It may be known from previous experience that the
relative error $\sigma_j/q(t_j)$ is fairly constant (homo-
scedastic), as in Fig. 3. Then the individual variances may be replaced by weightings $w_j = 1/q_j^2$ so
that Eqn. (5) becomes

$$S = \sum_{1}^{n}\left(1 - \frac{1}{q_j}\sum A_i\,e^{-\lambda_i t_j}\right)^2 \qquad (6)$$

(c) It may be known from previous experience that the
error is relatively constant, as in Fig. 1. Then the
weights w_j should be set to 1, so that Eqn. (5)
becomes

$$S = \sum_{1}^{n}\left(q_j - \sum A_i\,e^{-\lambda_i t_j}\right)^2 \qquad (7)$$

1.2 Minimization Procedure

To demonstrate the minimization procedure, a slightly
different form of equation will be used, namely

$$Y_j = A - B\,e^{-Ct_j} + \epsilon_j \qquad (8)$$

which is a simple form of the more general relation

$$Y_j = f(A, B, C, \ldots x_j) + \epsilon_j \qquad (9)$$

The change helps to indicate the steps in the procedure.
Equation (9) must be nonlinear in one or more of the
parameters A, B, C,..., otherwise estimates could be
made by ordinary linear regression.

The first step is to make good initial estimates a_1, b_1,
c_1 of the final estimates of A, B and C. This may be done
by curve-peeling, or by using one of the minimization
techniques such as simplex (see below), which are robust
enough to find a local minimum even from very bad
initial guesses.

Applying Taylor's theorem gives, in general, from
Eqn. (9):

$$f(A_1, B_1, C_1, \ldots x_j) \approx f(a_1, b_1, c_1, \ldots x_j)$$
$$+ (A - a_1)f_a' + (B - b_1)f_b' + (C - c_1)f_c' \qquad (10)$$

where f'_a, f'_b and f'_c are the partial derivatives of f with respect to A_1, B_1 and C_1 evaluated at the point a_1, b_1, c_1. In the particular instance, from Eqn. (8) we have

$$f'_a = 1$$
$$f'_b = -e^{-Ct_j}$$
$$f'_c = B\,e^{-Ct_j}.t_j \qquad (11)$$

These expressions may be readily evaluated if the form of the original equation is known, as in Eqn. (8). If $Y_j - (A - B\,e^{-Ct_j})$ is defined as the residual error e_j, the Taylor's theorem approximation to $f(A, B, C, \ldots x_j)$ may be used to write, from Eqns. (8, 11),

$$e_j = (A - a_1) + (B - b_1)X_1 + (C - c_1)X_2 + \epsilon_j \quad (12)$$

where $X_1 = -e^{-Ct_j}$ and $X_2 = B\,e^{-Ct_j}.t_j$. Equation (12) is a linear regression of e_j on the variates X_1 and X_2, but since Taylor's theorem provides only an approximation, the linearization is only approximate, and Eqn. (12) only provides second approximations a_2, b_2 and c_2, which are used to calculate new values for f'_a, f'_b and f'_c. The process is continued until the decrease in e_j and in a, b and c are considered small enough to be satisfactory. The generalization of this approach to include all the time points t_j, all unknown parameters $A, B, C, \ldots$, and possibly all the measured variables $Y_1, Y_2, \ldots$, may be deduced for this example. Whatever method is used to solve an equation such as Eqn. (12), clearly the user has to write a subroutine which will compute the partial differentials f'_a, f'_b appropriate to his equation set or model, and also sum the residuals e_j for all the time points.

1.3 Normal Equations

An alternative approach is to dispense with linearizing Eqn. (7), but simply to differentiate S with respect to each of the parameters to be estimated. In the case of Eqn. (8), this leads to three equations:

$$\frac{\delta S}{\delta A} = \sum^n Y_j - nA + B \sum^n e^{-Ct_j} = 0$$

$$\frac{\delta S}{\delta B} = \sum Y_j e^{-Ct_j} + A \sum e^{-Ct_j} - B \sum e^{-2Ct_j} = 0$$

$$\frac{\delta S}{\delta C} = \sum l_j e^{-Ct_j}.t_j + A \sum^n e^{-Ct_j}.t_j - B \sum e^{-2Ct_j}.t_j = 0$$

$$(13)$$

If these equations are set equal to zero, as above, they are called normal equations; their simultaneous solution will give the optimal values of A, B and C. This may seem a theoretically easier thing to do than to apply Taylor's theorem; in the particular model equation chosen (i.e., Eqn. (3)), there is only one nonlinear parameter C, and consequently Eqns. (13) can be solved easily. In general, however, there would be several such parameters (the λ's in Eqn. (7)), and the normal equations can only be solved with difficulty (by using variants of the optimization procedures described below). Thus

in the type of problem under consideration there is no advantage in using the normal equations.

1.4 Optimization

To produce estimates of parameters from equations such as Eqn. (6) or Eqn. (12), optimization, or search programs, described in the literature as "methods for the minimization of a function of several variables," are used. The objective function minimized is s or e; the variables are the A's and λ's. There are several good routines such as simplex (Nelder and Mead 1965), but Marquardt's method (Marquardt 1963) is currently very popular. It is expressly designed for least-squares problems. It is not necessary for the user to write a program, as the Oxford Numerical Algorithms Group have written and tested a set of programs on this (and other) topics, on both ALGOL and FORTRAN. A catalog is given in the NAG Mini-Manual (1980). An introductory discussion of optimization is given in Swann (1969).

1.5 Identifiability

For clarity in exposition, the examples so far discussed (Figs. 1, 3 and Eqn. (9)) have contained only one exponential term. Figure 2 illustrates problems that can arise when the model Eqn. (3) contains more than one exponential term (i.e., $i > 1$ in Eqn. (1)). The points on the graph are derived from:

$$y(t) = 50\,e^{-1.5t} + 20\,e^{-0.5t} + 30\,e^{-0.05t} \quad (14)$$

where t is in arbitrary time units. No error has been added to the data points, because it is not necessary for the argument. It may be noted, incidentally, that $y(0) = 50 + 20 + 30 = 100$.

If data collection is continued only until 10 time units have elapsed (full circles on graph), graphical curve-peeling leads to the identification of only two exponential terms, leading to a derived equation

$$y'(t) = 68\,e^{-1.11t} + 34.2\,e^{-0.064t} \quad (15)$$

where $y'(0) = 68 + 34.2 = 102.2$, near enough to the starting value of 100 (presumed known) to satisfy a quick check for validity. It is necessary to extend the data at least to 18 time units (open circles) to get an accurate estimate of the exponential term with slowest decay. Even then the curve-peeling technique if carelessly applied can lead to an equation (Fig. 2):

$$y'(t) = 38.5\,e^{-0.73t} + 30\,e^{-0.05t} \quad (16)$$

where only the check $y'(0) = 38.5 + 30 = 68.5$ might show that a term could be missing.

It might be thought that this fundamental inaccuracy could not occur if a graphical technique is replaced by nonlinear least squares estimation, but Finkelstein and Carson (1979) have shown that when error is present, a set of experimental data points similar to those of Fig. 2 could be fitted equally well by a two-way exponential or a three-exponential system. Thus it is extremely dangerous to identify a model solely by fitting an integrated set of rate equations to one set of experimental data points (Cornish-Bowden 1976).

There is also the problem of cost benefit. In rapid-reaction enzyme kinetic work, there may be no difficulty in extending the data collection period from 2 to 20 s, say, in order to make sure of identifying a slowly decaying transient. If the data are being collected from a set of experimental animals, which have to be kept not for 10 but for 20 days, or for a hospital patient whose stay must be similarly prolonged, the temptation to weigh the cost of prolonging the experiment against the accurate determination of one exponential term becomes very great. However, Fig. 2 shows that accurate estimation of this often apparently unimportant transient may affect very markedly the accuracy with which the more rapid transients are estimated; it may even provide the only suspicion of their existence. It is at this point that value judgements become important, as does planning of the experiment. Little can be said about this aspect here, except to point out that the most rapidly decaying transient may be badly characterized if the time points are equally spaced, as in Fig. 2. On the other hand, timing errors may creep in early in the experiment, particularly if mixing of drug or tracer is not instantaneous, which is assumed in straightforward compartmental models.

It has been assumed that the size of the model, or the number of differential equations (Eqn. (1)) is not precisely known. This aspect of identification is sometimes called validation of the model (Finkelstein and Carson 1979). There is another problem in which the model may be known or assumed, and yet it is not possible to make unique estimates of all the parameters. This problem arises very often in enzyme kinetic studies (Sect. 3). Here it will be illustrated by a simple example.

Figure 4 represents a three-compartment model, in which a drug is injected at zero time into compartment

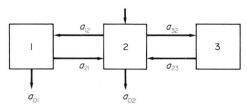

Figure 4
Three-compartment model with a possible ambiguity in defining the rate constants a_{ij}. The input is to compartment 2 (after Walter et al. 1980)

2. If measurements are made of drug concentration in compartments 1 and 2, it is not possible to estimate a_{01} and a_{21}, and a_{02} and a_{12}, uniquely. In the extreme case, either a_{01} or a_{02} could be zero, without the experimenter being able to tell. If, on the other hand, the secretion ratios from compartments 1 and 2 are measured, the

model becomes globally identifiable (i.e., in principle, estimates of all six rate constants can be made uniquely from one set of output observations).

It is easy to miss this ambiguity even in a simple model; in more complicated models it may be of great importance, particularly in prior planning of the experiment. Walter et al. (1980) have described a method by which all the possible minimal models with the same input and output behavior may be predicted, given the state Eqn. (1) and the initial conditions. They have produced a computer program, written in IBM FORTRAN, to implement the method.

It may be advisable for clinical and pharmacological work to employ a package model. Bloch et al. (1980) have prepared a program which simulates the response of a patient, called "MacDope," to drug administration, allowing analysis of the results. It has eight compartments simulating ingestion, excretion, dilution in plasma, metabolism and so on. Although this first version may have unsatisfactory features, there could be advantages in using a program on which some effort is being spent on maintenence and improvement.

1.6 Simultaneous Integration and Parameter Estimation

Most workers will be interested in Fig. 4, rather than simply in estimates of A's and λ's. As already stated, these are complex combinations of the primary constants k_i or k_i/V_i. It may be quite difficult to obtain the latter from the former. Thus, no directions have been given for estimating confidence limits for the A's and λ's, since they need not bear much relation to the confidence limits for the k's.

There is a strong case, therefore, for obtaining estimates of the rate constants directly, by integrating Eqn. (1), stopping the integration at each time point, and minimizing the overall sum of the squared deviations between computed and observed values. If more than one variable is being compared, weighting is an absolute necessity. This procedure has the advantage of forcing the user to provide a precise model, and of providing an unsurpassable feel of the behavior of the system. Although the integration does take time, and requires extensive storage space in the computer, matrix inversion routines do not produce results instantaneously; they also require storage space.

Two packages exist in which integration and parameter estimation are combined; CHEKMAT (Curtis and Chance 1974) and FACSIMILE (Kirby 1979). The latter is very large, requiring up to 1 Mbyte on an IBM 370 computer, but it can be used for very large problems, say in meteorology or nuclear engineering.

2. Steady-State Enzyme Kinetics

Affinity of an enzyme for its substrate is a characteristic of an enzyme reaction, which in it its simplest form has led to an understanding of enzyme kinetics. When substrate concentration is increased in the presence of an enzyme, the initial velocity V of the reaction increases to a maximum. The substrate concentration at half this maximum velocity can be shown to equate to the Michaelis–Menten constant K_m which is numerically low when the enzyme substrate affinity is high. Various mathematical treatments of data relating substrate concentration and velocity of reaction can be used to evaluate this constant.

One of the difficulties about the computation of parameters in Michaelis–Menten kinetics is a linearization, namely the Lineweaver–Burk plot, that has long been known to distort greatly estimates of K_m and V_{max} if they are obtained by simple linear regression. Of the alternative transformations, the so-called Hanes plot (s/v against v) has the advantage of showing up sigmoid (allosteric) kinetics as a distinct inversion (Gardiner and Ottaway 1969).

Cornish-Bowden (1976) introduced the idea of using the median of a set of estimates of K_m and V_{max}, derived from a single set of experimental data. The original method has been shown not to be free of bias, but an alternative way of estimating the median values is now available (Atkins and Nimmo 1980).

Both Atkins and Nimmo (1980) and Cornish-Bowden (1976) recommend the use of progress curves, rather than initial rates, for parameter estimation, although there is a problem with progress curves, due to the bias introduced by uncertainty about the true initial time point. Unless sophisticated technical methods are used, there is almost always a 10–15 s gap in enzyme assays, because of mixing problems, between adding the final reagent and getting the first reading.

Even if progress curves are ruled out, the problem of the true starting time, at which the estimate of initial rate should be performed, remains; it has been generally neglected in biochemistry. Unfortunately, the introduction of spectrophotometers with built-in microcomputers which can give a direct readout of "initial rate" is likely to make the neglect worse. There are two types of enzyme assay commonly in use. The first is the estimation of concentration of enzyme in a plasma or tissue sample. For this purpose, concentrations of reactants are adjusted so that the velocity of the reaction is large (if not maximum), and it may continue essentially linearly for a long time, perhaps several minutes. In these circumstances the initial time error is of negligible importance. This type of assay is, in numerical terms, overwhelmingly important because it is used in clinical biochemistry. Consequently the designers of fixed program microcomputers in commercial instruments tend to design only for these experimental conditions.

In the other type of assay, which is used for determining an enzyme mechanism, or parameters such as K_m and V_{max}, a wide range of substrate concentrations is (or should be) used. At low substrate concentrations it is inevitable that the reaction should not be linear for very long, simply because the substrate is rapidly used up. Experiment shows that estimates of the initial rate by construction of a tangent to a nonlinear progress curve

are almost always underestimates of the true initial rate. A systematic bias is therefore introduced into the measurements at the lower range of s values. It is well worthwhile to fit an empirical equation to the trace for each individual experiment, from which the error Δt may be estimated. With this information, a better estimate of the initial rate may be procured. The process implies extrapolation, which is usually undesirable, but even so reduces the systematic bias.

Let the observed elapsed time be T, so that $T = t - \Delta t$. It is not recommended to fit a polynomial in $t - \Delta t$ to the observed curve, because if the normal equations are derived by partial differentiation, as in Sect. 1.2, all the polynomial constants will occur only in one of the normal equations, so that they cannot be easily estimated. It is better to fit an equation of the type

$$q(T) = A(1 - e^{-\lambda(T + \Delta k)}) \tag{17}$$

where $q(T)$ is the amount of substrate disappearing, from which A, λ and Δt can then be separately determined. The estimate of the true initial rate is then $A\lambda$.

This method can be used with benefit even if the steady-state rate equation is more complex than that for a single-substrate process without inhibitors.

It would appear that almost all measurements nowadays are made on enzymes which are at least of this degree of complexity (i.e., they have two or more substrates), although strenuous efforts may be made to reduce the complexity to pseudo-single-substrate kinetics. There is, therefore, a good case for using the complete rate equation and fitting all parameters simultaneously.

Proper weighting of the raw data is essential for proper fitting of such equations, particularly when the object is to choose one of several mechanisms (Atkins and Nimmo 1980). Ottaway and Apps (1972) found it useful, when using a grid of concentrations for a two-substrate enzyme, to use the geographical mapping package SYMAP in order to visualize regions of apparently poor fit. No general rules can be given about an appropriate weighting. Each enzyme studied gives rise to its own error pattern. This implies that at least for a few substrate concentrations, enough replicate assays should be made for an estimate of the local variance to be formed. Anything more than this may be counterproductive, because complicated enzymes are often not stable enough *in vitro* to give reproducible results from beginning to end of an ambitious experimental design. The truth is probably that, just as in tracer–drug experiments, no single method of investigation will afford a complete picture of the properties of the model. Matters of judgement and experience must again be considered.

3. Non-Michaelis–Menten Enzyme Equations

3.1 Hill Plot

This is chiefly of interest because it may provide an estimate of the so-called Hill coefficient h. This in turn may provide a lower bound for the number of subunits in a ligand-binding protein (enzyme). Because an accurate estimate of h is unnecessary, the computation has often been done very badly.

Hill suggested an empirical equation:

$$Y = \frac{C[X]^h}{1 + C[X]^h} \qquad (18)$$

where $[X]$ is the concentration of ligand (substrate), Y is the fractional saturation of the protein with ligand, and h is an empirical constant. This equation fits many sigmoid initial velocity curves, or binding curves, quite well. Because the constant C has no physical meaning, the equation is usually transformed into

$$\log\left(\frac{Y}{1 - Y}\right) = \log C + h \log[X] \qquad (19)$$

from which an estimate of h is obtained by linear regression, neglecting the ends of the curve (Fig. 5).

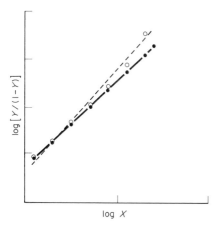

Figure 5
Plot of the linearized form of the Hill equation (Eqn. (19)): ●, data derived using the parameter values $V_{max} = 100$, $C = 13\,800$, $h = 1.8$; ○, the same data but with V_{max} reduced to 95. The dotted line has been fitted to these points by least squares; h is now 2.15

This procedure however neglects the fact that Y (usually assumed to be V/V_{max} for enzymes) demands a good estimate of V_{max}. Even if it could be used, a Lineweaver–Burk plot seriously underestimates V_{max} and estimation by visual inspection of the untransformed plot of v against s is even worse. At the very least, a Hanes transformation should be used, from which an estimate of $1/V_{max}$ can be obtained directly as the gradient of the plotted line for larger values of s (i.e., $[X]$).

What should be done is to fit the three constants V_{max}, C and h directly, using nonlinear regression. Probably, because C and h appear together in Eqn. (18), linearization by using a Taylor series expansion would be the ideal procedure. It is doubtful whether this has ever been done, but direct fitting of the three constants to the data by minimizing the sum of squares of deviations, using a good initial estimate of V_{max}, and less good estimates of C and h, has been carried out by a few workers.

3.2 Synthesis on a Nucleic Acid Template

Michaelis–Menten kinetics does not apply either to the synthesis of DNA or RNA on a DNA template, or to the synthesis of polypeptides on an RNA template, for several reasons. The reaction is not over quickly, but may require several minutes for completion: termination and more particularly initiation may take much more time than an individual elongation step. Most important, however, is the fact that several enzyme complexes (polymerase or ribosome) can be on the same template at one time. In addition, the complexes are large enough to occlude several nucleotides on the template. Estimation of progress rate therefore becomes a statistical problem in the stochastic sense; what is the probability that at its next move a given complex will find its progress halted by the tail end of the complex in front of it?

Recently, Mahon et al. (1980) have produced an equation called the termination model, for analyzing the linear phase, preceded by a lag, of incorporation of labelled nucleotides into RNA.

If D is the length of a scripton (template), V_p is the rate of propagation, and V_i is the number of polymerase molecules recognizing the initiation site per second, then no termination can occur before D/V_p seconds. Further, it was shown that, if $S(t)$ is the total number of nucleotides incorporated into RNA per scripton at any time t, then

$$S(t) = \int_0^t V_p N(t) \cdot t \qquad (20)$$

where $N(t)$ is the number of polymerases transcribing one scription at time t; it is a function of the three constants above. Integration gives

$$\left. \begin{array}{ll} S(t) = V_p V_i t^2/2 & t < t_{lag} \\ S(t) = V_i D[t - D/2V_p] & t > t_{lag} \end{array} \right\} \qquad (21)$$

From these equations V_i and V_p can be estimated if D is known; the method can be generalized to analyze systems with scriptons of varying length.

Mahon et al. used nonlinear least squares fitting, with a Taylor series expansion. They found that V_i is estimated more accurately than V_p, which is to be expected, since V_p needs an accurate estimate of the progress time t.

4. General Considerations

Most physical constants are only meaningful if they are less than zero. Advantage may be taken of this, just as the positive root of a quadratic is always chosen. However, it is not satisfactory to increase the objective function by a large numerical penalty whenever one constant goes negative in an optimization procedure. Some way of altering the vector of parameter estimates, in order to direct the search to a more productive region, is a better strategy.

A negative value for a parameter estimate may be informative even if it is incorrect. For instance, if an optimization consistently produces a reasonable fit, and credible parameter values, except for the one which is negative, the model is wrong, and should be changed. Conversely, a very small but negative value may mean that a parameter is zero (i.e., that it may be omitted from the model). Useful information of this kind in the construction of a valid model may be completely lost if a penalty function is written into a computer program so that it is unthinkingly applied whenever an estimate becomes negative.

See also: Mathematical Modelling in Biology

Bibliography

Atkins G L 1974 *Multicompartment Models for Biological Systems.* Chapman and Hall, London
Atkins G L, Nimmo I A 1980 Current trends in the estimation of Michaelis–Menten parameters. *Anal. Biochem.* 104: 1–9
Bloch R, Ingram D, Sweeney G D, Ahmed K, Dickinson C J 1980 MacDope: A simulation of drug disposition in the human body: Mathematical considerations. *J. Theor. Biol.* 87: 211–36
Colquhoun D 1971 *Lectures on Biostatistics: An Introduction to Statistics with Applications in Biology and Medicine.* Clarendon Press, Oxford
Cornish-Bowden A 1976 *Principles of Enzyme Kinetics.* Butterworth, London
Curtis A R, Chance E M 1974 *CHECK and CHEKMAT: Two Chemical Reaction Kinetics Programs,* AERE-R.7345. HMSO, London
Finkelstein L, Carson E R 1979 *Mathematical Modelling of Dynamic Biological Systems.* Research Studies Press, Forest Grove, Oregon
Gardiner W R, Ottaway J H 1969 Observations on programs to estimate the parameters of enzyme kinetics. *FEBS Lett.* 2: S34–38
Jacquez J A 1972 *Compartmental Analysis in Biology and Medicine.* Elsevier, Amsterdam
Kirby C R 1979 *Facsimile Implementation Manual,* AERE-R.9557. HMSO, London
Mahon G A T, McWilliam P, Gordon R L, McConnell D J 1980 The time course of transcription. *J. Theor. Biol.* 87: 483–515
Marquardt D W 1963 An algorithm for least-squares estimation of nonlinear parameters. *J. Soc. Ind. Appl. Math.* 11: 431–41
McMinn C L, Ottaway J H 1977 Studies on the mechanism and kinetics of the 2-oxoglutarate dehydrogenase system from pig heart. *Biochem. J.* 161: 569–81
Mowbray J, Ottaway J H 1973 The effect of insulin and growth hormone on the flux of tracer from labelled lactate in perfused rat heart. *Eur. J. Biochem.* 36: 369–79
Nelder J A, Mead J A 1965 A simplex method for function minimization. *Comput. J.* 7: 308–13
Numerical Algorithms Group 1980 *NAG Mini-Manual, Mark 8.* Numerical Algorithms Group, Oxford
Ottaway J H, Apps D K 1972 Some examples of the use of computer-produced contour plots in the fitting of enzyme rate equations to reaction-velocity measurements. *Biochem. J.* 130: 861–70
Shipley R A, Clark R E 1972 *Tracer Methods for In-Vivo Kinetics: Theory and Applications.* Academic Press, New York
Swann W H 1969 A survey of non-linear optimization techniques. *FEBS Lett.* 2: 39–55
Walter E, Lecourtier Y, Bertrand P, Gomeni R 1980 On the identification of explicative models. *Joint Automatic Control Conf.,* Vols. 1–2, Item WA6-H. American Society of Mechanical Engineers, New York

J. H. Ottaway

Biomechanics

Biomechanics is the branch of biology that applies engineering mechanics to the study of living organisms. The term is often applied specifically to studies of man but is more correctly used to include studies of all living organisms. Many different branches of mechanics have been applied to biological systems, and several are illustrated in this article.

1. History

The earliest book on biomechanics is Giovanni Borelli's *De Motu Animalium* (1680), which discusses the movements of mammals, birds and fishes. Borelli tackled several simple mechanical problems; for example, he used the principle of levers to estimate the forces which certain human muscles can exert, and Archimedes' principle in a discussion of the buoyancy of fish. A little later Stephen Hales published *Vegetable Staticks* (1727) and *Haemastaticks* (1733). The former describes a remarkable series of experiments on water movement in plants, and the latter some experiments using manometers to measure the blood pressure of farm animals. Little progress was made thereafter until the rise of the nineteenth century German school of physiologists. Johannes Müller (1801–58) carried out experiments with vocal cords from human corpses, relating his results to the physics of stretched strings. Human biomechanics has continued without interruption since then, but relatively little was done on the mechanics of other animals until the pioneering work of Sir James Gray (1891–1975) on mechanisms of locomotion.

2. Structural Properties of Materials

The principal structural materials of organisms are polymers (either proteins or polysaccharides), sometimes with inorganic crystalline fillers. Some, such as the mesogloea of jellyfish, are dilute aqueous gels. Others, such as the insect protein resilin, are amorphous elastomers with elastic and thermodynamic properties very like soft rubbers. Many biological materials are oriented semicrystalline polymers, for example, cellulose, a polymer of glucose and the principal constituent of plant cell walls and fibers, including cotton and hemp. Tendons consist of fibers of the protein collagen, and silk and wool are also protein fibers.

Many skeletal materials are composites. Insect cuticle, for instance, consists of polysaccharide fibrils in a protein matrix. Bone is a composite of 70% inorganic crystals (mainly calcium phosphate) and 30% collagen fibers. Mollusc shell consists of calcium carbonate crystals cemented together by protein. The composite structure of these materials makes them tough, like fiberglass and other man-made composites.

Study of biological materials has been greatly assisted by earlier work on rubber and plastics, but most biological materials are much more complex than man-made materials. Most man-made polymers are built from a single monomer but proteins consist of about 20 amino acids which polymerize in precisely controlled sequences to form long chains. The vast range of possible structures enables subtly different materials to evolve where slightly different properties are required. Structure at supramolecular level can also be very precisely controlled. One form of insect cuticle has microfibrils oriented in the manner of the molecules of a cholesteric liquid crystal. One of the principal types of bone consists of small cylindrical units, each built up of layers in which the fibers run alternately in left- and right-handed helices (see *Bone: Mechanical Properties*). The anisotropy of bone and other natural materials is often nicely matched to the stresses experienced in nature.

3. Statics

Some complex problems of statics arise in biomechanics. Consider, for instance, the forces the leg muscles must exert in human standing. One analysis has treated the body as an assembly of seven rigid segments (two feet, two shanks, two thighs and a trunk). Accordingly, the conditions for equilibrium in three dimensions gave 42 simultaneous equations. However, as there are 29 major muscles in each leg and there are also unknown forces at joints and ligaments, the problem is indeterminate. This is a recurring difficulty in biomechanics, which cannot easily be overcome by measuring internal forces in the body. Limited use has been made of transducers implanted in the bodies of men and animals to measure tension in tendons and strain in bones, but it is not feasible to implant many transducers in any one person or animal. Indirect evidence of muscle activity can be obtained from fine electrodes inserted into muscles, but it does not seem possible to calculate forces reliably from records of electrical activity.

4. Kinematics

The human hip is a simple ball and socket joint and the elbow is a hinge (with a second hinge between the two parallel bones of the forearm). Some other joints are more complex. The human knee behaves much like a hinge but consists of the spiral end surface of the femur rolling and sliding on the much flatter end of the tibia. The two are held together by ligaments, notably two ligaments which cross like an X in a gap in the middle of the joint. If slight play is ignored, there is just one degree of freedom of relative movement. The movement of the shin is approximately planar motion for most of its travel, but it rotates about a moving instantaneous center. As the knee approaches full extension, its movement becomes three-dimensional.

There are several interesting mechanisms in the skeletons of animals. Some of the most elaborate occur in certain fishes that extend the mouth forward as a tube when they feed. At least three radically different space linkages have evolved in different groups of fish, to give essentially the same movements.

5. Dynamics

Animal locomotion raises many problems in dynamics. Why, for instance, do men change from walking to running at a particular speed?

Consider first a man walking with speed u on legs of length l. If each leg is kept straight while its foot is on the ground the man's trunk advances in a series of circular arcs of radius l. This implies an acceleration u^2/l downwards, towards the supporting foot, but unless the man walks on adhesive ground his downward acceleration cannot exceed the acceleration of free fall, g. Hence u cannot exceed $(gl)^{1/2}$, or about $3 \, \text{m s}^{-1}$ for an adult man. The maximum speed will be less for a child (smaller l) and on the moon (smaller g). Observations agree quite well with these predictions. Men normally break into a run at $2.5 \, \text{m s}^{-1}$, and the faster racing walk involves a peculiar hip movement which reduces the vertical excursion of the center of mass.

More sophisticated models of human walking have been devised. Some are designed to explore the energy costs of different walking techniques, and seem to show that normal walking minimizes costs. Others are designed to estimate forces in individual muscles (see *Gait Analysis*).

Many investigations of animal dynamics require knowledge of the external forces acting on the body. Various types of transducers have been used to measure these forces, including force platforms set into the floor.

Elastic strain energy is important in many activities of animals. A flea taking off for a jump accelerates from rest to a speed of $1 \, \text{m s}^{-1}$ in a distance of 0.5 mm. To do this it must extend its legs in 1 ms. No muscle can contract so fast (except in a peculiar state of oscillation which enables small insects to beat their wings at high frequencies), but the flea has a built-in catapult made of the elastic protein resilin. Strain energy is stored in the resilin by a relatively slow muscular contraction and released rapidly by the trigger mechanism which initiates the jump.

In human running, the peak force on each foot is usually about 2 kN or three times body weight. Its moment about the ankle is balanced by a force of about 5 kN exerted by the muscles which are attached to the heel through the Achilles tendon. This stretches the tendon, storing elastic strain energy, while the kinetic and gravitational potential energies of the body are passing through a minimum. If the tendon were stiffer the muscles would have to do more positive work (increasing the total mechanical energy of the body) at one stage of the step, and negative work (converting mechanical energy to heat) at another. Both positive and negative work performance by muscle consume food energy.

6. Hydrostatics

Plants lose water by evaporation and this must be replaced by water from the soil. Such water as there is in dry soil is at a low potential because it is held by capillarity in narrow spaces or adsorbed to clay particles. If water is to move from soil to plant, the water potential in the plant must be even lower. Plant sap is a dilute aqueous solution and achieves a low water potential by having a low pressure. Indirect measurements have shown that in conditions of drought, the pressure of the sap of the creosote bush (a desert plant) falls to -80 atm. This pressure is apparently reached without cavitation, in the fine tubes which contain the sap.

Surface tension is important in some biomechanical problems. Lungs, for example, are spongy structures with air-filled pores and pockets of diameter 0.1 mm or less. Air is pumped in and out of the pockets, making them enlarge and contract as the animal breathes (see *Respiratory Function: Physiology*). If the inner surfaces of the lungs were coated with water, the pressure in the pockets would increase as their size decreased. They would tend to collapse completely, and very large pressures would be needed to reinflate them. This happens in a dangerous disease of newborn babies. The inner surface of the healthy lung is coated with a surfactant, which has a low surface tension when it forms a complete film, as in the contracted lung.

7. Hydrodynamics

The application of hydrodynamics can be illustrated by considering the motion of aquatic animals. An estimate of the power needed to propel a fish can be made by assuming the drag on the fish is the same as on a rigid body of the same size, moving at the same speed. Alternatively the power can be estimated from tail

movements shown in films, by calculating the rate at which kinetic energy is added to the wake. The latter method gives estimates about five times as high as the former. The discrepancy is believed to be due to thinning of the boundary layer by the undulation of the body, making the former estimate unrealistic.

Many microscopic protozoans swim by propagating transverse or helical waves along whip-like structures called flagella. The undulations of the flagella look rather like the swimming movements of fish such as eels, but the hydrodynamic mechanisms are quite different because of the difference of scale. The Reynolds number based on body length is about 10^5 for a medium-sized fish swimming moderately fast, but is of the order of 10^{-4} for a typical flagellate protozoan. Some bacteria swim by rotating a stiff helical flagellum about its own axis. This seems to be the only example known in biology of an axle capable of rotating indefinitely in the same direction: it is the nearest approach which nature has made to evolving a wheel.

Further examples of hydrodynamics in biology include the lubrication of joints and the flow of blood.

8. Aerodynamics

Aerodynamicists have studied the effects of wind on trees and the falling of thistledown and of propeller-like seeds such as sycamore and maple. More effort has been applied, however, to the study of animal flight.

Several species of birds and a species of bat have been trained to glide so as to remain stationary in inclined wind tunnels. The animals flapped their wings if the tunnel was too nearly level, so their minimum gliding angles could be determined for each of a range of air speeds. A small vulture and a fruit bat performed almost as well as a good model glider of similar size, but pigeons performed much less well.

Typical flapping flight of birds is difficult to analyze because it involves complex movements of flexible wings. More progress has been made in studying the hovering flight of hummingbirds and many insects. The hummingbird keeps its body inclined at a fairly steep angle and beats its wings forward and back horizontally. The wings generate lift like a helicopter rotor but their movement is reciprocating, not rotary.

To get enough lift, insect wings have to beat at high frequencies, 600 Hz in the case of a mosquito. Although the wings are light the power needed to drive them at such frequencies would be prohibitively high if their kinetic energy were degraded to heat at the end of each stroke. Instead, the wings and their attachments are built as a spring-mass system with a natural frequency corresponding to the wing-beat frequency. Hence muscular work is needed only to compensate for viscous losses and to do the necessary aerodynamic work. Some insects seem to have enough elastic compliance for this in their flight muscles but others have the compliance of the muscles supplemented by other materials, notably the elastic protein resilin.

Lift coefficients calculated from the weights and dimensions of insects and their wing-beat frequencies are sometimes as high as 3. Such values would be unattainable in a steady-state situation. One of the mechanisms that makes them possible is called the clap and fling. The wings are clapped flat against each other at the top of the upstroke. They separate again, starting with the leading edges, so that air flows round the leading edges into the expanding space between the wings. This generates circulation in the sense required to provide upward lift.

9. Acoustics

Hearing and sound production raise difficult problems in acoustics. The external ear, eardrum and ear ossicles of mammals serve as an impedance-matching device between the air and the fluid-filled sense organs of the inner ear (see *Ear Anatomy and Physiology*). Sounds of different frequencies are sensed by different parts of a helical sense organ (the cochlea), but the mechanism of this discrimination is uncertain.

Animals have many interesting mechanisms for producing sound. For example, crickets draw a plectrum on one wing along a file-like structure on the other, at the appropriate rate to excite a resonant panel on the wing. The wing is too small (compared to the wavelength of the sound) to be an efficient loudspeaker, but the male cricket improves its efficiency by singing in a burrow with openings shaped like exponential horns. Male crickets can be heard from a distance of 0.6 km.

See also: Biophysics; Cybernetics, Biological; Muscle

Bibliography

Alexander R McN 1982 *Locomotion of Animals*. Blackie, Glasgow
Alexander R McN 1983 *Animal Mechanics*, 2nd edn. Blackwell, Oxford
Carlsöö S 1972 *How Man Moves: Kinesiological Studies and Methods*. Heinemann, London
Currey J 1984 *The Mechanical Adaptations of Bones*. Princeton University Press, Princeton
McMahon T A 1984 *Muscles, Reflexes, and Locomotion*. Princeton University Press, Princeton
Preston R D 1974 *The Physical Biology of Plant Cell Walls*. Chapman and Hall, London
Vincent J F V, Currey J D (eds.) 1980 *The Mechanical Properties of Biological Materials*, Symposia of the Society for Experimental Biology Vol. 34. Cambridge University Press, Cambridge, UK
Vogel S 1981 *Life in Moving Fluids*. Willard Grant, Boston, Massachusetts
Wainwright S A, Biggs W D, Currey J D, Gosline J M 1976 *Mechanical Design in Organisms*. Arnold, London

R. McN. Alexander

Biometry

Biometry is the application of statistical techniques to analysis in biological research. Its origin dates back to the turn of this century when it was hoped that various statistical descriptive techniques would aid in the demonstration of evolutionary processes, especially natural selection. This hope was largely unfulfilled but in the process an elaborate body of statistical techniques was developed, which today is indispensable to the conduct and evaluation of biological research. The terms biological statistics and biostatistics are essentially synonymous with biometry, although biostatistics, at least in the USA, carries a connotation of biometric work related to medicine and public health. Statistical methods are important in biology because results of experiments are usually not clear-cut and therefore need statistical tests to support decisions between alternative hypotheses.

1. Analysis of Variance

The types of variables studied in biometry include measurement variables (e.g., length, height, width and concentrations of chemicals in tissues and body fluids); derived variables (such as percentages and ratios); rank-ordered variables (e.g., ranking of arrival times of animals at a food source, or maturation times of plants); and frequencies of qualitative characteristics (e.g., the number of animals surviving an experimental operation versus the number succumbing during the experiment). Each of these types of variables needs special methods for its analysis.

Biometry serves as a tool for the description of biological samples in terms of the location (magnitude) and dispersion (variation or scatter) of observed variables. The true characteristics or parameters of a population are rarely, if ever, known; hence they must be estimated from population samples. Descriptive biometry estimates statistics of location such as the mean, median or mode of a series of measurements or observations, or statistics of dispersion such as the variance or the standard deviation that estimate the inherent amount of variation. Such techniques are useful to characterize various populations of man or other organisms (e.g., the average height in a human population); to estimate inherent variability in such populations due to natural processes (e.g., variation in longevity in a human or animal population); or to estimate measurement error, such as in a series of blood sugar determinations made by a given technique. Estimation is not confined to single samples. By comparing statistics of location and of dispersion in two samples, conclusions can be reached about whether the samples differ in mean and variability of a given property. Thus, are blood cholesterol levels higher in people who eat meat than in vegetarians? Or, are fish raised under constant temperature less variable in length than those raised under fluctuating temperatures? Many studies are structured as repeated samples which lead to estimates of variation at different levels. Thus from samples of several human populations measured for metabolic rates, estimates of the variance of rate within populations and among populations may be obtained. Such estimates are usually obtained by a technique known as the analysis of variance, which permits the separation of different levels of variation. These estimates, known as variance components, are fundamental to the field of quantitative genetics in which the inheritance of measureable traits, such as height, weight, milk yield in animals, or protein content in plants, is studied. By carefully analyzing controlled crosses in such organisms, the variation attributable to paternal and maternal contributions, to the environment, and to other aspects of the design can be estimated and apportioned.

2. Distribution of Variables

Biometry is also concerned with describing the actual distribution of variables in biology and making inferences about the biological processes that have given rise to these distributions. Paramount among these distributions is the normal or Gaussian distribution manifested repeatedly in measurement variables of a great variety of biological phenomena. Variables determined by numerous factors that act in an additive fashion will be normally distributed. Exponential and log-normal distributions are examples of other distributions of measurement variables. Such distributions are used to predict the frequency of occurrence of deviant measurements in populations (e.g., of tolerances to toxic agents in populations). Among frequency variables the binomial and Poisson distributions are well-known examples in genetic and ecological research that test for randomness of spatial and temporal processes.

3. Testing of Hypotheses

Once the distributions and hence the probabilities of occurrence of different magnitudes of variables are known, inferences about the likelihood of various unusual observations are investigated. Such inferences in turn lead directly to the testing of hypotheses—perhaps the single most important aspect of biometry. In tests of hypotheses, scientists typically formulate a null hypothesis of no difference between the true values of two or more samples. For example, no difference is postulated between cholesterol levels of meat eaters and vegetarians. The mean levels for these populations from samples are then estimated and, based on a knowledge of statistical distribution theory, an inference is made whether a difference as large as the one observed in the two populations could have arisen by chance alone. If the odds on the origin of such a difference are less than one out of twenty, it is concluded that it is unlikely for the

observed difference to have arisen by chance alone and the difference is called significant. This conclusion is equivalent to the rejection of the null hypothesis. Hypothesis testing is central to the evaluation of the results of most medical, biological and agricultural experiments. Does the administration of a particular drug increase the life span of leukemia patients? Does crowding of fish in an aquarium decrease their growth rate? Does application of fertilizer increase the yield of a given crop? All such questions are answered by the formulation of an appropriate null hypothesis followed by a statistical test leading to acceptance or rejection of the null hypothesis. When combined with the analysis of variance such hypothesis testing has been elaborated into an important branch of the discipline called experimental design, in which the effects of various factors and of controls can be evaluated independently and jointly. Moreover, this technique promotes the efficient utilization of resources by specifying the minimal replication required to test specific hypotheses in a given experiment.

4. Bivariate and Multivariate Analysis

Bivariate analysis studies the interrelation between two variables. An estimate of the correlation between two variables establishes their association in nature but does not necessarily demonstrate a causal relation between the two variables. Regression analysis shows the variation of the dependent or criterion variable as a function of a second, predictor, regressor or independent variable. The functional relationships are fitted by least-squares methods, the most frequent being the linear function of the form $Y = a + bX$. In special cases bivariate data are fitted with curvilinear functions based on polynomial expressions of X or other special functional equations believed to represent the relationships between the variables concerned. An alternative method of estimation to least squares is the method of maximum likelihood, which in linear regression yields identical estimates.

Multivariate analysis is the extension of bivariate analysis to multiple predictor and criterion variables. Equations predicting a single criterion variable in terms of multiple independent predictors (multiple regression) is a common technique for allocating proportions of the determination of one variable to putative causes. Other important multivariate techniques include discriminant function analysis which uses multiple descriptive variables to separate efficiently two or more samples. Thus this method answers questions such as: What combination of morphological measurements can adequately discriminate between the two sexes or any two races in a collection of human skulls? Other approaches, known as numerical taxonomy, estimate the resemblance between species on the basis of numerous characteristics and, by a variety of multivariate techniques applied to a matrix of resemblances, arrange these species into classifications.

5. Nonparametric Statistic Methods

Since many biological variables have unknown distributions or are distributed in a manner unsuited to standard statistical tests, the application of nonparametric statistic methods has gained increasing popularity. In these methods the information concerning sample measurements is usually degraded, leading to less efficient estimates and hypothesis testing procedures. However, the interpretation is not as dependent on the standard assumptions of statistical analysis as are the more customary, so-called parametric methods. Thus two samples that should be compared in terms of their mean longevities might be pooled and the rank of the longevity of each individual determined in the pooled sample. Instead of comparing mean longevities the means of the ranks in each of the two samples can be compared. If the samples do not differ in longevity the ranks in the two samples should have the same means.

6. Computer Analysis

The increasing availability and speed of interactive computing is changing the practice of biometry. There is increasing emphasis on data analysis. Instead of estimating statistics of various samples under simple assumptions and employing standard statistical program packages for the analysis of the data, a variety of approaches can be employed under essentially experimental conditions. One may, for example, estimate the mean survival of a group of patients on the assumption that 10% of the original population comprised the most labile individuals who died early and would not have come to the attention of the investigator. Similarly, numerous displays of the same data set can be obtained using graphic terminals. For data for which the true distribution is unknown and for which no simple models can be postulated, hypothesis testing can be carried out by means of Monte Carlo simulation in which the distributions of various nonstandard statistical techniques can be evaluated by randomly generating numerous experimental cases on the computer and analyzing observed results in the light of the randomly obtained distributions.

See also: Biostatistics; Cancer Statistics; Statistical Methods in Medicine

Bibliography

Bliss C I 1967–1970 *Statistics in Biology*, Vols. 1, 2. McGraw-Hill, New York
Gill J L 1978 *Design and Analysis of Experiments in the Animal and Medical Sciences*, Vols. 1–3. Iowa State University Press, Ames, Iowa
Pimentel R A 1979 *Morphometrics: The Multivariate Analysis of Biological Data*. Kendall/Hunt, Dubuque, Iowa
Rohlf F J, Sokal R R 1981 *Statistical Tables*, 2nd edn. Freeman, San Francisco
Sokal R R, Rohlf F J 1981 *Biometry: The Principles and Practice of Statistics in Biological Research*, 2nd edn. Freeman, San Francisco

R. R. Sokal

Biophysics

Biophysics is the study of the physics of biological systems. Individual contributions to this field may emphasize structure or function, may focus on any scale from ion to ecosystem, and may be theoretical or experimental in approach. Concern with physical *ideas* is, however, the essential feature. Mere use of a physical instrument does not make the work biophysical; one has only to think of the microscope, or even, nowadays, the electron microscope, to agree that the *question being asked* must be physical too.

In contrast to the routine use of a physical instrument, its *development* in the biological context often requires penetration into the physics of the biological system concerned. For reasons of this kind, it is unfruitful to define the borders between biophysics and biological or medical physics, and between biophysics and bioengineering: these disciplines differ in mainstream emphases, but not at their interfaces.

1. History

Since the Renaissance, the feasibility of treating living matter as a physical system has become accepted piecemeal, often it seems only half-consciously, and with many alternations between advance and retrenchment. The least philosophically contentious perception has been that the macroscopic mechanics of organisms is fit subject matter for physical analysis. Probably the first rigorous application to biology of essentially modern physics was by Galileo, and his follower Borelli, in 17th century Italy. They considered to excellent effect the statics of such structures as trees and skeletons. In Britain, at much the same time, Harvey and Hales applied physical insights and measurements to the heart and the circulation of the blood.

Far more contentious was the matter of metabolism—"animal heat." In the 1770s and 1780s, Lavoisier in France and Crawford in Britain stated, with elegant experimental backing, the equivalence of respiration and combustion. This account of the "vital flame" aroused considerable opposition in 19th century Germany, under the influence of the Romantically based "nature philosophy," before the matter was resolved in favor of Lavoisier.

Over much the same period, the study of "animal electricity" developed from studies on frogs' legs by

Galvani (Italy, 1780s) through du Bois-Reymond's distinctly modern approach to the analysis of such phenomena (Germany, 1848), to the earliest recordings of action currents (nowadays less accurately called "action potentials"). Out of these came the first ionic theory of the origins of potential differences across biological membranes, formulated by Bernstein in 1868. The maturation of this subject, though contentious, was fairly steady; there was not the pattern of early insight, followed by an identifiable reactionary phase, which there was over animal heat.

Significant contributions to both the above topics were made by Helmholtz who also, by his early statement that energy was conserved in living as well as in inanimate systems, initiated biological thermodynamics. His later works in the 1860s on vision and on hearing, unequalled in their insight and scope, established the physics of sense organs on its modern footing.

2. Modern Biophysics

In the 20th century, investigation has continued—often at an explosively increasing rate—on all the topics whose beginnings have been outlined above. Hodgkin (1964) in Britain and Cole (USA) have made outstanding advances on nerve, while von Békésy (1960) in the USA and Katchalsky (Israel) have made pioneering contributions in hearing and the nonequilibrium thermodynamics of biological systems, respectively. However, many other topics have become truly biophysical only in this century. Perhaps the first of these was the study of muscle; this was drawn forward from the mainly morphological approach of previous generations, most notably by the combined mechanical and thermal approach of Hill (Britain) and by the mechanical and x-ray diffraction studies, respectively, of A F Huxley and H E Huxley (Britain).

Biophysical chemistry also emerged as a recognizable discipline in the first half of this century. Out of it has grown the largest single field of study in modern biophysics—that of macromolecules. Protein structure, with its implications for processes such as gas carriage and enzyme action, was the locus of the first major breakthroughs, notably by Pauling in the USA and Sanger and Perutz in Britain. Almost immediately after the initial work on proteins came the double-helical concept of the structure of DNA by Crick (Britain) and Watson (USA), and the consequent cracking of the genetic code. It was not until then that the theory of vitalism had ultimately fallen, for in cell replication it had been possible to conceive that some "life principle" was at work, despite the fact that metabolism, excitability, and force generation had long become accepted as explicable in totally physicochemical terms.

Genes carry information, but so do sensory structures, nerves and hormones. Also, both subcellular processes and the behavior of physiological systems are feedback controlled. Therefore some of the same theoretical concepts may be applied to the study on the one hand of molecules, and on the other of organs and whole organisms. Pioneered in the USA by Wiener (1961) and von Neumann, information and control theory—merging more recently into the general systems formalisms of modern engineering—have thus provided the intellectual tools for analyses of biological systems at both ends of the size range (see *Cybernetics, Biological*). The data for testing concepts have, however, been more obtainable in macroscopic cases. Consequently, the principal effect so far has been a counterweighting development of modern biophysics, at the other pole from molecular biology. Cardiovascular, endocrine, motor and perceptual processes—including the pathologies of each—have all been fruitfully considered in these terms.

3. Experimental Tools

X-ray diffraction was the principal experimental tool of the early molecular biophysicists, giving them information about protein, nucleic acid and virus structures in crystalline preparations (see *X-Ray Diffraction in Molecular Biology*). The technique has also been applicable to repeating membrane systems, and to the cytoplasmic structure of one living cell, that of striated muscle, which is itself a macro crystal. Nuclear magnetic resonance—perhaps particularly of protons for studies of water and of phosphorus for metabolites—may well succeed x-ray diffraction as the forefront tool in this area, for it gives information about the state of much smaller moieties when still within the intact, living body (see *Nuclear Magnetic Resonance Spectroscopy*).

More of the nature of physical tools for answering traditionally chemical questions, yet usually considered just within this subject, are the migration under centrifugal or electric fields and the light scattering of large, dissolved biomolecules which give information about their size and related properties. Similar caveats apply to the use of both absorption and emission spectroscopies over a variety of wavelengths to identify and quantify biological molecules (see *Spectroscopy*), and to the radical, modern extension of the electromagnetic emission concept, electron microprobe analysis, whereby the atomic constitution of cells and organelles can be partially assayed (see *Electron Microprobe Analysis*). A wealth of other spectroscopic methods, however, permit the analysis of biological questions having more definitively physical content, such as the redox state of respiratory systems, the fluidity of membranes, and photosynthetic and visual processes. Rather similarly, the effects of ionizing radiations on living matter may be either a tool (e.g., in genetics), or a subject of biophysical study in themselves (see *X Rays: Biological Effects*).

Membrane biophysics was founded on the intracellular electrode, as much as molecular biophysics was on x-ray diffraction, but continues to use it ever more widely. Initially recording, then controlling, the voltage across a cell membrane, the microelectrode now reaches down to the level of the single ionic channel by recording the current noise in submicrometer membrane pieces. Isotopic tracers are often used along with microelectrodes to study ion movement across barriers (cell membranes or whole tissues) and may be the only tool available when the migrating species is uncharged.

The biophysics of body systems and whole organisms (plant or animal) tends to use simpler instrumentation such as pressure or force transducers, or thermocouples. Flight is studied in wind tunnels and with stroboscopes; swimming is studied in flow tanks. Ecological physics, however, makes extensive use of both telemetry and radar; yet here, as at several points above, it is doubtful how often the question is anywhere near as physical as the instrumentation used to answer it.

4. Future Developments

With so much information available on the molecular make-up of cells, as well as simpler organisms such as viruses, physical theories are now being widely produced—though some are better founded than others—about what constitutes the living state (Ling 1984), the dynamical control systems which operate in it (Goodwin 1963), and the means by which it formed on the surface of this planet (Eigen and Schuster 1979). Yet the fact that some of the most fundamental processes are incompletely understood should not be overlooked. An example is energy transduction: the oxidative formation of ATP, the universal energy source of cells, is still the subject of competing theories (notably the "chemiosmotic" theory of Mitchell in Britain), and so is the means by which the scalar release of the energy of hydrolysis of ATP produces vectorial work in motile systems and in active solute transport (Hill 1977).

Another area of uncertainty, fundamental in importance if not in scale, is morphogenesis—arguably the least understood of all heavily researched processes in modern biology. It is not even clear whether the mechanisms by which genes are expressed in body shape (genotype converted into phenotype) are essentially chemical or physical; nor whether truly new mechanisms in either class await discovery, or if unperceived mathematical consequences of mechanisms already known are all that are involved.

Finally, the biophysics of brain, like its physiology and biochemistry, although immensely vigorous and exciting (see, for example, MacKay 1980 and Marr 1982), confronts problems which may be ultimately insuperable. One is simply that of scale and all the formal as well as experimental problems thence implied: earlier estimates of about 10^{10} neurons inside the cranium, making around 10^3 synapses (connections) each, are even being revised upward. Still more fundamental is the inescapable philosophical doubt of whether it is possible in principle for the human brain to formulate effective questions about itself.

5. Conclusion

In moving from physics to biophysics, the experimentalist finds inconvenient, messy material, frighteningly easy to disturb or destroy, with variables impossible to isolate; the theoretician finds never just a "many-body" but a "many-many-many-body" problem, with practically no phase homogeneous, and almost no relationship linear. Furthermore, organizational complexity means that the effects of scale are more severe: a particle physicist and an astrophysicist have far more of their intellectual armory in common than two biophysicists, one studying the avian genome and the other flapping flight. Yet we may think that Galileo, though he would have understood the methods of neither, would have approved the intentions of both.

See also: Artificial Membranes; Bioelectricity; Biomechanics; Cell Electrophoresis; Muscle; Thermodynamics, Classical

Bibliography

Alberts B, Bray D, Lewis J, Raff M, Roberts K, Watson J D 1983 *Molecular Biology of the Cell.* Garland, New York

Alexander R M 1983 *Animal Mechanics*, 2nd edn. Blackwell, Oxford

Caro C G, Pedley T J, Schroter R C, Seed W A 1978 *Mechanics of the Circulation.* Oxford University Press, Oxford

Casey E J 1962 *Biophysics: Concepts and Mechanisms.* Reinhold, New York

Eigen M and Schuster P 1979 *The Hypercycle: A Principle of Natural Self-Organization.* Springer, Berlin

Fung Y C 1981 *Biomechanics: Mechanical Properties of Living Tissues.* Springer, New York

Goodwin B C 1963 *Temporal Organisation in Cells.* Academic Press, London

Gregory R L 1981 *Mind in Science: A History of Explanations in Psychology and Physics.* Weidenfeld and Nicolson, London

Hill T L 1977 *Free Energy Transduction in Biology: The Steady-State Kinetic and Thermodynamic Formalism.* Academic Press, London

Hodgkin A L 1964 *The Conduction of the Nervous Impulse.* Liverpool University Press, Liverpool

Hughes W 1979 *Aspects of Biophysics.* Wiley, New York

Huxley A F 1980 *Reflections on Muscle.* Liverpool University Press, Liverpool

Katchalsky A, Curran P F 1965 *Nonequilibrium Thermodynamics in Biophysics.* Harvard University Press, Cambridge, Massachusetts

Lenihan J 1974 *Human Engineering: The Body Re-Examined.* Weidenfeld and Nicolson, London

Ling G N 1984 *In Search of the Physical Basis of Life.* Plenum, New York

MacKay D M 1980 *Brains, Machines and Persons*. Collins, London

Marr D 1982 *Vision: A Computational Investigation into the Human Representation and Processing of Visual Information*. Freeman, San Francisco

Nicholls D G 1982 *Bioenergetics: An Introduction to the Chemiosmotic Theory*, Academic Press, London

Rosen R M 1973 *Foundations of Mathematical Biology*. Academic Press, London

Talbot S A, Gessner U 1973 *Systems Physiology*. Wiley, New York

van Holde K E 1971 *Physical Biochemistry*. Prentice-Hall, Englewood Cliffs, New Jersey

von Békésy G 1960 *Experiments in Hearing*. McGraw-Hill, New York

Waterman T H, Morowitz H J 1965 *Theoretical and Mathematical Biology*. Blaisdell, New York

Watson J D 1976 *Molecular Biology of the Gene*, 3rd edn. Benjamin, Menlo Park, California

Wiener N 1961 *Cybernetics*. MIT/Wiley, New York

N. C. Spurway

Biostatistics

Biostatistics consists of that part of the theory and application of formal statistics that pertains to biological and medical problems. The range of subject matter is broad; techniques are drawn from data collection, sampling, experimental design, interval estimation, hypothesis testing, correlation and regression, probability theory, simulation, and stochastic modelling. Applications lie in the areas of vital statistics, clinical trials, bioassay, growth and development, agriculture, genetics, ecology, and epidemiology. The simplest techniques, descriptive procedures such as tabular display, graphical display, and the calculation of averages and accompanying measures of dispersion are very widely employed. Their aim is to represent a set of more or less complex data in such a way that its salient features are readily recognizable. Studies of association that compute measures of association without attempting generalization beyond the observed data also fall in this category.

On the other hand, inferential procedures are also routinely used in biostatistics. These draw probabilistic conclusions concerning general situations of which available data are representative. This implies that the data are collected by random sampling procedures based on the larger population for which the generalization is to be made. Failure to observe proper random choice of items for inclusion in a study may well lead to serious bias in the results. But if sample selection is proper, there are many well-known procedures either for comparing a sample statistic, such as a mean, median, standard deviation or correlation coefficient, with a predetermined standard or for comparing two or more such statistics from different samples. Common parametric tests, those

based upon strong distributional assumptions, include the t-test, the Fisher–Behrens test, the Newman–Keuls test, analysis of variance, the F-test for equality of variance, Duncan's multiple range test, and large-sample tests based on the normal distribution and the central limit theorem.

Nonparametric tests are those that do not rest upon the functional form of the underlying probability distribution. They may thus be used more broadly than parametric tests. There is a compensating loss of power, but this is sometimes small, and the alternatives to the use of nonparametric procedures are sometimes unattractive. Typical nonparametric procedures are the Wilcoxon test, the Mann–Whitney test, the run test, the Kolmogorov–Smirnov and chi-square tests for goodness of fit, the Friedman test, and the Kruskal–Wallace test. If the problem involves estimation rather than hypothesis testing, most of the procedures mentioned here, both parametric and nonparametric, can be used to set confidence limits.

In epidemiology, vital statistics and biological development, much work is based upon morbidity and mortality rates. These are simple proportions; however, the user must be alert to identify the population used in setting the denominator of the fraction. This population defines the extent to which generalization of the rate is allowable. In comparative studies, much emphasis is placed upon adjustment of morbidity and mortality rates for factors such as age, sex, and ethnic origin. With the effects of such factors removed, clearer comparisons among rates are sometimes possible. Probably the best adjustment procedure is cohort analysis, subdividing on the factor whose effects one wishes to remove from the data.

Probability theory forms the conceptual basis for inferential statistics and provides additional analytical tools for the biostatistician. For example, its central role in genetics has been recognized ever since science rediscovered the work of Mendel. But more generally, it provides a vehicle for describing or approximating a great variety of natural interactions, the outcomes of which are not deterministic. Many of the uses of statistics in biology involve collecting and analyzing data in attempts to check the reasonableness and perhaps the validity of a model that has been advanced as a possible explanation of some phenomenon. Among the areas where such modelling has been applied are the theory of learning, the spread of epidemic disease, and the induction of cancer.

Since important problems of biology and human health are known to be affected by many variables, multivariate methods are often called upon by biostatisticians. Multiple regression and correlation analysis, in various versions, form the backbone of most empirical studies of association among possible causal variables and final measurable outcomes. Factor analysis and principal component analysis can assist in the sorting-out process. Particularly in genetics, agriculture, and epidemiology, it is now common to encounter studies involving a large number of variables and requiring the use of modern electronic computers. Computers are also a necessary part of simulation studies which normally require huge numbers of re-computations.

Experimental design cuts across all areas of application. Good design assures that the estimates, comparisons and tests needed will be attained, and that their cost will not be greatly different from the minimal cost possible. On the other hand, bad design leads to flawed experiments where little or no information relevant to the problem under study eventuates or where costs mount and months may grow to years without completion or solution. Poor design may lead to a conclusion that real-world environmental data cannot yield statistically significant associations with appropriate dependent variables such as mortality rates, in cases where proper design would provide significant results. Experimental design was systematized in connection with agricultural research first by R A Fisher and his colleagues at the Rothamstead Experimental Station in the UK. A particular area of its concentrated use today is that of clinical trials and bioassay where experiments tend to be designed in advance according to lengthy protocols establishing step-by-step procedures, sometimes in minute detail.

Biostatistics has been used extensively in studies of physical quantities in biology and medicine. In the remainder of this article, research is cited that uses statistical techniques in the study of the interaction of biological systems with major physical factors: magnetism, sound, temperature, light, electricity, pressure, and ionizing radiation. Emphasis is placed upon the statistical techniques used, and an attempt is made to mention procedures employed most frequently. However, limited space allows only brief discussion. Related articles dealing with statistical methods in medicine and with mathematical models in biology can be found elsewhere in the Encyclopedia (see *Statistical Methods in Medicine; Mathematical Modelling in Biology*).

1. Magnetism

About thirty years ago, basic biostatistical procedures were used in studies of mouse leukocytes when the animals were maintained in magnetic fields (Barnothy et al. 1956). Percentage increases and decreases in numbers of segmented cells were computed along with associated confidence limits. Tests of hypotheses showed that several increases and decreases were significant in comparison to effects observed in control groups. In a related article (Barnothy and Barnothy 1958) graphical illustrations of variation in leukocyte count over time were shown for mice first kept in a magnetic field and then irradiated from a cobalt source. Again significant differences were found between treated mice and controls. More recently, tests for the significance of mean differences have been used in a study (Archer 1979) that used horizontal geomagnetic flux as a possible explanatory variable for anencephalus incidence. The several factors involved were stratified; mean values and their statistical errors were computed for each stratum as a basis for the tests.

Standard experimental design was used in a report (Krueger et al. 1974) dealing with the effects of electromagnetic fields on egg production. Hens were randomly assigned to a control group, and to five experimental groups each receiving exposure to a different electromagnetic field. Analysis of variance was used to check for significant differences among means. A simpler design was used in a double-blind assessment of human pain thresholds (Harper and Wright 1977). When commercially available "magnetic" bracelets were compared with placebo bracelets, no significant difference in pain threshold was found. Pain was generated by radiant heat on the back of the hand.

Also, Monte Carlo methods were used (Shih 1975) to simulate the effects of magnetic fields in high-energy electron radiotherapy. Accumulated electron energy deposition was exhibited in the form of contour plots based on the simulation results.

2. Sound

In recent years numerous studies of the effects of sound on biological systems have been carried out. For example, in a report concerning the effects of reverberation on speech discrimination (Gelfand and Hochberg 1976), analysis of variance was used after an arcsine transformation to test the significance of differences between normal and hearing-impaired subjects under several experimental conditions. Also, in a pattern analysis of auditory-evoked EEG potentials (Sayers et al. 1979), correlation coefficients were used as a basis for further statistical work; chi-square methods were employed to test distributions of phase spectra.

Various probability calculations based on Bayes's theorem (Lamore and Rodenburg 1980) were used in an attempt to study the relationship between the results of the short-increment sensitivity index (SISI) and auditory recruitment. At an earlier date, factor analysis (Harris 1963) was applied to a large number of DLI (difference limen for intensity) techniques. Harris found one set of factors relating to modulation techniques and another set relating to the memory method.

A very carefully designed community noise survey (Malchaire and Horstman 1975) involved sampling procedures based on Latin squares. Practical random sampling techniques were employed in the data-gathering phase, and both correlation and regression procedures were used to analyze relationships between percentiles of noise level distributions.

Extensive tables along with considerable analysis were used (Rop et al. 1979) in a recent study of hearing losses deriving from industrial noise exposure. Seven different noise levels were found to have significant negative effects on workers' hearing according to a likelihood ratio test. Age and sex distributions of expected percentage of workers with a total hearing loss greater than or equal to 90 dB were given separately for the right and left ears. Also, in a comparative examination of two methods for selecting hearing protectors (Waugh 1976), test of differences between standard deviations revealed that relatively small differences were highly significant.

3. Temperature

In a study of reduction in hamster liver size during heat exposure (Chayoth et al. 1977), confidence limits were set for average liver weight, and significance tests were run for differences between average liver weights in experimental and control animals. A paper comparing various physiological responses of girls and women to treadmill walking at elevated temperatures (Drinkwater et al. 1977) used both two and three factor analysis of variance and tested the main effects. It also went on to employ the Newman–Keuls test based on ordered means. In an experiment examining recovery after immersion in cold water (Harnett et al. 1980), subjects were placed in tanks at one of two temperature levels and later rewarmed by one of eight methods. Differences between treatment effects were tested pairwise by the Fisher–Behrens procedure because the equal variance assumption for analysis of variance was not satisfied.

A study of the relationships connecting atmospheric temperature and several biochemical and hemostatic variables (Bull et al. 1979) was centered on correlation coefficients. Such coefficients were computed between temperature differences, on the one hand, and differences between successive readings on the other. Tests for significance of the correlation coefficients were run, and the calculations were repeated for eight periods of lag.

Regression procedures were used in an examination (Miller and South 1978) of thermoregulation in the spinal cord of the marmot. The resulting linear models related maximum change in heat production in the cord to difference between cord temperature and preliminary hypothalmic temperature. Regression functions were again constructed in a report on the viability of rabbit kidneys preserved by hypothermic perfusion (Foreman 1975).

An example of probabilistic modelling (Curry et al. 1978) derived a mathematical theory of poikilotherm development from a minimal set of assumptions. It then justified developmental rate as a random variable and considered three special cases—all involving temperature dependence.

4. Light

An experiment (Peters et al. 1978) showed that sixteen hours of light daily increased weight gain and milk yield in Holstein cattle in comparison to control cattle. *t*-Tests and analysis of variance were used to test differences in mean effects. Confidence limits were also reported. In a study of photosensitivity of hatchling alligators (Kavaliers 1980), the diel pattern of the mean energy fluence rate was shown to be bimodal by graphical methods. *t*-Tests were used to identify significant differences in threshold fluence rates at different temperature levels and between scotophase and photophase. *t*-Tests were also used by investigators of the effect of light on porphyrin metabolism (Magnus et al. 1974). Confidence limits were set for mean ALA-S activity, ALA-S indicating hepatic S-aminolaevulinic synthetase, which was used as an index of porphyrin abnormality. Tests showed significant differences between rats exposed to light and those kept in the dark when porphyria had been induced.

The correlation coefficient between critical photoperiod and growing season was computed (Bradshaw 1976) for 22 populations of the mosquito *Wyeomyia smithii*. This coefficient was found to be highly significant. A multiple regression equation was established that linked growing season with latitude and altitude. Temporal changes in growth hormone, cortisol and glucose were examined in rhesus monkeys (Natelson et al. 1975). The distribution over time of secretory bursts was found to be nonrandom on the basis of an *F*-test. A chi-square test showed that significantly more bursts occurred during light than during dark. These investigators also employed cumulative binomial probabilities and Wilcoxon signed-rank tests as well as Spearman rank correlations.

The effects of automotive exhaust and carbon monoxide on chick embryos were investigated (Hoffman and Campbell 1978). Weight and biochemical data were analyzed by one-way analysis of variance and by Duncan's multiple range test. Chi-square was used to examine survival rate data. Some significant effects of both exhaust and carbon monoxide were identified. A chi-square test was also used in a study (Stahl et al. 1979) of inactivation of virus warts by light.

5. Electricity

The number of studies that consider some aspect of electricity in connection with biological phenomena is huge, and many employ statistical procedures. For example, in a well-designed study based on human EEG data (Giannitrapani 1979a), analysis of variance was used to identify significant differences in brain areas and frequencies between subjects differing in laterality preference. The same technique was employed (Giannitrapani 1979b) to analyze data on cross-spectra, phase-angle and coherence scores in normal and schizophrenic subjects.

In examining damage to the heart following defibrillator shock (Babbs et al. 1980), the investigators fitted dose–response curves to percentage data using linear regression following a probit transformation. In an epidemiologic survey of relationships between respiratory illness and the use of gas and electricity for cooking (Keller et al. 1979), multiple regression was used to relate incidence of disease to a variety of environmental factors.

Exponential functions were fitted to measures of average contraction velocity (Broman 1977) by least squares in an investigation of the influence of heavy contraction on a skin-derived single motor unit action potential. And the Poisson probability distribution was used in a study of electron energy losses in high-voltage electron microscopy (Jouffrey 1978) to model the probability of creating n plasmons.

Chi-square tests with Yates' correction and Fisher's exact test were both used to study the usefulness of 25 h Holter electrocardiovascular tape recordings in predicting coronary events after acute myocardial infarction (Cats et al. 1979). Chi-square procedures were also employed (Ivanova et al. 1980) to test for significance following application of the prognostic stratification method to various exercise and ambulatory monitoring data obtained from electrocardiographs.

6. Pressure

In a study of changes in arterial baroreceptor reflexes in the seal during diving (Angell-James et al. 1978), confidence limits for mean pulse interval and for blood pressure measurements were found before and after treatments and before and after diving. A student's t-test was used to evaluate differences between the means of paired data. Regression lines were computed to represent the association of blood pressure and pulse interval. Paired-data t-tests also appeared in a study of the influence of low- and high-pressure baroreceptors on plasma renin activity (Mark et al. 1978).

F-tests for significance of interactions and t-tests for differences in main effects appeared in a paper (Thomas et al. 1976) that examined the interaction effects of hyperbaric nitrogen and oxygen on rats. Analysis of covariance revealed significant differences in performance on cognitive test scores over time during a 17-day dry saturation dive (O'Reilly 1977). Analysis of variance and Newman–Keuls tests were also employed.

In a sophisticated examination of the relationships between environmental and genetic factors and blood pressure (Weinberg et al. 1979), the principal technique used was path analysis. Multiple regression procedures were also employed to adjust the observed data for age and sex. A chi-square test and a Kolmogorov–Smirnov test for normality were each used in different parts of the study. Also a study of blood pressure aggregation in families (Havlik et al. 1979) performed multiple regression using the residuals of observed data adjusted for sex and generation. Coefficients of correlation between systolic and diastolic blood pressure were computed.

7. Ionizing Radiation

Radiation effects at high levels of exposure have been studied extensively by statistical methods. Much of this work is based either on Japanese human data from Hiroshima and Nagasaki (Beebe et al. 1971), on data from weapons testing (Rallinson et al. 1974), or on laboratory exposure of animals (Anderson et al. 1979). The follow-up atomic bomb studies place great emphasis upon sampling and estimation procedures as do the weapons test analyses. Studies involving laboratory exposure use various types and levels of ionizing radiation. They often are concerned with estimation of a dose–response function. They also employ a broad range of inferential and descriptive procedures.

There is extensive literature concerning health effects of low-level ionizing radiation, such as natural background radiation. Much of this is summarized in the so-called BEIR-III report (Committee on the Biological Effects of Ionizing Radiations 1980). In particular, the use of models for the dose–response function is discussed there and in a later commentary (Fabrikant 1981). The difficult problem of obtaining empirical estimates of cancer risk from surveys has been discussed (Land 1980) from the point of view of sample size needed to obtain reasonable results in terms of precision and power. However, some such work has gone forward. One study (Jacobson et al. 1976) used correlation coefficients and simple linear regression in analyzing the relationship in US state data between background ionizing radiation, annual dose rate, and leukemia death rate. Others (Frigerio and Stowe 1976, Sanders 1978, Hickey et al. 1981) employed multiple regression techniques to relate natural background radiation level, along with certain atmospheric chemical pollutant factors and other variables, to mortality rates for a number of chronic diseases, mainly cancers.

8. Conclusion

Statistical procedures are useful throughout all branches of science; the fact that "biostatistics" is recognized as a word in the English language testifies to their importance in the fields of biology and medicine. This article has attempted to point out how widely statistical thinking is used among scientists working at the interface of physical phenomena and biological systems. Whether the processes involved are deterministic or stochastic, the biostatistician has a contribution to make. As the quantification of biological knowledge increases, the use of statistical methods will likewise increase. After all, the discussion concerns the data gathering, the accurate description, and the process of generalization that lie at the heart of the scientific method.

See also: Cancer Statistics; Statistical Methods in Medicine

Bibliography

Anderson E C, Holland L M, Prine J R, Smith D M 1979 Tumorigenic hazard of particulate alpha-activity in Syrian hamster lungs. *Radiat. Res.* 78: 82–97

Angell-James J E, Daly M deB, Elsner R 1978 Arterial baroreceptor reflexes in the seal and their modification during experimental dives. *Am. J. Physiol.* 234: 730–39

Archer V E 1979 Anencephalus, drinking water, geomagnetism and cosmic radiation. *Am. J. Epidemiol.* 109: 88–97

Babbs C F, Tacker W A, Van Vleet J F, Bourland J D, Geddes L A 1980 Therapeutic indices for transchest defibrillator shocks: Effective, damaging, and lethal electrical doses. *Am. Heart J.* 99: 734–38

Barnothy J M, Barnothy M F, Boszormengi-Nagy I 1956 Influence of a magnetic field upon the leucocytes of the mouse. *Nature* 177: 577–78

Barnothy M F, Barnothy J M 1958 Biological effect of a magnetic field and the radiation syndrome. *Nature* 181: 1785–86

Beebe G W, Kato H, Land C E 1971 Studies of the mortality of A-bomb survivors. IV. Mortality and radiation dose. 1950–1966. *Radiat. Res.* 48: 613–49

Bradshaw W E 1976 Geography of photoperiodic response in diapausing mosquito. *Nature* 262: 384–85

Broman H 1977 An investigation on the influence of a sustained contraction on the succession of action potentials from a single motor unit. *Electromyogr. Clin. Neurophysiol.* 17: 341–58

Bull G M, Brozovic M, Chakrabarti R, Meade T W, Morton J, North W R, Stirling Y 1979 Relationship of air temperature to various chemical, haematological and haemostatic variables. *J. Clin. Pathol.* 32: 16–20

Cats V M, Lie K I, VanCapelle F J, Durrer D 1979 Limitations of 24 hour ambulatory electrocardiographic recording in predicting coronary events after acute myocardial infarction. *Am. J. Cardiol.* 44: 1257–62

Chayoth R, Krauthammer N, Winikoff J, Sod-Moriah U A 1977 Control of liver size in heat-acclimated hamsters. *J. Appl. Physiol.* 43: 445–48

Committee on the Biological Effects of Ionizing Radiations, Assembly of Life Sciences, Division of Medical Sciences 1980 *The Effects on Populations of Exposure to Low Levels of Ionizing Radiation.* National Academy of Sciences, Washington, DC

Curry G L, Feldman R M, Sharpe P J 1978 Foundations of stochastic development. *J. Theor. Biol.* 74: 397–410

Drinkwater B L, Kupprat I C, Denton J E, Crist J L, Horvath S M 1977 Response of prepubertal girls and college women to work in the heat. *J. Appl. Physiol.* 43: 1046–53

Fabrikant J I 1981 The BEIR-III report: Origin of the controversy. *Am. J. Roentgenol.* 136: 209–14

Foreman J 1975 Prediction of viability of rabbit kidneys preserved by hypothermic perfusion. *Cryobiology* 12: 231–37

Frigerio N A, Stowe R S 1976 Carcinogenic and genetic hazard from background radiation. In: *Biological and Environmental Effects of Low-Level Radiation*, Vol. II. *Proc. Symp. IAEA and WHO, 1975.* International Atomic Energy Agency, Vienna, pp. 385–93

Gelfand S A, Hochberg I 1976 Binaural and monaural speech discrimination under reverberation. *Audiology* 15: 72–84

Giannitrapani D 1979a Laterality preference, electrophysiology and the brain. *Electromyogr. Clin. Neurophysiol.* 19: 105–23

Giannitrapani D 1979b Spatial organization of the EEG in normal and schizophrenic subjects. *Electromyogr. Clin. Neurophysiol.* 19: 125–45

Harnett R M, O'Brien E M, Sias F R, Pruitt J R 1980 Incidental treatment of profound accidental hypothermia. *Aviat. Space Environ. Med.* 51: 680–87

Harper D W, Wright E F 1977 Magnets as analgesics. *Lancet* ii: 47

Harris J D 1963 Loudness discrimination. *J. Speech Hear. Disord.* Suppl. 11

Havlik R J, Garrison R J, Feinleib M, Kannel W B, Castelli W P, McNamara P M 1979 Blood pressure aggregation in families. *Am. J. Epidemiol.* 110: 304–12

Hickey R J, Bowers E J, Spence D E, Zemel B S, Clelland A B, Clelland R C 1981 Low level ionizing radiation and human mortality: Multi-regional epidemiological studies. A preliminary report. *Health Phys.* 40: 625–41

Hoffman D J, Campbell K I 1978 Embryotoxicity of irradiated and non-irradiated automotive exhaust and carbon monoxide. *Environ. Res.* 15: 100–7

Ivanova L A, Mazur N A, Smirnova T M, Sumarokov A B, Nazarenko V A, Svet E A 1980 Electrocardiographic exercise testing and ambulatory monitoring to identify patients with ischemic heart disease at high risk of sudden death. *Am. J. Cardiol.* 45: 1132–38

Jacobson A P, Plato P A, Frigerio N A 1976 The role of natural radiations in human leucomogenesis. *Am. J. Publ. Health* 66: 31–37

Jouffrey B 1978 Electron energy losses with special reference to high-voltage electron microscopy. *Ann. N.Y. Acad. Sci.* 306: 29–46

Kavaliers M 1980 Circadian rhythm of extraretinal photosensitivity in hatchling alligators. *Alligator Mississippiensis. Photochem. Photobiol.* 32: 67–70

Keller M D Lanese R R, Mitchell R I, Cote R W 1979 Respiratory illness in households using gas and electricity for cooking. *Environ. Res.* 19: 495–503

Krueger W F, Giarola H J, Bradley J W, Shrekenhamer A 1974 Effects of electromagnetic fields on fecundity in the chicken. *Ann. N.Y. Acad. Sci.* 247: 391–400

Lamore P J, Rodenburg M 1980 Significance of the SISI test and its relation to recruitment. *Audiology* 19: 75–85

Land C E 1980 Estimating cancer risks from low doses of ionizing radiation. *Science* 209: 1197–203

Magnus I A, Janousek V, Jones K 1974 The effect of environmental lighting on porphyrin metabolism in the rat. *Nature* 250: 504–5

Malchaire J B, Horstman S W 1975 Community noise survey of Cincinnati, Ohio. *J. Acoust. Soc. Am.* 58: 197–200

Mark A L, Abboud F M, Fitz A E 1978 Influence of low- and high-pressure baroreceptors on plasma renin activity in humans. *Am. J. Physiol.* 235, Pt. 1: H29–33

Miller V M, South F E 1978 Thermoregulatory responses to temperature manipulation in the spinal cord of the marmot. *Cryobiology* 15: 433–40

Natelson B H, Holaday J, Meyerhoff J, Stokes P E 1975 Temporal changes in growth hormone, cortisol and glucose: Relation to light onset and behavior. *Am. J. Physiol.* 229: 409–15

O'Reilly J P 1977 Hana Kai II: A 17-day dry saturation dive at 18.6 ATA. VI. Cognitive performance, reaction time, and personality changes. *Undersea Biomed. Res.* 4: 297–305

Peters R R, Chapin L T, Leining K B, Tucker H A 1978 Supplemental lighting stimulates growth and lactation in cattle. *Science* 199: 911–12

Rallinson M L, Dobyns B M, Keating F R, Rall J E, Tyler F H 1974 Thyroid disease in children. A survey of subjects potentially exposed to fallout radiation. *Am. J. Med.* 56: 457–63

Rop I, Raber A, Fischer G H 1979 Study of the hearing losses of industrial workers with occupational noise exposure, using statistical methods for the analysis of qualitative data. *Audiology* 18: 181–96

Sanders B S 1978 Low-level radiation and cancer deaths. *Health Phys.* 34: 521–38

Sayers B M, Beagley H A, Riha J 1979 Pattern analysis of auditory evoked EEG potential. *Audiology* 18: 1–16

Shih C C 1975 High energy electron radiotherapy in a magnetic field. *Med. Phys.* 2: 9–13

Stahl D, Veien N K, Wulf H C 1979 Photodynamic inactivation of virus warts: A controlled clinical trial. *Clin. Exp. Dermatol.* 4: 81–85

Thomas J R, Burch L S, Banvard R A 1976 Interaction of hyperbaric nitrogen and oxygen effects on behavior. *Aviat. Space Environ. Med.* 47: 965–68

Waugh R 1976 Calculated in-ear A-weighted sound levels resulting from two methods of hearing protector selection. *Ann. Occup. Hyg.* 19: 193–202

Weinberg R, Shear C L, Avet L M, Frerichs R R, Fox M 1979 Path analysis of environmental and genetic influences on blood pressure. *Am. J. Epidemiol.* 109: 588–96

R. J. Hickey, A. B. Clelland and R. C. Clelland

Biotelemetry

Telemetry, as the word's roots suggest, is measurement at a distance. Biotelemetry is the discipline concerned with the measurement of living organisms at a distance. Though accurate, this inclusive definition requires two refinements to be useful since virtually all measurements in biology are made at a distance, especially in man: for example, the electrocardiogram is usually sensed at the chest wall and not directly from the heart. Distance in the telemetry context is presumed to be relatively great. By convention, hardware modes are understood to be excluded from biotelemetry though technically they qualify. Freedom of movement of the organism is implicit in biotelemetry. Biotelemetry is thus of profound significance and immense importance: profound because it permits study of life in natural environments; immense because of the vast number of situations in which it can be usefully applied.

1. Principles of Telemetry

Telemetry systems are special examples of communication systems whose purpose is to send messages from a source to a destination. The elements common to such systems may be represented as shown in Fig. 1.

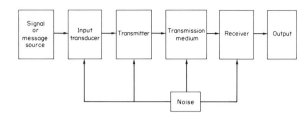

Figure 1

Block diagram of the segments of a communication system

The signal source in biotelemetry is the organism. The signal is sensed by an electrode which is connected to the transmitter. The transmitter conditions the signal and couples it to the transmission medium. The signal as contained in the transmission medium is detected by a receiver at the destination and displayed as desired or subjected to computations. Noise impinges on the system at several points, contaminating the signal in real communication systems.

Since the biological signals have time as a dependent variable, the convention $x(t)$ is adopted to represent an arbitrary signal at time t. There are many circumstances, however, when frequency-domain descriptions are convenient for communication systems.

In telemetry systems, the transmitter provides the transmission medium, most often a radiating electromagnetic wave, termed a carrier wave since it bears the information of interest. The carrier is represented by $A_c \cos \omega_c t$ in the case of a continuous-wave (cw) carrier where A_c is the amplitude of the unmodulated carrier. Since biological signals often require amplification, the transmitter has associated amplifiers and also subcarriers when multichannel capability is desired.

1.1 Modulation

The most important signal processing to be performed by the transmitter is modulation, which is the systematic alteration of the carrier by the signal of interest. Historically, amplitude modulation of a cw carrier was the first modulation mode achieved. This is the common AM standard radio broadcast mode. The modulation is represented by:

$$x_c(t) = A_c \cos \omega_c t + M x(t) A_c \cos \omega_c t$$
$$= A_c[1 + M x(t)] \cos \omega_c t \qquad (1)$$

where ω_c is the radian frequency, and M is a constant, the modulation index. The carrier frequency f_c is

$$f_c = \omega_c / 2\pi \qquad (2)$$

The envelope of the modulated carrier has the shape of the signal $x(t)$ when the carrier has a much higher frequency than the signal; otherwise the envelope is not recognizable. This modulation property is called frequency translation upwards, and it permits the use of antennae of manageable size to radiate the modulated carrier. This is important in biology since many signals are below 100 Hz and their direct radiation would require enormously large antennae. This carrier frequency requirement, which preserves the needed relationship between the signal $x(t)$ and the envelope, imposes the condition that

$$f_c \gg W' \qquad (3)$$

where W' is the upper frequency in the signal band width; and

$$M \leqslant 1 \qquad (4)$$

when, for mathematical convenience, the signal is normalized, that is, $|x(t)| \leqslant 1$. When $M = 1$, 100% modulation is achieved and the modulated amplitude varies between 0 and $2A_c$. When $M > 1$, overmodulation occurs, resulting in phase reversal and distortion of the envelope.

Spectral (Fourier) transformation of the time domain signal gives information about the transmission frequency bandwidth B_T required for an amplitude modulated signal. This transform is found to be:

$$X_c(f) = \frac{A_c}{2} [\delta(f - f_c) + \delta(f - f_c)]$$
$$+ \frac{MA_c}{2} [X(f - f_c) + (f + f_c)] \qquad (5)$$

where δ is the Dirac delta function. This is the translated message spectrum plus a pair of impulses at $\pm f_c$, the carrier frequency itself.

There are two properties of the AM spectrum worth observing: (a) there is symmetry about the carrier frequency; and (b) the transmission frequency bandwidth required for an AM signal is exactly twice the message bandwidth

$$B_T = 2W' \qquad (6)$$

Thus, if conservation is necessary, AM has its drawbacks. The double term is recognized by terming the spectrum above f_c as the upper sideband, and that below as the lower sideband, leading to the designation double-sideband modulation (DSB). Modifications of the frequency translated spectrum give rise to single-sideband modulation (SSB) and vestigial-sideband modulation (VSB). All are examples of linear modulation, and all have certain individual advantages.

Nonlinear modulation includes phase modulation (PM) and frequency modulation (FM) and less practical methods. Only frequency modulation will be considered here. First, a seeming paradox needs clarification. Frequency has implicit the concept of periodicity, but modulation connotes time variation. These are incompatible concepts; they can be resolved by considering the angular position of the phasor $\theta_c(t)$:

$$\theta_c(t) = 2\pi f_c t + \phi(t) \qquad (7)$$

The second term is the relative phase angle. The angular frequency is the time derivative of angular position. The instantaneous frequency $\mathscr{L}(t)$ in hertz is given by

$$\mathscr{L}(t) = \frac{1}{2\pi} \frac{d\theta_c}{dt} = f_c + \frac{1}{2\pi} \frac{d\phi}{dt} \qquad (8)$$

and the instantaneous frequency of an FM wave is

$$\mathscr{L}(t) = f_c + f_d x(t) \qquad (9)$$

where f_d is the frequency deviation constant.

The phasor $\theta_c(t)$ is related to the instantaneous frequency $\mathscr{L}(t)$ by

$$\theta_c(t) = 2\pi \int^t \mathscr{L}(t') \, dt' \qquad (10)$$

and the modulated waveform is

$$x_c(t) = A_c \cos \left[\omega_c t + 2\pi f \int^t x(t') \, dt' \right] \qquad (11)$$

Nonlinear modulation is also termed exponential or angle modulation and it differs from linear modulation in that the transmission bandwidth is much greater than twice the message or signal bandwidth. Further, the modulated spectrum is not simply the translated signal spectrum.

This latter point is emphasized as spectral analysis of FM signals comes under consideration. Of importance is the observation that $\mathscr{L}(t)$, the instantaneous frequency, is not identical to the spectral frequency f which describes the time-domain signal in the frequency domain in terms of fixed frequency sinusoidal components. Thus, there is no simple equivalence between $\mathscr{L}(t)$ and the FM spectrum, and the approach to the FM spectrum is by tone modulation in which the instantaneous frequency of an FM signal varies sinusoidally about the carrier frequency. Setting $x(t) = A_M \cos \omega_M t$ then

$$\mathscr{L}(t) = f_c + f_d A_M \cos \omega_M t$$
$$x_c(t) = A_c \cos \left(\omega_c t + 2\pi f_d \int^t A_M \cos \omega_M t' \, dt' \right)$$
$$= A_c \cos \left(\omega_c t + \frac{2\pi f_d A_M}{\omega_M} \right) \sin \omega_M t$$

The quantity $\beta = 2\pi f_d A_M / \omega_M = A_M f_d / f_M$ is termed the FM modulation index and is the maximum phase deviation in radians produced by the tone; we have

$$x_c(t) = A_c \cos(\omega_c t + \beta \sin \omega_M t) \qquad (12)$$

The further development of spectral analysis of FM signals involves the use of Bessel functions.

$$x_c(t) = A_c J_0(\beta) \cos \omega_c t$$
$$+ \sum_{n \, \text{odd}}^{\infty} A_c J_n(\beta) [\cos(\omega_c + n\omega_M)t$$
$$- \cos(\omega_c - n\omega_M)t]$$
$$+ \sum_{n \, \text{even}}^{\infty} A_c J_n(\beta) [\cos(\omega_c + n\omega_M)t$$
$$+ \cos(\omega_c - n\omega_M)t] \qquad (13)$$

where $J_n(\beta)$ are Bessel functions of the first kind, of order n and argument β, and Eqn. (13) represents a constant-amplitude wave whose instantaneous frequency is varying sinusoidally. It can be seen that the number of sidebands in frequency modulation is infinite.

Pulse-code modulation (PCM) is digital modulation as opposed to continuous-wave modulation and pulse modulation, which are analog signals. As will be seen, there are advantages to PCM with regard to noise contamination of the signal. If only a few discrete values are allowed for the modulated parameter, and if the separation between these values is great compared with perturbations, then it is easier for the receiver to determine which value was meant to be sent. In analog modulation, the modulated parameter can take on any value corresponding to the range of the message. Contamination by noise means that the receiver cannot determine the signal value.

In pulse-code modulation, the continuous signal $x(t)$ is sampled to get $x_s(t)$. The sample values are rounded off to the nearest preset discrete value. This step is quantization. The sampled and quantized signal $x_{sq}(t)$ is discrete in time and amplitude. If there are a finite number of quantum levels, then each level can be represented by a digital code of finite length. The most commonly used code is binary. Several coded pulses are required for each signal so the PCM bandwidth is considerably greater than the signal bandwidth. Quantized samples occur at a rate of $f_s \geqslant 2W$ samples per second so there must be $v f_s$ coded pulses per second, where v is the number of coded pulses transmitted per sample. Assuming no free distances in the coded signal, the permitted duration of any one pulse is $T_p = (v f_s)^{-1}$. Pulse resolution requires a bandwidth of at least $(2T_p)^{-1}$ and

$$B_T \geqslant (2v f_s)^{-1} \geqslant vW \qquad (14)$$

The baseband signal can be made to modulate a carrier (usually rf) for transmission. Carrier modulation can be amplitude-shift keying (ASK), frequency-shift keying (FSK) or phase-shift keying (PSK). Carrier modulation requires greater transmission bandwidth than Eqn. (14).

1.2 Multiplexing

While a single tracking telemetry channel can be useful in some biological applications, it is frequently desirable to send several channels over the same transmission system. This is accomplished by multiplexing. There are two fundamental methods: time-domain multiplexing (TDM) and frequency-domain multiplexing (FDM). TDM divides time and uses the divisions as slots and sequentially sends signals, one for each slot. The several input signals, all bandlimited in W, are sequentially sampled at the transmitter by a commutator. A set of such impulses is termed a frame; at the end of each frame, the commutator resets the transmission to the beginning of the next frame. A convention is required to identify the beginning and end of a sequence. One revolution is accomplished in $t_0 \leqslant W/2\,\text{s}$. If z is the number of inputs, the pulse-to-pulse spacing is

$t_0 z = (z f_0)^{-1}$; the spacing between successive samples from any one input is t_0. Frame synchronization is seen to be critical for TDM, and the commutator is usually electronic.

The concept in FDM is quite similar. The signals are provided with subcarriers which the individual signals modulate, one per channel. The separate outputs of the modulated subcarriers may be summed and then used to modulate the main carrier, thus sending the needed mixed signal as a single transmission. Subcarrier discriminators are required in these systems to separate the original signals at the receiver. It may be seen at once that avoidance of crosstalk, the situation in which one signal is coupled into another, is of importance in FDM. Proper choice of guard bands to separate the signals far enough in frequency, hence FDM, is essential. There are standardized schemes which have been calculated and tested. The White Sands, New Mexico, telemetry frequencies, standardized for use and accepted by NATO countries, are referenced.

2. System Limitations

2.1 Bandwidth

As has been seen throughout, the bandwidth concept is inevitable. If the signal must be sent in real time, then the output system must keep up with it and the signal speed is a determinant of the system. Even if the transmission rate is manipulable by the biotelemetry engineer, efficient use calls for minimizing transmission time. However, this cannot be manipulated arbitrarily without additional energy cost. If the bandwidth is insufficient, transmission time is slower than real time as signalling speed decreases. This intimate time–bandwidth limitation is real, although for reasons not developed here the signal cannot be band limited and time limited at the same time.

2.2 Contaminations

Noise is generally considered to be anything that is unwanted in the system. It is accurate and useful to distinguish between the contributors: attenuation, distortion, interference, thermal noise and shot noise.

(*a*) *Attenuation*. This is the reduction of signal strength with increasing distance. It is not as intractable as some other difficulties, but it is well to remember that the power law means that losses from deep space can be $10^{20}(200\,\text{dB})$.

(*b*) *Distortion*. An alteration of the signal due to imperfect response of the system to the signal is known as distortion. Theoretically, this may be corrected by improved system design, but in real physical systems some distortion must be tolerable. In theoretical distortionless transmission, the input signal $x(t)$ gives an undistorted output $y(t)$, which differs from the input only by a multiplication constant and a finite time delay:

$$y(t) = Kx(t - t_0) \tag{15}$$

and, in terms of the spectrum of the output

$$H(f) = K \exp(-j\omega t_0) \tag{16}$$

where $H(f)$ is the transfer function of the system. A network giving distortionless transmission must have a constant amplitude response (i.e., $|H(f)|$ is a constant) and a negative linear phase shift (i.e., $\theta(f) = -\omega t_0 \pm M\pi$).

The three major classifications of distortion are: (i) amplitude distortion, for which $|H(f)|$ is not constant; (ii) phase distortion, for which $\theta(f) \neq -\omega t \pm M\pi$; and (iii) nonlinear distortion, where the transfer function is not defined for nonlinear elements.

(*c*) *Interference*. This is contamination by extraneous signals whose form is similar to the desired signal. Interference comes from both man-made and natural sources. It can come from the line current (60 Hz in the USA, 50 Hz in Europe and elsewhere). Among the most common sources are automobile ignition, the atmosphere, solar-flare activity, and the center plane of the galaxy.

The simplest case of interference is probably that of an interfering cosine wave and an unmodulated carrier: if the cosine has amplitude A_i and frequency $f_c + f_i$, then the signal at the demodulator is

$$A_c \cos \omega_c t + A_i \cos(\omega_c + t\omega_i)t$$

If $A_i \ll A_c$ then the signal at the demodulator is:

$$y(t) = A_c(1 + M_i \cos \omega_i t) \cos(\omega_c t + M_i \sin \omega_i t) \tag{17}$$

where $M_i = A_i/A_c \ll 1$. Thus, the interfering wave amplitude modulates the carrier. When the interference is strong rather than weak relative to the carrier, as in this case, then the carrier can be considered to modulate the interfering wave.

(*d*) *Thermal noise*. This is always present in electrical systems. Its inescapability derives from kinetics. A particle at a temperature above absolute zero has thermal energy which imparts random motion. The random motion of an electron constitutes a random current. In a conducting medium, the random motion gives rise to a random voltage, termed thermal, resistance or Johnson noise. Particle–wave duality ensures thermal noise associated with electromagnetic radiation. Therefore, no electrical communication can take place without noise.

Noise is a nuisance to both the biologist and the communications engineer, but progress has been made in quantifying it, identifying its sources, and seeking both to reduce it where possible and to identify signals embedded in a noisy background.

From thermodynamics, it is seen that the power spectrum of thermal noise is constant over an extremely wide frequency range and thus contains all frequencies in equal proportions. Such a spectrum is termed white

by analogy with white light, and may be expressed as

$$G(f) = \eta/2, \qquad \eta = 4RkT \tag{18}$$

where R is the resistance, k is Boltzmann's constant, and T is the absolute temperature.

White noise power implicitly is infinite, or has infinite variance:

$$\int_{-\infty}^{\infty} G(f)\,df = \int_{-\infty}^{\infty} \frac{\eta}{2}\,df = \infty \tag{19}$$

which is physically not possible, and $G(f)$ cannot be flat over all frequencies. (The $\frac{1}{2}$ factor in Eqns. (18) and (19) emphasizes that half of the power is associated with positive frequency and half with negative frequency.) At radio frequencies, kT is very great compared with $h\omega$ and the spectral density is effectively flat; in the vicinity of $f_0 = kT/h$, the spectrum of thermal noise begins to decrease exponentially and f_0 is in range of 10^{13} Hz, the infrared band.

The central limit theorem and other studies indicate also that thermal noise has a Gaussian distribution since it is often the summation of large numbers of randomly moving electrons. Thus, at the receiver front end, the amplitude of thermal noise has a stationary Gaussian distribution with zero mean:

$$P(x) = \frac{1}{\sqrt{2\pi}\,\delta} \exp\left[\frac{-(x - M)^2}{2\delta^2}\right] \tag{20}$$

where δ is the standard deviation.

(*e*) *Shot noise*. While thermal noise is a property of particles above absolute temperature, impinging everywhere on the system and giving rise to the concept of the "noisy channel," there are other noise considerations. The discrete nature of charge carriers as current flows across a junction gives rise to statistical fluctuation which, though averaging zero for steady current, gives rise to noise termed shot noise, the mean square value of which is given by:

$$\overline{i_n^2} = 2qIB \tag{21}$$

where q is the electron charge, I is the current, and B is the noise bandwidth. Shot noise is also random and white in nature. Semiconductors have shot noise due to base current crossing the emitter junction.

Other noise formulations, such as spot noise, $1/f$ noise and pink noise are not considered here and are principally derived from the basic noise concepts treated above. Gaussian white noise adds to the receiver at the front end where the incoming signal is apt to be lowest. In summary, noise as a consequence of thermal considerations cannot be eliminated, is Gaussian, white and additive.

3. Receivers

The basic components of a receiver are tuning mechanisms, a demodulator and amplifiers. FM receivers are most often so-called superheterodyne receivers. There are three kinds of amplifiers in such a receiver: the radio-frequency (rf) amplifier which is tuned to the carrier frequency of interest; an intermediate frequency (IF) amplifier which provides most of the gain and is fixed-tuned; and the audio-frequency (AF) amplifier which brings the power up to the loudspeaker requirement. In AM signals, the receiver performs frequency translation downwards. The following illustration of this process provides an example of heterodyning.

Consider the wave process $x(t)\cos\omega_1 t$. Multiplication by $\cos\omega_2 t$ yields

$$x(t)\cos\omega_1 t \cos\omega_2 t = \tfrac{1}{2}x(t)\cos(\omega_1 + \omega_2)t$$
$$+ \tfrac{1}{2}x(t)\cos(\omega_1 - \omega_2)t$$

The product consists of sum and difference frequencies $f_1 + f_2$ and $f_1 - f_2$ each modulated by $x(t)$. If $f_1 \neq f_2$ then the signal spectrum has been translated to two new carrier frequencies. With selective filtering, the signal can be converted up or down in frequency. Converters can be represented schematically by Fig. 2. The operation is

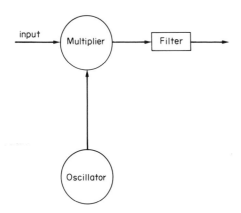

Figure 2
Frequency converter

called heterodyning. When the oscillator is exactly synchronized in both phase and frequency, the process is called synchronous or coherent detection. Communication systems use phase-lock loops to provide coherent detection; the loop effectively cleans up the input signal. The local oscillator in the phase-lock loop has its center frequency controlled by the voltage impressed upon it.

3.1 Signal-to-Noise at the Output

It has already been stated that the signal at the front end of the receiver may be at its lowest, and further corrupted by additive Gaussian white noise. The system usually further degrades performance by internally generated noise, and the ratio between the signal and noise at the output can be substantially less. The output signal power spectrum can be represented by

$$G_{so}(f) = |Hf|^2 G_{si}(f)$$

The output noise spectrum is given by

$$G_{no}(f) = |Hf|^2 G_{ni}(f) + G_{nx}(f)$$

where $G_{nx}(f)$ is the noise (referred to usually as excess noise) introduced by the system. The output ratio of signal to noise is then

$$\left(\frac{S}{N}\right)_o = \frac{\int_{-\infty}^{\infty} |H(f)|^2 G_{si}(f)\, df}{\int_{-\infty}^{\infty} |H(f)|^2 G_{ni}(f) + G_{nx}(f)\, df} \tag{22}$$

If, under the reasonable assumption that the input noise is white and the amplitude response of the system is constant over the frequency range of the input signal, $H(f) = H_o(S/N)$, then is simplified to

$$\left(\frac{S}{N}\right)_o = \frac{H_o^2 p_s}{H_o^2 \eta B_N + N_x} \tag{23}$$

where p_s is the input signal power, B_N is the noise-equivalent bandwidth, N_x is the output total excess noise power and $\eta = 4RkT$. The importance of temperature in unavoidable noise is again noted.

It can be shown that exponential modulation, in particular FM, can provide increased signal-to-noise ratios compared with amplitude or linear modulation. The cost is the otherwise unnecessarily wide transmission bandwidth needed. In biotelemetry, the particular problem must be studied and decisions made on what the requirements are: those of monitoring in the operating room during surgery may differ widely from those of a study of hibernation.

4. Biological Signals

These vary in complexity from what is not truly a biological signal, such as an artificial reference signal placed on an animal for tracking and location purposes, to a highly complicated signal, such as the electroencephalogram which consists of largely aperiodic waves, complex in nature, and mere microvolts in signal strength. The definition of what is living and what is not will not be exercised here save to say that it is simple only superficially. A safe beginning may be established by considering what medicine has established as vital signs in man: pulse, blood pressure, temperature and respiration. For comprehensiveness and utility, the electrocardiogram may be considered as the pulse source for telemetry purposes. The electrocardiogram is a time history of the electrical activity that governs the mechanical pumping cycle of the heart. It is a millivolt signal and is periodic. Since it is already electrical in nature it can be sensed directly using an electrode with the aid of conducting paste. Telemetry of this signal in the individual at rest provides minimal difficulty. The problems of movement are discussed with other artefacts in Sect. 5.

Temperature, a much sought-after signal in many biological studies, may be converted to the electrical domain by thermistors which do not need a reference temperature. A possible disadvantage is that the thermistor has a nonlinear transfer function.

Blood pressure, a vital sign of considerable interest to measure and transmit remotely, has been measured mechanically by the principle of tonometry—the lateral pressure against the arterial wall "bumping" the sensor. Mechanical inputs may be converted into electrical outputs by piezoelectrical crystals and by microphone detection of the Korotkov sounds.

Respiration has been measured by gas-sensitive thermistors at the nostril and from movement of the chest wall by impedance pneumography. Molecular inputs may be converted to electrical outputs by polarographic techniques.

The electroencephalogram, besides being a useful tool in the diagnosis and treatment of epilepsy, has in recent years, under certain conditions, been regarded as the hallmark of useful full life, and its absence—the isoelectric state—as cerebral death. Because it is an electrical signal it can be sensed directly, with the aid of proper conducting paste, from the scalp. It is of very low voltage and its aperiodic complex nature makes it one of the most difficult of all signals to telemeter.

5. Surface Electrodes in Active Subjects

The ability to study organisms by telemetry can be severely restricted if the signal at the point of the sensor is so disturbed that it is not recognizable as the transmitted signal. This may frequently be the case when the interface between the electrode, electrolyte, and tissue is perturbed as the subject moves.

The ability of electrons to flow is fundamental to migration of charge carriers including those at junctions. The function of temperature is again important: the probability function of an electron occupying a temperature-dependent energy state is given by the Fermi–Dirac function:

$$f(E) = \{1 + \exp[(E - E_F)/kT]\}^{-1} \tag{24}$$

where $f(E)$ is the probability that a specific energy level E is occupied at a specific temperature T, and E_F is the Fermi level. The Fermi level is that energy level for which the probability of occupation is 0.5. Most levels above the Fermi level are empty; most below are filled. When sufficiently excited, electrons can cross into the conduction band and participate in charge migration. The simplest model for such charge migration, which may be presumed to occur when the electrode, electrolyte and tissue come into contact, is that of a single electron initially at rest at a distance from the solid electrode conductor, and thus regarded as being at infinity from it. With motion, the electron comes to a distance x from the metal, and a positive charge is induced on the metal surface. There is a force of attraction between the electron and the induced positive charge. This is equivalent to the force which would exist

between the electron and an equal positive charge located at a distance $-x$ from the metal surface. The force is termed the image force and the formulation the method of images. The image force is given by

$$F = \frac{-q^2}{4\pi(2x)^2 \epsilon_0} \tag{25}$$

where F is the image force, q is the electron charge, and ϵ_0 is the permittivity of free space $= 10^9/4\pi c^2 = 8.854 \times 10^{-12}$ F m^{-1}, where c is the velocity of light in free space. The work done by an electron at infinity to arrive at x is

$$\phi(x) = \int_\infty^x F \, dx = \frac{q^2}{16\pi\epsilon_0 x} \tag{26}$$

where ϕ is the work function.

As the distance of the electron from the metal surface goes to zero, this model expectedly fails because the real surface is not a smooth mathematical plane but a collection of atoms with interatomic spacing a_0. There is evidence that the force becomes constant as the electron comes close to the surface and is of the order $-q^2/[4\pi(2a_0)^2 \epsilon_0]$. The total work done in the transfer of the electron in its free state to its bound state at the surface of the metal where $x = 0$ is given by

$$\phi = \int_0^{a_0} \frac{-q^2}{4\pi(2a_0)^2 \epsilon_0} \, dx + \int_{a_0}^\infty \frac{-q^2}{4\pi(2x)^2 \epsilon_0} \, dx$$

Hence

$$\phi = \frac{-q^2}{16\pi a_0 \epsilon_0} - \frac{q^2}{16\pi a_0 \epsilon_0} = \frac{-q^2}{8\pi a_0 \epsilon_0}$$

In a typical solid, $a_0 \approx 3$ Å and a rough estimate of the work function is 3 eV, the same order of magnitude obtained by photoelectric techniques (1–5 eV). At the metal electrode, the signal is conducted through the solid. Radiation or heat assists the electrons to cross from the valence band to the conduction band and the conductivity of the metal crystal lattice is a function of incident light and temperature—again there is internal thermal noise.

Closely related to the work function is the contact potential between conductors. If two conductors are in contact there is a potential difference measurable between each point just outside of the two conductors. With the condition that electrons can flow from one conductor to the other, the nature of the junction does not matter. This contact potential is usually a slowly varying dc potential which reaches 1.5 V, a value not negligible in the biological context. Contact potential is related to work function, in particular to the expression $q^2/(8\pi a_0 \epsilon_0)$, and depends on the work functions of the conductors, which are in turn dependent on their surface state and not on geometry. It may be expressed by:

$$V_{AB} = (\phi_A - \phi_B)/q \tag{27}$$

This is the consequence of the requirement for electrons from two conductors to equilibrate at the same kinetic energy. Problems in biotelemetry electrodes emerge when there is change in the conductors' surface states as a consequence of temperature, time and physicochemical change.

A third interface phenomenon involves the transference of electrons when there is an apparently insurmountable potential barrier to such electron transfer. The classical physics view of an electron as a particle with momentum and energy cannot provide an explanation. This is provided by consideration of the wave characteristics of the electron: wave number and angular momentum. Particle-wave duality provides the following (Planck) relationship:

$$\Delta E = h\nu$$

where ΔE is the energy difference between two different electron orbits, h is Planck's constant, and ν is the frequency of radiation emitted by the atom when the electron shifts from one orbit to another in which its energy is less. Further relationships of use are

$$P = h/\lambda$$

which is the de Broglie relation, where P is the momentum and λ the wavelength, and

$$E = \hbar\omega$$

where $\hbar = h/2\pi$ and ω is the angular frequency ($2\pi f$); and

$$P = \hbar k$$

where $k = 2\pi/\lambda$ and is the propagation constant. These relationships are involved in the following considerations of the wave behavior of electrons. For simplicity, the assumption is made that the electron which evades the apparently insurmountable barrier is in a vacuum.

If the probability amplitude of the electron is denoted as Ψ, then the chance of finding it is its absolute square $|\Psi_2|^2$ (with the subscript 2 identifying the vacuum).

The probability amplitude Ψ varies with time and space:

$$\Psi \sim \exp\left\{-\frac{j}{\hbar}\left[\left(\frac{P^2}{2m} + E_p\right)t - Px\right]\right\} \quad (28)$$

in a region where E_p is the potential energy, t is the time, x is the distance and m is the electron mass.

Substituting V_2 to denote the potential energy E_p in the vacuum and uniformly using subscript 2 to denote a vacuum,

$$\Psi_2 \sim \exp\left\{-\frac{j}{\hbar}\left[\left(\frac{P_2^2}{2m} + V_2\right)t - P_2 x\right]\right\}$$

Since the electron transfer is radiationless, its energy E_2 in the vacuum must equal the energy it had in the metal, namely E_1, and the momentum mass expression $P_2^2/2m$ is negative. Since 2 and m are positive, P_2^2 must be negative. This is possible if the quantity is imaginary and equal to a real quantity multiplied by j. Substituting jP_2' for P_2:

$$\Psi_2 \sim \exp\left\{-\frac{j}{\hbar}\left[\left(\frac{(jP_2')^2}{2m} + V_2\right)t - jP_2' x\right]\right\}$$

and by restricting the argument to the space variation of Ψ_2, that is, the distance x from the interface:

$$\Psi \sim \exp\left[\left(-\frac{j}{\hbar}\right)(-jP_2'x)\right] = \exp\left(\frac{-P_2'x}{\hbar}\right)$$

The quantities P_2', x and $\hbar$ in the exponent are real; therefore, the space dependence of Ψ_2 in the vacuum is real. The probability of finding the electron in the vacuum is then

$$|\Psi_2|^2 = \exp(-2P_2'x/\hbar) \quad (29)$$

and the further away from the interface, the exponentially less chance there is of finding Ψ_2. This quantum mechanical wave behavior, in which there is a finite probability $|\Psi_2|^2$ of finding the electron in the vacuum, is called quantum mechanical tunnelling.

These dynamic events at the interface in stable equilibrium when the subject is at rest permit clean sensing of the signal. When the subject moves, however, the system is so perturbed that the signal is virtually completely lost and the transmission is practically that of noise alone. This movement artefact is the curse of biotelemetry from superficial electrodes; the promise is defeated by the very motion sought. Various efforts have been made to stabilize the interface, but the problem is far from solved.

6. Depth Electrodes

In some circumstances, it is essential to implant electrodes. In certain cases of epilepsy, electrodes are implanted in the brain and anchored to bone. In such cases, electrode and interface move in the same direction simultaneously and relationships are relatively undisturbed, so the signal is transmitted and received with little interference. Much valuable seizure data have been obtained this way. The use of depth electrodes is not without risk, however.

7. Biotelemetry Studies

Thus far, biotelemetry studies have ranged from transmitting the temperature of birds' eggs to the monitoring of man and animals in space flight. In exobiology, attempts have been made to detect life on Mars. Man has been monitored in health and disease; probably the most common telemetered signal is that of the ECG in coronary care units and in post-myocardial infarction exercise. Man has also been studied in a most difficult transmission environment: the sea. Animals have been tracked more accurately and the navigation of migratory birds has been studied. Organ systems inside the body have been studied, in healthy and diseased states.

Equipment has changed from the 40 kg backpack once used to transmit the ECG from a cyclist to present systems which exploit advances in complementary-symmetry metal oxide semiconductors and other electronics which weigh a few grams, permitting the use of telemetry in situations rarely conceived even in relatively recent times. Microminiaturization has allowed concealment of telemeters in animals whose group behavior could be perturbed by apparatus visible on one member. Practical consideration of tissue-electronics effects is essential. Physiological correlations between mother and fetus have been so studied, as have fetal aortic flow dynamics, and successful solutions of the attendant problems open many avenues for biologists.

Studies of herd phenomena which do not require implantation have been made using interferometry, in which antennae are used together with other essential system components to track and locate animals of commercial importance. The principle of this method is based on the difference in phase angle between two received signals from the same source; at the receiving point there is a pair of receivers and a set of paired antennae which provide a directional angle, thus giving a fix on the transmitter. Accuracy depends on the phase resolution, and can be more exact than some satellite determinations. Nonetheless, there is increasing use of satellites in the monitoring of important biological resources on land and in the sea.

Miniaturization and thus unobtrusiveness will also permit greater use of telemetry sensing devices in those who are handicapped in fundamental biological functions. Although not used sufficiently at present, coherent light and optical character recognition aids to orientation and distant recognition—of danger and street signs, for example—for the sightless have much promise, as does sonic recognition of obstacles. A robot guide which operates with remote navigational and sensing aids and is in ultrasonic telemetric contact with its sightless user has been introduced. Here, biotelemetry has made cautious beginnings in what could be extraordinary advances in the discipline.

The full exploitation of telemetry from surface sensors awaits the solution of the electrode interface problem which has scarcely been touched by the explosion in technology at the electronic circuitry elements of the system. Useful solutions of the electrode problems could lead to fulfilment of the enormous promise of telemetry in the study of man and other animals in their environments in health, in sport, and in remote and emergency health care.

Bibliography

Adey W R, Hahn P M 1971 Biosatellite III Results. *Aerosp. Med.* 42: 273–336

Adey W R, Zweizig J R 1972 Physiological considerations in implantable underwater telemetry. *Proc. Int. Telemetering Conf.*, Vol. 3. International Foundation for Telemetering, Woodlands Hills, California, pp. 474–86

Brabyn J A 1982 New developments in mobility and orientation aids for the blind. *IEEE Trans. Biomed. Eng.* 29: 285–89

Cheeseman C L, Mitson R B (eds.) 1982 *Telemetric Studies of Vertebrates*. Academic Press, London

Davis B R, Willcocks M C 1982 Application of interferometry to the monitoring of sheep and cattle behavior in arid zone paddocks. *Biotelemetry Patient Monit.* 9: 185–204

Geddes L A 1972 *Electrodes and the Measurement of Bioelectric Events*. Wiley–Interscience, New York

Hanley J 1976 Telemetry in health care. *Biomed. Eng.* 11: 269–72

Hanley J, Broughton R 1976 Data acquisition in recording from special environments. In: Remond A (ed.) 1976 *Handbook of Electroencephalography and Clinical Neurophysiology*, Vol. 3. Elsevier, Amsterdam

Institute of Electrical and Electronics Engineers 1965 *Radio Spectrum Utilization: A Program for the Administration of the Radio Spectrum*. Institute of Electrical and Electronics Engineers, New York

Mackay R S 1961 Radio telemetering from within the body. *Science* 134: 1196–202

Panter P F 1965 *Modulation, Noise, and Spectral Analysis*. McGraw-Hill, New York

Pauley D J, Reite M 1981 A microminiature hybrid multichannel implantable biotelemetry system. *Biotelemetry Patient Monit.* 8: 162–72

Power J H, McCleave J D 1980 Riverine movements of hatchery-reared atlantic salmon (*Salmo salar*) upon return as adults. *Environ. Biol. Fishes* 5: 8–14

Schmidt-Koenig K 1979 *Avian Orientation and Navigation*. Academic Press, London

Sinberg-Eriksen P, Gennser G, Lindström K, Benthin M, Dahl P 1985 Pulse-wave recording—Development of a method of investigating foetal circulation *in utero*. *J. Med. Eng. Technol.* 9: 18–27

Smith R F, Stanton K, Stoop D, Brown D, Janusz W, King P 1977 Vectorcardiograph changes during extended space flight (M093): Observations at rest and during exercise. In: Johnson R S, Dietlein L F (eds.) 1977 *Biomedical Results from SKYLAB*. National Aeronautics and Space Administration, Washington, DC

Tachi S, Tanie K, Komoriya K, Abe M 1985 Electrocutaneous communication in a guide dog robot. *IEEE Trans. Biomed. Eng.* 37: 461–69

Telemetry Working Group, Inter-Range Instrumentation Group 1966 *Telemetry Standards*, Document 106-66. Secretariat Range Commander's Council, White Sands Missile Range, New Mexico

Winters S R 1921 Diagnosis by wireless. *Sci. Am.* 124: 465

J. Hanley

Blood Cell Analysis: Automatic Counting and Sizing

Since the 1950s there has been a rapid deployment of increasingly sophisticated instruments for counting cells automatically, impetus coming from the increasing clinical demand for comprehensive, rapid blood examinations. A useful historical review of developments in this area has been given by Melamed and Mullaney (1979).

Instruments developed for blood-cell counting and sizing are based on one of three main principles: (a) cells flowing through an illuminated chamber are recorded photoelectrically, (b) cells in a chamber are scanned through a "window" and the presence of each cell is detected photoelectrically, and (c) cells flowing through a narrow aperture through which an electrical current is passing produce an alteration in resistance which may be detected and quantified. Simple instruments employing manually prepared diluted blood samples incorporate one such principle, combined with a relatively straight-forward mechanism for aspirating a predetermined volume of diluted blood, and with electronic detection and analysis of impulses generated. More automated instruments can obtain and process, differentially, aliquots of whole blood and the related electronic data, providing an almost simultaneous display of the results of various estimations. In this article, some of the more commonly available instruments are described.

1. Flow Cytometry with Photoelectric Detection

Moldavan (1934) proposed a flow-through system using a capillary tube and photoelectric detection; he noted uniformity of tube size, focusing, maintenance of cells in single-cell suspension and sensitivity of photoelectric detection as difficulties. Some of these were overcome by the development of a device for "hydrodynamic focusing" in which cells in suspension were slowly introduced into a faster-flowing stream of fluid which provided a laminar sheath and which centered the particles in a 10 μm wide core where they could be accurately focused and enumerated using dark-field illumination. This device was incorporated into a blood-cell counter which enjoyed a short vogue and which was capable of producing erythrocyte counts and (after lysis of erythrocytes with detergent) leukocyte counts with coefficients of variation (CV) of about 2 and 9%, respectively (Crosland-Taylor et al. 1958). In this instrument, coincidence—the registering of only a single pulse when two or more particles pass through the counting device simultaneously—was insufficient to be clinically important.

In recent years, other instruments have been developed incorporating a flow cell with dark-field illumination and photoelectric detection of particles. Some combine cell counting with measurement of other parameters, such as photoelectric hemoglobinometry. An automated system for erythrocyte count and hemoglobin determination involving automatic blood sampling, pumps, mixing systems for reagents and samples, colorimetry, photoelectric particle detection and enumeration, and recording, was developed by Technicon Instruments (Sturgeon and McQuiston 1965). This instrument could assay 40 specimens per hour with a CV of about 0.5% for hemoglobin estimation and about 4% for erythrocyte counts. Subsequent addition of leukocyte

counting (on an aliquot of blood in which erythrocytes were lysed) and hematocrit estimation by electrical conductivity of whole blood permitted the delivery of four directly determined parameters and the calculation of the mean cell volume (MCV), mean corpuscular hemoglobin (MCH) and mean corpuscular hemoglobin concentration (MCHC) (Nelson 1969). The conductivity technique for hematocrit estimation proved to have too many artefacts and is no longer considered useful. In a further development, platelet counting became available, using whole blood with lysis of erythrocytes by 2 M urea. This had the advantage of eliminating the prior preparation of platelet-rich plasma, a time-consuming measure with potential for error. Precision was good (CV about 4%) but artefactual fluctuations (e.g., due to the presence of paraproteins) were a disadvantage. In the Hemalog 8 system, a further advance in the Technicon line, the conductivity method for hematocrit is replaced by an automated microhematocrit technique (Lewis and Ward 1975). Platelet counting using urea lysis is also incorporated. Throughput is about 60 specimens per hour and performance meets clinically acceptable standards. A more recent instrument in this line, the H6000, combines the principles of the Hemalog system with automated differential white-cell counting (see *Blood Cell Analysis: Morphological and Related Characteristics*). Erythrocytes and platelets are counted in a single flow cell, with differentiation based on detection of differences in cell size. The hematocrit is derived from the size and number of pulses detected during erythrocyte counting, a silicon photovoltaic sensor being employed.

The Hemac 630L manufactured by Ortho Diagnostic Instruments is a counting system based on the detection of the light scattered when particles in a flow cell pass through a laser beam. The instrument carries out determinations of hemoglobin concentration and erythrocyte and leukocyte counts. The hematocrit is derived from the summation of red cell pulse size; the MCV and MCHC are calculated. The instrument meets clinically acceptable standards (Lewis and Bentley 1977). A later model, the ELT8, performs platelet and erythrocyte counts on the same dilution with differentiation based on cell size. Evaluations of the ELT8 have been described by Merle-Béral et al. (1981) and Raik et al. (1982).

Acridine Orange–RNA complex, which emits orange fluorescence on exposure to uv light, has been used to facilitate conventional visual reticulocyte counts (Vander et al. 1963). Based on this principle, automated reticulocyte counting has been performed using fluorescent flow cytometry (Seligman et al. 1983, Tanke et al. 1983).

2. Photoelectric Scanning of Cell Suspensions in a Chamber

The principles of this approach have been described by Lagercrantz (1952) and Wolff (1950). A chamber containing a suspension of erythrocytes is moved beneath a rectangular window under dark-field illumination. When a cell appears in the window, the light impulse produced is photoelectrically recorded. An apparatus based on this approach was devised which could count up to 10^4 cells in 2 min with a CV of 2–6% (Macfarlane et al. 1959). The instrument had limited capability however, and the principle has not been widely applied.

3. Instruments Based on Electrical Impedance in a Narrow Aperture

Coulter (1953, 1956) observed that particles in dilute suspension passing through a narrow orifice through which an electrical current is also passing, produce a detectable change in electrical resistance related to the size and number of particles.

Practical utilization of this phenomenon is influenced by several factors. The position of the particle in the aperture is important; variations in the speed of passage and in the current density in different zones of the aperture influence the size and duration of the pulses detected. Thus particles not passing through the center of the aperture produce "atypical" pulses. This anomaly can be corrected by hydrodynamic focusing.

Rigid particles produce a greater pulse per unit volume than deformable particles. Thus, fixed red cells produce electrical pulses representing 1.5 times their actual volume, and rigid latex spheres of accurately known volume generate pulses indicating a substantially greater volume. These factors have to be considered when standardizing instruments for clinical use. Kachel (1979) and Jones (1982) have reviewed electrical-impedance particle counting in detail.

Early reports of erythrocyte counting by electrical-impedance change indicated good precision and correlation when compared with standard manual methods. Differences were regarded as being largely due to variability in manual counts (Brecher et al. 1956, Mattern et al. 1957). The latter also described leukocyte counting following lysis of erythrocytes by detergent—the basis of leukocyte counting in subsequent instruments of this type. Platelet counting required the removal of erythrocytes by settling or differential centrifugation, since threshold settings were insufficiently sensitive to discriminate between them, and counts had subsequently to be corrected for hematocrit. Correlation with manual counts was close, with sample preparation the greatest source of variation.

The next generation of instruments (represented by the Coulter Model S) incorporates an automatic diluting system, which divides the sample into aliquots for erythrocyte count and direct MCV determination in one channel, and leukocyte count and hemoglobin determination in the other. Hemoglobin is determined in a photoelectric colorimetric device after several seconds exposure of blood to a cyanide-containing lysing reagent which converts the hemoglobin to cyanmet-hemoglobin. However, such conversion is not essential as estimations

are read at the isobestic point of hemoglobin (525 nm). This approach avoids the need to wait for the formation of cyanmet-hemoglobin, prior to estimation of hemoglobin concentration, and permits the use of non-cyanide-containing reagents. The other indices (hematocrit, MCH and MCHC) are computed from the directly measured parameters and the results delivered by digital printer. In practice, the instrument produces up to 120 sets of results per hour with good reproducibility and accuracy. Carry-over from one sample to the next is 2–3%. Linearity is good in clinically relevant ranges and the instrument is in general mechanically and electronically reliable and efficient (Pinkerton et al. 1970). Subsequent modifications, based on the observation that lymphocytes and nonlymphoid leukocytes can be distinguished on the basis of size, allow rapid determination of the proportion of lymphocytes in the leukocyte channel, with the majority of the remaining cells usually being, by inference, neutrophils; the results agree well with conventional differential counts (England et al. 1975). After various minor modifications, a major practical advance occurred with the introduction of the Coulter Model S Plus described by Rowan et al. (1979). The new instrument has improved particle flow, hydrodynamic focusing, diminution in sample carry-over, improved instrument monitoring displays and other practical advantages. Platelet counts are performed and estimates of variation in erythrocyte size (red-cell-distribution width, RDW), mean platelet volume (MPV), platelet-size distribution (PDW) and plateletcrit are made. The conventional seven parameters measured by the Coulter Model S are also measured by the S Plus and the results delivered by digital printout at a rate, in practice, of 70–80 samples per hour. Accuracy and reproducibility of conventional parameters are similar to those for the Model S (correlation coefficient $r > 0.985$ for all parameters), and for platelet counts $r > 0.98$ on comparison with manual and semiautomated methods. Platelet counts are determined in the same channel as the erythrocyte count by counting particles in the range 2–20 fl and deriving mathematically the particle number in the size range 0–70 fl, assuming log-normal platelet volume distribution. The diagnostic value of the new parameters (RDW, MPV, PDW and plateletcrit) remains to be fully determined; however, preliminary assessment of the red-cell-distribution width and mean platelet volume has shown useful clinical correlations (Bessman et al. 1981, 1983, Jones 1982, Rowan and Fraser 1982). A modification to the Model S Plus allows rapid determination of the proportion of lymphocytes on the basis of size, and the results correlate well with conventional differential counts (England et al. 1982).

Other instruments based on electrical impedance measurement are represented by the Toa counters manufactured by Toa Medical Electronics. Particle detection is based on a change in capacity which occurs when a particle with a dielectric constant different from the suspending solution passes a transducer which consists of two electrodes mounted on opposite sides of a capillary through which the particles pass. Changes in the phase and amplitude of the detected signal are processed electronically to give a measure of the number or size of particles. An instrument incorporating this technology with hydrodynamic focusing (Toa CC-120 Microcell Counter) provides estimates of hemoglobin, erythrocyte count, MCV, hematocrit and leukocyte count with good correlations with Coulter S values on the same samples, and reproducibility within clinically acceptable limits (Inaoka et al. 1980).

The Ultra-Flow 100 manufactured by Clay Adams employs electrical impedance to count platelets. The system is adjusted to count particles of 3.25–40 fl and simultaneously—by amplitude discrimination thresholding—determine the erythrocyte count in the same diluted blood sample. Counts between 1.75–3.25 fl and 40–43 fl are used to monitor excessive background interference in the low ranges and overlap with microcytic erythrocytes in the upper ranges. The instrument (with the absolute erythrocyte count dialled in) counts 3.25–40 fl particles, compares this with the erythrocyte count in the diluted sample and computes the absolute platelet count. It is useful only where an instrument determining the erythrocyte count is available. The results produced are clinically acceptable (Guthrie et al. 1980).

A comprehensive and detailed account of the principles and methods used in blood cell counting and the clinical implications of blood cytometry can be found in van Assendelft and England (1982).

See also: Blood Cell Analysis: Morphological and Related Characteristics; Clinical Biochemistry: Automation

Bibliography

Bessman J D, Williams L J, Gilmer P R 1981 Mean platelet volume: The inverse relation of platelet size and count in normal subjects and an artifact of other particles. *Am. J. Clin. Pathol.* 76: 289–93

Bessman J D, Gilmer P R, Gardner R H 1983 Improved classification of anemias by MCV and RDW. *Am. J. Clin. Pathol.* 80: 322–26

Brecher G, Schneiderman M, Williams G Z 1956 Evaluation of electronic red blood cell counter. *Am. J. Clin. Pathol.* 26: 1439–49

Coulter W H 1953 Means for counting particles suspended in a fluid. US Patent No. 2,656,508

Coulter W H 1956 High speed automatic blood cell counter and cell size analyser. *Proc. National Electronics Conf.*, Vol. 12. National Electronics Consortium, Oak Brook, Illinois, pp. 1034–40

Crosland-Taylor P, Stewart J W, Haggis G 1958 An electronic blood-cell-counting machine. *Blood* 13: 398–409

England J M, Bashford C C, Hewer M G, Hughes-Jones N C, Down M C 1975 A semi-automatic instrument for estimating the differential leucocyte count. *Biomed. Eng.* 10: 303–4

England J M, Chetty M C, de Silva P M 1982 Estimation of lymphocyte percentage and number on the Coulter Counter Model S Plus Phase II. *J. Clin. Pathol.* 35: 1194–99

Guthrie D L, Lam K T, Priest C J 1980 The Ultra-Flo 100 platelet counter: A new approach to platelet counting. *Clin. Lab. Haematol.* 2: 231–42

Hastings J W, Sweeney B M, Mullin M M 1962 Counting and sizing of unicellular marine organisms. *Ann. N.Y. Acad. Sci.* 99: 280–90

Inaoka Y, Bando S, Tominaga M, Katayama Y 1980 Routine usage of the CC-120 microcellcounter. *Sysmex J.* 3: 6–12

Jones A R 1982 Counting and sizing of blood cells using aperture impedence systems. In: van Assendelft and England 1982, pp. 49–72

Kachel V 1979 Electrical resistance and pulse sizing (Coulter sizing). In: Melamed M R, Mullaney P F, Mendelsohn M L (eds.) 1979 *Flow Cytometry and Sorting*. Wiley, New York, pp. 61–104

Lagercrantz C 1952 On the theory of counting individual microscopic cells by photoelectric scanning. An improved counting apparatus. *Acta Physiol. Scand.* 26 (Suppl. 93): 1–135

Lewis S M, Bentley S A 1977 Haemocytometry by laser-beam optics: Evaluation of the Hemac 630L. *J. Clin. Pathol.* 30: 54–64

Lewis S M, Ward P G 1975 An evaluation of the Hemalog. *Lab. Pract.* 24: 13–18

Macfarlane R G, Payne A M M, Poole J C F, Tomlinson A H, Wolff H S 1959 An automatic apparatus for counting red blood cells. *Br. J. Haematol.* 5: 1–16

Mattern C F T, Brackett F S, Olson B J 1957 Determination of number and size of particles by electrical gating: Blood cells. *J. Appl. Physiol.* 10: 56–70

Melamed M R, Mullaney P F 1979 An historical review of the development of flow cytometers and sorters. In: Melamed M R, Mullaney P F, Mendelsohn M L (eds.) 1979 *Flow Cytometry and Sorting*. Wiley, New York, pp. 3–9

Merle-Béral H, Rémy F, Rahaël M, Lesty C, Binet J L 1981 Evaluation des performances de l'ELT8. *Nouv. Rev. Fr. Hematol.* 23: 61–66.

Moldavan A 1934 Photo-electric technique for the counting of microscopical cells. *Science* 80: 188–89

Nelson M G 1969 Multichannel continuous flow analysis on the SMA-4/-7A. *J. Clin. Pathol.* 22 (Suppl. 3): 20–25

Pinkerton P H, Spence I, Ogilvie J C, Ronald W A, Marchant P, Ray P K 1970 An assessment of the Coulter Counter Model S. *J. Clin. Pathol.* 23: 68–76

Raik E, McPherson J, Barton L, Hewitt B S, Powell E G, Gordon S 1982 An evaluation of the ELT-8 hematology analyzer. *Pathology* 14: 153–64

Rowan R M, Fraser C 1982 Platelet size distribution analysis. In: van Assendelft and England 1982, pp. 125–41

Rowan R M, Fraser C, Gray J H, McDonald G A 1979 The Coulter Counter Model S Plus: The shape of things to come. *Clin. Lab. Haematol.* 1: 29–40

Seligman P A, Allen R H, Kirchanski S J, Natale P J 1983 Automated analysis of reticulocytes using fluorescent staining with both acridine orange and an immuno-fluorescent technique. *Am. J. Hematol.* 14: 51–66

Sturgeon P, McQuiston D T 1965 A fully automated system for the simultaneous determination of whole blood red cell count and hemoglobin content. *Am. J. Clin. Pathol.* 43: 517–31

Tanke H F, Rothbarth P H, Vossen J M J J, Koper G J M, Ploem J S 1983 Flow cytometry of reticulocytes applied to clinical hematology. *Blood* 61: 1091–97

van Assendelft O W, England J M (eds.) 1982 *Advances in Hematology Methods: The Blood Count.* CRC Press, Boca Raton

Vander J B, Harris C A, Ellis S R 1963 Reticulocyte counts by means of fluorescence microscopy. *J. Lab. Clin. Med.* 62: 132–40

Wolff H S 1950 An apparatus for counting small particles in random distribution with special reference to red blood corpuscles. *Nature (London)* 165: 967

P. H. Pinkerton and K. C. Carstairs

Blood Cell Analysis: Morphological and Related Characteristics

Examination of patient blood samples in the clinical laboratory begins with the measurement of hemoglobin concentration and the enumeration of one or more of the formed elements of blood: white cells (leukocytes), red cells (erythrocytes) and platelets. The next step is the microscopy of a stained blood film which is prepared by spreading a small drop of blood along the central portion of a 7.5×2.5 cm glass slide. This is the classic or "wedge" blood film, which is dried, fixed in methanol and stained so that the various elements are colored differing shades of red, blue or purple. The wedge film has the disadvantage that only a small portion of the film is suitable for examination and, within this area, different types of leukocytes are unevenly distributed. The standard method of determining proportions of types of white cells is the "100 cell differential" wherein each of 100 consecutively found leukocytes is assigned to its class—neutrophil, lymphocyte, monocyte, eosinophil, or basophil—and then each class is reported as a percentage of the total. If reactive, precursor or malignant variants are found, these too are reported. Normal human blood has approximately 4000–11 000 white cells per microliter, and the accuracy and reproducibility of the standard differential are poor (Rumke et al. 1975). Nevertheless, much useful clinical information is provided by this apparently simple procedure. Millions of blood films are examined daily, a task requiring hundreds of thousands of hours of labor: an average technologist reports twelve to fifteen slides per hour. The skilled morphologist does much more than a leukocyte differential and will report on other features such as an increase or irregularity of background color (representing abnormality of quantity or quality of protein in the plasma); red cell arrangement (agglutination, indicating abnormal anti-red-cell antibodies); red cell size, color and shape (there are many possible patterns, each of which may indicate the type or cause of an anemia); and red cell inclusions (which may indicate the type or cause of an anemia, or indicate an infection). Platelet number and size can also be assessed with sufficient accuracy to be clinically useful. The human observer automatically makes allowances for the uneven distribution of cells, the considerable variations in stain colors that routinely occur, and recognizes the many artefacts due to dirt, stain deposit and failures in technique of slide preparation. Moreover, the good morphologist, having detected an abnormality, may then proceed to extend the examination, searching for other features to provide further evidence for a particular disease process.

Because blood-film microscopy is so clinically useful, but also very expensive and dependent on the variable skill of the morphologist, much time and money have been expended in an attempt to automate the process. Much has been achieved in the area of leukocyte recognition, but automated reporting of the whole blood film is still in its very early stages.

This article deals with the following aspects: (a) white cell differential counting by examination of fluid blood; (b) automation of microscopy of stained blood films; and (c) research tools designed to isolate or detect special subsets of leukocyte classes.

1. Examination of Fluid Blood

The first equipment commercially available for this purpose, using differential counts by continuous flow cytological analysis, was the Hemalog D series made by Technicon Instrument Corporation of the USA (Mansberg et al. 1974). It provided a differential count based on 10 000 cells, thus giving the most reproducible differential of any automated device. It was also excellent for detecting small numbers of certain abnormal cells.

Early models processed samples at the rate of one per minute, but the effective rate was considerably less when the time required for setting up and routine maintenance was considered. The later model, the D/90, processed samples at 90 per hour, and it is this model that is described below. A simplified version is incorporated into Technicon's H6000 model. This machine consists of the differential counting module, an automated hemoglobinometer and cell counter, and an automated blood film maker and stainer. Bar-coded patient and sample identification technology is provided.

1.1 Technicon Hemalog D/90

(a) *Principle of operation.* Anticoagulated blood is automatically aspirated into the machine and is divided into three portions, each of which follows a separate pathway. Red cells are lyzed and the types of leukocytes have their internal constituents differentially colored. Each portion eventually passes through a sheathed-stream flow cell where leukocytes are classified and counted by complex photoelectrical analysis.

(b) *Detailed operation.* Each blood specimen on a 40 sample turntable is successively and automatically positioned, mixed and a sample aspirated. Each sample is then divided into three portions, diluted and the red cells lyzed. The first portion provides a total white cell count and peroxidase-containing leukocytes are stained (neutrophils stain shades of grey, eosinophils black and lymphoid cells are unstained). In the second portion, monocytes (containing esterases) are stained red. A few monocytes stain only slightly and are classified under "remainder" in the results print-out. In the last portion, heparin-containing basophils are deeply colored by Alcian blue. Each of the three portions, now diluted, free of intact red cells and containing specifically stained leukocytes, enters a separate sheathed-stream flow cell where cell number, cell size and cell staining are determined.

(*c*) *Cell identification and counting.* A single tungsten halogen lamp acts as a common light source for the special-purpose optical system for each flow cell. In the first two channels, a beam splitter divides the collected light from the flow cell for the measurement of forward scattered light (which corresponds mainly to cell size) and of light loss (which corresponds mainly to the intensity of cell staining). The basophil channel has two light-scatter detectors operating at differing wavelengths. Data from 10 000 cells are displayed on an oscilloscope screen in the form of xy plots, with each cell represented as a single point. Light scatter is on the y axis and absorption on the x axis. On the screen are high and low threshold displays for scatter and absorption, plus two moving thresholds which bracket the neutrophil representation. The screen is divided by the axes and threshold limits (set by calibration procedures) into ten rectangular areas of unequal size in which normal and abnormal white cell distributions form characteristic patterns.

(*d*) *Printed results.* The data are transformed into a traditional leukocyte differential count of the five main classes of leukocytes, plus "large unstained cells" (usually atypical lymphocytes or primitive cells), "high-peroxidase leukocytes" and a "remainder" (usually monocytes with low enzyme activity). This type of apparatus has been evaluated by Bain et al. (1980a,b) and Urmston et al. (1980).

The number of classes of leukocytes that the Hemalog D series positively identifies is limited. It is, however, unrivalled at detecting small percentages of abnormal cells because of the large number of leukocytes screened. The manufacturers do not lay claim to identifying any abnormalities in other cell lines.

All the more complex automated cell counters made by Coulter Electronics, Ortho and TOA-Sysmex can now routinely provide a "limited differential," dividing leukocytes into two (lymphoid, others) or three (lymphoid, monocytes, others) classes. Because Ortho equipment uses measurement of light scatter rather than volume determination, it may easily have the potential for identifying further subclasses by identifying cell coloration (after the addition of dyes) or the degree of heterogeneity of structures within the cell.

2. Computerized Pattern Recognition Devices (*PRDs*)

The first attempts to produce devices which could replace the human eye and a portion of brain function were stimulated by a committee of the National Coal Board of the UK formed to investigate the possibility of making "a machine to replace the human observer." Much of the initial work was performed at the Radcliffe Infirmary, Oxford, and University College, London. Image detection and recognition prototypes were made; by 1955 a flying-spot microscope was described for the sizing and counting of red cells (Causley and Young 1955).

Almost simultaneously, work was being carried out in the USA, and by 1961 the Perkin Elmer Corporation had built the CELLSCAN (Izzo and Coles 1962), which slowly processed (by off-line computer) manually positioned microscope images. From this was developed the faster CELLSCAN–GLOPR system (Preston 1973), which eventually became the basis for PRDs manufactured by Coulter Electronics.

The first commercially available PRD was Corning's LARC (Megla 1973). It used a spinner to make the blood film and a complex, controlled-environment staining device. It operated at the rate of approximately 30 slides per hour. Image identification was by analysis of a series of histograms of density for each color across each cell. Other manufacturers followed. In the USA, Abbott Laboratories (ADC 500), Coulter Biomedical Research Corporation (Diff 2, Diff 3-50 and the Diff 4), and Geometric Data (a range of models culminating in the 590), and in Japan Omron Electronics (Microx) and Hitachi (Automated Blood Cell Differential Analyzer) have all produced PRDs. At the time of writing, both Japanese firms are still manufacturing but in the USA, only Geometric Data's range of models survives. The LARC was the first to disappear, followed by the ADC 500; Coulter's Diff 3 was rapidly superseded by the 3-50 and followed by the short-lived Diff 4. However, many of these machines are still in use and three different models will be described briefly below. Japanese PRDs are not marketed in Europe or the USA and apart from information available in sales brochures, little is known about the internal design of the equipment. However, the larger models are more automated than their American counterparts, even spun slide preparation, staining and loading being fully automated.

All PRDs are superficially similar in that a stained blood film on a glass slide is examined by means of a computer-controlled microscope. White cells, red cells and platelets are analyzed, one by one, by color, density, size and shape, with the analog data being digitized, compressed and cell-identification features extracted by computer(s). In reality, however, each manufacturer uses very different methods for slide preparation, slide staining, illumination, data collection and data processing.

In an attempt to eliminate some of the many problems of the wedge blood film, the ADC 500, the Diff 3-50 and

Japanese models employ a spinner to prepare blood films. This device produces a monolayer of cells over the entire surface of the slide. Blood is placed at the centre of the slide which is then positioned in the horizontal plane inside the device and rotated at high speed. The resulting monolayer gives an even distribution of cell types with little or no overlapping of cells. The spinner needs very careful design, otherwise it produces its own set of artefacts. Modern spinners are complex, capable of very high rates of acceleration and deceleration, produce little air turbulence and automatically stop as soon as the optimal thickness of the blood layer has been achieved. Spinners may produce potentially infective aerosols; either virus filters or a water curtain can be used to protect the operator. The Hematrak series uses a semiautomatic spreader for making wedge blood films with a much enlarged area suitable for microscopy. A spinner may also be used.

For the staining of blood films, the ADC 500 and the Coulter Diff 3-50 have differing versions of the Ames Company sequential stainer in which slides are successively transported face down over a flat steel platen and exposed to a sequence of reagents, held in place between the platen and slide by surface tension. The platen has a pattern of grooves to produce mixing of reagents, and appropriately placed drainage channels. Geometric Data, Omron and Hitachi use proprietary batch stainers. Each company provides stains modified to suit its machines' performance characteristics.

The speed of operation of these machines is dependent on mechanical factors, computer efficiency and the frequency of slide reviews. Each slide has to be moved many times on the xy axes for each differential; similarly, either the slide or objective has to move in the z axis to maintain focus. Each movement takes a finite amount of time, and additional time is required for microscope components to settle and become vibration-free. Thus, all manufacturers provide modified microscope components; movements are carried out by viscous-damped stepper motors (up to 8000 steps per second) on the xy axes and linear conductor motors or piezoelectrical devices for z axis movement.

The computer has to control the finding and framing of cells, the acquisition of scene data being sorted according to wavelength, density, area and shape; it then compresses and presents these data to the cell-identification module. The more these activities can proceed in parallel, the higher the potential rate of reporting. In addition, the manufacturer may choose to "hardwire" certain of the functions, rather than using software, thus gaining speed at the expense of system flexibility.

For any given batch of slides, the percentage that is reviewed depends not only on the complexity of the abnormality, but the degree of accuracy that the user believes the PRD to possess and the extent to which the user is willing to accept the tendency of a particular PRD to emphasize or ignore certain features. Thus, the degree to which the manual review process slows productivity

may be as much the user's responsibility as the manufacturer's.

2.1 ADC 500

The ADC 500 was introduced in 1977 (Green 1979a,b). Films are made using a spinner, with a virus filter to protect the environment. A sequential stainer is used, which also mounts each slide on a special holder which can be imprinted with a machine- and human-readable identification code. The PRD itself consists of three compact modules which can be arranged to suit the space available. One module contains the power supply, another the computer and the third contains the microscope, illumination systems, printer and instruction keyboard. The ADC 500 accepts stacks of up to 50 slides which are then automatically and sequentially analyzed and reported. The microscope is proprietary and (as in all PRDs) the user may use a standard binocular head for slide review. There is no CRT monitor. The objective lens is a Leitz $\times 40$ NA 1.0 planapochromat, driven in the z axis by a linear induction motor. The light source is a 150 W xenon arc lamp. After light has traversed the condenser, specimen and objective, it is split into six beams: three for the three separate 50×50 element solid-state photodiode high-resolution scanner arrays (operating at 412, 525 and 560 nm, respectively), one for the cell-finder array and two for the autofocus arrays. (Focus is maintained by comparing the outputs of two 64 element photodiode arrays placed slightly out of focus on each side of the image plane. Focusing is a closed-loop system, operating independently of the computer and data capture.) Once a cell has been detected by the finder array, it is framed and image data are obtained by each of three high-resolution ($\sim 0.5\,\mu m$) arrays. The data are digitized into 64 density levels. Cell features are extracted sequentially by a special hard-wired processor, each element of the image being assigned to background, red cell, leukocyte nucleus or leukocyte cytoplasm, according to color and density. Density histograms are produced for each scene for each wavelength. After the processor has completed its analysis of image data, the Nova 2 minicomputer completes the cell analysis. The basic method for data transformation (Green 1976) is a multicolored generalization of the monochrome technique of Prewitt and Mendelsohn (1966).

In the ADC 500, focusing is independent of data capture, and at the same time one cell is registered, the digitized image of a second is processed and a third is being classified. The machine recognizes six normal types of leukocytes, and six abnormal or atypical categories, grades eight abnormalities of red cells and gives an estimate of platelet sufficiency. Results are printed on the machine's own paper roll and identified according to the number on the slide holder. Slides for review can be automatically reentered and any type of cell detected can be recalled for viewing (but not necessarily the identical cell already classified). The manufacturer claims 45 differentials (500 white cells each) per hour, that is 22 500 white cells per hour. The machine can be adjusted to produce differentials based on samples less than 500, and the rate of analysis increased to 70–90 per hour.

2.2 Diff 3-50

The Coulter Diff 3-50 was introduced in late 1980 to supersede the Diff 3. Films are made using diluted blood and a spinner fitted with a water curtain. A modified wedge film can also be used. Staining is by a slightly modified sequential stainer. The Diff 3-50 is a free-standing module containing the microscope and automatic magazine loader, keyboard, twin printers, a CRT display for report formatting and machine control, a second CRT for cell display, plus power supply and computer. The objective lens is a Zeiss $\times 40$ NA 1.0 planapochromat, and the light source is a 60 W tungsten bulb. Focusing is carried out by applying results of image-histogram analysis to a coarse (mechanical) and fine (piezoelectrical) focusing mechanism, that is, focusing is sequential to image capture. Cell location is performed using solid-state sensors in the optical system, but cell image data are sequentially captured by a plumbicon operating in two spectral regions centered at 510 and 580 nm. The digitized image data are manipulated in a software-controlled preprocessor to extract cell features by means of a parallel/serial implementation of a binary transform technique (Golay 1977). A simplified explanation of the Golay transform is given by Ingram and Preston (1970). The technique permits spatial information concerning any image point and its six surrounding points to be reduced to one of 28 descriptive numbers. Further processing provides 42 measures of cell characteristics. Cell classification is carried out by the Nova 4 minicomputer which utilizes multiple pairwise comparisons of linear discriminant functions containing some or all of the 42 cell characteristic measurements. The Diff 3-50 recognizes 10 types of white cells, and 6 kinds of red cell abnormality, as well as platelet sufficiency. Slides for review can be identified and rescanned automatically with the observer using the microscope or CRT display. Any type of nucleated cell can be relocated and displayed. The maximum speed for differentials (with no review, red cell or platelet analysis) is 58 per hour. In the standard operating mode, the rate is 40–48 slides per hour.

2.3 Hematrak Series

There are several models in this series, the most advanced of which is the Hematrak 590. Slides are made by a semi-automatic device making a modified wedge film. Spun films may also be used. Staining is by a proprietary batch stainer. The model 590 is a single free-standing unit incorporating an automatic slide loader, optics, keyboard for report modification and machine control, printer, a CRT display for cell display and report review, power supply, computers and floppy disk pack. An auxiliary printer is available which prints a summary of all slide results and quality control data as an aid to editing. Bar-coding of samples is available.

The objective lens is a $\times 40$ NA 1.0 planapochromat and moves in the z axis. Both surfaces of the slide are automatically oiled; this is required because image data capture is by means of a flying spot from a cathode ray tube which passes downwards through the objective lens and condensor to three filtered photomultipliers (operating at 425, 530 and 595 nm). For visual use of the microscope facility, substage illumination is used. Cell finding is done within a field of $\sim 320 \times 300\,\mu m$; when a cell is found, scan parameters are changed to produce a high-resolution field $32 \times 20\,\mu m$ with $0.25\,\mu m$ spacings, which moves to the cell in question, thus saving on mechanical movements. Photomultiplier outputs are digitized into six density levels. Cell features are extracted from the raw data by a hard-wired preprocessor which applies a series of virtual masks line by line to the image data at each of the six levels. The number of image elements above each density level for each cell (but not equally for each spectral region) is accumulated. The result is that for each cell, 94 counts are stored relating to cell features. The computer then performs its classification function by a series of sequential binary decisions at several hundred decision nodes based on the 94 feature counts and the derived secondary features. Miller has described further details of this n-point transformation methodology (Miller et al. 1974, Miller 1976).

The Hematrak 590 can print out brief reports on its auxiliary printer as an aid to the user in deciding which slides require review. Simple plots of cell size can also be printed. Individual abnormal cells can be relocated, and their identities edited into the report. Ten types of leukocytes and six kinds of red cell abnormality can be identified, as well as estimates of the platelet count. Only abnormal or unclassified cells can be retrieved for examination by the operator.

The speed of operation of the machine is determined by the type of report that is required, the degree of abnormality of the blood film and the type and quality of the preparation. Speed can be maximized by setting the machine to count only 100 white cells, 0 red cells and omitting the platelet count. The speed of reporting is affected by the number of manual reviews that are carried out: some reviews are mandatory, some are optional. The maximum rate is ~ 100 slides per hour; in hospital practice, the reporting rate is probably between 25 and 60 per hour.

Currently available is a modification for counting reticulocytes (newly formed red cells) and shortly a package may be available for the identification and enumeration of lymphocyte subsets.

2.4 PRDs for Red Cell Morphology

Research equipment has been extensively applied to the problem of diagnosing various anemias by applying computerized pattern recognition techniques to red cell morphology (Bacus and Weens 1977, Bacus 1980, Westerman et al. 1980). It appears that the diagnostic power of such devices may exceed that of many morphologists.

3. Research Equipment

Various advanced and complex flow cytometry and cell-sorting devices now exist which can identify differing types of cells flowing in a controlled stream by carrying out multiple measurements of each cell, either utilizing the cell's intrinsic properties (e.g., uv absorption by nucleic acids), protein content and size, or by induced characteristics produced by fluorescent dyes or fluorescent antibodies. Various classes of cells within a heterogeneous population can then be separated from the others and collected for further examination.

The power of these machines and their associated computers is now so great that possibly, for the first time, technology has far outstripped the current needs of the researcher.

Melamed et al. (1979) have provided a comprehensive review of the subject. Among the manufacturers in this field are Ortho, Becton-Dickinson, and Coulter Electronics.

See also: Blood Cell Analysis: Automatic Counting and Sizing; Clinical Biochemistry: Automation

Bibliography

Arndt-Jovin D J, Jovin J M 1978 Automated cell sorting with flow systems. *Annu. Rev. Biophys. Bioeng.* 7: 527–58

Bacus J W 1980 Quantitative morphological analysis of red blood cells. *Blood Cells* 6: 295–314

Bacus J W, Weens J H 1977 An automated method of differential red blood cell classification with application to the diagnosis of anemia. *J. Histochem. Cytochem.* 25: 614–32

Bain B J, Scott D, Scott T J 1980a Automated differential leucocyte counters: An evaluation of the Hemalog D and a comparison with the Hematrak, II: Evaluation of performance on routine blood samples from hospital patients. *Pathology* 12: 101–9

Bain B J, Neill P J, Scott D, Scott T J, Innis M D 1980b Automated differential leucocyte counters: An evaluation of the Hemalog D and a comparison with the Hematrak, I: Principles of operation; reproducibility and accuracy on normal blood samples. *Pathology* 12: 83–100

Causley D, Young J Z 1955 Flying spot microscope. *Science* 109: 371–74

Golay M J E 1977 Analysis of images. US Patent No. 4,060,713

Green J E 1976 Method and apparatus utilizing colour algebra for analyzing scene regions. US Patent No. 3,999,047

Green J E 1979a A practical application of computer pattern recognition research, the Abbott ADC 500 differential classifier. *J. Histochem. Cytochem.* 27: 160–73

Green J E 1979b Rapid analysis of hematology image data, the ADC 500 pre-processor. *J. Histochem. Cytochem.* 27: 174–79

Herzenberg L A, Sweet R G, Herzenberg L A 1976 Fluorescence-activated cell sorting. *Sci. Am.* 234: 108–17

Horan P K, Wheeless L L 1977 Quantitative single cell analysis and sorting. *Science* 198: 149–57

Ingram M, Preston K 1970 Automatic analysis of blood cells. *Sci. Am.* 223: 72–82

Izzo N F, Coles N 1962 Blood cell scanner identifies rare cells. *Electronics* 35: 52–57

Mansberg H P, Saunders A M, Grower W 1974 The Hemalog D white cell differential system. *J. Histochem. Cytochem.* 22: 711–24

Megla G K 1973 The LARC automatic white blood cell analyzer. *Acta Cytol.* 17: 3–14

Melamed M R, Mullaney P F, Mendelsohn M L (eds.) 1979 *Flow Cytometry and Sorting.* Wiley, New York

Miller M N 1976 *IEEE Trans. Biomed. Eng.* 23: 400–5

Miller M N, Levine M S, Partin E 1974 Color separation for discrimination in pattern recognition systems. US Patent No. 3,827,804

Onoe M, Preston K, Rosenfeld A (eds.) 1980 *Real-Time Medical Image Processing.* Plenum, New York

Preston K 1973 Automated microscopy for cytological analysis. In: Barer R, Cosslett V E (eds.) 1973 *Advances in Optical and Electron Microscopy*, Vol. 5. Academic Press, New York, pp. 43–93

Preston K, Onoe M (eds.) 1976 *Digital Processing of Biomedical Images.* Plenum, New York

Prewitt J M, Mendelsohn M L 1966 The analysis of cell images. *Trans. N.Y. Acad. Sci.* 128: 1035–53

Rumke C L, Bezemier P D, Kuik D J 1975 Normal values and least significant differences for differential leukocyte counts. *J. Chron. Dis.* 28: 661–68

Urmston A, Hyde K, Gowenlock A H, Maciver J E 1980 Evaluation of the Hematrak differential leucocyte counter. *Clin. Lab. Haematol.* 2: 199–214

Westerman M P, O'Donnell J, Bacus J W 1980 Assessment of the anemia of chronic disease by digital image processing. *Am. J. Clin. Pathol.* 74: 103–6

K. C. Carstairs and P. H. Pinkerton

Blood Flow: Invasive and Noninvasive Measurement

Accurate measurement of blood flow to organs or tissues is necessary in many areas of medical and biological research. The clinical value of flow measurements, particularly in cardiology and in cardiac and vascular surgery, is well recognized. The ability to detect blood flow, even if it cannot be precisely measured, has also proved invaluable to obstetricians for the measurement of fetal heart rate and to other clinicians for investigation of the circulation.

Many methods for measuring blood flow are available. These include:

(a) venous occlusion plethysmography, whereby the rate of increase in volume of a part of the body is measured as blood flows into it while the outflow (venous return) is temporarily cut off;

(b) techniques using ultrasound and the Doppler effect in order to derive blood velocity from the change in frequency (due to the Doppler effect) which occurs when a beam of ultrasound is scattered by moving red cells in the blood;

(c) electromagnetic induction techniques, in which the blood flow is determined from the voltage induced when blood (an electrical conductor) flows through a magnetic field;

(d) indicator dilution methods, in which an indicator is added and its concentration is measured after thorough mixing with the flowing blood; and

(e) indicator clearance methods, whereby blood flow is determined from the rate at which an indicator is carried away from the site of measurement by the flowing blood.

1. Venous Occlusion Plethysmography

Venous occlusion plethysmography is one of the oldest methods of blood-flow measurement. It is noninvasive and can yield highly accurate results (Barendsen 1980). The limb (or part of a limb) is enclosed in a rigid container filled with warm water, and an inflatable cuff encircles the limb close to where it enters the container at a waterproof seal (Fig. 1). The cuff is inflated rapidly

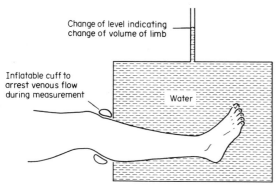

Figure 1
Venous occlusion plethysmograph for measuring blood flow in a limb or part of a limb

to a pressure which just exceeds the pressure in the veins, preventing blood from leaving the limb, but not restricting the arterial inflow. During each measurement, which takes a few seconds, arterial blood continues to flow into the limb, distending the veins, and displacing an equal volume of water from the container. The rate at which water is displaced is measured electronically and recorded. After each measurement, the cuff is deflated to allow the veins to return to their normal volume.

The original method was rather cumbersome, and an air-filled plethysmograph applied to a segment of a limb or to a single digit is now often preferred. Alternatively the rate of increase in the circumference of the limb is measured during venous occlusion and the rate of increase in volume calculated on the assumption that the

limb is a cylinder of fixed length (Whitney 1953). The change in circumference is measured using a strain gauge consisting of a fine, mercury-filled elastic tube which encircles the limb. It is assumed that the volume of the limb is directly proportional to l^2 where l is the circumference of the limb. Since the area of cross section of the tube is reduced as its length is increased, the electrical resistance of the strain gauge is directly proportional to l^2. Hence the rate of change of limb volume is directly proportional to the rate of change of resistance (see *Plethysmography*).

2. Use of Backscattered Ultrasound and the Doppler Effect

The use of ultrasound and the Doppler effect for the investigation of circulation has increased rapidly in the last decade and research is in progress on many aspects of this technique. A beam of ultrasound at a frequency between 2 and 10 MHz is directed obliquely across a blood vessel (Fig. 2). The red cells in the blood scatter

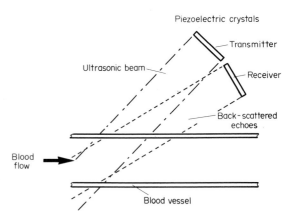

Figure 2
Measurement of blood flow by ultrasound. A probe in which the crystals are mounted is applied to a blood vessel at operation, or held against the skin for transcutaneous measurements on an underlying vessel

the ultrasound in all directions. Some of the backscattered ultrasound impinges on a receiver crystal which is adjacent to the transmitter crystal. In practice, both crystals are often fixed in juxtaposition in a single probe which can be held against the surface of the skin in the vicinity of a blood vessel. A gel or oil is used as an ultrasound coupling medium, filling the space between the crystal faces and the skin. Owing to the Doppler effect, movement of the blood towards, or away from, the crystals causes the received sound to be at a higher, or lower, frequency than the transmitted sound, the change in frequency being directly proportional to the component of the blood velocity in the direction of the ultrasonic beam (see *Doppler Blood Flow Measurement*).

Since normal blood velocities give rise to changes in frequency which are within the human audible range, many simple inexpensive instruments have been produced which allow the operator to hear a signal which varies in pitch with the velocity of the blood. Such instruments are in common use for the detection and qualitative assessment of arterial and venous flow, and the confirmation of pregnancy by the detection of the fetal pulse (see *Fetal Monitoring*). More elaborate machines have been developed for the imaging of blood vessels and the monitoring of fetal heart rate during labor.

In order to obtain a quantitative measurement of blood flow, rather than an indication of blood velocity, it is necessary to know the area of cross section of the vessel and the mean velocity over that area. Two methods have been used to determine the velocity distribution across the vessel. One approach is to analyze the frequency spectrum of the Doppler signals, since the proportion of the received signal within a certain frequency band indicates the proportion of blood flowing at a corresponding range of velocities. Alternatively, if the transmitter output consists of short ultrasonic pulses, signals from different parts of the blood vessel can be separated according to the time interval between transmission and reception of a pulse. This is known as a range-gated Doppler system. Whichever method is used, it is also necessary to know the angle between the ultrasonic beam and the blood vessel, and the cross-sectional area. Ultrasonic instruments which combine pulse–echo techniques for imaging the vessel and the Doppler effect for velocity measurements have recently been developed.

3. Electromagnetic Induction

An electromagnet (Fig. 3) can be placed so that it creates a uniform magnetic field perpendicular to the axis of the blood vessel. Since blood is an electrical conductor, a potential difference is induced along the diameter of the vessel at right angles to both the magnetic field and the direction of blood flow. The potential difference E is given by $E = Bdv$, where B is the magnetic flux density, d is the diameter of the vessel, and v is the mean velocity. Hence the volume of blood $\dot{Q}$ flowing in unit time is given by

$$\dot{Q} = \pi Ed/4B$$

A major advantage of this method is that the flowmeter is unaffected by changes in the velocity profile across the vessel, provided that the blood velocity is symmetrical about the axis of the vessel (Shercliff 1962).

An electromagnetic flowmeter consists of electronic circuits which energize the magnet and amplify the induced voltage picked up by a pair of electrodes in the probe. The probe may be cannulating or cuff type. The former consists of a rigid tube with a pair of electrodes fixed on the inside wall, and an electromagnet on the outside; the latter (Fig. 4) is placed round the intact blood vessel, so that the electrodes make contact with the vessel wall, and the integral magnet provides the transverse magnetic field (Wyatt 1971).

In order to eliminate errors due to a potential difference between the electrodes caused by electrolytic

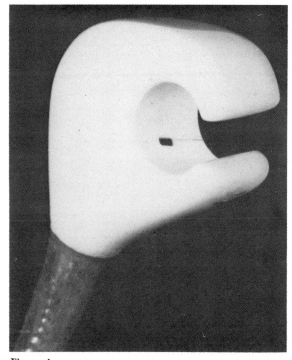

Figure 4
A cuff-type electromagnetic flowmeter probe for use on blood vessels of 10 mm diameter. One of the electrodes can be seen within the lumen of the probe (after Terry 1972. © United Trade Press, London. Reproduced with permission)

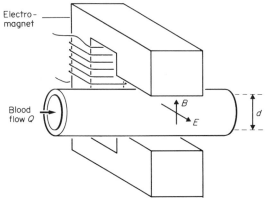

Figure 3
Principle of the electromagnetic flowmeter

effects, the electromagnet is energized by alternating current at a frequency of up to 1000 Hz. The current waveform may be sinusoidal, rectangular or trapezoidal. In each case, the change in the magnetic field causes an unwanted voltage to be induced between the electrodes, and this must be rejected by the electronic circuits, so that only the voltage which is in phase with the magnetic field is accepted and amplified. For maximum accuracy the zero level should be adjusted with the vessel momentarily occluded. In some flowmeters, the magnet current is zero for a short time during each cycle, which allows the zero level to be corrected automatically without the need to occlude the vessel.

Electromagnetic flowmeters are normally calibrated by noting the output for known rates of flow. In a cannulating flowmeter, a change in the electrical conductivity of the blood has little, if any, effect on the output, and the errors due to all causes can be less than 5%. The output of a cuff-type flowmeter depends on the electrical conductivity of the blood and the vessel wall. Furthermore, large errors can occur if the blood flow is not symmetrical about the vessel axis, which commonly occurs if the vessel is curved, branching or narrowed by atheroma. It is therefore advisable to calibrate the probe at the site of measurement if an accurate measurement is required.

4. Indicator Dilution Methods

The use of indicators for measurements on the circulation was introduced by Stewart (1897) and developed for the measurement of blood flow by Hamilton et al. (1928). Many different methods based on this technique are now in use. They can be divided into two groups, depending on whether the indicator is injected at a constant rate (for long enough for a steady state to be reached) or in the form of a rapid injection of a single bolus of the indicator.

If an indicator is added to the flowing blood at a known, steady rate (Fig. 5) and its concentration is measured after thorough mixing has occurred, the indicator will reach the site of measurement at the same rate at which it is injected, i.e., $\dot{Q}C = X$, where $\dot{Q}$ is the blood flow, C is the indicator concentration, and X is the rate of injection of indicator. Hence the blood flow is calculated from the indicator concentration and the injection rate.

The second group of methods is based on the principle that all the indicator, injected in a bolus, will reach the site of measurement, after dilution and mixing with the blood. Hence,

$$\int_{t=0}^{\infty} \dot{Q}C(t)\,\mathrm{d}t = Y$$

where $\dot{Q}$ is the blood flow, $C(t)$ is the concentration of the indicator at time t, Y is the total quantity of indicator injected, and t is the time from the start of the injection.

If the blood flow is constant, the equation can be rearranged as follows:

$$\dot{Q} = \frac{Y}{\displaystyle\int_{0}^{\infty} C(t)\,\mathrm{d}t}$$

This is known as the Stewart–Hamilton equation.

Blood flow through the heart is commonly measured by dye dilution, for example, by measuring the concentration of a dye such as indocyanine green in the blood photometrically. Thermal dilution, using cold saline, can also be used. Here, saline is injected through a catheter into the blood entering the heart. A temperature sensor at the tip of a second catheter measures the temperature of the blood after mixing in the heart. Equipment is available, using each of these methods, for both measuring and calculating the blood flow automatically.

5. Indicator Clearance Methods

A number of very different methods of flow measurement are based on the simple principle that the rate at which an indicator is removed from a point or a region in the body is proportional to the flow, provided that certain conditions are met. For example, if a bolus of the radioisotope ^{133}Xe, dissolved in saline, is injected into tissue, the perfusion of the tissue can be determined from the rate of decay of radioactivity at the site of the injection. In another method, known as thermal clearance, a thermistor is embedded in tissue and maintained at a temperature a few degrees above body temperature; the electrical power supplied, which is directly proportional to tissue perfusion, is measured.

A similar technique, hot-film anemometry, is used to determine the velocity profile in the larger blood vessels. A thin-film resistor, a fraction of a millimeter across, is mounted at the tip of a needle which is inserted in a blood vessel. The varying blood velocity is determined from the power supplied, as the needle is advanced across the vessel.

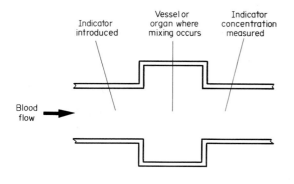

Figure 5
Principle of flow measurement by indicator dilution

6. Choice of Method

Plethysmography and Doppler ultrasound techniques have the important advantage of being noninvasive. While plethysmography provides an accurate measurement of mean flow, it has the drawback of requiring the subject to remain still, in a suitable position, while measurements are made. Ultrasonic methods provide clinically valuable information on the velocity and pulsatility of blood flow, but elaborate equipment is necessary for precise measurements.

If a vessel is exposed during surgery, the choice lies between the ultrasonic and electromagnetic methods. Further, either technique can be used for long-term flow measurement using a surgically-implanted probe placed round an artery, ultrasonic flowmeters having the advantages of lower power consumption in the transducer and freedom from baseline drift.

Blood flow through the heart and the major vessels can be measured by indicator dilution techniques, if a needle or a catheter can be inserted into the appropriate artery or vein. A great variety of techniques, based mainly on dilution or clearance principles, are available for flow measurements in the various organs and tissues of the body. The available methods have been reviewed by Cobbold (1974), and Woodcock (1975) and Mathie (1982). Many clinical applications have been described by Woodcock (1976).

See also: Hemodynamics; Blood Pressure: Invasive and Noninvasive Measurement; Cerebral Blood Flow: Regional Measurement; Cardiac Output Measurement

Bibliography

Barendsen G J 1980 Plethysmography. In: Verstraete M (ed.) 1980 *Methods in Angiology: A Physical-Technical Introduction*. Martinus Nijhoff, The Hague, pp. 38–92

Cobbold R S C 1974 *Transducers for Biomedical Measurements: Principles and Applications*. Wiley, New York

Hamilton, W F, Moore J W, Kinsman J M, Spurling R G 1928 Simultaneous determination of the pulmonary and systemic circulation times in man and of a figure related to the cardiac output. *Am. J. Physiol.* 84: 338–44

Mathie R T (ed.) 1982 *Blood Flow Measurement in Man*. Castle House Publications, Tunbridge Wells

Shercliff J A 1962 *The Theory of Electromagnetic Flow-Measurement*. Cambridge University Press, Cambridge, UK

Stewart G N 1897 Researches on the circulation time and on the influences which affect it. IV. The output of the heart. *J. Physiol. (London)* 22: 159–83

Terry H J 1972 The electromagnetic measurement of blood flow during arterial surgery. *Biomed. Eng.* 7: 466–72

Whitney R J 1953 The measurement of volume changes in human limbs. *J. Physiol. (London)* 121: 1–27

Woodcock J P 1975 *Theory and Practice of Blood Flow Measurement*. Butterworth, London

Woodcock J P (ed.) 1976 *Clinical Blood Flow Measurement*. Sector, London

Wyatt D G 1971 Electromagnetic blood-flow measurements. In: Watson B W (ed.) 1971 *IEE Medical Electronics Monographs* 1–6. Peter Peregrinus, London, pp. 181–243

H. J. Terry

Blood Gas Analysis

The first commercial equipment for blood gas analysis was based on the design of Professor Paul Astrup and manufactured by the Radiometer Company of Copenhagen in the 1960s. Since the early days, there have been considerable innovations in technique, mainly in the development of direct reading electrodes sufficiently small to be combined in a single cuvette, and with these it has become possible to perform a complete analysis on a single small blood sample.

1. The Blood Sample

There are few estimations in clinical chemistry where correct sample handling is more critical than in blood gas analysis. It is probably impossible to transport blood from the patient's blood vessel to the cuvette of the analyzer without some degree of change, but this can be minimized with careful technique. Arterial blood, by far the most satisfactory sample, is usually taken from the radial artery. Simple arterial puncture produces only minor discomfort to the patient and has been shown to be a safe procedure (Cole and Lumley 1966). Venous blood, from the back of a prewarmed hand, can be used for pH and p_{CO_2} estimations, particularly during anesthesia. Capillary blood, obtained by heel prick after a ten minute warming period and collected in heparinized glass capillaries, has traditionally been used for blood gas estimations in neonates and small children. Unfortunately, in infants with impaired cardiopulmonary function, correlation with arterial values is poor and it is often such infants who most need blood gas analysis (Gandy et al. 1964).

1.1 Collection and Storage

Although it has been recommended in the past that blood samples should be collected into non-interchangeable all-glass syringes, under most clinical conditions plastic syringes appear to be satisfactory. Changes in pH or gas tensions are similar in both types of syringe, and are due to metabolism of the blood itself rather than losses through the syringe walls. This results in significant falls in p_{O_2} and rises in p_{CO_2} and pH. These effects can be minimized by storage of the sample in iced water if it cannot be analyzed immediately. Even under these conditions the time between sampling and analysis should not exceed three hours (Siggaard Andersen 1961).

2. Measurement of Hydrogen Ion Concentration

The determination of the hydrogen ion concentration of blood plasma is an integral part of acid–base estimation. Traditionally, the acidity has been expressed as a pH value, that is, as the negative logarithm of the hydrogen ion concentration.

In all modern blood gas analyses pH is measured by means of an electrode based on the original design of MacInnes and Dole (1929) using "pH sensitive" glass. This glass is composed of a three-dimensional lattice of oxygen atoms held in irregular chains by silicon atoms. The spaces in the lattice are filled by cations of different energy levels and varying freedom of movement. If a membrane of this glass separates two solutions, one of known pH (i.e., a buffer) and one of unknown pH, the resultant exchange of ions between the glass surface and the solutions results in an emf being produced across the glass membrane, the magnitude of which is given by:

$$E = K + \frac{2.303\ RT}{F}\ \text{pH}$$

where R is the gas constant, T is the absolute temperature and F is the Faraday constant. Thus the change in emf from pH_1 to pH_2 is:

$$E_1 - E_2 = \frac{2.303\ RT}{F}\ (\text{pH}_1 - \text{pH}_2)$$

If $\text{pH}_1 - \text{pH}_2 = 1$ pH unit, the emf for unit pH change, ΔE, is given by:

$$\Delta E = \frac{2.303\ RT}{F} = 61.7\ \text{mV at 38 °C}$$

Assuming 98% efficiency, one pH unit will be approximately equal to 60 mV.

To detect this emf, a silver/silver chloride anode dips into a known buffer on one side of the glass membrane, and on the other side the circuit is completed by a salt bridge (usually potassium chloride solution) in contact with both the test solution and the cathode. This cathode is usually in the form of a calomel reference electrode which itself produces a known stable emf against which that of the glass electrode is compared (Fig. 1). The resultant emf is amplified by means of a high-impedance amplifier, and displayed in terms of H^+ concentration or pH units. The whole electrode assembly is encased in a chamber thermostatically controlled to better than ± 0.1 °C. It is calibrated by means of a pair of buffers that cover the range of pH to be measured (i.e., around 7.4 at 37 °C). The values usually chosen are 6.84 and 7.38. These two points determine the slope of the calibration plot of pH against emf, and when they have been fixed only a single point value is required just before the estimation.

Although standard buffers are accurate to ± 0.005 of a pH unit, blood pH measurement in the average laboratory does not approach this level of precision. One study reported a mean difference of 0.053 pH units between identical whole blood samples in different labo-

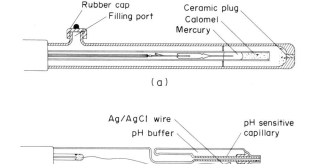

Figure 1

(a) Calomel reference electrode, and (b) capillary pH electrode. In actual use they are joined by a solution of potassium chloride (the salt bridge)

ratories. In the past, chilling of the electrodes by the insertion of cold samples and poor control of bath temperature have both been responsible for measurement errors. These have now largely been eliminated by the use of prewarming coils and thyristor control of thermostats. Variations in the potential developed across the liquid junction (the junction potential) are still an important source of error. Some modern electrodes have a liquid junction incorporating a cellophane membrane saturated with KCl solution. Contamination of this membrane with blood or protein can produce significant errors which are often extremely difficult to detect. The glass electrode itself may become contaminated with blood protein and must be regularly cleaned with solutions of hypochlorite or proteolytic enzymes.

3. Measurement of Carbon Dioxide Concentration

If a blood sample is exposed to a strong acid, CO_2 is expelled and may be measured volumetrically or manometrically—this is the basis of the Van Slyke technique; the total CO_2 content is measured. However, blood carbon dioxide is now nearly always expressed as partial pressure (p_{CO_2}), which can be determined indirectly or directly.

3.1 Indirect Method

This is the equilibration or "Astrup" technique (Astrup 1959) and relies on the fact that a straight line is obtained when whole blood pH is plotted against log p_{CO_2}. The blood sample is divided into three parts. The pH of the first is measured and the remaining two parts are placed in the two arms of a microtonometer and equilibrated to a known p_{CO_2} with predetermined mixtures of CO_2 in oxygen. Their new pH values are then measured and plotted on a Siggaard Andersen nomogram (Fig. 2), and a straight line drawn between them. The intersection of the original pH with this line indi-

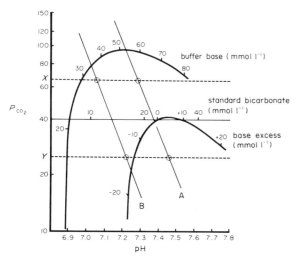

Figure 2
Siggaard Andersen nomogram. The points of intersection of the calculated p_{CO_2} values for the sample after tonometry (X and Y) with their measured pH's, provide the buffer line. Line A represents the normal, line B a patient with a metabolic acidosis

cates the p_{CO_2} of the original sample. Furthermore, the slope of this line is an indication of the buffering power of the blood and the acid–base state of the patient may be derived from the intersection of this line with pre-drawn curves on the nomogram.

3.2 Direct Method

Direct measurement is achieved using an electrode similar to that described by Stow et al. (1957) and later modified by Severinghaus and Bradley (1958). It consists of a combined pH and reference electrode mounted behind a selectively permeable membrane, originally of PTFE but now more usually of silicone rubber (Fig. 3). Between the membrane and the tip of the electrode is a thin spacer of nylon mesh containing "carbon dioxide electrolyte"—usually a solution of bicarbonate and potassium chloride. The carbon dioxide but not the hydrogen ion in a blood sample diffuses through this membrane into the electrolyte and the following reaction

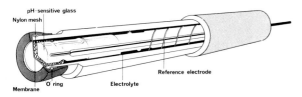

Figure 3
Carbon dioxide electrode

occurs:

$$CO_2 + H_2O \rightleftarrows H_2CO_3 \rightleftarrows H^+ + HCO_3^-$$

The change in the H^+ concentration of the electrolyte is measured by the electrode, amplified and displayed directly as p_{CO_2}. The modern version is essentially similar but the reference electrode usually takes the form of a silver/silver chloride wire wound in a spiral around the measuring electrode and both are immersed in an electrolyte which is continuous with that covering the tip. The whole assembly can be calibrated with humidified gas mixtures containing known concentrations of carbon dioxide.

The pH/log p_{CO_2} relationship is linear to within 0.002 pH units from 1–100% CO_2 in the calibrating gas (i.e., from 1 to 93 kPa) although most electrodes read only to 26 kPa. The response time is between 30 s and 3 min, mainly depending on the thickness and nature of the membrane. The electrode responds more rapidly to an increase than to a decrease in p_{CO_2}.

4. Measurement of Oxygen Concentration

Blood oxygen concentration can be expressed in three ways: (a) as an oxygen tension (p_{O_2}), (b) as an oxygen saturation (s_{O_2}), or (c) as an oxygen content (c_{O_2}). At present, oxygen tension measurements are the most common method of estimating blood oxygen concentration, but oxygen saturation and content are still required for some purposes, particularly those associated with respiratory physiology.

4.1 Oxygen Tension

Oxygen tension is almost always determined by means of the polarographic electrode (Fig. 4) originally described

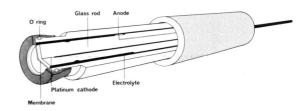

Figure 4
Polarographic oxygen electrode

by Clarke (1956). It consists of a platinum cathode 10–25 µm in diameter sealed into the end of a glass tube with the tip of the platinum exposed at its surface. A polypropylene membrane covers the glass tip which is roughened to allow a thin film of electrolyte (buffered potassium chloride) between the membrane and the tip. The anode is silver/silver chloride which also dips into the electrolyte. A constant polarizing voltage of about 0.6 V is applied between the anode and cathode. Gases from the blood sample diffuse through the membrane and oxygen is reduced at the tip of the cathode according to the reactions:

$$O_2 + 2H_2O + 2e^- \rightarrow H_2O_2 + 2OH^-$$

$$H_2O_2 + 2e^- \rightarrow 2OH^-$$

Thus, four electrons are required for the complete reaction. The current flow, which is proportional to the p_{O_2} of the blood sample, is measured, amplified and displayed directly as an oxygen tension using the appropriate instrumentation. The zero point of the oxygen electrode is set by introduction of an oxygen-free gas mixture into the cuvette; the "span" (the gain of the amplifier) may be set with humidified room air or a gas mixture containing a known concentration of oxygen.

The method is subject to error due to:

(a) The blood gas factor—the current output of the electrode is lower for blood than for a gas mixture of similar oxygen tension (Rhodes and Moser 1966).

(b) Nonlinearity—if the blood sample has an oxygen tension much greater than the upper calibration value its measured p_{O_2} can be considerably lower than the true value.

(c) Instability—oxygen electrodes are inherently unstable and drift of up to 1–2% of the reading over a period of 5 min is not uncommon. For maximum accuracy, these electrodes *must* be calibrated *immediately* before use.

(d) Interference—both halothane and nitrous oxide interfere with the operation of the oxygen electrode in such a manner that their presence in the blood sample increases the apparent p_{O_2}. Errors of up to 25% of the reading may be produced in this way (Dent and Nettar 1976).

4.2 Oxygen Content

For many years, the Van Slyke manometric technique was the only accepted direct method of measuring blood oxygen content. Although accurate in experienced hands, it was technically difficult and very time consuming. A reliable alternative technique has now been developed which utilizes a galvanic cell as the oxygen detector. This is commercially available as the "Lex O$_2$ Con" (Fig. 5), which can give excellent results provided it is operated correctly (Adams and Cole 1975).

To determine oxygen content, an oxygen-free gas mixture containing 1% CO, 2% H$_2$ and nitrogen is bubbled through distilled water in a scrubber unit until equilibrium is reached. A 20 μl sample of blood is injected which then immediately hemolyzes—its oxygen being liberated by the CO. The liberated oxygen passes in the gas flow to a fuel cell where a current is generated in proportion to the oxygen content of the original sample. Calibration of the instrument is normally car-

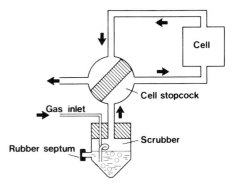

Figure 5
Fuel cell analyzer for oxygen content

ried out with room air but iced water equilibrated with 100% oxygen can also be used.

4.3 Oxygen Saturation

Oxyhemoglobin saturation can either be derived from measurements of oxygen content and of hemoglobin (Hb):

$$s_{O_2} = \frac{c_{O_2}}{(\text{Hb} \times 1.39) + (p_{O_2} \times 0.003)}$$

or may be estimated directly using some form of oximeter—a dedicated spectrophotometer which measures the amount of light transmitted through or reflected from a layer of blood in a cuvette (Cole and Hawkins 1967). Oximeters are thus of two types:

(a) reflection oximeters, which measure the ratio of light reflected from the surface of a whole blood sample at two different wavelengths; and

(b) transmission oximeters, which measure the optical absorbance of a hemolyzed blood sample at two or more wavelengths.

The wavelengths are chosen such that one indicates total hemoglobin (the isobestic point), and the other indicates the value for oxyhemoglobin (Figs. 6 and 7). In modern instruments, the necessary calculations are performed on a microcomputer and the results are displayed digitally. The zero point can be set with dithionated blood and the 100% point with nonsmoker's blood tonometered with oxygen.

Present day oximeters are both accurate and precise (Dennis and Valeri 1980). Once calibrated, they can remain stable for long periods and their only limitation is that their readings may be affected by the presence of abnormal hemoglobins (e.g., carboxyhemoglobin and methemoglobin). By measuring the molar absorbance of the sample at other appropriate wavelengths, however, some instruments yield values for these hemoglobins.

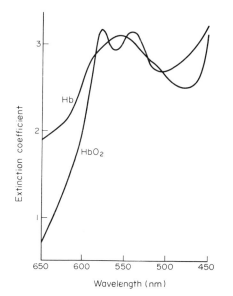

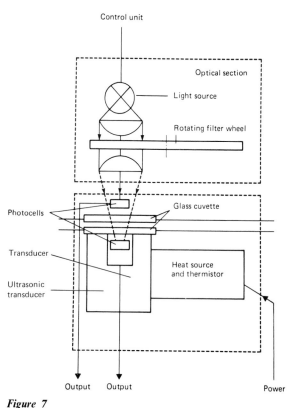

Figure 6
Absorption curves of hemoglobin and oxyhemoglobin. At 505 nm the absorbencies are equal—this is an isobestic point; at 600 nm they are widely different

Figure 7
Optical system of a commercial transmission oximeter (Radiometer, OSM2)

5. Measurement of Hemoglobin Oxygen Affinity

The degree of ease with which hemoglobin takes up oxygen in the lungs and unloads it to the tissues is known as its oxygen affinity. An increased affinity is known as a "shift to the left," a decreased affinity—a "shift to the right." Oxygen affinity is usually expressed as the p_{50}—the oxygen tension at which hemoglobin is 50% saturated. The normal value is around 3.7 kPa. Methods of measurement are outlined below.

5.1 Curve Generating Machines

Curve generating machines are complex and expensive: they attempt to produce a complete dissociation curve when, in most cases, only the p_{50} is required. (Even with a pronounced shift, the hemoglobin dissociation curve is essentially of the same shape.) These machines are also unreliable and extremely slow.

5.2 Two-Point Equilibration

Here, the blood sample is successively equilibrated in a tonometer maintained at 37 °C with two gas mixtures each containing an accurately known concentration of oxygen in 5.6% CO_2 and nitrogen. The oxygen concentrations are chosen so that the resultant saturations lie on each side of the 50% point. After measurement of the oxygen saturations and correction of the p_{O_2} to a pH of 7.4, a straight line of s_{O_2} against p_{O_2} is drawn through the points (see Fig. 8). This technique avoids measurement of p_{O_2} and the inherent inaccuracy of the oxygen electrode (Astrup et al. 1965).

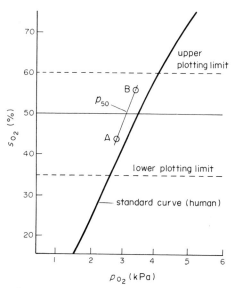

Figure 8
Two-point equilibration. A and B represent oxygen saturations at known oxygen tensions and should lie within the upper and lower plotting limits

5.3 One-Point Technique

In this case, the s_{O_2}, p_{O_2} and pH of the blood sample are measured and the values entered into a computer program which represents the corrected hemoglobin dissociation curve. The program then derives the tension which produces 50% saturation on that particular curve. If the sample is equilibrated to a known oxygen tension before the measurement of saturation, similar results are obtained but the inaccurate p_{O_2} estimation is avoided (Aberman et al. 1975). These techniques are particularly suitable for large screening programs for abnormal p_{50}'s and are mainly used for the detection of abnormal hemoglobins in a special population.

6. Commercial Blood Gas Analyzers

All commercial analyzers for the bench determination of blood gas and acid–base state are basically similar in that they first measure the pH, p_{CO_2} and p_{O_2} of the blood sample. The differences lie in their mode of calibration, sampling, washing, calculation of derived values, necessary corrections, monitoring of electrode performance and diagnostic service routine. The equipment may conveniently be divided into three types: manual, semiautomated and fully automated.

6.1 Manual Machines

In these instruments (still in commercial production), all calibrating, sample-insertion and cuvette-rinsing procedures are carried out by the operator. They often have a combined cuvette for the O_2 and CO_2 electrodes, but a separate capillary electrode for the pH estimation. Their use requires the services of a skilled technician. The instruments are capable of more accuracy and precision than any other type when correctly set up and calibrated. They have the undoubted advantage that the electrodes can be preconditioned by insertion of one half of the sample, the actual reading being taken after the second half is inserted. They can be calibrated appropriately to minimize errors due to nonlinearity of electrodes.

6.2 Automated Machines

It is not possible to differentiate clearly between fully and semiautomated machines. In general, the semiautomated machines (e.g., Instrumentation Laboratories, IL 1304 and the Corning 168) have a limited number of automated functions (usually confined to sampling, calibration and washing), whereas the automated machines (e.g., the Radiometer ABL3, the IL 1312 and the Corning 178) are in complete control of all functions.

In a typical automated measuring cycle (Fig. 9), the sample is inserted or aspirated, automatically measured and the results are displayed. Derived values are calcu-

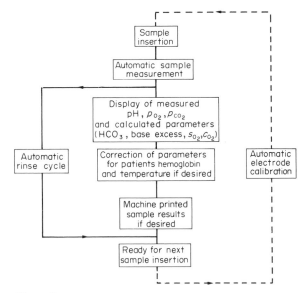

Figure 9
Flow diagram of a fully automated blood gas analyzer

lated and the necessary corrections for hemoglobin and temperature are made. The instrument then prints out the results, and rinses and prepares for the next sample. Single- or two-point calibrations usually occur at predetermined intervals (the actual time depending on the individual manufacturer) so that, theoretically at least, the instrument should be in calibration and always ready for use. Unfortunately, it is possible for electrode drift to occur between calibrations, especially when the time interval is long—in some instances 2 h. This implies that fully automated machines are inherently less accurate than those which must be calibrated immediately before use. Automated machines have either a fixed measuring time, i.e., they make the measurements a fixed time after insertion of the sample (Radiometer ABL3); alternatively the rate of change of the electrode output is monitored by the instrument computer and displayed only when this change is below a predetermined level, i.e., a plateau is reached. The fixed time-measuring cycle can present problems when measuring high p_{O_2} values which tend to fall constantly after reaching a peak—possibly due to loss of oxygen by diffusion into the cuvette wall. As a plateau is never reached with these samples, the second method is also invalid. However, the response time of modern electrodes is extremely short, and these problems are minimized.

7. Quality Control

The ideal quality-control material for blood gas analysis is whole blood of known pH, p_{O_2} and p_{CO_2}. It is possible to produce blood of a known p_{O_2} and p_{CO_2} by equi-

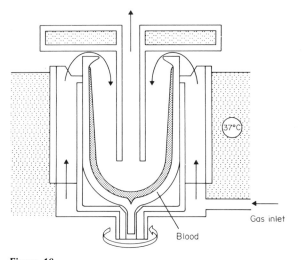

Figure 10
Commercial thin film tonometer (Instrumentation Laboratory, Model 237)

libration with gas mixtures of known concentrations in a tonometer. Tonometers are of two types:

(a) Bubble tonometers. In bubble tonometers, the blood sample is exposed to a stream of fine bubbles of a humidified gas mixture under controlled conditions. These instruments are extremely efficient and able to deal with sample volumes of up to 20 ml. They are difficult to clean, however, and frothing and hemolysis can create problems (Adams and Morgan-Hughes 1967).

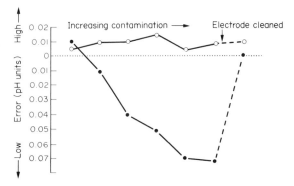

Figure 11
The failure of a commercial quality-control material to demonstrate malfunction of a pH electrode which was clearly shown by tonometered calf serum: ●, readings with serum; ○, readings with commercial quality-control material

(b) Thin-film tonometers. These expose a thin film of the blood sample in a rotating cup or oscillating cuvette to a stream of the gas mixture, again under controlled conditions of humidity and temperature. Commercial thin-film tonometers include the IL Model 237 (Fig. 10), which accepts up to 8 ml of blood, and the "Astrup microtonometer" (Radiometer's BMS 2) which uses disposable cuvettes and is suitable for extremely small samples of the order of 100 μl.

A whole blood sample of known pH is, however, impractical and a serum sample is usually prepared. A

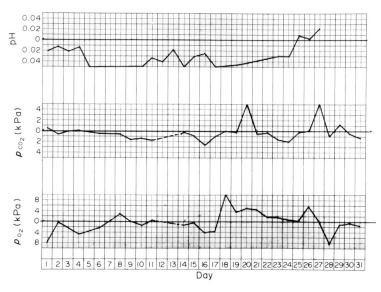

Figure 12
Quality control charts. Lines at zero represent true value for tonometered sample and continuous lines that produced by instrument

500 ml bottle of calf serum is divided into 5 ml aliquots. One of these is tonometered to three known p_{CO_2} values and a buffer line is constructed. This will hold for the remaining samples for about two years if the samples are stored at $-20\,°C$. Defreezing and tonometry to a known p_{CO_2} produce a sample of accurately predictable pH and if this differs materially from the expected value it is a good indication of electrode malfunction (Bird and Henderson 1971).

7.1 Commercial Standards

Several commercial materials are available for quality control. These standards are essentially gas-equilibrated buffer solutions although one does contain preserved red cells. When used carefully, they can demonstrate problems with O_2 and CO_2 electrodes but their usefulness in the case of pH electrodes is doubtful (Fig. 11). Quality control figures can be recorded most clearly on a chart similar to that shown in Fig. 12 which nowadays can be computerized (Shrout 1982). Acceptable limits of accuracy have been proposed by Cotlove et al. (1970) who suggest that the standard deviation of the analytical error should be less than half that of the reference range.

See also: Blood Gas Tensions: Continuous Measurement

Bibliography

Aberman A, Cavanilles J M, Weil M H, Shubin H 1975 Blood p_{50} calculated from a single measurement of pH, p_{O_2} and s_{O_2}. *J. Appl. Physiol.* 38: 171–76

Adams A P, Morgan-Hughes J O 1967 Determination of the blood-gas factor of the oxygen electrode using a new tonometer. *Br. J. Anaesth.* 39: 107–13

Adams L, Cole P V 1975 A new method for the direct estimation of blood oxygen content. *Cardiovasc. Res.* 9: 443–46

Astrup P 1959 Ultra-micro-method—for determining pH, p_{CO_2} and standard bicarbonate in capillary blood. In: Woolmer R F (ed.) 1959 *A Symposium on pH and Blood Gas Measurement: Methods and Interpretation.* Churchill, London, pp. 81–100

Astrup P, Engel K, Severinghaus J W, Munsun E 1965 The influence of temperature and pH on the dissociation curve of oxyhemoglobin of human blood. *Scand. J. Clin. Lab. Invest.* 17: 515–23

Bird B D, Henderson F A 1971 The use of serum as a control for acid–base determination. *Br. J. Anaesth.* 43: 592–94

Clark L C 1956 Monitor and control of blood and tissue oxygen tensions. *Trans. Am. Soc. Artif. Intern. Organs* 2: 41–47

Cole P V, Hawkins L H 1967 The measurement of the oxygen content of whole blood. *Biomed. Eng.* 2: 56–63

Cole P V, Lumley J 1966 Arterial puncture. *Br. Med. J.* 1: 1277–78

Cotlove E, Harris E K, Williams G Z 1970 Biological and analytic components of variation in long term studies of serum constituents in normal subjects. Physiological and medical implication. *Clin. Chem. (Winston-Salem, N. C.)* 16: 1028–32

Dennis R C, Valeri C R 1980 Measuring percent oxygen saturation of hemoglobin, percent carboxyhemoglobin and methemoglobin and concentrations of total hemoglobin and oxygen in blood of man, dog and baboon. *Clin. Chem. (Winston-Salem, N. C.)* 26: 1304–8

Dent J G, Nettar K J 1976 Errors in oxygen tension measurement caused by halothane. *Br. J. Anaesth.* 48: 195–97

Gandy G, Grann L, Cunningham N, Adamsons K Jr, James L S 1964 The validity of pH and p_{CO_2} measurements in capillary samples in sick and healthy newborn infants. *Pediatrics* 34: 192–97

MacInnes D A, Dole M 1929 A glass electrode apparatus for measuring the pH value of very small volumes of solution. *J. Gen. Physiol.* 12: 805–11

Rhodes P G, Moser K M 1966 Sources of error in oxygen tension measurement. *J. Appl. Physiol.* 21: 729–34

Roberts M J, Cole P V 1982 Simple and rapid method for the assessment of the oxygen affinity of haemoglobin. *Ann. Clin. Biochem.* 19: 354–57

Severinghaus J W, Bradley A F 1958 Electrodes for blood p_{O_2} and p_{CO_2} determination. *J. Appl. Physiol.* 13: 515–20

Shrout J B 1982 Controlling the quality of blood gas results. *Am. J. Med. Tech.* 48: 347–51

Siggaard Andersen O 1961 Sampling and storing of blood for determination of acid–base status. *Scand. J. Clin. Lab. Invest.* 13: 196–204

Stow R W, Baer R F, Randall B F 1957 Rapid measurement of the tension of carbon dioxide in blood. *Arch. Phys. Med. Rehabil.* 38: 646–49

Weisbrot I M, Kambli V B, Gorton L 1974 An evaluation of clinical laboratory performance of pH blood gas analyses using whole blood tonometer specimens. *Am. J. Clin. Pathol.* 61: 923–35

P. V. Cole

Blood Gas Tensions: Continuous Measurement

The measurement of arterial oxygen and carbon dioxide partial pressure (p_{O_2} and p_{CO_2}) is important in the management of critically ill patients. These variables are recognized as useful clinical indicators for the early detection of respiratory problems. The measurements are particularly important in preterm (premature) infants who are especially vulnerable to respiratory illnesses, either because of immaturity of the lung or inadequately developed mechanisms for the control of breathing. The most important consequences of a respiratory illness are that the amount of oxygen in the arterial blood supplying the body falls and the amount of carbon dioxide rises. If the amount of oxygen falls to a very low level, brain-damage or death may follow. Avoidance of too low a level of oxygen in the blood is achieved by giving the baby extra oxygen to breathe and, when necessary, assisting breathing by mechanical means (see *Neonatal Intensive Care Equipment*). A particular problem with oxygen therapy in preterm infants is that if the oxygen level in arterial blood goes too high, damage to the eyes (retrolental fibroplasia), which may progress to blindness, can occur.

Carbon dioxide is produced in the body. The only way that it can be removed effectively is through the lungs. If the lungs are diseased, the amount of carbon dioxide in the blood builds up and mechanical ventilation (artificial respiration) may have to be used to bring it down. Blood flow to the brain is controlled to a large extent by the amount of carbon dioxide in arterial blood. If the carbon dioxide level is high, cerebral blood flow is greatly increased. If the blood oxygen level is low at the same time, often the case in respiratory illnesses, cerebral blood flow increases still more.

Respiratory illnesses, usually accompanied by cerebral hemorrhage or infarction (death of brain tissue), are the commonest causes of early neonatal death in infants born in the UK. Hemorrhage into the cerebral ventricles, or "intraventricular hemorrhage," is found in over three-quarters of all small preterm infants who die. The available evidence points strongly to the conclusion that the rupture of vessels in the germinal layer (a highly vascular layer of cells where intraventricular hemorrhage originates) can happen as a consequence of fluctuations in cerebral blood flow, caused partly or largely by alterations in the partial pressures of oxygen and carbon dioxide in the blood.

For many years the only way of measuring arterial p_{O_2} and p_{CO_2} was by the analysis of blood samples withdrawn from the patient. Unfortunately this technique only gives information at one point in time and it is now known that large variations in these gas levels can occur over the course of a few minutes. In the case of the infant, this intermittent sampling technique can also result in an undesirable loss of blood. Continuous measurement of oxygen and carbon dioxide—without the need for repeated blood sampling—is clearly preferable. There are essentially two approaches to continuous blood-gas measurement—invasive and noninvasive monitoring.

1. Invasive Blood-Gas Monitoring

Both p_{O_2} and p_{CO_2} catheter-tip sensors have been developed but the continuous measurement of blood oxygen tension has become by far the more widespread. This may partly reflect clinical need but also indicates the technical difficulties inherent in incorporating a Severinghaus p_{CO_2} sensor in the tip of a catheter. In this respect the potential fragility of the pH glass bulb, the very high electrical impedance of the sensor and the requirement of a two-point, sterile *in vitro* calibration, have left the development of a clinically usable catheter-tip p_{CO_2} sensor essentially unrealized.

1.1 Catheter-Tip Sensors

The most widely used catheter-tip p_{O_2} sensor for neonatal monitoring is shown in Fig. 1. The cathode is the tip of a 180 μm diameter Trimel-coated silver wire. The reference electrode is a silver cap, and the contact to the

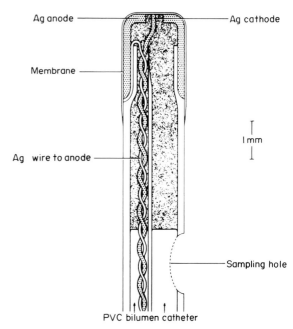

Figure 1
A catheter-tip p_{O_2} sensor for neonatal monitoring

reference electrode is a Trimel-coated silver wire sealed with conducting epoxy resin. The silver cap is held onto the catheter with epoxy resin, while the space between the cathode and the reference electrode and the space behind are filled with epoxy resin of negligible water absorption.

The catheter is a 1.65 or 1.33 mm diameter poly(vinyl chloride) tube of bilumen cross section. An entry port in the catheter wall enables blood samples to be drawn through one lumen for comparative p_{O_2} analysis as well as p_{CO_2}, pH measurements and other studies. The second lumen carries the cathode and reference-electrode connecting wires.

After shaping, the tip is dipped in a mainly KCl electrolyte and hot-air dried. The tip is then dipped in a solution of polystyrene in toluene and left to dry in air, forming a diffusion membrane. At this stage the sensor is completely dry and inactive and in this condition has a long shelf life. The sensor is sterilized by γ-irradiation.

These catheter-tip sensors are introduced into the aorta of the newborn baby via an umbilical artery. A solution of sodium bicarbonate, dextrose and heparin is infused through the catheter at about 65 ml kg^{-1} per day. With the sensor in contact with blood, water vapor diffuses through the diffusion membrane, dissolves the KCl crystals and a conventional polarographic p_{O_2} sensor is formed. The time taken for the output current to stabilize can vary from 10 to 45 min. The zero current of the sensor is assumed to be negligible so that the device can be calibrated against an arterial blood sample. The response time of these sensors varies, depending on the membrane thickness, but typically is in the range 20–50 s (to reach 95% of the eventual value).

An alternative structure for a polarographic p_{O_2} sensor is a monopolar device in which the cathode is covered with a hydrophilic diffusion membrane and the anode is placed on the opposite side of the membrane—usually on the skin. The advantage of this approach is the small physical size which is achievable because of the simplified construction. Monopolar sensors as small as 0.3 mm diameter have been reported for insertion in the radial artery of critically ill patients (Watanabe 1973).

A catheter-tip sensor has been reported (Coon et al. 1976) which measures p_{CO_2} and pH simultaneously *in vivo*. The sensor consists of a PdO hydrogen ion-sensitive electrode and an Ag/AgCl reference electrode surrounded by a thin layer of bicarbonate solution and enclosed with a gas-permeable silicone polycarbonate copolymer membrane, the tip of which is pH sensitive. For p_{CO_2} measurement, the Severinghaus technique is used with the PdO surface being used to detect the change in the pH of the bicarbonate solution. For pH measurement, an additional reference electrode outside the sensor but in contact with the blood is required. The measured potential between the Pd/PdO wire and the external reference is a linear function of the pH of the solution.

The pH-sensitive membrane is made up of copolymer elastomers containing about 60% polysiloxane and 40% poly(bis-phenol-A) carbonate. A mobile H^+ "carrier," *p*-octadecyloxy-*m*-chlorophenylhydrazone mesoxalo-nitrile (OCPH), is added to provide the H^+ selectivity. These catheter-tip p_{CO_2} sensors are approximately 0.6 mm in diameter.

2. Noninvasive Blood Gas Monitoring

A noninvasive method of continuously estimating the partial pressure of arterial blood gases transcutaneously without the need to withdraw arterial blood samples has been developed. The technique is now used extensively to estimate the partial pressures of arterial oxygen and carbon dioxide, these measurements being most commonly written as tcp_{O_2} and tcp_{CO_2}, respectively. The transcutaneous technique is based on the observation that the diffusion of oxygen and carbon dioxide through intact human skin is substantial and can be readily detected by a p_{O_2} or p_{CO_2} sensor placed on the skin.

2.1 Transcutaneous p_{O_2} Monitoring

The p_{O_2} at the surface of skin at normal temperature is near zero. However, if skin temperature is increased, the p_{O_2} measured at the surface of the skin can approach the arterial value quite closely. This was demonstrated as early as 1951 when Baumberger and Goodfriend showed that when a finger was immersed in an electrolyte at 45 °C, the p_{O_2} of the solution approached that of arterial blood. The vasodilation which accompanies this temperature increase is a crucial factor in ensuring good

correlation between the skin surface p_{O_2} and arterial p_{O_2} (Huch et al. 1969).

The estimation of arterial p_{O_2} by the transcutaneous technique (Huch et al. 1972) relies on the correct choice of skin temperature. At this temperature a fortuitous cancellation of errors occurs, resulting in a measurement of p_{O_2} at the surface of the skin which reliably reflects arterial p_{O_2}. Heating the skin causes several effects:

(a) vasodilation of the dermal capillaries, thereby "arterializing" the capillary blood;

(b) a rightwards shift of the oxyhemoglobin dissociation curve, i.e., the p_{O_2} increases at the electrode site; and

(c) an increase of oxygen diffusion through the skin.

What happens is that heating causes vasodilation, and by effect (b) above, results in a capillary p_{O_2} below the electrode, which is greater than arterial p_{O_2}. However, by a judicious choice of temperature, this overreading can be arranged to balance the reduction in p_{O_2} caused by tissue metabolism. Thus, on the further condition that the consumption of oxygen by the transcutaneous sensor is low, the p_{O_2} measured at the skin surface will reflect arterial p_{O_2}. With newborn infants a sensor temperature of 44 °C has been found to be optimal for reliable monitoring of arterial p_{O_2}. (It is estimated that at this sensor temperature the skin temperature will be about 43 °C.)

A transcutaneous p_{O_2} sensor is shown in Fig. 2. Essentially, it is a polarographic oxygen sensor of low

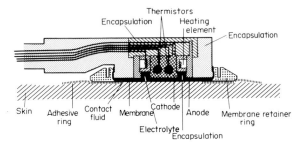

Figure 2
Transcutaneous p_{O_2} sensor

oxygen consumption in which is incorporated an electrical heater to maintain the skin at the necessary temperature. Typically, the cathode is 20 μm diameter platinum fused in glass and surrounded by a silver anode. The electrolyte is contained by a 25 μm Teflon diffusion membrane. The current output of such a sensor at 44 °C is in the range $3–4 \times 10^{-9}$ A at a p_{O_2} of 150 mmHg. A thermistor is used to indicate sensor temperature and a second thermistor acts as a safety cutout to prevent overheating of the sensor in case of a primary thermistor malfunction.

In the newborn and adult patients, attachment of the transcutaneous sensor to the skin is achieved by the use of a double-sided adhesive disk similar to those employed in ECG electrodes. It is clearly important to achieve a gas-tight seal around the circumference of the sensor.

The sensor is calibrated in air (20.9% oxygen) and the zero of the system determined by exposing the sensor to nitrogen.

Figure 3 shows the results of a comparison of tcp_{O_2} and arterial p_{O_2} in newborn infants over a period of six

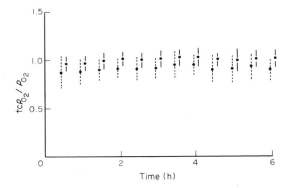

Figure 3
Graphical comparison of tcp_{O_2} (dotted lines) and arterial p_{O_2} (solid lines) plotted against time

hours. The tcp_{O_2} was measured with a transcutaneous sensor and p_{O_2} with an intravascular p_{O_2} sensor. The ratio of tcp_{O_2}/p_{O_2} against time is shown for the transcutaneous sensor calibrated both *in vitro* and against a blood sample (Pollitzer et al. 1979).

In order to minimize the possibility of skin damage, the transcutaneous monitoring period in practice is now generally limited to four hours.

2.2 Transcutaneous p_{CO_2} Monitoring

The noninvasive estimation of arterial p_{CO_2} is similarly achieved by the use of a transcutaneous p_{CO_2} sensor attached to the skin of the patient. The sensor is based on the Stow–Severinghaus principle, and incorporates a heater for raising the skin temperature (Fig. 4).

The transcutaneous measurement of p_{CO_2} (tcp_{CO_2}) differs in theory from the noninvasive p_{O_2} measurement, although the measurement technique is very similar. In the case of tcp_{CO_2} measurement, there is no balance-of-errors equation. In this case, heating the sensor has the following effects on the tcp_{CO_2} readings:

(a) the solubility of CO_2 decreases, thus increasing p_{CO_2};

(b) metabolism of tissue beneath the sensor site increases, causing an increase in CO_2 production; and

(c) the diffusion of CO_2 through the stratum corneum increases.

All these effects give positive errors and since the p_{CO_2} sensor does not consume gas, the tcp_{CO_2} measurement at the skin surface will be higher than arterial p_{CO_2}. The tcp_{CO_2} measurements in the newborn infant are made at 44 °C, as is the case with oxygen monitoring.

The relationship between tcp_{CO_2} measured at a sensor temperature of 44 °C and arterial p_{CO_2} can be expressed as:

$$tcp_{CO_2} = mp_{CO_2} + C$$

Many differing values have been reported for m and C, but the values reported by Eberhard of $m = 1.22$ and $C = 6.6$ appear to have gained more general acceptance (Eberhard et al. 1980). However, it is likely that the relationship above is dependent on the temperature profile in the skin and may therefore be dependent on the geometry of the particular sensor.

A combined tcp_{O_2}/p_{CO_2} sensor has been developed (Parker et al. 1979), which is capable of measuring both gases at the surface of the skin. The single sensor is essentially a tcp_{CO_2} device with a cathode incorporated in it to measure p_{O_2}. Since the current output of a tcp_{O_2} sensor is necessarily small, the production of hydroxyl ions by the cathode is small resulting in a correspondingly small change in the pH of the electrolyte. The sensor thus uses a common electrolyte for both p_{O_2} and p_{CO_2} measurement in addition to a common diffusion membrane and anode. The sensor is shown in Fig. 5. The main advantages of this sensor are its convenience of

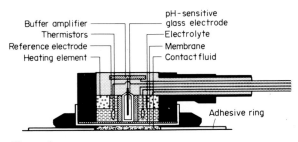

Figure 4
A tcp_{CO_2} sensor

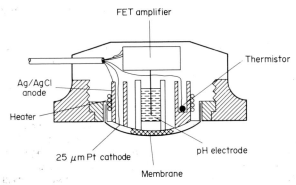

Figure 5
A combined tcp_{O_2}/p_{CO_2} sensor

use, lower cost and the minimizing of skin damage by monitoring on a single skin site.

See also: Blood Gas Analysis

Bibliography

Baumberger J P, Goodfriend R B 1951 Determination of arterial oxygen tension in man by equilibration through intact skin. *Fed. Proc., Fed. Am. Soc. Exp. Biol.* 10: 10–11

Coon R L, Lai N C J, Kampine J P 1976 Evaluation of a dual function pH and p_{CO_2} *in vivo* sensor. *J. Appl. Physiol.* 40: 625–29

Eberhard P, Mindt W, Schäfer R 1980 *A Sensor for Noninvasive Monitoring of Carbon Dioxide.* Hoffman La Roche, Basle

Huch A, Huch R, Lübbers D W 1969 Quantitative polarographische Sauerstoffdruckmessung auf der Kopfhaut des Nuegebornen. *Arch. Gynaekol.* 207: 443–51

Huch R, Lübbers D W, Huch A 1972 Quantitative continuous measurement of partial oxygen pressure on the skin of adults and newborn babies. *Pflugers Arch.* 337: 185–98

Parker D, Soutter L P 1975 *In vivo* monitoring of blood p_{O_2} in newborn infants. In: Payne J P, Hill D W (eds.) 1975 *Oxygen Measurements in Biology and Medicine.* Butterworth, London, pp. 269–83

Parker D, Delpy D T, Reynolds E O R 1979 A single electrochemical sensor for transcutaneous measurement of p_{O_2} and p_{CO_2}. *Birth Defects* 15(4): 109–16

Pollitzer M J, Reynolds E O R, Morgan A K, Soutter L P, Parker D, Delpy D T, Whitehead M D 1979 Continuous comparison of *in vitro* and *in vivo* calibrated transcutaneous oxygen tension with arterial oxygen tension in infants. *Birth Defects* 15(4): 295–304

Watanabe H 1973 Disposable sensors for the continuous *in vivo* monitoring of arterial p_{O_2}. In: *Proc. 8th Meeting of the Association for the Advancement of Medical Instrumentation.* Association for the Advancement of Medical Instrumentation, Arlington, Virginia

D. Parker and D. T. Delpy

Blood Pressure: Invasive and Noninvasive Measurement

Blood pressure measurement can be classified as either (a) invasive or direct, or (b) noninvasive or indirect. In invasive blood pressure measurement, a catheter is inserted through an artery or a vein and positioned in the region in which the pressure is to be measured. A pressure transducer is employed to convert pressure changes to proportional electrical signals which are amplified and displayed on a pressure monitor. The transducer may be mounted at the tip of the catheter, or more commonly, an external transducer is used, which is connected to a saline-filled catheter. The saline serves to transmit the pressures generated at the distal end of the catheter to the transducer. A three-way, disposable

stopcock is fitted into the system at the transducer. The stopcock provides a means of isolating the patient from the measuring system, and of calibrating the system by venting the transducer to the atmosphere for zero pressure, and pressurizing it using a sphygmomanometer to obtain calibration points. The transducer incorporates a diaphragm which flexes with changing pressure. A strain-gauge element (forming the arms of a Wheatstone bridge) is bonded to the diaphragm, and the signals from the bridge are amplified. The transducer is balanced at calibration and also before use, by venting the transducer to the atmosphere as described, and applying an offset voltage so that a zero signal is registered at atmospheric pressure. Any subsequent vertical movement of the transducer in relation to the distal end of the catheter makes rebalancing necessary because of the change in hydrostatic pressure introduced.

Noninvasive techniques of blood pressure measurement currently in vogue are sphygmomanometry and oscillometry. Sphygmomanometry is much more commonly used, but modern technology has made oscillometry an attractive technique which can be employed to measure and display blood pressure parameters continuously.

This article reviews blood pressure transducers and their measurement characteristics, including nonlinearity, hysteresis, drift and dynamic response. Noninvasive blood pressure measurements are also briefly discussed.

1. Principle of Pressure Measurement

Pressure is most commonly measured using the principle of motion balance in relation to the applied pressures acting on a spring element. Thus, if the restoring force from the spring element is a linear function of the deflection, there is a linear relation between the applied pressure and the deflection. The most commonly used spring elements for converting pressure to a proportional deflection are diaphragms and bourdon tubes. A spring element has two sides to which pressure can be applied; thus, in principle, every pressure measuring device is a differential device and it is the difference between the two acting pressures which determines the deflection of the spring element.

In measuring the pressure inside a pressure vessel, the parameter of interest is usually the difference between the pressure inside the vessel and the outside atmospheric pressure. Here, pressure in the vessel is on one side of the spring element and the atmospheric pressure is on the other side. A pressure measuring device arranged in this way is termed a gauge-pressure measuring device.

In other instances it may be desirable to measure a pressure with respect to zero pressure. In this case one side of the spring element must be evacuated. A pressure measuring device which measures pressures with respect to a vacuum is termed an absolute-pressure measuring device. When it is not possible to vent one side of the spring element to atmosphere, the applied pressure is measured with respect to a closed gas volume which has a pressure near atmospheric pressure.

The various forms of pressure measurement are illustrated in Fig. 1.

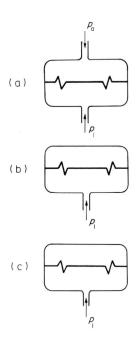

Figure 1
Types of pressure measurement: (a) gauge pressure measuring $p_1 - p_a$, where p_a is the atmospheric pressure and p_1 is the applied pressure; (b) absolute pressure measuring $p_1 - p_0 = p_1$, where p_0 is the zero pressure (vacuum) and p_1 is the applied pressure; (c) nonstandard pressure measuring $p_1 - p_R$, where p_1 is the applied pressure and P_R is the pressure in the reference chamber. $p_R = p_S T/T_S$, where p_S is the atmospheric pressure at the time when the chamber was sealed, T_S is the absolute temperature when the chamber was sealed and T is the absolute temperature at the time of measurement

2. Invasive Measurement Using Transducers

As described, the motion of a pressure transducer diaphragm needs to be detected and converted into an electrical signal. The position of the diaphragm can be sensed by various methods. The main methods used in the measurement of pressure involve the use of several types of gauge.

(*a*) *Strain gauges.* These are probably the most common type of gauge used. They depend upon a change of length of a wire producing a change in the wire's resistance. The diaphragm is coupled to an unbonded strain gauge—via a post—to an array of wires connected to form a Wheatstone bridge. Motion of the diaphragm changes the stress in the wires and provides an electrical output. Semiconductor materials have been used recently as they also show change in resistance with strain. Alternatively, the diaphragm may be made of silicon with a strain gauge integrated into it forming a more sensitive gauge (Fig. 2).

(*b*) *Inductive gauges.* These sense the position of the moving diaphragm by detecting the subsequent change in the inductance of a coil or the inductive coupling between two coils.

(*c*) *Optical methods of pressure measurement.* These have been in use in physiology for many years. A sensitive form of transducer measures the change in position of a diaphragm by measuring the deflection of a beam of light incident thereon. Catheter-tip devices using fiber-optic paths have also been investigated.

(*d*) *Capacitance gauges.* A fixed plate held close to a diaphragm constitutes a capacitor, the value of which changes with the distance between the two plates. Such devices are difficult to construct and are very sensitive to external interference. Thus they are seldom used today.

3. Pressure Transducer Characteristics

The relationship between applied pressure and output signal of a pressure transducer is not absolutely linear. Further, the transducer is affected by external influences, for example, temperature and acceleration. The magnitude of these imperfections determines the quality of the transducer.

3.1 Nonlinearity

Nonlinearity is defined as the maximum deviation from a best straight line, fitted through the calibration points, with rising pressure. Normally a good approximation for the best straight line is a line with slope given by the end points of the calibration curve and located such that the deviations between the calibration points and the line are equal on both sides of the line. Nonlinearity is expressed as a percentage of full scale input.

Linearity is of great importance and modern gauges are linear to within 1% over the working range of -7 to 40 kPa. If, however, the gauge has been stressed above this range then it may well exhibit appreciable nonlinearity. It will also show a zero offset. There is always a small zero offset when no pressure is applied to a gauge. This is minimized in the construction and setting up of the gauge during manufacture. It must be offset by balancing the transducer. Because some amplifiers can offset large amounts of imbalance, and the transducer may be distorted outside its linear range, it is important to check this from time to time. Modern amplifiers, with an automatic balancing arrangement, have a smaller range and are unable to balance a poor transducer. The

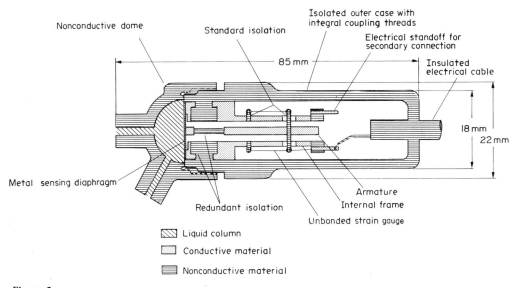

Figure 2
Strain-gauge transducer

pressure range of capture of a digital system is usually ±5 kPa.

The best method of checking the linearity of a transducer is by (a) venting the transducer to the atmosphere and balancing, (b) applying a series of pressures to the transducer from say 0 to 27 kPa in at least ten stages always raising the pressure and plotting the output from the transducer, and then (c) decreasing the pressure, again in the ten stages, plotting the output. The points produced when the pressure is raised should be joined as should the points obtained when the pressure is decreased. The two lines should be linear and coincident. Any deviation from linearity can be measured and the hysteresis determined.

3.2 Hysteresis

Hysteresis is defined as the maximum difference in signal output for the same pressure input when approached from opposite directions. Combined nonlinearity and hysteresis should not be more than 1% over the working range. The transducer if linear can now be calibrated by applying two pressures of say 0 and 27 kPa, and noting the output.

3.3 Drift

Only a small amount of drift in both baseline and sensitivity is acceptable. When zero pressure is applied to a transducer, there is still an electrical output caused by thermal effects which has to be balanced out by the signal processing electronics. This change in zero point at zero pressure is known as thermal zero shift. It is expressed as a percentage of the full scale per °C. Thermal sensitivity-shift indicates the change in sensitivity under varying temperature. It is expressed as percentage error per °C. The effect on sensitivity is usually insignificant, but transducers do tend to have an unacceptable amount of zero or baseline drift after switching on. Thermal equilibrium must be reached (this usually takes about half an hour) and the effect of changes in environmental temperature on transducer drift must be checked. Calibrating errors in measuring static pressures should be small provided the transducers are allowed to reach equilibrium.

The use of continuous flush devices to prevent the blockage of catheters has introduced another problem. As a result of the continuous flow down the catheter there is a pressure drop along its length as well as a major pressure drop at the controlling site. In normal circumstances the effect is negligible and does not significantly change the pressure measured. When there is a restriction in the catheter, however, the pressure drop across the controlling mechanism decreases and there is a greater pressure drop along the catheter, thus increasing the mean pressure at the transducer. The opposite effect is caused by a leaking stopcock. Here the flow is in the opposite direction and the mean pressure falls.

3.4 Dynamic Response

The dynamic response of a transducer–catheter system is much more difficult to test than, and cannot easily be predicted from, the static characteristics of a pressure transducer. It depends mainly on the length and bore of the catheter and the material from which it is made, and also on the volume displacement of the pressure transducer.

Harmonic analysis of a pressure recording made with a catheter-tip manometer reveals the harmonic content of the signal. Fourier's theorem states that a complex waveform can be broken down into a mean term and a sum of sine or cosine waves of appropriate magnitude and phase. The transducer–catheter system must be able to respond accurately to the mean term and to the highest frequency contained within the waveform without distortion. Since a high heart rate of 120 beats min^{-1} corresponds to a frequency of 2 Hz, 15 harmonics will easily contain all the information required to reproduce that waveform. The transducer–catheter must respond therefore without distortion up to a frequency of 30 Hz. This is easily achieved with catheter-tip devices which respond to frequencies of up to several thousand hertz and which can record heart sounds as well. To achieve a 30 Hz response with a catheter and external transducer requires a study of the theory of pressure transmission. Numerous papers on this subject have been published (Gabe 1972, Fry 1960, Hansen 1949, MacDonald 1974). The theory is very complex, but the basic facts can be appreciated by considering a simple system.

Consider a needle of length L and radius r connected to a pressure transducer. The only characteristic of the transducer to be known is the volume displacement (ΔV) of the transducer for a given applied pressure (Δp), the density and viscosity of the fluid filling the transducer being ρ and μ, respectively. The system can then be considered analogous to a mass coupled to a spring which will resonate provided the damping is less than critical. It can be shown that the natural undamped frequency f of the system is given by

$$f = \frac{1}{2\pi}\left(\frac{\pi r^2 E}{\rho L}\right)^{1/2}$$

where $E = \Delta p / \Delta V$. The damping factor is given by

$$\beta = \frac{4\mu L}{r^3 \pi E}$$

The response of the system can be tested by applying a sudden change of pressure, such as that produced by bursting a balloon to release the pressure. The theoretical results are shown in Fig. 3, which has been normalized for amplitude and natural damping frequency. The critical case $\beta = 1$ shows no overshoot; the undamped case would oscillate for ever. Figure 4 shows how this would affect the amplitude frequency response graph, again normalized for amplitude and frequency. It can be seen that a constant frequency response is impossible. To achieve a flat response within a few percent of 30 Hz would require either a high natural frequency or careful damping with a lower natural frequency. It is very difficult to hold a system to a constant damping during use and thus a high natural frequency with low damping is the best choice, since any increase in damping can only help the response characteristics, and changes in resonant frequency are easier to see in a pressure trace. The resonant frequency should be at least twice the required frequency response. The

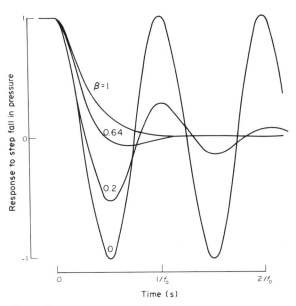

Figure 3
Theoretical response of a step change in input to a system, with different damping

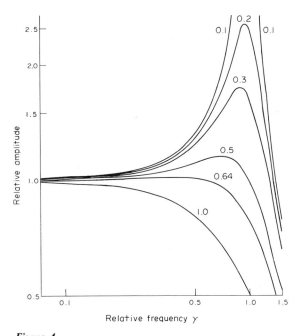

Figure 4
Amplitude–frequency plot normalized for amplitude and frequency

response of a Statham pressure transducer connected to a needle in line with a cardiac catheter is shown in Fig. 5.

The amplitude frequency response is not the only consideration in achieving a faithful reproduction of the

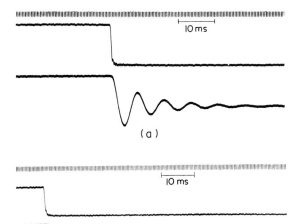

(a)

(b)

Figure 5
(a) Actual step response of a needle connected to a Statham P23Gb pressure transducer. The needle's internal diameter = 0.495 mm and its length = 31 cm; f_D = 132 Hz. (b) Response of a catheter to a step reduction in pressure using a Statham P23Gb transducer and no. 7 cardiac catheter 125 cm long, internal diameter 0.0177 cm; f_D = 83 Hz

pressure. For a zero delay there should be no phase lag between the harmonics. This cannot be achieved. Thus, a constant delay is required entailing a linear phase–frequency response dependent upon the damping of the system. Figure 6 shows the phase–frequency

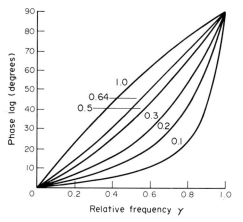

Figure 6
Phase–frequency response of a system showing that a linear change in phase (constant time delay) is achieved at a damping of 0.6

response at different damping factors. The only linear response over the entire range is at a damping of 0.6. This corresponds to a constant time delay. However, if the resonant frequency is high enough, then a linear response up to 0.3 of the natural frequency can be achieved.

3.5 Distributed Systems

A more exact analysis of the transducer–catheter arrangement can be made if it is considered as a distributed system, and transmission line theory used. The practical benefits of this approach have not been obvious (Ranke 1952, van der Tweel 1957, Latimer 1968).

3.6 Practical Considerations

The frequency response of the pressure measurement system is dependent upon the volume displacement of the transducer and on the catheter material, size and length. Great care must be taken to avoid leaks in the system. High-quality stopcocks are recommended and all efforts should be made to eliminate air trapped in the system. This can be extremely difficult. Woven Dacron catheters are especially difficult to make air-free. It is much easier to obtain a good response with disposable polyurethane catheters.

Catheter-tip devices overcome all the problems of frequency response. In these, a diaphragm is forced to move by the blood pressure. The displacement, transformed into an electrical signal, is fed to a suitable amplifier. Most invasive catheter-tip devices are miniature versions of external devices and are either semiconductor gauges or thin-film strain gauges. The problem associated with their use, apart from their cost, is that zero pressure is not easily effected with the catheter in the patient. If zero pressure is set before the catheter is inserted, the catheter should be at body temperature, since thermal shift cannot be corrected after the transducer has been inserted. Because the catheter measures pressure relative to atmospheric pressure, it must be vented. A tube from the back of the diaphragm to the outside, usually within the electrical connector, is required. This must not become blocked or restricted as the transducer will then drift significantly and become useless.

The amplifiers in use to measure the pressure excite the Wheatstone bridge with either a direct current (dc) or alternating current (ac). The ac variety are termed carrier amplifiers. These amplifiers are less likely to drift than dc amplifiers and are suitable for use with all semiconductor gauges as well as wire and thin-film gauges. Some semiconductor gauges become polarized with dc energization and cannot thus be used. If intravascular heart sounds are to be recorded, the carrier frequency must be at least twice the highest frequency contained within the heart sounds.

4. Noninvasive Measurements

The traditional ausculatory or Riva–Rocci method for measuring blood pressure with a sphygmomanometer uses a pressure cuff placed around the arm above the elbow and inflated to a pressure above systolic pressure level. The cuff is slowly deflated and a stethoscope is placed above the brachial artery to detect the eventual flow of blood in the arm.

Systolic pressure is the highest pressure at which successive sounds are heard. Diastolic pressure is obtained by deflating the cuff still futher until the sounds become dull and muffled.

Nearly all the noninvasive methods of measuring blood pressure are based upon this latter method. Various techniques may be used to detect blood flow as the cuff is deflated—microphonic, ultrasound and plethysmographic techniques being the most common. These methods have met with varying degrees of success. They are subject to motion artefact, recording only systolic and diastolic pressure, and are limited to measuring peripheral pressures. Mean arterial pressure can be measured by oscillometry, a technique in which pressure pulses are detected by a transducer incorporated in an inflatable cuff. The cuff is automatically and repeatedly inflated at selected regular intervals, and within these intervals the pressure in the cuff is reduced in controlled steps. The apparatus is microprocessor-controlled, and values are indicated on an LED display. At cuff pressures above systolic, the pressure pulses detected are of constant amplitude, and increase by some 30% in amplitude as the cuff pressure approaches systolic pressure. Peak pulse amplitude, however, occurs at mean arterial pressure, and thus this pressure can easily be detected by electronic techniques.

Central pressures are much more difficult to record noninvasively and no satisfactory method has yet been found. An ingenious method, although not used practically yet, has been proposed, namely injecting microbubbles of gas into the circulation which will resonate at a frequency proportional to a power of the pressure when excited by ultrasound.

See also: Physiological Measurement

Bibliography

Fry D L 1960 Physiologic recording by modern instruments with particular reference to pressure recording. *Physiol. Rev.* 40: 755–88
Gabe I T 1972 Pressure measurement in experimental physiology. In: Bergel D H (ed.) 1972 *Cardiovascular Fluid Dynamics*, Vol. 1. Academic Press, London, pp. 11–50
Hansen A T 1949 Pressure measurement in the human organism. *Acta Physiol. Scand.* 19 (Suppl. 68): 1–230
Latimer K E 1968 The transmission of sound waves in liquid-filled catheter tubes used for intravascular blood pressure recording. *Med. Biol. Eng.* 6: 29–42
MacDonald D A 1974 *Blood Flow in Arteries*, 2nd edn. Edward Arnold, London
Ranke O F 1952 Registrierung laufender Wellen als Registrierprinzip. *Arch. Kreislaufforsch.* 18: 99–107
van der Tweel L H 1957 Some physical aspects of blood pressure, pulse wave, and blood pressure movements. *Am. Heart J.* 53: 4–17

C. J. Mills

Blood Viscosity Measurement

Over half of the deaths in the Western world are due to a cardiovascular cause and most of these are due ultimately to failure of blood flow to a vital organ, in particular the heart, brain or limbs. A common cause of reduction in blood flow is the formation of obstructive lipid-rich masses in the arterial wall (atherosclerosis), often complicated by thrombosis. It is now appreciated, however, that an increase in blood viscosity with a reciprocal reduction in flow may produce similar tissue anoxia and damage. The study of blood viscosity and its determinants has come into prominence recently because of advances in technology which make the accurate measurement of blood viscosity possible.

1. Definition of Viscosity and Its Determinants

Viscosity is a liquid's resistance to flow, set up by shear stresses within the flowing fluid. When flow is streamlined, i.e., when adjacent layers of the liquid are moving parallel to each other, the difference in velocity between these layers is a measure of the shearing within the flowing fluid. This velocity gradient is the shear rate (velocity difference divided by distance between layers) and is expressed in s^{-1}. The tangential force per unit area measured in pascals (Pa) or millipascals (mPa) applied to the fluid which produces shearing and flow is termed the shear stress. The viscosity of a fluid is defined as the shear stress divided by the shear rate and is measured in mPa s. Thus

$$\text{viscosity} = \frac{\text{shear stress}}{\text{shear rate}}$$

Rheology is the study of the deformation and flow of matter; hemorheology is the study of the flow properties of blood.

Blood is a complex fluid consisting of a fluid part (plasma), which forms just over half of the blood volume, and a cellular part composed mainly of red blood corpuscles, with a variety of different white cells and platelets. When anticoagulated blood is rapidly centrifuged, the red cell layer (hematocrit) occupies about 42–48% of the volume of blood. In certain diseases this may be greatly increased (polycythemia) or decreased (anemia).

Blood plasma is a solution of salts, proteins, minute amounts of hormones, and trace elements. The important proteins which determine its viscosity properties are fibrinogen (which is also important in forming the fibrin framework of clots), α_2-macroglobulin (a protease inhibitor), various immunoglobulins, and other globulins of high molecular weight. Plasma viscosity is dependent on temperature, a fall in temperature causing a rise in viscosity. At a constant temperature the viscosity of plasma is constant, i.e., shear stress and shear rate are proportional. Plasma is therefore a Newtonian fluid and its viscosity can be easily measured in a capillary-tube viscometer, a constant force (gravity) producing at a constant temperature a rate of plasma flow which is inversely related to the viscosity of the plasma. At 37 °C, normal plasma viscosity is 1.25 ± 0.15 mPa s. For clinical purposes a variety of semiautomated machines enable plasma viscosity to be rapidly measured in large numbers of samples at a variety of temperatures (Harkness and Whittington 1971, Harkness 1981). Some additional information can be obtained by measuring the serum viscosity. Here, blood is allowed to clot on being withdrawn from a vein. This causes the complete removal of the fibrinogen. The viscosity of the resultant serum is now independent of fibrinogen and allows the contribution of fibrinogen to plasma viscosity to be assessed.

Whole-blood viscosity is complex. Because the interaction of red cells and plasma varies according to the shear rate, blood viscosity also varies with shear rate (non-Newtonian behavior). Another feature of the non-Newtonian behavior of blood is that it possesses a yield stress, which has to be exceeded before stationary blood will begin to flow.

Blood viscosity must be measured at a variety of defined shear rates and a variety of rotational viscometers are now available (Dormandy 1981). The sample of anticoagulated blood is placed in a cup which is surrounded by a water jacket to maintain a constant temperature, usually 37 °C (Fig. 1). A suspended conical or cylindrical bob is lowered into the blood. An electric motor attached to either the cup or the bob produces a constant speed of rotation (shear rate); the torque (shear stress) transmitted to the blood sample is measured. Alternatively, a constant torque is applied and the resultant rate of rotation (shear rate) is measured.

The major determinants of blood viscosity are the hematocrit and plasma viscosity. The shear rate dependence of blood viscosity is due partly to the aggregation of red cells at low shear rates by fibrinogen and other high molecular weight globulins, and partly to red-cell deformation at high shear rates. At high shear rates ($100-200$ s^{-1}) the red-cell aggregates are dispersed and the cells are deformed to form ellipsoids with their long axes aligned with the flow streamlines. Normal blood viscosity at such shear rates is 5 ± 1.5 mPa s (at 37 °C). At shear rates below 10 s^{-1}, blood viscosity rises steeply because the shear forces no longer deform the red cells and red-cell aggregation starts to occur. For

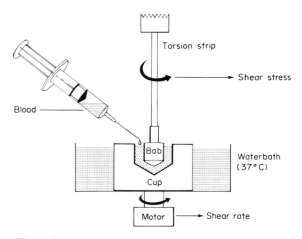

Figure 1
Rotational viscometer

example, the viscosity of normal blood at shear rate 1 s^{-1} is 17 ± 5 mPa s (at 37 °C).

2. Blood Viscosity and Blood Flow

The theoretical basis of the relationship between viscosity and blood flow depends on the Poiseuille–Hagen formula, which relates the volume rate of steady flow Q of a simple liquid in a straight, rigid tube to the pressure gradient along the tube $p_1 - p_2$, the length of the tube l, the internal radius r and the fluid viscosity η:

$$Q = \frac{(p_1 - p_2)\pi r^4}{8\, l\eta}$$

However, the situation in the intact animal circulation is more complex as blood is not a simple fluid, blood vessels are not straight or rigid, and blood flow is pulsatile. If, however, Poiseuille's law is applied to the intact circulation, an increase in blood viscosity should produce a decrease in blood flow rate with the important proviso that the vessels may increase in diameter and/or there may be an increase in cardiac output. Studies of isolated, vasodilated animal organs perfused with bloods of different viscosities have confirmed that increase in blood viscosity is associated with decreased flow. However, the fall in blood flow is less than that predicted by the measurement of blood viscosity even when flow is reduced to produce low shear rates. The probable explanation is that in the major resistance vessels (the arterioles and capillaries), the hematocrit and viscosity fall below venous hematocrit and viscosity. This fall in hematocrit and blood viscosity in tubes with diameter less than 300 µm is known as the Fåhraeus–Lindqvist effect (Fåhraeus and Lindqvist 1931). The fall in hematocrit and viscosity progresses down to a vessel diameter of 15 µm, at which point whole-blood viscosity approaches plasma viscosity. In such small diameter vessels the red cells migrate towards the center of the vessels (axial migration) resulting in an increase in the thickness of the outer zone of cell-free, low-viscosity plasma.

Since the microcirculatory hematocrit is different from venous-sample hematocrit, comparative measurements of blood viscosity should be made not only at native venous hematocrit, but also corrected to a standard hematocrit (e.g., 45%). Such values are obtained best by calculation (Dormandy 1981). Such a correction also enables an assessment to be made of the contribution of other determinants, such as plasma viscosity, red-cell aggregation and deformability, to whole-blood viscosity.

3. Clinical Relevance of Blood Viscosity

With the advent of new machines that can measure blood viscosity and its determinants rapidly and accurately, the relevance and importance of such measurements have become apparent. Reviews of clinical conditions associated with increased blood viscosity have been published (Lowe et al. 1981, Lowe and Forbes 1981).

Overt hyperviscosity syndromes are found in polycythemia, leukemias and paraproteinemias which are due to elevation of the hematocrit, white-cell count and plasma globulins, respectively. In addition, it is now appreciated that increased blood viscosity may play a part in the production of ischemic symptoms in arterial disease in peripheral, coronary and cerebral arteries; it may also play a role in venous thrombosis and in circulatory shock, where the microcirculation to vital organs is disturbed.

However, viscosity can be reduced, in the short term, by lowering the hematocrit by venesection (removing a pint of blood at a time, the fluid content being replaced much more quickly than the cellular content), by lowering plasma viscosity by plasma exchange, or lowering fibrinogen levels by the administration of certain snake venoms and other drugs. These are exciting clinical developments but as yet they have no established place in routine patient management.

See also: Blood Flow: Invasive and Noninvasive Measurement; Hemodynamics

Bibliography

Dormandy J 1981 Measurement of whole blood viscosity. In: Lowe et al. 1981, pp. 67–78
Fåhraeus R, Lindqvist T 1931 The viscosity of blood in narrow capillary tubes. *Am. J. Physiol.* 96: 562–68
Harkness J 1981 Measurement of plasma viscosity. In: Lowe et al. 1981, pp. 79–87
Harkness J, Whittington R B 1971 The viscosity of human blood plasma: Its change in disease and on the exhibition of drugs. *Rheol. Acta* 10: 55–60
Lowe G D O, Barbenel J, Forbes C D (eds.) 1981 *Clinical Aspects of Blood Viscosity and Cell Deformability.* Springer, Berlin
Lowe G D O, Forbes C D 1981 Blood rheology and thrombosis. *Clin. Haematol.* 10: 343–67

C. D. Forbes and G. D. O. Lowe

Bone: Mechanical Properties

The bones of the skeleton play an important part in the support and movement of the body, and also provide mechanical protection for vital organs. These functions are possible because bone is both stiff and strong. Bones and bone material are not homogeneous but have clearly defined structures. In the typical long bone, such as the femur, the central portion or shaft consists of a cylinder of compact cortical bone, but the ends are filled with a spongy, cancellous bone consisting of a three-dimensional lattice made up of bony trabeculae.

The basic microstructural unit of adult human bone is the Haversian system or osteon, which forms a roughly cylindrical elongated structure (Fig. 1). The

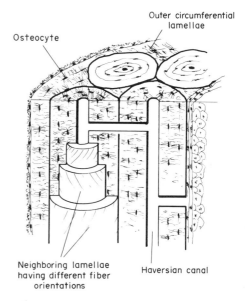

Figure 1
The structure of compact bone: each Haversian system may contain up to 30 lamellae, and have a diameter of up to 0.1 mm

osteon has a central Haversian canal containing blood vessels, lymphatics and nerves. Concentric lamellae surround the Haversian canal, and the outer limit of the osteon is delineated by a cement line of calcified mucopolysaccharide ground substance.

The lamellae contain both organic and mineral constituents, with about 60–75% of the dry weight being made up of inorganic material. Approximately 95–99% of the organic material is collagen fibers. Scanning electron microscopy suggests that each lamella consists of several layers of collagen fibers, separated by a thin layer of ground substance. Each lamella has a predominant fiber orientation, which differs in adjacent lamellae. The major mineral component is hydroxyapatite, most of which occurs as crystals having plate or needle shapes. The crystals are deposited at specific sites on the collagen (White et al. 1977) and their growth is aligned with the fiber, the crystallographic *c* axes being parallel to the fiber. Bone from which organic material has been removed retains its shape and some of its strength, which suggests that the mineral phase is continuous. Bone is a living material and cells, called osteocytes, are found in lacunae between the lamellae.

In the shaft of long bones the osteons are packed in parallel array, with their long axes approximately directed along the shaft of the bone. The lamellae on the outer and inner surfaces of the shaft do not form Haversian systems but are parallel to the bone surface. The Haversian canals of adjacent osteons are connected by lateral Volkmann's canals. The canals and the lacunae contain fluid which transports nutrients and waste products, and are also free to move in response to stress. The open-textured cancellous bone at the ends of the long bones consists of trabeculae made up of rods and plates.

The mechanical properties of bone have been extensively investigated. The aims have been to relate the mechanical properties to the composition of bone and structure of bone in health and disease, to provide basic information on the mechanics of skeletal support and locomotion, and to provide parameters for the analysis of the effects of abnormal loads, and the presence of implants.

1. Mechanical Properties

Bone, like other biological materials, shows wide variation in composition and structure, and this is reflected in the range of values of the mechanical parameters reported in the literature. The conditions under which the specimens are prepared and tested are important, for example, dry bone has a significantly higher strength than wet bone (Evans 1957).

Bone shows a linear load–deformation response, from which a modulus of elasticity may be calculated; yield typically occurs at strains of $\sim 1\%$. After yield, the bone undergoes plastic deformation, followed by failure. Time-dependent behavior is also observed.

1.1 Load–Deformation Behavior

The structural alignment of osteons in compact bone is reflected in anisotropy of mechanical properties. The modulus of elasticity in the longitudinal direction is about 1.5–2 times that measured tangentially. There is less information on the radial modulus, but it probably lies between the values for the other directions. It has been suggested that the anisotropy is such that the bone can be considered as orthotropic (Ashman et al. 1984).

The value of the longitudinal modulus for slow tensile tests on wet specimens commonly falls in the range 15–20 GPa. Compression tests show a much wider variation; Fung (1981) ascribes the variability to the nonhomogeneous, anisotropic composite structure of bone, but it may be due to the difficulty inherent in compression tests (Swanson 1971).

Anisotropy is also apparent in the shear modulus. Specimens in torsion along the axis of the bone have a shear modulus of about 5–6 GPa, but tangential specimens are about twice as stiff (Bonfield and Li 1967). The load–deformation response of cancellous bone consists of an initial linear phase, followed by deformation at almost constant load and finally an increasingly stiff response (Gibson 1985). It is both weaker and less stiff than cortical bone.

1.2 Time Dependence

Strain-rate dependence has been demonstrated in both tension and compression. The elastic modulus becomes higher with increasing strain rate, the modulus being a logarithmic function of the rate. Ultrasonic methods have been used to demonstrate anisotropy and produce values of the modulus of elasticity which may be in excess of 20 GPa.

Both stress relaxation and creep have been demonstrated. The data of Lakes et al. (1979) suggest that the isochronous relaxation modulus is strain dependent, implying nonlinear viscoelastic behavior. The dynamic response also shows time-dependent effects. Specimens tested over the frequency range 0.002–100 Hz showed progressive changes in modulus and phase angle.

1.3 Failure

The values of the failure parameters reported in the literature show a wide variation.

The stress at failure appears to be directionally dependent. For wet bone the stress at failure for both tension and compression is about 100 MPa in the longitudinal direction, but the radial and tangential strength is less. The strength is rate dependent, with the stress at failure increasing with increasing strain rate.

Impact strength has been investigated using Charpy-type tests in which the energy lost by a weighted pendulum in fracturing the specimen is measured. The energy absorbed per unit cross-sectional area is about 14 kNm m^{-2} (Tsuda 1957). There is also evidence that longitudinal specimens are tougher than tangential specimens. Bone is highly notch sensitive (Swanson 1971), although the root radius of the notch appears to have little effect.

2. Models of Bone

The mechanical properties of bone depend on its composition and structure, and strength and stiffness increase with increasing mineralization. Attempts have been made to model the elastic behavior of bone using composite theory.

The modulus of bone has been calculated using Voigt (equal strain in each component) and Reuss (equal stress) models. The resulting predictions gave upper and lower bounds of the modulus–mineral content relationship which are widely separated, the experimental results falling between them. A combined model produces better agreement between experiment and prediction, but this may be due to the presence of an additional adjustable parameter (Piekarski 1978). A microstructural model based on a near-hexagonal array of osteons leads to predictions of the angular variation of modulus which closely simulates the experimental behavior of bone (Katz 1980).

Bone is a living material, which will respond to force by modifying its composition. A model incorporating such adaptation has been developed and applied to the problem of the response of bone to a medullary pin (Cowin and Van Buskirk 1979), but experimental validation is still lacking.

3. Stress Analysis

Bones often have a complicated shape and structure which makes analytical stress analysis difficult, but they may be incorporated into finite element models. Analyses include the determination of stresses in intact bones and joints, and the effect of fixation devices and implants (Gallacher et al. 1982).

See also: Soft Connective Tissues: Mechanical Behavior; Biomechanics

Bibliography

Ashman R B, Cowin S C, Van Buskirk W C, Rice J C 1984 A continuous wave technique for the measurement of the elastic properties of cortical bone. *J. Biomech.* 17: 349–61

Bonfield W, Li C H 1967 Anisotropy of nonelastic flow in bone. *J. Appl. Phys.* 38: 2450–55

Cowin S C, Van Buskirk W C 1979 Surface bone remodelling induced by a medullary pin. *J. Biomech.* 12: 269–76

Evans F G 1957 *Stress and Strain in Bones: Their Relation to Fractures and Osteogenesis.* Thomas, Springfield, Illinois

Fung Y C 1981 *Biomechanics: Mechanical Properties of Living Tissues.* Springer, New York

Gallacher R H, Simon B R, Johnson P C, Gross J F 1982 *Finite Elements in Biomechanics.* Wiley, New York

Gibson L J 1985 The mechanical behaviour of cancellous bone. *J. Biomech.* 18: 317–28

Katz J L 1980 The structure and biomechanics of bone. In:

Vincent J F V, Currey J D (eds.) 1980 *The Mechanical Properties of Biological Materials.* Cambridge University Press, Cambridge, pp. 137–68

Lakes R S, Katz J L, Sternstein S S 1979 Viscoelastic properties of wet cortical bone—1. Torsional and biaxial studies. *J. Biomech.* 12: 657–78

Piekarski K 1978 Structure, properties and rheology of bone. In: Ghista D N, Roaf R (eds.) 1978 *Orthopaedic Mechanics: Procedures and Devices.* Academic Press, London, pp. 1–20

Swanson S A V 1971 Biomechanical characteristics of bone. In: Kenedi R M (ed.) 1971 *Advances in Biomedical Engineering,* Vol. 1. Academic Press, London, pp. 137–87

Tsuda K 1957 Studies on the bending test and impulsive bending test on human compact bone. *Kyoto Furitsu Ika Daigaku Zasshi* 6: 1001–15

White S W, Hulmes D J S, Miller A, Timmins P A 1977 Collagen–mineral axial relationships in calcified turkey leg tendon by x-ray and neutron diffraction. *Nature (London)* 266: 421–25

J. C. Barbenel

Brachytherapy

Brachytherapy is radiotherapy using radioactive sources placed close to or inserted into the patient. It may be subdivided into three types of treatment: interstitial, intracavitary and surface. In interstitial therapy the sources are introduced surgically into the tissues, whereas in intracavitary therapy they are placed in natural body orifices. If the disease is superficial, surface therapy may be used, the sources being distributed over a surface applicator which is applied to the area to be treated.

One advantage of brachytherapy is that, because of the short source-to-tumor distance, the tumor is treated with the minimum irradiation of surrounding normal tissue, thus enabling a high tumor dose to be given without risk of morbidity. Further, the biological effectiveness of the radiation is enhanced because it is delivered at a constant low dose rate.

The method has its limitations in that some tumor sites are inaccessible. Another disadvantage is that personnel who manipulate the sources are exposed to radiation, although this hazard can be significantly reduced by using afterloading techniques (see *Radiotherapy: Afterloading Techniques*).

1. Sealed Sources

The sealed sources currently in common use are ^{226}Ra, ^{137}Cs and ^{90}Sr. ^{198}Au, ^{182}Ta and ^{192}Ir (in solid form) are similarly employed. A further isotope used in a sealed source gaining popularity is ^{125}I. The iodine is incorporated in a resin and encapsulated in titanium. The low γ-ray energy it emits (27–35 kV) simplifies protection procedures, even the use of protective gloves being feasible with this energy. Because of high attenuation in tissue the dose is also highly localized, and iodine seeds can be used in permanent implants even though the half-life is 60 days. The physical properties of the isotopes used in brachytherapy are summarized in Table 1.

Table 1
Physical properties of sealed sources commonly used in brachytherapy

Isotope	Half-life	Radiation	Energies[a] (MeV)
^{226}Ra	1620 years	β, γ	0.65–1.17
^{137}Cs	30 years	β, γ	0.66
^{198}Au	2.7 days	β, γ	0.41
^{182}Ta	115 days	β, γ	1.12
^{192}Ir	74 days	β, γ	0.32
^{90}Sr	28 years	β	0.54, 2.3
^{125}I	60 days	γ	0.027

a Only the energies of the principal, therapeutically useful γ rays are given, except for ^{90}Sr which emits only β particles

1.1 Radium-226

Radium was for many years the principal isotope used in brachytherapy (see *Radium in Medicine: Early History*). Sealed in metal housings, radium sources are emitters of γ radiation, and are available as needles or tubes. Needles are employed primarily for interstitial therapy and have a sharp point at one end to facilitate insertion and an eyelet at the other to which a suture is attached for retrieval, usually a few days after insertion. Tubes have slightly rounded or flat ends and may have an eyelet.

The main parameters which describe a radium source are:

(a) activity, the total radioactive content of the source, usually expressed in milligrams of radium;

(b) total length, the overall length of the source;

(c) active length, the length occupied by the radioactive material;

(d) linear activity (sometimes called line density), the radium content per unit active length, normally expressed in milligrams per centimeter; and

(e) filtration, the total wall thickness of the source.

Radium decays to a gas, radon (half-life 3.8 days), which in turn decays through a series Ra A (^{218}Po), Ra B (^{214}Pb), Ra C (^{214}Bi), etc., until a stable isotope of lead is reached. Of this series only Ra B and Ra C emit γ rays of sufficient energy to be useful in therapy. Thus radium sources really contain radium in radioactive equilibrium with its decay products. To preserve this equilibrium the source must be sealed. Since radon is expelled during the manufacturing process, the disintegration products Ra A, Ra B, Ra C (half-lives 3.05, 26.8, 19.7 min, respectively) are lost and it is therefore necessary for radioactive equilibrium to be reestablished before the source is calibrated and brought into use. A period of

approximately 35 days is required; after this time the radon and Ra C are present to 99.8% of their equilibrium value.

The radium used in needles and tubes is in the form of radium sulfate mixed with zirconium phosphate powder. The mixture is doubly encapsulated in iridium–platinum, the inner capsule being termed the "cell" and the outer capsule the "sheath" (Fig. 1). The cells are

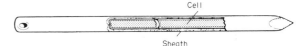

Figure 1
Structure of a radium needle

sealed by brazing and the sheaths by welding. The finished sources are rigorously tested after manufacture to ensure the absence of surface contamination and radon leakage. Uniformity of distribution is checked by autoradiography. A certificate of measurement is issued, stating the content in milligrams of radium. Each appliance is engraved with an appropriate identification and serial number.

Dividing the radium in cells has two advantages. First, there is less change in the overall active length as the contents settle. Second, in the event of breakage the spillage of radioactive material is reduced. The high toxicity of radium combined with its long half-life and gaseous decay product represent a serious hazard, one reason for the declining popularity of radium.

Types of needles approved for use in UK hospitals have elongated countersunk eyelets, and trocar points in 25% iridium–platinum. The cells are of 10% platinum and have a wall thickness of 0.2 mm; all needle sheaths are of 20% iridium–platinum. The nominal content of needles varies from 0.5 mg to 5.0 mg radium, distributed in one, two, three or four cells over an active length of 15–50 mm. Common values of linear activity are 0.33 and 0.66 mg cm^{-1}.

North American types of needles have bodkin eyelets, and trocar points in 25% iridium–platinum. Their cell walls are 0.2 mm, 10% iridium–platinum and their sheath 0.3 mm, 20% iridium–platinum. Their loadings vary from 0.25 mg cm^{-1} to 10 mg cm^{-1}.

1.2 Cesium-137

^{137}Cs has largely replaced radium in brachytherapy. As a fission by-product, ^{137}Cs is cheap compared with radium, which is difficult to extract from limited natural resources. Cesium is much less toxic than radium and has no gaseous decay products. Having a half-life of about 30 years its activity decreases by 2.3% per annum. Since its main use is as a radium substitute, the sources, encapsulated in iridium–platinum or, latterly, stainless steel, are constructed in exactly the same way. Activities are normally specified in terms of milligrams of radium equivalent. The slightly lower energy of the γ rays emitted by cesium makes little difference, therapeutically, although less lead is needed for shielding, greatly reducing the radiation protection problem.

1.3 Gold-198

Point sources of ^{198}Au are useful for interstitial treatments of relatively inaccessible sites such as the bladder, since its short half-life allows the sources to be permanently implanted. Two types of sources are available in the UK, grains (cylinders of length 2.5 mm and diameter 0.8 mm) and seeds, roughly twice as large. Both are encapsulated in 0.15 mm of platinum to absorb the β radiation and are produced by the neutron irradiation of stable ^{197}Au in a nuclear reactor. Activities of up to 50 mCi are available. In the USA and Canada, similar sources are available of length 2 mm and diameter 0.5 mm. ^{198}Au superseded radon for short half-life, permanent implants many years ago.

1.4 Tantalum-182

The use of ^{182}Ta is decreasing in radiotherapy. Since its half-life is much longer than the usual treatment time (Table 1), it is never used as a permanent implant. The usual form is a "hairpin" shape made from flexible wire, which can be introduced into the bladder and removed at the end of the treatment via a catheter.

1.5 Iridium-192

^{192}Ir is used in the same way as ^{182}Ta, and preferred by some radiotherapists because the energy of its γ rays is high enough for bone absorption and tissue attenuation of the radiation not to be a problem, but low enough to allow easier shielding than for tantalum.

1.6 Strontium-90

^{90}Sr is a pure β emitter and is useful in surface treatments where only very shallow penetration of radiation is required, for example in treating the cornea of the eye. Sources are usually made by bonding the ^{90}Sr to silver foil which is mounted on the concave side of a spherical surface applicator. The foil is then covered with enough polythene to remove the low-energy (0.54 MeV) electrons which are also produced in the decay mode. Polythene is chosen because it has a low atomic number, x-ray production in the polythene due to electron interaction thus being minimal. For a typical applicator, the dose to the lens of the eye is less than 20% of that to the cornea.

2. The Paterson–Parker Rules

The Paterson–Parker rules answer two vital questions: "How should sources be distributed to produce a uniform dose?" and "What dose is obtained?"

For surface irradiations the sources (usually radium or cesium tubes, gold grains, or iridium wire) are mounted at a short distance from the skin surface. (The treating distance h is usually 0.5–1.5 cm.) The rules describe the arrangement of sources for different ratios of treated area to treating distance so that the most uniform dose distribution is obtained. Dose homogeneity of better than $\pm 10\%$ can be achieved. The sources are fixed rigidly in position by mounting them on plastic (sticking plaster for gold grains) accurately molded to the body part to be treated and of a thickness equal to h. Low atomic-weight materials are used in order that absorption and scatter of radiation will be similar to that in tissue. Such a construction is known as a mold, and is circular or rectangular in shape. The rules also take into account curvature of the treated area. A typical mold applied to the back of the hand is shown in Fig. 2.

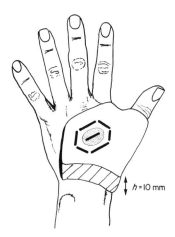

Figure 2
Typical mold for the back of the hand with treating distance $h = 10$ mm. Six sources are arranged to form a circle with one source at the center according to the Paterson–Parker rules

Associated with the distribution rules are families of curves which show, for various values of h, how the dose delivered to the treated surface is related to the area treated, the amount of radium or other isotope on the mold, and the time for which it is applied (Fig. 3). In practice, the dose required and the treatment time are specified first, and the source activity is then calculated. Much clinical experience acquired over the years using the original röntgen unit (R) has made radiotherapists reluctant to allow the introduction of SI units into brachytherapy.

The dose to tissue from a simple planar mold decreases with depth, but the effect may be lessened and greater penetration achieved by increasing the treating distance h. Where access can be gained to both sides of the tissue to be treated (e.g., the pinna or the lip), then a more uniform dose may be obtained by placing sources on both sides, the so-called sandwich mold.

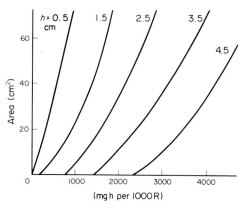

Figure 3
Dosimetry data for planar implants and molds

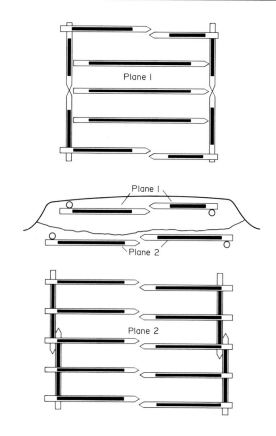

Figure 4
Arrangement of needles in a typical implant

Intracavitary sources are used for treating the body and cervix of the uterus, and the vagina. Cesium or radium tubes are normally arranged as three line sources; special Paterson–Parker rules provide details of the distribution and dose delivered.

In interstitial therapy, needles are inserted directly into the tissues to be treated. A complete array of needles is known as an implant. A single planar implant will treat a slab of tissue 10 mm thick. Lesions up to 25 mm thick can be treated with two parallel planes, as shown in Fig. 4. Larger tumors are treated by distributing the sources throughout the treated region—a volume implant. Two shapes are commonly used, a cylinder (with cesium or radium needles) and a sphere (obtainable only with point sources such as gold grains). Again, particular Paterson–Parker rules give the required distribution of sources and allow dose calculation.

Post-implantation radiographs enable the actual distribution of sources to be found. Unlike molds, practical difficulties preclude the arrangement of sources exactly as specified by the rules. Modern computer techniques allow detailed dose distributions around actual implants to be calculated.

The dose close to each implanted source is of course very high, thus the overall dose variation is much greater than with molds. These local "hot spots" seem to have little effect clinically, however, and are ignored in assessing the effective dose.

Some sites such as the breast require large areas to be implanted with flexible sources. Iridium wire is used to afterload the implant. Usually only one line density of wire is employed and so distributions slightly different from those for the common type of implant are needed.

It should be noted that there are many alternative dosage systems for cervical treatments other than the Paterson–Parker (Manchester) system (see, for example, Fletcher 1966, Walstam 1954), and other systems have been developed to meet the requirements of afterloading interstitial techniques, for example, that of Pierquin et al. (1978) for iridium wire.

3. Future Developments

The use of molds will probably decline as electron therapy becomes more widely available. Intracavitary treatments are well established and remain unchallenged by other therapeutic modalities. Implants will remain for the forseeable future, but with greater emphasis placed on afterloading to allow more accurate positioning and less radiation exposure to the staff. Another step toward the improvement of interstitial brachytherapy might be the use of neutron sources such as ^{252}Cf which produce radiation of different relative biological efficiency compared with γ rays, and which should offer some therapeutic gain.

See also: Radionuclides: Clinical Uses; Radiotherapy: Beta Particles; Radiotherapy: Treatment Planning

Bibliography

Fletcher G F H (ed.) 1966 *Textbook of Radiotherapy*. Lea and Febiger, London

Hilaris B S (ed.) 1975 *Handbook of Interstitial Brachytherapy*. Publishing Sciences Group, Acton, Massachusetts

Meredith W J 1967 *Radium Dosage: The Manchester System*, 2nd edn. Churchill Livingstone, Edinburgh

Pierquin B, Dutreix A, Paine C H, Chassange D, Marinello G, Ash D 1978 The Paris system of interstitial radiation therapy. *Acta Radiol. Oncol., Radiat. Phys., Biol.* 17: 33–48

Walstam R 1954 The dosage distribution in the pelvis in radium treatment of carcinoma of the cervix. *Acta Radiol.* 42: 237–50

A. M. Perry

Brain-Stem Electric Response Audiometry

Brain-stem electric response (BSER) audiometry is a form of electric response audiometry (see *Electric Response Audiometry*) employing response components originating in the auditory nerve and brain stem. The BSER may be recorded from noninvasive electrodes on the mastoid and vertex, and its principal components occur within 10 ms of the onset of the acoustic stimulus. The technique has applications both in the estimation of the behavioral audiometric threshold, which can be achieved with a degree of accuracy comparable with other available electric responses, and in the determination of the site and nature of a lesion in the auditory pathway. The ability to elicit responses from the cochlear nerve and nuclei in the auditory brain stem has important applications in the diagnosis of brain-stem disorders, particularly in neurotology.

See also: Audiometers; Electroencephalic Audiometry; Objective Audiometry

Bibliography

Gibson W P R, 1978 *Essentials of Clinical Electric Response Audiometry*. Churchill Livingstone, London

S. Gatehouse

C

Cancer Statistics

Although medical physicists are more usually associated with the fields of radiotherapy, nuclear medicine, bioengineering, ultrasonics, electronics and computer science, they are also able to extend their sphere of influence to include cancer statistics. In particular, with their experience of radiotherapy physics they are well placed to give constructive advice on, for example, methods of calculating survival rates following radiotherapy, surgery or chemotherapy, and on the design of clinical trials to study the possible differences between two treatments. Statistical problems are also present in nuclear medicine and radiology when diagnostic images have to be assessed, in the production of meaningful and reliable management information statistics such as radiotherapy treatment workload figures, and in the design and operation of hospital cancer registries. Routine enquiries concerning the use of the chi-squared tests and t-tests and on the meaning of $p < 0.05$ will also be encountered within a medical physics department, although not related only to cancer statistics. Standard explanations of these basic tests to determine statistical significance can be found in Bourke and McGilvray (1975), Mould (1976) and Swinscow (1980).

1. Patient Data

Collection, storage and retrieval of data must be considered thoroughly before embarking on any statistical study, whether it be a retrospective survey, planned prospective clinical trial, or the setting up of a cancer patient registry. Data presentation should also be well thought out, and the physicist should be aware that as well as graphs and tables as aids to presentation, there are also pie diagrams, pictograms, bar charts, histograms and dot diagrams. An example of the latter is given in Fig. 1 for 56 patients treated for disseminated melanoma. It is seen that in a relatively small space a large amount of information can be displayed for instant viewing. The data include treatment group, sex, survival time in months, stage of the disease, whether patients are alive or dead and if alive, whether the disease is still present. By the use of further symbols, and of color, the dot diagram in Fig. 1 can be extended to include more patient details, providing a very convenient method of data presentation.

Data collection methods vary widely throughout the world and within individual countries. Some hospitals have departments of biostatistics, epidemiology and community medicine, with trained staff to abstract patient data from case notes and to undertake patient follow-up studies. Others may rely on a peripatetic clerk who visits at regular intervals from a central bureau, or on a single permanent member of staff. Data storage and retrieval systems vary just as widely, from the most elegant and sophisticated computer methods to a simple exercise-book system inherited from the 1930s and in existence with little change for some 50 years. The optimum method for data storage and retrieval depends

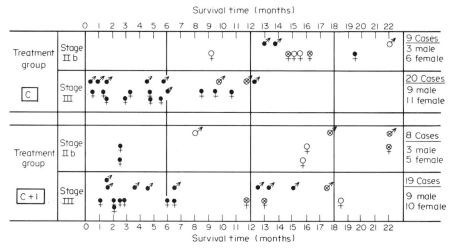

Figure 1

Dot diagram for 56 cancer patients: ●, dead; ⊗, alive with melanoma present; ○, alive with no sign of recurrence (after Newlands et al. 1976)

on the statistical study being undertaken, whether less than 100 patient records are to be analyzed, 500–3000 new patients per year are to be registered at a given hospital, or whether several thousands of case histories are to be stored annually for a regional or national system. If there is to be anything other than a rather elementary analysis of the data required, a computer database is a prerequisite. This is not to say that a mainframe or a minicomputer is always necessary, since in recent years the capability of the microcomputer has increased and such a machine is usually available in a hospital. Figure 2 shows the storage and retrieval system, based on a Hewlett-Packard 85 microcomputer, which is used for the Westminster Hospital cancer registry (Mould 1983b). Data analysis programs are also available. Figure 2 is given only as an example because the criteria for data storage and retrieval systems vary with the type of study, the funds available and the existing computer facilities in the hospital.

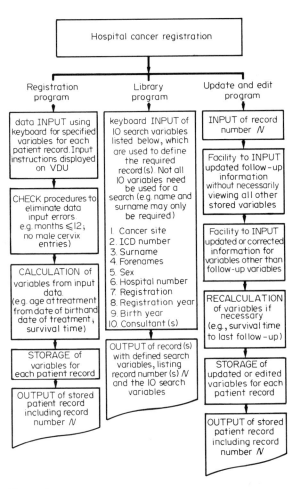

Figure 2
Flow diagram of a patient data storage and retrieval system

Table 1

Patient data requirements which must be considered for a hospital cancer registry

Data type	Requirement
Name and status and personal details	Surname Forenames Sex Maiden name—if female Date of birth Country of birth Occupation
Address	House/flat number Street/road Town/city and region/country Postcode/zipcode Does patient live within local catchment area?
Tumor description	Cancer site ICD (international classification of disease) number Histology Was histology verified? Degree of differentiation Stage
Tumor history	Symptoms Symptom duration Family cancer history Was patient previously treated? Was there a previous independent primary cancer? Details of previous cancer, if any
Treatment at hospital of registration	Was treatment given? Treatment details including board grouping of radiotherapy and/or surgery and/or chemotherapy Consultant(s) Anniversary date (date of first planned treatment) Age at anniversary date Hospital from which patient was referred, if any Hospital to which patient now referred, if any
Patient reference numbers	Hospital case note number National cancer registration number NHS number (in UK)
Follow-up	Date of follow-up and patient state (retain a series of follow-ups for estimate of recurrence, if one occurs) Date of death Cause of death Details of any subsequent independent second primary cancers Survival time, apparent disease-free time and terminal disease time, if appropriate Was a post mortem undertaken? Metastatic details

The specification of a patient data bank also varies, but there is a certain basic data block which is always required—name and address, cancer site, patient reference numbers, description of treatment and date of birth, for example. Table 1 lists data items to be considered when devising a hospital cancer registry, although not all form the basic data block. Whether the histology was confirmed and whether the patient was treated are most important items of information since it is not unknown for publications to quote survival data for a series of patients containing treated and untreated cases, in which case any meaningful comparison with another series is impossible. Without proper histological confirmation of cancer, a proportion of those patients followed-up and found to be alive with no sign of recurrence might not have had cancer in the first instance. The bias this would cause is obvious. Reference numbers are most important for tracing purposes and in the UK the National Health Service Number is especially useful (Mould 1980). It is also most important to determine whether or not the patient was previously treated, because survival analysis of a group of patients containing an unknown proportion of cases whose survival time is calculated from the date when a recurrence was treated will underestimate the efficacy of the treatment. Occupation is important for epidemiological studies, but unfortunately in many instances, particularly in the UK, this information is not stated in the case notes. Further, an individual may have several different occupations during his lifetime, thus the specification of a single occupation may be misleading. Although the stage of disease is an important prognostic factor in cancer and although international tumor-staging classifications have been published, many patients are not staged, and there are also a large number of different hospital staging systems. This makes comparison between various centers rather difficult. Therefore, if stage is to be stored, it must be unequivocally defined.

2. Cancer Incidence, Prevalence and Mortality

An incidence rate refers to the number of cases of disease diagnosed in the population during a given period of time (usually per annum). A prevalence rate refers to the number of cases at a given point in time or during a given period who are alive with disease present. A mortality rate refers to the number of cases in the population who died during a given period of time (usually per annum). Very few prevalence rates are published but incidence and mortality rates are regularly published in reports of the Office of Population Censuses and Surveys (OPCS) in London, of the National Cancer Institute (NCI) in Bethesda and of the International Agency for Research on Cancer (IARC) in Lyon. In particular, the IARC series of Cancer Incidence in Five Continents (Waterhouse et al. 1976) provides a wealth of data on incidence rates.

Patterns of incidence and mortality differ markedly for different cancer sites and different countries. For further details see the publications of the OPCS, NCI and the IARC and Mould (1983a). It should be noted that a high incidence of disease does not automatically imply a high mortality. For example, skin cancer has a high incidence, but since this cancer (excluding melanoma of the skin) can be successfully treated, the mortality rate is low. Since mortality rates and incidence rates are usually quoted per 10^5 or per 10^6 population, these rates are calculated from regional or from national data and not from a hospital cancer registry where the sample sizes are too small.

Mortality and incidence rates are sometimes expressed as standard rates, the rates being standardized to a given registration year. For England and Wales, using 1968 as the standard registration year, the standardized registration ratio (SRR) for lung cancer in males will be by definition $SRR_{1968} = 100$. The increase in male lung cancer between 1962 and 1977 can then be demonstrated in terms of the SRR values, showing a steady increase of $SRR_{1962} = 87$ to $SRR_{1977} = 115$. For female lung cancer in England and Wales the increase is more dramatic with $SRR_{1968} = 100$ rising to $SRR_{1977} = 166$. This data is published in the OPCS Series MB1 reports for which No. 1 is Cancer Statistics Registrations (1971) and No. 8 is Cancer Statistics Registrations (1977).

Age-standardized incidence rates and age-standardized mortality rates are also published in which the rates are expressed relative to a defined standard population (Armitage 1977, Mould 1983a). Standard populations for the World, Africa and Europe are to be found in the IARC publication by Waterhouse et al. (1976).

Annual incidence rates for some of the major cancers in England and Wales are given in Table 2. This demonstrates the high incidence of lung cancer in males in England and Wales, 118.5 per 10^5 population and the fact that the incidence for some cancers is markedly different between the sexes (see data for bladder cancer). Lung cancer in males is also high in other parts of the world although not as high as in England and Wales. As

Table 2

Crude annual incidence rates, England and Wales, 1977

	Incidence rate per 10^5 population	
Cancer site	Males	Females
All sites	405.7	368.2
Lung	118.5	30.5
Breast	0.9	85.6
Stomach	30.7	20.2
Cervix uteri		15.9
Prostate	32.2	
Bladder	24.9	9.0
Skin[a]	43.4	35.9

a Excludes melanoma of skin   Source: Office of Population Censuses and Surveys 1982 Series MB1, No. 8

examples, the maximum crude annual incidence rate for any cancer site is that for lung cancer in the populations of San Francisco, Detroit and Denmark, the Singapore Chinese and New Zealand Maori. For Japanese male populations, the site with the highest cancer incidence is the stomach; for males in Sweden and Norway it is the prostate gland (Waterhouse et al. 1976). For females in the USA, England and Wales, Scandinavia and Israel, the most commonly occurring cancer is breast cancer, whereas in Bombay, Ibadan, Bulawayo (Africa) and Jamaica it is cancer of the cervix uteri (Waterhouse et al. 1976). This briefly illustrates the wide variations in incidence between cancer sites and between countries.

The differences in incidence are due to many factors, not all of which have been identified. In Hirayama et al. (1980) there is an excellent summary of cancer risks for the different sites by histology; prognosis; patterns of occurrence (time trends, international variations, variation within countries, migration); host factors (sex, age, genetic predisposition, precancerous lesions, predisposing morbid conditions, multiple primary neoplasms); and by environmental factors (socio-economic status, tobacco, drugs, alcohol, diet, radiation, occupation, air pollution, sex life and pregnancy, biological agents).

Two of the most well-known cancer risk factors are tobacco and ionizing radiation. The former is well documented in papers by Doll and others, particularly for the survey on British doctors. The latter is dealt with in many articles on radiation protection, and in the follow-up reports on the atomic bomb survivors of Hiroshima and Nagasaki. Data have been summarized by Mould (1983a) for leukemia mortality statistics for American radiologists, British radiologists and atomic bomb survivors, and for the distribution of induction periods for radiation-induced cancer in man for 34 cases of histologically proven cancer which were reported in the literature before 1957. The mean time lapse between irradiation and diagnosis of cancer was 30 years. It was found that there was a similar distribution of induction periods for asbestos-induced lung cancer in man.

Mortality from cancer in England and Wales, in comparison with mortality from other selected causes of death, is given in Table 3. Cancer is the second most common cause of death after heart disease. Occupational mortality in England and Wales is given in a supplement every ten years, following a decennial census, and the latest of these supplements for 1970–1972 (Series DS No. 1) was published by the OPCS in 1978. Annual mortality statistics for England and Wales are also published by the OPCS and are presented for cause of death (Series DH2) and for areas of England and Wales (Series DH5). Whereas survival-rate figures usually refer to the patients who have been treated and are usually grouped by treatment method, mortality statistics include all persons whether treated or untreated. Table 4 presents cancer mortality rates for the selected cancer sites in Table 2.

Table 4
Crude mortality rates, England and Wales, 1980

Cancer site	Mortality rate per 10^6 population	
	Males	Females
All sites	2899	2416
Lung	1117	332
Breast		482
Stomach	267	178
Cervix uteri		82
Prostate	210	
Bladder	123	51
Skin	8	7

a Excludes melanoma of skin   Source: Office of Population Censuses and Surveys 1982 Series DH2, No. 7

3. Cancer Survival

One of the most important factors affecting survival is the stage of the disease at presentation for treatment, since the earlier the stage the better the patient's prognosis. This is demonstrated for cancer of the cervix uteri in Fig. 3. Five-year survival rates for England and Wales for patients registered in 1975 and for the selected cancer sites in Tables 2 and 4, are given in Table 5.

It is important to avoid bias when quoting statistical rates. It would (for example) be incorrect to assume that one group of patients given treatment A who are all under 30 years of age have a better survival rate than another group of patients given treatment B who are all over 75 years of age, simply because the first group underwent treatment A. The age difference between groups must be taken into account. This is done by calculating a relative (age-corrected) survival rate, which is 100 multiplied by the ratio of the uncorrected (crude) survival rate and the expected survival rate in the normal population for a group of people with the same age and sex distribution as the treated group. Since the age distribution for the different cancers is not always similar, it is always better to consider the relative survival

Table 3
Crude mortality rates, England and Wales, 1980

Cause of death	Mortality rate per 10^6 population	
	Males	Females
Heart disease	5965	5832
Cancer	2899	2416
Respiratory disease[a]	1765	1626
Road accidents	181	71
Suicide	110	67
Diabetes mellitus	83	110
Tuberculosis	17	8
Multiple sclerosis	11	18

a Excludes lung cancer   Source: Office of Population Censuses and Surveys 1982 Series DH2, No. 7

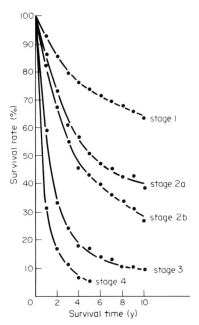

Figure 3
Survival rates for cancer of the cervix calculated using a life-table method (after Mould and Staffurth 1979)

Table 6
Crude and relative five-year survival rates, male population, England and Wales, 1975

Cancer site	Five-year survival rate	
	Crude	Relative
Tongue	24.1	31.9
Prostate	21.8	35.1
Bladder	39.0	54.2
Testis	63.9	67.3
Brain	10.9	12.2
Lung	5.3	7.0

Source: Office of Population Censuses and Surveys 1982, Series MB1, No. 9

rates than the crude rates. Table 6 compares crude and age-corrected survival rates for selected cancers. Because cancers of the tongue, prostate and bladder occur at a greater mean age than those of the testis, brain and lung, there is a larger difference between the crude and relative rates for the first three cancer sites.

The calculation of a survival rate for a group of cancer patients is one which a physicist might be expected to make. The method of calculation used for many years was a direct method in which the T-year percentage survival rate was expressed as 100 multiplied by the ratio of the number of patients surviving at least T years to

the number of patients entering for treatment at time $T = 0$. Thus to calculate a 5-year survival rate a follow-up period of five years was essential for the whole group of patients. If some patients had been lost to follow-up between 0 and 5 years subsequent to treatment, these cases were discarded from the analysis. Such a method loses useful information if some of the patients are known to have survived three years before being lost to follow-up. The life-table (actuarial) method of survival-rate calculation described by Mould (1976) and others, and the Kaplan–Meier method described by Schwartz et al. (1980) and others, do not have this disadvantage and use all the available information. The Kaplan–Meier method uses the precise individual survival times of the patients, whereas the life-table method uses survival time intervals, usually annual, and groups the data. In this method, the probability p_i of survival in each annual interval i is computed, where $p_i = (1 - q_i)$ and q_i is the probability of dying in the ith interval. The probability of surviving to the end of the ith interval (T is the end-point of the ith interval) is given by the product of the p_i values. Thus:

$$P_T = (1 - q_1).(1 - q_2).(1 - q_3)....(1 - q_i)$$

The percentage T-year survival rate is therefore $100\ P_T\%$.

Prediction models have also been proposed in which the T-year survival rate can be estimated without any patients being observed to year T. This enables long-term survival rates such as 10-year and 15-year rates to be predicted, and also indices of "cure"—such as in the lognormal model (Boag 1949). However, the several parametric statistical models described in the literature have not been fully tested for all cancer sites and they should be used with caution. One exception is the lognormal model for cancer of the cervix, which has been exhaustively tested using staged data (Mould and Boag 1975). This model has the basic assumption that the distribution of survival times of carcinoma cervix patients who die with their disease present, primary or metastatic, can be represented by the lognormal distribu-

Table 5
Crude five-year survival rates, England and Wales, 1975

Cancer site	Five-year survival rate	
	Males	Females
All sites	22.2	34.5
Lung	5.3	5.5
Breast	47.3	49.1
Stomach	5.0	5.4
Cervix uteri		46.7
Prostate	21.8	
Bladder	39.0	36.4
Skin[a]	71.8	75.6

a Excludes melanoma of skin   Sources: Office of Population Censuses and Surveys 1982 Series MB1, No. 9

tion. The model has three parameters: M and S of the lognormal distribution

$$Y = \frac{1}{\sqrt{2\pi}\,t}\,e^{-x^2/2}$$

where t is the survival time and $x = (\log t - M)/S$; and a third parameter C, which represents the proportion of patients "cured," that is, those patients who do not die with their disease present. The parameter values M, S and C can be solved by maximum likelihood, but if the value of S can be fixed a priori, as it can for cancer of the cervix, the model becomes more stable than if all three parameters are unknown. Typical values of the index of cure C for series of cancer of the cervix patients obtained from the Middlesex, Royal Marsden, Chelsea Hospital for Women, University College, Hammersmith and Christie Hospitals and from the Norwegian Radium Hospital, Oslo, are given in Table 7.

Table 7
Index of cure for cancer of the cervix calculated using the lognormal model with a constant $S = 0.35$

Disease stage	Number of series of patients	Range of estimates of C
1	7	0.57–0.71
2	9	0.36–0.43
3	6	0.14–0.27

Source: Mould and Boag 1976

For many years the five-year survival rate was considered to be the equivalent of a cure rate, but since many patients survive for longer than five years subsequent to treatment and still die with the disease present, a five-year survival rate is clearly not always a good indication of cure. Alternatives are the statement of longer term survival rates such as the 15-year rate, a predicted index of cure such as in Table 7, or the Easson and Russell (1968) definition. They define the occurrence of a cure as "the time after treatment when the annual death rate from all causes is similar to that of a normal population group of the same age and sex distribution." The method involves a graphical representation of the survival experience of the treated group and of the normal population group to determine the time point at which the death rates from both groups become equivalent. It is perhaps best to quote T-year survival rates, which can be stated unequivocally rather than try to attempt to define what is meant by cure since it is unlikely that there will ever be full international agreement on the meaning of this term.

What is of more importance than stating an index of cure, is a quantitative statement of quality of life subsequent to treatment as opposed to only stating the quantity of life—the survival time. This is achieved by dividing the survival times into an apparent disease-free

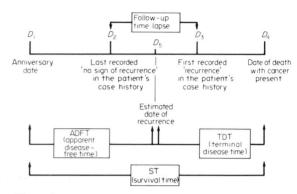

Figure 4
Definition of terms ADFT and TDT for a patient who experiences a remission before recurrence of the cancer and eventual death. The anniversary date D_1 is the date of first treatment. D_5 is the mid-date between D_2 and D_3

time plus a terminal-disease time, where the dividing point is the estimated date of recurrence (Fig. 4). The estimated date of recurrence can be calculated from the last recorded "no sign of recurrence" (NSR) and the first recorded "recurrence" (REC), but care must be taken that the follow-up time lapse is not so large that the estimate is inaccurate. A maximum time difference of 12 months is recommended. Results for cancer of the cervix showed that for stage 1 some 20% of cases experienced an $\text{ADFT} = 0$, that is, no apparent disease-free time subsequent to treatment, that for stage 2 there were some 30% $\text{ADFT} = 0$ cases and for stage 3 there were some 60% $\text{ADFT} = 0$ cases. This data refers to patients treated at several London hospitals, 1944–1962. Quality of life subsequent to treatment can also be expressed using patient-performance scale, of which one example is the Karnofsky status, Table 8.

4. Clinical Trials

Clinical trials in general are discussed in many sources (Schwartz et al. 1980, Pocock 1982, Armitage 1975, Hill 1977). Those specifically designed with respect to cancer have been discussed by Mould (1979).

When planning a clinical trial, unequivocal answers must be available to many questions including those below:

(a) What population is to be studied?

(b) What treatment methods are to be investigated?

(c) What randomization methods are to be investigated?

(d) How is the criterion of successful treatment to be defined and measured and for what improvement in this criterion is it considered worthwhile organizing the trial?

Table 8
Karnofsky status for patient performance

Score	Performance status	
100	normal, no complaints no evidence of disease	able to carry on normal activity; no special care; special care is needed
90	able to carry on normal activity; minor signs of symptoms of disease	
80	normal activity with effort, some signs or symptoms of disease	
70	cares for self, unable to carry on normal activity or do active work	unable to work, able to live at home, cares for most personal needs, a varying amount of assistance is needed
60	requires occasional assistance but is able to care for most needs	
50	requires considerable assistance and frequent medical care	
40	disabled, requires special care and assistance	unable to care for self, requires equivalent of institutional or hospital care; disease may be progressing rapidly
30	severely disabled, hospitalization indicated, although death is not imminent	
20	very sick, hospitalization necessary, active supportive treatment necessary	
10	moribund, fatal processes progressing rapidly	
0	dead	

(e) What level of statistical significance is acceptable when analyzing the results?

(f) Given the number of patients available each year, what is the likely duration of the trial?

Other features which require consideration when planning clinical trails include: design of a written protocol, method of controlling the trial, double blind techniques, end-point evaluation, crossover trial techniques, prognostic factors, trial form design for data entries, statistical analysis, ethical considerations, multicenter trial organization, procedures to adopt in the event of protocol violations, monitoring of the progress of the trial, funding, and publication.

In addition, the trial designer may have to consider whether it is appropriate to use historical controls. This decision should be taken with great care since the possibility of bias from historical data is very great. The organization of the clinical trial office which administers the trial should be well thought out, particulary if the trial is multicenter. It is also most important that patient coding forms for the treatment and follow-up data are uncomplicated, otherwise incomplete data entries may result, ruining the planned final analysis of the trial.

Clinical trials in cancer may consider a T-year survival rate as the indication of success, and the Logrank test for analyzing the results. This latter test is described by Peto et al. (1976, 1977). There are, however, many statistical tests of significance. Those which are nonparametric are well described by Siegel (1956). Sequential analysis designs are described by Armitage (1975) and charts such as that in Fig. 5 are to be found in *Documenta Geigy*.

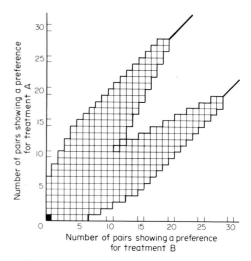

Figure 5
Sequential analysis chart (after Bross I 1952 *Biometrics* 8: 188–205, courtesy Ciba-Geigy Ltd.)

An example of their use is given in Bakowski et al. (1978). Sequential analysis designs can sometimes be appropriate when the success criterion is tumor regression which can be assessed in a relatively short time. The advantage of a sequential trial design is that the trial end-point is determined not by the entry of a fixed number of patients, but by the point in time when one of the chart boundaries is crossed. For a description of the stopping rules see Mould (1979).

5. Future Trends

Physicists' involvement in cancer statistics in the future is likely to include advice on obtaining and analyzing management information statistics. These may include cost-benefit analyses and patient through-puts on a linear accelerator or telecobalt unit which can be used for forward planning for the purchase of capital equipment. Computer-aided diagnosis involving the use of decision trees will also become more widely used in medicine. Collaboration with community medicine specialists and other medical colleagues on features of epidemiological studies (Alderson 1976, Friedman 1974, Lilienfeld 1976) could also develop. Collaboration with medical and administrator colleagues in the field of cancer statistics is to be recommended, since there are many areas (e.g., drug evaluation, survival analysis, mathematical modelling, epidemiology and computer applications) where the physicist with his scientific and mathematical background can make a significant contribution.

See also: Biostatistics; Statistical Methods in Medicine

Bibliography

Alderson M 1976 *An Introduction to Epidemiology.* Macmillan, London

Armitage P 1975 *Sequential Medical Trials*, 2nd edn. Blackwell Scientific, Oxford

Armitage P 1977 *Statistical Methods in Medical Research.* Blackwell Scientific, Oxford

Bakowski M, Macdonald E, Mould R F, Cawte P, Sloggem J, Barrett A, Dalley V, Newton K A, Westbury G, James S E, Hellmann K 1978 Double blind controlled clinical trial of radiation plus razoxane (ICRF 159) versus radiation plus placebo in the treatment of head and neck cancer. *Int. J. Radiat. Oncol. Biol. Phys.* 4: 115–19

Boag J 1949 Maximum likelihood estimates of the proportion of patients cured by cancer therapy. *J. R. Stat. Soc., B* 11: 15–53

Bourke G, McGilvray J 1975 *Interpretation and Uses of Medical Statistics.* Blackwell Scientific, Oxford

Easson E C, Russell M H 1968 *The Curability of Cancer in Various Sites: 4th Statistical Report of the Christie Hospital and Holt Radium Institute, Manchester.* Pitman, London

Friedman G D 1974 *Primer of Epidemiology.* McGraw-Hill, New York

Hill A B 1977 *A Short Textbook of Medical Statistics.* Hodder and Stoughton, London

Hirayama T, Waterhouse J, Fraumeni J 1980 *Cancer Risks by Site.* International Union Against Cancer, Geneva

Lilienfeld A M 1976 *Foundations of Epidemiology.* Oxford University Press, New York

Mould R F 1976 *Introductory Medical Statistics.* Pitman, London

Mould R F 1979 Clinical trial design in cancer. *Clin. Radiol.* 30: 371–81

Mould R F 1980 The importance of the NHS number. *Br. J. Radiol.* 53: 512

Mould R F 1983a *Cancer Statistics.* Hilger, Bristol

Mould R F 1983b The Westminster Hospital microprocessor cancer registry. *Br. J. Radiol.* 55: 897–904

Mould R F, Boag J W 1975 A test of several parametric statistical models for estimating success rate in the treatment of carcinoma cervix uteri. *Br. J. Cancer* 32: 529–50

Mould R F, Staffurth J 1979 Carcinoma of the cervix uteri at the Royal Marsden Hospital, London, 1962–1970: Survival results. *Br. J. Radiol.* 52: 157–58

Newlands E S, Oon C J, Roberts J T, Elliott P, Mould R F, Topham C, Madden F J F, Newton K A, Wesbury G 1976 Clinical trial of combination chemotherapy and specific active immunotherapy in disseminated melanoma. *Br. J. Cancer* 34: 174–79

Peto R, Pike M C, Armitage P, Breslow N E, Cox D R, Howard S V, Mantel N, McPherson K, Peto J, Smith P G 1976 Design and analysis of randomized clinical trials requiring prolonged observation of each patient. I. Introduction and design. *Br. J. Cancer* 34: 585–612

Peto R, Pike M C, Armitage P, Breslow N E, Cox D R, Howard S V, Mantel N, McPherson K, Peto J, Smith P G 1977 Design and analysis of randomized clinical trials requiring prolonged observation of each patient. II. Analysis and examples. *Br. J. Cancer* 35: 1–39

Pocock S 1982 Statistical aspects of clinical trial design. *Statistician* 31: 1–18

Schwartz D, Flamant R, Lellouch J 1980 *Clinical Trials.* Academic Press, London

Siegel S 1956 *Nonparametric Statistics for the Behavioral Sciences.* McGraw-Hill, New York

Swinscow T D V 1980 *Statistics at Square One*, 7th edn. British Medical Association, London

Waterhouse J, Muir C, Correa P, Powell J (eds.) 1976 *Cancer Incidence in Five Continents*, Vol. 3, IARC Scientific Publication No. 15. International Agency for Research on Cancer, Lyon

Waterhouse J, Muir C, Correa P, Powell J (eds.) 1982 *Cancer Incidence in Five Continents*, Vol. 4, IARC Scientific Publication No. 42. International Agency for Research on Cancer, Lyon

R. F. Mould

Cardiac Catheterization

Cardiac catheterization is used in the invasive assessment of cardiac disease. The catheters used are plastic tubes, up to 125 cm in length, and 1.5–2.5 mm in diameter. The two most commonly used types of catheter are the hollow catheter and the electrode catheter (Fig. 1). The hollow catheter has at least one lumen connecting a hole at or near the tip of the catheter to an

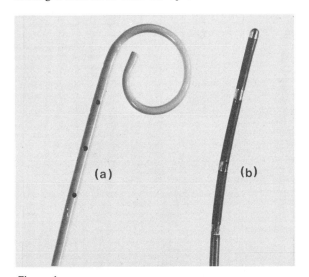

Figure 1
Two main types of cardiac catheter: (a) a hollow "pigtail" catheter used for sampling blood pressure from, and injecting radiopaque dye into, the left ventricle, and (b) an electrode catheter with four platinum electrodes

external port, through which blood samples may be withdrawn, pressures may be measured, and fluids may be injected. The most common fluids used are drugs and radiopaque dyes. The electrode catheter usually has no lumen, and has one or more platinum electrodes on its surface near to the tip, through which the heart's electrical activity may be recorded, and electrical stimuli transmitted. Insertion of a catheter allows direct access to the heart, which means more precise diagnosis and more direct treatment of cardiac disease.

1. Cardiac Function

The heart is a four-chambered pump consisting of two ventricles, the main pumping chambers, which are supplied by two atria, the reservoirs which collect blood returning to the heart and supply it to the ventricles at the correct time to assure maximum efficiency (Fig. 2). Blood returning from the body's veins passes first into

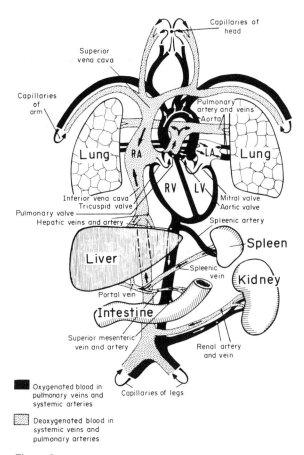

Figure 2
Human circulatory system

the right atrium via the two great veins, the superior vena cava and the inferior vena cava. The right atrium pumps the blood through the tricuspid valve into the right ventricle, which supplies the lungs with blood via the pulmonary valve and pulmonary artery. The blood passes through the network of fine blood vessels (pulmonary capillary bed) in the lungs, and comes into contact with inspired air in the small lung chambers (alveoli). Hemoglobin in the blood is oxygenated by diffusion, and the blood then returns to the left atrium via the pulmonary veins, passing through the mitral valve into the left ventricle. This highly muscular chamber is the "boiler house" of the heart: it must generate enough pressure to supply the whole body with blood. It contracts in synchrony with the right ventricle, sending blood out through the aortic valve into the aorta (the main artery of the body) and then to the systemic circulation. The blood gives up its oxygen on contact with body tissues, and returns to the heart through the venous system (see *Hemodynamics*).

The ventricles of a normal, adult heart can hold about 150 ml of blood, though only some 95 ml of this (the stroke volume) is ejected during each beat (see *Cardiac Output Measurement*). The ratio of stroke volume to total volume is called the ejection fraction, and is an easily measured and sensitive parameter used to describe the pumping efficiency of the heart. In the normal heart, the ejection fraction at rest is about 65%, increasing on exercise to 80%. However, following a heart attack or due to exercise-induced angina (a condition caused by narrowing of the heart's own coronary arteries, resulting in an insufficient blood supply during exercise), ejection fraction may be reduced—in some cases falling to less than 30%, and will further reduce when the subject exercises. At rest, the heart rate is usually around 70 beats per minute, and as 95 ml of blood is ejected during each beat, the cardiac output is in the region of 6.5 l min^{-1}. On strenuous exercise, this may be increased four-fold to cope with the body's increased oxygen demand. The increase is brought about mainly by an increase in heart rate to 180–200 beats per minute, but also by an increase in stroke volume (ejection fraction) as previously stated. Trained athletes usually have larger-than-normal heart chambers, giving them an increased stroke volume. Since their oxygen requirements at rest are unchanged, they have a lower resting heart rate than normal (down to 40 beats per minute) while still maintaining the same cardiac output.

The electrical conduction system of the heart governs the heart rate and the effective sequencing of the atrial and ventricular pumps. The heart's natural pacemaker is the sinus node, a small area near the top of the right atrium at its junction with the superior vena cava. To understand the mechanism by which the sinus node controls heart rate, the situation in the cells of muscle tissue elsewhere in the heart must first be considered. These cells rest in a polarized state, i.e., the interior of the cell is negatively charged with respect to extracellular fluid. This electrical imbalance is maintained by an ionic imbalance (extracellular sodium concentration is much higher than intracellular), and will remain undisturbed unless an adjacent cell undergoes depolarization. This triggers a biochemical process which affects the membrane surrounding the cell, previously a barrier to free exchange of ions, and allows sodium to rush into the cell, cancelling the negative potential inside. This rapid electrical discharge causes the cell to contract by a process known as excitation–contraction coupling. A complex interchange of ions maintains this contraction for some 200–350 ms, after which the cell repolarizes, returning to its former state. The ionic imbalance is restored by an "ion pump," which forces ions to cross the cell membrane against the concentration gradient (see *Bioelectricity*).

The important characteristic of the sinus node pacemaker is its self-triggering ability, whereby a "trickle charge" of ions during the resting phase causes self-depolarization when a critical (threshold) potential is reached. This depolarization process is continuous, and is conducted to the muscle cells, leading to the regular beating of the normal heart. All of the right and left atria are activated and contract within 100 ms of the initial wave of depolarization leaving the sinus node, but to allow time for the ventricles to be efficiently filled by the pumping of the atria, a delay is introduced before ventricular contraction begins. The atria are insulated from the ventricles by a ring of nonconducting fibrous tissue, through which there is one gap. Before the electrical wave of depolarization can pass through this gap, it must traverse the atrioventricular node, which conducts the impulse very slowly, introducing a 100 ms delay. After leaving this node, the impulse is very rapidly conducted to and through the ventricular mass by further specialized conducting tissues: the His bundle, which divides into a right and left bundle branch, then splits up into a diffuse network of small fibers (the His–Purkinje network). This finely balanced system of conduction is subject to many faults, due to congenital abnormalities and also to acquired heart diseases, and many of the techniques in the catheter laboratory aim to explore and treat these faults.

2. Cardiac Abnormalities

Cardiac abnormalities can be broadly separated into two categories: congenital (inherited) and acquired (due to aging or disease processes).

Congenital defects usually arise because at some stage during the development of the fetal heart, a malformation takes place. Generally, the earlier in fetal development that this occurs, the more serious the defect. The commonest defect arises because the fetus does not use its lungs, and therefore does not pump the blood from the right ventricle through the pulmonary circulation. Instead, the pulmonary artery is connected directly to the aorta, allowing both ventricles to supply the systemic circulation. When the child is born, this communication between the great arteries usually closes within several hours, but it can remain open (patent ductus arteriosus). This connection between the normally high-pressure left-side of the heart and the low-pressure right-side leads to shunting of the left-heart output into the pulmonary circulation. This means that far more blood is passing through the lungs than is going out through the aorta to the body, and will eventually lead to deterioration in the function of the ventricles.

A similar functional situation arises in the so-called "hole-in-the-heart" syndrome, where a small hole exists either between right and left atria (atrial septal defect) or between right and left ventricles (ventricular septal defect). This defect is usually the result of a malformation in the interatrial or interventricular wall (septum), at an early stage in fetal development. Other, more complex, abnormalities can arise whereby the pressure in the right side of the heart is abnormally high. In this situation, a septal defect will lead to blood being shunted from the right side to the left side of the heart. Hence deoxygenated venous blood enters the systemic circulation, producing the "blue baby" syndrome. Catheterization is an invaluable tool in the assessment of congenital abnormalities, and is mandatory before surgical correction is attempted.

Much more commonly seen in the catheter laboratory is the problem of acquired heart disease. Disease of the heart's own circulation (the coronary arteries) is the single largest cause of death in the Western world. The heart has a right and left coronary artery, the left splitting into two branches shortly after its origin at the aorta. An occlusion of any of these three vessels, due to furring of the arteries (atheroma) or to a blood clot (thrombosis), leads to the symptoms of a heart attack, 50% of first heart attacks being immediately fatal. If the patient survives the attack, many problems can develop as a result of the occlusion. The area of heart muscle previously supplied by the blocked artery dies (myocardial infarction) and is replaced by scar tissue, which cannot contract. This affects the heart's pumping efficiency, and usually leads to a drop in ejection fraction. If the scar is large, it may become thin and bulging due to the constant pumping of the blood in the ventricle against it. This condition is known as an aneurysm, and can drastically reduce pump efficiency. Cardiac catheterization provides the best means of diagnosing it.

The other most common problem is heart-valve defect whereby one of the valves may fail to open properly (stenosis) or fail to close properly (incompetence) (see *Heart Valves*). Stenosis of a valve causes an increased pressure in the chamber of the heart which is pumping blood through the valve, leading to the enlargement of that chamber and to a decrease in the pumping efficiency. Incompetence causes the stroke volume to increase, since part of the output is regurgitated through the valve which has failed to close properly. The severity of these conditions is assessed by catheterization before surgery.

3. Diagnosis and Assessment of Cardiac Disease

The objective of cardiac catheterization is to establish the existence and exact location of cardiac disease or defect and to quantify its severity. All quantitative assessment during catheterization is instrumentation-based, and also highly invasive, since access to the central circulation is a prerequisite. The improvement in diagnosis resulting from the use of these techniques must always be balanced against the physical damage and mental trauma they produce, and for this reason a wide array of noninvasive techniques exist which seek to replace the invasive measures employed during catheterization. These noninvasive techniques include phonocardiography, radionuclide scanning, echocardiography (see *Echocardiography*) and electrocardiography (see *Electrocardiography*) sometimes using signal averaging techniques to identify low amplitude conduction-system potentials. Despite continuing advances in these fields, the precise diagnostic tests performed in the catheterization laboratory usually precede surgical treatment.

4. Catheterization Techniques

Cardiac catheters are inserted into the heart via veins and arteries in the arms, neck or groin, where the vessels lie close to the surface of the skin. Insertion is either percutaneous, through a short plastic introducer (cannula), which preserves the vessel after the catheter is removed, or via a cut-down, the vessel being tied off using cat gut to stop the flow of blood, and then cut open to allow access for the catheter. Following arterial cut-down, the vessel is repaired and its function restored; veins are more numerous and therefore relatively expendable, and are tied off permanently after cut-down. A catheter inserted into a vein can be manipulated into the right atrium, right ventricle and pulmonary arteries. An arterial catheter enters the heart retrogradely (against the flow of blood), and can be manipulated into the aorta, left ventricle and coronary arteries. The left atrium is generally not accessible, but two methods are employed to assess left atrial function. First, a catheter may pass into the left atrium through a defect in the atrial septum or though a sharp needle passed down the catheter and pushed through the septum (the hole thus formed closes completely when the needle is removed due to the overlapping network of muscle fibers in the heart). Secondly, the catheter may be advanced as far as possible into the pulmonary artery. When it can go no further, it is said to be "wedged." The pressure in this position (the pulmonary capillary wedge pressure) is very close to the left atrial pressure, because the resistance of the pulmonary capillary network is normally very low, and so very little pressure drop occurs between wedge position and left atrium. In order to block the pulmonary artery behind the tip more efficiently, a latex balloon is sometimes provided which, when inflated, effectively blocks the pulmonary circulation by preventing blood from flowing around the catheter.

5. Pressure Measurements

To measure intracardiac pressures, the catheter is filled with saline and connected, externally to the patient, to a pressure transducer, commonly a strain gauge in a Wheatstone-bridge configuration. The electrical signal produced—as a result of the displacement of the strain gauge diaphragm—is amplified, displayed and recorded for further analysis. The frequency components of the intracardiac pressure waveforms do not generally exceed 40 Hz, but this figure is outside the capabilities of the measuring systems used in most catheterization laboratories. The use of extension plastic tubing for convenience between the catheter and the transducer, and the relatively high compliance of the catheter material markedly reduces frequency response, and also causes the system to be resonant at a frequency of between 10 and 50 Hz. Such resonance can cause severely underdamped responses ("ringing") on fast-rising pressure waveforms. This underdamping can sometimes be compensated by hydraulic overdamping using fine capillary tubing or needles, or by using a compensatory electronic filter to smooth the responses. However, these techniques are not usually satisfactory. The remedy really lies in the use of wide-bore stiff-walled catheters directly connected to low-compliance transducers. Catheters with transducers at their tip (Millar catheters) obviate most of the problems associated with fluid-filled recording systems but they are expensive and fragile, and most laboratories use them for research only.

6. Coronary Arteriography

Disease of the coronary arteries causes the common condition of angina pectoris, a crushing pain in the chest brought on typically by exercise. Coronary artery disease usually takes the form of atheroma, in which fatty, granular deposits develop in the arterial wall, subsequently narrowing the artery and leading to hardening and/or occlusion of the vessel—an event known as a myocardial infarction, or more commonly, as a coronary thrombosis or heart attack. The pain is a result of the tissue supplied by the vessel being starved of blood. The tissue subsequently dies (necrosis) and is replaced by scar tissue (fibrosis) which cannot contract, hence impairing the pumping performance of the ventricle. In addition, the blood supply to the remaining viable tissue may be reduced (ischemia), and it is the condition of this tissue which can be improved by restoring the blood supply, usually by coronary artery bypass grafting, where a vein taken from the leg is first grafted onto the diseased vessel beyond the block and then attached to the aorta. In order to locate the damage to coronary arteries precisely, a coronary arteriogram is carried out. A small amount of radiopaque dye is injected into the coronary arteries through a catheter, a film taken simultaneously enabling the areas of narrowing and/or block to be located. The Greek word atheroma literally means gruel and the lumpy appearance of a diseased coronary artery in the arteriogram explains immediately why the word came to be used in this context.

Exciting recent advances in catheterization techniques have been aimed at unblocking coronary arteries which are stenosed or occluded. Both mechanical and chemical methods of unblocking are available. The mechanical method uses a very fine catheter with a balloon at its tip. The catheter is placed at the site of narrowing, the balloon is inflated, and the arterial walls are dilated, restoring the blood supply. This technique is known as percutaneous transluminal coronary angioplasty, or PTCA, and has given promising results in the treatment of acute and chronic coronary artery lesions. Preliminary work on the use of laser irradiation, delivered via a fiber-optic catheter, to unblock coronary arteries, has suggested that this technique may be of future value. Chemical methods involve infusion of a clot-dissolving (thrombolytic) drug, e.g., streptokinase, either into a peripheral vein or, more commonly, directly into the affected artery. When used in acute heart attack victims, this technique can restore the blood supply and decrease the amount of damaged tissue.

7. Angiography

The functional condition of the chambers of the heart and the great vessels connected to them may be examined by angiography, a technique in which a high-pressure injection of x-ray contrast medium into the chamber or vessel is recorded on cine film and/or video tape. A left ventricular angiogram involves the injection of about 50 ml of radiopaque contrast material over a period of 3–4 s. A high-pressure injector is needed to overcome the resistance of the narrow-bore catheter. The angiogram gives information on the contractility of the ventricle and outlines areas of dead muscle present as a result of a heart attack. Incompetence of the inlet valve is shown by regurgitation of the dye during the ejection phase, when the valve should be closed.

The use of angiography in assessment of ventricular function is limited by the effects of the dye injection, which include induction of premature irregular beats (ectopics), and decrease in contractility, due to the high viscosity and ionic content of the contrast medium. The recent introduction of nonionic contrast media has improved the situation, and the use of digital background subtraction, with digital subtraction angiography (DSA), has allowed lower injectate volumes, thus minimizing physiological upset, while maintaining picture quality.

Since both angiography and coronary arteriography cause the coronary circulation to fill with contrast medium instead of oxygenated blood, there is a significant risk of inducing cardiac arrest, thus resuscitation equipment is always on hand during these investigations. The mortality associated with these procedures is very low (around 1 in 1000 cases).

8. Detection of Intracardiac Shunts

Intracardiac shunting of blood is usually due to a defect either in the wall between the atria or the wall between the ventricles. It can often be surgically corrected, but its size and position must first be assessed. In the catheter laboratory two methods are used to detect shunts: indicator dilution and oximetry. In the former method, a bolus of the indicator is injected into the heart, either centrally or into a peripheral vein, and its passage through the heart is monitored by a detector. Dye, radioisotope and ascorbic acid (vitamin C) are the most commonly used types of indicator. The dye, which is usually green, is monitored by the measurement of the absorption of red or infrared light by the blood, either by means of direct arterial blood sampling or by using an ear lobe as an *in vivo* cuvette. A radioactive indicator is detected by a radioactivity counter. Ascorbic acid undergoes a redox reaction with a platinum detecting electrode, generating hydrogen ions and thus changing the electric potential on the electrode. A proportion of the injected indicator will cross the defect in either the left-to-right or right-to-left direction, depending on which side of the defect maintains the higher pressure, producing an abnormal "hump" in the clearance curve recorded. The position of the indicator sensor and the relative positions of the shunt peak on the curve and the main peak can indicate the site and direction of the shunt. The ascorbate and radionuclide methods described above are sometimes carried out by single-breath inhalation of hydrogen or a radioactive gas, respectively, the gases bypassing the venous circulation and quickly reaching the left side of the heart. A sensor in the right side does not record any activity for at least 10 s (body circulation time) unless a left-to-right shunt exists.

Oximetry is a simple technique in which a blood sample taken from one of the heart's chambers or great vessels is analyzed for oxygen content. A normalized figure is usually obtained, called the oxygen saturation. Fully oxygenated arterial blood has an oxygen saturation of 100%. Deoxygenated blood in the right atrium has a saturation of around 80%, unless a left-to-right shunt causes oxygenated blood to mix with the venous blood, producing a rise in saturation. The sampling site at which this rise is noted gives an indication of the shunt position, and the magnitude of the rise gives the shunt size (usually expressed as the ratio of pulmonary blood flow to systemic blood flow, which is greater than 1:1 if a left-to-right shunt exists). An oximeter measures the blood's transmission or reflection of one or more particular wavelengths of light. A commonly used oximeter measures the ratio of light diffusely backscattered by the blood sample at two light frequencies (600 nm and 900 nm). This method becomes inaccurate at low oxygen saturations; the transmission method is inaccurate at the high saturations.

Some of these methods may also be used to measure the cardiac output (see *Cardiac Output Measurement*). The ratio of blood pressure to cardiac output is called the vascular resistance (analogous to Ohm's law, where electrical resistance is equal to voltage/current), and is a useful parameter to describe the workload on the heart. The heart is analogous to a current generator, and an increased vascular resistance without a change of "current" (i.e., cardiac output) will lead to increased "voltage," or blood pressure. High blood pressure (hypertension) is often treated by attempting to decrease the vascular resistance.

9. Conduction System

Disorders in the heart's electrical conduction system, secondary to myocardial infarction, to congenital heart disease, or existing without underlying heart disease (idiopathic), can be accurately assessed in the catheter laboratory, and their correct treatment decided upon. Drug treatment or pacemaker implantation (see *Cardiac Pacemakers, Implantable*) are the alternatives for most disorders of this type, but in some serious rhythm disturbances (arrhythmias), surgery may be required to avoid life-threatening attacks of tachycardia (very rapid heart rate).

The simplest form of electrical treatment is the insertion of a temporary pacemaker, a bipolar catheter being positioned at the apex of the right ventricle (see *Cardiac Pacemakers, Temporary*). The catheter is connected to a pacing box, which delivers regular electrical stimuli at a rate of 60–100 per minute, depending on the needs of the patient. Patients given permanent pacemaker systems under general anesthetic always have a temporary pacemaker inserted beforehand, as the anesthetic may cause a slowing in heart rate. In some centers, this problem is avoided by inserting permanent pacemakers under local anesthetic in the catheter laboratory.

A patient with a complicated rhythm disorder may have up to eight catheters inserted simultaneously. The femoral veins in the groin are large vessels and can accept up to four catheters each. These catheters are used to record characteristic electrical signals from the different elements of the conduction system, and to transmit externally generated electrical stimuli. An attempt is made to restart the arrhythmia from which the patient suffers by different protocols of stimulation, such as rapid atrial or ventricular pacing (at up to 200 beats per minute), and the introduction of between one and four extra beats (ectopics) at various cardiac sites. The acute effect of drugs in preventing the induction of arrhythmias can be studied in the laboratory.

In some patients, the ring of fibrous tissue which electrically insulates the atria and the ventricles (the annulus fibrosus) is broken not only by the normal atrioventricular node and His bundle, but also by a second accessory pathway, which may be sited anywhere on the ring. During normal rhythm, the presence of such a pathway is often indicated by a characteristic surface-electrocardiogram pattern. The condition is usually benign, but in certain circumstances an arrhythmia may be initiated where conduction passes down the normal pathway and returns to the atria by the accessory pathway, setting up a very fast heart rate (up to 250 beats per minute). Such heart rates do not allow time for the ventricles to fill properly and cardiac output is consequently reduced, sometimes to life-threatening levels. Catheter diagnosis of this condition, known as the Wolff–Parkinson–White syndrome, permits the correct use of drugs or pacemakers, or can direct a surgeon to destroy the abnormal pathway, by cutting or freezing it using a cryothermal probe.

Catheter ablation is the most recent advance in the treatment of arrhythmias. When abnormal tissue can be identified during catheterization, the same catheters can be used to deliver very high energies (typically a 5000 V, 50 A pulse of 5 ms duration) directly to the tissue, with the patient under general anesthetic. These high energies damage the tissue, and in some cases can provide a permanent cure without the need for drugs, pacemaker, or open heart surgery. The technique is still in its infancy, and improved circuitry and catheter design may lead to its use as first-line treatment of some cardiac arrhythmias.

10. Computer Applications

The advent of dedicated microprocessors has seen an upsurge in computer involvement in the catheter laboratory (see *Computers in Cardiology*). Analysis of angiograms and coronary arteriograms yields quantitative data on ventricular wall motion and the degree of coronary artery narrowing. Pressure tracings may be signal-averaged to remove measurement artefacts caused by variations in heart position produced by respiration. Indicator dilution methods of measuring cardiac output are all readily automated, removing the tedious job of planimetering the area under a clearance curve. Electrical stimulation protocols may be computer controlled, allowing the clinician to concentrate on interpreting the results. Fully automated catheter laboratory systems, requiring the minimum of operator interaction and thus minimizing operator error, are now available, and will contribute to the clinical results obtainable from catheterization.

See also: Cardiac Function: Noninvasive Assessment; Heart Valves

Bibliography

Hurst J W, Logue R B 1978 *The Heart*, 4th edn. McGraw-Hill, New York

A. D. Cunningham

Cardiac Function: Noninvasive Assessment

An initial assessment of a patient's cardiac status can be made at clinical examination, mainly from the presence or absence of symptoms. Such symptoms, however, are too unspecific or too insensitive to be a reliable basis on which to decide the patient's subsequent management and treatment. Dyspnea, for example, although of great relevance in that it limits the patient's lifestyle and capacity for effort, is obviously a nonspecific system—it may be due to respiratory rather than cardiac disease. On the other hand, the clinical signs of ventricular dysfunction (e.g., gallop rhythm on auscultation) are relatively insensitive although fairly specific; if these signs are detected at clinical examination, then it is likely that ventricular function is severely impaired. Fortunately, additional investigative means of assessing cardiac function are available. These are classified as invasive methods (e.g., catheterization and angiography) and noninvasive methods. In this article several noninvasive procedures are described, each of which, in particular clinical circumstances, has an important contribution to make. Other noninvasive techniques are discussed elsewhere (see *Echocardiography; Dynamic Cardiac Studies*).

It should be emphasized that all tests of cardiac function as well as the interpretations that are put on them should be examined and analyzed with respect to their sensitivity, specificity and accuracy.

1. Chest Radiography

Usually, a posteroanterior radiograph of the chest is taken, but a left lateral view can often give additional information. Right anterior oblique and left anterior oblique views are sometimes used with a barium swallow to help assess individual cardiac-chamber enlargement. The most common radiographic measurement made is the cardiac transverse diameter, defined as the quotient of the maximal cardiac width × 100% and the transverse thoracic diameter. (Its value should be less than 50%, except in young children.)

In cardiac disease, the cardiac size is a function of the duration as well as the severity of the pathology. In acute pathologies, the cardiac size can remain near normal despite gross hemodynamic abnormalities which are the evidence of acute heart failure.

The appearance of the lung field radiographs can be very useful in assessing pulmonary venous hypertension but again, sensitivity and specificity are lacking; moreover, technical factors influence the appearances on the radiograph. It is recognized that the chest x ray will fail to show diagnostic features in as many as 25% of patients who have left-heart failure complicating acute myocardial infarction. The earliest radiographic evidence of a raised pulmonary venous or capillary pressure is usually a diversion of blood flow to both upper zones of the lung fields. The most severe changes detected by radiography are caused by acute pulmonary edema; exudative changes are seen spreading out from the hilar zones, in a "butterfly" pattern. However there is often a lag between radiographic appearances and measured pulmonary capillary pressure. Thus radiographic findings are of limited value in terms of patient management; their prognostic value is more definite.

2. Standard Electrocardiography

Standard 12-lead electrocardiography (with surface electrodes on a resting patient) is a useful investigation (see *Electrocardiography*) but it is well recognized that the procedure has serious limitations in respect of both sensitivity and specificity for the identification of ventricular enlargement (approximately 70% true positives for left ventricular hypertrophy and perhaps 5% false positives using widely accepted criteria), and for the identification of cardiac ischemia in patients presenting with angina (a true-positive rate of 50% when patients with concomitant aortic valve disease or systemic hypertension are excluded). The recognition of these diagnostic limitations of the resting ECG has led to the introduction of exercise electrocardiography.

3. The Exercise Tolerance Test and Exercise Electrocardiography

Exercise tolerance testing is a standard procedure undertaken in patients thought to have coronary heart disease. Disposable ECG electrodes are attached to the chest wall and the patient's electrocardiogram is observed on an oscilloscope both during and after exercise. Changes in heart rate and the ECG waveform are monitored, particularly ST segment depression and T-wave depression (ST–T depression). The ECG is also recorded for later examination. The design of such tests has changed markedly since Master described a two-step exercise test in 1942 in which the patients made circular trips over a low wooden stile of set dimensions.

There are two main types of dynamic exercise testing: multistage treadmill effort testing (Fig. 1), and graduated work load (bicycle ergometry) (Fig. 2). The treadmill test is more readily undertaken by older patients and is perhaps less leg-dependent than bicycle ergometry. Bicycle ergometry has the advantage of permitting other investigations such as exercise radionuclear ventricular function studies. (In treadmill exercising, the patient moves forwards and backwards; this makes the fixed positioning of a radiation detector impossible.) Improvements in the electrical record of physiological parameters with analog-to-digital conversion and data processing of signals have greatly enhanced the quality of the electrocardiogram in treadmill studies. Commercial recording systems are now available which use modern microtechnology and which offer a choice of exercise protocols (Bruce and Hornsten 1969, Nagle et al. 1965, Sheffield

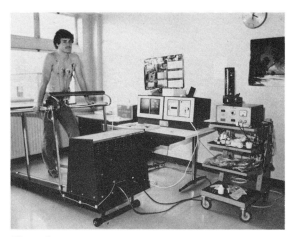

Figure 1
Treadmill effort test

Figure 2
Bicycle ergometer exercise test

1965); treadmill speed and incline are changed automatically (Table 1). Some systems provide assisted ST–T depression and slope analysis, and comment on their diagnostic significance (Simoons et al. 1981).

Table 1
Multistage treadmill effort tolerance test:
Bruce protocol

Stage	Duration (min)	Treadmill speed (mph)	Incline (%)
I	3	1.7	10
II	3	2.5	12
III	3	3.4	14
IV	3	4.2	16
V	3	5.0	18
VI	3	5.5	20
VII	3	6.0	22

Source: Bruce and Hornsten 1969

4. The Value of an Exercise Tolerance Test

4.1 Quantitation of Effort Capacity

It is possible to compare the total exercise time (the actual duration of the exercise) with that predicted on the basis of the person's age, and express this as a measurement of functional aerobic impairment (FAI), or by its complement, the measured effort capacity, as a percentage of the predicted value.

In a stable clinical situation it is customary nowadays to use a symptom limited test. Previously it was common practice to undertake a "maximum effort tolerance test" to a target heart rate of, usually, 220 beats min^{-1} minus the subject's age. If this target heart rate were achieved without pain, or without changes caused by ischemia appearing on the electrocardiogram, the test was judged "normal."

4.2 Definition of the Limiting Symptom

The particular symptom which actually limits an exercise test is itself an important factor in diagnosis. The limiting symptom in some patients is not, as expected, the presenting symptom. In other instances, it becomes clear that the exercising patient has clearly developed angina although there is no evidence of cardiac ischemia on the electrocardiogram. Most cardiologists would classify this test result as positive evidence of ischemic heart disease (Table 2).

Table 2
Assessment of observations in exercise tests for the diagnosis of ischemic heart disease

Observation	Sensitivity (%)	False negative (%)
ST segment depression alone	54	46
Angina alone	68	32
ST segment depression and angina	88	12

4.3 Electrocardiographic Evidence of Underlying Cardiac Ischemia

As previously indicated, the limitations of the resting electrocardiogram in identifying patients suffering from angina are well recognized. Exercise testing, however, increases the number of true positives identified. Modifications of ECG techniques, particularly the use of multiple-lead systems (Fig. 2), has also increased the yield of true positives—from 60% to 90% without loss in specificity (Table 3). The classic appearance of cardiac

Table 3
Sensitivity in exercise tests using multiple lead systems for the diagnosis of ischemic heart disease

Number of leads	Sensitivity %
V5	60
V2, AVF, V5	70
16 lead multiple precordial lead ST mapping	90

ischemia on the electrocardiogram is ST segment depression of 1 mm or more and flat bottomed (square wave) for 0.08 s (2 mm) at a standard recording sensitivity of 10 mm mV^{-1} and paper speed of 25 mm s^{-1} (Fig. 3).

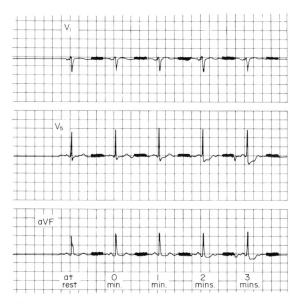

Figure 3
Ischemic ST segment depression developing during exercise testing

Down-sloping ST segment depression is also a possible indication of cardiac ischemia. Up-sloping ST depression gives an increased yield of true positives but at the cost of a significant increase in false positives.

4.4 Assessment of Ventricular Function

It is customary to monitor heart rate continuously during the test and to check blood pressure every three minutes by sphygmomanometry. The heart rate and the arterial blood pressure are two of the major determinants of left ventricular work, and the double product [(Maximal heart rate) × (maximal systolic blood pressure)/100] correlates well with myocardial consumption of oxygen (MVO_2). Heart rate and blood pressure are clearly influenced by drug therapy. Beta-blocking drugs modify the effect of exercise on the heart rate and blood-pressure of normal patients and patients with ischemic heart disease. A low double product may indicate either underlying impairment of left ventricular function or, simply, poor cooperation on the part of the patient.

4.5 Identification of Exercise-Induced Arrhythmias

Effort-induced cardiac ischemia can cause ventricular instability which may become manifest during an exercise test as frequent or multifocal extrasystoles. If these occur in short runs (ventricular tachycardia) the test should be terminated and the patient observed until a stable rhythm returns. Occasionally, exercise induces intraventricular conduction defects (bundle branch block) and, more rarely, second or third-degree atrioventricular block. These induced defects are manifestations of serious underlying cardiac pathology.

4.6 Evaluation of the Therapeutic Effect of Drugs

A patient can assess his own response to a drug regime, subjectively, by noting changes in his symptoms. The physician, also, can assess an angina sufferer's response to drug therapy by checking on the number of glyceryl trinitrate pills which he takes of his own volition in response to symptoms. An objective assessment of the regime can be made by serial exercise testing. In a number of patients a discrepancy may be found between the subjective and objective assessments of response to therapy. Similarly, the exercise testing of patients before and after coronary vein graft surgery is a useful objective means of evaluating the benefit of the surgery.

4.7 Identification of High-Risk Groups

For angina patients, the greater the degree of ST-segment depression, especially in the early stages of the exercise test, the greater the number of main coronary arteries likely to be affected by disease, and the lower the survival rate at five years after medical therapy (European Coronary Surgery Study Group 1982). There is, however, an overlap between groups (categorized according to the number of arteries involved), and it is usual to proceed to coronary angiography to define unambiguously the degree of anatomical involvement. It is, however, clear that functional testing of cardiac status (response to exercise testing) could well define "at risk" subsets more clearly than the angiographic studies which are relied upon at present. Studies have been undertaken of survivors of acute myocardial infarction using a low-level exercise test at an early date after infarction (Theroux et al. 1979, De Feyter et al. 1982). The test is terminated at a maximal heart rate of 130 beats min^{-1} or at the development of angina or of ischemic ST segment depression. The development of these latter symptoms indicates a high-risk subset in whom angiographic and other studies should be considered.

5. Indications for Terminating an Exercise Test

Symptom-linked exercise testing should not be undertaken in patients with unstable angina or who have experienced a recent onset of chest pain. It should be undertaken with caution in patients who have aortic valve stenosis. The test should be terminated if there is a fall in systolic blood pressure during exercise. The test should also be terminated if the patient develops any neurological symptoms such as light-headedness or staggering. Runs of ventricular extrasystoles and gross ST segment depression are other good reasons for termination. Given these precautions, the mortality rate associated with symptom-limited, multistage, treadmill exercise testing is estimated at 1 in 100 000.

6. The Exercise Tolerance Test versus Radionuclide Ventriculography

The important diagnostic role of the multistage effort test in ischemic heart disease is acknowledged, particularly when its limitations are recognized. A positive test is very likely to be true positive but caution is required with a negative test. Evidence suggests that radionuclide ventriculography may be a more sensitive test than exercise testing (see Table 4; Berger and Zaret 1981).

Table 4
Comparison of radionuclide ventriculography and exercise tolerance testing in ischemic heart disease

	Sensitivity (%)	Specificity (%)
Radionuclide ventriculography	87	92
Exercise tolerance testing	64	98

Source: Berger and Zaret 1981

The combination of the two procedures may in the future reduce the need for cardiac catheterization and angiography.

7. 24-Hour Ambulatory ECG Monitoring

In 24 h ambulatory ECG monitoring, the ECG signals are recorded on a tape recorder carried by the patient.

Initially described as Holter monitoring, this technique is in widespread clinical use. For optimal results the skin under the electrodes should be meticulously prepared and the electrodes positioned most carefully. Amplitude modulation (AM) recording systems are in most common use. They offer a great deal of information on intermittent dysrhythmias in symptomatic and asymptomatic patients. The problem remains that a patient who has a paroxysmal dysrhythmia may not have an attack during the monitored period. Repeated tests may have to be undertaken before typical symptoms are detected. Caution is required in assessing asymptomatic sinus and atrioventricular (AV) junctional bradycardia since these rhythms occur in a large number of normal subjects, as do supraventricular and ventricular extrasystoles. Ambulatory monitoring is a very useful aid in detecting intermittent high-grade AV block or intermittent natural pacemaker dysfunction (Fig. 4). More recently, frequency modulation (FM) recording systems

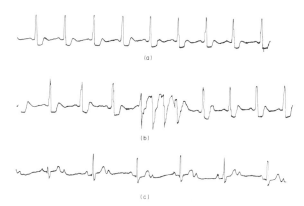

Figure 4
Electrocardiographic changes during 24 h ambulatory monitoring: (a) ischemic ST segment depression, (b) paroxysmal ventricular tachycardia, and (c) 2:1 atrioventricular block. (Every second sinoatrial signal is blocked at an atrioventricular node. Normal QRS complex appears after every second P wave)

have been introduced. These systems have an enhanced role to play in the detection and quantitation of ST segment depression and elevation and of T wave changes over a 24 h period of normal activity (Fig. 5). During the monitored period the patient keeps a diary of symptoms to relate the electrocardiographic appearances to the symptoms. Nonetheless, it is evident from 24 h monitoring that episodes of cardiac ischemia often occur without angina and sometimes at lower heart rates than might have been anticipated. Furthermore, episodes of a variant of angina (Prinzmetal's angina) due to coronary spasm and associated with acute ST segment elevation can sometimes be diagnosed only by this

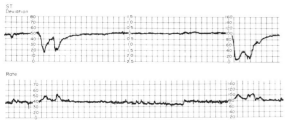

Figure 5
Quantitative measurement of ST segment depression
in mV from isoelectric baseline with simultaneous
heart-rate recording using a frequency-modulated
ambulatory monitoring system

method as the attacks are not effort related. The technique has been used, too, to monitor post-myocardial
infarction patients to detect those with ventricular irri-
tability; such patients may be a "high risk" subset and
need special treatment. The technique is also used to
monitor the efficacy of antiarrhythmic and antianginal
drug therapy.

Commercial systems are available which scan the
recorded rhythms on the cassette tapes at 60–120 times
the recording speed. Arrhythmias are analyzed automatically; graphic and tabular printouts of the recorded
dysrhythmias are provided. Nonetheless, supervision by
medical, nursing, and technical staff is usually mandatory. The procedure remains rather time-consuming and
there are technical problems with the equipment which
have not yet been completely solved.

8. Systolic Time-Intervals Measurement

Systolic time intervals are measured from a simultaneous
recording of the phonocardiogram, the carotid artery
pulsation, and the electrocardiogram. Their measurement provides a noninvasive means of assessing left
ventricular function. The following measurements are
taken (see Fig. 6):

(a) LVET (left ventricular ejection time), the time
    interval between the onset of carotid upstroke and
    the incisura of the dicrotic notch;

(b) $Q - S_2$ (electromechanical systole), the time interval
    between the onset of QRS complex and the aortic
    component of the second heart sound; and

(c) PEP (preejection period), obtained from $(Q–S_2) -$
    LVET.

All of these measurements have to be corrected for heart
rate. However, the ratio PEP/LVET is not greatly
influenced by heart rate—its normal value is 0.42 or less.

In practice, systolic time-interval measurement is subject to equipment-induced variations; the findings are
nonspecific. The method has been shown to be useful in

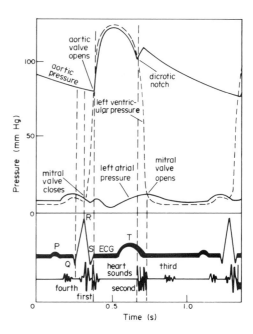

Figure 6
Correlation of the four heart sounds with the
electrical and mechanical events of the cardiac cycle

the difficult differentiation between constrictive pericarditis and constrictive cardiomyopathy (Ghose et al.
1976)—normal values of the PEP/LVET ratio (~ 0.41)
in constrictive pericarditis and high values (~ 0.62) in
constrictive cardiomyopathy. In recent years the technique has tended to be replaced by radionuclide
ventriculography and by sector scanning, real-time
echocardiography. To confirm constrictive pericarditis it
is usual to undertake a pericardial biopsy. The biopsy
can be performed through the central tendon of the
diaphragm using an extraperitoneal approach and
avoiding the older technique of thoracotomy. A histological diagnosis of cardiomyopathy can be made by endo-
myocardial biopsy at cardiac catheterization.

9. Conclusion

The role of noninvasive methods of assessing cardiac
status is likely to increase. Some studies have suggested
that invasive studies are unnecessary in many instances
in selecting patients for valve-replacement studies
(Sutton et al. 1981, Hall et al. 1983). Other studies
suggest that it is no longer necessary to subject those
patients to angiography who present with atypical chest
pain and who have a normal thallium scintigraphy
study. Some other recent studies have indicated that in
a group of patients with ischemic heart disease, either an
exercise tolerance test or radionuclide ventriculography,
can identify those individuals who are particularly at

risk (Dewhurst and Muir 1983, Theroux et al. 1979, European Coronary Surgery Study Group 1982). However, the final position has not been resolved as other studies (Jennings et al. 1984, Rozanski et al. 1983) have not confirmed these findings. An eventual resolution of the matter could be of major prognostic importance. It could give an indication as to whether surgical treatment or medical treatment is preferable for the individuals at high risk.

See also: Cardiac Output Measurement; Monitoring Equipment in Coronary and Intensive Care

Bibliography

Berger H J, Zaret B L 1981 Nuclear cardiology. *N. Engl. J. Med.* 305: 855–63
Bruce R A, Hornsten T R 1969 Exercise stress testing in evaluation of patients with ischemic heart disease. *Prog. Cardiovasc. Dis.* 11: 371
Dagenais G, Rouleau J R, Christen A, Fabia J 1982 Survival of patients with a strongly positive exercise electrocardiogram. *Circulation* 65: 452–56
De Feyter P J, van Eenige M J, Dighton D H, Visser F C, de Jong J, Roos J P 1982 Prognostic value of exercise testing, coronary angiography and left ventriculography 6–8 weeks after myocardial infarction. *Circulation* 66: 527–36
Dewhurst N G, Muir A L 1983 Comparative prognostic value of radionuclide ventriculography at rest and during exercise in 100 patients after first myocardial infarction. *Br. Heart J.* 49: 111–21
European Coronary Surgery Study Group 1982 Long term results of prospective randomised study of coronary artery bypass surgery in stable angina pectoris. *Lancet* ii: 1173–80
Ghose J C, Mitra S K, Chhetri M K 1976 Systolic time intervals in the differential diagnosis of constrictive pericarditis and cardiomyopathy. *Br. Heart J.* 38: 47–50
Hall R J C, Kadushi O A, Evemy K 1983 Need for cardiac catheterisation in assessment of patients for valve surgery. *Br. Heart J.* 49: 268–75
Jennings K, Reid D S, Hawkins T, Julian D J 1984 Role of exercise testing early after myocardial infarction in identifying candidates for coronary surgery. *Br. Med. J.* 288: 185–87
Nagle F J, Balke B, Naughton J P 1965 Graduational step tests for assessing work capacity. *J. Appl. Physiol.* 20: 745
Rozanski A, Diamond G A, Berman D, Forrester J S, Morris D, Swan H J C 1983 The declining specificity of exercise radionuclide ventriculography. *N. Engl. J. Med.* 309: 518–22
Sheffield L T 1965 Graded exercise in the diagnosis of angina pectoris. *Mod. Concepts. Cardiovasc. Dis.* 34: 1
Simoons M L, Hugenholtz P G, Ascoop C A, Distelbrink C A, de Laud P A, Vinke R V M 1981 Quantitation of exercise electrocardiography. *Circulation* 63: 471–75
Sutton M G St J, Oldershaw P, Sacchetti R, Panelli M, Lennox S C, Gibson R V, Gibson D G 1981 Valve replacement without preoperative cardiac catheterisation. *N. Engl. J. Med.* 305: 1233–38
Theroux P, Waters D D, Halphen C, Debaisieux J-C, Mizgala H F 1979 Prognostic value of exercise testing soon after myocardial infarction. *N. Engl. J. Med.* 301: 341–45

J. A. Kennedy

Cardiac Output Measurement

Cardiac output is the volume of blood ejected either from the left ventricle into the systemic circulation or from the right ventricle into the pulmonary circulation. It is conventionally measured in liters per minute, while in physiological studies the concept of cardiac index is employed. This permits a comparison of output measurements in individuals of different body weight and is defined as

$$\text{Cardiac index} = \frac{\text{cardiac output}}{\text{body surface area (m}^2)}$$

Stroke volume is the volume of blood ejected per heart beat:

$$\text{Stroke volume} = \frac{\text{cardiac output}}{\text{heart rate}}$$

and

$$\text{Stroke index} = \frac{\text{stroke volume}}{\text{body surface area (m}^2)}$$

Cardiac output can vary considerably in healthy individuals. Resting levels in an adult man are around 5 liters per minute. This falls during sleep and can rise to more than 20 liters per minute during exercise. Changes in cardiac output occur during many disease states, with a decreased flow to kidney, brain and other tissues, impairing organ function.

Earliest studies of cardiac output were based on the Fick principle and required the direct measurement of oxygen consumption and both arterial and venous blood sampling for measurement of oxygen content. Subsequently, indicator dilution techniques obviated much of this need for sampling. In addition, radioisotopes and ultrasound provide further methods for measurement of cardiac output.

1. The Fick Principle

The Fick principle states that blood flow through organs can be estimated if a measurable substance is either removed or added as blood flows through them. Thus cardiac output f can be measured using the equation

$$Q/t = f(C_A - C_V)$$

where Q is the subject's oxygen consumption measured by the spirometric collection of expired air, C_A is the oxygen concentration of a systemic arterial blood sample, and C_V is the oxygen concentration of a mixed venous blood sample from the pulmonary artery.

The oxygen content of arterial blood is greater than the oxygen content of venous (pulmonary arterial) blood due to oxygen inhaled and transferred to the blood via the lungs.

The concentration of oxygen in arterial blood is also relatively uniform, but this is not true of venous blood: each tissue may utilize varying quantities of oxygen, and the blood oxygen content in the superior vena cava thus differs from that in the inferior vena cava. Also, mixing is incomplete in the right atrium and only in the pulmonary artery can a truly mixed sample be obtained. This method therefore requires right-heart catheterization, arterial blood sampling and meticulous expired air collection. It is likely that data will have an error of at least 10%.

For example, if an oxygen consumption of 300 ml min^{-1} results in a systemic arterial blood oxygen concentration of 19 ml oxygen per 100 ml blood, and the venous oxygen concentration is found to be 14 ml per 100 ml, then $(C_A - C_V) = 5$ ml oxygen per 100 ml blood. Thus

$$f = \frac{Q}{t(C_A - C_V)} = \frac{300}{5} \times \frac{100}{1} \text{ ml min}^{-1}$$

$$= 6 \text{ l min}^{-1}$$

The Fick principle was the first to allow repeated and reasonably accurate measurement of cardiac output. However, it has now been superseded by other techniques.

2. Indicator Dilution Methods

The volume of fluid in any container can be calculated if a known quantity of indicator (e.g., dye, cold saline or radioisotope) is added and the concentration of indicator measured after it has been uniformly dispersed throughout the fluid. This technique was developed and refined by Stewart and Hamilton.

A suitable indicator is one which can be detected and measured, and which remains within the circulation during at least its first circuit. Subsequent removal from the circulation is an advantage.

2.1 Theory

Consider a tube with a mixing chamber and a steady flow of water of Q ml s^{-1} (Fig. 1). Let fluid be collected

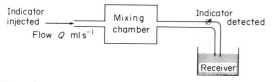

Figure 1
Basic principles of indicator dilution technique

in the receiver until all the indicator has passed out of the mixing chamber over time t. If I (mg) is the amount of indicator injected and $\bar{C}$ is the mean concentration of indicator in the receiver, then

$$\bar{C} = I/Qt$$

Thus

$$Q = I/\bar{C}t \text{ ml s}^{-1} \quad (60 \, I/\bar{C}t \text{ ml min}^{-1})$$

The concentration of indicator distal to the mixing chamber is recorded as a time–concentration curve.

Such a model oversimplifies the situation with regard to cardiac output. A known quantity of indicator injected into the venous side of the circulation passes through the pulmonary circulation, returns to the left ventricle and is distributed by the arterial system (where the blood is sampled and concentration measured) before returning in the venous circulation to the right side of the heart. Circulation time varies as the blood passes through the various organs such as liver, kidney and brain. Some blood therefore returns to the heart more quickly than the rest and consequently recirculation interrupts the time–concentration curve (Fig. 2). This,

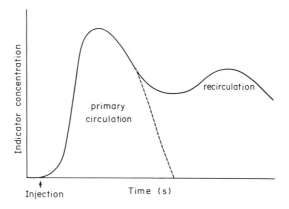

Figure 2
Semilog plot of indicator time concentration curve as used to calculate cardiac output

however, can be allowed for and corrected since the initial downslope of the curve is monoexponential and can be replotted on semilogarithmic paper as a straight line. Extrapolation of the downslope enables a reasonable approximation to be made of recirculation. The area under the curve represents $\bar{C}t$ and can be measured in various ways including planimetry.

2.2 Indicator Substances

A variety of indicator substances are used. Dyes such as Coomassie blue and indocyanine green were initially used. Indocyanine green has a peak spectral absorption at 800 nm, the wavelength at which the absorption of both reduced hemoglobin and oxyhemoglobin is identical, and so changes in their concentration as found in venous and arterial blood do not affect detection of the indicator. Additional indicators include radioisotopes such as ^{131}I-labelled human serum albumin and more recently the use of cooled saline has proved particularly useful.

In a standard method for the clinical measurement of cardiac output, cannulas are inserted into an artery and vein and a bolus of known volume and concentration of indocyanine green is then injected as centrally and as rapidly as possible. Arterial blood is then withdrawn at a constant rate through a densitometer cuvette. A time–concentration curve of indicator in the withdrawn blood is recorded with the system being calibrated by the serial passage through the cuvette of indicator samples of varying concentration. Analog computers have been developed for cardiac output measurements with an immediate digital printout.

3. Thermodilution

The thermodilution technique involves the injection of a solution, usually dextrose–saline cooled to less than body temperature, into the venous system (preferably the right atrium) using a multichannel catheter. The resultant blood temperature change is proportional to flow and is detected by a thermistor incorporated into the distal end of the catheter which is positioned in the pulmonary artery. Computer systems are available for curve integration, output calculation and rapid digital display.

Cardiac output is given by the equation

$$Q = \frac{AV(T_b - T_c)}{ST_b d_t} \times \frac{D_c S_c}{D_b S_b} - B$$

where A and B are constants from the thermodilution/Fick regression equation of Branthwaite and Bradley (1968), V is the volume of saline ejected less the dead space of the catheter, T is the temperature (°C), D is the density, S is the specific heat, t is the time (s) and $ST_b d_t$ is the interval of change in blood temperature with time, the subscripts b and c representing blood and injectate, respectively.

Good correlation has been found between thermodilution and Fick methods (Fegler 1954) and between dye and thermodilution (Goodyer et al. 1959). Results are also reproducible in a wide range of clinical conditions and in different species. The advantages of thermodilution include reproducibility and the feasibility of repeated frequent measurements. The method can be readily used and is invaluable in intensive care. Potential inaccuracies have to be recognized although they rarely pose a problem in practice. These include heat loss in the lungs and spontaneous variation of blood temperature which can occur during both stable and unstable hemodynamic situations. As with all indicator dilution techniques, the accuracy depends on adequate mixing and normal distribution. In situations where the anatomy is disturbed, such as intracardiac shunts and valve incompetence, absolute values are much less accurate.

The development of the balloon-tipped flow-directed catheter (Ganz et al. 1971) has added to the advantages of the thermodilution method. This catheter can be readily "floated" from a peripheral vein into the pulmonary artery and also allows measurement of pressure and taking of samples as well as cardiac output measurement.

4. Radionuclide Techniques

Early methods using radionuclides involved their detection by means of an external detector placed over the heart. Recent developments include the use of gamma cameras, improved detector collimation and a wider choice of radiopharmaceuticals for injection (see *Radionuclides: Clinical Uses*).

Most simply, an external NaI(Th) crystal detector is placed over the fourth left interspace at the sternal edge. It is linked to a ratemeter and a recorder. Two peaks of radioactivity are recorded after injection (Fig. 3). (The

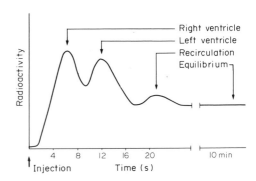

Figure 3
Time–concentration curve during precordial counting after injection of ^{131}I serum albumin

same principle is employed by a recently developed nuclear stethoscope.)

The first peak is due mainly to right ventricular activity and the second mainly to left ventricular activity. The "left" ventricular washout is exponential and is replotted to provide the area under the curve of primary circulation through both cardiac chambers. A venous sample is taken after 10 min when mixing is complete for measurement of blood radioactivity and a simultaneous equilibrium value for precordial radioactivity is obtained.

Cardiac output in liters per minute is calculated from the formula

$$Q = \frac{60I}{A} \times \frac{1000}{C/d}$$

where I refers to the amount of radioactivity injected, A to the area under curve of primary circulation, C to the blood or plasma concentration of radioactivity and d to the equilibrium value.

Gamma camera techniques have largely replaced other isotopic techniques. Isotopes such as ^{99m}Tc are used to assess left ventricular function and allow derivation of cardiac output. Two techniques are commonly used: the first pass method and the dynamic recording method.

4.1 First Pass Method

A small-volume bolus of isotope is injected peripherally and its transit through the heart recorded. Data are acquired by a gamma camera, usually positioned in the left anterior oblique position to obtain optimal right and left ventricular separation, and with caudal tilt in an attempt to separate left atrial activity from that of the left ventricle. Quantitative information can be derived from time–activity curves from selected regions of interest (usually the left ventricle). Computer linkage is required.

4.2 Dynamic Recording Method

Data acquisition is triggered by another signal, usually the ECG, so that "gated" activity is obtained during ventricular systole and diastole. Estimations of end systolic and end diastolic volumes are made and the ejection fraction calculated (Fig. 4). The disadvantages of this technique include the influence of high background counts and uncertain delineation of the border of the left ventricle, although techniques are being developed to improve computer delineation of cardiac boundaries.

The accuracy of gamma camera methods has been assessed in several ways:

(a) by comparison of isotope and contrast left ventriculography using the area/length method, which has provided a reasonable correlation (Strauss et al. 1971), and

(b) by analysis of fluctuations of radioactivity over the left ventricle during a cardiac cycle. Thus:

$$\text{Ejection fraction} = \frac{c_d - c_s}{c_d}$$

where c_d is the end diastolic count and c_s is the end systolic count. Correlations with angiography have again been good ($r = 0.84$–0.87). This technique avoids any geometrical assumptions as to the shape of the left ventricle and, as the half-life of ^{99m}Tc is 6 h, measurements can be repeated before and after any intervention, and during exercise.

Both methods provide a measurement of ventricular volume from which calculations of cardiac output can be made:

End diastolic volume − end systolic volume
$$= \text{stroke volume}$$

Cardiac output = stroke volume × heart rate

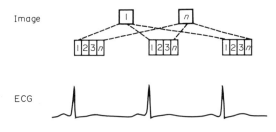

Image   ECG

Figure 4
Gamma camera acquisition is gated to the patient's electrocardiogram. Counts are stored in frames where frame 1 represents the total counts obtained during the first portion (1–n) of all cardiac cycles recorded

These techniques are promising and are being increasingly employed. However, it must be remembered that, although doses are low, the patient is exposed to radioactivity. There are statutory regulations limiting the total dosage administered to each patient.

5. Measurement of Stroke-Volume Changes by Ultrasonic Doppler (Transcutaneous Aortovelography)

The continuous-wave ultrasonic Doppler technique is used to measure aortic blood velocity transcutaneously. An ultrasonic beam is directed towards the aorta from a transducer placed in the suprasternal notch. The highest instantaneous blood velocity in the aorta observed in the ultrasonic field is measured. Its time integral varies directly with aortic flow and hence with stroke volume (Fig. 5). This "systolic velocity integral"

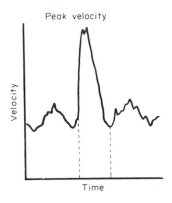

Peak velocity

Velocity

Time

Figure 5
Schematic representation of the tracing of aortic blood velocity. Value of peak velocity is indicated. The area equals the systemic velocity integral. Velocity is zero at the abscissa. Nonzero velocity during diastole is caused by moving blood and tissue interfaces

(SVI) is equivalent to the area under the time–velocity curve for each ejection. Good correlations are found in both animals and man, at rest and on exercise, between this method and blood flow measured by invasive techniques (Light 1976, Sequeira et al. 1976, Light et al. 1976).

Recordings can be obtained simply and quickly in up to 90% of subjects and, at power levels less than $100\ \mathrm{mW\ cm^{-2}}$, the method is safe. Although accuracy depends on the assumption that little change occurs in aortic dimensions during the cardiac cycle, other non-invasive techniques suggest that any such changes are not clinically significant (Goldberg 1971, Olson and Shelton 1972). This method appears useful in measuring sequential changes in acutely ill patients although only relative volume flow is obtained. Calculation of the absolute volume flow requires the use of a "calibration" measurement using another standard technique.

6. Echocardiography

Medical ultrasound uses piezoelectric material to generate and detect high-frequency ($> 20\ \mathrm{kHz}$) sound waves. Such high-frequency sound waves can be transmitted through biological tissue. If media of different acoustic impedance are encountered perpendicular to the path of the sound beam, the reflected sound travels back to the transducer and is reconverted to electrical activity. The distance between the transducer and echo source can be calculated knowing the speed of sound through the biological tissue (average value $1540\ \mathrm{m\ s^{-1}}$ in human tissue (Goldman and Jueter 1956)) and the time elapsed

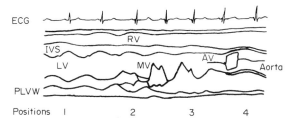

Figure 7
Schematic representation of the ultrasonic patterns obtained in each of the four standard ultrasonic beam positions. Key: RV, right ventricle; PLVW, posterior left ventricular wall; MV, mitral valve; AV, aortic valve; IVS, interventricular septum

from transmission of the sound to the reception of the echo.

Ultrasonic signals may be displayed on a cathode-ray tube as an amplitude modulated display (AMD), presenting time from sound transmission to echo return as a measurement of distance along the x axis and the amplitude of the returning sound (intensity of the echo) on the y axis. In echocardiography the pattern of motion is the characteristic used to identify certain intracardiac structures and accordingly an M-mode or time–motion display is recorded (see *Echocardiography*).

In the normal human there is a large area of the anterior chest which allows waves from an ultrasonic transducer to enter the soft tissues without encountering bone or air-containing lung tissue, the "ultrasonic window" (Fig. 6). Structures within the heart are recognized by patterns of motion and the anatomical relationship between structures with specific patterns (Fig. 7). Measurement of cardiac output is determined from echoes derived from the left ventricle. The borders of the left ventricle are the interventricular septum (IVS) anteriorly, and posterior left ventricular wall posteriorly. The cavity of the chamber is bordered by the left side of the interventricular septum and the endocardial surface of the posterior left ventricular wall (Fig. 8). By measuring the distance between the left side of the IVS and posterior left ventricular endocardium during the cardiac cycle the left ventricular internal dimension (LVID) can be obtained during systole and diastole. The LVID systole is measured at the onset of the QRS as end diastole may be difficult to determine using other criteria. The LVID is measured at the nadir of septal motion in patients with normal septal movement (Fig. 8).

The echocardiographic left ventricular (LV) dimensions have been correlated with angiographic LV volumes (Feigenbaum et al. 1969) and many similar studies have correlated echocardiographic dimensions with angiographic ventricular indices, with stroke volume measurements determined by indicator dilution, and with Fick cardiac outputs, all with good statistical relationships (Popp and Harrison 1970, Teichholz et al. 1972). Caution must be exercised, however, as these

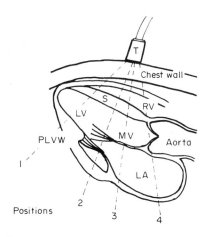

Figure 6
Schematic representation of the ultrasonic transducer at the ultrasonic window showing the four basic beam positions and the anatomical structures encountered in each position. Key: T, transducer; RV, right ventricle; S, septum; LV, left ventricle; MV, mitral valve; LA, left atrium; PLVW, posterior left ventricular wall

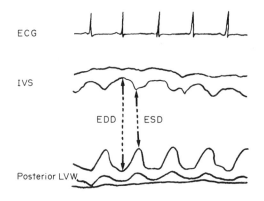

Figure 8
Schematic representation of the left ventricular echogram showing the measurement of the end diastolic diameter EDD and the end systolic diameter ESD. Key IVS, interventricular septum; LVW, left ventricular wall

original studies attempted to show that the echocardiographic measurements had a relationship to volume and could provide useful estimates of ventricular volume and cardiac output. Despite these limitations, formulae for the calculation of ventricular volumes have been derived.

The rationale for echocardiographic estimation of left ventricular volume is based on the following assumptions: (a) LV can be described as a prolate ellipse; (b) the two short axes (d_1 and d_2) are equal; (c) the long axis l is twice the short axes d_1 and d_2; (d) the LVID and the echocardiogram approximates the short axes (d_1 and d_2); and (e) the LV walls contract uniformly. The calculations for these assumptions are as follows:

Formula for volume of prolate ellipse (assumption (a))

$$V = \frac{4\pi}{3} \times \frac{d_1}{2} \times \frac{d_2}{2} \times \frac{l}{2}$$

Assumption (b)

$$V = \frac{4\pi}{3} \times \frac{d}{2} \times \frac{d}{2} \times \frac{l}{2} \text{ or } V = \frac{d^2\pi}{3} \times \frac{l}{2}$$

Assumption (c)

$$V = \frac{\pi d^2}{3} \times \frac{2d}{2} \text{ or } V = \frac{\pi d^3}{3}$$

Assumption (d)

$$V = \frac{\pi}{3}(\text{LVID})^3 \text{ or } V = 1.047(\text{LVID})^3$$

$$\text{or } V = (\text{LVID})$$

Equation for error of assumption (c)

$$V = \frac{7.0}{2.4 + \text{LVID}}(\text{LVID})^3$$

This formula is applicable to ventricles which have the general shape of a prolate ellipse but in diseased

ventricles the long axis is not always twice the size of the short axis. As the ventricle dilates the two dimensions approximate each other and the use of a correction factor has been suggested.

The recent introduction of two-dimensional echocardiographic techniques (Bom et al. 1973, Carr et al. 1979) has allowed a more accurate determination of ventricular volumes in patients with cardiac aneurysm and regional left ventricular wall movement abnormalities, giving improved correlation with directly measured ejection fraction and stroke volume (Weyman et al. 1976). The two-dimensional echocardiogram, however, underestimates ventricular volume when compared with those measured from angiography. Further technical refinements may allow more accurate measurement.

Thus, echocardiography is an easily repeatable, reproducible assessment of cardiac output and useful in assessing changes due to physiological challenges, pre- and postoperative assessment, and studying the sequential influences of pharmacological interventions.

7. Impedance Plethysmography

Impedance plethysmography is a noninvasive technique which can be used to monitor changes in cardiac output (see *Plethysmography*). It is an indirect measurement allowing changes in blood flow to be followed qualitatively. The values, however, may differ markedly from the absolute values obtained by invasive methods.

7.1 Principles and Method

Kubicek et al. (1974) introduced the use of four disposable band electrodes of self-adhesive mylar tape (Fig. 9). A constant current of 4 mA VMS at 100 kHz is fed across the thorax via electrodes 1 and 4. The current is

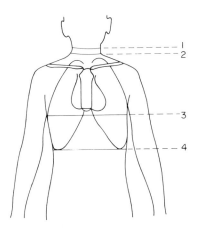

Figure 9
Schematic representation of the sites of application of the four disposable band electrodes for impedance plethysmography

supplied by a high-quality isolated transformer. The impedance of the adult thorax lies in the range 20–35 Ω with a fall during systole of 0.2 Ω. The impedance cardiograph tracing measures the change in thoracic impedance Z_0 (measured between electrodes 2 and 3), the impedance change during the cardiac cycle ΔZ, and its first derivative dZ/dt. Simultaneous phonocardiography and carotid pulse wave tracings are often taken simultaneously and calculations made as defined by Kubicek (Fig. 10). Good correlations have been

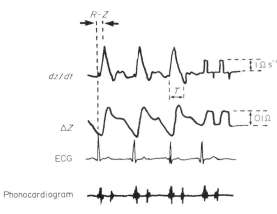

Figure 10
Schematic representation of the simultaneous tracings obtained during impedance plethysmography. The diminution in the thoracic impedance during systole is recorded as an upward deflection as is the corresponding negative first derivative

found in both animals and humans with normal hearts (Pate et al. 1975, Williams and Caird 1980). Problems may occur in subjects with slow ventricular ejection, and in patients with valvular lesions or right bundle branch block (Williams and Caird 1980).

8. Choice of Techniques

There has been great expansion in the methods available for measurement of cardiac output. Depending on facilities, any method may be appropriate. It seems likely, however, that noninvasive methods will become increasingly popular. For limited clinical studies where access to sophisticated nuclear medicine facilities or hemodynamic facilities are not available, echocardiography would appear the method of choice. In intensive-care situations where measurement of intravascular pressures is necessary, thermodilution techniques employing the "floating" Swan–Ganz catheter would seem suitable.

See also: Dynamic Cardiac Studies; Doppler Blood Flow Measurement; Cardiac Function: Noninvasive Assessment

Bibliography

Bom N, Lancee C T, Van Zwieten G, Kloster F E, Roelandt J 1973 Multiscan echocardiography: Technical description. *Circulation* 48: 1066–74

Branthwaite M A, Bradley R D 1968 Measurement of cardiac output by thermal dilution in man. *J. Appl. Physiol.* 24: 434–38

Carr K W, Engler R L, Forsythe J R, Johnson A D, Gosinck B 1979 Measurement of left ventricular ejection fraction by mechanical cross sectional echocardiography. *Circulation* 59: 1196–206

Fegler G 1954 Measurement of cardiac output in anesthetised animals by a thermodilution method. *Q. J. Exp. Physiol.* 39: 153

Feigenbaum H, Wolfe S B, Popp R L, Haine C L, Dodge H T 1969 Correlation of ultrasound with angiocardiography in measuring left ventricular diastolic volume. *Am. J. Cardiol.* 23: 111

Ganz W, Donoso R, Marcus H, Forrester J S, Swan H J C 1971 A new technique for measurement of cardiac output by thermodilution in man. *Am. J. Cardiol.* 27: 392–96

Goldberg B B 1971 Suprasternal ultrasonography. *J. Am. Med. Assoc.* 215: 245–50

Goldman D E, Jueter T F 1956 Tabular data of the velocity and absorption of high-frequency sound in mammalian tissues. *J. Acoust. Soc. Am.* 28: 35–38

Goodyer A V N, Huvos A, Eckhardt W F, Ostberg R H 1959 Thermal dilution curves in the intact animal. *Circ. Res.* 7: 432–41

Griffith J M, Henry W 1974 A sector scanner for real time two-dimensional echocardiography. *Circulation* 49: 1147–52

Kubicek W G, Kottke F J, Ramos M V, Patterson R P, Witsoe D A, Labree J W, Remole W, Layman T E, Schoening H, Garamala J T 1974 The Minnesota impedance cardiograph—Theory and applications. *Biomed. Eng.* 9: 410–16

Light L H 1976 Transcutaneous aortovelography: A new window on circulation. *Br. Heart J.* 38: 433–42

Light L H, Cross G, Hanson P L, Brotherhood J, Hanson G C, Peisach A R, Sequeira R F 1976 Pilot evaluation of transcutaneous aortovelography. In: Woodcock J (ed.) 1976 *Clinical Blood Flow Measurement*. Pitman, London, pp. 107–14

Olson R M, Shelton D K 1972 A nondestructive technique to measure wall displacement in the thoracic aorta. *J. Appl. Physiol.* 32: 147–51

Pate T D, Baker L E, Rosborough J P 1975 The simultaneous comparison of the electrical impedance method for measuring stroke volume and cardiac output with four other methods. *Cardiovasc. Res. Cent. Bull.* 14: 39–52

Popp R L, Harrison D C 1970 Ultrasonic cardiac echography for determining stroke volume and valvular regurgitation. *Circulation* 41: 493–502

Sequeira R F, Light I H, Cross G, Raftery E B 1976 Transcutaneous aortovelography: A quantitative evaluation. *Br. Heart J.* 38: 443–50

Strauss H W, Zaret B L, Hurley P J, Natarajan K, Pitt B 1971 A scintiphotographic method for measuring left ventricular ejection fraction in man without cardiac catheterization. *Am. J. Cardiol.* 28: 575–80

Teichholz L E, Krenlen T H, Herman M V, Gorlin R 1972 Problems in echocardiographic volume determinations:

Echoangiographic corrections. *Circulation, Suppl. II* 46: 75

Weyman A E, Peskoe S M, Williams E S, Dillon J C, Feigenbaum H 1976 Detection of left ventricular aneurysms by cross sectional echocardiography. *Circulation* 54: 936–44

Williams B O, Caird F I 1980 Impedance cardiography and cardiac output in the elderly. *Age Ageing* 9: 47–52

A. R. Lorimer, W. S. Hillis and A. C. Tweddel

Cardiac Pacemakers: Computerized Data Handling

The successful treatment of cardiac dysrhythmias using cardiac pacemakers has resulted in an increasing number of pacemaker patients (Fig. 1). Many major implanting

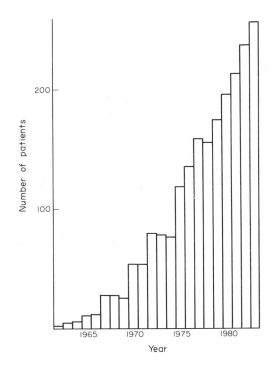

Figure 1
A histogram of the number of new patients receiving cardiac pacemakers in Glasgow during the period 1962–1983

centers now keep records of more than 1500 patients; data from these large numbers of patients must be stored for periods of 10–20 years.

In addition to precise patient record-keeping on a local scale, the proliferation of implanted pacemakers, using a variety of power sources, intended for different pacing modes (see *Cardiac Pacemakers, Implantable*)

and produced by more than 20 manufacturers, necessitates an objective method for analyzing their performance. Thus although manufacturers maintain records of the performance of their own hardware, groups of implanting centers have established their own pacemaker performance databanks, pooling their data. Although there is a proliferation of systems, their common objective is to provide accurate follow-up records for individual patients and to make independent estimates of the reliability and longevity of pacemakers from representative samples.

1. Data Collection

Data collected by manufacturers rely mainly on the factory analysis of faulty generators which have been returned from implanting centers during a warranty period. Clinical teams are interested also in pacemaker malfunctions which, though not necessarily technical, may be related to design. Examples are extrusion of the generator through the skin, displacement of the lead from the position chosen at implant, and increase in stimulation threshold (see *Cardiac Pacemakers, Temporary*). Several centers have established systems of data handling which accommodate both technical and clinical problems.

A system of data handling, established by Green (1975), uses a series of codes to define major faults. Failure of a conducting lead is defined as a break in the conductor or as faulty insulation. Generator failure is classified in two modes: the first caters for predicted changes in generator parameters that are consistent with depletion of the power source, while the second failure mode evidences unexpected (and therefore potentially more dangerous) changes.

Data are entered into the databank at the time of the initial implant and at all subsequent pacemaker operations. At the initial implant, details of the patient's history and serial numbers of the implanted generator and lead are recorded. The reasons for subsequent surgical interventions and the findings at these interventions are summarized by implant lifetime codes and subcodes for both the generator and the lead. Implant lifetimes are calculated by the computer to provide a clear record of pacemaker performance. Patients who have died and patients who have been lost to follow-up are entered into the record to ensure that survival analysis is accurate.

2. Data Processing

The principal advantage of computer storage of patient data is that updated summaries of each patient's pacemaker history as well as current implant lifetimes are readily available. This information can be presented as a print-out produced at monthly intervals, or can be displayed on a visual display unit (VDU) in "on-line"

Table 1
A survival table for a pacemaker generator. The example shows part of the table (the first year) for a Medtronic 5950 generator

No. of months after implantation (x)	Implants at start of interval (l_x)	Withdrawn during interval (u_x)	Failed during interval (d_x)	Effective number exposed to failure (l'_x)	Monthly failure rate (q_x)	Proportion surviving (p_x)	Cumulative survival (P_x)	Standard error (s_x)
1	405	32	0	389.0	0.000	1.000	1.000	0.000
2	373	11	1	367.5	0.003	0.997	0.997	0.003
3	361	10	0	356.0	0.000	1.000	0.997	0.003
4	351	10	0	346.0	0.000	1.000	0.997	0.003
5	341	8	0	337.0	0.000	1.000	0.997	0.003
6	333	3	1	331.5	0.003	0.997	0.994	0.004
7	329	10	0	324.0	0.000	1.000	0.994	0.004
8	319	9	0	314.5	0.000	1.000	0.994	0.004
9	310	3	0	308.5	0.000	1.000	0.994	0.004
10	307	8	1	303.0	0.003	0.997	0.991	0.005
11	298	1	1	297.5	0.003	0.997	0.988	0.006
12	296	1	0	295.5	0.000	1.000	0.988	0.006

$$l'_x = l_x - \tfrac{1}{2}u_x; \quad q_x = d_x/l'_x; \quad p_x = 1 - q_x; \quad P_x = p_1 p_2 p_3 \ldots p_x; \quad s_x = P_x \left\{ \sum_{k=1}^{x} [q_k/(l'_k - d_k)] \right\}^{1/2}$$

systems. Such technical summaries are useful at pacemaker follow-up clinics and particularly at emergency admissions of patients.

Of the several methods of pacemaker data analysis that have been proposed, the cumulative survival table (Schaudig et al. 1975) is the most common. This method permits processing of data from patients who report for follow-up observation at different intervals. In addition to patients who have died or who have been lost to follow-up, patients with incomplete observation periods can also contribute to the data analysis—for the duration of those periods. The method follows the same procedure used in the analysis of patient death rate after the diagnosis of a disease. The effective number (l'_x) exposed to risk of failure or death in successive intervals x is the number entering the interval minus half the number withdrawn during the interval. Thus the failure or death rate (q_x) is the number failing in the interval x divided by the effective number exposed. The proportion surviving, p_x, is $(1 - q_x)$ and the cumulative proportion surviving, P_x, to the end of interval x is $P_x = p_1 p_2 p_3 \ldots p_x$. A typical survival table is shown in Table 1.

Application of this type of analysis permits reliability predictions for pacemaker generators. A follow-up regime can be established for each type of pacemaker, together with a planned policy of replacement, benefiting both the patient and the implanting center by minimizing premature and emergency surgery.

3. Databank Systems

There are basically three types of pacemaker databank systems;

(a) off-line, mainframe computer systems,

(b) on-line systems on a dedicated minicomputer with disk storage of data, and

(c) on-line systems on a dedicated minicomputer or microcomputer with floppy-disk storage.

The first two types can readily provide statistical analyses on a national basis. Although the batch processed mainframe option is relatively cheap, the dedicated minicomputer linked by telephone modems to large implanting centers offers greater flexibility since it also allows off-line batch processing for smaller centers.

The floppy-disk storage system can store pacemaker data for only one hospital or one group of hospitals but it can be used for other cardiological applications, e.g., cardiac catheterization data analysis. At follow-up clinics, a program of operator–computer interaction can be devised. After checking that correct capture and sensing are observed on the electrocardiogram (ECG), changes in the parameters of the pacemaker can be compared with statistical predictions to determine the probability of the pacemaker surviving until the next programmed visit to the clinic. The computer can produce lists of patients overdue for appointment and those at risk, in addition to generating letters to each patient's family practitioner.

See also: Computers in Cardiology

Bibliography

Green G D 1975 *The Assessment and Performance of Implanted Cardiac Pacemakers.* Butterworth, London
McGregor D C, Covvey H D, Noble E J, Smardon S D, Wilson G J, Goldman B S, Wigle E D 1980 Computer-assisted reporting system for the follow-up of patients with cardiac pacemakers. *Pace* 3: 568–88

Schaudig A, Zimmermann M, Thurmayr R, Kreuzer E, Reichest B 1975 Computer applications for monitoring of pacemaker patients. In: Schaldach M, Furman S (eds.) 1975 *Advances in Pacemaker Technology*. Springer, Heidelberg

G. M. Brewster and A. L. Evans

Cardiac Pacemakers, Implantable

The development of small, high-capacity electrical power sources and the increasing sophistication and reliability of electronic components have accelerated the use of implanted cardiac pacemakers in treating many types of cardiac dysrhythmias. The ultimate aim is to design pacemakers which, once implanted, will function throughout the lifetime of the patient with no need for further surgery.

Natural contractions of the chambers of the heart are controlled by the conduction of electrical impulses through specialized muscle fibers. The electrical signal originates at the sinoatrial (SA) node of the right atrium and spreads throughout the atria (Fig. 1). The only

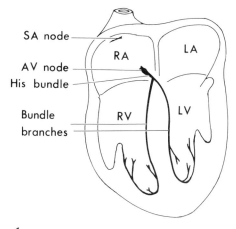

Figure 1
The natural conduction system of the heart: LA, left atrium; RA, right atrium; LV, left ventricle; RV, right ventricle

conducting tissue joining the atria to the ventricles is the atrioventricular (AV) node. The AV node, the His bundle and its bundle branches form a slow conduction path which delays the initiation of ventricular contraction. This delay allows the ventricles to be filled with blood pumped from the atria. Defects in the conduction system cause abnormal heart rhythms and inefficient cardiac functions. An artificial cardiac pacemaker provides a more appropriate sequence of electrical stimuli to the heart. The pacemaker consists of a pulse generator and an electrode–lead system for conveying the pulses from the generator to the heart wall.

1. Indications for Artificial Cardiac Pacing

The most common indications for the need to implant a cardiac pacemaker are dizziness or periods of unconsciousness caused by a diminished cerebral blood flow resulting from a slow ventricular rate. This reduction in heart rate is most often caused by a conduction block at the AV node which prevents atrial signals reaching the ventricles. In this situation, ventricular rate is controlled by a slower focus of electrical activity on the ventricular conduction system; intrinsic ventricular rates are typically 30–40 beats per minute. Abnormal activity of the SA node with normal atrioventricular conduction also causes slow or variable heart rates which can be controlled by the use of an implantable pacemaker.

2. Techniques for Pacemaker Implantation

Artificial cardiac pacing was first achieved by opening the chest and supplying stimuli via electrodes sewn to the outer surface of the heart or inserted into the wall of the heart. Now, with the development of cardiac catheterization techniques (see *Cardiac Catheterization*), flexible leads are passed through a major vein to introduce electrodes into the right side of the heart (Fig. 2). Typically a lead is passed, under radiographic control,

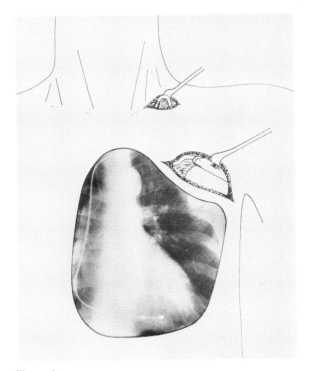

Figure 2
A composite diagram showing the generator site and the entry of the lead into the vein. The location of the tip of the lead in the apex of the right ventricle is shown in the radiograph

into the right atrium or ventricle and manipulated so that its tip lodges in a stable position in the chamber to be paced. Pacing threshold voltages (see *Cardiac Pacemakers, Temporary*) are accurately measured because a slow progressive increase in stimulation threshold is experienced with long-term pacing electrodes over a period of years, and this may lead to loss of pacing. The endocardial electrogram is measured through the implanted lead system to ensure adequate detection of the signal by the generator. For most generator designs, the detected signal should be greater than 2 mV.

The lead is usually connected to the generator by a screw on a pin-and-socket fitting. The generator and lead are then implanted beneath the skin or beneath a layer of muscle—usually in the pectoral region.

3. Cardiac Pacemakers

A pacemaker consists of an electrical power source and pulse generator circuit together with an electrode–lead system to transfer electrical signals to and from the heart.

3.1 Electrical Power Source

The ideal pacemaker power source should be small and light and should maintain output for the natural lifetime of the patient. The first implanted pacemakers used mercury–zinc cells, giving a generator lifetime of some 30–36 months. Longer-lived cells using radioisotopes have also been used as power sources, for example, 5 Ci of ^{238}Pu (half-life 87 years). The radioisotope serves as a source of heat which is converted to electrical energy by the Seebeck (thermoelectric) effect using bismuth telluride elements. The weight of shielding around the radioactive source and the need for its recovery after the removal of the generator from the patient have limited the use of this type of power source.

Most generators now in use have power cells with lithium anodes. Lithium forms a strongly reactive cation which yields high cell voltages with a number of cathode materials. Two types of such cells are shown in Fig. 3. The most common solid-state cell used for cardiac pacemakers has a lithium anode and a cathode of molecular iodine bonded to poly(vinyl pyridine). No electrolyte is included in the cell construction because a crystalline lithium iodide electrolyte forms between the anode and the cathode during the chemical reaction. As the thickness of the electrolyte layer increases during the cell lifetime, the internal resistance of the cell rises by 50–100 Ω per month, producing a corresponding drop in cell voltage. When the iodine in the cathode is depleted, the emf of the cell falls rapidly.

The cell developed by the Société des Accumulateurs Fixes et de Traction (SAFT) in France is an example of a lithium cell employing a liquid electrolyte in the form of lithium perchlorate dissolved in propylene carbonate. The electrolyte is absorbed in a polypropylene separator between the anode and the cathode. The anode is a

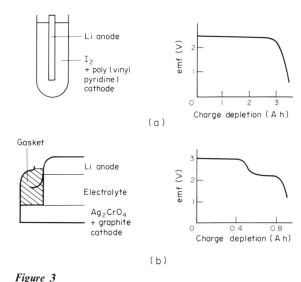

(a)

(b)

Figure 3
Two examples of lithium anode cells used for pacemaker generator power sources, and their corresponding discharge curves: (a) is a lithium iodide solid state cell, and (b) is a SAFT cell which has a liquid electrolyte

lithium disk and the cathode is a pellet of silver chromate and graphite. The cell voltage remains constant during the reduction of silver chromate to lithium chromate, subsequently decreasing to a lower plateau during the further reduction of the chromate ion to more stable chromium oxide. This double plateau serves as an indicator of cell depletion.

3.2 Pulse Generator Circuit

The pulse generator circuit typically includes an amplifier, an oscillator and a voltage doubler (Fig. 4).

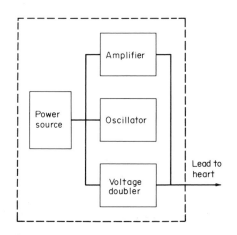

Figure 4
A schematic diagram of the function components of a pacemaker

These components may be hermetically sealed within a single hybrid package. The oscillator determines the desired pacing rate and pulse duration. The amplifier senses the endocardial electrogram and defines a refractory period to provide protection against electrical interference. Following a pacemaker pulse or a sensed endocardial signal the refractory circuit prevents the amplifier from sensing for a predetermined period, typically 325 ms. In the presence of strong continuous electrical interference the amplifier constrains the oscillator to emit pulses at a fixed rate. The voltage doubler boosts the oscillator output to provide pulses of about 5 V amplitude at the pacemaker electrodes.

The use of modern, integrated circuitry results in a lower current drain from the power source of some 5–25 μA. When the generator output is inhibited by intrinsic cardiac signals the current drain is lower (some 4–8 μA).

3.3 Electrode–Lead System

Either a unipolar or a bipolar electrode system is used to stimulate the heart. In a unipolar system the electrical stimulus is applied between a single electrode in contact with the heart wall and a second electrode on the pulse generator. In bipolar stimulation the pulse is applied between two electrodes on the heart wall. Bipolar systems are less susceptible to interference from external electromagnetic sources. Typical impedances presented to the generator are 400 Ω for unipolar pacing and 600 Ω for bipolar systems.

Leads are usually constructed of helical spring coils made of platinum–iridium or Elgiloy (a high fatigue-strength alloy) and sheathed in silicone rubber. (Polyurethane insulation has been investigated as a means of manufacturing leads of smaller diameter but its use is not common.) A lead must withstand flexing some 100 000 times per day for the remainder of the patient's life.

Electrodes must not suffer electrolytic degradation or corrosion as a result of polarization effects. The most widely used electrode material is, again, platinum–iridium. An ideal electrode design would produce a high current density in tissue with a low current density at the metal–tissue interface. Electrodes currently used have surface areas of 8–12 mm^2.

3.4 Pacemaker Function

Early pacemakers emitted pulses at fixed time intervals bearing no relation to the patient's natural heart rate (Fig. 5a). To avoid the complication of possible competition between the artificial and the natural cardiac pacemaker, the demand generator has been developed (Fig. 5b). This type of generator, which today is used in more than 90% of pacemaker implants, continuously detects any natural endocardial electrical signals which are present and outputs pacing pulses only if these natural signals are insufficient to maintain heart function. In the absence of an adequate natural heart rate these generators pace the heart at a fixed (asynchronous)

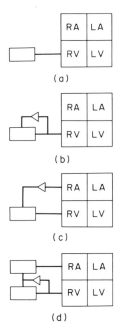

Figure 5
Functional diagrams of different type of pacemakers: (a) an asynchronous ventricular pacemaker delivering pulses at a fixed rate, (b) a demand pacemaker which can be inhibited by the endocardial signal, (c) a synchronous model which senses atrial activity and paces the ventricle at the same rate, and (d) an atrioventricular sequential pacemaker which senses ventricular activity and can pace both the atrium and the ventricle. Rectangles and triangles represent oscillators and amplifiers, respectively

rate. Such generators cannot increase the heart rate in response to physiological demands but this problem can be overcome for patients with normal SA node function by using generators which detect the atrial signal and pace the ventricle synchronously (Fig. 5c). These atrial synchronous pacemaker systems require the placement of electrodes in both the atrium and the ventricle. In the absence of an atrial signal the ventricle is paced asynchronously.

The closest approach to the artificial production of a "natural" heart rhythm is the atrioventricular sequential pacemaker (Fig. 5d). The generator detects ventricular signals but can pace the atrium and the ventricle, with a preset delay between the pacing impulses. When the ventricular rate is too slow, a pacing pulse is transmitted to the atrium. The generator then waits for this signal to be conducted to the ventricle. In the absence of normal AV conduction, a subsequent pacing impulse is transmitted to the ventricle to produce the natural atrioventricular sequence of contraction. The AV sequential pacemaker also requires electrode placement in both the atrium and the ventricle.

Although the atrial synchronous and atrioventricular sequential pacemakers offer more efficient pacing of the heart, their widespread use is presently limited by the need to implant more than one electrode–lead system. The majority of pacemaker patients are over 60 years of age and are able to maintain (for their lifestyle) an adequate cardiac output at a heart rate of 70 beats per minute.

Some early pacemakers were designed so that the pulse rate, and in some cases the pulse amplitude, could be altered by a form of magnetic coupling. A more recent related design involves the transmission of digital pulses by inductive coupling between an external programmer and a receiving antenna in the implanted pacemaker, making it possible to change the rate, amplitude and duration of the pacing pulse, the sensitivity, the refractory period, and even the mode of pacing, with no discomfort to the patient. Thus throughout the lifetime of the implant the lifetime of the generator can be maximized by reprogramming the output to minimize the current drain on the power source while retaining an adequate margin of safety for the patient. The programmed instructions to the generator are preceded by an identification code to prevent accidental reprogramming by external sources.

See also: Cardiac Output Measurement; Cardiac Pacemakers, Temporary; Cardiac Pacemakers: Computerized Data Handling; Electrocardiography; Cardiac Function: Noninvasive Assessment

Bibliography

Camm J, Ward D 1983 *Pacing for Tachycardia Control.* Telectronics, Sydney
DeGraff A C, Frieden J (eds.) 1977 Appraisal and reappraisal of cardiac therapy: Cardiac pacing and pacemakers. *Am. Heart J.* 94: 115–24, 250–59, 378–86, 517–28, 658–64, 795–804
Furman S, Escher D J W 1970 *Principles and Techniques of Cardiac Pacing.* Harper and Row, New York
Nieveen J (ed.) 1981 *Arrhythmias of the Heart.* Excerpta Medica, Amsterdam

A. L. Evans and G. M. Brewster

Cardiac Pacemakers, Temporary

It is often necessary to control a patient's heart rate, quickly in some circumstances, by means of an external cardiac pacemaker. Electrical impulses are applied by the pacemaker to the heart wall using an electrode–lead system introduced into the right side of the heart via a superficial vein.

A need for temporary pacing may be indicated in the following circumstances:

(a) a slow heart rate produced either by sinus-node bradycardia or by heart block (an abnormality in the electrical conduction system of the heart);

(b) tachycardia, which may be controlled either by "overdrive" pacing, the heart being driven by the pacemaker at a rate faster than the underlying natural rate, or by interruption pacing in which a reentry-type arrhythmia is stopped by an appropriately timed stimulus; and

(c) post-surgical management, particularly after open-heart surgery, to maintain a sufficient output of blood from the heart.

1. Pacing Techniques

Although it is possible to pace the heart via external electrodes positioned on the chest or via electrodes applied to the myocardium by percutaneous needle puncture, these methods are a traumatic or painful alternative to the simple transvenous approach. In this method, the veins at the root of the neck or the brachial or the femoral veins are used to introduce a lead; the vein is either exposed by dissection or entered by a needle percutaneously. The pacing lead is passed into the vein and advanced, usually under radiographic control, into the right ventricle (or into the right atrium for atrial pacing).

When the lead is in a stable position in the apex of the ventricle, electrical tests are performed to ensure that the electrodes are making good contact with the heart muscle. The negative output terminal of the pacemaker generator is connected to the electrode situated at the tip of the lead and the positive terminal to the proximal electrode which is usually some 2 cm from the tip. The application of a 5 V pulse of 2 ms duration is usually

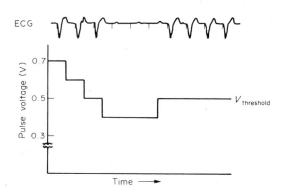

Figure 1
Determination of the threshold stimulation voltage for pacing. The pulse voltage is reduced in steps until capture is lost

sufficient to ensure "capture"—the spread of depolarization from the point of contact of the electrode. The pulse voltage may now be reduced (Fig. 1) until capture is just lost, and then increased to the capture threshold voltage. The generator output is eventually set at some three times the threshold value to allow for possible short-term change in heart-muscle response. The lead is then tied securely to the vein at the point of entry and a dressing is placed over the wound. Temporary pacing leads are commonly left in situ for ten days. Care must be taken to ensure that the lead is not displaced from its position in the heart by, for example, a vigorous arm movement when the lead has been inserted via a brachial vein.

2. The Pulse Generator

Physiological parameters affecting the design of the generator are the heart rate, which is usually maintained at 70 beats per minute in simple heart block, and the threshold voltage, the value of which decreases as the pulse width increases (Fig. 2), although there is no

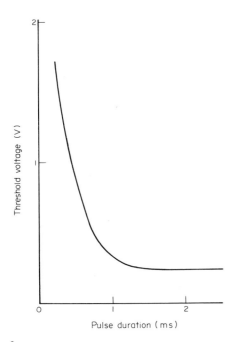

Figure 2
Variation in the threshold voltage with the duration of the pacemaker pulse

advantage in using a pulse width greater than 2 ms. In addition, if the generator is to supply an electrical pulse only when the natural beat-by-beat heart rate falls below its normal rate (i.e., when the period between two heart beats exceeds 857 ms), the generator must be able to detect the heart's own electrical activity transmitted via

the electrodes and lead. Such pacemaker generators are termed demand generators; their mode of operation and pertinent signals are illustrated in Fig. 3.

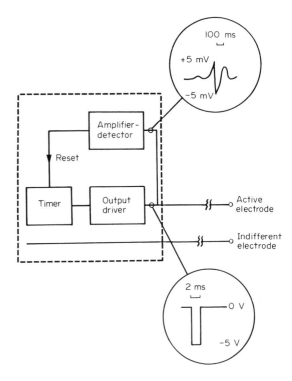

Figure 3
Schematic diagram of a demand pacemaker generator showing typical input and output signals

Bipolar ventricular electrograms exhibit an intrinsic deflection (corresponding to the passage of the cardiac depolarization wave) of 11.75 ± 6.02 mV amplitude and 2.82 ± 0.81 V s^{-1} slew rate (Furman et al. 1977). When the pacemaker's amplifier–detector system senses the presence of an electrical signal in the heart, the timer is reset, and will then generate a pulse every 857 ms unless its function is interrupted by the sensing amplifier. The output driver presents this pulse to the heart in a suitable form. Demand generators have a refractory period, after a pacing stimulus and after a sensed R-wave signal, during which the amplifier–detector system is incapable of sensing a signal, thus preventing the timer from being reset too soon.

3. Pacing Problems

Most difficulties encountered with temporary pacing systems are due either to faulty positioning of the electrodes (or a displacement of the lead) or to a change in the threshold level of stimulation. Satisfactory pacing is restored by repositioning the lead or by increasing the pulse voltage.

External pacemaker generators operating in demand mode are more susceptible to electrical interference than implanted generators. Although the units are designed to revert to a fixed rate mode when there is continuous electrical interference, pulsed interference signals at a repetition rate near that of the pacemaker can cause complete inhibition of the pacemaker. The problem is resolved either by removing the source of interference or by switching to fixed-rate pacing.

Pacemaker generators are battery powered: since they are directly connected to the heart, their design must ensure that there is no chance of hazardous leakage currents passing through the heart. The accepted upper limit for such leakage currents, at frequencies below 1 kHz, is 10 μA (see *Medical Electrical Equipment: Safety Aspects*).

See also: Cardiac Pacemakers, Implantable; Cardiac Pacemakers: Computerized Data Handling

Bibliography

Furman S, Escher D J W 1970 *Principles and Techniques of Cardiac Pacing.* Harper and Row, New York
Furman S, Hurzeler P, de Caprio V 1977 Cardiac pacing and pacemakers—III. Sensing the cardiac electrograms. *Am. Heart J.* 93: 794–801
Nieveen J (ed.) 1981 *Arrhythmias of the Heart.* Excerpta Medica, Amsterdam
Schaldach M, Furman S 1975 *Advances in Pacemaker Technology.* Springer, Berlin

A. L. Evans and G. M. Brewster

Cell Electrophoresis

Particles which are electrically charged with respect to a suspending medium will move under the influence of an electric field. Such particles are said to possess an electrophoretic mobility, the magnitude of which depends on their surface charge, the physical properties of the medium, and the magnitude of the applied field. The source of the charge in cells, resultant electrokinesis, its measurement, and some applications of the technique are outlined.

1. Origins of Surface Charge

The charge on the surface of a particle may arise from:

(a) Ionization, as occurs, for example, with the carboxyl and amino groups of proteins. The degree of ionization depends closely on pH so that the net charge may be positive at low pH, negative at high pH, and zero at the isoelectric point. The charge on mammalian cells is due largely to the carboxyl groups of sialic acid, thus all these cells exhibit a negative charge at physiological pH 6.8–7.4.

(b) Ion adsorption in aqueous solution. Cations have a greater degree of hydration than anions, and so tend to remain in solution, whereas smaller anions may be adsorbed onto particles, thus imparting a positive charge.

(c) Ion dissolution of sparingly soluble compounds. This is often not equal for the pair of ions, and may be adjusted by the concentration of one of the pair of ions. Heavy-metal oxides and hydroxides are sensitive to the hydrogen-ion concentration which thus affects surface charge.

Most surfaces in a medium thus possess a surface charge and hence a surface potential termed the ψ potential, which is usually negative. The charged surface attracts counterions from the medium which in turn attract further ions of opposite charge, forming a diffuse double layer. The inner zone of adsorbed ions and the associated counterions forms a region of high viscosity and decreasing potential. The limit of this is the Stern plane beyond which is a region of shear. In an electrophoretic measurement the expressed charge or ζ potential is that at the surface of shear. The relationship of the ψ and ζ potentials is shown in Fig. 1. Factors which affect this are defined by the Smoluchowski equation for surfaces

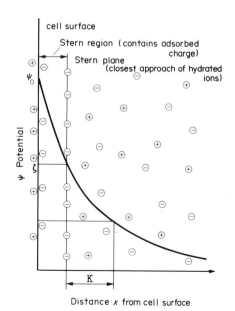

Figure 1
Schematic representation of the electric double-layer to show the relationship of ψ and ζ potentials in an ionic medium

with a large Ka, the ratio of the hydrodynamic "thickness" to the particle radius. (This thickness is usually termed $1/K$, the distance over which the potential decreases by an exponential factor at low potentials.) The electrical mobility is given by

$$U = \frac{\epsilon \zeta}{\eta}$$

where U is the mobility in $m^2\,s^{-1}\,V^{-1}$, ϵ is the dielectric constant, and η is the viscosity in $kg\,m^{-1}\,s^{-1}$. This can be related to intrinsic (σ_0) and adsorbed (σ_β) charges by:

$$\sigma_0 + \sigma_\beta = K \cdot 2kT \sinh(ze/2kT)$$

where $K = 2n_0 e^2 z^2/kT$, n_0 is the ionic concentration, e is the electronic charge, z is the valency, k is Boltzmann's constant, and T is the absolute temperature.

2. Electrokinetic Measurement

There are a number of conditions which arise due to the relative motion of a fluid and a charged surface, including:

(a) electrophoresis—the movement of a charged particle relative to a fluid due to an applied electric gradient, and

(b) electroosmosis—the movement of a fluid relative to a charged surface due to an applied electric gradient.

The converse of these can occur and give rise to:

(c) streaming potential—an electric field created by the movement of a liquid relative to a charged surface (e.g., flow through a glass tube), and

(d) sedimentation potential—an electric field created by the settling of charged particles in a fluid.

Both electrophoresis and streaming potential may be readily measured, and electroosmosis is an important complication in the practical measurement of electrophoretic mobility.

3. Measurement of Electrophoretic Mobility

The earliest and simplest techniques use microscopic viewing of the particles with manual timing of particles across a fixed distance, usually in two opposite directions to compensate for drift. Then

$$U = v/E$$

where v is the velocity in $m\,s^{-1}$ and E is the gradient in $V\,m^{-1}$. Conventionally these units are quoted $\times 10^8$ to give values of U in $\mu m\,s^{-1}\,V^{-1}\,cm$. Limitations on the particle velocity are imposed by Joule heating effects taking into account the voltage gradient and conductivity. Typical limiting values in physiological media (0.150 M) are about $3\,kV\,m^{-1}$ in 1 mm diameter

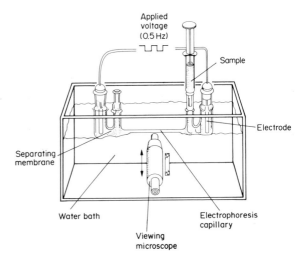

Applied voltage (0.5 Hz)
Sample
Electrode
Separating membrane
Water bath
Viewing microscope
Electrophoresis capillary

Figure 2

A simple microelectrophoresis apparatus for cells and particles greater than 10^{-6} m, showing essential movements for stationary-layer location. The electrodes may be nongassing and enclosed in the sample volume, or separated by sinters, or may be gassing and separated from the sample by semipermeable membranes

chambers and $1\,kV\,m^{-1}$ in 2 mm chambers. Figure 2 shows a typical layout of a microelectrophoresis apparatus. Positive counterions adsorbed onto the negatively charged glass move under the influence of the applied field setting up electroosmotic flow. The velocity profile of particles in this system is parabolic and true electrophoretic velocity occurs only at the stationary layer where electroosmotic velocity is balanced by the return flow (Fig. 3). In a cylindrical system the stationary layer is a coaxial layer positioned at s, where:

$$s = 0.293 \times radius \text{ (from either inner surface)}$$

In a rectangular container where height $h \gg$ wall separation d, the position of the stationary layer is found by

$$s = 0.177d \qquad h/d = 5$$
$$s = 0.211d \qquad h/d = \infty$$

Methods exist which reduce electroosmotic flow to a negligible amount. A coating on the glass (or container) wall such as that of a protein at its isoelectric point (e.g., myoglobin at pH 6.9) can give effective neutrality. Creation of a region of high viscosity at the glass surface by use of cross-linked methyl cellulose is an effective method of reducing electroosmosis to low levels. Closely spaced electrodes at some distance from the container wall are free from electroosmosis at the point midway between the electrodes. Unless the stationary layer can be located with precision, such techniques are necessary for many automatic methods.

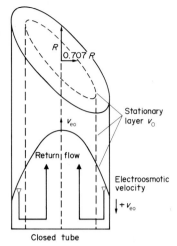

Figure 3
The electroosmotic generation of a parabolic flow
profile in a tube results in a coaxial stationary layer.
It is only in this plane that particle velocity is true
electrophoretic mobility and is normally selected by
depth-of-field limitations

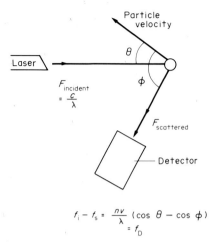

$$f_i - f_s = \frac{nv}{\lambda} (\cos\theta - \cos\phi)$$
$$= f_D$$

Figure 4
Schematic of the relationship between particle angular
velocity with respect to incident light direction and
viewing direction, and the Doppler shift for a given
wavelength. In a medium of refractive index n, the
effect on the wavelength must be included

Automatic techniques employ grating systems or are
based on the Doppler (frequency) shift. With grating
systems, an image of particles can be projected onto a
vidicon and velocity can be determined by the rate of
cutting the scan lines. The use of a moving grating to
obtain signals from moving particles and stationary
reference objects allows velocity and direction (polarity)
to be measured.

There are a number of techniques based on the
Doppler (frequency) shift of light scattered from moving
particles. Although coherence is not essential for some
of these methods, a laser is a convenient source of
high-intensity light and does allow the use of coherent
detection. The frequency shift f obtained is determined
by the angular velocity θ of the particle with respect to
the source of light and with respect to the viewing
direction ϕ, related (Fig. 4) by the general Doppler
equation:

$$f = \frac{nv}{\lambda} (\cos\theta - \cos\phi)$$

where λ is the wavelength, n is the refractive index, and
v is velocity. Frequency is proportional to velocity and
can be obtained by heterodyning the scattered light with
an unshifted reference beam at the detector surface.

A special case of the Doppler method relies on
particles crossing two intersecting beams which are at
symmetrical angles with respect to the particle movement (Fig. 5). Here positive and negative Doppler-
shifted signals are generated, which can then be combined to give a Doppler difference signal which is twice
the frequency of the heterodyne product and which is
independent of viewing direction. This has the advantage

of allowing a high solid angle of light collection but the
disadvantage of signal degradation if there are many
scatterers in the measurement volume. Thus the reference technique is suitable for high concentrations of
small particles, and the differential (or incoherent)
method for small numbers of particles.

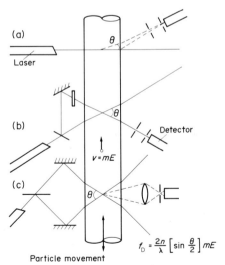

$$f_D = \frac{2n}{\lambda} \left[\sin\frac{\theta}{2} \right] mE$$

Figure 5
Three Doppler systems used in cell electrophoresis;
(a) parasitic heterodyne, using scatter from the
container walls as a reference ($\theta_{max} = 20°$),
(b) reference mode which uses an attenuated portion
of the light on the axis of the detector ($\theta_{max} = 60°$),
(c) Doppler difference technique heterodynes the two
beams in the measurement volume and thus allows a
large solid angle of light collection (m is unit mobility)

4. Applications

Since electrophoretic mobility is an expression of the net charge density on the particle surface, alteration of surface groups can be monitored by means of cell electrophoresis. This is of practical application in the study of aggregation behavior of particulate material in aqueous and nonaqueous media. Flocculation occurs in solid–water suspensions at decreasing ζ potential which itself depends on pH and ionic concentration. Suspensions will deflocculate at high pH and low salt concentrations due to hydration and electric double layer repulsion. Deflocculating agents can be studied by cell electrophoresis and are of importance in soil structure, borehole drilling, water purification, and provision of stable building substrates. Trivalent cations are efficient at lowering ζ potentials and are used to flocculate industrial wastes. High ζ potentials are necessary to maintain flow in paints, aerosols and coolants. The measurement of electrophoretic mobility is a direct guide to the behavior of such substances.

The biological action of inorganic particles is related to their surface charge. Hemolysis of erythrocytes by quartz particles is effected by ionized silanol groups and can be reduced by metal cations and by proton-accepting polymers such as poly(vinyl pyridine-N-oxide). The presence of trace elements affects the lytic ability of quartz as well as its ζ potential, thus having implications in the likelihood and prevention of silica-related industrial disease.

The carcinogenicity of asbestos dusts appears to be related in a certain way to their surface charge as well as to their crystal shape. The mechanism of the particle–cell interaction is not fully understood, however, since in the presence of tissue proteins particles take on the charge of the adsorbed material. Thus crocidolite which in 150 mM NaCl has a mobility of 2.5–3.0 units has a mobility of only 0.75 in quite low concentration of serum.

Crystals generated *in vivo* such as hydroxyapatite, urate or brushite are associated with the inflammatory diseases of rheumatoid arthritis, osteoarthiritis and gout. The inherent ζ potential of these particles is directly related to the inflammation they cause, and may be modified by a number of agents. Coating of the crystals to reduce the ζ potential also reduces inflammation, as does any physical treatment of the crystals which produces a lower charge.

Certain aspects of bacterial growth have been studied by cell electrophoresis. Bacterial cells in log phase exhibit a higher ζ potential than at any other phase and there are characteristic pH/mobility curves which reflect surface groups (Shaw 1969). The presence of particular substances in the bacterial cell wall or membrane can be characterized by the effect of specific enzymes on the mobility of the whole particle. For example, a lack of response of electrophoretic mobility to the enzymes trypsin, ribonuclease and lipase suggests that the substrates for these enzymes are absent whereas they are present in mammalian membranes. The binding of detergents, disinfectants, and other bacteriocidal agents to the bacterial wall can be demonstrated by a change in electrophoretic mobility, as can the adaptation of bacteria to these agents by a change in the bacterial cell wall.

Fungal cells, like bacteria, exhibit surfaces with ionizable groups, except, for example, the sporangia of a phycomycete. The pure cellulose nature of these sporangia imparts zero charge at any pH. However, the spores carry a charge due to carboxyl groups in their structure. The structure exhibits binding of antifungal agents, and the mobility of the spores can be reduced or even reversed by surfactants such as dodine (*n*-dodecylguanidine acetate), but this mobility change is probably not the basis of the antifungal action.

Mammalian red blood cells owe their charge to the carboxyl groups of acylated neuraminic acid (sialic acid) and the resultant mobility is characteristic of the species. Values obtained in 150 mM saline at pH 7.2 are: rabbit, -0.5; guinea pig, -1.01; human, -1.07; rat, -1.29; and dog, -1.25 units ($10^{-8}\,m^2\,s^{-1}\,V^{-1}$). These are normally stable values. Bimodal distributions are found in the rat, due to mobility differing with cell age; unimodal distributions occur in humans and rabbits with a standard deviation of about 3–4%. Variations do occur in disease, probably due to factors adsorbed from the plasma, but real differences have been demonstrated in the membrane fatty acid binding, for example in multiple sclerosis.

Platelets are cellular blood components which normally have a mobility of 0.78 units, but this can alter in response to adrenaline and to adenosine diphosphate. The shape of the dose–response curve is diagnostic for cardiovascular disease related to clot formation.

Cell electrophoresis is also used extensively to study components of the lymphoid system. Lymphocytes show two or three subgroups; monocytes, one major group; and granulocytes, two or three subgroups of different mobility. In leukemia, cell electrophoresis demonstrates the presence of a single distribution of a predominant cell type corresponding to the type of disease, be it B, T or null cell. There have been reports of direct lymphocyte responses to antigen challenge in the form of a 20% increase in mobility, and also demonstrations of granulocyte and macrophage decrease in mobility in response to lymphocyte products. It appears that it is possible to demonstrate an *in vitro* model of an *in vivo* function using cell electrophoresis.

See also: Bioelectricity

Bibliography

Drain L E 1980 *The Laser Doppler Technique*. Wiley, London

Preece A W, Sabolovic D 1979 *Cell Electrophoresis: Clinical Applications and Methodology*. Elsevier–North Holland, Amsterdam

Shaw D J 1969 *Electrophoresis*. Academic Press, London
Sherbet G V 1978 *The Biophysical Characterisation of the Cell Surface*. Academic Press, London

A. W. Preece

Cell Population Kinetics

The tissues of higher plants and animals are composed of cells and the products of these cells. Within the body of a complex organism, such as man, cells are constantly being produced by division, undergoing differentiation and eventually dying. Cell population kinetics is aimed at understanding and quantifying these dynamic events in the life history of a cell population.

The process of mitosis, which is the normal pattern of nuclear division in both plant and animal cells, has been known for many years and is usually closely linked temporally to cell division. Despite the long-standing knowledge of mitosis, it is only relatively recently that the concept of a division or cell cycle has been recognized. Mitosis achieves the separation (into two daughter cells) of identical sets of genetic information which are contained in the DNA of the chromosomes. In 1953 Howard and Pelc discovered that new DNA was synthesized not during mitosis, as had been supposed, but during a separate discrete period prior to mitosis. This led to the concept of a cell cycle, in which a cell goes through a series of distinguishable phases between one mitosis and the next.

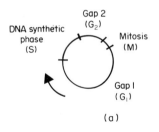

(a)

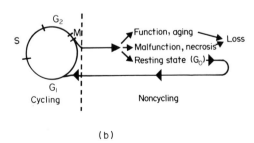

(b)

Figure 1
The cell cycle: (a) simple model illustrating the four phases of the cycle, (b) more complex model incorporating cell loss and the concept of growth fraction

The period of DNA synthesis is generally referred to as the S phase. The interval between mitosis and the S phase is Gap 1 (or G_1) while the period between S and mitosis is Gap 2 (or G_2) (Fig. 1a). Although these two phases were originally designated "gaps" in the proliferative cycle, it is becoming increasingly clear that many biochemical events, vital to cell proliferation and its control, occur during these phases.

1. Basic Cell Kinetic Parameters

The time taken for a cell to pass from one mitosis to the next is known as the cell cycle time t_C. In most real cell populations the kinetic situation is more complicated than depicted in Fig. 1a. Most populations are a mixture of proliferative and nonproliferative cells. This observation leads to the concept of the growth fraction or proliferative index I_P, which is defined as

$$I_P = N_C/N$$

where N_C is the number of proliferative cells and N the total number of cells in the population. The rate of cell production (new cells/unit number cells per hour) in a population depends on both the growth fraction and the average cycle time of the cells which comprise this fraction. Figure 1b illustrates a more complex cell kinetic model into which a noncycling fraction of cells has been incorporated. As is also shown, cells which leave the cycle do not necessarily share the same fate. They may differentiate irreversibly, function and eventually die; they may undergo immediate necrosis; or they may enter a resting phase in which they retain the ability to reenter the proliferative cycle under suitable conditions or stimuli. This resting but fertile state has been termed G_0.

The overall growth of a cell population depends on the balance between cell production and cell loss due to death or to migration from the population. Cell loss is often expressed as a fraction of cell birth and this cell loss factor Φ is defined as (Steel 1968):

$$\Phi = k_L/k_B$$

where k_L is the rate of cell loss and k_B is the birth or cell production rate. Most tissues in adult animals such as skin or intestinal epithelium retain a fairly constant size, which does not imply a static situation, but rather a dynamic equilibrium between cell birth and loss. Such tissues are said to be in a kinetic steady state (i.e., $\Phi = 100\%$).

Growing cell populations are manifestly not in a steady state. The extreme case of growth is that represented by an exponential increase in cell number. In a pure population of exponentially growing cells, cell number doubles in a period equal to t_C. It should be noted that not all expanding populations grow in a simple exponential fashion as will be discussed later in relation to tumors.

2. Cell Kinetic Techniques

The methodology of cell kinetics has undergone a dramatic expansion in the past 25 years. For this reason, it is impossible to do more than mention a few of the commonly used techniques. For a more detailed description of these methods, their applications and drawbacks, the reader is referred to Aherne et al. (1977).

Perhaps the oldest technique to yield hard data involves the use of metaphase-arrest (or stathmokinetic) agents. Principal amongst these are colchine, colcemid and the Vinca alkaloids, vincristine and vinblastine. When proliferating cells are exposed to one of these agents under optimal conditions, all cells which enter mitosis are arrested in metaphase. By following the time course of metaphase accumulation the rate of entry of cells into mitosis and, thereby, the cell production rate can be measured.

However, the biggest boost to cell kinetics came with the synthesis of tritiated thymidine (^{3}H-TdR) in 1957 and subsequently ^{14}C-thymidine (^{14}C-TdR). Thymidine is incorporated specifically into DNA by cells in S; when radioactively labelled thymidine is administered to a cell population, any cells in S will become labelled and can subsequently be recognized. Thus by exposing a proliferating cell population to ^{3}H-TdR for a short interval (pulse labelling), the fraction of cells in S can be estimated; this is known as the labelling index (I_S):

$$I_S = N_S/N$$

where N_S is the number of cells in S and N is the total number of cells.

In addition, there are a number of techniques using ^{3}H-TdR or ^{14}C-TdR which yield quite detailed kinetic information. Primary amongst these is the fraction labelled mitosis (FLM) method, in which the cell population is pulse labelled with ^{3}H-TdR and the fraction of mitoses which are labelled is determined in samples taken at various times thereafter. Under ideal conditions, peaks of labelled mitoses are generated, as shown in Fig. 2. This method enables the duration of the cell cycle and its phases to be determined and, provided that

at least two clear peaks are obtained, the growth fraction can also be calculated. For details of this and other techniques using labelled nucleotides see Aherne et al. (1977).

Another method is flow cytometry, a technique which enables a rapid analysis of the cellular content of compounds, such as DNA, to be made. Individual cells stained with a fluorescent dye which binds quantitatively to DNA, are passed through a laser beam in a flow cytometer. The intensity of fluorescence of many thousands of individual nuclei or cells per minute can be measured and recorded. By this means DNA histograms can be compiled which allow the proportion of cells in the various cell cycle phases to be calculated (Fig. 3).

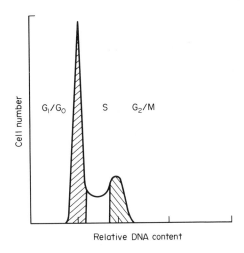

Figure 3
A typical DNA histogram for a proliferating cell population. From this curve the fraction of G_1/G_0 cells, G_2/M cells, and S phase cells can be calculated

Multiparameter analysis, in which RNA content or cell size, for example, is measured simultaneously with DNA content, is also possible and may allow, at least in some limited cases, proliferating cells to be distinguished from nonproliferating cells (Darzynkiewicz et al. 1979). This rapid technique does, however, have the drawback that suspensions of single cells or nuclei are necessary. Thus information from the spatial distribution of cells and from the architecture of the tissue is lost.

As stressed above, cell kinetics is a quantitative field of investigation. Unfortunately, the biological systems to which it is applied are complex and often poorly understood. The results obtained are usually averages which are often difficult to place within statistical confidence limits. For these reasons mathematical models play an important role in cell kinetics (Steel 1977), and computers have been widely used in the application of these models.

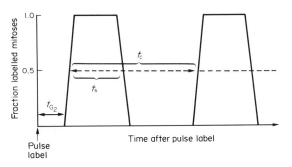

Figure 2
Two idealized peaks of labelled mitoses. In biological systems the peaks are damped due to variation in phase durations resulting in less sharply defined peaks

3. Cell Populations

Before carrying out kinetic experiments the cell population must be defined. This may be done on the basis of both functional and morphological criteria or by location within a tissue or organ. Further, using similar criteria, it may be possible to recognize various compartments or subpopulations of cells within the population of interest.

To clarify the situation, consider an example of a real cell population, the small intestinal epithelium. It is clear that this tissue has at least two distinct compartments: the fingerlike villi which project into the lumen of the intestine and carry out the digestive and absorptive functions, and the pit-like depressions (the crypts) where new cells are produced to replace those lost from the villi. Thus, the villi can be considered as the functional compartment and the crypts as the proliferative compartment. In fact, it is possible to recognize within the crypts several distinct cell kinetic compartments (Fig. 4).

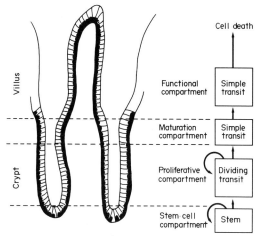

Figure 4
A diagrammatic representation of the kinetic compartments of the small bowel epithelium

It has been known for some time that mitoses and labelled cells are restricted to the lower two-thirds of the crypt. This suggests that the cells at the top have already left the proliferative cycle and embarked upon the differentiation pathway leading to functional cells on the villi. These cells therefore constitute a maturation compartment. A further compartment—the stem-cell compartment—has been suggested to exist, probably at the base of the crypts. The stem-cell concept is an important one. Stem cells are capable both of long-term self renewal and of supplying cells to other compartments. Only stem cells are capable of renewing a tissue after physical, chemical or radiation damage.

In Fig. 4, the various compartments of the small intestinal mucosa are analyzed in terms of their cell kinetic characteristics. The stem-cell compartment has

no input of cells but does show proliferation and an output of cells into the proliferative compartment. The proliferative compartment is a dividing transit compartment having input, further proliferation and eventual output of cells. Both the maturation and the functional compartments are simple transit compartments with input and output but no proliferation.

Cells are finally lost from the villi into the intestinal lumen. Other tissues can be analyzed in a similar manner to that described above, although the situation may be more complex and less well understood.

A detailed review of the cell kinetics of a wide range of epithelial tissues may be found in Wright and Alison (1984).

4. Tumor Cell Kinetics

A major stimulus to cell kinetic research has been the relevance it might have to an understanding of tumor growth and possibly to the improvement of therapy. A parameter often quoted for tumors is the doubling time t_D, which may be determined by measuring the increase in tumor size with time and thus constructing a growth curve. If tumors grew exponentially, their doubling time would remain constant with increase in tumor size. The slope of the curve plotted on semilog paper would give the growth rate k_G. Then

$$t_D = \ln 2 / k_G$$

In fact, most tumors do not maintain exponential growth, rather their growth rate slows as they age. A number of mathematical equations have been developed in an attempt to fit a wide variety of experimental data. Perhaps the most widely used equation which meets this criterion is the Gompertz equation:

$$M_t = M_0 \exp\{A/r[1 - \exp(-rt)]\}$$

where M is the mass of the population (M_0 at the start of observation and M_t at time t later), A is a growth rate constant, and r is a constant that governs the decrease in growth potential as tumors enlarge.

When values for parameters such as t_D or t_C are quoted for tumors they are generally average values for all tumor cells. In fact, tumors are known to be kinetically very heterogeneous. It has already been shown that within "normal" tissues it may be possible to discern compartments. In tumors, signs of such organization are generally obscure or lacking, depending on the degree of differentiation. However, in a given area, a tumor may comprise proliferative cells, resting G_0 cells, differentiating, end state, and dying cells. Of particular importance is the existence of cells with stem-cell properties. These cells are of vital importance in determining the regrowth of tumors after chemo- or radiotherapy. Tumor stem cells are impossible to recognize in situ and their existence and response to therapy can only be demonstrated indirectly. This is an area of great interest in current cell kinetic research.

Another area of great interest is the application of flow cytometry in the field of cancer research. This technique enables DNA content and many other characteristics of tumor cell populations to be determined rapidly. Flow cytometric data may aid in the diagnosis of disease and in determining the prognosis and treatment of choice of individual patients (Barlogie et al. 1983).

See also: Radiosensitizers; Radiation Carcinogenesis; Chemical Carcinogenesis; Cervical Cytology: Automation

Bibliography

Aherne W A, Camplejohn R S, Wright N A 1977 *An Introduction to Cell Population Kinetics.* Arnold, London

Barlogie B, Ruber M N, Schumann J, Johnson T S, Drewinko B, Swartzendruber D E, Göhde W, Andreeff M, Freireich E J 1983 Flow cytometry in clinical cancer research. *Cancer Res.* 43: 3982–97

Darzynkiewicz Z, Traganos F, Andreeff M, Sharpless T, Melamed M R 1979 Different sensitivity of chromatin to acid denaturation in quiescent and cycling cells as revealed by flow cytometry. *J. Histochem. Cytochem.* 27: 478–85

Howard A, Pelc S R 1953 Synthesis of deoxyribonucleic acid in normal and irradiated cells and its relation to chromosome breakage. *Heredity* 6 (Suppl. 2): 261–73

Steel G G 1968 Cell loss from experimental tumours. *Cell Tissue Kinet.* 1: 193–207

Steel G G 1977 *Growth Kinetics of Tumors: Cell Population in Relation to the Growth and Treatment of Cancer.* Clarendon Press, Oxford

Wright N, Alison M 1984 *The Biology of Epithelial Cell Populations.* Clarendon Press, Oxford

R. S. Camplejohn

Centrifuges: Principles and Applications

The development of centrifuges and centrifugation techniques has been driven by the need to study and manipulate the various classes of biological components from cells to macromolecules, which have been identified in biological research, and by the need to apply this knowledge for therapeutic purposes. Of particular importance in this development was the realization that cells and their products consist of macromolecules of discrete size and that larger structures consist of assemblies of such macromolecules. This important intellectual advance was indeed, in part, a result of the work of Svedberg and co-workers who developed the ultracentrifuge, and were thus able to demonstrate that proteins were diverse molecules of discrete size (Svedberg and Pedersen 1940).

1. Theory of Centrifugation

Particles of appropriate size will sediment under the influence of the earth's gravitational field. A centrifuge exposes particles to accelerations in excess of those applied by the gravitational field by placing them in a field of force (the so-called centrifugal field) generated by rotation. The particles are normally in solution, but centrifuges have also been developed which operate on particles in the gaseous state.

The factors determining the behavior of a spherical particle in a medium in a centrifugal field may be derived from Stokes' law. The sedimentation rate of the spherical particle is represented by

$$v = \frac{2r^2 g'(\rho_s - \rho_m)}{9\eta}$$

where v is the sedimentation rate, r is the radius of the sphere, g' is the force acting on the sphere, ρ_s is the density of the sphere, ρ_m is the density of the medium, and η is the viscosity of the medium. The sedimentation rate is thus proportional to the square of the particle radius, proportional to the force acting on the particle, proportional to the difference in density between the particle and its surrounding medium, and inversely proportional to the viscosity of the medium. If the densities of the particle and the surrounding medium are equal the sedimentation rate is zero.

The force acting on the particle is generated by rotation. The equation relating the particle's acceleration to its angular velocity ω and its distance r' from the center of rotation is:

$$g' = \omega^2 r'$$

This acceleration is normally expressed relative to the gravitational field. The behavior of particles in the centrifugal field is frequently described by the parameter s, the sedimentation coefficient, which is the rate (in seconds) of sedimentation in a field of unit centrifugal force, and is given by

$$s = (1/\omega^2 r')(dr'/dt)$$

Many macromolecules have values of s in the range 10^{-13} to 10^{-11} s. These values are normally presented in terms of Svedberg units (S) which have a value of 10^{-13} s. This means that macromolecules will frequently have S values in the range 1–100. The value of s for a particle depends on the solvent and its temperature. To allow standardization of sedimentation coefficients, values are frequently quoted in terms of $S_{20,w}$; this is the theoretical value which would be obtained if the measurement had been carried out using water as the solvent at 20 °C. This correction can be made provided the density ρ and viscosity η of the solvent and partial specific volume $\bar{v}$ of the particle are known using the equation:

$$s = \frac{1 - \bar{v}\rho_{20,w}}{1 - \bar{v}\rho_{t,s}} \frac{\eta_{t,s}}{\eta_{20,w}}$$

A further correction will also be required to allow for

nonideal effects at finite concentrations of the particle being examined. Thus the value of $S_{20,w}$ is measured at various concentrations and the value extrapolated to infinite dilution. These concepts of sedimentation underlie the uses of centrifugation both where the rate of movement of particles in the centrifugal field is studied or, alternatively, where the particles are separated from other particles by virtue of differences in sedimentation rate. An application of particular significance (historically) is the study of the sizes and shapes of macromolecules using the analytical ultracentrifuge.

Isopycnic or density equilibrium centrifugation is another important way of using centrifugation to study the behavior of classes of particles or to achieve their separation. As shown, the net force acting on a particle in a centrifugal field is zero when the particle has the same density as the surrounding medium. It is thus possible to determine the density of particles in a medium by sedimenting them to their isopycnic point in a density gradient. It is also possible to separate classes of particles differing in density.

There are two theoretical aspects of the use of isopycnic centrifugation which merit discussion. The first relates to the formation of a density gradient by centrifugation of an initially uniform solution of a solute such as a salt. The second concerns the time taken for another class of particle—normally a macromolecule—to reach density equilibrium in a density gradient of solute formed during the centrifugation or, alternatively, in a density gradient formed prior to initiation of centrifugation. The formation of density gradients by centrifugation has been intensively studied. A sedimentation–diffusion equilibrium results from an appropriate combination of solution, time and centrifugal field. The denser component in the solution sediments, but the nonuniform concentration generated by the centrifugation results in back diffusion. At equilibrium the concentration of solute increases with distance from the center of rotation and thus gives rise to a density gradient. The density gradient depends only on the angular velocity, distance, temperature and various parameters describing the behavior of the solute in the solvent system used. The density gradient at equilibrium is described by the equation:

$$d\rho_0/dr' = \omega^2 r'/\beta_0$$

(This equation neglects the effect of terms dependent on hydrostatic pressure.) The thermodynamic basis for this equation was described by Vinograd and Hearst (1962). The density gradient proportionality constant β_0 is determined for a solute in a given solvent. It has been found that certain salts, particularly those of cesium, generate gradients suitable for isopycnic banding of nucleic acids under conditions readily produced by modern high-speed centrifuges.

The time taken for particles, typically macromolecules, to reach equilibrium in a system consisting of an initially homogeneous solution of particles and solute can in principle be calculated using an equation

derived from that of Meselson et al. (1957) by Fritsch (1975):

$$t = \frac{\beta_0(\rho_p - \rho_m)}{\omega^4 r_{eq}^2 s}\left(1.26 + \ln\frac{r_b - r_t}{\sigma}\right)$$

where ρ_p is the particle density, ρ_m is the medium density, ω is the angular velocity, r_{eq} is the distance from the center of rotation of the particle band at equilibrium, s is the sedimentation coefficient, r_b is the distance to the bottom of the gradient, r_t is the distance to the top of the gradient, and σ is the diffusion coefficient of the particle. A related equation has been derived for the case in which a narrow zone of particles is introduced at the top of a preformed density gradient:

$$t = \frac{\beta_0(\rho_p - \rho_m)[\log(r_{eq} - r_t)/r_t + 4.61]}{\omega^2 r_{eq} s(d\rho/dr')}$$

where $d\rho/dr'$ is the slope of the density gradient. The difficulty in using these equations is obvious as some of the parameters required to solve them are difficult to determine, e.g., diffusion coefficients. However, practical approaches to approximate solutions which are very useful in designing experiments have been described (Birnie and Rickwood 1978).

2. Practical Aspects of Use

The applications of centrifugation in biomedical research and clinical practice fall into two broad areas: (a) the analysis of the properties of particles, and (b) the manipulation of particles. The two crucial particle parameters may be loosely described as size and density. There are thus both analytical and preparative applications of the centrifugation techniques based on these properties. The type of equipment required to carry out these centrifugation techniques is dictated largely by the size of the particles and the scale of the application. The particles range from cells to macromolecules via subcellular particles, i.e., S values from greater than 10^7 to less than 10. The density of the particles ranges from less than 1 to about 2. The scale of the application varies from the analysis of volumes of less than 1 ml to industrial scale application where hundreds of liters may be processed. The performance of the equipment is naturally related to the scale of the application; equipment capable of handling large volumes of material will not be capable of achieving the highest centrifugal fields.

A number of types of centrifuge have been designed to carry out various applications. They range from small bench centrifuges capable of achieving a centrifugal field of perhaps a few hundred g to ultracentrifuges capable of achieving 500 000g. There are specialized instruments such as flow-through centrifuges which are designed to process large volumes of material, and at the other end of the spectrum there are analytical ultracentrifuges designed to make high precision measurements on extremely small volumes of sample.

The greatest area of development and use of centrifuges in the research field has involved the use of centrifugation on what is essentially a preparative scale. The same equipment has also been used to provide analytical insights, but the techniques and equipment have been developed with preparation in mind. Three techniques are of particular importance: differential centrifugation, rate-zonal centrifugation and isopycnic centrifugation. The first two are techniques for the separation of particles on the basis of their sedimentation coefficient. Differential centrifugation involves separation of an initially homogeneous mixture of particles under conditions where all the particles of one class will reach the side or bottom of the container in which they are being centrifuged during the run, but other classes of particle will not be completely cleared from the medium. It is obvious that the approach will never achieve the complete purification of one class of particle from another as the particle distribution is initially homogeneous. The efficiency of purification can be improved by the use of repeated cycles of differential centrifugation. It is an extremely useful approach for large-scale purifications and also for concentrating particles from large volumes. Rate-zonal centrifugation involves the centrifugation of particles which are initially present in a thin zone. The zone is introduced onto a density gradient prior to centrifugation. The density gradient subsequently stabilizes zones of particles as they sediment in the centrifugal field. This approach will in principle achieve an essentially complete separation of particles which differ sufficiently in sedimentation coefficient. It should be noted that while the density gradient will have an effect on the sedimentation of the particles because of the varying density and viscosity of the medium surrounding the particles, the role of the density gradient is primarily to stabilize the sedimenting zones against convective disturbances. Isopycnic centrifugation is used to separate particles on the basis of their buoyant density in a given medium. Thus, in principle, particles differing sufficiently in density can be separated in an appropriate density gradient. The separation is in theory independent of the particle size, but practically it must be possible to move the particle to its equilibrium position in a reasonable time in the centrifugal field obtainable with a centrifuge of appropriate capacity. It should also be noted that the density of a macromolecule is a function of its partial specific volume and its degree of hydration in a given solvent. Thus DNA will have a density of approximately 1.3 in dilute aqueous sucrose solution, but 1.7–1.8 in concentrated CsCl solution. The ability to separate density classes of particle will therefore be dependent in part on the availability of suitable density gradient media.

The widespread use of techniques of preparative centrifugation is reflected in the range of rotors available for use in preparative centrifuges. There are five types in common use: fixed-angle, vertical, swinging-bucket, zonal and flow-through. The fixed-angle rotor is one in which the rotor pockets are at a fixed angle to the vertical axis of the rotor. This angle can vary from approximately $15°$ to $40°$, depending on the particular application for which the rotor is designed. Fixed-angle rotors are typically used for differential and for isopycnic centrifugation employing density gradients generated during centrifugation. Particles will move only a relatively short distance before encountering the wall of the tube; thus material is pelleted relatively rapidly. The short effective pathlength also means that density gradients formed in the rotor will be quite shallow and will be formed relatively rapidly. Vertical rotors are essentially an extreme form of fixed-angle rotor in which the angle of the pocket to the vertical axis is zero. The pathlength is thus limited to the width of the tube. These rotors have found particular application in isopycnic centrifugation. Thus self-forming density gradients are formed rapidly and are extremely shallow. Swinging-bucket rotors consist of a yoke assembly with a series of so-called buckets suspended from it. These buckets hang vertically while the rotor is at rest, but pivot to a horizontal position when under acceleration. The pathlength is typically long compared with fixed-angle rotors of similar capacity. These rotors are typically used for rate-zonal and isopycnic separations. In the case of isopycnic centrifugation using this type of rotor, a preformed density gradient and a rapidly sedimenting particle will be required in order to obtain a separation in a reasonable time. The disturbances caused by reorientation of bucket contents during acceleration and deceleration of this type of rotor are relatively minor, making it well suited to rate-zonal separations. Zonal rotors are a specialized type of rotor designed for large-scale preparation work under optimal conditions. They are used for rate-zonal or isopycnic separations. The cylindrical or bowl-shaped rotor, in which the volume increases with the square of the distance from the center of rotation, is typically divided into sector-shaped compartments. To avoid the problems of reorientation the rotor is normally loaded with density gradient and sample while it is rotating, and unloaded in the same way. Static loading and unloading, however, are possible using specially designed zonal rotors. The sector-shaped cells in the rotor provide the ideal solution to the problem of convective disturbances caused by wall effects. Flow-through rotors are a highly specialized type used for large-scale preparation work. The medium containing the particles to be separated is pumped continuously through what is essentially a zonal rotor. Particles are either pelleted or banded isopycnically. In the latter case the rotor will normally be unloaded while rotating.

3. Applications

The applications of centrifugation range from the preparation of various classes of blood cells to the purification of DNA molecules. Between these two extremes of size there has been a tremendous amount of work on separation of subcellular organelles. It is difficult to imagine that the present knowledge of cellular and subcellular processes could have been gained without the use of centrifugation techniques. This knowledge has been crucial in the development of certain areas of medicine.

The direct use of techniques based on centrifugation in clinical practice is more limited. Two principal areas of application are apparent. The first is the use of techniques involving centrifugation for the analysis of pathological specimens. This group of applications includes the measurement of packed red-cell volume (hematocrit) by centrifugation of a blood sample in a capillary tube under defined conditions, and the use of cytocentrifuges to produce microscope slides suitable for cytological examination of cellular elements in body fluids. There is also a whole group of applications based on the use of centrifugal techniques in rapid analyzer systems. High-speed or analytical ultracentrifugation is not a technique well suited to routine clinical analysis, but it has found limited application in a number of conditions. Immunoglobulin levels in pathological sera have been studied by analytical ultracentrifugation (Stanworth 1973), and hormone receptor assays using high-speed centrifugation have been developed to assist in assessing the hormone dependence of mammary carcinomas.

The second class of clinical applications of centrifugation is concerned with the preparation of blood components. The preparation of packed red cells for therapeutic use represents a relatively simple application in this area. Other purified elements (e.g., platelets and leukocytes) are also prepared by centrifugation techniques. These cellular elements can be purified by sophisticated continuous-flow centrifugation techniques. The removal of malignant leukocytes from the circulation can be achieved in some cases by continuous-flow centrifugation (Goldman and Lowenthal 1975).

See also: Clinical Chemistry: Physics and Instrumentation

Bibliography

Birnie G D (ed.) 1972 *Subcellular Components, Preparation and Fractionation.* Butterworth, London
Birnie G D, Rickwood D (eds.) 1978 *Centrifugal Separations in Molecular and Cell Biology.* Butterworth, London
Fritsch A 1975 *Preparative Density Gradient Centrifugations.* Beckman Instruments International, Geneva
Goldman J M, Lowenthal R M (eds.) 1975 *Leucocytes: Separation, Collection and Transfusion.* Academic Press, London
Meselson M, Stahl F N, Vinograd J 1957 Equilibrium sedimentation of macromolecules in density gradients. *Proc. Natn. Acad. Sci. USA* 43: 581–88
Rickwood D (ed.) 1984 *Centrifugation: A Practical Approach,* 2nd edn. Information Retrieval Ltd., London
Schachman H K 1959 *Ultracentrifugation in Biochemistry.* Academic Press, London
Stanworth D R 1973 Ultracentrifugation of immunoglobulins. In: Weir D M (ed.) 1973 *Handbook of Experimental Immunology,* 2nd edn., Vol. 1. Blackwell, Oxford, Chap. 9
Svedberg T, Pedersen K O 1940 *The Ultracentrifuge.* Clarendon Press, Oxford
Vinograd J, Hearst J E 1962 Equilibrium sedimentation of macromolecules and viruses in a density gradient. *Prog. Chem. Org. Nat. Prod.* 20: 372–422

T. D. K. Brown

Cerebral Blood Flow: Regional Measurement

A visual examination of an anatomical cross section of the human brain confirms the great spatial complexity of cerebral structures. Within the cortical territory in particular, gray matter is seen to be convoluted with white matter, and there are islands of gray matter surrounded by white matter, deep in the brain. Each gray matter structure has a specific neurological task to perform and to this end receives nutrients which diffuse from the local cerebral vasculature. Per milliliter of tissue, the quantity of blood flowing (blood flow) is three to four times greater in the gray matter than in white matter. Thus, within the brain there is a complex distribution of blood flow which relates to the distribution of neurological function.

Technological developments in radionuclide imaging (for example, single photon emission tomography and positron emission tomography) have provided the means of measuring the blood flow in localized regions of the brain: one is no longer confined to measurements of mean whole brain blood flow or hemisphere blood flow which are inherently inadequate for the measurement of regional flow.

The first attempts at measuring regional cerebral blood flow (rCBF) were effected by causing a bolus of radiolabelled tracer to pass through the cerebral vasculature. The tracer diffuses into the tissue and the rate at which it clears from the tissue is measured—a rate which is blood-flow dependent. The first experimenters used discrete, collimated radiation detectors to measure the clearance of the tracer. Today, one might effect the measurement by taking a series of tomographic images of the brain. These measurements of rCBF are based on the use of tracers which diffuse freely into, and exchange freely with, the cerebral tissues, the residence of tracer being solely dependent on nutrient blood flow. This prerequisite distinguishes rCBF measurements from studies in which the tracers remain within the vasculature and therefore give little information about the access of diffusible nutrients to the cerebral tissues.

Steady-state tomography is another, more elegant, method of measuring rCBF. It is experimentally less demanding than serial tomography. In one form of the technique, the patient is continuously infused with tracer. The tracer builds up in the cerebral tissues and reaches a steady (equilibrium) concentration level which is blood-flow dependent. The blood flow is measured from tomograms that can be taken unhurriedly.

Ideally, for all these methods of measuring rCBF, the tracer used should be inert. Further, and perhaps obviously, the radiation emitted must be sufficiently penetrative to pass through brain, skull and scalp tissue to be easily detected externally.

The history of the measurement of regional cerebral blood flow and the details of the various techniques of measurement, their advantages and disadvantages, are discussed in this article.

1. History

Ingvar and Lassen (1961) were the first to report a method of measuring rCBF in man. They injected a saline bolus containing ^{85}Kr, a 0.67 MeV β emitter, into the carotid arteries of patients who were undergoing neurosurgery. The skulls of the patients being temporarily removed, it was possible to record the local clearance of the ^{85}Kr using a β-sensitive end-window Geiger counter. Obviously, such an approach to measurement had severe limitations, and so the search continued for an inert, diffusible tracer which emitted γ rays that could be recorded through the intact skull. Then, in 1963, Glass and Harper showed that ^{133}Xe, an emitter of 81 keV and 35 keV photons, could be detected in the brain through the skull. The ^{133}Xe was administered by injecting a saline solution into the carotid artery. Also in 1963, Mallett and Veall drew attention to the fact that inhaled ^{133}Xe gas is sufficiently soluble in blood to partition into the pulmonary circulation, thus offering a noninvasive means of transporting tracer to the peripheral tissues. The clearance of ^{133}Xe from the brain, after the patient had stopped inhaling the gas, provided data that were blood-flow dependent.

These three papers written in the early 1960s formed the platform upon which methods of measuring rCBF were developed over the following 20 years. Both the intracarotid and inhalation methods of administering ^{133}Xe were pursued. Sveinsdottir et al. (1977) went on to build arrays of up to 254 radiation detectors to view a single cerebral hemisphere. On the basis of the regional clearance rate curves that were obtained, Scandinavian groups were able to define, in terms of blood flow, the detailed territories of the brain that were associated with particular cerebral functions (Ingvar and Lassen 1977). Sveinsdottir's group injected the ^{133}Xe tracer into the carotid artery. Although attractive because of its noninvasiveness, the inhalation method gives less accurate results because the clearance rate measurements are adversely influenced by radiation from tracer that has recirculated. Also, ^{133}Xe gas in the upper airways is an unwanted source of background radiation as is tracer in

the scalp tissue overlying the skull. Veall and his co-workers identified these sources of error and formulated a procedure to correct for recirculation (Veall and Mallett 1966) and for tracer in the overlying tissue (Crawley and Veall 1971). This was taken further by Obrist et al. (1967). Both groups measured the concentration of ^{133}Xe in the air expired by the patients to monitor the time course of the activity in the recirculating arterial blood.

The more invasive intracarotid injection method of introducing the tracer is now less used, and a variation on the ^{133}Xe inhalation method has been introduced. A saline solution of ^{133}Xe is injected intravenously (Thomas et al. 1979). Again, a correction is necessary for recirculation. The single feature that emerges, however, in respect of noninvasive ^{133}Xe methods of measuring regional cerebral blood flow in which discrete detectors are used, is that the spatial accuracy of the measurements is rather poor. Many workers have therefore advisedly restricted the use of this methodology to the recording of global values of central blood flow. The clinical applications of such measurements are to be found in papers by Thomas et al. (1977), Wade et al. (1981) and Brown and Marshall (1982). They have avoided using the method for pathological conditions in which focal cerebral effects are known to exist.

2. Single Photon Emission Tomography

A discrete detector viewing the head records a superimposition of signals that have their sources in different cerebral structures at various depths. Given the complex distribution of the cerebral vasculature, it is essential to have a tomographic readout of the distribution of the blood flow tracer within the brain, if regional blood flow is to be measured accurately. Lassen et al. (1981) used a scanner that was designed specifically to record the clearance of ^{133}Xe from the cerebral tissue tomographically. While this methodology in conjunction with the inhalation procedure has produced definable transaxial tomographic distributions of regional blood flow, the quality of the data obtained is far from ideal. There are a number of reasons for this:

(a) The 81 keV photons emitted by ^{133}Xe are heavily absorbed and scattered by the cerebral tissues, thus the recorded signals, particularly those from structures deep in the brain, are appreciably attenuated. The registering of scattered photons represents a source of noise in the rCBF data.

(b) Scattered photons from tracer within the airways of the head are also a source of noise.

(c) Serial tomographic scans are necessary to define regional cerebral clearance. This is not conducive to good measurement statistics and yet tomography is a procedure which is severely demanding, statistically.

(d) Although corrections may be made for errors introduced by the recirculation of tracer using curves obtained by monitoring expired air, this latter procedure is yet a further source of noise, degrading the regional cerebral blood-flow tomogram.

(e) A value for the blood–tissue partition coefficient of xenon is needed to compute the regional cerebral blood flow. However, xenon is a highly lipophilic element, hence the partition coefficient value is greatly dependent on the fat content of the tissue. O'Brien and Veall (1974) have reported that the partition coefficient in cerebral tumors is appreciably different from that of normal brain tissue.

Recently, Von Kummer et al. (1983) reported another source of error in using inert gas clearance methods, following experiments in which they used fine electrode probes to measure the clearance of hydrogen from discrete regions in cat brain. The inaccuracies arise from the diffusion of tracer from regions of low blood flow, where tracer concentration is high, to adjacent high flow regions where the tracer concentration is lower.

A steady-state tomographic procedure is required to measure regional cerebral blood flow accurately. A radionuclide is needed that is less lipophilic than ^{133}Xe and which emits more energetic γ rays. Kuhl et al. (1982) have shown that the normal brain tissue uptake of amphetamine labelled with ^{123}I (160 keV x rays), intravenously injected, is closely related to regional cerebral blood flow. However, discrepanices have been noted between the cerebral uptake of the tracer and the regional cerebral blood flow in certain pathologies, for example, brain tumors, because amphetamine is not inert and therefore its uptake is related to the metabolic integrity of the tissues and not solely to cerebral blood flow. A possible solution is to use the inert gas ^{81m}Kr as tracer. It has a half-life of 13 s and emits a 190 keV x ray. It is a decay product of ^{81}Rb, half-life 4.7 hours. The ^{81m}Kr daughter product can be eluted off from the parent nuclide either in saline solution or as a gas. The steady-state distribution of the tracer in the cerebral tissue represents the balance between its delivery to the tissue (blood flow), its radioactive decay and, to a lesser extent, its clearance from the tissue by blood flow. However, it is not possible to deliver sufficient tracer into the peripheral tissues of adults simply by inhalation as the solubility of the isotope is too low, although some success has been achieved in young children (Arnot et al. 1970). In adults, it has been continuously infused in solution through an arterial catheter the distal end of which is located either in the aortic arch or in the carotid artery. Fazio et al. (1980) and Kanno et al. (1981) have reported the use of this invasive technique. So far only qualitative results have been obtained.

3. Positron Emission Tomography (PET)

A noninvasive, steady-state tomographic measurement of regional cerebral blood flow has been developed, based on the use of the inert tracer, ^{15}O-labelled water. The patient inhales tracer amounts of $C^{15}O_2$ continuously and the ^{15}O transfers rapidly to water molecules in the lungs (West and Dollery 1962). A continuous arterial supply of $H_2{}^{15}O$ is presented to the peripheral tissues. Figure 1 shows how the labelled water builds up in the brain for a cerebral blood flow of 50 ml blood/100 ml of tissue/min. The steady-state level of the tracer in the tissue is very much less than that in the arterial blood. A simple, regional-blood-flow-dependent relationship

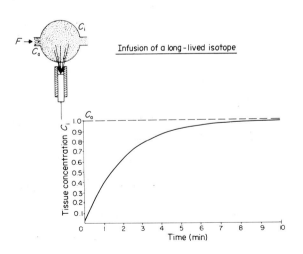
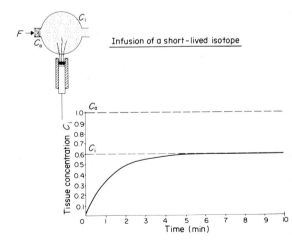

Figure 1
The steady-state radioisotopic measurement of regional blood flow in tissue. A comparison is made between the constant arterial infusion of a long-lived isotope with a half-life of the order of hours, and water labelled with ^{15}O which has a half-life of 2.1 min. Blood flow = 0.5 ml/ml of tissue/min

exists between the arterial concentration of the tracer (C_a) and the cerebral tissue concentration (C_i):

$$C_i = C_a(F/V + \lambda) \qquad (1)$$

where λ is the decay constant of ^{15}O ($0.335\ \text{min}^{-1}$), F (ml min^{-1}) is the blood flow in tissue and V is the volume of tissue. The concept of this form of measurement was introduced by Jones et al. (1976). It is possible, using a PET scanner, to obtain a tomographic image of the absolute tissue concentration of a positron-emitting isotope such as ^{15}O. When related to the corresponding concentration of an arterial blood sample, measured in a well counter, the tracer Eqn. (1) can be solved and the regional cerebral blood flow F/V obtained. Frackowiak et al. (1980) have shown that absolute values of regional cerebral blood flow can be obtained by this approach. Intravascular bolus injections of H_2^{15}O have also been used to measure rCBF with a PET scanner using either the first pass uptake of tracer (Raichle et al. 1983) or the clearance rate from the cerebral tissue (Huang et al. 1983). The use of water as the diffusible blood flow tracer is attractive in that it is not lipophilic. Further, regional cerebral blood flow thus measured can be related to PET measurements of regional cerebral function using other positron emitting tracers (see *Positron Emission Tomography*). For example, the regional cerebral metabolic rate for oxygen and glucose utilization can be measured by inhaling molecular ^{15}O or by injecting ^{18}F-labelled deoxyglucose intravenously. Such combined studies of regional cerebral blood flow and metabolism (nutritional supply and demand) and also blood volume (vasculature) are now beginning to advance the understanding of the pathophysiology of the human brain (Frackowiak et al. 1981, Wise et al. 1983a, Wise et al. 1983b, Rhodes et al. 1983, Gibbs et al. 1984). The major drawback of such measurements is that a cyclotron is required to produce these positron-emitting tracers, and a PET scanner to record their distribution within the brain.

See also: Single Photon Emission Tomography; Blood Flow: Invasive and Noninvasive Measurement; Radionuclide Brain Imaging; Radionuclide Cisternography; Radionuclide Imaging

Bibliography

Arnot R N, Glass H I, Clarke J C, Davies J C, Schiff D, Pictor Warlow C G 1970 Methods of measurement of cerebral blood flow in the newborn infant using cyclotron produced isotopes. *Radioakt. Isot. Klin. Forsch.* 9: 60–75

Brown M M, Marshall J 1982 Effect of plasma exchange on blood viscosity and cerebral blood flow. *Br. Med. J.* 284: 1733–36

Crawley J C W, Veall N 1971 Gamma spectrum subtraction technique for measurement of activity in body organs and its use for cerebral blood flow studies. *Proc. IAEA Symp. Dynamic studies with Radioisotopes in Clinical Medicine and Research.* International Atomic Energy Agency, Vienna, pp. 585–90

Fazio F, Fieschi C, Collice M, Nardini M, Banfi F, Possa M, Spinelli F 1980 Tomographic assessment of cerebral perfusion using a single-photon emitter (Krypton-81m) and a rotating gamma camera. *J. Nucl. Med.* 21: 1139–45

Frackowiak R S J, Lenzi G L, Jones T, Heather J D 1980 Quantitative measurement of regional cerebral blood flow and oxygen metabolism in man using ^{15}O and positron emission tomography: Theory, procedure and normal values. *J. Comput. Assist. Tomogr.* 4: 727–36

Frackowiak R S J, Pozzilli C, Legg N J, du Boulay G H, Marshall J, Lenzi G L, Jones T 1981 Regional cerebral oxygen supply and utilization in dementia: A clinical and physiological study with oxygen-15 and positron tomography. *Brain* 104: 753–78

Gibbs J M, Wise R J S, Leenders K L, Jones T 1984 Evaluation of cerebral perfusion reserve in patients with carotid-artery occlusion. *Lancet* i: 310–14

Glass H I, Harper A M 1963 Measurement of regional blood flow in cerebral cortex of man through intact skull. *Br. Med. J.* 265: 593

Huang S C, Carson R E, Hoffman E J, Carson J, MacDonald N, Barrio J R, Phelps M E 1983 Quantitative measurement of local cerebral blood flow in humans by positron computed tomography and ^{15}O-water. *J. Cereb. Blood Flow Metabol.* 3: 141

Ingvar D H, Lassen N A 1961 Quantitative determination of regional cerebral blood-flow in man. *Lancet* ii: 806–7

Ingvar D H, Lassen N A 1977 (eds.) *Cerebral Function, Metabolism and Circulation*, Acta Neurol. Scand. Suppl. 64 Vol. 56, 262–88

Jones T, Chesler D A, Ter-Pogossian M M 1976 The continuous inhalation of Oxygen-15 for assessing regional oxygen extraction in the brain of man. *Br. J. Radiol.* 49: 339–43

Kanno I, Uemura K, Miura S, Miura Y 1981 HEADTOME: A hybrid emission tomograph for a single photon and positron emission imaging of the brain. *J. Comput. Assist. Tomogr.* 5: 216–26

Kuhl D E, Barrio J D, Huang S-C, Selin C, Akerman R F, Lear J L, Wu J L, Lin T H, Phelps M E 1982 Quantifying local cerebral blood flow by *N*-Isopropyl-*p*-(^{123}I)-Iodoamphetamine. *J. Nucl. Med.* 23: 196–203

Lassen N A, Henriksen L, Paulson O B 1981 Regional cerebral blood flow in stroke by 133-Xenon inhalation and emission tomography. *Stroke* 12: 284–88

Mallett B L, Veall N 1963 Investigation of cerebral bloodflow in hypertension, using radioactive-xenon inhalation and extracranial recording. *Lancet* i: 1081–82

O'Brien M D, Veall N 1974 Partition coefficients between various brain tumours and blood for ^{133}Xe. *Phys. Med. Biol.* 19: 472–75

Obrist W D, Thompson H K, King C H, Wang H S 1967 Determination of regional cerebral blood flow by inhalation of 133-Xenon. *Circulation Res.* 20: 124–35

Raichle M E, Martin W R W, Herscovitch P, Mintun M A, Markham J 1983 Brain blood flow measured with intravenous H$_2$ ^{15}O, II: Implementation and validation. *J. Nucl. Med.* 24: 790–98

Rhodes C G, Wise R J S, Gibbs J M, Frackowiak R S J, Hatazawa J, Palmer A J, Thomas D G T, Jones T 1983 *In vivo* disturbance of the oxidative metabolism of glucose in human cerebral gliomas. *Ann. Neurol.* 14: 614–26

Sveinsdottir E, Larsen B, Rommer P, Lassen N A 1977 A

multidetector scintillation camera with 254 channels. *J. Nucl. Med.* 18: 168–74

Thomas D J, Du Boulay G H, Marshall J, Pearson T C, Ross Russell R W, Symon L, Wetherley-Mein G, Zilkha E 1977 Cerebral blood-flow in polycythaemia. *Lancet* ii: 161–63

Thomas D J, Zilkha E, Redmond S, Du Boulay G H, Marshall J, Ross Russell R W, Symon L 1979 An intravenous ^{133}Xenon clearance technique for measuring cerebral blood flow. *J. Neurol. Sci.* 40: 53–63

Veall N, Mallett B L 1966 Regional cerebral blood flow determination by ^{133}Xe inhalation and external recording: The effect of arterial recirculation. *Clin. Sci.* 30: 353–69

Von Kummer, Herold S, Von Bries F 1983 Inaccuracies in the calculation of CBF from inert gas clearance. In: Hartman A, Hoyer S (eds.) 1983 *Proc. Int. Symp. Measurement of Cerebral Blood Flow and Cerebral Metabolism in Man.* Springer, Berlin

Wade J P H, Pearson T C, Ross Russell R W, Wetherley-Mein G 1981 Cerebral blood flow and blood viscosity in patients with polycythaemia secondary to hypoxic lung disease. *Br. Med. J.* 283: 689–92

West J B, Dollery C T 1962 Uptake of oxygen-15-labeled CO_2 compared with carbon-11-labeled CO_2 in the lung. *J. Appl. Physiol.* 17: 9–13

Wise R J S, Bernardi S, Frackowiak R S J, Legg N J, Jones T 1983a Serial observations on the pathophysiology of acute stroke: The transition from ischaemia to infarction as reflected in regional oxygen extraction. *Brain* 106: 197–222

Wise R J S, Rhodes C G, Gibbs J M, Hatazawa J, Palmer T, Frackowiak R S J, Jones T 1983b Disturbance of oxidative metabolism of glucose in recent human cerebral infarcts. *Ann. Neurol.* 14: 627–37

T. Jones

Cervical Cytology: Automation

Since the development of the Papanicolaou smear technique by Papanicolaou and Traut (1943), population cervical smear screening has become widely established as a means of detecting premalignant changes in early carcinoma of the uterine cervix in women. The major cost in such programs is that of screening the smears, and for this reason much effort has been devoted to the development of instrumentation to make the screening process either fully or partially automatic. This article outlines the requirements that must be met by such instrumentation, describes the various principles and techniques that have been applied, and assesses the current state of the art.

1. Basic Requirements and Approaches

Cervical cytology involves the microscopic examination of cells exfoliated from the epithelium of the uterine cervix in order to find any which show abnormalities,

indicating the presence of a malignant or premalignant lesion (Koss 1968). Cells are collected by scraping, wiping or washing the surface of the cervix (e.g., with a wooden or plastic spatula), and smearing the resulting material onto a glass slide. In conventional screening, the cells are fixed, stained and examined visually. Smears usually contain very large numbers of cells (1500–100 000), but in some malignant or premalignant specimens only very few (< 10) of these will be abnormal (Boddington 1968). The detection of abnormal specimens is made even more difficult by the fact that normal specimens vary widely in appearance owing to factors such as hormonal changes during the menstrual cycle, recent pregnancies, or nonmalignant infections. Thus conventional visual screening is a tedious, time-consuming but nevertheless highly skilled process, and is by far the most costly part of a population screening program. It is for these reasons that much effort has been devoted to automation in cervical cytology screening.

The earliest attempts at automation were aimed at the complete elimination of visual screening, but these were not successful because the required diagnostic accuracy was not attained. For this reason most subsequent work has been directed towards the "prescreener"—a device which assists the cytotechnician by removing the unequivocally normal material, and separates out doubtful material for visual inspection and diagnosis. This can be performed either by selecting doubtful specimens for complete visual screening, or by "marking" individual doubtful cells for subsequent relocation and diagnosis by the cytotechnician (Tolles 1971).

Although the human screener obtains much useful diagnostic information from the relationships between different cells in a specimen, automated prescreening systems have so far been based solely upon the measurement and classification of individual cells. The cell parameters used must be able to separate normal cells from diagnostically malignant or premalignant (henceforth called abnormal) cells, and must be suitable for measurement by the instrumentation used in the system. Some parameters which have been used, or proposed for use, for cell classification in different systems include: nuclear diameter/chord intercept, nuclear area, nucleo/cytoplasmic area ratio, nuclear peak/mean density, nuclear DNA content, and cell volume. Several studies have been carried out to determine typical parameter values for samples of the various normal and abnormal cell types encountered in cervical cytology specimens (Koss 1968).

The classifications assigned to the individual cells in a specimen are put together in a specimen classifier to give an overall specimen result. The simplest specimen classifier gives a direct separation between normal and abnormal specimens from the number of objects classified as abnormal cells; however, other more complex procedures may be required in practical laboratory systems to give more detailed cytological information, such as the presence of infection. In addition, the specimen classifier must be able to detect any specimens that are unsuitable for analysis because of errors in specimen collection or preparation.

One of the most fundamental requirements of automated cytology systems is that their accuracy should be at least as high as that of conventional screening. The absolute accuracy required is still under debate, but recent results suggest that abnormal specimen detection rates of 95–98% will probably be necessary for clinical acceptability. A second requirement is that the overall analysis rate and cost of a system must be such as to give a useful cost benefit in comparison with visual screening.

Some of the most serious problems in cervical cytology automation stem largely from the nature of the specimen. The earliest attempts at automated pre-screening showed that the conventional Papanicolaou-stained smear was not well suited to automated analysis. The accurate classification of single cells is most easily performed on even monolayers or suspensions of isolated cells, but the conventional scrape specimen contains many tissue fragments (cell clumps), leukocyte clusters and mucus strands in addition to single exfoliated cells. For this reason special preparation techniques have been developed for cervical cytology specimens for automated analysis. Some of these techniques are outlined in Sect. 4.

Most prescreeners that have so far been developed are based on one of two technologies: image analysis and flow cytometry systems. These will be discussed separately because they present very different characteristics and problems; nevertheless, many of the achievements, techniques and difficulties in the two systems are very similar.

2. Image Analysis Systems

Figure 1 shows a block diagram of a typical image analysis system for cervical cytology automation. Specimens are prepared for scanning on a conventional glass microscope slide or similar substrate, and ideally consist of an even dispersion of single cells. The slide is moved in a "raster" pattern under a microscope by a motorized stage, so that the whole slide is covered as a series of field images, most of which will contain many cells.

In order to analyze the specimen, each field image is first converted into electrical signals by a scanner, such as a photoelectric cell/Nipkow disk, a television camera or a solid-state light-sensitive diode detector. The signal is processed by the analysis subsystem, which first segments (or thresholds) the data to delineate the boundaries of the objects to be analyzed (i.e., cells and/or cell nuclei). The scanner data from the relevant object areas are then processed to calculate parameter values for each object, and these are used by the cell classifier to obtain a classification for the object (e.g., normal cell, abnormal cell or debris). This process is repeated for each cell in the specimen, and on completion of the scan the specimen classifier derives an overall specimen result from the total population of cell classes obtained.

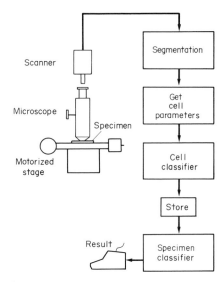

Figure 1
Basic image analysis system for cytology automation

Image analysis systems can be used to give a wide range of cell parameters. Most scanners break down the image into a regular array of small points, or "pixels," with a number for each which represents its density level. Pixels belonging to one object are linked together, and various object parameters can be calculated from the resulting chains (Rosenfeld and Kak 1976). Object diameters can be obtained from the maximum length of pixel lines across the object. Similarly, the area of an object is obtainable from the number of pixels within the object. Peak/mean density and integrated optical density (which is proportional to DNA content under suitable staining conditions) can be obtained from densitometric values of pixels. Nucleo/cytoplasmic ratios may be calculated using two suitable thresholds, or by colorimetric techniques. Finally, cell morphology parameters can be measured from the geometric relationships between object boundary pixels.

The earliest practical instruments designed for cervical cytology automation utilized relatively simple hardware-implemented analysis techniques to achieve the necessary high speeds. However, it was found that these gave unacceptably high error rates, mainly because of the large numbers of noncell artefacts (leukocyte clusters, cell overlaps, debris) mistaken for abnormal cells. More recently, the steady decrease in the cost of digital computing equipment has led to more and more use of software analysis techniques; their greater logical flexibility has led to improvements in several areas (Husain et al. 1976). Pattern recognition techniques have been used successfully in many systems for separating abnormal cells and noncell objects. Improved classifiers have been developed which give more reliable cell and specimen classifications from the available cell parameters (Duda and Hart 1973). Mathematical and statistical models for the analysis processes have been developed which have given new insight into the optimal design of automated cytology systems.

In spite of these advances, several problems remain to be solved. Firstly, the problem of high data rates. Although the cost of computing has fallen dramatically, the basic speed of computers has not changed significantly. Some of the scanners that have recently been developed are capable of continuous scanning at high data rates and so some hardware data reduction technique is required to compress the data for input to a computer. This at present limits the application of software techniques in certain critical areas such as segmentation. Secondly, some of the parameters utilized in recent systems are highly dependent on the quality of image data, so that more critical demands are placed upon the accuracy of scanning, focusing and segmentation. The problems of optimizing specimen classifiers to cope adequately with the very large inter-specimen variations that occur in cervical cytology specimens remain largely unexplored.

3. Flow Systems

The two basic flow cytometry systems that have been applied to automated cervical cytology are shown in Figs. 2 and 3. The specimen to be analyzed is suspended in a fluid medium, and is passed in a narrow stream through some form of measuring device. In the Coulter counter, this consists of a narrow orifice across which a voltage is set up by means of electrodes in the main chambers. By measuring the change in current which occurs as a cell passes through the orifice, the cell volume can be obtained. The second type of flow cytometry

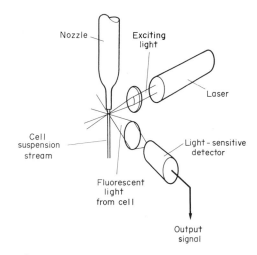

Figure 3
Fluorescence-activated flow cytometry system. The secondary fluorescence generated as cells pass through the laser beam can be detected and measured to give cell parameters (depending on the stain used) such as total DNA content

system involves a light beam passing through the cell suspension stream. Various cell properties can be measured by means of direct changes in this beam of light as a cell passes through, or by fluorescent emission from suitably stained cells. Facilities for sorting the cells on the basis of the parameter measurements are also incorporated in many systems.

The flow system can measure several cell parameters which are potentially valuable for cervical cytology automation (Van Dilla and Mendelsohn 1979). Cell volume can be measured using the Coulter principle outlined above. The nuclear DNA content can be measured either directly by the ultraviolet light absorption of the cell or by the total fluorescence obtained after staining with a DNA-specific fluorescent dye. Similarly, total RNA content can be obtained by using an appropriate fluorescent dye. Another parameter which has been found to be useful in cell classification is the light scattered as the cell passes through the light beam; this can provide information concerning the size, homogeneity, and refractive index of the cell. Lastly, the diameter of the cell or nucleus can be estimated in flow systems with a narrow beam of exciting illumination by measuring the length of the object fluorescence pulse and the fluid flow rate.

Several different methods of using such systems for cervical cytology prescreening have been proposed (Herman et al. 1979). Perhaps the most basic of these is a direct analysis of the population of parameter values obtained by passing the specimen through the system, usually displayed as a histogram in one-parameter systems or as a scatter plot in two-parameter systems. The specimen is then classified as suspect if a significant

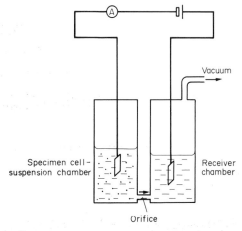

Figure 2
Principle of the Coulter flow cytometry system. Current pulses caused by the change in resistance between the electrodes as the cells pass through the orifice can be related to cell volume

number of signals fall outside the normal parameter regions. Another approach utilizes the cell sorter to separate any cells with unusual parameter measurements from normal cells. This enriched sample of abnormal cells is then prepared for visual screening.

As with image analysis systems, several basic problems which indicate the difficulty of using flow cytometry systems for cervical cytology automation have become apparent—the accuracy of the system, in particular. Some abnormal specimens contain very few abnormal cells in an otherwise normal cell population, so that the overall effect of these cells on total population characteristics is very small. A second related problem is that of false abnormal cell signals caused, for example, by small cell clusters and debris. In combination, these problems make adequate discrimination between normal and abnormal specimens difficult in systems which rely upon whole-specimen histogram or scatter plot analysis. The apparent advantages of cell sorting systems in overcoming this problem have proved difficult to realize because of the technical and economic difficulty of reliably capturing and preparing the sorted cell samples for visual screening. However, recent advances in morphological analysis using multiple-slit scanners, and in multiparameter analysis and sorting, may well help overcome these problems.

4. Specimen Preparation

Cervical cytology specimens for automated cytology should ideally consist of even, well-dispersed monolayers or suspensions of cells, with defined concentration, and with cells stained to demonstrate the specific cell components required for measurement (Wheeless and Onderdonk 1974). To obtain samples which approach this ideal, specimens must be processed in the cytology laboratory before laying on a microscope slide for screening. The most common technique for transporting specimens is to suspend cells (collected by the conventional scrape, swab or wash techniques) in a mild preservative fluid such as 20% alcohol in balanced salt solution. The first stage of the laboratory procedure usually consists of disaggregating cell clusters, either by mechanical means (e.g., by pumping through a hypodermic syringe needle or by ultrasonic techniques), or chemical means (e.g., enzymic agents such as trypsin and hyaluronidase). The resulting suspension contains mostly single cells, but with a cell concentration that varies widely between different specimens. Various methods for standardizing the concentration have been used, such as visual inspection/dilution, or more complex measurement methods based on Coulter counting or light scatter in a column of the suspension.

Two methods of dispersing cells on a microscope slide for image analysis systems have been used. Cells can either be laid by pipetting the suspension onto the substrate, or by centrifuging the cells directly onto the slide in a slide centrifuge. Adhesion to the slide can be aided by precoating the slide with an agent such as gelatin, albumin or polylysine. With polylysine this stage can be used to give a more even dispersion by washing off excess cells.

Many different stains have been used in automated cytology. Most image analysis systems utilize absorption stains. The conventional Papanicolaou staining method has been used, but its variegated cytoplasmic coloration and limited nucleo/cytoplasmic differentiation causes difficulty during analysis. Nuclear stains that have found use include hematoxylin, thionin, acriflavine, and gallocyanin chrome alum. Cytoplasmic counterstains include eosin, congo red and tartrazine. For flow analysis, the various fluorescent stains in use include acridine orange, ethidium bromide and propidium iodide.

5. Current Status

It will be evident from the preceding sections that the problem of automating cervical cytology has proved to be far more difficult than was originally envisaged. In spite of extensive research efforts over many years, no successful cervical cytology automation system has yet appeared. Nevertheless, extensive research continues in many countries, and several common trends may be identified in these projects. Advances in image analysis hardware and software have led to the possibility of using powerful and complex techniques, previously only available in expensive research equipment, in relatively low-cost laboratory systems. These improvements have been brought about by several factors, notably the reduction in cost of computing power, the introduction of improved software techniques for artefact rejection and classification, and the introduction of new scanners such as the Chalnicon camera and solid-state light-sensitive diode arrays. Specimen preparation techniques developed for automated cytology now give excellent results, but more research and development is necessary to automate these to meet the cost and throughput requirements of laboratory systems. Finally, efforts must continue to reach agreement amongst cytopathologists as to the exact requirements to be met by automated systems before they become acceptable for clinical use.

See also: Blood Cell Analysis: Automatic Counting and Sizing; Blood Cell Analysis: Morphological and Related Characteristics; Clinical Biochemistry: Automation

Bibliography

Bacus J W, Wiley E L, Gabraith W, Marshall P, Willbanks G D, Weinstein R S 1984 Malignant cell detection and cervical cancer screening. *Anal. Quant. Cytol.* 6: 121–30

Boddington M M 1968 Scanning area required for screening cervical smears. In: Evans D M D (ed.) 1968 *Cytology Automation.* Livingstone, Edinburgh, pp. 14–23

Duda R O, Hart P E 1973 *Pattern Classification and Scene Analysis.* Wiley, New York

Herman C J, Bunnag B, Cassidy M 1979 Clinical cytology

specimens for cancer detection. In: Melamed M R, Mullaney P F, Mendelsohn M M (eds.) 1979 *Flow Cytometry and Sorting*. Wiley, New York, pp. 559–72

Husain O A N, Tucker J H, Page Roberts B A 1976 Automation in cervical cancer screening. *Biomed. Eng.* 11: 161–66, 204–10

Koss L G 1968 *Diagnostic Cytology*, 2nd edn. Lippincott, Philadelphia

Papanicolaou G N, Traut H F 1943 *Diagnosis of Uterine Cancer by the Vaginal Smear*. Commonwealth Fund, New York

Rosenfeld A, Kak A C 1976 *Digital Picture Processing*. Academic Press, New York, pp. 333–403

Tolles W 1971 Development of standards. *Acta Cytologica* 15: 243–45

Van Dilla M, Mendelsohn M L 1979 Introduction and resumé of flow cytometry. In: Melamed M R, Mullaney P F, Mendelsohn M M (eds.) 1979 *Flow Cytometry and Sorting*. Wiley, New York, pp. 11–37

Wheeless L L, Onderdonk M A 1974 Preparation of clinical gynecologic specimens for automated analysis: An overview. *J. Histochem. Cytochem.* 22: 522–25

J. H. Tucker

Chemical Carcinogenesis

About one-fifth of all deaths in Western Society are now attributable to cancer, and detailed epidemiological studies have concluded that as many as 80% of human cancers are the result of exposure to carcinogenic factors present in the environment (Higginson 1969). The latter include agents as diverse as radiation, viruses, and a substantial number of chemicals that have accumulated as an unfortunate consequence of the industrialization process which has taken place during the past century. Any hope for a solution to the cancer problem must therefore involve (a) a concerted effort to identify potential carginogens in order that they may be removed from the environment, and (b) fundamental studies on the mechanisms of tumor induction with the aim of understanding and, ultimately, intervening to prevent the progression of the disease. The purpose of this article is to illustrate the advances which have been made using these two approaches and to demonstrate how the combined efforts of chemists, biochemists, biologists and genetic engineers offer a realistic prospect of attaining the goal of understanding cancer (as opposed to the much more nebulous and distant ideal of finding the long-sought-after "cure") within the foreseeable future.

1. Identification of Environmental Carcinogens

Since the classical observations that chimney sweeps exposed to soot exhibited an abnormally high incidence of skin cancers, a great deal of effort has gone into the identification and classification of carcinogenic agents present in the environment. Epidemiological studies

have been dramatically successful in a few cases in showing that certain groups of workers are particularly prone to specific types of cancer as a result of occupational exposure to chemicals. Examples of this are the lung cancers contracted by workers in the asbestos industry, and the watch-face painters of Japan who developed leukemias and bone tumors as a result of licking brushes contaminated with fluorescent radium-based paint. Such obvious cases are, however, relatively rare since most human cancers cannot be traced back to a specific event or exposure to a particular chemical. A notable exception to this is the now indisputable evidence linking lung cancer to the smoking of cigarettes. This will be discussed in more detail later.

The list of chemicals which have carcinogenic activity in animals has now reached alarming proportions, and hundreds more are being tested each year. Some of the classes of compounds which have carcinogenic activity in animal systems are shown in Table 1. A more complete list of compounds tested in both bacterial and animal systems is given in Bartsch et al. (1980).

Early studies on the detection of environmental carcinogens were hampered in that cancer is essentially a disease of old age. The steep rise in cancer incidence with age is normally interpreted in terms of models in which normal cells go through a series of independent genetic events (mutations) before becoming fully malignant cancer cells. Thus individuals exposed to certain industrial chemicals today may not develop tumors for several decades, by which time the possibility of tracing and identifying the offending chemical is extremely small. Even with animal model systems using mice or rats, which have a much shorter life span than humans, the positive identification of a carcinogen may take years. In view of the many thousands of chemicals in use and the hundreds of new ones added yearly, the logistics of keeping track of potential carcinogens by testing in animals are clearly horrendous. For this reason, a great deal of effort has gone into the development of short-term tests designed to detect chemicals that can cause mutations, i.e., changes in DNA—the large molecule present in all cells which contains the information that controls each cell's appearance and behavior. The most successful and widely used of these short-term tests is the Ames test, by which the effects of chemicals on the growth characteristics of bacteria in culture (*Salmonella typhimurium*) are quantified to serve as a measure of mutation rates.

2. Smoking and Cancer

The UK (Scotland, in particular) is top of the league table of countries with a high incidence of lung cancer which accounts for almost half of all cancer deaths in men. Death from lung cancer can be particularly drawn out and painful. The prognosis is very poor since the tumors do not in general respond favorably to treatment with cytostatic drugs in the same way as, say, some of

Table 1
Compounds with carcinogenic activity in animals

Aromatic amines and nitro compounds
2-Acetylaminofluorene
N-Acetoxy-*N*-acetyl-2-aminofluorene
N-Acetoxy-*N*-myristoyl-2-aminofluorene
N-Hydroxy-*N*-acetyl-2-aminofluorene
N-Hydroxy-*N*-myristoyl-2-aminofluorene
N-Myristoyloxy-*N*-acetyl-2-aminofluorene
N-Myristoyl-*N*-myristoyl-2-aminofluorene
2-Nitrofluorene
4-Nitroquinoline 1-oxide

Alkylating agents
Epichlorohydrin
Epoxide 201
    (4-Methyl-7-oxabicyclo[4.1.0]heptane-3-carboxylic acid)
Ethylmethane sulfonate
Glycidaldehyde
Methylmethane sulfonate
N-Nitroso-*N*′-nitro-*N*-methylguanidine
N-Nitrosoethylurea
N-Nitrosomethylurea
N-Nitroso-*N*-methylurethane
1,3-Propane sultone
β-Propiolactone

Pharmaceutical and industrial chemicals
Benzene
Diethylstilboestrol (DES)
N,*N*′-Dinitrosoethambutol
Phenacetin (Acetophenetidin)

Haloalkenes, haloalkanes and related halo compounds
Chloroethylene oxide
1,4-Dichlorobut-2-ene
Tetrachloroethylene
Trichloroethylene
Vinyl chloride
Vinylidene chloride

Mycotoxins
Aflatoxin B₁
Patulin
Penicillic acid

N-Nitrosamines
N-Nitroso-*N*-(acetoxy)methyl-*N*-butylamine
N-Nitroso-*N*-(acetoxy)methyl-*N*-ethylamine
N-Nitroso-*N*-(acetoxy)methyl-*N*-methylamine
N-Nitroso-*N*-(acetoxy)methyl-*N*-propylamine
N-Nitroso-*N*-butyl-*N*-(3-carboxypropyl)amine
N-Nitroso-*N*-butyl-*N*-(4-hydroxybutyl)amine
N-Nitroso-*N*-(butyloxy)methyl-*N*-methylamine
N-Nitroso-*N*,*N*-di(2-acetoxypropyl)amine
N-Nitrosodibutylamine
N-Nitrosodiethylamine
N-Nitroso-*N*,*N*-di(2-hydroxypropyl)amine
N-Nitrosodimethylamine
N-Nitroso-*N*,*N*-di(2-oxopropyl)amine
N-Nitrosodipentylamine
N-Nitrosodiphenylamine
N-Nitrosodipropylamine
N-Nitroso-*N*-(2-hydroxypropyl)-*N*-propylamine
N-Nitroso-*N*-methyl-*N*-phenylamine
N-Nitroso-*N*′-methylpiperazine
N-Nitroso-*N*-methyl-*N*-propylamine
N-Nitrosomorpholine
N-Nitrosonornicotine
N-Nitroso-*N*-(2-oxopropyl)-*N*-propylamine
N-Nitrosopiperidine
N-Nitrosopyrrolidine

Organochlorine compounds and pesticides
1,1-Dichloro-2,2-bis(*p*-chlorophenyl)ethylene (DDE)
4,4′-Dichloro-α-(trichloromethyl)benzhydrol (Kelthane)
1,1-Dichloro-2,2-bis(*p*-chlorophenyl)ethane (DDD)
1,1,1-Trichloro-2,2-bis(*p*-chlorophenyl)ethane (DDT)
1-Chloro-(2,4,5-trichlorophenyl)vinyl-*O*-*O*-dimethylphosphate
    (Gardona)

Polycyclic aromatic hydrocarbons
Benz[*a*]anthracene
3,4-Dihydro-3,4-dihydroxybenz[*a*]anthracene
Benz[*a*]anthracene 5,6-oxide
Benzo[*a*]pyrene
Benzo[*a*]pyrene 4,5-oxide
7,8-Dihydro-7,8-dihydroxybenzo[*a*]pyrene
7*β*,8α-Dihydroxy-9α,10α-epoxy-
    7,8,9,10-tetrahydrobenzo[*a*]pyrene
7,12-Dimethylbenz[*a*]anthracene
7-Methylbenz[*a*]anthracene
3-Methylcholanthrene
15,16-Dihydro-11-methylcyclopental[*a*]phenanthren-17-one

Triazenes, hydrazines and azides
1-(4-Chlorophenyl)-3,3-dimethyltriazene
1-(4-Methylphenyl)-3,3-dimethyltriazene
1-(4-Methoxyphenyl)-3,3-dimethyltriazene
1-(3-Nitrophenyl)-3,3-dimethyltriazene
1-Phenyl-3,3-dimethyltriazene
1-Phenyl-3-methyltriazene
1,1-Dimethylhydrazine
1,2-Dimethylhydrazine

the leukemias. It is all the more surprising, therefore, that it has taken so long for the general public to accept the now overwhelming evidence for a direct link between lung cancer and cigarette smoking, and even longer for environmental agencies and government departments to react positively to discourage the habit. Although some people who smoke 50 cigarettes a day live to the age of 90, it is indisputable that smoking cigarettes increases the risk of dying of lung cancer by a factor of 10–50. This eminently preventable disease, with its enormous impact on medical resources, not to mention its cost in terms of human suffering, could largely be eliminated by the abolition of cigarette smoking.

What types of chemicals are present in cigarette smoke? Not surprisingly, among the many chemicals which have been identified in tobacco-smoke condensate are several which have been shown to be carcinogenic in animal systems, and mutagenic in the Ames test. These are mainly in the polycyclic aromatic hydrocarbon class shown in Table 1. Interestingly, another class of compound (cocarcinogens) has also been found in tobacco smoke which is not included in Table 1, since compounds of this class are not by themselves carcinogenic: they can, however, greatly enhance the cancer-causing activity of other agents. The presence of such cocarcinogens or tumor promoters is probably the explanation of the extremely striking evidence that particularly high lung-cancer death rates occur among both asbestos workers and uranium miners who are also cigarette smokers.

3. Stages of Chemical Carcinogenesis

Epidemiological studies have shown that the development of cancer in humans involves the interplay of multiple factors, both genetic and environmental. The early interpretation that the production of a malignant cell is the culmination of a series of distinct steps has received support from experimental model systems for tumor induction in animals. The most widely studied of these is the mouse-skin model originally described by Berenblum (1941) whereby tumor development could be operationally divided into two distinct phases, termed initiation and promotion (Fig. 1). It was shown by Berenblum that a single small dose of a primary carcinogen is by itself insufficient to cause the development of tumors in mouse skin. However, if the single treatment with the carcinogen was followed by applications two or three times weekly of an oil extracted from a plant (*Croton tiglium*, one of the Euphorbiaceae) then benign tumors (papillomas) developed after 2–3 months, some of which progressed to form malignant carcinomas. No tumors were observed if the mice were treated with croton oil alone, or if the oil was applied before the carcinogen (see Fig. 2). The conclusion from this and many other experiments was that the carcinogen, or initiator, can cause basic defects in a small proportion of the normal skin cells—probably by inducing a mutation in the DNA. The promoting agent present in the croton

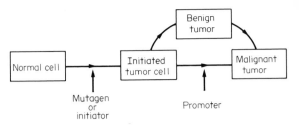

Figure 1
Stages of chemical carcinogenesis. An initiated tumor cell is obtained after a critical interaction between a normal cell and an initiator, which is usually a mutagen. Under the influence of a promoter, the initiated cell may progress to form a benign tumor and subsequently become malignant

oil, which has subsequently been isolated and identified (a derivative of the diterpene phorbol, see Table 2), appears to enable the expression of the mutated gene by some as yet poorly understood mechanism. The end result is that the "initiated" or "dormant" tumor cell is stimulated to grow by the promoter, probably undergoing further genetic changes en route to the final malignant tumor.

The molecular events accompanying these sequential changes are poorly understood and are the subject of intense research effort, but the implications for the study of human tumor development are obvious and far-reaching. Everyone reading this article will have been exposed to minute doses of carcinogens at some time in their lifetime, and will almost certainly have a small proportion of initiated tumor cells. The likelihood that a particular individual will develop a tumor may therefore depend on the degree of exposure to promoters. The validity of the tumor-promotion model as a major factor in human cancer is now well documented. In addition to

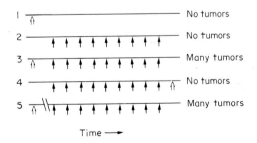

Figure 2
Experimental subdivision of chemical carcinogenesis into stages of initiation and promotion using a mouse-skin model system. The symbol ⇑ denotes a single treatment with an initiator and ↑ a single treatment with a promoter. Usually, promotion is started about 1 week after initiation. Example 5 shows that this time lag can be increased up to about 1 year without reducing the number of tumors formed

Table 2
Compounds with tumor-promoting activity in animals

Compound	Tissue in which tumors are induced
Phorbol ester derivatives	
12-*O*-tetradecanoylphorbol-13-acetate (TPA)	skin, lung, liver, bone marrow, mammary gland, forestomach
Phenols	
anthralin	skin
Fatty acids and derivatives	skin
methyl 12-oxonital-*trans*-10-octadecenoate	forestomach
methyl sterculate	liver
Sweeteners	
saccharin	bladder
cyclamate	bladder
Bile acids	
taurodeoxycholic acid	colon
lithocolic acid	colon
sodium deoxycholate	colon
sodium cholate	colon
sodium chenodeoxycholate	colon
Industrial chemicals and pesticides	
butylated hydroxytoluene (BHT)	liver, lung
DDT	liver
polychlorinated biphenyls (PCBs)	liver
Miscellaneous	
cigarette-smoke condensate	skin
phenobarbital	liver
mechanical irritation	skin, palate
uv light	skin
hormones	mammary gland

the previously cited example of promoters in cigarette smoke giving rise to lung cancer, substantial evidence suggests that promoters play a role in the development of cancers of the colon, breast, liver and upper alimentary tract. A list, which is by no means comprehensive, of compounds with promoting activity in experimental systems is shown in Table 2.

4. Reversibility of Carcinogenesis

The process whereby a normal cell becomes initiated appears to be irreversible: initiated tumor cells can lie dormant for a large proportion of an animal's life span, only to become activated when exposed to a suitable promoting influence. There is no known method of removing initiated cells or reversing the (presumably mutational) event which caused their appearance. Some aspects of promotion, however, do appear to be reversible. If a normal initiation/promotion experiment in mouse skin is halted before the appearance of the malignant carcinomas, many of the benign papillomas regress and the animal may return to what appears, superficially at least, to be a "normal" state. Similar conclusions have been reached from epidemiological studies on people who have given up smoking. Their risk of contracting lung cancer decreases year by year until it reaches the same level as that of individuals who have never smoked at all. This recognition that some aspects of promotion are reversible has prompted the search for substances which may accelerate regression or in some way interfere with the activity of promoters. A number of such substances have now been found. Among the most active of these are the retinoids—synthetic derivatives of naturally occurring vitamin A (Slaga et al. 1978). Studies on the prevention of cancer in animals have been sufficiently promising that clinical trials have now been initiated to determine the efficacy of retinoids in the prevention of human epithelial tumor development.

5. Is There a "Cancer Gene"?

Cancer researchers have for decades been confronted with the problem of devising a hypothesis which could explain how so many diverse chemical and physical agents can exert the same effect in transforming a normal cell into a tumor cell. Tumors are known to arise as a result of exposure to radiation, chemical carcinogens, viruses, and even mechanical injury. One elegant solution to the problem was proposed more than ten years ago by Robert Huebner and George Todaro of the National Cancer Institute in the USA. It had previously been shown that a type of virus which has RNA as its basic genetic material (called a retrovirus) can directly induce normal cells to form tumors. The reason that such viruses can act in this way is that they carry specific "cancer genes" which, when introduced into a cell, make it grow in an uncontrolled fashion. Huebner and Todaro proposed that such viral genes, now known as oncogenes, might be part of the genetic material of every normal cell, perhaps as the result of an earlier infection with a virus. The oncogenes were thought to remain quiescent within normal cells but be capable of activation by a suitable agent, for example, radiation or a chemical carcinogen. Although this theory fell into disrepute because of the lack of any convincing evidence that viruses are a major factor in human cancer, most recent results obtained using the techniques of gene transfer and genetic engineering have provided a profusion of data in its favor. Several research groups have been able to isolate—using special techniques which allow the transfer of genes from tumor cells to normal cells—the oncogenes which are activated in human lung, colon and bladder tumors. The researchers were amazed to find that the genes they had isolated were very similar, but not identical, to oncogenes which had previously been found in some viruses. It is important to note that these discoveries do not implicate viruses as the cause of cancer, but rather demonstrate that all normal cells contain the complete information required to make a tumor. The data obtained up to now represent only the tip of the iceberg. Many other human and animal oncogenes are presently being studied. A fascinating pattern is emerging: the oncogenes which are switched on in a tumor may depend on the type of tissue in which the tumor arises, and on the state of differentiation, or maturity, of the particular cell which suffers the mutation. An added bonus from this type of study may well be that we will learn not only about the way tumors are formed, but also about the mechanisms which control normal cell growth and development.

6. How Are Oncogenes Activated?

A normal cell may be required to switch on or off hundreds or even thousands of genes during its lifetime. Most, but by no means all, of these switches are accomplished by mechanisms which do not involve changing the primary structure of the cell's DNA. How then might a chemical carcinogen, which is known to cause specific point mutations, induce the activation of a cellular oncogene? One possibility is that the mutation might have occurred within a region of the gene which controls its activity. Indeed, sequencing of activated oncogenes from a variety of human and animal tumors has demonstrated the existence of point mutations which have this effect. In some animal tumors induced by treatment with chemical carcinogens, the particular mutations found may correlate with the type of carcinogen used (for a review see Balmain 1985). Another possibility is that mutations could occur not inside the gene but in regions, called promoters or enhancers, which are located on the DNA adjacent to the gene itself. A more plausible hypothesis, however, is that a rearrangement of the DNA sequence has taken place, possibly as a consequence of the initial mutation, resulting in a new, highly active enhancer being inserted near the oncogene. Such rearrangements could occur on a relatively small scale (enhancer sequences are normally only about 50–100 bases long) or on a large scale, involving the translocation of large pieces of individual chromosomes. These notions are not without precedent. Classical work carried out with viruses has provided an example of how a promoter sequence from a virus can, when inserted into normal DNA at the appropriate position, switch on a cellular oncogene. While it has not yet been demonstrated that a similar "promoter insertion" model applies also to chemical carcinogenesis, such data may be forthcoming from comparative sequence studies on active and inactive cellular oncogenes. Similarly, there is a great deal of cytological evidence that large-scale chromosomal translocations are associated with many forms of cancer. The most celebrated example of this is the Philadelphia chromosome, which is observed in most patients with chronic granulocytic leukemia. In the Philadelphia translocation, a piece of human chromosome 22 is moved to a new site on chromosome 9 and vice versa. It has recently been shown that this movement results in the relocalization of a cellular oncogene in a different chromosomal position, with a resultant change in its pattern of expression. Application of the

techniques of genetic engineering and gene transfer to these questions should result in their successful resolution within the foreseeable future.

See also: Cancer Statistics; Radiation Carcinogenesis; Cell Population Kinetics

Bibliography

Balmain A 1985 Transforming *ras* oncogenes and multi-stage carcinogenesis. *Br. J. Cancer* 51: 1–7
Bartsch H, Malaveille C, Camus A-M, Martel-Planche G, Brun G, Hautefeuille A, Sabadie N, Barbin A, Kuroki T, Drevon C, Piccoli C, Montesano R 1980 Bacterial and mammalian mutagenicity tests: Validation and comparative studies on 180 chemicals. In: Montesano R, Bartsch H, Tomatis L (eds.) 1980 *Molecular and Cellular Aspects of Carcinogen Screening Tests.* International Agency for Research on Cancer, Lyon, pp. 179–241
Berenblum I 1941 The carcinogenic action of croton resin. *Cancer Res.* 1: 44–48
Higginson J 1969 Present trends in cancer epidemiology. *Proc. Can. Cancer Res. Conf.* 8: 40–75
Slaga T J, Sivak A, Boutwell R K (eds.) 1978 *Mechanisms of Tumor Promotion and Cocarcinogenesis.* Raven, New York

A. Balmain

Chromatography

Chromatography is a technique for the resolution of a mixture by separation of certain or all of its components in concentration zones or in phases different from those in which they are originally present, irrespective of the force or forces causing the components to move from one phase to another. The technique is used widely in biology and medicine, for example, in the study of biological fluids and metabolic pathways and in the examination of complex molecules such as carbohydrates, proteins and nucleic acids. It is also used in forensic medicine (see *Forensic Applications of Chromatography*).

In 1906, Tswett used a column of finely divided calcium carbonate to separate a solution of green plant pigments, washing them on to the column with light petroleum. He observed that a series of bands of different pigments began to separate out as they moved down the column. This was due to the differing adsorption of the pigments to the calcium carbonate. The technique is hence known as adsorption chromatography. In 1941, Martin and Synge introduced partition chromatography. The medium packed in the columns they used contained a certain amount of water and the separation of substances depended on the distribution or partition of the substances between the water and the solvent flowing through the column. Martin and

Synge later developed paper chromatography, in which separation of mixtures was achieved on sheets of filter paper. Other types of chromatography which have since been introduced include thin-layer chromatography, gel-filtration chromatography, gas–liquid chromatography and most recently high-performance liquid chromatography.

Basically all chromatography systems consist of two phases: a stationary phase and a mobile phase. Stationary phases can be solid, liquid or solid–liquid mixtures, whereas mobile phases can be liquid or gaseous and flow over or through the stationary phase. The nature of these phases depends on the type of chromatography being employed and the compounds to be separated.

1. Adsorption Chromatography

This technique involves separation between a mobile phase (usually referred to as a solvent) and a stationary solid phase. Separation is achieved in a column packed with stationary phase (silica gel, alumina, calcium carbonate or magnesium oxide) or on a plate spread with a layer of stationary phase (thin-layer chromatography).

The mobile phase may be a single solvent or a mixture of up to three or four different solvents. The choice of solvents and adsorbent depends on the components to be separated. If a particular component of a mixture has a high affinity for the adsorbent it will move slowly through the column, whereas another component with less affinity will move more rapidly. The most active form of an adsorbent is produced by heating it strongly to remove all surface contaminants. The amount of sample applied to the column is important as the adsorptive power of the surface decreases rapidly with increasing sample load. It is therefore necessary to have the adsorbent-to-sample ratio as high as possible.

Adsorption chromatography is limited to interactions involving hydrogen bonding and weaker electrostatic forces. If interactions are ionic, the process is known as ion-exchange chromatography (see Sect. 5).

2. Thin-Layer Chromatography

Thin-layer chromatography (TLC) is useful for the separation of very small amounts of sample. Here the absorbent is spread uniformly as a thin film on an inert support (a glass or plastic plate). The adsorbents contain a binding agent such as calcium sulfate to enable binding to the support. A slurry of adsorbent in water is spread on the plate and then dried in an oven at about 100 °C in order to remove the water and leave the adsorbent bound to the plate. The sample is applied to a corner of the plate in small amounts of solvent to prevent spreading, and the plate is introduced into a glass tank containing the developing solvent to a depth of about 1.5 cm. The solvent gradually travels up the plate, separating the mixture. If further resolution is required, the plate can be rerun in the opposite direction using a second solvent system. This process is known as two-dimensional chromatography.

Substances separated by TLC can be detected by a number of different techniques. Some adsorbents contain a fluorescent dye to enable a plate to be examined under ultraviolet light after development: compounds may show up as blue, green or black spots. Another useful method of detection involves developing the plate in a tank with a few crystals of iodine. The iodine accumulates on the plate, compounds showing up as dark brown spots on a yellow background. Both detection methods are nondestructive. Corrosive reagents (e.g., 50% sulfuric acid) or powerful oxidizing agents can be used for location of compounds. The plates may also be sprayed with color reagents specific to certain compounds (e.g., ninhydrin for amino acids, orcinol for pentoses). Wherever possible, standards should be run on the chromatogram with the sample for ease of identification. Where this is not possible, identification is usually carried out by calculating R_F values for a particular solvent system. The R_F value, a constant for a particular compound under standard conditions, is defined as the ratio of the distance moved by the solute to the distance moved by the solvent front.

The main advantage of TLC over column chromatography is the time involved in the separation. Most plates are developed in 30–90 min. Quantitative estimation of the separated components is possible by use of a scanning densitometer or by scraping off the areas equivalent to their spots and eluting from the adsorbent using a suitable solvent. TLC is still used in a limited number of applications in medicine although it has largely been superseded by gas–liquid chromatography and high-performance liquid chromatography because of difficulty in accurate quantitation. It is, however, still used to detect abnormal sugars in biological samples in inborn errors of metabolism. Aminoacidurias can be investigated by running urine on a two-dimensional system. A semiquantitative separation of lecithin and sphingomyelin in amniotic fluid by TLC is widely used as an index of fetal lung maturity.

3. Partition Chromatography

This technique is based upon the principle of a compound partitioning itself between two liquid phases. Paper chromatography is carried out on sheets or strips of filter paper absorbing water, which is regarded as the stationary phase. The nonaqueous solvent (the mobile phase) is allowed to travel by capillary action along the paper. When it reaches a compound this will distribute itself between the two solvents according to its distribution coefficient. The more soluble it is in the mobile phase, the further along the paper it will travel.

Two-dimensional chromatography may be used as for TLC. Solvent systems vary according to the nature of compounds to be separated, e.g., for amino acids, butanol–acetic acid–water (40:10:50) is used; for mono- and disaccharides, butanol–pyridine–water (50:28:22); and for chlorophylls and carotenoids, chloroform–petroleum ether (30:70). Detection methods are also similar to those for TLC, with the exception that corrosive reagents are not used as they would cause the paper to disintegrate. Development times are, however, a lot longer, some taking overnight to run. The applications of paper chromatography are in the main similar to those of TLC and again have largely been replaced by more advanced techniques.

4. Gas–Liquid Chromatography

Gas–liquid chromatography (GLC) is a special form of the generalized technique of partition chromatography in which the sample to be analyzed is introduced into an inert gas stream (usually nitrogen or argon) and carried through a column. The components of the mixture distribute themselves between the mobile phase and a liquid phase (stationary phase) held on an inert granular solid, and are eluted at different rates. At the column outlet, the presence of the separated components in the carrier gas is monitored by a suitable detector and the response of the detector is recorded against time. The column is maintained in an oven at an elevated temperature which volatilizes the compounds to be analyzed.

Since the introduction of GLC in the mid-1950s a vast amount of research has been carried out on its uses in almost every field of biology and medicine. Progress has been made in column technology, detection systems, the automation of sample injection and data processing systems. This has led to increased efficiency, sensitivity and versatility of GLC as an analytical tool. It is important that the solid support should be inert to the sample. The most commonly used support is Celite (diatomaceous silica), and because of the problem of sample–support interaction this is often silanized to modify the hydroxyl groups. The stationary liquid phase should be thermally stable at the temperature used for analysis. The phase is usually a high boiling point organic compound and can be loaded on the support in concentrations of 1–25 wt%. The choice of liquid phase depends upon the class of compound to be analyzed. The detection system most commonly used is the flame-ionization detector, which detects almost all organic compounds. Other systems include the electron-capture detector, which responds only to compounds that capture electrons (e.g., halogen compounds), and the nitrogen detector sensitive only to organophosphorus and nitrogen-containing compounds. Analysis can be either at constant temperature or by temperature programming (gradually increasing column temperature). The quantity of material present may be determined by measuring the area or height of a peak obtained on a recorder. Quantitation involves the use of an internal standard. The standard should have physical properties

as close as possible to the test compounds and elute at a rate near to but distinct from them.

An increasingly useful feature of sample preparation which has extended the scope of GLC is the chemical conversion of the compounds to be separated into less polar, more volatile, or more stable derivatives before chromatography. Derivatives chosen depend on the substance concerned, the complexity of the procedure, and the chromatographic behavior of the reaction products other than the desired derivative, e.g., fatty acids may be converted to methyl esters, carbohydrates may be silanized, or amino acids may be silanized at their carbonyl group or acetylated at their amino group. Biological fluids can therefore be analyzed for: (a) single components (e.g., ethyl alcohol, drugs, and individual steroids), or (b) multicomponents (e.g., drugs and their metabolites, fatty acids, amino acids, hydroxy acids, keto acids, sugars, and steroids).

5. Ion-Exchange Chromatograpy

This technique is used for the separation of ionic substances which range from simple inorganic ions to polyelectrolytes such as enzymes, proteins, hormones and nucleic acids. Three types of ion-exchange materials are available: ion-exchange resins, ion-exchange gels and ion-exchange celluloses. The difference between these materials is mostly in their individual microstructure. Ion-exchange resins can be further subdivided into cation and anion exchangers. Cation exchangers have negatively charged groups in the resin and attract positively charged molecules. The pore size of resins is much smaller than that of gels and celluloses and thus resins are used for the fractionation of small molecules such as inorganic ions and amino acids. In column chromatography, a solution containing the ions to be separated is introduced at the top of the column. The conditions are such that the ions in the solution are rapidly exchanged with those of the resin, giving a narrow band of sample ions at the top. Different ions are then displaced one by one, either by slowly changing the pH of the eluting liquid, or by increasing the temperature. This principle has been adopted commercially to produce analyzers for automatic amino acid analysis. The sample is introduced via an injection system and is pumped onto a strongly acidic cation exchange resin column. The column is developed by pumping a buffer through the column, the pH and ionic strength of which are gradually increased by mixing buffers of different pH values and ionic strengths. The effluent from the column is allowed to mix with the ninhydrin color reagent, the color produced by heating in an oil bath being read colorimetrically (see *Colorimetry*). Colorimeters are connected to a recorder to give a continuous trace of the effluent leaving the column showing peaks representing the different amino acids present. Ion-exchange chromatography papers are primarily cellulose, and these have been modified to produce diethylaminoethylcellulose (strongly basic) and carboxymethylcellulose (weakly acidic) papers as well as some other less-used types. Papers are used to separate the same types of compounds as resin columns.

6. Permeation Chromatography

This technique involves the separation of molecules on the basis of their molecular size and shape. The most commonly used materials are a group of organic compounds possessing a three-dimensional network of pores which confer gel properties upon them. Gel filtration is the separation of molecules of varying molecular size utilizing these gel materials. A column of gel particles is in equilibrium with a solvent suitable for the molecules to be separated. Large molecules which are completely excluded from the pores will pass through the interstitial spaces, whereas smaller molecules will be distributed between the solvent inside and outside the molecular sieve and will pass through the column at a slower rate. Gels which have been used include cross-linked dextrans, agarose, polyacrylamide and polystyrenes. The usual way of characterizing the various types of gel is by means of their water regain values. This is the amount of water taken up in the completely swollen gel granules by 1 g of the dry gel grains. The gels with low water regains have the smallest pore size and are used for the fractionation of small molecules, whereas gels with high water regains are used for the fractionation of high-molecular-weight compounds. Gel filtration has a number of applications: its main use is in the purification of biological macromolecules such as viruses, proteins, enzymes, hormones, antibodies, nucleic acids and polysaccharides. Mixtures of lower-molecular-weight compounds can also be separated (e.g., peptides and oligonucleotides). The molecular weights of certain proteins have been determined by gel filtration. The effluent volume of a substance is approximately a linear function of the logarithm of the molecular weight, over a considerable molecular weight range. Solutions of compounds of high molecular weight may be desalted. The high-molecular-weight substances move with the void volume, whereas the low-molecular-weight components move more slowly. This method is faster and more efficient than dialysis.

7. Affinity Chromatography

Affinity chromatography uses the ability of enzymes and some other proteins to bind substrates, coenzymes and other ligands specifically and reversibly. The ligand is covalently bound to an insoluble matrix and the support packed into a chromatographic column. A solution of the impure enzyme or protein is then applied to the column and the enzyme is selectively retained. Impurities which are not bound are eluted and the enzyme is subsequently displaced by elution with a solution of the substrate in a medium of different pH or ionic strength. Successful application of affinity chromatography depends largely on how closely the chosen conditions permit the ligand–macromolecule interaction which is characteristic of the components in free solution. Careful consideration must be given to the nature of the inert matrix and to the steric restrictions generated by immobilization of the ligand. The nature, mode of attachment and concentration of the ligand greatly influence the adsorption and subsequent elution of the complementary macromolecule. In considering the choice of matrix the following conditions must be satisfied:

(a) it must possess suitable and abundant chemical groups to which the ligand may be coupled covalently, and must be stable when attached;

(b) it must be stable during binding of the macromolecule and its subsequent elution, and

(c) it must not interact with other macromolecules to minimize nonspecific adsorption.

In practice, the most commonly used matrices are agarose and cross-linked dextrans. The choice of ligand depends solely on the specificity of the macromolecule to be purified. It must also possess a suitable chemical group which will not be involved in the specific binding of ligand to macromolecule, but which can be used to link it to the matrix.

Affinity chromatography has been used with great success to purify a number of enzymes and proteins. In addition, a number of antibodies, antigens, nucleic acids, lipids, cells and viruses have been purified. The technique also provides a means of studying hormone and drug receptors and other complex cellular structures.

8. High-Performance Liquid Chromatography

High-performance liquid chromatography (HPLC) or high-pressure liquid chromatography is the latest development in the field of chromatography. Its great advantage is that separation and detection can be carried out in one operation and it is particularly suited to trace analysis. The main reason for the high performance is the use of a small particle size resulting in an increase in pressure to obtain adequate flow rates of mobile phase. The method is used extensively for analytical techniques, and high resolution in a very short time can be obtained with small samples. The sample volume injected can be higher than for GLC; sample collection is simple.

The essential components of an HPLC system are a solvent delivery system, an injector, a column and a detection system. The solvent delivery system is a pump which should be capable of providing high pressures (up to 40 MPa) giving a pulseless flow at low, constant, easily controlled flow rates. Some systems provide for the continuous change of solvent composition. Introducing the sample is more difficult in HPLC than in GLC because of the higher pressure inside the system. In

HPLC, the sample is usually introduced using a sample valve in which the sample is contained in a core or a loop that can be rotated into and out of the solvent stream. Detection systems should be sensitive, versatile and have a linear response over a wide concentration range. The two most common types are ultraviolet detectors and refractive index detectors. Others which have been used for certain applications are fluorescent detectors, electron-capture detectors, electrochemical detectors and reaction detectors. All modes of separation can be employed: adsorption, partition, ion-exchange and molecular exclusion. Appropriate methods are selected by considering the polarity, ionizability and molecular weight range of the components of the sample. The advantages arising from the use of very small particles, bonded stationary phase and paired-ion chromatography have led to an increase in publications of methods for the examination of biological fluids by HPLC. It appears that HPLC will continue to expand, and methods will be modified to give greater speed and sensitivity.

See also: Clinical Chemistry: Physics and Instrumentation

Bibliography

Dixon P F, Gray C H, Lim G K, Stoll M S 1976 *High Pressure Liquid Chromatography in Clinical Chemistry*. Academic Press, London
Jones A R 1970 *An Introduction to Gas–Liquid Chromatography*. Academic Press, New York
Simpson C F 1982 *Techniques in Liquid Chromatography*. Wiley, New York
Smith I, Seakins J W T 1976 *Chromatographic and Electrophoretic Techniques*, Vol. 1, *Paper and Thin Layer Chromatography*, 4th edn. Heinemann, London
Williams D L, Nunn R F, Marks V (eds.) 1978 *Scientific Foundations of Clinical Biochemistry*, Vol. 1, *Analytical Aspects*. Heinemann, London

E. C. Jamieson

Chromosome Analysis, Automatic

Automatic chromosome analysis is of critical importance in genetic counselling and in mass screening for genetic defects. Amniocentesis and chorion villous sampling (CVS) can enable fetal chromosome defects to be diagnosed. Such procedures are used in: (a) the screening of high-risk women (e.g., women over the age of 35 who desire children); (b) screening newborns by means of the amnion, or birth membrane, or by CVS; (c) giving genetic counselling to individuals with family histories of mental retardation and other known cytogenetic abnormalities; and (d) monitoring the effects of low levels of radiation and environmental pollutants on human chromosomes, especially those of children and industrial workers.

It has been observed, using only the classical staining and karyotyping methods, that one person in 200 has a major chromosome abnormality. This figure increases significantly to 1 in 50 people when banding techniques are used. This latter technique often enables translocations (i.e., the transfer of a segment of a chromosome from one to another site on the same or another chromosome) to be precisely specified with regard to the origin and final location of each translocated fragment. Such information is vital in evaluating the significance of a translocation in terms of mental retardation and other abnormalities.

Most cytogenetic laboratories are being overloaded because of the number of patients seeking genetic counselling. If chromosome analysis is to be made available to all those desiring genetic counselling and for mass screening, then automatic chromosome analysis is essential. At the present time, there are too few qualified technicians for the great number of individuals desiring chromosome analysis. Automating the procedure with interactive capabilities promises to assist significantly in enabling existing and future genetic counselling laboratories to cope with the increased work load.

Chromosome analysis can be divided into two distinct phases: the first being that of finding metaphase spreads on a glass-slide preparation (Fig. 1), the second being the

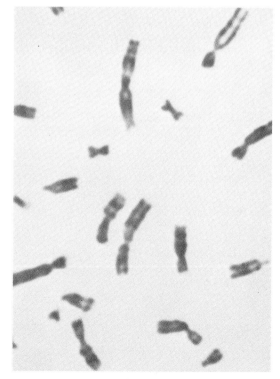

Figure 1
Human metaphase chromosome spread as seen using a 100 × microscope objective.

production of the karyotype or classification of the chromosomes of a metaphase spread (Fig. 2). In manual laboratory techniques much of the time spent by the technician is in locating the metaphase spreads on the

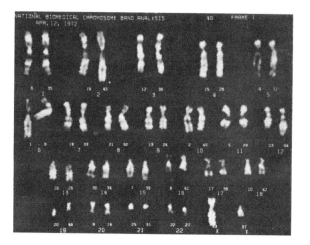

ATIONAL BIOMEDICAL CHROMOSOME BAND ANALYSIS   45   FRAME I
APR, 12, 1972

Figure 2
A human karyotype produced in 1971 by the automatic chromosome-analysis facility developed by the authors

glass slide and photographing the spreads in preparation for karyotype analysis. Automating this first phase alone could probably quadruple the number of patients that could be handled by the same number of technicians. Performing the karyotyping utilizing a light pen on a graphics monitor TV screen without having first to make a photograph will also improve technicians' throughput. Automating this second phase would also reduce the time required for training new cytogenetic technicians.

1. Summary of the Flow of Processing

After the patient material (e.g., blood sample) with patient identification is presented to the facility, the cells must be cultured and the chromosome spreads prepared and stained for banding analysis. The automatic computer analysis is then performed and the resulting report delivered to the proper recipient (e.g., the patient's physician or the screening supervisor). At any time, the statistical databank can be interrogated by means of a database programming system (to be discussed later). We shall now consider the automatic computer-processing steps. It should first be noted, however, that the procedure has a number of variations (mostly concerned with the order of performing the processing components), all of which the facility can accomplish. These variations somewhat complicate the discussion;

therefore they will not be detailed until after the description of basic processing.

The processing flow diagram begins with the microscope slide (Fig. 3). After the patient identification is entered into the computer, the corresponding microscope slide is placed on the automatic focusing stage of

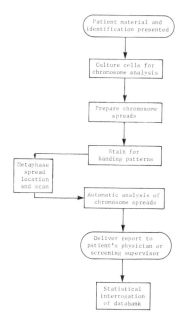

Figure 3
Flow of processing for an automated chromosome-analysis facility

the automated microscope. The microscope will then automatically select a good chromosome spread and display this on the TV monitor. The operator either approves or rejects the display, depending on its quality, the number of overlapping chromosomes, and so on. It is expected that at least 80% of the chromosome spreads selected by the automated microscope will be approved by the operator.

If the operator rejects the spread by pressing the appropriate key on the keyboard, the automated microscope will continue to search the slide for another good spread. Several end points of this process may be identified: first, when the entire area of the slide has been scanned; second, when the requisite number of chromosome spreads have been located; and third, when the requisite number of "good" spreads have been found. When the operator approves a spread, the "approved" key is pushed, and the computer records the coordinates of this good spread for future reference. During the scanning for a good metaphase spread, a $20 \times$ microscope objective is used; the selected spread is then observed under a $100 \times$ microscope objective. Then, using a light pen, the operator interactively edits the chromosome spread image, pointing out to the computer where chromosomes are touching and overlapping and where there may be artefacts.

On the completion of editing, the operator pushes the "analysis" key: this directs the computer to digitize the image and put it into the memory of the computer. The computer then performs the analysis of all chromosomes of that spread, determining whether the spread is normal or abnormal, and displays the karyotype on the TV monitor. If the spread is abnormal, the computer asks the operator if the analysis is to be repeated. The computer actually tells the operator why the spread is abnormal, and if it is felt that there is an error, the operator can interactively correct the karyotype. It is generally found that little if any editing is required, so only very infrequently will the operator have to do this step. However, for 100% accuracy, it is necessary to include this option. The complete report on the cell is then produced by the computer (Fig. 4).

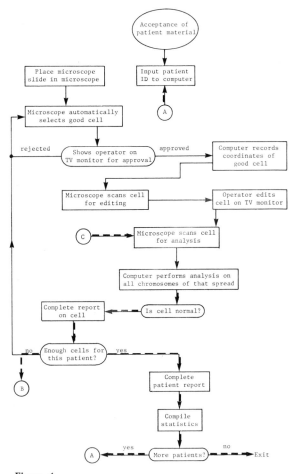

Figure 4
Flow diagram of the automatic computer-processing steps

If the requisite number of cells for that patient (an input parameter) has been analyzed, the complete patient report is compiled and printed. At the same time, the computer enters these results into the databank, producing whatever statistical results are required. Otherwise the search for the next cell for that patient is initiated by the automated microscope.

Many variations in the order of the procedures just described are possible. For example, it may be more efficacious to search the microscope slide for the requisite number of good spreads all at once, recording the coordinates of all the good cells. Then at a later time the microscope slide can be replaced into the automated microscope and the computer will locate each cell for performing the analysis. In this system, all the searching is done at once, and all the analysis is done afterwards, whereas in the first-mentioned system, the search-and-analysis steps were interleaved. A very important variation in the order of the procedures is the capability of concurrently overlapping the editing and analysis steps. The editing of one cell can be accomplished the same time that the computer is carrying out the analysis of the previous cell. For example, in current systems the classical chromosome analysis takes about 40 s; it is found that this is ample time for the operator to perform the editing of the next cell to be analyzed. Similarly, all interrogations of the statistical files can be concurrently overlapped in time with the automated microscope searching of microscopic slides, further enhancing the utilization of the facility.

2. Computer Analysis of Chromosomes

In the following section, methods of automatic computer analysis developed by the present authors are described, but all other methods to a large degree closely relate to those discussed here. The syntax-directed pattern-recognition approach for computerized image analysis of chromosomes was conceived and developed by Ledley who, in the early 1960s, applied these methods to classical automatic chromosome analysis together with Ruddle. Later numerous researchers worked in the field, including Mendelsohn, Rutovitz and Neurath. By the late 1960s, the authors had developed their own SPIDAC computer-controlled microscope and vidicon scanning system, and the MACDAC interactive image-editing graphics console for automatic chromosome analysis, because low-cost digital image scanners and displays were not yet commercially available. In the early 1970s, Ledley worked with Lubs and Ruddle to demonstrate clearly the efficacy of the automatic analysis, including banding analysis (Ledley et al. 1972). However only since the early 1980s, with the extensive interest and growth of genetic counselling and subsequent economic pressures, has there been significant commercial interest in the utilization of the algorithm methods developed for automating cytogenetic laboratories.

2.1. Scanning the Glass Slide for Chromosome Spreads

The purpose of scanning for good metaphase spreads is to locate the coordinates of such spreads and to grade the spread with respect to the qualities needed for automatic karyotyping. Figure 5 shows a block diagram

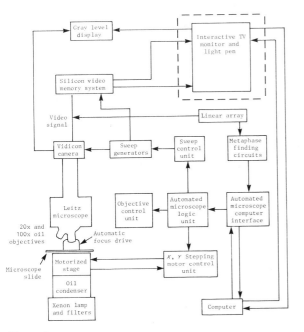

Figure 5

Block diagram of SPIDAC system, including the associated parts of the MACDAC and computer hardware and software

of the automated computer and microscope components for this process. To use the automated microscope, a thick oil is placed on the microscope slide, and the slide is placed in the holder. The $20 \times$ oil microscope objective is automatically focused by an instantaneous automatic high-focus device. With the $20 \times$ objective, the motorized stage moves the slide so that successive strips $525 \, \mu m$ in length will be scanned in the field of view of the linear array. The stage moves at a velocity of about $0.16 \, cm \, s^{-1}$ covering the area of the slide at a rate of $1 \, cm^2$ in approximately two minutes. During the scanning, the information is sent into special metaphase-finding circuits, and the output of these circuits is delivered via the computer interface to the computer's memory. The time to detect all chromosome spreads on a $2 \, cm^2$ area of the microscope slide is about four minutes.

During the time the stage is moving, the information from the linear array is transmitted to the metaphase-finding circuits. These circuits rapidly execute a logical procedure that utilizes an algorithm dependent on the characteristic shapes of the chromosomes and their

proximity to one another in a chromosome spread. The number of objects in a spread is also roughly calculated, and based on these criteria, the occurrence of a chromosome spread is detected and a "good" or "no good" decision is made. When a good chromosome spread is detected the computer determines the coordinates of the spread and records these for later use. Depending on the order of processing desired, the automatic microscope will continue to look for other chromosome spreads or will proceed with the chromosome karyotyping analysis.

The movement of the stage is under the control of the computer, which can position the microscope slide to within $1.25 \, \mu m$ (i.e., the minimum motor-step size is $2.50 \, \mu m$). This choice of step size is limited only by the overall speed with which the scan can be made; the mechanical setup is accurate to within a fraction of a micrometer. For analysis, the computer positions the stage so that the center of the chromosome spread, the coordinates of which have been previously determined, is directly under the objective, and the $100 \times$ objective is used. A change of filter is required so that no adjustment of the vidicon camera will be necessary. The $20 \times$ and $100 \times$ objectives are parfocal, so that the automatic focus device will keep the cell in focus for both objectives. A scan is made with the TV camera, and the image is presented to the operator for editing.

The logical circuits for metaphase-spread detection recognize chromosome spreads during a search of the glass slide. The technique employed in the logic for recognition of chromosome spreads uses a triplet of conceptual cursors. As the slide is scanned and digitized by the linear array, the triplet of conceptual cursors presents a flag when 010 occurs, being the identification of an object of small width. A so-called "cover" is initiated when such a flag occurs. If, as the scan proceeds, another small object is identified within a prescribed distance from the first, then the cover is continued; otherwise the cover is terminated. In this manner a cover is continued along a scan line until there are no more small objects within the prescribed distance of the last object. The number of objects is also counted. Thus a large object (such as a nucleus of a cell) or an isolated object is not detected.

The information concerning the covers is transmitted to the computer. The computer then performs an analysis of the covers. If a long enough cover appears at approximately the same place on a sequence of scanned lines, and if the total number of objects identified is large enough, then the collection of covers is said to be a chromosome spread, and its location is computed as the geometric center of the collection of segments making up these covers. Of course objects are not actually being counted, only "hits" in the manner described above. It turns out, however, that a good chromosome spread will generally have a sufficiently large number of objects to present enough hits to make a good collection of covers. Figure 6 illustrates the covers, and the detection of two different chromosome spreads in the same field. The presence of a nucleus near a spread will not interfere

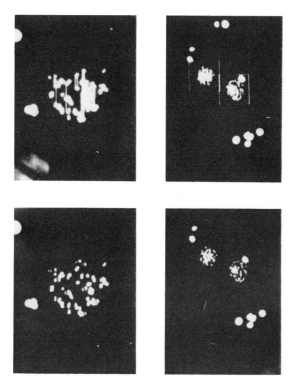

Figure 6
Illustration of method used to recognize chromosome spreads during a search of the glass microscope slide. The technique employed uses, in effect, a triplet of conceptual cursors in the computer's memory. When a properly spaced 010 triplet occurs, an object of small width is identified and a "cover" is initiated (see vertical lines on the chromosome spread). If, as the scan proceeds, another small object is identified within a prescribed distance from the first, then the cover is continued; otherwise the cover is terminated. A collection of covers, properly placed, is said to be a chromosome spread, and its location is that of the geometric center of the collection of segments making up these covers. The figure shows squares around two chromosome spreads that were so detected, where the center of each square is at the center of the chromosome spread as detected by the computer. The pictures were taken from the MACDAC display scope face

with the detection of a good spread. A poor spread will not present a sufficient number of objects.

2.2 Interactively Editing a Chromosome Spread

The editing of a chromosome spread is carried out through the computer by means of a light pen and a TV monitor. The editing of a chromosome spread consists of identifying artefacts in the image on the monitor that are not chromosomes but are similar in size, and also in "separating" any touching chromosomes in the picture.

Overlapping chromosomes are also pointed out to the computer by means of the light pen. For touching chromosomes, a line is drawn along which the chromosomes will be separated. For overlapping chromosomes the area of overlap is indicated to the computer with the light pen. A "zoom" feature is also included that will display an enlargement of any section of the picture for more detailed study. The area to be zoomed is pointed out to the computer by means of the light pen. Before the editing of the chromosome spread, the precise positioning of the spread can be adjusted through keyboard directions to the computer so that all of the chromosomes of the spread will appear under the microscope objective.

The interactive program used for chromosome analysis has the following options:

(a) Adjust the position of the microscope stage: (i) move up, (ii) move down, (iii) move right, (iv) move left.

(b) Enter the edit code and execute program CLEANUP: (i) erase an object, (ii) partially fence to eliminate artefacts, (iii) separate chromosomes, (iv) connect two objects, (v) indicate overlapping chromosomes, (vi) completely fence to eliminate artefacts, (vii) erase last code inserted.

(c) Count the number of objects: (i) present the number of objects found and indicate on the display which objects were counted, (ii) go back to edit, (iii) go back to main program.

2.3 Classical Computer Analysis of Chromosomes

The classical analysis of chromosomes involves four steps, which are carried out by the computer after the digitized and edited picture has been stored in the computer's memory. The steps are: (a) searching, (b) bounding, (c) locating arm ends and centromere, and (d) syntax analysis. These four steps are repeated for each chromosome being analyzed. Finally, using the results obtained, the process of karyotyping or classifying the chromosomes is carried out after all the chromosomes in the spread have been analyzed.

(a) Searching. In processing a frame, the program processes one object at a time by means of an internal, programmed scan of the picture, starting at the top left-hand corner and continuing row by row. This internal scan is accomplished by a "bug," which looks for a picture spot that has a gray level greater than the "cutoff" gray level. The cutoff gray level is defined such that the interior points of any object will have gray-level values greater than that level. After each object is found, it is processed. In this way the objects are sequentially processed, until finally—when the scanning bug reaches the lower right-hand corner of the picture—all of the objects have been processed and the program proceeds to the analysis applicable to that frame.

(*b*) *Bounding*. When an object has been found, its processing is carried out in terms of a boundary analysis. A bug is moved around the boundary of the object in such a direction that the interior of the object is kept to its right. The next boundary point is determined by looking clockwise around the present boundary point, starting from the previous boundary point. When a certain number N of boundary points (that is, a certain boundary length) has been traversed, a segment is defined. The segment is then characterized by the coordinates of its center point, the components of a leading vector, and the components of a trailing vector. The length chosen for the segment must be short enough that the angle between the leading and trailing vectors is approximately a measure of the curvature of the segment. Then the vector sum of the leading and trailing vectors is approximately the tangent to the segment at its center point and gives a measure of the direction of the segment. There are three user parameters associated with a segment, varied to suit the particular problem under consideration. These are the segment length, the

arrow length, and the distance between centers of successive segments. A boundary characterization list is constructed as each boundary segment is analyzed successively, until the original boundary entry point is reached again. As the boundary is traversed, each boundary point is flagged, so that the object will not again be "found" by the search program.

(*c*) *Locating arm ends and centromere*. The boundary list for each chromosome is used to find the perpendicular axis of symmetry of the chromosome. Next, a best-fit parabola is found for the chromosome as a whole (Fig. 7). Then, in a parabolic coordinate system with the symmetry parabola as one axis, the ends of the arms are found as the extremes of a set of intersections of a certain family of curves with the boundary of the chromosome. The centromere is found as the narrowest region of the chromosome, where the boundary of the chromosome is nearest to the parabolic symmetry axis. Special cases arise for the achrocentric D and G groups of chromosomes. Particular programs identify chromosomes of

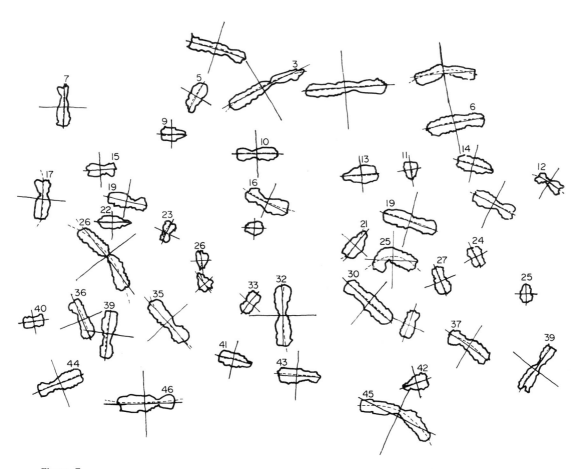

Figure 7
Parabolic axes of chromosomes. The dashed line shows the best parabolic axis for each chromosome

these types, and for the determination of the arm ends and the centromere, special programs are called which depend on local curvature concavities for the centromere location.

The arm lengths of the chromosomes are the distances between the center of the centromere and the ends of the arms. The total length of the chromosome is the sum of the average lengths of the long and short arms. The total area is the number of points within the perimeter. The arm area is the number of points contained in an arm pair. Using these parameters, the classical karyotype or classification of chromosomes into ten groups is carried out, where the groups are denoted as A1, A2, A3, B, C and X (the female chromosome), D, E16, E17-18, F, G and Y (the male chromosome).

(d) Syntax analysis. The above description presents only the barest outline of the automatic computer program used by the authors for analyzing a chromosome and making measurements on it. Other methods have also been developed, one of the most interesting of these being the curvature-syntax method. Here the curvatures of the boundary segments are determined, and these data, together with the relative positions of the edge segments around the periphery of the chromosome, are used in a syntactical analysis to identify the provisional ends of the chromosome arms. The provisional arm ends are then used to calculate the center of gravity of the

chromosome. Following this calculation, the centers of the boundary segment situated midway between each pair of ends of the chromosome arms are identified, permitting the division of the chromosome into four quadrants. Each provisional arm end is moved to the point of maximum distance from the center of gravity within the limits of its quadrant. The proper arm pairs are then identified by pairing the ends that are closest together. Next, starting at the ends of an arm pair, the narrowest portion of the chromosome is determined to be the centromere (see Fig. 8).

Both systems of classical analysis can be utilized in the same program. The parabolic symmetry method can be used first, and the results of this method given a confidence test to assess the correctness of the results. If the confidence level is too low, then the computer reanalyzes the chromosome using the syntax method. In practice, less than 1–2% of the chromosomes require such a reanalysis. Other researchers have described various systems for accomplishing the same tasks of finding the arm ends and centromere of a chromosome. But most methods are related to either the parabolic axis technique or the syntax methods.

2.4 Banding Computer Analysis of Chromosomes

(a) Finding the profile. The first step in chromosome-banding analysis is to find the gray-level profile along the axis of the chromosome by integrating gray levels on successive lines perpendicular to the axis. Since the chromosomes can be bent, a curvilinear parabolic coordinate system is used. The principles involved in finding the profile of a chromosome are simple: the principal axes of the chromosome are found, and modified to parabolic axes. The longer axis is then divided into convenient segments of equal length. The chromosome is examined spot by spot, and the gray-level value of each spot is added to a tally for the longer-axis segment corresponding to the long-axis coordinate of the spot, as a tally is kept of the number of spots contributing to each segment. When the examination of the chromosome is completed, the gray-level tally for each longer-axis segment is divided by the corresponding spot-count tally to find the average gray level for that segment.

The spot-by-spot examination of the chromosome requires some programming complexity. Since only those spots belonging to the chromosome under consideration can be used, the use of all the spots within a "smallest" rectangle around a chromosome will result in errors in cases where a piece of another chromosome may also be included in the rectangle.

(b) Karyotyping with banding. The second step of banding analysis is the karyotyping or identification of each individual chromosome. This proceeds within the groups determined by the classical analysis. That is, the analysis of the banding patterns is accomplished after the classical analysis has been completed. Since the classical analysis results in ten groups of chromosomes (i.e., A1, A2, A3, B, C and X, D, E16, E17-18, G

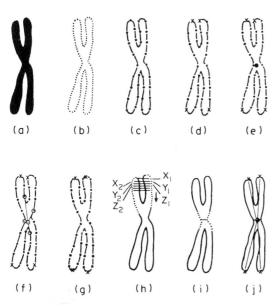

Figure 8
FIDAC analysis of chromosomes showing: (a) a chromosome silhouette, (b) boundary points, (c) segment center points, (d) arm ends, (e) average of arm ends, (f) four quadrants, (g) adjusted arm ends, (h) beginning search for centromere location, (i) adjusted centromere location, and (j) measurement distances

and Y), the banding patterns will be used: (i) to distinguish between chromosomes within the same group (i.e., within groups B, C and X, D, E17-18, F, G and Y), and (ii) to verify the consistency of the banding patterns with the classical chromosome-analysis results (i.e., particularly in the cases of A1, A2, A3 and E16).

Of course, the banding, or gray-level profile curves are used, as illustrated in Fig. 9. These profile curves are

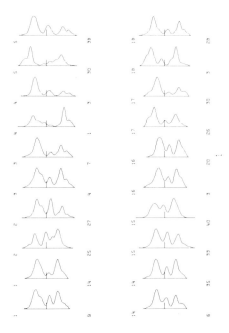

Figure 9
Actual banding profiles of some chromosomes

produced by averaging gray-level values of sampled points taken in strips perpendicular to the long axis of the chromosome, whereas this long axis is a best-fit parabola as discussed above. The curves represent the results of smoothing the raw data to make the identification of such critical points of the curve as maximum, minimum and inflection points more reliable.

Working with these curves, four techniques of identification are used, namely: (i) Fourier-analysis coefficients, (ii) differences from a "standard" curve for each of the 22 pairs and X and Y, (iii) the use of a discrimination function and Bayesian approaches, and (iv) the application of pragmatic and ad hoc methods. Often the best results are obtained with this latter approach, i.e., to follow the logic or reasoning of the cytogeneticist in classifying chromosomes. This method has the added advantage of being better understood by the cytogeneticist, who must, in any event, be the final judge of the success of the system. Thus the cytogeneticist can significantly aid in suggesting methods for improving the computer program, even if he himself is

not directly involved in the programming act. A disadvantage of this method is that the words used by the cytogeneticist are often not sufficiently precise, or may in fact not express his intended meaning.

The computer then displays on the TV monitor the resulting chromosome karyotype, with the chromosomes rotated and aligned standing vertically (Fig. 2). Since the chromosomes are presented with their banding pattern as scanned by the TV camera, the skilled cytogeneticist can review the computer results and interactively make adjustments if required. Thus chromosomes can be interchanged in their classification position and questionable chromosomes can be indicated. This karyotype can then be photographed and a hard copy made. The hard copy can then be included in the patient's chart.

2.5 The Chromosome Databank

(*a*) *The patient's report*. The operator can then direct the computer to compile a summary report for the patient. The summary report will include the patient identification, the number of cells examined for that patient, the results of the examination, the results of the karyotyping and a picture of one or more karyotyped cells.

The entire objective of the system is to detect and identify abnormal cases. Of course, the detection of abnormalities is based on the ability to identify normal chromosome spreads as an essential prerequisite. There are four types of abnormalities that can occur: (i) abnormalities in the number of chromosomes in a spread, (ii) abnormalities in the measurements of a chromosome in a spread, (iii) abnormalities in the morphology of a chromosome in a spread, and (iv) abnormalities in the banding pattern of a chromosome in a spread. Combinations of these types of abnormalities frequently occur. Breaks, rings, etc., are all classified under abnormalities of morphology.

(*b*) *Statistical analysis*. The DOCS (display of chromosome statistics) programming system is used for statistical analysis of the chromosome databank. This is an interactive system in which the scientist can communicate "on-line" with the disk memory of the computer, from the computer console. The interaction enables the evaluation and display of large masses of data in a file. For instance, analyses of variance, *t*-tests, and other statistical tests can be executed on data selected from the file; or histograms, least-squares lines, scattergrams, etc., can be displayed by the computer (Fig. 10). The DOCS system guides the user in a step-by-step dialogue, so that all of the detailed specifications needed for the desired statistical analysis or data display can be obtained.

Examples of types of displays available are as follows:

(*i*) *Histograms*. The distribution frequency for any measurement and patient, upon identification of the information, is given in a histogram (patient number, measurement, type of chromosome). The maximum and minimum of the appropriate parameters

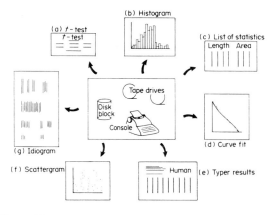

Figure 10
Illustration of interactive and graphic capabilities of
the DOCS (display of chromosome statistics)
programming system

are called from the disk memory. The histogram
may be drawn on the printer or plotter.

(ii) Idiograms. The computer idiogram is an arrangement of the entire chromosome complement of a
cell into the ten recognizable types, with each
chromosome represented as a stick figure.

(iii) Scattergrams. Upon identification of the parameters
to be plotted on the *x* and *y* axes (patient number
and two measurements), the maxima and minima
are obtained from the disk, and the axes are drawn
on the printer or plotter.

(iv) Lists. The contents of the disk or tape or basic
statistics derivable from the accumulations are
printed out.

(v) t-Tests. The statistical mean of a given measurement of any one chromosome for one patient is
compared with that of another patient, or a sub-
population, or the entire population, or any combination. The *t*-statistic is given with the number of
degrees of freedom.

Fragments from an interactive dialogue are shown in
Fig. 11 in order that the reader may get some feel for the
techniques involved (the letters and numbers in the
parentheses give the FORMAT specification for the
reply, e.g., (14) means a four digit integer).

3. Summary

In this article the algorithmic methods used for automatic chromosome analysis have been briefly outlined,
including those first used in automatically locating chromosome spreads on the glass slide, and then those used
in performing both classical and banding analysis for
karyotyping. The problem of automatic chromosome

```
       WHAT PATIENT, WHAT CELL (214)
00000030
       NO PATIENT   0 IN PATIENT LIST
       WHAT PATIENT, WHAT CELL (214)
15200010
       WHAT MEASUREMENTS? (213)
000007
       1520     10  2  46
       DISPLAY  (A4)
hist
       INDIVIDUAL CELLS (0)
       AVERAGE OF CELL MEANS (1)
       AVERAGE OF INDIVIDUAL MEAS (3)
       AVERAGE OF PATIENT MEANS (9)   FORMAT 12
9
       WHAT MEASUREMENTS? (13)
       HOW MANY STRIPES ON X-AXIS (BETWEEN 5 AND 50),   (12)
15
       DISPLAYS (A4)
hist
       INDIVIDUAL CELLS (0)
       AVERAGE OF CELL MEANS (1)
       AVERAGE OF INDIVIDUAL MEAS (3)
AVERAGE OF PATIENT MEANS (9)   FORMAT 11
1
       WHAT PATIENT   (14)
0000
       WHAT MEASUREMENTS? (13)
015
       HOW MANY STRIPES ON X-AXIS (BETWEEN 5 AND 50),   (12)
10
       WHAT TYPE DO YOU WISH (USE 2 TO 21), '13
       IF YOU WISH ALL TEN TYPES, TYPE 0
000
       DISPLAY? (A4)
          STOP.  0
FA83A INT REQ OCA
```

Figure 11
Fragments from an interactive dialogue, DOCS
programming system

analysis played an important role in the initial development of the field of automated image pattern recognition, and led to the development of syntax-directed
image analysis by Ledley. Only recently, with the greatly
increasing need for clinical cytogenetics, has there been
extreme commercial interest in these techniques. Indeed,
with the rapidly increasing patient load, automatic chromosome analysis may well be essential if the clinical
demand is to be satisfied.

See also: Clinical Biochemistry: Automation; Genetic Code;
Genetic Engineering; Cervical Cytology: Automation

Bibliography

Aggarwal R K, Fu K S 1974 Automatic recognition of
   irradiated chromosomes. *J. Histochem. Cytochem.* 22:
   561–68
Bender M A, Kastenbaum M A, Lever C S, Pelster D R
   1972 Combination Bayesian and logical approach to
   analysis of normal and abnormal chromosome spreads.
   Comput. Biol. Med. 2: 151–66
Bishop R P, Young I T 1977 The automated classification
   of mitotic phase for human chromosome spreads. *J.
   Histochem. Cytochem.* 25: 730–40

Bruschi C, Tedeschi F, Puglisi P P, Marmiroli N 1981 Computer-assisted karyotyping system of banded chromosomes. *Cytogenet. Cell. Genet.* 29: 1–8

Caspersson T, Castleman K R, Lomakka G, Modest E J, Moller A, Nathan R, Wall R J, Zech L 1971 Automatic karyotyping of quinacrine mustard stained human chromosomes. *Exp. Cell Res.* 67: 233–35

Castleman K R, Melnyk J, Frieden H J, Persinger G W, Wall R J 1976 Computer-assisted karyotyping. *J. Reprod. Med.* 17: 53–57

Eaves G N 1967 Image processing in the biomedical sciences. *Comp. Biomed. Res.* 1: 112–23

Fleischmann T, Gustafsson T, Hakansson C H 1971 Computer-display of the chromosomal fluorescence pattern. *Hereditas* 68: 325–28

Gallus G, Montanaro N, Maccacaro G A 1968 A problem of pattern recognition in the automatic analysis of chromosomes: Locating the centromere. *Comp. Biomed. Res.* 2: 187–97

Gallus G, Neurath P W 1970 Improved computer chromosome analysis incorporating preprocessing and boundary analysis. *Phys. Med. Biol.* 15: 435–45

Gilbert C W 1966 A computer programme for the analysis of human chromosomes. *Nature (London)* 212: 1437

Gilbert C W, Muldal S 1971 Measurement and computer system for karyotyping human and other cells. *Nat. New Biol.* 230: 203–7

Green D M, Bogart J P, Anthony E H 1980 An interactive, microcomputer-based karyotype analysis system for phylogenetic cytotaxonomy. *Comput. Biol. Med.* 10: 219–27

Green D K, Cameron J 1972 Metaphase cell finding by machine. *Cytogenetics* 11: 476–87

Ladda R, Atkins L, Littlefield J, Neurath P, Marimuthu K M 1974 Computer-assisted analysis of chromosomal abnormalities. *Science* 185: 784–87

Ledley R S 1964 High speed automatic analysis of biomedical pictures. *Science* 146: 216–23

Ledley R S 1966 Computer aids to medical diagnosis. *J. Am. Med. Assoc.* 196: 933–43

Ledley R S 1969 Automatic pattern recognition for clinical medicine. *Proc. IEEE* 57: 2017–35

Ledley R S 1973 Pattern recognition with interactive computing for a half dozen clinical applications of health care delivery. *AFIPS Conf. Proc.* 42: 463–77

Ledley R S, Ing P S, Lubs H A 1980 Human chromosome classification using discriminant analysis and Bayesian probability. *Comput. Biol. Med.* 10: 209–18

Ledley R S, Lubs H A, Ruddle F H 1972 Introduction to chromosome analysis. *Comput. Biol. Med.* 2: 107–28

Ledley R S, Ruddle F H 1969 Chromosome analysis by computer. *Sci. Am.* 221: 87–95

Ledley R S, Ruddle F H, Wilson J B, Belson M, Albarran J 1968 The case of the touching and overlapping chromosomes. In: Cheng G C (ed.) 1968 *Pictorial Pattern Recognition*. Thompson, Washington, DC, pp. 87–97

Low D A, Selles W D, Neurath P W 1971 Bayes' theorem applied to chromosome classification. *Comp. Biomed. Res.* 4: 561–70

Lubs H A, Ledley R S 1973 Automated analysis of differently stained human chromosomes—A review of goals, problems and progress. *Nobel Symp.* 26: 61–67

Lundsteen C, Bjerregaard B, Granum E, Philip J, Philip K 1980 Automated chromosome analysis, I: A simple method for classification of B- and D-group chromosomes represented by band transition sequences. *Clin. Genet.* 17: 183–90

Lundsteen C, Gerdes T, Granum E, Philip J, Philip K 1981 Automatic chromosome analysis, II: Karyotyping of banded human chromosomes. *Clin. Genet.* 19: 26–36

Marimuthu K M, Selles W D, Neurath P W 1974 Computer analysis of giesa banding patterns and automatic classification of human chromosomes. *Am. J. Hum. Genet.* 26: 369–77

Mendelsohn M L, Conway T J, Hungerford D A, Kolman W A, Perry B H, Prewitt J M S 1966 Computer-oriented analysis of human chromosomes, I: Photometric estimation of DNA content. *Cytogenetics* 5: 223

Mendelsohn M L, Hungerford D A, Mayall B H, Perry B H, Conway T J, Prewitt J M S 1969 Computer-oriented analysis of human chromosomes, II: Integrated optical density as a single parameter for karyotype analysis. *Ann. N.Y. Acad. Sci.* 157: 376

Mendelsohn M L, Kolman W A, Perry B, Prewitt J M S 1965a Computer analysis of cell images. *Postgrad. Med.* 38: 567–73

Mendelsohn M L, Kolman W A, Perry B, Prewitt J M S 1965b Morphological analysis of cells and chromosomes by digital computer. *Methods Inf. Med.* 4: 163–67

Mendelsohn M L, Mayall B H 1972 Computer-oriented analysis of human chromosomes, III: Focus. *Comput. Biol. Med.* 2: 137–50

Neurath P W, Ampola M G, Low D A, Selles W D 1970 Combined interactive computer measurement and automatic classification of human chromosomes. *Cytogenetics* 9: 424–35

Neurath P W, Bamford S B, Mitchell G W Jr 1968 Computer-aided measurements of neutrophil appendages in sex chromosome aberrations. *Med. Res. Eng.* 7: 17–20

Oosterlinck A, Van Daele J, DeBoer J, Dom F, Reynaerts A, Van Den Berghe H 1977 Computer-assisted karyotyping with human interaction. *J. Histochem. Cytochem.* 25: 754–62

Paton K 1969 Automatic chromosome identification by the maximum-likelihood method. *Ann. Hum. Genet.* 33: 177–84

Perry J 1969 System for semi-automatic chromosome analysis. *Nature (London)* 224: 800–3

Piper J, Mason D, Rutovitz D, Ruttledge H, Smith L 1979 Efficient interaction for automated chromosome analysis using asynchronous parallel processes. *J. Histochem. Cytochem.* 27: 432–35

Ruddle F, Smith S, Ledley R S, Belson M 1969 Replication-precision study of manual and automatic chromosome analysis. *Ann. N.Y. Acad. Sci.* 157: 400–23

Rutovitz D 1968 Automatic chromosome analysis. *Br. Med. Bull.* 24: 260–67

R. S. Ledley and H. A. Lubs

Clinical Biochemistry: Automation

The first automated clinical biochemistry analyzers (autoanalyzers) were introduced in the early 1950s to increase the precision of some routine estimations, and to increase sample throughput rate to meet the increasing demands that were being made on clinical laboratory services. The ensuing years have witnessed a huge increase in their capacity, complexity, and cost. A modern multichannel autoanalyzer is capable of performing up to 30 tests per hour on each of 240 samples. From this extreme, right down to the simple single-channel autoanalyzer, there is sure to be an instrument available on the market which will meet any laboratory's particular requirements.

In this article, the principles of continuous flow and discrete autoanalyzers are described. Some examples are given of how automation can be enhanced by interfacing the analyzers to laboratory computers and by their control using microprocessor technology. The fundamental principles of the analytical measurements involved are described elsewhere (see *Clinical Chemistry: Physics and Instrumentation*).

There are essentially two classes of chemical determinations performed by automatic analysis, namely end-point and kinetic determinations. In end-point analysis, sample and reagent are mixed and the reaction is allowed to proceed to completion prior to measurement. In kinetic analysis, concentration is determined by the initial chemical reaction rate.

1. Continuous-Flow Autoanalyzers

1.1 The Basic System

The heart of the system (Fig. 1) is a proportioning pump in which flexible tubing is clamped between a smooth platten and a roller assembly. The pump is designed to produce a smooth, steady flow of liquid through the system. The volume of liquid flowing is proportional to the cross section of the tubing. The stream of liquid is segmented by the introduction of air bubbles which have a washing effect on the internal walls of the tube. The sample probe might typically remain in the sampling position for 50 s, move into a wash tube for 10 s and then move into the following sample cup for a further 50 s, and so on. Successive samples thus pass to the pump along the same length of tubing, separated by a small

volume of water. Diffusable constituents are separated by dialysis into the reagent stream. The sample and reagent mixture is heated to the required temperature for a time long enough to ensure that the chemical reaction is complete, but not too long, otherwise the time delay would be inconvenient and cross-contamination from one sample to the next would be unnecessarily large. With the reaction complete, the bubbles are removed so that a steady signal is obtainable from the measuring device—typically a colorimeter, fluorimeter or nephelometer. Each signal is output in analog form, traced on a pen recorder. The trace consists of a rise curve, a steady-state plateau region of amplitude proportional to the serum concentration being measured, and a fall curve. For optimum throughput, the rise and fall curves should be as steep as possible; the plateau region need only be sufficiently long for a reading to be taken. The parameters affecting the performance of this type of system have been extensively studied and are more or less optimized in standard commercial packages (e.g., the Technicon AA II autoanalyzer). Further increases in throughput rate can be obtained by incorporating curve regenerators into the system between the colorimeter and the recorder. These devices perform a real-time analog deconvolution on the raw data and effectively increase the slope of the rise and fall curves, enabling the plateau to be reached earlier. A 50% increase in throughput can be obtained by curve regeneration. Some users are unhappy with curve regeneration, however, because it can mask anomalies in the raw data.

1.2 Multichannel Systems

In single-channel continuous flow autoanalysis the pump acts as a "heart," pumping fluids through the system. The first attempt to introduce a "brain" came when a programmer unit was incorporated to create a multichannel instrument, the programmer controlling the switching among several analytical channels by means of an array of relays. With the Technicon SMA 6/60, for example, each one of the instrument's six chemistry channels is multiplexed through to the recorder for a period of 10 s. The operation is phased to ensure that the

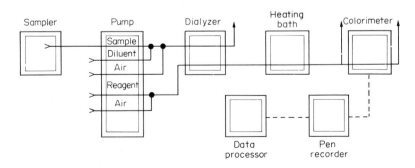

Figure 1
Basic continuous flow autoanalyzer system: ———, fluid pathways; –––, electrical connections

particular chemistry channel appearing on the recorder is in the correct plateau region at the appropriate time. This is accomplished by inserting or removing coiled lengths of tubing to lengthen or shorten the distance travelled by the mixture.

The programmer type of brain is limited in comparison with recent microprocessor-controlled auto-analyzers such as Technicon's SMA II and SMAC. These highly sophisticated "biochemist simulators" have the ability to run more than 10 tests, effectively in parallel, by converting the analog signals from the colorimeter into digital form suitable for storage and processing by the microprocessor control unit. Each curve is mathematically analyzed to assess its integrity and to enable the plateau reading to be made with considerable accuracy. The results from each channel are stored in arrays under microprocessor control. Phasing is therefore no longer necessary. Drift and carry-over corrections can be made with ease.

The continuous-flow autoanalyzers have undoubtedly served as the leading workhorses for routine tests in laboratories over the past two decades. The technology is well proven and generally reliable.

2. Discrete Analyzers

In a discrete analyser (Northam 1969), the sample is transferred to a cup and transported through the system (Fig. 2). Diluent and one or more reagents can be added

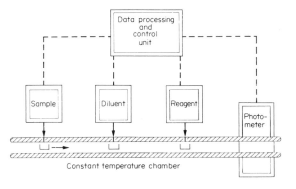

Figure 2
A discrete analyzer

by pumping them from appropriate reservoirs. In the more recent instruments a microcomputer controls the selection of the reagent, and the volumes of sample, diluent and reagent, by setting the number of pulses that drive a stepping motor. It also takes measurements from the photometer and translates them into meaningful concentration units. Results are normally output on a printer and/or on an interfaced laboratory computer.

Cross-contamination between samples is virtually nonexistent in discrete analyzers. Hence high throughput

rates can be achieved—generally between 100 and 300 samples per hour. Discrete analyzers are normally used for end-point reactions, although—by incorporating two photometers into the system—kinetic measurements are possible on some instruments.

3. Centrifugal Analyzers

Centrifugal analyzers are discrete analyzers in which sample and reagent are mixed by centrifugal force. A disk of up to 40 radial segments is loaded by means of an automatic pipetter with sample and diluent in one compartment of a segment and reagent in another (Fig. 3). When the disk is spun, typically at 1000 rpm, the

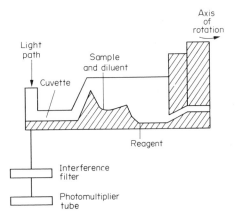

Figure 3
Cross section of part of the rotor of a centrifugal analyzer

sample and reagent mix and are transferred by centrifugal force to the cuvette. The rotor assembly is located within a chamber maintained at constant temperature. Photometer measurements of optical density are digitized and multiplexed by the microprocessor control unit. The contents of all segments of the rotor can be studied in parallel. Both end-point and reaction-rate determinations can be made. Calculations to convert optical density units into chemistry concentration units are performed under microprocessor control.

4. Computer Interfacing

The conveyer-belt approach to clinical analysis is incomplete without a laboratory computing system, effectively a large databank in which results are stored against a patient's identification code. There are three ways of inputting data:

(a) manually, by reading results from a chart recorder or a printer;

(b) automatically, but off-line by the production by the autoanalyzer of an ASCII-coded paper tape which is subsequently read by the laboratory computer; and

(c) automatically and on-line, by direct link between the laboratory computer and the analyzer.

In mode (c), data are normally transferred between the analyzer and the computer in real-time by the computer's input/output bus. The interfacing system adopted in a particular instance depends on the capacity of the laboratory computer, the number of instruments which it serves, and on the throughput of the analyzer. When more than one analyzer is involved, priority of access to the computer is another determining feature in choosing the interfacing system.

5. Economic Considerations

The availability of finance is an often desirable constraint on automation in the laboratory. The demand for clinical information can be virtually endless. It is always advantageous to have information, relating say to serum electrolytes, even though the probability of any abnormality being present is small. Thus, in the absence of financial constraints, the demand for laboratory services tends to increase in proportion to the hospital throughput and to the tests available. The imposition of a financial upper limit forces the number and availability of tests to be rationalized.

The equation governing the choice of a laboratory autoanalyzer has indeed a large financial component. Variables include the purchase price of the analyzer, the number of operators required, the cost of reagents, the cost of maintenance contracts and the ability to select tests. Appropriate weighting factors operate, depending on local priorities. In general, the equation favors the discrete, selective analyzer because on such a system a test which is not requested need not be performed, thus saving on reagents. Continuous-flow analyzers cannot be truly selective. Printout of results can be restricted to the requested tests, but all tests are actually performed because lines cannot be shut off without adversely affecting the dynamic equilibrium of the instrument.

6. Conclusions

The fully automatic, modular, multichannel, selective, programmable, microprocessor-controlled clinical chemistry analyzer is now an accepted part of laboratory life. Commercial interest in the field is vast and developments are rapid. Up-to-date detailed information on a particular instrument is best obtained from the manufacturer's sales literature. Articles in journals are useful in the study of details of a specific chemical method applied to a particular instrument or instruments. Information on specific features and performances of auto-analyzers in journal articles and in text books is often out of date before it appears in print.

The range of autoanalyzers is well documented by Cook (1979a, 1979b, 1980). Details of the principles of autoanalyzers are well presented in Foreman and Stockwell (1975).

See also: Computers in Clinical Biochemistry; Microcomputers: An Application in the Clinical Laboratory

Bibliography

Cook I J Y 1979a Multiple channel analysers: Increasing use by chemical laboratories. *Med. Technol.* 9 (11): 44–49
Cook I J Y 1979b Multiple channel analysers: Increasing use by chemical laboratories. *Med. Technol.* 9 (12): 27–31
Cook I J Y 1980 Multiple channel analysers: Increasing use by chemical laboratories. *Med. Technol.* 10 (1): 15–18
Foreman J K, Stockwell P B 1975 *Automatic Chemical Analysis.* Ellis Horwood, Chichester
Northam B E 1969 Discrete analysis systems. *J. Clin. Pathol.* 22 (3): 42–50

D. J. Wyper

Clinical Chemistry: Physics and Instrumentation

There are two aspects to the role of clinical chemistry in patient care: the analysis of a specimen (usually blood), and the interpretation of an analytical result in terms of the diagnosis of an illness or monitoring the progress of therapy. This article will be concerned only with the contribution of physics to analytical procedures.

The widespread use of mechanization of sample handling, instrumental methods of analysis and automatic data processing enable a clinical chemistry laboratory to carry out as many as one million tests per annum. These modern methods depend increasingly on complex instruments and expensive "black boxes," the principles of which are not always well understood by those who use them. Physicists contribute to instrumental methods in clinical chemistry in several ways; for example, in the development of new instruments and techniques, in the education of instrument users, and in the servicing and quality control of instrument performance.

1. Analytical Principles

Many analytical methods are based on the measurement of physical parameters such as mass and volume. These two properties are measured in the classical procedures of gravimetric and volumetric analysis. The absorption and emission of radiation are the bases of several instrumental methods of analysis (e.g., flame emission spectrometry, colorimetry and radiochemical methods).

In addition, the electrical properties of materials are used as measurement parameters (e.g., ion-selective electrodes for measuring pH, p_{CO_2}, p_{O_2}, potassium and ammonium). However, thermal properties have so far been little used in clinical chemistry other than in the determination of osmolarity by freezing-point depression.

In any system of analysis the aim is to make a measurement that is unique to the substance to be determined. Such specificity may be achieved in one of two ways: either the physical signal generated is unique to the substance being determined (e.g., the spectrum of an atom), or the substance is separated from the matrix by techniques such as dialysis, chromatography, fractional distillation, solvent extraction or precipitation before it is measured by a nonspecific method. Once the required signal has been obtained, some form of calibration is necessary in order that the signal may be interpreted in terms of amount of the analyte. Some methods (e.g., pressure measurement and enzyme assays) are absolute in the sense that the amount of the analyte can be estimated by knowledge of and computation from instrumental parameters. More commonly a prepared standard material will be passed through the analytical procedure and the signal it generates compared with that from the sample. When this form of calibration procedure is used, great care is required during the evaluation of the analytical procedure to ensure that no interference effects exist or develop which may lead to a difference in response of the system to standard and sample material. The accuracy of the system may be continuously monitored by quality-control procedures in which the response of the system to standard solutions is recorded day by day and the concentration of the analyte in a reference material with the same matrix is measured with each batch of sample. The signals from the standard solutions and the observed concentration of the analyte in the reference material should remain within defined limits to allow results from one day to the next to be compared in a meaningful way.

2. Analytical Instrumentation

Analytical methods in clinical chemistry fall into two general categories. One category contains methods which involve the analysis of a large number of samples for several components (e.g., the electrolytes, sodium, potassium, chloride and bicarbonate). This requirement justifies investment in elaborate automated equipment dedicated to those analyses. The second category covers those methods where a smaller number of samples require selective tests that may necessitate either special instrumentation or, more likely, a flexible instrument and a skilled analyst. Instruments in the clinical chemistry laboratory fulfill two functions: to enable an analysis to be made which is not otherwise possible and to enable analyses to be carried out faster, more accurately,

on smaller quantities or more cheaply than by alternative methods. Present-day instruments generate electrical signals which may be measured by a meter, a chart recorder or a digital display. Instrumentation in this sense began in the 1940s with the advent of flame photometers and colorimeters. These instruments were inexpensive and simple, and required little in the way of maintenance services. The simpler instruments tend to require greater operator skills and are more labor intensive than the advanced systems in which the sample is fed to the instrument automatically and the signal processed by electronic circuits. At this level of sophistication, instrument maintenance becomes more demanding. With the increasing reliance on instrumentation for the successful operation of the clinical chemistry laboratory there is a corresponding demand for immediate on-site servicing to minimize the "downtime" of the instrument.

The variety of analytical methods and the corresponding branches of physics upon which they are based are so great that within the confines of this article only the briefest mention of them is possible. This information is summarized in Tables 1 and 2. The most commonly used techniques in routine clinical chemistry are presented in Table 1 in approximate decreasing order of frequency of use. Additional minor tests totalling less than 1% of the laboratory throughput include turbidimetry, osmometry, spot card tests and titrimetry. The practice in individual laboratories will often differ from that given in Table 1 according to local needs and resources. Nevertheless, two general conclusions may be drawn: firstly, over 95% of tests depend on electronic instruments of one kind or another and secondly, of those instruments, the colorimeter or spectrophotometer makes the greatest single contribution.

The second place occupied by analytical atomic spectroscopy reflects the large number of determinations of sodium and potassium by emission flame photometry rather than a demand for the determination of many different elements as may be achieved by atomic absorption spectroscopy. In view of the importance of colorimetry and spectrophotometry a brief description of them will be given here.

Photoelectric colorimeters and spectrophotometers are used to measure the absorption of electromagnetic radiation in the wavelength range 200–800 nm by molecules, radicals or ions in solution. The wavelength absorbed is characteristic of an electronic transition within the absorbing species. The width of the absorption band (0.5–50 nm) is much greater than that of an atomic line, due to the fine structure of the energy levels arising from molecular vibration and rotation. The wavelength and intensity of the band may be affected by the solvent and other components of the solution. The absorption spectrum of a compound or its colored derivative provides a rapid and sensitive method for qualitative and quantitative analysis. In a complex mixture, the width of absorption bands frequently results in band overlap. The principle limitations are set by the

Table 1
Routine techniques of clinical chemistry

Technique	Principle	Application	Advantages
Colorimetry/spectro-photometry	measurement of the optical density of solutions in the visible and ultraviolet region of the spectrum at wavelengths characteristic of the compound to be determined	qualitative and quantitative determination of a wide range of compounds or their colored derivatives (e.g., urea, creatinine and glucose); over 30 tests in total	sensitive, rapid analysis, simple instrumentation, readily automated
Analytical atomic spectroscopy	emission, absorption or fluorescence of radiation in the visible and ultraviolet region of the spectrum by a free atom	quantitative determination of most elements and in particular sodium, potassium, calcium, magnesium, copper, zinc and lead	specific, sensitive $(10^{-12} g)$, accurate, simple instrumentation
Ion-selective electrodes	an electrical potential difference is developed across a membrane owing to its selective permeability to, or reactivity with, a particular ionic species	quantitative determination of ions, gases, and (with enzyme coated membranes) organic molecules, e.g., O_2, CO_2 H^+, Ca^{2+}, glucose, urea	in some circumstances ionic activity is of greater significance than the total concentration of the analyte
Immunoassay	a form of saturation analysis in which a known amount of a labelled compound (e.g., an antigen) is in competition with an unknown amount of the same, or reactively similar, compound in a reaction with a limited amount (i.e., insufficient to satisfy all binding sites) of a binding reagent, e.g., an antibody. The label may be a radioactive isotope, a fluorescent molecule or an enzyme	qualitative and quantitative determination of hormones (e.g., insulin, testosterone, oestrogen, thyroxine)	specific, sensitive $(10^{-12} g)$
Electrophoresis	separation of charged compounds of biological interest by their migration in a medium under the influence of an electric field. The resulting distribution of compounds is identified by staining, ultraviolet fluorescence, radioactive labelling, antigen (antibody reaction), quantitative analysis of the medium	identification of compounds (e.g., proteins, amino acids, lipoproteins)	simultaneous screening for the presence or absence of several compounds
Reaction-rate analysis	the rate of product formation, or reagent depletion, is used as a measure of catalytic activity and is usually determined colorimetrically	quantitative determination of enzyme activity (e.g., lactate dehydrogenose, creatine, phosphokinase)	specific, sensitive, simple, rapid

chemical nature of the sample, and many developments are directed at improving the specificity and sensitivity of the reagents used to develop colored compounds. Quantitative analysis is achieved on the basis of the Beer–Lambert law which states that the absorbance (logarithm to the base 10 of the intensity of the incident radiation divided by the transmitted intensity) is directly proportional to the concentration of the absorbing species and the absorption path length. The basic components of spectrophotometers are a light source, wavelength selector, absorption cell (cuvette) and photodetector. Colorimeters or absorptiometers commonly use nondispersive wavelength selection (a filter with bandwidth 4–40 nm) and solid-state or simple phototube detectors, while spectrophotometers employ a prism or grating monochromator (with bandwidth down to 0.2 nm) and a photomultiplier. Colorimeters are inexpensive and suited to repetitive measurements of

Table 2
Special techniques of clinical chemistry

Technique	Principle	Application	Advantages
Chromatography	mixtures of compounds are separated by means of their differing distribution coefficients between contiguous moving and stationary phases. The difference in distribution may result from molecular weight, charge or shape, singly or in combination	separation and identification of compounds (e.g., ethyl alcohol, amino acids, drugs, metabolites and steroids)	relatively simple technique for the separation of complex mixtures
Fluorimetry and phosphorimetry	reemission of absorbed energy (usually radiation) by a molecule; fluorescence occurs within 10^{-8}–10^{-3} s and phosphorescence after 10^{-3} s	quantitative determination of natural products and drugs (e.g., flavins, purines, tyrosine and their luminescent derivatives, steroids, and vitamins). May be used as a nonradioactive tracer technique by attaching a fluorescent group to the molecule of interest	highly specific and sensitive
Ultrafiltration	solute flow through a semipermeable membrane is enhanced by a pressurized flow of solvent through the membrane	separation and concentration of high-molecular-weight ($>10^4$) solutes (e.g., proteins), removal of salts, and separation of "unbound" from "protein bound" species	simple, rapid technique when high resolution is unnecessary
Centrifugation	separation of large molecules or particles in a medium of controlled density on the basis of their density or size under the influence of an applied centrifugal force	separation of cells, cell fragments and large proteins by sedimentation. Lipids are separated by flotation	most suitable for separating the larger particles of biological interest with minimum damage to the material
Turbidimetry and nephelometry	scattering of radiation on passage through a transparent medium containing a particulate second phase	quantitative determination of protein in urine and cerebrospinal fluid, and immunoglobulins and lipids in serum	relatively simple and rapid technique for quantitation of precipitates
Activation analysis	radioactivity is induced in the sample by irradiation with neutrons, charged particles or γ rays	determination of most elements ($\sim$70) *in vitro*, some ($\sim$6) *in vivo*	sensitive and accurate quantitative analysis
Electron spectroscopy	release of an electron from a sample by a photon or electron of known and relatively low energy	investigation of protein structure, determination of elemental content of surfaces and films to a depth of $\sim$5 nm; examination of the structure of complex molecules in gaseous or solid state	high sensitivity, easily interpreted spectra, nondestructive of sample
Electron spin resonance	resonant absorption of energy from a microwave beam by unpaired electrons of atoms, ions, molecules or molecular fragments in a sample located in a strong, stable, slowly varying magnetic field	study of transition-metal complexes, reactions involving free radicals, and "spin labelling" of drugs	suitable for study of radicals
Infrared spectroscopy	measurement of the absorption, emission or polarization of radiation arising from molecular vibrations	structural and qualitative analysis, and quantitative analysis of gases	relatively simple and generally available instruments

Table 2 *continued overleaf*

Table 2 *cont.*

Technique	Principle	Application	Advantages
Mass spectroscopy	separation of ions and charged molecular fragments in a particle beam according to their mass-to-charge ratio	determination of trace elements. Measurement of molecular weights. Identification of compounds (in combination with gas chromatography)	mass spectra are usually simpler than optical spectra. The technique can be used for structural, qualitative and quantitative analysis
Mössbauer spectroscopy	modulation of the resonant absorption of low-energy γ rays in bound nuclei with split nuclear energy levels by relative acceleration of source and sample	study of molecules containing transition metals (e.g. iron in hemoproteins)	complementary to single-electron-energy spectrometry
Nuclear magnetic resonance spectroscopy	resonant absorption of energy from a radiofrequency beam by nuclei with residual magnetic moments and located in a strong, uniform, slowly varying magnetic field	examination of the physical environment of nuclei and of molecular structure particularly utilizing the proton and ^{13}C as a structural tracer isotope	a means for studying the structure of molecules by its effect on selected nuclei
Raman spectroscopy	measurement of the wavelength change and intensity of monochromatic radiation scattered by a molecule	structural studies of biological macromolecules; quantitative analysis	Raman spectra are simpler than those obtained by infrared absorption
X-ray fluorescence spectroscopy	x-ray emission is stimulated in the sample by an intense beam of primary x rays, or electrons if a vacuum system is used. The fluorescent radiation is characteristic of the elements in the specimen	determination of major elements. Microanalysis of solid samples. Distribution of elements in solid samples	suitable for simultaneous multielement analysis; nondestructive of sample

absorbance at a fixed wavelength. The more expensive spectrophotometer can also fulfill this function, but its main purpose, by virtue of its accurate and variable wavelength control, is the measurement of absorption spectra.

Table 2 lists those techniques that are used to give more selective clinical information, and for research purposes. The first five methods—chromatography, fluorimetry/phosphorimetry, ultrafiltration, centrifugation and nephelometry—are established techniques in clinical biochemistry and can provide valuable diagnostic information, particularly from chromatographic separations. However their applications are limited to relatively few samples in comparison with the large numbers of electrolyte determinations. The remaining techniques, arranged in alphabetical order, are primarily research tools used particularly for the study of the structure and function of molecules.

With the aid of modern instrumentation and technology, virtually any analysis becomes feasible, at a price, and hence the cost-benefit of all investigations must be considered. Computer-aided or controlled, automatic analytical systems can analyze specimens faster, cheaper and more precisely than a human analyst. It is therefore imperative to consider what information will really benefit the patient if the chemist and clinician

are not to be overwhelmed by the flood of data modern instruments can provide. The physicist can provide the tools for the analysis of clinical samples but it is the responsibility of the clinical biochemist to decide how they shall be used.

See also: Chromatography; Colorimetry; Fluorimetry; Spectroscopy; Neutron Activation Analysis; Radioimmunoassay; Nuclear Magnetic Resonance Spectroscopy

Bibliography

Bauer H H, Christian G D, O'Reilly J E 1978 *Instrumental Analysis.* Allyn and Bacon, Boston, Massachusetts
Skoog D A, West D M 1971 *Principles of Instrumental Analysis.* Holt, Rinehart and Winston, New York
Williams D L, Nunn R E, Marks V (eds.) 1978 *Scientific Foundations of Clinical Biochemistry*, Vol. 1: *Analytical Aspects.* Heinemann, London

J. B. Dawson

Clinical Temperature Measurement

Measurement of body temperature has a long-established tradition in medicine. Clinical use of the mercury-in-glass thermometer dates back to 1776 and is still the most popular choice when occasional measurements of body temperature are required. Today there is a range of more sophisticated techniques which permit continuous display and recording of temperature from the major body orifices and sites on the skin surface. Simple robust instruments are also used to indicate the temperature of fluids and gases which are preheated before being administered to a patient.

1. Clinical Temperature Range

Thermodynamically, the human body can be regarded as a central core (comprising trunk, head and legs) closely controlled about a temperature of 37 °C (± 2 °C) and surrounded by a thermally labile shell, the temperature of which varies between 32 and 35 °C under normal conditions. Body temperature generally refers to the core temperature. When a patient's core temperature rises above 41 °C he or she is said to be hyperthermic, and when below 30 °C, hypothermic.

2. Nonelectrical Thermometers

The clinical mercury-in-glass thermometer consists of a mercury-filled reservoir bulb, between 2 and 5 mm in diameter, connected to a stem about 100 mm long. There are three models, covering the normal range (35–42 °C), the subnormal range (25–40 °C), and the ovulation range (35–38 °C). Although the accuracy for the normal-range thermometer is specified as ± 0.1 °C, it has been found that the error can be as much as ± 0.5 °C (Knapp 1966). Rigidity and fragility confine the application of the mercury-in-glass thermometer to measuring mouth, axilla and rectal temperatures.

In an attempt to overcome the disadvantages of the mercury-in glass thermometer, mylar strips impregnated with microencapsulated cholesteric liquid crystals which produce temperature-dependent color changes have been used to indicate trends in temperature from the skin surface.

For specialized applications where accurate tissue temperatures are required in the presence of electromagnetic fields or radiofrequency radiation, a fluoro-optic instrument has been developed. It utilizes the property of certain phosphors to fluoresce (when excited by ultraviolet light) with an intensity which is dependent on temperature.

Two types of dial thermometers are used in medical equipment such as blood warmers, humidifiers and autoclaves. In one type the differential expansion of two dissimilar metals displaces a pointer over a graduated scale. The other type has a gas-filled temperature sensor. The gas-pressure variations are converted into a mechanical movement by a Bourdon pressure gauge.

3. Electrical Thermometers

Thermocouple thermometers rely on the Seebeck effect in which the emf produced across the measuring and reference junctions of two dissimilar metals or alloys (typically copper and constantan) is dependent upon their temperature difference (Fig. 1). Thermocouple

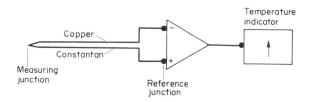

Figure 1
Thermocouple circuit

instruments are accurate to ± 0.1°C and probes are available for a variety of body sites.

Thermistors are semiconductor compounds manufactured from heavy metals such as cobalt and manganese. Their resistance R is defined by the relationship: $R = A \exp (b/T)$, where A and b are constants, and T is the temperature (°C). This characteristic can be linearized over the range 0–50 °C by shunting the thermistor with a resistor equal to the thermistor resistance at 20 °C (Hill and Dolan 1976). In the thermistor thermometer, the thermistor forms part of a Wheatstone bridge circuit (Fig. 2). A pocket-sized thermistor thermometer (Fig. 3)

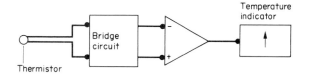

Figure 2
Thermistor circuit

can be used for intermittent temperature taking. Thermistor probes can be damaged by excessive heat and should not be autoclaved. Temperature-dependent rf oscillator circuits in which a thermistor varies the carrier frequency have been made in the form of a pill which can be swallowed. The body temperature of mobile patients can thus be measured by detecting the emitted signal externally.

4. Zero-Temperature-Gradient Thermometers

An estimate of core temperature can be obtained from surface sites on the body by creating a zone of zero heat flow across the shell. This method uses a flat probe

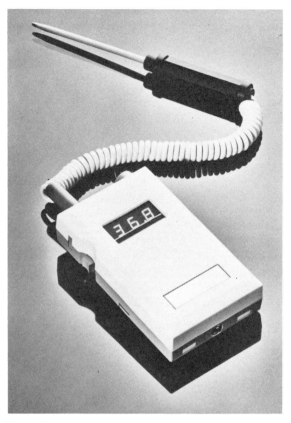

Figure 3
Thermistor thermometer

placed on the body surface within which a heater and two thermistors are separated by a thin thermal insulating layer (Fig. 4). The temperature difference across the layer is held close to zero by controlling the power supplied to the heater. The signal from the skin thermistor then measures core temperature (Fox et al. 1973).

5. Sites for Temperature Measurement

Core temperature can be obtained from the axilla (35.3–36.7 °C), rectum (36.9–37.7 °C), esophagus (36.9–

37.7 °C), mouth (35.9–37.2 °C), tympanic membrane of the ear (36.9–37.7 °C) and nose. Temperatures in parentheses are the normal ranges (Holdcroft 1980).

Surface sites of clinical importance include the great toe, the temperature of which depends on peripheral perfusion and is useful in assessing blood flow (Jolly and Weil 1969); abdominal skin temperature in the neonate which provides a good indicator of body cooling (see *Neonatal Intensive Care Equipment*); and forehead temperature which has been suggested as a trend indicator for core temperature.

In intensive care units, it is customary to have patient-monitoring systems which include two temperature channels capable of displaying core temperature and core–surface temperature difference ΔT. ΔT is useful in determining hypovolemic and shocked states (see *Monitoring Equipment in Coronary and Intensive Care*).

To obtain an accurate estimate of mean skin temperature as many as 15 sites have been used (Mitchell and Wyndham 1969). However a four-site arrangement is regarded as adequate using the formula:

$$T \text{ (average skin)} = 0.3\, T \text{ (nipple + arm)}$$
$$+ 0.2\, T \text{ (thigh + calf)}$$

In determining mean body temperature, it is usual to combine core and average skin temperatures as follows:

$$T \text{ (body)} = xT \text{ (average skin)} + yT \text{ (core)}$$

where x and y are constants whose values range from 0.8 to 0.9 and from 0.1 to 0.2, respectively.

See also: Thermography

Bibliography

Benzinger M, Benzinger T H 1972 Tympanic clinical temperature. In: Plumb H H (ed.) 1972 *Temperature: Its Measurement and Control in Science and Industry.* Instrument Society of America, Pittsburgh, pp. 2089–102

Crocker B D, Okumura F, McCuaig D I, Denborough M A 1980 Temperature monitoring during general anaesthesia. *Br. J. Anaesth.* 52: 1223–29

Fox R H, Solman A J, Isaacs R, Fry A J, MacDonald I C 1973 A new method for monitoring deep body temperature from the skin surface. *Clin. Sci.* 44: 81–86

Hill D W, Dolan A M 1976 *Intensive Care Instrumentation.* Academic Press, London

Holdcroft A 1980 *Body Temperature Control in Anaesthesia, Surgery and Intensive Care.* Baillière Tindall, London

Jolly R H, Weil H M 1969 Temperature of great toe as an indication of the severity of shock. *Circulation* 39: 131–38

Knapp H A 1966 Accuracy of glass clinical thermometers compared to electronic thermometers. *Am. J. Surg.* 112: 139–41

Mitchell D, Wyndham C H 1969 Comparison of weighting formulas for calculating mean skin temperature. *J. Appl. Physiol.* 26: 616–22

K. B. Carter

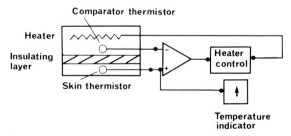

Figure 4
Zero-temperature-gradient thermometer circuit

Color Blindness

Color blindness or color-vision deficiency is the inability to distinguish certain spectral mixtures, particularly those easily distinguished by normal observers. All people fail to discriminate between different colors at night when vision is mediated solely by rods. However, rare groups of individuals fail to discriminate certain colors even when vision is (normally) mediated by cones. Color blindness is a consequence of either inherited anomalies (congenital dyschromatopsias) or pathological disturbances (acquired dyschromatopsias) of the visual system.

1. Color-Vision Tests

Regardless of whether or not the color-vision disorder is inherited or acquired, specific diagnosis is invariably based on the individual's performance in color-vision tests. The majority of these tests function qualitatively, although some also function quantitatively. Three basic types of test are used for clinical examination: pseudoisochromatic plate tests, arrangement tests and anomaloscope tests.

Plate tests require the observer to identify some figure, letter or other symbol "embedded" in a background. Both the figure and background are made up of clusters of various sized dots or spots differing systematically only in color. The colors are chosen to capitalize on the color confusions characteristic of different types of deficiency. In the most basic type of plate, the vanishing-figure type, the figure is seen by the normal observer but not by the defective observer who confuses the figure and background colors. (The colors are pseudoisochromatic in the sense that only the defective observer fails to distinguish them.) There are three variants of this basic design: the hidden digit type, containing a figure seen by the defective observer but not by the normal observer; the qualitatively diagnostic type, containing two separated figures, both seen by the normal observer but only one of which is seen by the defective observer (depending upon his type of defect); and the transformation type, containing two superimposed figures, one seen by the normal observer, the other by the defective observer.

Arrangement tests differ from plate tests in design and procedure but rely on the same pseudoisochromatic confusions. They require the observer to organize a set of color samples of constant (or near-constant) luminous reflectance according to a specified hue or saturation order. Color-defective observers make characteristic inversions of the order, the errors being concentrated in confusion zones peculiar to the defect. Such tests employ pigment colors chosen from color order systems such as Munsell papers of fixed value but variable chroma or hue (see *Colorimetry*). Like pseudoisochromatic plate tests, arrangement tests are designed and standardized either for natural daylight or for Commission Internationale de l'Eclairage (CIE) standard illuminant C

conditions. Other illuminants change the apparent colors of the test elements. Colors confused by the defective observer under standard illuminant C may not be confused under other illuminants.

Anomaloscope tests require the observer to make color matches between a spectral (or nearly spectral) test light and two spectral (or nearly spectral) primary lights. The lights are chosen so that in general only two photopigments are active in the match, their relative contributions being proportional to the amounts of the primaries required to match the test light. For the normal observer, an acceptable match is obtained over a very limited range of fixed ratios of the two primaries. For the defective observer, however, either the midpoint of this range is displaced from the normal or the range of acceptability is abnormally wide.

There are four types of matches or equations used in anomaloscopes: the Rayleigh equation, involving matching a yellow (585–600 nm) test light to a mixture of red (> 640 nm) and green (545–555 nm) primary lights; the Engelking–Trendelenburg equation, matching a blue–green or blue (490 nm) test light to a mixture of blue (470 nm) or blue–violet and green (517 nm) primary lights; the Pickford–Lakowski equation, matching a white test light to a mixture of blue (470 nm) and yellow (585 nm) primary lights; and the Moreland equation, matching a mixture of yellow (579 nm) and blue–green (480 nm) test lights to a mixture of blue (439 nm) and green (499 nm) primary lights.

The tests described above are not infallible so it is sound practice to use a battery of them in any serious investigation or selection of color-defective observers. Pokorny et al. (1979) suggest the following minimal requirements for evaluating color vision: a pseudoisochromatic plate screening test (preferably one including plates for tritan defects, see Sect. 2), the Farnsworth–Munsell 100-hue test, the appropriate illuminant for pigment tests (standard illuminant C), and an anomaloscope. A complete diagnosis of defective color vision requires fairly elaborate spectroscopic equipment for measuring photopic relative luminous efficiency curves, color-matching functions, and wavelength discrimination.

2. Congenital Color-Vision Deficiencies

When the deficiency originates from hereditary factors, it can be classified according to which and how many of the cone systems are affected and according to whether the cone systems are actually missing (a reduced color-vision system) or altered in their sensitivity (an anomalous color-vision system). The salient properties of the various congenital color defects discussed below are summarized in Tables 1 and 2.

We know from trichromacy (see *Color Vision*) that persons having normal color vision require three degrees of freedom to quantify their color vision, one degree of freedom corresponding to each of the independent cone signals. Persons having anomalous color vision (i.e., anomalous trichromats) also require three degrees of freedom, but the ratios of their cone signals must differ from those of the normal observer (a consequence of univariance). This results in a discrimination loss compared with the normal trichromat, a loss most likely not originating from an anomalous screening pigment or from some other prereceptor cause but from an anomalous cone pigment, i.e., one having either a different action spectrum or a wavelength maximum displaced from the normal. The exact nature of the loss depends on whether the anomalous pigment is in the L cones (protanomalous trichromacy), M cones (deuteranomalous trichromacy) or S cones (tritanomalous trichromacy). For the protanomalous and for the deuteranomalous observer, the discrimination loss can be expressed by the Rayleigh matching equation—matching a spectral yellow test light by adjusting the proportion of spectral red and green primary lights. Only two primaries are required because just two photo-

pigments (the M and L) are active in this part of the spectrum. (The contribution of the S cones is relatively insignificant for most observers.) One class of protanomalous trichromat (the simple protanomalous type) require in their matches a higher proportion of the red primary than do normal trichromats (i.e., a midpoint displaced to the red) but retain a fairly narrow red–green matching range. To the normal trichromat the mixture field at the protanomalous match appears orange compared to the yellow test field. A second class, the extreme protanomalous type, has a wide matching range that includes the simple protanomalous match and the normal match and may extend to the red primary. Both classes show a luminosity loss for long wavelengths. Simple deuteranomalous trichromats, on the other hand, require a higher proportion of the green primary. Their match appears greenish to the normal observer. The more severe form, the extreme deuteranomalous trichromat, has a wide matching range including the simple deuteranomalous match and the normal match, and

Table 1
Inheritence, incidence and classification of congenital color-vision defects

| Type | Classification | | | | Incidence[b] (%) | |
	Color vision	Discrimination deficit	Mechanism	Inheritance[a]	Male	Female
Protanomaly	trichromatic	red to yellowish-green (various)	alteration	X-linked recessive	1.0	0.02
Deuteranomaly	trichromatic	red to yellowish-green (various)	alteration	X-linked recessive	5–6	0.38
Tritanomaly	trichromatic	greenish-blue to blue (various)	alteration	autosomal dominant(?)		
Protanopia	dichromatic	red to green	reduction	X-linked recessive	1.0	0.02
Deuteranopia	dichromatic	red to yellowish-green	reduction	X-linked recessive	1.0	0.01
Tritanopia	dichromatic	greenish-blue to blue	reduction	autosomal dominant	0.002–0.007	0.001
Complete achromatopsia (with normal visual acuity)	monochromatic	total	reduction		0.000 001	0.000 001
Complete achromatopsia (with reduced visual acuity)	monochromatic	total	reduction	autosomal recessive	0.003	0.002
X-linked incomplete achromatopsia	dichromatic/monochromatic	red to green/total	alteration	X-linked recessive		
Autosomal recessive incomplete achromatopsia	dichromatic/monochromatic	various/total	alteration	autosomal recessive		
Normal	trichromatic	none			91.6	99.0

a Not known for complete achromatopsia (with normal visual actuity)   b For white Europeans. Not known for tritanomaly, X-linked incomplete achromatopsia and autosomal recessive incomplete achromatopsia

Table 2
Properties of congenital color defects

Type	Degrees of freedom	Altered cone systems	Absent cone systems	Reduced luminous efficiency	Relative luminous efficiency function (peak)	Confusion[a] point	Spectral neutral point
Protanomaly	3	L		long wavelengths	540 nm		
Deuteranomaly	3	M		middle wavelengths	560 nm		
Tritanomaly	3	S		no	555 nm		
Protanopia	2		L	long wavelengths	540 nm	$x_p = 0.7365$ $y_p = 0.2635$	494 nm
Deuteranopia	2		M	middle wavelengths	560 nm	$x_d = 1.4000$ $y_d = -0.4000$	499 nm
Tritanopia	2		S	no	555 nm	$x_t = 0.1710$ $y_t = 0.0000$	570 nm, 400 nm
Complete achromatopsia (with normal visual acuity)	1		S and either L or M	long or middle wavelengths	540 nm or 560 nm		all wavelengths
Complete achromatopsia (with reduced visual acuity)	1		S, M and L	long wavelengths (rod system)	507 nm		all wavelengths
X-linked incomplete achromatopsia	2/1		L and M	long wavelengths	445 nm or 507 nm		
Autosomal recessive incomplete achromatopsia	2/1		S and either L or M	long or middle wavelengths	540 nm or 560 nm		
Normal	3			no	555 nm		

a Commission Internationale d'Eclairage (CIE) chromaticity

extending to the green primary. Both classes of deuteranomalous observer however, unlike their protanomalous counterparts, have a normal or nearly normal luminosity function.

The discrimination loss for tritanomalous trichromats cannot be expressed by the Rayleigh equation. Their red–green ratios lie within normal limits. The loss must be expressed by an equation in which the S cones play a part, for example, the Engelking–Trendelenburg equation—the match of a blue–green test light to a mixture of blue and green primary lights. The match is not as diagnostic as the Rayleigh match since all three cone photopigments are activated by the test and primary lights, and since the inert yellow pigments of the eye lens and retina have an important effect on the ratio. In general, however, tritanomalous trichromats require more blue primary than the average normal trichromat. Their luminosity function is normal.

Congenital color defectives whose color vision is completely quantified by two degrees of freedom are called dichromats. They have only two classes of cone pigment and hence only two independent cone signals.

Like anomalous trichromats, dichromats are identified according to the affected (in this case missing) photopigment type. There are three types. Protanopes have normal S and M cones but lack L cones. In terms of the Rayleigh equation, they match the test light to the green primary, the red primary, and all possible mixtures of the primaries. (Since their S pigment does not contribute to the match, they have only one degree of freedom in this situation.) As a consequence of the missing photopigment, chromaticness discrimination is absent from about 520–700 nm (green to red) and there is a luminosity loss for long wavelengths. In addition, protanopes have a diagnostic spectral neutral point or zone and a confusion or copunctal point. The spectral neutral point defines the spectral wavelength (494 nm) that they exactly match to an equal-energy (achromatic) white. It is the wavelength balancing the quantum catches in the two remaining photopigment systems. The copunctal point defines the intersection in the CIE trichromatic chromaticity diagram (see *Colorimetry*) of the straight lines representing stimuli of the same dichromatic chromaticity (i.e., those which the protanope confuses and which he can match to each other by adjusting only luminance). Its chromaticity coordinates are approximately 0.7365 and 0.2635.

Deuteranopes lack M cones. Like protanopes they match the yellow test light to the green primary, the red primary and all possible mixtures of the two, but typically require different luminance settings than protanopes. For them, total discrimination is absent from 530 to 700 nm (greenish-yellow to red), but there is no appreciable luminosity loss of the middle wavelengths. Their neutral point is about 499 nm and the chromaticity coordinates of their copunctal point are 1.4000 and −0.4000.

Tritanopes lack S cones. On the Rayleigh equation the tritanope makes a unique match that falls within the distribution of matches made by normal trichromats. This is predictable because they have normal M and L cones (i.e., those active in the match). If a satisfactory Engelking–Trendelenburg equation could be devised involving the activity of just the S and M cones then the tritanope could be diagnosed as matching a violet light to a green light (they would also match it to a blue or blue–green light). Unfortunately, the equations presently available do not lie on tritan confusion lines and hence do not rule out the participation of the L cones. As a result, tritanopes do not usually accept a full range of matches (which would be diagnostic of dichromacy) and hence cannot always be distinguished from tritanomalous trichromats. For tritanopes, total discrimination is absent from about 445 to 480 nm (greenish-blue to blue). There is no appreciable luminosity loss. They have two neutral points, one near 570 nm, the other near 400 nm. Thus there must be two wavelengths in the spectrum that produce the same ratio of absorptions in the M and L cones as does white light. The chromaticity coordinates of their copunctal point are 0.1710 and 0.0000.

Congenital color defectives whose color vision has only one degree of freedom (complete achromatopsia) are either cone or rod monochromats. Cone monochromats have one class of cones, L or M, but not both. Rods and S cones are totally absent or nonfunctioning. If the M cones are the sole ones functioning, there is a sensitivity loss at long wavelengths (similar to protanopia). On the other hand, if L cones are functioning, the sensitivity loss is at middle wavelengths (similar to deuteranopia). Residual color discrimination may occur if the field of view is sufficiently large to involve parafoveal photoreceptors.

Rod monochromats, as their name implies, lack all three functioning cone types (they are also called complete achromats with reduced visual acuity). Like cone monochromats, they have no chromaticness discrimination, being totally color blind. They match all color stimuli to an achromatic white or any spectral wavelength simply by varying luminance. Unlike cone monochromats, however, their color-perception deficiency is overshadowed by other ocular defects. They have very reduced acuity in foveal vision. In the normal trichromat

the foveal area is populated almost exclusively by cone receptors. In the rod monochromat, however, it is populated mainly by rods or by morphologically abnormal cones. Presumably the cones that exist contain a rhodopsin-like pigment because the spectral response curve of the rod monochromat resembles the scotopic (rod) luminous efficiency function, peaking at 510 nm. This interpretation is borne out by the facts that rod monochromats show no Purkinje shift and no cone–rod transition in dark adaptation. Besides having poor visual acuity, rod monochromats are often without protective iris pigment (albinism), are subject to photophobia and pendular nystagmus, and are unable to maintain foveal fixation.

In addition to the complete achromatopsias, there are a number of incomplete achromatopsias which are always accompanied by reduced visual acuity and occasionally by other ocular defects as well (e.g., photophobia, pendular nystagmus, and abnormal distribution of macular pigment). Individuals in this category seem to have one class of cones, plus rods. Nevertheless, despite having only one functioning cone class, they retain some dichromatic color vision. This must involve the active participation of the rods. Normally, ratios of rod–cone signals do not provide the basis for color vision because the two types of signals are not simultaneously available for comparison. Rods and cones operate in what are largely nonoverlapping luminance regimes (the region of overlap is called the mesopic luminance region).

When the S cones are the sole class functioning with rods, the incomplete achromat has dichromatic color vision at mesopic and low photopic levels (the luminous efficiency function resembles a linear combination of the spectral responses of S cones and rods). He is, however, monochromatic at low mesopic/scotopic levels (where the luminous efficiency function resembles the spectral response of the rods) and high photopic levels (where it resembles the spectral response of the S cones). A Purkinje shift occurs, but spectral sensitivity is depressed for wavelengths above 550 nm. This condition is often misnamed blue monocone monochromacy or π_1 cone monochromacy, the misleading implication being that the absence of L and M cones precludes color-vision processing (the terms blue and π_1 refer to the S cone process). Given the residual dichromatic color vision present, the condition is better described as X-chromosome-linked incomplete achromatopsia. This emphasizes that its genetic origin is distinct from that of the other (autosomal recessive) incomplete achromatopsias in which only M or only L cones function with the rods.

When the M cones are the sole class functioning with the rods, the dichromatic color vision occurs at low and moderate photopic luminances and is most obvious when large stimulus fields are used. The photopic luminosity function is protanopic and Rayleigh matches are displaced toward the red primary. A neutral point may occur in the blue–green region of the spectrum.

When the L cones function alone with the rods, dichromatic vision occurs at low and moderate photopic luminances. There may be a reduced high-luminance scotopic response. The photopic luminosity function is deuteranopic and Rayleigh matches are displaced toward the green primary. A neutral point may occur in the yellow–green region of the spectrum.

The overall frequency of congenital disorders differs between racial and cultural groups. It is approximately 8.4% in European whites, slightly higher in North American whites (10%), but lower for Mongols (4–5%), Africans (1.5–4%), Australian aborigines (1–2%) and American Indians (1–2%). The frequency of specific disorders depends upon whether the defect is inherited as an X-chromosome-linked recessive trait (protan and deutan defects and some forms of incomplete achromatopsia), or as an autosomal dominant trait (tritan defects), or as an autosomal recessive trait (rod monochromacy and some forms of incomplete achromatopsia). The incidences of the first are predictably higher in males than in females. Table 1 lists the incidences reported for white European males and females.

3. Acquired Color-Vision Deficiencies

Deficiencies of color vision also originate from diseases or injuries affecting the refractive media (lens, vitreous and aqueous humor), the receptor apparatus of the eye, the optic nerve, or its connections and terminals in the brain. They may also originate from drug intoxication and hypovitaminosis. Such acquired or pathological forms of color blindness are less typical than their congenital counterparts. (They are termed acquired color-vision deficiencies even though in some cases, such as in the primary hereditary optic atrophies, the optic-nerve disorder causing the defect is inherited.) Unlike congenital disorders, acquired disorders are usually progressive (deteriorating with time although in some cases amenable to treatment), often dissimilar in the two eyes, and often accompanied by other visual complaint (reduced visual acuity or field loss). These characteristics make them difficult to classify. Cases resembling anomalous trichromatism, dichromatism and monochromatism are found; however, the disorders cannot be associated with specific malfunctions of one or more of the three cone receptor mechanisms, except perhaps in the case of the S cones. Instead, classification is most frequently made according to the major axes of chromatic discrimination loss (Pokorny et al. 1979). There are three basic patterns: a nonspecific loss (no axis), a loss, the axis of which corresponds to the red–green axis of discrimination, and a loss with axis corresponding to the blue–yellow axis of discrimination. (Redness–greenness and blueness–yellowness are two independent dimensions of perceptual color space. They probably reflect the way signals generated by the visual photopigments are processed by the opponent color systems.)

There are two broad categories of red–green defect (types I and II) and one general category of yellow–blue defect (type III).

The type I (red–green) acquired defect is observed in retinal diseases primarily affecting the macular cones and in intoxications of digitalis, quinine, oral contraceptive agents, phenothiazine, salazosulfapyridine and furaltodone. It is associated with a major loss of visual acuity and eccentric fixation. As the disease progresses, the chromatic discrimination on the red–green axis progressively deteriorates with an accompanying parallel loss in visual acuity. In advanced stages there is total color blindness in the affected visual field. The usual outcome resembles congenital achromatopsia with reduced visual acuity (rod monochromacy).

The type II (red–green) acquired defect is observed in diseases affecting the optic nerve (including optic neuritis, retrobulbar neuritis, optic atrophies), and in optic-nerve intoxications, malfunctions of the optic disk and tumors of the optic nerve or chiasm. Drugs and other chemicals causing optic-nerve intoxications and hence related type II defects are: the oral antidiabetics, chlorpropamide and tolbutamine; the quinoline derivatives, especially clioquinol; the antipyretics, ibuprofen and phenylbutazone; the monoamine oxidase inhibitors; the sulfonamides; the tuberculostatics, dihydrostreptomycin, ethambutol, isoniazide, paraminosalicyclic acid, rifampicin and streptomycin; the nitrofurane derivatives, furaltodone and nalidixic acid; thallium; disulfiram; cyanide; and tobacco. In early stages, most colors appear reduced in saturation. As the disease progresses, chromatic discrimination deteriorates along the red–green axis and to a lesser extent along the blue–yellow axis. A neutral zone of the spectrum emerges in which colors appear gray at above 500 nm. In advanced stages severe losses in visual acuity occur and the neutral zone widens to include the whole midspectral region. The relative luminous efficiency curve remains normal, but the Rayleigh match is displaced toward the green primary.

The type III (blue–yellow) acquired defect is the most common. In conventional nomenclature it is described as a blue–yellow defect, but in some cases it is more correctly described as acquired tritanopia (loss of S cone sensitivity). In other cases, however, the defect cannot be ascribed to a specific mechanism. It occurs in many choroidal, pigment epithelial, retinal and neural disorders, including brown nuclear cataracts, chorioretinal inflammations and degenerations, vascular disorders, senile macular degeneration, papilledema, glaucoma, retinitis pigmentosa, autosomal dominant optic atrophy, and prechiasmal optic-nerve disorders. Synthetic antimalarial agents (especially the chloroquine derivatives), indomethacin, erythromycin, phenothiazine derivatives, and oral contraceptive agents may all give rise to toxic retinopathy or retinal vascular disturbances with a predominant type III acquired defect. Methyl alcohol poisoning, chronic alcoholism, or vitamin B_{11} (folic acid) or B_{12} (cobalamin) deficiency can also result in a type III disturbance of color discrimination. Aging may lead to a reduced discrimination of blue, culminating in an acquired tritan-type defect. This is partly due to senile changes of the retina and pigment epithelium, but mainly to senile discoloration (yellowing) of the lens. In the early stages the defect is characterized by mild or moderate loss of discrimination in the blue and yellow spectral regions. Confusions are made between blues and greens and between yellows and violets. The discrimination is normal elsewhere and there is no neutral point in the spectrum. Visual acuity is hardly affected. In advanced stages, color vision becomes dichromatic and a neutral zone emerges at about 550 nm. This ordinarily indicates the loss of function or impairment of the S cones. The spectral luminous efficiency usually remains normal.

Bibliography

Alpern M 1974 What is it that confines in a world without color? *Invest. Ophthalmol.* 13: 648–74
Boynton R M 1979 *Human Color Vision.* Holt, Rinehart and Winston, New York
Grützner P 1972 Acquired color vision defects. In: Jameson D, Hurvich L M (eds.) 1972 *Visual Psychophysics: Handbook of Sensory Physiology* VII/4. Springer, Berlin, pp. 643–59
Lakowski R 1969a Theory and practice of color vision testing: A review, Part I. *Br. J. Ind. Med.* 26: 173–89
Lakowski R 1969b Theory and practice of color vision testing: A review, Part II. *Br. J. Ind. Med.* 26: 265–88
Pokorny J, Smith V C, Verriest G, Pinckers A J L G (eds.) 1979 *Congenital and Acquired Color Vision Defects.* Grune and Stratton, New York
Wright W D 1946 *Researches on Normal and Defective Color Vision.* Kimpton, London
Wyszecki G, Stiles W S 1982 *Color Science: Concepts and Methods, Quantitative Data and Formulas*, 2nd edn. Wiley, New York

L. T. Sharpe

Color Vision

This article deals with the neurophysiological and perceptual aspects of color vision. The emphasis is on man. Only vertebrates are considered and then only insofar as their physiology and behavior are likely to be valid for man.

1. General Considerations

Human color vision is trichromatic, that is, color sensations can, under carefully controlled conditions, be completely quantified with three degrees of freedom. There are several ways to expand this statement. The most fundamental is to say that any four colors, A, B, C and D, are always linearly dependent (all four cannot be vectors orthogonal to each other in color-perception space). A match expressed by the relation $a\mathrm{A} + b\mathrm{B} + c\mathrm{C} + d\mathrm{D} = 0$ must hold in which the scalar multipliers a, b, c, d are not all zero. Some of the scalars must, however, be negative.

Quantitative systems of color matching are all based on this relationship (see *Colorimetry*). Three of the colors, say A, B and C, are chosen as fixed or reference primaries. Their choice is constrained by the requirement that none of them can be matched by a mixture of the other two. Given this, an unknown color D may then be matched completely either by mixing A, B and C in suitable proportions (i.e., $a\mathrm{A} + b\mathrm{B} + c\mathrm{C} = d\mathrm{D}$), by mixing it first with one of the primaries then matching the combination by suitable proportions of the other two (e.g., $a\mathrm{A} + b\mathrm{B} = d\mathrm{C} + d\mathrm{D}$), or by mixing it with two of the primaries then matching the combination by adjusting a single primary (e.g., $a\mathrm{A} = b\mathrm{B} + c\mathrm{C} + d\mathrm{D}$).

The principle of trichromacy, as far as it applied to mixing colored pigments and paints, was known early in the eighteenth century, but it was not until the nineteenth century that it was recognized as a fact of physiology. In 1801, Thomas Young reconciled the infinity of colored lights with the trichromacy of color perceptions by suggesting:

> Now as it is almost impossible to conceive each sensitive point of the retina to contain an infinite number of particles, each capable of vibrating in perfect unison with every possible undulation, it becomes necessary to suppose the number limited; for instance, to the three principal colors, red, yellow and blue....

Young's speculation can be brought up to date. What he called particles are most likely electrons in the π orbitals of chromophores. Chromophores are supplied by the retina and combine with proteins (often called opsins) to form photopigments (photolabile pigments whose structure changes upon the absorption of light). The chromophore in each photopigment absorbs the light quanta but the opsin determines which quanta are preferentially absorbed. (The sole effect of light absorption upon a photopigment as far as is known is to isomerize the chromophore, thereby initiating a series of thermal reactions.) Photopigments are not distributed evenly throughout the retina, but are anchored to lamellae in the outer segments of different classes of photoreceptors. Each photoreceptor contains only one photopigment type.

Young argued for three photopigment types, but there are in fact four; all have the 11-*cis* isomer of vitamin A aldehyde (11-*cis*-retinal) as their prosthetic chromophore group but differ in their opsins. As far as color vision is concerned, however, Young was correct: only three of the photopigment types, the cone photopigments, are responsible for color vision. The fourth, the rod photopigment, subserves colorless low-luminance vision and plays no role in normal trichromatic color vision. It may, however, play a role in the color vision of incomplete achromats (see *Color Blindness*).

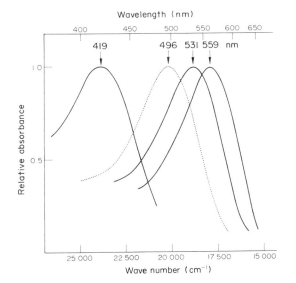

Figure 1

The relative absorbance spectra of the four photopigments of the normal human retina as a function of wavenumber (reciprocal wavelength). The ordinate is log (intensity of incident light/intensity of transmitted light). The solid curves are for the three classes of cones, the dotted curve for the rods. These curves are based on microspectrophotometric measurements of 137 receptors from seven human retinae

Figure 1 shows, for the cone photopigments, how the quantity of log (incident light/transmitted light) varies with wavelength. The peaks of the three classes lie in the violet (at a wavelength of approximately 419 nm), in the green (approximately 531 nm), and in the yellow–green (approximately 559 nm) regions of the electromagnetic spectrum. For convenience they can be referred to as short-wavelength absorbing (S), middle-wavelength absorbing (M), and long-wavelength absorbing (L), although the description is correct only insofar as it refers to the relative peaks of their spectra. The sensitivities of the pigments overlap considerably: all three pigments absorb short, middle and long wavelengths of light.

Besides having different peaks of maximum absorption, the absorption spectra of the cone photopigments have different shapes: the bandwidth (in cm^{-1}) of their absorption spectra decreases as the wavenumber or frequency of maximum absorption decreases. The S pigment is broadest.

Although the different cone types cannot be distinguished morphologically, they can be distinguished functionally. The S cone system in particular seems to have response characteristics setting it apart from the M and L cone systems. These include poor quantum efficiency, low absolute sensitivity, poor differential sensitivity, low spatial acuity, large summation area, low temporal resolution, long time constants, long critical duration, vulnerability to retinal diseases, apparent

saturation under equilibrium conditions, and sudden loss of sensitivity following long-wavelength adaptation (also called transient tritanopia). Such basic sensitivities are probably not abnormalities of the S cones themselves. They more likely result from the low representation of the S cones throughout the retina and from the post-receptoral neurons recoding and regulating S cone response (see below).

The curves in Fig. 1 were obtained from microspectrophotometric measurements of human-cone outer segments. They could have been estimated by other means such as analyzing dichromatic and normal color matching (the König–Helmholtz approach), measuring the absolute and incremental threshold sensitivity of the eye (the two-color threshold approach), or measuring the reflectivity of the fovea (the reflection densitometry approach). Such relative absorbance curves do not characterize the cone photopigments for the specific pigment concentrations of the retina. In the retina, the photopigment molecules are closely packed in receptors and exhibit significant optical density. As a result the spectral absorption of the pigments will widen due to self-screening. This is a direct consequence of the increased optical path length. To characterize the absorption properties of the cone photoreceptors themselves, an estimate of the effective optical density of the photopigments in the cones is needed. This can be determined from the absorbance spectrum and the specific absorbance (approximately 0.015 μm^{-1}) of the pigments, assuming that the greatest possible density is along the axial length of the cone outer segments, approximately 35 μm for the longest fovealar cones and decreasing to 25 μm for the shortest parafoveal cones. The absorbance spectra of Fig. 1 must also be corrected for the selective filtering of radiant energy by the eye lens and the vitreous humor, which shifts the maximum of the cone sensitivities to higher wavelengths.

Young assumed that his particles would be evenly distributed throughout every sensitive part of the retina. Actually, the six million or so cones containing photopigment are heavily concentrated in an area 1500 μm in diameter (5.2° of visual angle) about the central visual pole, roughly coextensive with the retinal pit or fovea. Even within this area, cones are not evenly spread; the very central island or bouquet (50–75 μm in diameter or 0.17°–0.24°) probably lacks S cones. Indeed S cones are probably sparsely represented throughout the entire central retina. This central area is demarcated by a yellow, nonphotolabile carotenoid—the macula lutea or yellow spot which is associated phenomenologically with the Maxwell spot (a shaded spot seen centrally when observing a flickering uniform field of deep blue light). The density of this pigment varies considerably. It is most intensive on the slopes and in the margin of the inner fovea (an annular zone approximately 70–100 μm wide). Outside the foveal pit the pigmentation is less dense, diminishing with retinal eccentricity. Inside the fovea proper, especially the foveola (an area approximately 280–300 μm wide or 1.4° of arc), pigmentation is very slight (it may be absent) compared with the foveal slopes. Thus to characterize fully the absorption properties of the cones at the retina the absorbance curves of Fig. 1 have to be corrected not only for the optical densities of the photopigment and the preretinal screening filters but also for the varying optical density of the macular pigmentation. In other words, cones residing in different retinal locations will have different spectral sensitivities.

Regardless of their spectral sensitivity all cones respond to light in the same way—by initiating electrical signals in the nerve cells of the visual system. These signals are independent of photon wavelength or energy, and depend only upon the rate at which photons are absorbed. All that varies with wavelength is the probability that an individual photon will be absorbed (the absorbance spectra of Fig. 1 reflect this changing probability). Cone output being limited to a single dimension of change is fundamental to trichromacy. It means that there can be three and only three color signals—the relative outputs of the three cone classes.

The signals generated by the three cone classes pass through the neural layers of the retina and the dorsal lateral geniculate nucleus (dLGD) on their way to the striate cortex. These structures contain the neural machinery needed to compare the outputs of the different cone classes; the neural machinery needed for color discrimination. The first comparisons (so far identified in the primate) take place in the color opponent ganglion cells of the retina. These cells are most common near the fovea and have small receptive fields with corresponding small dendritic spreads (generally less than 10 µm in diameter). Such cells combine inputs from different cone types. Their firing rate increases (above a spontaneous level of activity) when one spectral region (say long-wavelength light) stimulates their receptive field and decreases when another region (say middle-wavelength light) does. (The receptive field of a cell is any portion of the receptor mosaic where appropriate stimuli from the visual field evoke or modify the cell's responses.) Other ganglion cells have broadband or nonopponent characteristics, responding in the same way to all wavelengths. They are believed to be involved in processing luminance (not color) information.

In primate ganglion cells, color opponency may or may not go with spatial opponency. Cells with spatial opponency have a center-surround receptive-field arrangement; the center receives input from a different cone type (or types) than the concentric surround. (The field center probably receives input from very few cones.) Depending on whether the cell is an on or off unit (see *Visual Fields and Thresholds*), stimulation of the center can either increase or decrease the firing rate from its "spontaneous" baseline value. Spatially nonopponent cells, on the other hand, have no simple spatial differentiation in their receptive fields. The balance of excitatory and inhibitory influences produced by a stimulus does not change with the position of the stimulus in the receptive field.

The nature of the input to color opponent ganglion cells, whether they are spatially opponent or not, varies. Some cells are trichromatic, receiving inputs from all three cone types, with two inputs being opposed to the remaining one. Others are bichromatic, receiving opposed inputs from only two cone types—the L and M cones. S cones do not provide input to bichromatic cells, only to trichromatic ones—and then only excitatory input.

Almost all ganglion cells in the monkey project directly to the dorsal lateral geniculate nucleus of the thalamus, a multilayered topologically organized structure. The units found here are similar to those of the retina in terms of the spatial and chromatic organization of their receptive fields. The same three basic types are found: those having both color and spatially opponent organization (roughly 77% of the cells found in the parvocellular layers of the macaque dLGD; 61.5% of those found in the combined magno- and parvocellular layers), those having a color opponent organization in a spatially undifferentiated receptive field (about 7% parvocellular layers; 7.5% combined layers) and those having spatially organized receptive fields but lacking any spectrally opponent organization (about 16% parvocellular layers; 31.5% combined).

All afferents from the color opponent dLGD project exclusively to a region of the occipital cortex known variously as area 17, primary visual cortex, or striate cortex. This region has a very specific topological relation to areas of the dLGN and consequently also to the retina and visual field. In the monkey visual cortex, four predominant types of spatial arrangement of receptive field types have been reported: those with concentric receptive fields, those with simple receptive fields, those with complex receptive fields, and those with hypercomplex receptive fields. Almost all receive inputs from M and L cones (S cones input is rare) and all save the hypercomplex type have double-opponent color properties. In double-opponent color cells, one delineated area of the receptive field is excited by, say, long-wavelength light and inhibited by middle-wavelength light, whereas another area or areas responds in the opposite manner. Such cells probably receive inputs from two or more dLGD units of the same opponent color type but opposite polarity. In the concentric type, the opposite opponent color systems occupy the center and surround; in the simple type, they occupy the central rectangular area and the parallel antagonistic flanks; in the complex type they are spatially undifferentiated. The hypercomplex cells, on the other hand, have single-opponent color properties. A central activating area is flanked on either side by silent antagonistic regions having the same spectral sensitivities. The orientation selectivity these cells display is due to spatial antagonism between cones with the same spectral properties rather than to opponent color antagonism between cones with different spectral properties. (This is also true of simple and complex cells.)

The highest cortical areas to which color analysis has

been traced both anatomically and physiologically are two visual areas (V4 and V4A) in the prestriate cortex. The first of these lies on the anterior bank of the lunate sulcus; the second on the lateral part of the posterior bank of the superior temporal sulcus. They seem to be areas of great specialization. Unlike units in the striate cortex, the units here are selective primarily for color. Over 50% are color opponent, some having half-bandwidth responses as narrow as 10 nm.

2. Absolute and Differential Sensitivity of the Eye

Using psychophysical procedures, the eye's absolute and differential threshold sensitivity to various stimulus patterns can be probed. Experiments of this sort reveal that the human eye is not equally sensitive to all wavelengths of light. When rods mediate detection, a monochromatic light of 507 nm is the most efficient light stimulus (less energy is required to see at that wavelength), but when cones take over, a light of 555 nm becomes most efficient (see *Vision*). This is a consequence of the different spectral sensitivities of the rod and combined cone systems. The transition from rod to cone vision produces a shift in the relative brightnesses of the various spectral regions (called the Purkinje shift). Whereas the rod system is five times more sensitive to monochromatic radiations of 507 nm than to radiations of 430 nm or 571 nm, the cone system is five times more sensitive to 555 nm than to 489 or 637 nm.

At absolute threshold for all but the long wavelengths (> 630 nm), rods are more sensitive than the combined cone systems (depending of course upon the spatial and temporal parameters of the stimulus). The greater sensitivity of rod vision results in there being no color sensation for most spectral lights even at luminances well above the threshold for light detection. The interval between the threshold of light detection (the achromatic rod response) and the threshold of color detection (the chromatic cone response) is known as the photochromatic interval. When expressed as a rod–cone threshold ratio it is about 1000:1 (3 log units) at 500 nm, slightly less than at 400 nm, and virtually nonexistent at long wavelengths (> 630 nm).

Not only does the ability to detect monochromatic lights change throughout the spectrum, but the ability to distinguish between monochromatic lights λ and $\lambda + \Delta\lambda$ (wavelength discrimination) also changes. In the violet and red regions, below 430 nm and above 650 nm, large differences $\Delta\lambda$ introduce no perceptible difference to the color of a test light. On the other hand, in the blue–green (490 nm), orange–yellow (590 nm), and blue (440 nm) regions, wavelength differences as small as 1 nm can be detected on the basis of their different hues. (The change in ability to discriminate is related to, but not wholly explained by, the changing slopes of the photopigment curves, greatest discrimination occurring where the rate of change of the difference in absorptions in the different

cone classes is largest.) Saturation discrimination, the amount of monochromatic light L_λ which has to be added to a white light L_w to make the mixture noticeably colored, also changes as a function of wavelength. The smallest proportion of L_λ to L_w occurs for blue wavelengths, increasing steadily to a maximum value in the yellow and then decreasing again in the red. Because of this relation, reds and blues are said to have greater saturation than yellows and greens. This is confirmed in two ways. Firstly, there are more discriminable steps between white and spectral reds and blues than between white and spectral greens and yellows. Secondly, spectral blues and reds appear more saturated than spectral yellows of equal brightness.

Decreasing the saturation of a light while holding its wavelength constant (i.e., increasing the ratio of L_w to L_λ) concomitantly changes the apparent hue. This apparent hue shift (Abney effect) is variable for different wavelengths. With desaturation, hues correlated with wavelengths greater than 578 nm appear yellower, those correlated with wavelengths less than 578 nm appear greener (the blue–green wavelengths) or yellower (the yellow–green wavelengths). There are two exceptions—the yellow of 578 nm and the purple complementary (see *Colorimetry*) of 559 nm appear invariant with desaturation.

The hue of a spectral light also changes as a function of brightness (the Bezold–Brücke effect). At high retinal illuminances above 1000 trolands (1000 trolands $\equiv$ 318.3 cd m^{-2}, viewed through a 2 mm diameter pupil), hues correlated with wavelengths greater than 500 nm (yellow–greens, yellows, and reds) appear yellower, while those correlated with wavelengths less than 500 nm (blue–greens, blues, and violets) appear bluer than at lower retinal illuminances (100 trolands). There is a similar though less pronounced variation in apparent hue at low retinal illuminances near the color threshold (below 10 trolands). Hues correlated with wavelengths between 470 and 578 nm (blue–greens, greens, and yellow–greens) tend to appear greener, while hues correlated with wavelengths above 578 nm (oranges and reds) appear redder. There are, however, three spectral hues and a fourth extraspectral hue (a purple) which are invariant (see *Colorimetry*). Showing no Bezold–Brücke shift are a blue at about 474 nm, a green at about 505 nm, a yellow at about 578 nm and a reddish purple with a complementary wavelength at about 559 nm. The blue, green and yellow invariant hues agree closely with the so-called psychological primaries or pure hues, but the reddish purple hue is judged perceptibly different from psychological red (which is also extraspectral).

Both the Abney and Bezold–Brücke effects demonstrate what a risky business it is to assign precise hue names to spectral wavelengths. Explanations of these phenomena usually depend on assumptions about the saturation of the receptor processes or color opponent processes most strongly affected by the light. For instance, a deep red light (670 nm) activates the cones containing the L photopigment most strongly and the cones containing the M photopigment less strongly (see Fig. 1). When the light is very bright, the L pigment approaches bleaching and does not respond strongly to further increases in intensity. The relative increase in response of the M cones, which are far from saturating, is accordingly stronger. As a result, there is an apparent alteration in color towards the yellow. This would explain some but by no means all aspects of the Bezold–Brücke effect; explaining the Abney effect is more difficult.

3. Subjective Color Phenomena

The activities of the striate and pre-striate cortical units (described above) are unlikely to be the immediate precursors of color perception. In fact, the way neurophysiological events correspond to subjective experience has still not been explained. Still awaiting physiological explanation are a number of well-known "subjective" color phenomena. They are called subjective because they are instances where different physical stimuli (defined as different by their spectral flux distributions) produce the same psychological percept (e.g., color constancy) or conversely where the same physical stimuli produce different or new percepts (e.g., afterimages, simultaneous and successive contrast, and spatiotemporal effects). Admittedly this is a loose way of grouping together some of the most interesting color phenomena.

Three examples will be given. Consider, first, color constancy. This refers to the perceptual approximate constancy of the color of objects when the objects are viewed under different illuminants. For instance, white paper appears white under both natural daylight and artificial incandescent illumination. The constancy is surprising because changing illuminants necessarily changes the local spectral flux reaching our eye from the paper and hence the ratios of absorption in the three cone classes; under artificial light the paper should appear more yellow than under natural light. There are limits, of course, to this phenomenological constancy. Constancy fails if a monochromatic source replaces a broad-band one or if the spectral reflectance of the color object is near monochromatic. Nor does it apply to the perceived brightness of objects; the relative brightness of different parts of the spectrum persists. How the visual system achieves color constancy is still a matter of great speculation. It cannot be explained solely in terms of the quantum catches of the cone pigments. One suggestion made by Edwin Land is that each of the three cone systems, given the appropriate neural wiring, forms an independent image of the world in terms of a lightness scale or more correctly in terms of the reflectance ratios at color boundaries (i.e., object edges or contours). Thus there is one lightness scale for the L pigment system, a second for the M pigment system, and a third for the S pigment system. Comparison of these separate images, perhaps in color-opponent cells, would provide the basis

for color sensation. So long as the ratios of differences between the three independent scales (which are in themselves ratios) remain constant, the perceived color would remain constant. Only after gross changes in the actual spectral flux distribution (such as caused by changing from broad-band to monochromatic light) would the lightness images and hence the comparison ratios be significantly altered. Still to be worked out, though, is whether or not spectral reflectance analysis need depend upon a high degree of cone independence (from the retina to the cortex) and whether absolute lightness judgements are made within each cone system. Neither the first or second seems to be a characteristic property of the primate visual system.

A different sort of subjective color phenomenon is that evoked by certain spatiotemporal patterns of illumination, even though the light falling on the retina is either wholly achromatic or wholly monochromatic. The necessary condition for such illusions appears to be the alteration, at a frequency (4–6 Hz) a little below that which abolishes flicker, of patterns differing only in total reflectance. One such pattern is Benham's disk or top. It is made up of a uniformly dark sector and a light sector, the latter containing four groups of arc. When the disk is rotated clockwise so that the same retinal area is successively stimulated by the different levels of reflected light, a pattern of rings is perceived on a uniform background. The position of the rings is determined by the radial placing of the arcs. The outermost ring appears dark blue, the next olive green, the third brownish, and the innermost red. The colors are not very saturated. Changing the speed of the disk, direction of rotation, or the luminance or spectral composition of the light used to illuminate it, alters the appearance of the colors, but even in monochromatic light the apparent colors of different arcs differ conspicuously. A wide variety of explanations has been offered to account for the appearance of these subjective colors including afterimages, chromatic aberration of the eye, specific wavelength-dependent phase differences, phase-dependent lateral inhibition in the retina, and wavelength-dependent frequency modulation of nerve impulses. None of these is wholly satisfactory.

Finally, consider the McCollough effect—a relatively small but long-lasting biasing of the visual system. After regularly alternating exposure to, say, red vertical and green horizontal gratings, the observer perceives color after effects contingent on the orientation of achromatic grating counterparts. The after effects are approximately complementary to the hues of the exposure lights. Thus in the above example, vertical gratings appear greenish and horizontal gratings appear pinkish. The effects are highly robust and extremely persistent. In some cases they may be repeatedly elicited even weeks after the initial exposure. Explanations of these effects usually appeal to the selective adapting or fatiguing of cortical neurons with orientation and color preferences (e.g., the simple cells with dual opponent-color-receptive fields). The decay of such effects however may have a longer time course of recovery than the supposedly responsible neural units. This prompts speculation that the McCollough effect and its long-lived relatives—such as the after effects produced by adaptation to colored gratings of different spatial frequency or by wearing bipartite color spectacles or prismatic lenses—involve neuronal learning or conditioning. (There is manifest evidence for short-term plasticity in cortical connections.) Largely ignored in such explanations, however, is the role that receptoral and neural mechanisms at more peripheral levels of visual processing might play. Long-lasting after effects might build up not only as the result of learned contingencies in the cortex but as the result of serial contingencies in the color-processing hierarchy.

It would be instructive to end by returning to the opening definition of trichromacy: any color sensation can be completely quantified with three degrees of freedom. This definition has an important qualification, often overlooked—it applies only under carefully controlled experimental conditions. Behind this qualification lies much of the psychology of color vision. The sensation of any particular color depends not only on the ratios of excitation of the three cone systems produced by that color, but also on a complex of other factors. These include the mode of appearance of the color (whether it is a self-luminous light or a reflecting or transmitting surface), the illuminating and viewing conditions (if it is a reflecting or transmitting surface), and the previous and concurrent exposure to other colors in the visual field. Such factors involve the processes of the retina and visual pathways but also the higher processes, the cognitive functions, of association cortex. As a final example, the color brown is often described as a dark, desaturated, yellow or orange. This is quite inaccurate, since there are light browns and tans. Brown is in fact a completely new experience of hue, similar in this respect to some olive greens and maroons. The sensation of brown cannot be elicited by merely reducing the brightness of a yellow, orange, or any other spectral light (this has its own effect, the Bezold–Brücke effect). Nor can it be elicited by mixing yellow or orange with other spectral lights (this has its own effect too, the Abney effect). Such changes carried to the extreme will of course produce black, not brown. For a brown sensation to be experienced, the yellow or orange light must be surrounded with a brighter annulus of white light or the observer must be preexposed to a white light. This "blackens" the yellow or orange light by simultaneous or successive contrast. This example succinctly demonstrates that the explanation of color perception is beyond current physiological knowledge.

See also: Visual Cortical Neurophysiology

Bibliography

Boynton R M 1979 *Human Color Vision.* Holt, Rinehart and Winston, New York

De Valois R L, De Valois K K 1975 Neural coding of color. In: Carterette E C, Freedman M P (eds.) 1975 *Handbook of Perception*, Vol. 5. Academic Press, New York

Hurvich L M 1981 *Color Vision.* Sinauer Associates, Sunderland, Massachusetts

Le Grand Y 1968 *Light, Colour and Vision*, 2nd edn. Chapman and Hall, London

McAdam D L (ed.) 1970 *Sources of Color Science.* MIT Press, Cambridge, Massachusetts

Marriott F H C 1976 Colour vision. In: Davson H (ed.) 1976 *The Eye*, Vol. 2A. Academic Press, New York

Mollon J D 1982 Colour vision and colour blindness. In: Barlow H B, Mollon J D (eds.) 1982 *The Senses.* Cambridge University Press, Cambridge

Mollon J D, Sharpe L T (eds.) 1983 *Colour Vision: Physiology and Psychophysics.* Academic Press, New York

Stiles W S 1978 *Mechanisms of Colour Vision.* Academic Press, London

Wyszecki G, Stiles W S 1982 *Color Science—Concepts and Methods, Quantitative Data and Formulae*, 2nd edn. Wiley, New York

Young T 1802 On the theory of light and colours. *Philos. Trans. R. Soc. London, Ser. A* 92: 12–48

Zrenner E 1983 *Neurophysiological Aspects of Color Vision in Primates.* Springer, Berlin

L. T. Sharpe

Colorimetry

Colorimetry is any technique for measuring the color of objects and lights. The technique may be qualitative or quantitative, visual or photoelectric, direct or indirect.

1. Color-Order Systems

The simplest and most straightforward techniques of color measurement involve direct visual comparisons. The observer compares an unknown color with physical standards appropriate to the color dimension being evaluated. For specific purposes, the range of standards need only be large enough to encompass the unknown color, and the individual standards can be realized in any convenient form (e.g., colored paints, papers, filters, or solutions). However, for general purposes, the choice of standards is decided by additional considerations. They should (a) have permanent color characteristics, (b) represent the widest range of possible colors, and (c) be organized according to some color plan or set of principles allowing the precise specification of any color chosen at random. Since it is impossible to produce an infinite variety of standards, most matches will not be perfect. Some logical scheme must be introduced, therefore, to organize the standards so that visual interpolation of the exact match is possible. Visual interpolation is easiest when the standards are arranged in a series of small, approximately equal intervals.

Sets of color standards having the properties outlined above are called color-order systems. Plans for scaling color standards which are based on the principles of color perception are called color-appearance systems. These are the most useful systems provided that their scales are used under carefully controlled conditions. This proviso is necessary because the appearance of colors changes with lighting and viewing conditions. An explanation of the perceptual or psychological attributes of color will help illustrate what uniform scales of color appearance are.

Perceptually, color may be conceived as varying along three separate dimensions: hue, saturation and brightness (lightness). Hue is the color dimension corresponding to the isolated sections of the visible electromagnetic spectrum: violet (wavelength ~ 390–455 nm), blue (455–492 nm), green (492–577 nm), yellow (577–597 nm), orange (597–622 nm) and red (622–770 nm). Purple is also a hue although it comprises energy from the two opposite ends of the spectrum (violet plus red). The inclusion of purple enables hue to be treated as a continuous dimension running from red to orange, yellow, green, blue, violet, and back again through the purples to red. White, grays and black are missing from this continuum because they are not hues. They are achromatic colors, i.e., colors without a hue assignable to a specific region of the spectrum. They either have energy simultaneously present from more than one section of the spectrum (grays and white) or none at all (black). Whites may also be formed by mixing complementary wavelengths (see Sect. 2). In these special cases, the white will have energy present from two isolated regions or one spectral region plus purple.

Saturation is the color dimension corresponding to the vividness of hue. It represents the proportion of perceived hue to perceived achromatic content in the total perception of a color. On the one hand, white, black, and grays have only a perceived achromatic content and are fully desaturated. On the other hand, the spectral colors have only a perceived hue content and are fully saturated even though they are not themselves seen as being equal in saturation (spectral blues and reds appear more saturated than spectral greens and yellows). Fully saturated purples have two perceived hue contents (spectral red plus spectral blue). All other colors have both a perceived hue and perceived achromatic content and are desaturated. The ratio of their perceived hue component (specified as a pure monochromatic spectral content) to their combined hue and achromatic components is a measure of their colorimetric purity.

The combined hue and saturation dimensions are called the chromaticness attribute of a color as distinguished from its brightness or lightness attribute. Brightness or luminosity refers to self-luminous colors (lights). It is the attribute permitting a color to be judged similar to one of a series of achromatic colors ranging from dim to dazzling bright. Generally, a color appears brighter as radiant energy is increased. Lightness, on the other hand, refers to non-selfluminous or object colors

which are either light-reflecting (surface colors) or light-transmitting (volume colors). It is the attribute permitting an object color to be judged similar to one of a series of achromatic object colors ranging from black to white (surface colors) or from black to perfectly clear (transparent volume colors).

Geometrically, these attributes constitute a three-dimensional space of color perception with hue and saturation defining a polar-coordinate system orthogonal to a vertical axis of brightness or lightness (Fig. 1). Horizontal cross sections of this space correspond to color planes of constant brightness (lightness). The

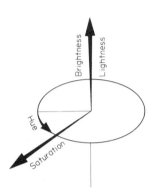

Figure 1
Geometric representation of the organization of color-perception space in terms of hue, saturation, and brightness or lightness

origin or central point of each cross section represents the associated achromatic color, the length of the radius vector represents the degree of saturation, and the polar angle represents the hue of a color. Vertical half-sections of this space (i.e., planes intersecting at the lightness or brightness axis) correspond to color planes of constant hue.

The best known color appearance system based on principles of color perception is the Munsell system. It uniquely classifies colors according to three dimensions: hue, chroma and value. Hue as defined in the Munsell system requires no further explanation, its definition being the same as the one given above. Chroma corresponds closely to the saturation of a color: low chroma colors appear achromatic, high chroma colors appear saturated. Similarly, value is related to the lightness of an object, although it is actually a compressive function of an object color's reflectance. Low-value samples have low reflectance and appear equally dark; high-value samples have high reflectance and appear equally light.

The Munsell system comprises several hundred colored chips arranged on pages of constant hue. Chips displayed along horizontal rows have the same value; chips displayed along vertical columns have the same chroma. Because of the limitations of pigment availability and permanence, the number of chroma and value steps varies with the hue. Although the intent is to make the steps between adjacent chips perceptually uniform in spacing (this is an essential property of a perceptually uniform color space), the Munsell system fails to satisfy this property. Steps within a single psychological dimension are not equivalent. Most observers find, for instance, that colors of constant high value differ perceptually more than their low-value equivalents. What is more, the differences in step size are even greater between the psychological dimensions: for Munsell chips of moderate chroma, one unit of value roughly equals two units of chroma and three units of hue.

Aware of these shortcomings, a select committee of the Optical Society of America (OSA) has produced its own series of uniform color scales (referred to as the OSA UCS system). After much deliberation, the committee chose to arrange and specify its color standards as the lattice points of a regular cuboctahedron. This geometric figure has two advantages: it optimizes the chances of including a near match for any color chosen at random and it helps color differences in all three psychological dimensions to be visualized. The basic unit of the OSA UCS system is a cluster of 12 colors corresponding to the 12 vertices of a cuboctahedral shell. Within the shell, each color is equally different from a central color and from each of its nearest neighbors (only 6 neighbors are possible in Munsell space). Given this basic octahedral unit, the system grows as each color located at a vertex becomes in turn the center of another cuboctahedral cluster. The network of clusters is extended until the lattice points include all colors that can be produced within the gamut of available colorants.

The actual color standards corresponding to the lattice points are realized as 555 two-inch square (25.8 cm²) painted color cards. Each card is described by a trichromatic notation according to a lightness scale (L) and two hue scales: yellowness (j) and greenness (g). All three scales have both negative and positive values. The L scale ranges from $+5$ to -7 with 0 perceptually equal to a gray of 30% reflectance under CIE standard illuminant D_{65} (see Sect. 3). The j and g scales range from $+10$ to -10 with neutral grays corresponding to the coordinates $j = g = 0$. Positive values of j correspond to yellow hues, negative values to blue hues; positive values of g correspond to green hues, negative values to red or red–purple hues. Four hundred and twenty of the standards sample this space at two unit intervals (i.e., the unit distance in the cuboctahedral clusters equals two units of either L, g or j). This means that only even values of j and g occur for even values of L; only odd values for odd values of L. An additional 134 intermediate standards are included, however, so that the much used near-neutral central-lightness regions are sampled at single unit intervals. For this limited region, both odd and even values of j and g occur for odd and even values of L.

In terms of the cuboctahedron model, the three scales can be depicted as in Fig. 2. The lattice points of the

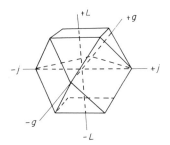

Figure 2
OSA UCS cuboctahedron space showing location of the lightness (L), yellowness (j) and greenness (g) axes through its center

cuboctahedron provide three sets of square grids and four sets of triangular grids (the seven different pairs of parallel faces). If the lightness axis L is perceived as running vertically through the centers of a pair of square faces, then horizontal cleavage planes (i.e., those orthogonal to the L axis) will correspond to constant lightness charts (i.e., standards only differing in their j and g values), and vertical cleavage planes (i.e., those parallel to the L axis) will correspond to charts of constant or opposite hues (i.e., j or g scales).

2. Trichromatic Colorimetry

As well as being matched to physical standards, unknown colors may be specified by metameric color matching. This technique exploits a fundamental fact of human physiology: of the four types of photon absorbing pigment in the normal eye, only three—those residing in the cone photoreceptors—are responsible for color vision. These cone pigments are maximally sensitive to different regions of the electromagnetic spectrum, but their sensitivities overlap. For convenience they are referred to as short-wavelength, middle-wavelength, and long-wavelength absorbing, although this description is correct only insofar as it refers to their relative wavelengths of maximum absorption (see *Color Vision*).

The sensitivities of the cone pigments differ; for example, short-wavelength (high-energy) photons are most likely to be absorbed by short-wavelength absorbing cones and least likely to be absorbed by long-wavelength absorbing cones. The reverse is true for long-wavelength (low-energy) photons. However, once a photopigment absorbs a photon all information about its wavelength and energy is lost. Thus, even though a cone's input is multivariant, its output is univariant, and depends solely upon how many quanta are caught by its pigment, not upon which quanta are caught (this assumes that quantum efficiency is constant). The limitation of cone output to a single dimension of change is fundamental because it means that any color can be specified by only three values: the relative quantum catches of the three cone classes. Obviously this will change with the wavelength characteristics of the color.

At present, quantum catches cannot be counted. To do so, it is necessary to know exactly the fractional absorption spectrum of the cones, as well as the optical density of several ocular screening pigments. Fortunately these shortcomings can be bypassed by invoking a corollary of cone univariance (i.e., metamerism). Metameric colors are matching colors with dissimilar spectral energy distributions. They may be contrasted to isomeric colors, i.e., matching colors with identical spectral energy distributions. Metamerism between colors holds only for a given illuminant and a given observer. Despite these restrictions, it holds under a wide variety of experimental conditions that may alter the appearance of metameric colors, e.g., preexposure to bright adapting lights, the introduction of strongly colored surrounds, or concomitant changes in the radiance (irradiance) or the spectral composition of both colors. The reason that metamerism occurs (i.e., why physically distinct colors can appear identical) is physiologically straightforward: their disparate spectral energy distributions produce identical quantum catches. (This is only true if the distributions have at least three crossovers, one in each area of maximum response of the cones.) Metamerism thus provides a way of completely specifying an unknown color by three values other than the cone outputs. Any color can be matched in appearance, but not in physical composition, by known amounts of arbitrarily chosen color stimuli or primaries—color stimuli whose mixture produces the same quantum catches as the unknown color. Only three independent ways of varying these primaries are needed so that the quantum catches of the three cone classes can be altered independently.

Independence may be achieved in a number of ways. In additive colorimetry, the unknown color is matched by superposing three colored lights, usually red, green or blue. Their choice is arbitrary but is constrained by linear independence: it must not be possible to match any one of the lights by an additive mixture of the other two (this ensures that the quantum catches of the three cone classes are altered independently). A second constraint is that the amounts of the three primaries (the tristimulus values of the unknown color) must be permitted sometimes to be negative or zero. In practice this is the same as adding one or two of the lights to the color being matched. This condition arises because combining lights necessarily lowers the saturation of the mixture. Therefore, although the hue of an unknown color can be matched by additive mixture of primaries, the saturation often cannot. The mixture color will appear less saturated than the unknown color. Consequently, an exact match can only be obtained by desaturating the unknown color (i.e., by adding one of the primaries to it). This action balances the ratios of quantum catches produced by the unknown and matching colors.

In subtractive colorimetry, the unknown color is matched by color-filtering white light. The three independent variables are the densities of the primary filters or wedges (usually yellow, cyan and magenta). In (Maxwell) disk colorimetry, the unknown color is matched by varying the relative areas of (not more than four) colored pieces of papers mounted on a sector disk. One of the areas must be achromatic or white. The disk is then rotated at a rate faster than the flicker resolution of the eye, fusing the disk sectors to give the appearance of a uniformly colored surface. The three independent variables are the angles occupied by the colored sectors of the disk. (Four sectors are required to give independent control over three; the area of the fourth is always determined by the combined area of the other three.) In Nutting colorimetry, the unknown color is matched by adding a monochromatic wavelength to white light, then adjusting the relative proportion of the two in the mixture. The three independent variables are the spectral wavelength, its brightness, and the brightness of the white light. This method follows directly from the definitions of hue, saturation and brightness. Choosing the spectral wavelength allows the unknown color's hue to be matched, adding the achromatic component matches its saturation, and adjusting the absolute amounts of the two matches its brightness (or lightness).

3. Indirect Colorimetry

Visual colorimetry is elegant, straightforward and precise. It reduces the vast array of color appearance to three dimensions of input. However, its generality is limited: different observers make different metameric matches for the same unknown color. In fact, the variability between two observers is often larger than their individual standard deviations. For visual colorimetry to have validity for a large population the color matches of many observers must be averaged, often an impractical and time-consuming task. A different but related approach is to develop a system of color measurement based upon the immutable characteristics of a standard or hypothetically ideal observer. The standard observer can be defined by a set of color-matching functions: three independent functions of wavelength which weight the quantal flux of an unknown color in the same way as the cone spectral sensitivity functions. Conveniently, these functions can be derived by using the principles of univariance and metameric color matching. Three primary colors are chosen, preferably of monochromatic wavelengths spectrally separated as much as possible, and then the amounts (which may be positive or negative) required to match the wavelength of an equal-energy spectrum are determined, in radiometric or photometric units. These amounts are called the spectral tristimulus values, from which the amounts required to match any heterogeneous mixture of wavelengths (i.e., the amounts needed to match any unknown color) can be computed. This is done by first multiplying the unknown color's relative spectral distribution curve $S(\lambda)$ by the spectral tristimulus values then by integrating these products over the whole spectrum:

$$X = k \int_\lambda S(\lambda)\bar{x}(\lambda)\,d\lambda$$

$$Y = k \int_\lambda S(\lambda)\bar{y}(\lambda)\,d\lambda$$

$$Z = k \int_\lambda S(\lambda)\bar{z}(\lambda)\,d\lambda$$

where X, Y and Z are the tristimulus values of the unknown color, $\bar{x}(\lambda)$, $\bar{y}(\lambda)$ and $\bar{z}(\lambda)$ are the color-matching functions (or spectral tristimulus values), and k is a normalizing factor usually set equal to

$$100 \Big/ \int_\lambda S(\lambda)\bar{y}(\lambda)\,d\lambda$$

The appearance of colored objects depends not only on the source irradiating them, but also on their spectral reflection (or transmittance) characteristics. So, when specifying the tristimulus values of colored objects, two new terms need to be introduced. For light-diffusing surfaces, $S(\lambda)$ in the above equations is replaced by $\rho(\lambda)S(\lambda)$, where $\rho(\lambda)$ is a spectral reflectance factor; and for light-transmitting surfaces $S(\lambda)$ is replaced by $\tau(\lambda)S(\lambda)$, where $\tau(\lambda)$ is a spectral transmittance factor.

Changing the illuminant changes the spectral distribution of the incident flux and consequently changes the spectral distribution of the reflected (or transmitted) flux. The Commission Internationale de l'Eclairage (CIE), the international authority responsible for standardizing lighting conditions, therefore recommends four precisely defined sources for making color measurements. These standard illuminants designated by the letters A, B, C and D_{65} represent the following spectral energy distributions:

A   is a gas-filled coiled-tungsten filament lamp operated at a color temperature of 2856 K;

B   is standard source A filtered and operated at a correlated color temperature of 4874 K (corresponding to direct sunlight);

C   is standard source A filtered and operated at a correlated color temperature of 6774 K (corresponding to light of an overcast sky); and

D_{65} is a theoretical source which cannot be realized and is only approximated by a filtered xenon arc source operated at a correlated color temperature of 6507 K (corresponding to light of an overcast sky). (D_{65} can also be approximated less exactly by certain fluorescent lamps having correlated color temperatures of 6500 K.)

Obviously a color's tristimulus values will depend upon the individual color-matching functions chosen to weight $S(\lambda)$. Many sets of color-matching functions exist; others can be algebraically derived from the original set of color primaries by homogeneous linear transformations. For widespread exchange of color information, however, the tristimulus values of a color must refer to some internationally accepted set of color-matching functions. The CIE recommends sets that approximate color-matching functions of a standard or hypothetically ideal observer. They are derived by averaging many sets of color-matching functions measured under standard conditions for different observers.

The CIE defines, necessarily, the properties of its standard observer for two different viewing conditions because color matches depend on the field of view, there being a continuous change in the amounts of the primaries required for the color match as field size is increased. The reasons for this are physiological: the amount of intrusion by the rods (the fourth type of photoreceptor process mostly subserving twilight or low brightness vision) and the density of macular pigment (an inert pigment screening the cones) changes across the retina. Accordingly, the CIE (1931) specifies a small field standard observer for color-matching in fields of view up to four degrees in angular subtense and a large field or supplementary standard observer (CIE 1964) for color matching in larger fields.

For each of these standard observers, the CIE provides tables of two different but equivalent (linearly related) sets of color-matching functions (Wyszecki and Stiles 1982): one set based on real color primaries red, green and blue and the other on imaginary primaries X, Y and Z. Their choice of unreal primaries has several advantages, the main one being that the tristimulus values and chromaticity coordinates designating any color (monochromatic or otherwise) are never negative. Negative numbers are a necessary complication of using real primary colors because it is impossible to match all colors by strictly additive mixtures. The choice of imaginary primaries also allows the introduction of other convenient properties. Thus in the case of the 1931 CIE standard observer, the transformation from the real primaries is defined so that the $\bar{y}$ color-matching function equals the CIE photopic luminous efficiency function. (Besides the CIE function there are other small field color-matching functions, but these mainly interest vision scientists attempting to derive the color sensitivity functions of the cones from color-matching behavior.)

Associated with each set of color-matching functions $\bar{x}$, $\bar{y}$ and $\bar{z}$ is a set of chromaticity coordinates x, y and z. Chromaticity coordinates are the ratio of each tristimulus value to their sum:

$$x = X/(X + Y + Z), \quad y = Y/(X + Y + Z)$$

and

$$z = Z/(X + Y + Z)$$

By reason of their derivation, they disregard the brightness (lightness) quality of a color and specify only its chromaticness (hue plus saturation). Since color-matching functions are the tristimulus values of the spectral colors, their chromaticity coordinates correspond to the chromaticities of the spectrum colors.

Therefore, plotting any one of the chromaticity coordinates against any other (usually x and y) in a two-dimensional chromaticity diagram defines the locus of spectrum colors. All physically possible colors must fall inside the region bounded by this locus and the straight line joining its violet and red extremities (the line of nonspectral purples). Only imaginary colors such as the CIE primaries can fall outside it.

The chromaticity diagram associated with the 1931 CIE observer can be used to illustrate a number of useful properties of these diagrams (Fig. 3). The point with

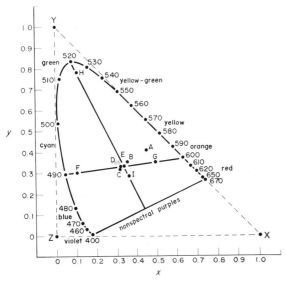

Figure 3
CIE 1931 (x, y) chromaticity diagram. The solid line connects the chromaticity points of the spectral colors and the nonspectral purples. This defines the gamut of all physically possible colors

chromaticity coordinates $x = y = 1/3$ is the equal-energy (E) or white point of the diagram (i.e., where the tristimulus values are equal). This point allows the dominant wavelength, excitation purity, and complementary color of a color (actually a colored light) whose chromaticity coordinates are known to be determined. For colored objects, the equal-energy point is replaced by the phenomenological white point: the chromaticity of the illuminant irradiating the object (usually A, B, C or D_{65}).

For reflecting objects, the white point corresponds to the chromaticity coordinates of a diffusing surface that does not absorb light of any wavelength and is therefore perfectly white when irradiated (i.e., $\rho(\lambda) = 1.0$) by the various standard sources. For transmitting objects, it corresponds to the chromaticity coordinates of a perfectly clear transmitter (i.e., $\tau(\lambda) = 1.0$). A magnesium oxide layer deposited from burning magnesium ribbon approximates a perfect reflector; fused quartz glass approximates a perfect transmitter.

The complementary wavelength λ_c of a color is found at the point of the spectrum locus where the line joining its chromaticity point and the white point intersects the opposite side of the diagram. It is the spectral wavelength which, when mixed in suitable proportions with the color, matches the white point. (This is a necessary consequence of additive mixing: all mixtures of two lights must lie on a straight line drawn between them.) It can be seen from Fig. 3 that many greens do not have spectral complementaries. Their complementaries are nonspectral purples. In Fig. 3, F and G, like H and I are pairs of complementary colors (lights) with respect to E. The dominant wavelengths of F, G, and H are 490, 600 and 520 nm, respectively. I is a purple and has no dominant wavelength in the spectrum; it is specified by the dominant wavelength of its complementary H (i.e., 520 nm). The excitation purities of F, G, H and I are 0.81, 0.53, 0.90 and 0.22, respectively.

The dominant wavelength λ_d of a color is found where the line joining its chromaticity point and the white point intersects the spectrum locus on the same side of the diagram. It corresponds loosely to the predominant hue quality of the color under neutral adaptation. Nonspectral purples lack dominant wavelengths and must be specified by their complementary wavelengths $-\lambda_d$ or λ_c.

The excitation purity of a color is the ratio of the distance between the white point and its chromaticity point and the distance between the white point and the spectrum locus (or the nonspectral purple line). It corresponds loosely to the saturation of a color, being equal to zero when the chromaticity point of the color coincides with the white point and equal to unity when the chromaticity point of the color coincides with the chromaticity point of the color's dominant wavelength.

A color's dominant wavelength and excitation purity do not correspond exactly to its hue and saturation because the 1931 CIE chromaticity diagram is not perceptually uniform. This is a fault of all chromaticity diagrams plotted in Cartesian coordinates. A fixed linear distance does not represent a fixed perceptible color difference throughout the diagram. No linear or projective transformation of these diagrams will correct their anomaly. This fact has two consequences: one theoretical, the other practical. Firstly, the geometry of color space is probably elliptic, not linear; Riemannian, not Euclidean. Secondly, the calculation of the color differences between two colors specified by their tristimulus values or chromaticity coordinates cannot be based on a simple algebra. So, not only does the relative size of the small color differences between colors vary throughout the diagram, the equality of two small differences breaks down if the differences are doubled or quadrupled.

Despite these problems there are many formulae that attempt to calculate the magnitude of color differences. Since the relation between chromaticness and brightness (lightness) differences is nonlinear, such formulae normally include separate factors for these two values. Most color-difference formulae are based on projective or curvilinear transformations of the 1931 CIE (x, y) chromaticity diagram. The CIE recommends two such spaces and their associated color formulae to promote uniformity of practice in evaluating color differences: the CIE LAB 1976 color-difference formulae often used to establish color tolerances in the color printing and textile-dyeing industries; and the CIE LUV 1976 formulae used to specify color differences in the lighting and color television industries. More information about these and other color difference formulae can be found in Wyszecki and Stiles (1982).

4. Photoelectric Colorimetry

This form of colorimetry replaces the visual observer with three photodetectors with spectral sensitivities which reproduce closely some suitable set of color-matching functions. (In principle the photodetector sensitivities should exactly equal the cone sensitivities of an ideal or standard observer.) The outputs of these detectors, assuming a linear relation to light input, are then directly proportional to the tristimulus values of the colors being measured.

The relative spectral response curves of physical detectors can be modified either by color filtering or template masking. In filter colorimetry, a single photocell or photomultiplier is used successively to measure the light passing through colored glass or gelatin filters whose transmission curves duplicate closely the chosen color-matching functions. (Using the photocell successively is equivalent to having three separate photocells.) These filters are placed either serially or adjacently between the color being measured and the photodetector. In template colorimetry, the filters are replaced by wavelength apertures positioned in the spectral plane of a double monochromator or other spectroscopic system. This arrangement offers greater accuracy and flexibility in approximating the color-matching functions when the spectrum into which the templates are placed is large. However, it is generally less light efficient than a photocell filter combination.

The coupling of small computers to spectrophotometers and spectroradiometers has reduced the need for photoelectric instruments. The computer stores and integrates color-matching functions automatically with spectral distribution curves as they are being measured. Thus there is no great advantage, in terms of accuracy or time, offered by photoelectric colorimeters which read tristimulus values directly. They are useful nevertheless as color-difference meters for measuring or establishing conformity to industrial color tolerances.

5. Colorimetric Analysis

This is a generic term in analytical chemistry encompassing the many techniques of absorption spectrometry. These methods exploit the phenomenon that chemical substances in solution preferentially absorb (and therefore selectively transmit) light. Absorption is produced by the small vibrations or resonances of atoms and molecules under the influence of light. In liquids and solids the energy levels of atoms and molecules are expanded into broad energy bands, with the result that the resonances are spread over broad ranges of frequency. When the resonances produce selective absorption in the visible region the substance appears colored. Thus the color of a homogeneous transparent solution provides a means of uniquely identifying its constituent substance. And the optical density of the solution provides a method of determining the substance's concentration. The concentration is calculated by applying the two laws governing absorption: Bouguer's or Lambert's law relating the amount of light absorbed to the thickness of the absorber and Beer's law relating the amount of light absorbed to the concentration of the absorber in solution. When combined, these laws express the important relationship: the ratio of incident to transmitted intensity of light is directly proportional to the concentration of the absorbing solution and the distance travelled through it. This allows the intensity of an unknown concentration of a substance to be directly compared to the intensity of a known or standard concentration of the same absorbing substances: for when equal fractions of light intensity are absorbed, the concentration of the unknown solution (C_2) must equal the concentration of the known solution (C_1) multiplied by their thickness (d) ratio (i.e., $C_2 = C_1 d_1/d_2$). This type of analysis is not limited to substances which in themselves appear colored. The process works equally well if the substance in solution reacts with inorganic and organic reagents to produce an identifiable hue.

The simplest applications of analytical colorimetry involve direct color comparison. An experimenter prepares color standards corresponding to known concentrations of the substance he wishes to detect. The standards may be colored glasses or sealed cells (Nessler tubes) containing the solution or a stable solution of similar hue. More sophisticated forms of colorimetric analysis use spectrophotometry or spectroscopy instead of visual comparisons because of their greater precision and particularly because they lend themselves to automatic measuring and recording procedures. With such techniques it is possible to measure most chemical elements and many organic compounds at concentrations less than one part per hundred million. There is another advantage in using spectrophotometric techniques: they extend the principles of colorimetric analysis into the adjacent ultraviolet and infrared regions of the spectrum, regions where visual comparisons are impossible. The optical properties of materials in these regions are similar to those in the visible region in their basic relation to the quantum-mechanical interactions of electrons. They differ only in detail.

See also: Clinical Chemistry: Physics and Instrumentation

Bibliography

Billmeyer F W Jr, Saltzman M 1966 *Principles of Color Technology*. Wiley, New York

Burnham R W, Hanes R M, Bartleson C J 1963 *Color: A Guide to Basic Facts and Concepts*. Wiley, New York

Commission Internationale de l'Eclairage 1970 *CIE Document on Colorimetry* (*Official Recommendations of the CIE*), Publication 15. Commission Internationale de l'Eclairage, Paris

Committee on Colorimetry of the Optical Society of America 1953 *The Science of Color*. Optical Society of America, Washington, DC

Judd D B, Wyszecki G 1975 *Color in Business, Science and Industry*, 3rd edn. Wiley, New York

Wright W D 1969 *The Measurement of Color*, 4th edn. Hilger, London

Wyszecki G 1978 Colorimetry. In: Driscoll W G, Vaughan W (eds.) 1978 *Handbook of Optics*. McGraw-Hill, New York

Wyszecki G, Stiles W S 1982 *Color Science—Concepts and Methods, Quantitative Data and Formulae*, 2nd edn. Wiley, New York

L. T. Sharpe

Communication Aids for the Physically Handicapped

Although the inability to communicate is frequently only one of a number of interrelated disabilities suffered by a severely physically handicapped individual, it can often be regarded as the most disabling of all. It can leave the sufferer who may well have no accompanying or associated mental impairment without any means of conveying thoughts, needs, or emotions to others. The challenge which confronts the technologist attempting to provide a solution to this problem is formidable, and no less taxing upon the available resources than is the drain on the emotional stamina of the disabled person.

1. Current Technology

Considerable progress has been made in recent years in the development and provision of electronic communication aids for the vocally handicapped. The ready availability and decreasing cost of modern electronic and microprocessor technology means that a promising future, with genuine prospects of useful and gainful employment, is becoming available to large numbers of severely physically handicapped people. However, an electronic communication aid will never completely cure a vocal disability, but will merely provide some enhancement of inherent communicating ability, an enhancement which will be tempered by the physical capabilities of the patient in using the aid.

Improvements and rationalizations of methods of communication and the introduction of new media of expression, such as the Bliss system and the Makaton vocabulary (which have developed in parallel and have been shown to complement each other with remarkable effect), go hand in hand with technological progress. Although the majority of electronic aids provide painfully slow rates of communication—an often unavoidable drawback occasioned by the poor control ability of the disabled person—the inherent "intelligence" offered by microprocessor technology can improve the situation significantly. In the situation where the rate of information transfer between the user and his equipment is very low, great improvements in the rate of communication can be achieved through the use of features such as anticipation and by means of internally stored libraries of words, phrases, sentences or larger units of text which can be generated by the aid in response to a highly abbreviated user effort.

The range of purpose-built or dedicated electronic communication aids now available is very wide, in terms of both capability and price. It is vital, therefore, that great care be taken in assessing the communication ability, physical capabilities and posture of the potential recipient as they relate to the operation of a communication aid.

In addition, with the "home computer" being readily and cheaply available, with widespread service and software backup, it is becoming more difficult to justify the design, construction, and probable additional cost of dedicated aids. It is relatively simple to incorporate the same software and interfacing to the user in an "off the shelf" device at a considerably lower price. The costs involved in hardware and software interfacing of the patient to a specially designed system are little different from those for a general-purpose, mass-produced unit; moreover, the expenses that would have been incurred in hardware development can be ignored, given equipment already available on the mass consumer market.

Dedicated communication aids are now appearing on the market, the prices of which reflect a family association with other equipment produced for a much larger consumer population but which require relatively minor modification in production methods to make them suitable for such a specialized market. There are also other aids which are manufactured at a loss, but are nevertheless made available to disabled users as a result of the philanthropic and charitable policies of some of the larger manufacturing organizations.

While it could be argued that a commercially available microcomputer system may exhibit features that would never be of use or accessible to the disabled user, these are provided at no additional cost, and still generally at a lower price than a purpose-built machine. Often the only real justification for a dedicated aid is that of self-contained power and portability, as this, until recently, has not been an attribute of many of the commercially available microcomputers. Also, the most commonly used microcomputer output media—television screens and electronic printers—are not necessarily the most appropriate if the aid is to be used for face-to-face communication, and both lack true portability.

If a general-purpose piece of equipment such as a home computer is to be used as a communication aid, a prerequisite is that it should have the ability to run the appropriate program as soon as it is switched on. To some extent this may limit the choice of equipment, particularly in the case of computers which require programs to be loaded from tape.

The provision of a commercially available microcomputer system purely as a communication aid, particularly for the younger user who may well have been nonvocal since birth, ignores the wider potential of such a machine to provide a range of other functions concerned with education, recreation, environmental control and mobility. The impact of the new technology has been no less apparent in schools for the physically disabled than in other educational establishments, and there is a need to provide physically handicapped children in special schools with access to the ever expanding range of educational and recreational programs. There will also be the requirement, as these children progress through school and prepare themselves to earn an independent living, to allow them to use business and word-processing software which is highly sophisticated and not capable of simple modification to suit the individual physical capabilities of such users.

This implies the development of purpose-built pieces of equipment that are not merely aids to communication, but which can be used to convert whatever control signals the user can produce into a form which appears to an unmodified microcomputer system as conventional keyboard input. This is the concept of a "transparent" or "surrogate keyboard," where the user's communication aid, which may well be microprocessor based, can also be used as a keyboard substitute for another computer system. Such equipment already exists, but the potential for expanding it to provide a universal control interface between a physically disabled user and a wide range of other equipment including wheelchairs, telephones and environmental controllers requires further study.

2. Types of Equipment

All but a few of the currently available aids make use of one, or perhaps a combination, of three modes of user input known as direct selection, encoding and scanning.

In a direct selection aid, the user can point directly, sometimes with appropriate electronic or mechanical assistance in the form of a head pointer or mouth stick, to letters, words or pictures presented on a "menu" before him. The act of pointing would typically result in a specific switch closure which would indicate which of the offered items was required. With direct selection aids there is a one-to-one mapping between the user action

and the system response; the highest possible communication rate is attained using this selection method.

In an aid which makes use of an encoded input, the number of items on the device menu is greater than the number of available input switch channels, and simultaneous activation of a combination of some of these input lines forms a code which can uniquely identify the item required. The main advantage of an encoded system over a direct selection system is that the motor demand on the user is less. This advantage may take the form of either a smaller required range of motion (because there are fewer keys), or may allow the use of larger keys which make less demand on accuracy. Direct selection is in fact a special case of encoding where the number of input lines is equal to the number of items presented; both methods offer similar potential rates of communication.

For those individuals whose motor control is very restricted and who therefore present the fewest available control sites, scanning aids frequently provide the only, and of necessity the slowest, solution. Access to the menu is provided through any motor act which can be transduced to produce a switch closure. In the most common arrangement, the menu is represented on an optoelectronic display or television screen as a two-dimensional matrix. Scanning proceeds with the indicator light or cursor moving under user control, first along a row, until the column containing the required item is encountered, and then each element in the selected column is scanned until the item is reached and final selection takes place. This type of scanning typically employs one or two control switches, although in other situations four switches or a joystick type of control may be used. The arrangement of the menu is normally designed to optimize communication rate by locating frequently used elements close to the origin of the scanned matrix. This means that such elements would typically be encountered in the scanning of the first row or column and selected without the necessity of the user having to scan a number of other rows or columns before being able to select them.

The output from a communication aid can take several forms depending on the purpose, price and complexity of the aid. This ranges from a simple visual indication of a selected item, or a cyclic, sequential presentation of a number of items previously selected, to the generation of text on a television screen or other optoelectronic display. A most desirable feature is the provision of the facility for correcting and editing a message before its final output, as there is nothing more frustrating for the disabled user than to have his every mistake highlighted by his communication aid.

Output in the form of synthetic human speech is now becoming commonplace, although currently available aids which provide this feature require considerable improvement in respect of voice quality. Flexibility of vocabulary, portability, economy of power consumption and intelligibility are, however, all improving steadily.

3. Matching the User to the Equipment and the Types of Transducer Available

The range of sensing devices available for detecting and using residual physical function to control communication aids is vast. The simplest form of input is via an appropriately mounted microswitch if the patient has some reliable movement and is able to exert at least a small force. For patients with feebler movements, capacitive touch switches or infrared-beam breakers are available. Some switches can be triggered by reproducible noises such as grunts or whistles, and there are also pneumatically operated switches which are activated by sucking or blowing down a tube. For patients with voluntary eyelid control, blink-sensing switches can be used, and it is now possible with fairly simple electronics to employ eye motion tracking to operate a communication or other aid. In yet more severe cases where the patient is incapable of perceptible movement, it may be possible to monitor residual muscular activity electronically and use these myoelectric signals for control.

One of the most neglected aspects of electronic communication aids concerns their control by disabled users. Most equipment is manufactured and sold on the assumption that some sort of patient-operated actuator, which can be represented by one or more switches, a joystick or other input device, is available. The exact nature of this actuator, which is fundamental to the patient's efficient or optimal use of the equipment, is seldom given the degree of consideration it deserves.

It has frequently been the case that an otherwise well-conceived and engineered design has been discarded by a user after a short period of frustration merely because the type of input and the mode in which the equipment was tested have proved inappropriate to the patient's particular capabilities. The actual means of input to the aid depend on a characteristic of the user, and frequently has to be specially constructed. Where appropriate, user specific input software may have to be written to match the aid to the user in the most efficient manner. Adequate training in the use of the aid is, of course, also necessary.

4. Provision of Communication Aids

It is unfortunately true that much innovative technology, which can be demonstrated to be a means of providing maximal benefit for the physically disabled, nonvocal patient, nevertheless fails to progress beyond a reasonably successful prototype stage. One of the major contributing factors to this state of affairs is the broad range of physical capabilities encountered in the disabled population, and the day-to-day variation in their characteristics. While the microcomputer offers the potential for great flexibility in the type of user input, and the capability of self-adaptation to changing circumstances, most aids suffer from inherent inflexibility. Setting-up procedures tend to be cumbersome and time consuming, and this results in a basically workable design becoming uneconomic to produce for a wider disabled population, and is therefore abandoned immediately, or after an economically disastrous initial production run. Many of the aids which do appear on the market tend to lack sophistication, limiting their usefulness, or are overpriced, making them accessible only to small numbers of potential users. Service backup and repair facilities are often inadequate.

In countries where there is some provision for the central funding and supply of communication aids (e.g., the UK), the range of aids offered is generally limited and there are inadequate facilities for training, assessment, prescription and personalization. In addition, there is usually very little follow-up once a communication aid has been issued. This situation is being improved in the UK by the establishment of regional communication aids resource centers.

See also: Microcomputers; Mechanical Devices in Medicine and Rehabilitation

Bibliography

Arnott J L, Pickering J A, Swiffin A L, Battison M 1984 An adaptive and predictive communication aid for the disabled exploits the redundancy in natural language. *Proc. 7th Annual Conf. Rehabilitation Engineering*, Vol. 4. Rehabilitation Engineering Society of North America, Bethesda, pp. 349–50

Bolton M P, Taylor A C 1984 A computer interface for tetraplegics using optimised encoding. *Proc. 7th, Annual Conf. Rehabilitation Engineering*, Vol. 4. Rehabilitation Engineering Society of North America, Bethesda, pp. 39–40

Easton J, Enderby P 1983 *The Acquisition of Aids for the Speech Impaired.* Assistive Communication Aids Centre, Frenchay Hospital, Bristol

Friedman M B, Kiliany G, Dzmura M 1984 An eye gaze controlled keyboard. *Proc. 7th Annual Conf. Rehabilitation Engineering*, Vol. 4. Rehabilitation Engineering Society of North America, Bethesda, pp. 446–47

Goldberg G 1984 Application of an optical keyboard for job-site access to standard computer systems. *Proc. 7th Annual Conf. Rehabilitation Engineering*, Vol. 4. Rehabilitation Engineering Society of North America, Bethesda, pp. 41–42

Latham C 1980 *Communication Systems.* Graves Medical Audiovisual Library, London

Perkins W J 1983 *High Technology Aids for the Disabled.* Butterworth, London

Poss J 1984 Life with the adaptive firmware card. *Proc. 7th Annual Conf. Rehabilitation Engineering*, Vol. 4. Rehabilitation Engineering Society of North America, Bethesda, pp. 117–18

Rosen M J, Goodenough-Trepagnier C 1982 Communication systems for the non-vocal motor handicapped: Practice and prospects. *Eng. Med. Biol. Mag.* 1 (4): 31–35

Vanderheiden G C, Grilley K 1975 *Non-Vocal Communication Techniques and Aids for the Severely Physically Handicapped.* University Park Press, Baltimore, Maryland

A. W. S. Brown

Computer-Aided Diagnosis

Although the computer is a relatively modern development, mankind has always been interested in external aids to diagnosis. In ancient times the positions of the stars in the heavens, the entrails of chickens, and small polished bones were all utilized from time to time in an attempt to discover what ailments afflicted patients. More recently, the medical profession has an enviable record of accepting innovation and technology; thus the advent of the electronic digital computer in the 1950s, although greeted with scepticism by some, was generally welcomed. It was envisaged—optimistically—that this device would allow accurate diagnoses and "rational" decisions to be made, perhaps for the first time in medicine.

This expectation has not, to date, been fulfilled. Many computer systems were produced in the 1960s to carry out medical diagnosis but these were mostly unwieldy, expensive and impractical. Unfortunately they were also accompanied by wild claims as to their value which were not fulfilled. In short, they promised the moon, cost the earth and in practical terms delivered very little.

In the 1970s however, with the advent of the microprocessor, smaller, cheaper and similar systems began to appear, designed to assist doctors rather than replace them. These smaller systems (as opposed to the unwieldy systems of the 1960s) may well have a valuable part to play in clinical medicine.

1. Basic Concepts

Before describing any computer system in detail, it may be helpful to consider some of the basic concepts which arise when computers are introduced into medical diagnosis. Perhaps the best dissertation on this point is that set out by Professor Lee Lusted (1968), formally Professor of Radiology at the University of Illinois, Chicago and a pioneer in the field.

Professor Lusted pointed out first and foremost that talking of "computers doing diagnoses," or "computer diagnosis" is not at all helpful at the present time—and what is really meant when we discuss the use of computers in this field is computer-*aided* diagnosis. In most successful systems the computer has been used as an adjunct to medical knowledge and experience, rather than as a diagnostic system in its own right. This has an important consequence: the clinician is likely to remain the central pivot of the diagnostic process, at least in the near future. From this it follows that clinicians (as well as computer scientists) should be very closely concerned with the design and implementation of computer-aided diagnostic systems.

Lusted also pointed out that many systems have in the past failed due to lack of preparation—and it is worth adding that such lack of preparation has usually been on the medical rather than the computing side. In particu-

lar, the cardinal sin involves acquiring a (large and expensive) computer and then trying to find a use for it! Ideally, preparation should involve specifying what the computer needs to accomplish in the diagnostic process—and then obtaining a specific computer system to meet these requirements.

Finally, many computer systems have been produced which work quite well in the mathematical laboratory but fail miserably when applied to patients. It should go without saying that the testing of any system must be carried out in a real-life situation; in medicine, this must be at the bedside or on the wards with real patients.

2. Computer Systems

It is both undesirable and impossible to describe every system in detail simply because both hardware and software used in the various experiments around the world have varied so widely. Thus computer-aided diagnostic systems have been implemented upon almost all types of computer, from large mainframes such as the IBM 370 and ICL 1906 series, through a variety of minicomputers, to microsystems such as the PET, Commodore and Texas SR59. As regards software, virtually every major language has been used for computer-aided diagnostic systems ranging from high-level languages (such as FORTRAN, ALGOL and BASIC) to machine code. There is thus a wide variety of systems and languages in use, but perhaps in summarizing the present position two points may be relevant.

First, the variety of systems in use stems partly from the different medical and statistical approaches adopted by the various research teams which have tackled the problem. Briefly these approaches can be divided into two: the determinist approach, and the probabilist approach. Examples of the former approach are seen in the wide variety of systems which invoke the use of diagnostic "rules" (e.g., "if the patient has pain and tenderness in the right lower abdominal quadrant, diagnose appendicitis"), and/or "trees." Interestingly as regards the second (probabilist) approach, most proponents have used a theorem published as long ago as 1763 by an English clergyman, the Reverend Thomas Bayes.

The other major cause for the wide variation in those systems which have been used for computer-aided diagnosis has been the way in which the computing field itself has developed in recent years, particularly in relation to hardware. In the mid-1960s, users were singularly fortunate to encounter (in even quite a large machine) a central core of more than 16K. Nowadays, with the advent of microsystems such an amount of core store can be bought for tens rather than hundreds of dollars. It is not difficult to appreciate that this revolution in the availability of cheap core store has produced—over the years—totally different computing systems to cope with similar problems.

3. Examples of Current Systems

A surprising amount of work is currently being carried out in relation to diagnosis and computers. Much of this relates to the use of computerized axial tomography (see *Computerized Axial Tomography*), of which the chief system is popularly known as the EMI Scanner. The EMI Scanner differs from most current computer-aided diagnostic systems in that it analyzes a complex set of visual and/or electronic images. Most computer-aided diagnostic systems do not do this, but concentrate rather upon a detailed analysis of the patient's complaints (symptoms) and the patient's features when physical examination is carried out (physical signs).

Examples of this type of system are many and varied. In Leeds, UK, systems have been devised to deal with patients suffering from acute abdominal pain, dyspepsia, alteration of bowel habit, gynecological problems and breast cancer. Elsewhere in the UK, workers (particularly in Glasgow and London) have devised computer-aided systems for the diagnosis of dyspepsia, thyroid disease and jaundice. In Europe, workers in Copenhagen, Paris, and a variety of other centers have also produced systems in which the computer is able to assist with the diagnosis of patients suffering from jaundice and thyroid disease.

In the USA, as well as studies similar to those already mentioned, probably the most interesting work has been carried out by Lusted and his colleagues in Chicago who have undertaken a number of highly interesting studies of the way humans and computers analyze data from x-ray investigations.

These types of computer-aided study are now becoming truly worldwide. For various gastroenterological complaints (such as patients with acute abdominal pain or inflammatory bowel disease) the World Organisation of Gastroenterology has now undertaken studies involving computer-aided diagnosis in approximately 20 countries, in Europe, North America and elsewhere (Bouchier and de Dombal 1979).

4. A Typical Current System

It may be helpful to consider one current system in a little more detail. The system chosen is one in use in a wide variety of centers. It was developed in the early 1970s in Leeds to assist with the diagnosis of patients suffering from acute abdominal pain. For clarity, a typical case will be described in which the diagnosis consists of three stages. The case is hypothetical (to preserve confidentiality) but it is both realistic and typical of many thousands of cases processed in various centers.

Stage 1. The patient concerned arrives in hospital complaining of abdominal pain of less than one week's duration. The patient is seen by the surgical resident on call, interviewed and examined and the data are recorded on a special form, an example of which is shown in Fig. 1.

Stage 2. Data are then entered (by a prespecified code) into the computer—either by a doctor or by one of the computing team, more usually the latter. The computer compares details of the case in Fig. 1 with several thousand similar cases in its memory bank calculating (from the work of Bayes) the probability of various common diseases and making a prediction as to the diagnosis.

Stage 3. This information is then displayed to the doctor, in the format of Fig. 2. Some points are worth noting. No patient is identified to the computer by name. If computers are to be used in medicine the public has a right to expect that this information is as confidential as that given to a doctor. Next follow the symptoms and signs obtained by the doctor. Finally the computer makes its diagnostic prediction (given the information available). Note that the computer is not strictly "making a diagnosis." It is merely selecting from amongst the common causes of acute abdominal pain (given that the information is accurate and given that the patient has

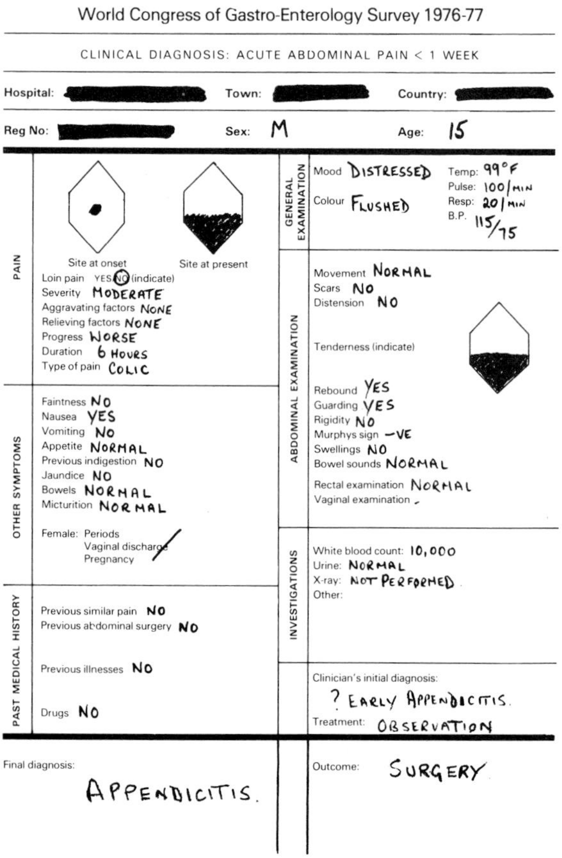

Figure 1
The case history of a patient with acute abdominal pain. Format is that used in World Organisation of Gastroenterology studies

CASE REF   ENC 1

MALE	NO PREV. ABD. PAIN
AGE 10–19	NO PREV. ABD. SURGERY
SITE ONSET—CENTRAL	NOT TAKING DRUGS
SITE PRES—LOWER 1/2	MOOD—DISTRESSED
NO AGGRAV. FACTORS	FLUSHED
NO RELIEV. FACTORS	ABD. MVT NORMAL
PAIN BECOMING WORSE	ABD. SCAR ABSENT
DURATION—12 HRS	NO ABD. DISTENSION
PAIN NOW COLICKY	TENDERNESS—LOWER 1/2
MODERATE PAIN	REBOUND TENDERNESS
NAUSEA	GUARDING PRESENT
NO VOMITING	NO RIGIDITY
NORMAL APPETITE	NO ABD. SWELLING
NO INDIGESTION	MURPHY SIGN NEGATIVE
NO JAUNDICE	BOWEL SOUNDS NORMAL
NORMAL BOWELS	NORMAL RECTAL EXAM
NORMAL MICTURITION	

APPEND	DIVERT	PERFDU	NONSAP
99.20	0.00	0.00	0.73
CHOLE	SMBOBT	PANC	R.COLIC
0.00	0.00	0.00	0.00

Figure 2
Computer-aided diagnostic printout of same case as Fig. 1

one of the causes listed) the most likely cause in this particular case.

5. Diagnostic Results

The results of a wide variety of computer-aided diagnostic systems have been described in a large series of publications. These are discussed at length by de Dombal (1979). Briefly the results can be summed up as follows. First, many computer-aided diagnostic systems have proved to be between 10% and 25% more accurate (in consecutive, prospectively collected, large series of patients) than the doctors who actually saw the patients. This was shown as early as 1971, when in a series of 552 patients in Leeds, suffering from acute abdominal pain, the doctors accuracy ranged from 42% to 81% (depending on the grade and seniority of doctor) while the accuracy of the computer-aided system in the same patients was 91% (de Dombal et al. 1972).

Far more interesting, however, has been a second, generally noted effect, namely that, during the operation of computer-aided diagnostic systems, the clinicians own (unaided) diagnostic performance has generally improved, so that by the end of each trial, the diagnostic performance of the clinician approximately matched that of the computer. Wilson and his colleagues (1975) showed that the accuracy of quite junior doctors rose— in dealing with patients suffering from acute abdominal pain—from just over 70% to between 80% and 90% after the introduction of a computer-aided diagnostic and decision support system. Similar findings have been

noted since then, in other centers and in other areas of clinical medicine.

6. Tasks and Prospects for the 1980s

It is clear that computer-aided diagnostic systems are no panacea in clinical medicine. The search for a diagnostic "Rosetta stone" has proved to be valueless. However, some useful features of computer-aided diagnosis have already emerged. It has been shown in a number of areas that the clinician aided by a computer is more accurate and more effective as a diagnostician than he is without this aid. The consequences for clinical medicine (and society) of an "across the board" improvement of even 10% in diagnostic performance would be very considerable; this is indeed the prospect for the 1980s.

Prior to the attainment of such a situation, however, some further tasks have to be performed, and these relate largely to clinical (as opposed to the computing) aspects of the problem. First and foremost there is a great need to clarify medical terminology and to conduct large-scale studies to distinguish between symptoms and signs which are merely interesting and those which—in a particular situation—are of crucial importance. At present this is not yet done at all well either by doctors themselves or by the medical text-books: there is considerable scope for computer-aided investigation in this respect.

Perhaps the interim verdict on computer-aided diagnosis in medicine should be that it has proved to be quite useful in restricted areas of medicine as an adjunct to the doctor's own diagnostic process. In this adjunctive role we may be confident that the use and usefulness of the computer in clinical diagnosis will become more widespread over the next ten years.

See also: Microcomputers

Bibliography

Bouchier I A D, de Dombal F T (eds.) 1979 Studies co-ordinated by the research committee of the organisation mondiale de gastroenterologie. *Scand. J. Gastroenterol.* 14: Suppl. 56
Cloe L E 1976 Health planning for computed tomography: Perspectives and problems. *Am. J. Roentgenol.* 127: 187–90
de Dombal F T 1979 Computers and the surgeon: A matter of decision. *Surg. Annu.* 11: 33–57
de Dombal F T, Leaper D J, Staniland J R, McCann A P, Horrocks J C, 1972 Computer-aided diagnosis of acute abdominal pain. *Br. Med. J.* 2: 9–13
Lusted L B 1968 *Introduction to Medical Decision Making.* Thomas, Springfield, Illinois
Wilson P E, Horrocks J C, Young C K, Lyndol P J, Page R E, de Dombal F T 1975 Simplified computer-aided diagnosis of acute abdominal pain. *Br. Med. J.* 2: 73–75

F. T. de Dombal

Computerized Axial Tomography

Computerized axial tomography (CAT) is a radiographic technique in which x-ray images representing slices through a patient are reconstructed by means of computer programs. An x-ray tube moves in a circular path with the patient on the axis of rotation and detectors either rotate simultaneously about the same axis or are fixed in position around the plane of rotation. Earlier systems used a complex movement called rotate–translate in order to economize on the number of detectors needed, but they have been mainly superseded because of their slower scan times and patient throughput. Electronic detectors are used to collect arrays of x-ray absorption data; the much wider linear dynamic range of electronic detectors compared to film enables a higher but still low and clinically acceptable dose to be given to the patient which can be converted into vastly improved detection of soft-tissue lesions. The additional advantage of tomography is that overlaying organ structure from the images, which would otherwise obscure the subtle low-contrast differences, is removed.

1. Historical Application

The first practical clinical demonstration of CAT was in 1971 (Hounsfield 1973). The superior imaging of cerebral lesions revolutionized neuroradiological practice and the CAT scanner rapidly became a standard item of equipment worldwide in all major hospitals which had neuroradiological departments. The inventor, Godfrey N Hounsfield, has received over 40 honors and decorations for his contribution to medical diagnostic imaging, which he worked out while studying pattern recognition and information transfer systems at the Central Research Laboratories of EMI Ltd (now Thorn-EMI).

The neuroradiological application was extended from the brain to the spinal cord as soon as whole-body systems were available that had a large enough scanning aperture, and disposed of the "water bag," which earlier systems used for coupling the head to the detectors.

The need for improved soft-tissue discrimination in oncology (cancer studies) has led to the CAT scanner being used for staging cancer disease progression and managing patient treatment. An important benefit of CT images is that the use of high-intensity beams in radiotherapy can be accurately planned using the digital data which comprise the image. Indeed, a variety of programs and specialist display systems now available allow the radiotherapist to plan multiple-beam treatments which optimize radiation to the tumor site while minimizing damage to neighboring sensitive tissues.

2. Technology

The key technology elements necessary for the CAT scanner are:

(a) high-output x-ray tubes;

(b) compact, reliable electronic x-ray detectors;

(c) cost-efficient digital computers;

(d) cost-efficient digital storage computer peripherals; and

(e) clinically oriented image-display systems.

The ready availability of all these in the 1970s allowed numerous manufacturers to offer systems to the world healthcare market.

A variety of system design concepts were developed which differed mainly in the data gathering and coding method. These were named first, second, third and fourth generation designs by their manufacturers as they were announced. The key differences between the designs are given in Table 1.

The minimum clinical requirements which are acceptable for current applications are given in Table 2.

Table 1
Design differences between first to fourth generation CT systems

Feature	Generation			
	First	Second	Third	Fourth
Typical number of detectors	1 (+1)	10–30	250–700	500–2000
Scan mechanism	rotate–translate	rotate–translate	rotate only	rotate–stationary
Typical scan time (s)	240	20–60	3–10	2–30

Table 2
Minimum clinical requirements for current CT systems

Area of study	Minimum scan time (s)	Resolution (mm)	Contrast discrimination (%)
Brain	10–20	1–2	0.5
Spinal cord	10–20	0.5–1	0.5
Chest	2–5	1–2	2
Abdomen	1–5	2–4	5
Heart (gated multiple scans)	ideally 0.01	0.5–2	0.5

System performance within these boundaries is now met by all manufacturers, although perfect spinal-cord imaging has not yet been achieved. Heart imaging is difficult and many compromises have to be made. Gating from ECG signals and addition of multiple scans remains the only method of data gathering since a single scan is not technically possible within one heart beat at the resolution or contrast-discrimination levels required, because of x-ray tube power limitation. Cardiac research work is proceeding, however, to determine the optimum compromise for clinical validity in the resulting images. It is important that heart muscle is imaged clearly as well as the major blood vessels supplying the heart wall.

3. Principle of Image Reconstruction

3.1 Data Sampling

If an angular array of projections of x-ray beam attenuation readings is taken over at least 180° around an object such that each part of the object is sampled in two orthogonal dimensions, then it is possible to reconstruct an image of the object represented by the map of attenuation in the plane of the sampled readings. The mathematical basis for this was first documented by Radon (1917).

Figure 1 shows a single detector being used to measure the x-ray attenuation of an object such that only the central element is common to all the sample readings taken—only this element can thus be determined accurately. If an offset is used and the experiment repeated, a second element is now common to all readings. Repeating this for the full diameter of the object enables all the diameter elements to be determined. If this is repeated for all the angles necessary to fill the circumference with picture elements, then every element is fully determined—with central elements significantly oversampled but the periphery correctly sampled.

When discrete sampling theorem rules are observed, the minimum number of samples necessary to determine

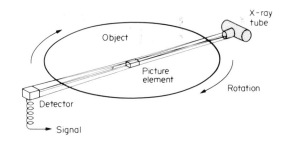

Figure 1
Single detector sampling

an object diameter D, with resolution d, is:

$$D/2 \text{ (angles)} \times 2D/d \text{ (views)} = D^2/d \text{ (readings)}$$

A 500 mm diameter object thus requires 785 398 readings for a 1 mm resolution accuracy.

3.2 Methods of Data Sampling

The single-detector systems first developed used a linear translation of the framework holding the x-ray tube and detector to create samples across the diameter (Fig. 2a).

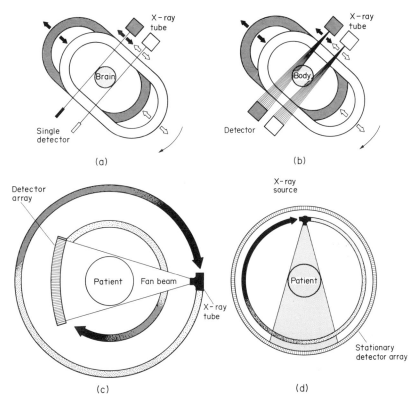

Figure 2
Systems used for data sampling: (a) first generation, single detector, rotate–translate; (b) second generation, fan beam, rotation only; (c) third generation, fan beam, rotation only; (d) fourth generation, fan beam, rotation only

The framework then rotated a small angle (1°) before translating in the reverse direction. A total of 180 angular steps was taken with diametric sample intervals determined by a fine graticule on the translation framework. Due to the masses involved, there was a mechanical speed limit of 3–4 min for a 180° rotation.

The second generation of CAT scanner systems (Fig. 2b) added additional detectors so that multiple angles were recorded simultaneously. The speed improvement was in direct proportion to the number of detectors, although a resolution improvement was usually introduced by increasing the number of detectors per degree. The practical scan time limit for body-sized apertures was about 20 s per 180° scan.

Third generation systems use a bank of detectors large enough to include the whole patient in the subtended angle (Fig. 2c). Over 200 detectors were initially used, but 500–700 are now common. The cost of each detector channel including the necessary electronics has to be considerably reduced over the first and second generation designs for this type of system to be practical.

No movement other than rotation is necessary, since the required diametric sample interval is given by the detector arc itself—each detector viewing one particular radius only.

A much-improved level of detector stability is necessary with this design concept, and many manufacturers withdrew from the market due to technical difficulties in achieving this. The scan speed is limited by x-ray tube flux output and computer data-transfer rates. Scan times as low as 1 s are possible.

Fourth generation designs (Fig. 2d) were developed in order to overcome the need for the high level of stability in third generation system detectors. Only the x-ray tube rotates, with a stationary detector bank occupying the full 360° scan plane. A much larger number of detectors are necessary for this type of system, other things being equal, but not in direct proportion to that apparently needed based on the third generation since the diametric sample interval is now obtained from the number of readings taken as the x-ray tube rotates, while the angular samples are determined by the detector geometry (the opposite to third generation). The scan speed is again limited by x-ray flux output and computer data-transfer rates, the latter tending to dominate in this design. Scan times as low as 1 s are possible.

3.3 Image Reconstruction

The matrix of readings from all types of system can be arranged as a rectangular set of views of the diameter versus angles around the object. The number of readings necessary to determine the object is represented by the area enclosed in the rectangle. The locus of a given point within the object mapped onto this matrix describes a sinusoidal curve with an amplitude proportional to radial distance from the center and phase shift as a function of angular position from the starting point of sampling the data (Fig. 3). This map is called a sinogram. The differences between each type or generation

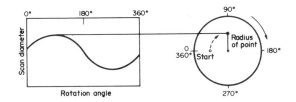

Figure 3
Locus of a point within a scanned object projected onto the data matrix (the sinogram)

of system can be seen by plotting the locus of detector views onto the matrix (Fig. 4).

The critical stability required for the detectors in the most cost-effective and popular fast scanner design (third generation type) is demonstrated by the locus diagram (Fig. 4c), where instability results in noise on a

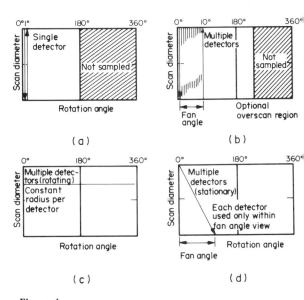

Figure 4
Loci of detector views on CAT scanner data matrices: (a) first generation, (b) second generation, (c) third generation, (d) fourth generation

single channel only—represented by a constant radius and therefore assigned to a circular ring pattern in the final image. Given that stability equals or is better than other sources of noise, such a system is capable of excellent fast imaging and most major manufacturers have selected this principle.

The other systems average individual channel noise over a larger and more diffuse area of the image, and are thus favored by some designers since detector stability is less critical.

3.4 Back Projection

The data matrix is back projected in order to create a tomographic image (Budinger and Gullberg 1974). Simple back projection produces a blurred image owing to natural smearing (Fig. 5a).

In order to give acceptable sharpness, an edge-enhancing filter has to be used (Fig. 5b). The method of

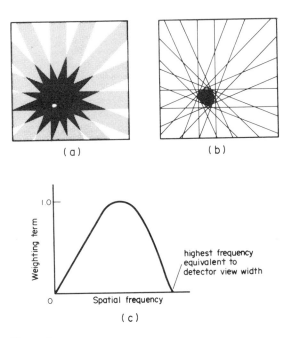

(a)   (b)

(c)

Figure 5
Diagrammatic representation of back projection:
(a) simple, (b) filtered, (c) filter function curve

reiteration employed in the first scanner systems accomplished this, but was rather slow. The most common method used now involves convolving the data matrix with a filter function chosen to give the "correct" image. The filter function approximates to a positive going ramp where dc is suppressed to zero amplitude but with a high-frequency roll-off applied. The upper cutoff point frequency limit is set by both the sampling theorem and the effective beam width determined by the detector aperture and the x-ray tube focal spot (Fig. 5c). Several filter functions are usually available to the radiologist in order to assist in the visual assessment of the image. Alternatively, a two-dimensional Fourier transform method may be used to filter the data in the frequency domain rather than in the sampling-space domain. The designer's choice depends upon the processing costs to produce an acceptable image in a reasonable time. The convolution method tends to be much faster, but offers less flexibility to the designer in changing subtle aspects of image quality than does the Fourier transform method.

3.5 Attenuation Coefficients

Although the prime requirement for diagnosis is a sharp, clear image which represents the anatomy directly, the method of data collection inherently produces a measurement of absolute absorption related to the physical and chemical nature of the object. The gray-scale dimension of the image is therefore a representation of the absorption coefficient for each individual picture element. Because the x-ray beam is not monochromatic, an absolute value is impossible to achieve, but if the "effective" energy of the beam is known, then it is possible to relate the data readings to the absorption coefficient. The scale used for image creation assigns, arbitrarily, zero value to an attenuation equivalent to water (for any beam energy). An attenuation which is effectively zero (the air path with no object in the scanner) is assigned to −1000 units, which results in an object with twice the attenuation of water such as bone giving +1000 units. It is possible to detect subtle disease differences such as those between tumors or abscesses from this absolute scale of "CT numbers" or Hounsfield units, as they are now called, in honor of the inventor. The system must, however, give stable and consistent readings for this to be possible since the differences are of only a few units. The use of a water coupling medium in the first scanner was intended to give such a stable reference, but the clinical need for easy patient handling and the ability to scan any part of the body led to the total abandonment of water coupling. Modern designs come remarkably close to the original performance in spite of the engineering compromises which have to be made.

3.6 Artefacts

The images produced have a variety of errors or artefacts within them. The most common artefacts in slow scanners were streaks caused by movement of the patient. Since the image is built up from many individual projections, displacement results in failure to overlap correctly at the back-projection stage and a streak is the common resulting visual defect.

Artefacts do not affect diagnosis in the majority of cases, since they are nowadays well controlled. Fast scanning has almost eliminated movement problems whilst other errors can be reduced by mathematical corrections during or after computation of the image.

Each system has its own inherent error types which have to be understood and should be eliminated during the design stage. Assessment of image quality is finally made by scanning known objects and comparing the results. Since it is important to be able to image patients, the design of these "phantoms," as they are called, simulates as far as possible the characteristics of tissue and bone, and allows an assessment of the potential detectability of real lesions.

4. X-Ray Generation

X-ray beams for CT are produced by bombarding a tungsten surface with high-energy electrons in a vacuum. The beam is predominately Bremsstrahlung radiation with a wide energy spectrum up to the maximum energy of the electron beam.

In spite of the advances in power rating and reliability of x-ray tubes, the output of x rays per kilowatt of input beam energy has remained relatively constant. The upper power level, and therefore maximum flux constraint, is determined by the melting point, or more correctly the critical vapor pressure point, of tungsten. The beam is filtered by metal windows, typically aluminum or copper, to produce the best dose to information transfer relationship. Since the image information is directly linked to the number of detected photons, and noise levels are predominantly photon quantum noise limited, image quality can be improved only by increasing either collection efficiency or the incident total x-ray flux. Collection efficiency depends on the photon losses due to absorption in detector windows, poor conversion within the detector and inefficient collection geometry. The latter is usually the major element, but overall, modern designs convert 50–70% of the available photons, which means that little further improvement is possible.

4.1 X-Ray Flux

At a mean effective energy of 70 keV as used in most CT scanners, the flux is approximately 3×10^{12} photons $s^{-1} kW^{-1} sterad^{-1}$. It is this that limits the maximum speed of modern CT machines, since image quality is directly proportional to flux. A CT image requires 10^9–10^{11} detected photons after absorption of 100–1000 fold by the patient's body. The required input flux is therefore 10^{11}–10^{12} photons into approximately 1×10^{-2} steradians for a theoretically perfect system.

Rotating anode tubes suitable for CT are capable of a power rating of 30 kW at a 4 s scan time or 15 kW at a 10 s scan time, but require significant derating if the power supply is pulsed. The energy needed per scan for a clinically valid image is 10–100 kJ or even more at the anode surface. The mass of the rotating anode restricts the heat capacity to 300–1000 kJ and therefore only a limited number of high-quality scans can be carried out before the critical temperature is reached, resulting in an extended cooling period, which can be conveniently before the next patient examination. The input energy is dissipated by emission and limited to only about 1.6 kW which determines the duty cycle of the system.

The flux utilization of the best systems is very high and the efficiency is considerably higher than conventional x-ray film used with screens in common practice. There could be only marginal improvement in the future since approximately 50% of the theoretical limit is now being achieved.

5. X-Ray Detectors

Either solid-state scintillators or gas ionization detectors can be used, each with its own advantages or disadvantages. The requirement is that they are small, cheap, and perform reliably with low noise and high stability. The system design implications are more important than comparisons of the merits of each detector in isolation.

5.1 Solid-State Detectors

Solid-state scintillators give a pulse of light in the visible spectrum from each x-ray photon absorbed. This light pulse is then converted into an electrical signal by a photomultiplier or silicon photodiode. The afterglow of the scintillation pulse must be low enough to prevent signal distortion since pulse counting is not possible due to the high rates involved (1×10^6 Hz) per channel and direct integration into charge is used. The linear dynamic range must be sufficient to record accurately both the highest count rate as well as the lowest—the latter in particular since reciprocals of logarithms are taken during computation and a small error is therefore magnified. Sodium iodide crystals doped with thallium were used in the first scanners since these give a high light output for each x-ray photon and are readily available. They have to be totally sealed since the material is hygroscopic and cost effective designs with large arrays of multiple detectors are not possible. The afterglow is also rather high which restricts the dynamic range.

Bismuth germanate and cesium iodide became favored materials as they could be obtained in commercial quantity. Cadmium tungstate, now available in quantity, is an ideal material for use with silicon photodiodes since the light output is high, of the correct spectral color to match the response of silicon, and it has excellent mechanical properties as well as low afterglow. All these materials are easily fabricated into small detector elements of the required size and cost in large arrays.

5.2 Gas Ionization Detectors

These are favored for third generation or "rotate-only" systems. Xenon gas at high pressure is used. In the presence of a high voltage across the detector cell, ionization caused by collision of x-ray photons with xenon molecules results in a pulse of current due to the primary and secondary ionization events. When correctly designed, these detector systems have excellent performances although their apparent simplicity disguises many design problems such as microphony and high-impedance amplifier stability. For rotate-only systems they are unsurpassed. The dimensions and cost per channel requirements are easily met for rotate-only systems.

6. System Design

All functions are controlled by digital computer technology, and although the block diagram shown in

Fig. 6 shows only a single computer, microcomputers with their own intelligence are distributed around the system for local operations. The operator uses the control keyboard to perform any activity required. It is common practice to use separate television display monitors for diagnosis purposes which are driven by their own computer and are isolated from the scanner itself. The images are then transferred by magnetic tape or floppy disk.

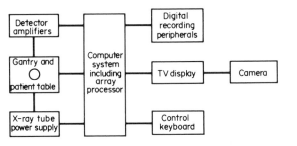

Figure 6
Block diagram of a CT system

The functions involve: x-ray tube operating conditions, patient location, detector calibration, scanning, data correction, image calculation, data storage—both image and raw data if required, image display and manipulation, system operation display, and photography of images. Image processing is now normally carried out either wholly or partly in a special front-end-array processor which differs in design architecture from the normal minicomputer. Although this device is still under software control from the main control computer, the speed of computation is many times faster than possible with even the most efficiently coded standard computer. The vast increase in the amount and rate of data has made this mandatory in order to achieve fast computation times for large-area, high-resolution images.

Data can be stored digitally on disk—either "floppy" or "hard" multiple platter—or, of course, on magnetic tape. The latter is used as archival storage, but tends to be inconvenient, especially if searching for images of a particular patient examination. Floppy disks are convenient and tend to be used as is x-ray film—the patient diagnosis file is associated with a particular disk dedicated only to that patient. They are not treated as archival data, however, since they are easily corrupted.

The multiple-platter hard disk gives very rapid access to any examination from hundreds stored on the same unit; this is important during the initial diagnostic stage when fast comparisons have to be made. Since each image contains 10^6 bits or more of information, even diskpaks up to 200 Mbyte (1.6×10^9 bits) can only store one or two weeks of patient data even with data compression techniques, and cannot be archival due to cost and size constraints.

The digital optical disk, derived from the laser video disk, promises a very cheap and reliable means of storing image data when such systems become readily available. Each disk can hold 10^{10} bits and is no larger than an LP record. This medium could be archival, rapid and extremely cost effective and will revolutionize image storage and access in CT as well as other digital imaging techniques in use in hospitals.

The manipulation of image data on the screen is important to the radiologist in extracting the maximum information content. The most important operation is "windowing," whereby only the gray-scale range relevant to the tissue densities of interest are displayed. More dense tissue is shown white, and less dense black, preventing this information confusing the diagnostician, and allowing the best use to be made of the television screen's dynamic range, which is more limited than film. An analogous operation is used in conventional radiography when selecting the exposure time and high voltage, since these affect the tissue-density range which can be recorded on film and can be considered as a windowing procedure.

The view required of the organ under examination may not be at right angles to the patient axis, and a rearrangement of the data from several images at an alternative angle may be preferred. The coronal (front) or sagittal (side) are common view angles since these correspond to conventional x ray, or an oblique angle may be necessary for instance to follow a curving vessel. These alternative views are all possible by reconstruction from sets of images of contiguous slices of the patient.

Quantification is also a powerful diagnostic tool, and comparison of tissue-density differences can be easily carried out—the area of concern being drawn on the screen by the operator using an electronic cursor. In some cases the area can be calculated automatically without operator intervention. Measurement of linear dimensions is another simple but powerful diagnostic facility.

Radiotherapy planning uses the full information of a CT image—density, position and organ delineation by the operator, in order to calculate dose levels from proposed radiation treatment. It is possible to experiment on the screen with several such plans, almost in real time, to optimize the dose to the tumor, while minimizing that to healthy organs.

See also: Computerized Image Analysis in Radiology; Positron Emission Tomography; Single Photon Emission Tomography

Bibliography

Brooks R A, di Chiro G 1976 Principles of computer assisted tomography (CAT) in radiographic and radioisotopic imaging. *Phys. Med. Biol.* 21: 689–732
Budinger T F, Gullberg G T 1974 Three-dimensional reconstruction in nuclear medicine emission imaging. *IEEE Trans. Nucl Sci.* NS-21 (3): 2–20

Hounsfield G N 1973 Computerized transverse axial scanning (tomography). *Br. J. Radiol.* 46: 1016–51

Radon J 1917 Über die Bestimmung von Funktionen durch ihre Integralwerte längs gewisser Mannigfaltigkeiten. *Ber. Verh. Saechs. Wiss. Leipzig Math. Phys. Kl.* 69: 262–77

J. N. A. Ridyard

Computerized Image Analysis in Radiology

Image analysis with a computer, or automated image analysis, involves the development of complex models intended to simulate and improve upon human analytical performance. There is a need for such efforts, since the frequency of human error in medical decision-making, including the interpretation of many classes of radiant images (Lehr et al. 1976), is astonishingly high. A major element of this error lies in the failure of the reader to detect the patterns of disease which so often are embedded in a background of structured and statistical noise (Kundel and Revesz 1976). Yet another source of human error is the matching of information once perceived in the image with the appropriate diagnostic classification. Here the problem is not one of perceiving the lesion, but rather a problem of recognizing the signs and patterns of disease.

1. Advantages and Disadvantages of Automation

When the scope and range of a diagnostic problem have been narrowly defined, automated analysis has been shown either to improve on or not to be conspicuously different from human accuracy (Lodwick et al. 1979a). However, it has not been possible to duplicate the range and versatility of human perception and decision-making, in part because we are not yet certain of how humans make decisions from images. There is also a problem with developing a database of necessarily global proportions.

Reader variation and error is a serious problem in human performance (Hessel et al. 1978). Automated feature extraction based upon quantifiable data should provide greater consistency. Furthermore, a system which can separate abnormals from normals with a high level of accuracy will permit humans to concentrate their diagnostic efforts on abnormal cases.

In evolving the basic techniques necessary for feature identification, we gain a better understanding of how humans perform. This feedback appears to be an important factor in improving automated diagnosis.

It should be noted that investigators tend to underestimate the complexity of diagnosis from images and often assume that solutions can be rapidly achieved. Real accomplishments have been modest, and experience has shown that the most successful approach to automating image analysis is to combine the respective talents of humans and machines.

2. Investigative Arena

While the need for machine analysis of images is not restricted to medicine, only the narrow area of radiological imaging will be considered here. While the 1963 report of computer-aided diagnosis of bone tumors appears to be the first effort to automate diagnosis from x rays (Lodwick et al. 1963), the report of Becker et al. (1964) on extracting and measuring the cardiothoracic ratio of the chest x ray and that of Lehr et al. (1970) on automated detection of lesions in brain scintiscans are the early pioneering efforts in total automation. The chest, however, comprising 40% of all radiographic examinations, continues to be the major target for the development of automated diagnostic models (Hall et al. 1971). However, the chest film has been a difficult subject because of a wide variability from image to image in terms of density, contrast, detail and background fog. These variations have required extensive preprocessing of each image to accomplish a level of standardization necessary for feature extraction. In spite of the excellence of quality control in the modern radiographic department, variation in patient size and exposure factors will continue to adversely affect analog–digital conversion of film information. Computed tomographic images, nuclear scintiscans and ultrasound images which have direct data-input–computer interfaces have made possible primary data acquisition, digitization and automated conversion of digital information into images. Such direct data acquisition has enhanced the usefulness of automated analysis (Henschke et al. 1979). Similar direct digital readouts of radiographic images on image amplifier tubes are being extensively employed for digital subtraction technology.

In the future all radiographic images are likely to have the option of direct digital readout at directly quantifiable gray levels.

3. Definition of the Problem

If computer analysis of medical images is to be of value, the technology must solve problems that need solutions, generally those which are aimed at improving the management of the patient. For example, a program designed to determine accurately only the cardiothoracic ratio will be of limited value, since cardiothoracic ratio is only one of several measurements useful in determining the possibility of cardiac enlargement. On the other hand, a program which can discriminate between several textural patterns in the lung will be of greater value especially if these patterns are characteristic of diseases such as pneumonia, pulmonary edema or lymphogenous spread of cancer. These diagnoses are of great significance in deciding appropriate therapy and are difficult for humans to classify properly. The end point must be a specific, reliable solution to a difficult problem of some frequency, with the possibility in the long run that the solution will be cost effective.

4. Target Areas for Automation

The following criteria have been of value in the selection of appropriate targets for automated diagnosis:

(a) Those organ systems, the images of which are subject to the highest rates of human error: Lehr et al. (1976), in a study using a test library of 2145 cases, found images of certain organ systems subject to higher rates of error than others (see Table 1).

Table 1
A comparison of the error rates associated with the radiological image analysis of certain organ systems

Organ system	Error rate (%)
Vascular	37.5
Bone	35.7
Skull	35.0
Abdomen	33.1
Chest	33.0
Extremities	31.0
Urograms	28.8
Gastrointestinal	25.5
Biliary	15.9

(b) Types of examinations which are of high frequency; these will maximize the value of the iterative process (i.e., the chest or hip).

(c) Examinations where the target area is easily found; these will be managed with greater reliability.

(d) Imaging techniques where the information is initially acquired in digital format, such as computed tomography; these will have more quantitative data requiring less preprocessing (McCullough 1977).

(e) If the accessible medical intelligence content of an imaging technique is high, as with bone; this will indicate a better chance of success.

(f) Problems of great public-health interest; these will have the dual virtues of importance and of high probability of funding. Typical examples are early detection of cancer, and diagnosis and staging of pneumoconiosis.

5. Automated Diagnostic Models

Automated models vary substantially in what they are designed to accomplish. Some typical functions of a model follow.

5.1 Image Segmentation or Field Analysis

This is the process of becoming oriented in the image in order to find the parts of interest, a problem roughly equivalent to human perception. Establishing the logic to find the areas of interest is perhaps the most difficult task in image analysis. This is particularly true of radiographic image analysis where accurate outlines of an object are needed but where segments are not too clearly defined in the original image. Detection is a major source of error for humans, and since the way in which humans perceive a lesion from the background of structured and statistical noise is poorly understood (Kundel and Revesz 1976), we lack a good model to build upon. Further, no two radiographs are alike, each is very complex with much detail and with a large number of superimposed objects. Subtle findings may yield significant diagnostic information which is difficult for the nonradiologist or the machine to distinguish from noise. The literature on image analysis segmentation methods has been well summarized by Harlow and Conners (1979). Boundary formation is an important part of segmentation. Region formation developed somewhat later than boundary formation although the purpose of both is to locate the objects of interest in the image. In region analysis, one attempts to locate areas of the image that share some uniform property such as gray levels, shade or texture. The most promising results relate region growing to functional approximation. The analysis of the scene is usually from the top down in that one first attempts to locate large objects and later to recognize smaller details. Tree structures are often used, such as those of Ballard and Sklansky (1973) for finding tumors in chest radiographs. Because in some instances, such as with pulmonary infiltrates or with breast thermography (Ziskin et al. 1975), machine perception may be impossible, investigators simply bypass this step by human identification of the fields of interest on the image. This is usually done interactively.

5.2 Finding Edges and Shapes

These basic and primitive operations of image analysis are employed for object identification as well as boundary detection. Edge detection and shape analysis are fundamental to the recognition of certain diagnostic classes. Shape analysis has been extensively studied for the diagnostic classification of congenital heart disease, where both Fourier and Fourier–Walsh shape descriptors have been effectively used (Cook et al. 1979). Automated edge detection is of central significance in volumetric determinations, especially cardiac volume (Chow and Kaneko 1973), where the objective is to record changes in cardiac volume through the entire cycle of the heart beat, and in computed tomography (CT), where Steiner et al. (1975) found that the volume of a hematoma which induced coma was rather specific. When the edge to be detected is discontinuous, as happens where overlapping shadows create a complex image, interactive human tracing of edges has been a useful solution.

5.3 Texture Analysis

This is another basic tool of medical image processing. The most frequent application is to match a particular texture to a texture class. Another application is to determine whether two textures are visually distinct, a problem particularly important in image segmentation. A third problem is to divide a group of textures into subgroups which are visually similar; this is also useful for region finding (Harlow and Conners 1979).

Texture analysis has been most extensively employed in computer classification of pneumoconiosis. Several methods of texture analysis have been used. In one, an optical computer is used to calculate the Fourier spectrum within a circle placed by hand in the lung field. In others, texture measurements are calculated based on the point-to-point variations in the density of the lung (Jagoe and Patton 1975). Conners and Harlow (1980) compared four texture classification algorithms, with the conclusion that the spatial gray-level dependence method (SGLDM) is the most powerful and versatile. Another application of texture analysis has distinguished between Paget's disease, metastasis in bone and normal bone (Lodwick et al. 1979a).

5.4 Other Essential Elements of the Model

These include a sound database, systems hardware, and the analytic software with a monitor for controlling the preprocessing, segmentation, feature extraction and classification procedures. Each of these components has its unique characteristics and problems. Special emphasis must be given to a carefully selected and verified library of cases. Assembly of a large series of radiant images is invariably a costly, difficult and time-consuming task. Each diagnosis to be considered must be represented by a sample of cases large enough to be statistically significant. The scope of an image-analysis program is usually restricted by lack of representative material in the database. While collection of new case material is important and may be a continuous process, care must be taken that for any sequence of experiments designed to assess progress, the database must remain a fixed variable.

Systems hardware varies widely from project to project and, with rapid advances in technology, quickly becomes outmoded. Pictorial information is often described by digitized arrays, syntactic and systematic strings and high-dimensional trees or graphs. Because most image-processing tasks require only repetitive Boolean operations or simple arithmetic operations defined over extremely large arrays of pixels, the resources of large computers are wasted. The information densities of large radiographic images, such as chest films, are representative of this problem and require a high level of processing capability. To meet specialized needs, systems are being designed consisting of hundreds of LSI bit-/microprocessors with large numbers of shared-memory modules and flexible networks for efficient image processing and pattern recognition applications.

For scanning, the effective image dimensions at which radiologists examine an image are significantly higher than a 2000×2000 matrix. Unfortunately, the state of the art in computer hardware presently precludes the use of digital images with dimensions of 2000×2000 pixels. Practically speaking, the scanner height and optics must remain constant. Hence, the choice of spatial resolution of which images are to be scanned is important. Often, to circumvent the limitations of digital image dimensions and yet obtain satisfactory spatial resolution, only small areas of the radiograph are scanned. Scanning only small areas permits one to distribute gray levels over a very small range in density values allowing the digital image to accurately represent small fluctuations. For texture analysis, the problem is to obtain enough information from the image to allow the computer to perform discrimination accurately. If spatial resolution is too low, the computer is not given enough detail; if too high, the processing time is too long because the dimensions of a digital image will have to be large to cover a large enough area to reflect the patterns characteristic of the various disease processes. For texture in bone, a spatial resolution where every pixel corresponds to a $0.2\,\mathrm{mm}^2$ area on the radiograph has been found satisfactory (Lodwick et al. 1979a). For boundary determination, scanning rasters of much coarser dimensions are satisfactory and time saving.

Analytic software includes a monitor for controlling operations of preprocessing, segmentation, feature extraction and recognition. For the latter, discriminant functions are widely employed. A new Bayesian modified maximum likelihood decision rule (MMLDR) (Choy 1976) has been found useful in multiclass problems. Depending upon the size of the database, both training and testing samples are useful. With a database of limited size, the 10% jackknife procedure is often employed for evaluating training and testing results.

6. Representative Problems in Automated Analysis

Problem 1: to diagnose five classes of congenital heart disease from a data set which includes normal chest films (six-class problem).

Millions of chest radiographs are analyzed yearly by radiologists. The magnitude of the task makes it difficult to maintain a consistent and reliable standard of judgement. A computer system for analyzing radiographs could allow the radiologist to handle more patients and make his diagnoses more consistent and repeatable. Such a system also has the potential of objectively or numerically describing the heretofore subjective information which resides in a chest film.

Successful attempts have been made to analyze the chest film for indications of rheumatic heart disease. For this task, size measurements have been the principal source of diagnostic descriptors. By contrast, the automated evaluation of congenital heart disease involves films from many age groups ranging from infants to adults. Because the images are nonstandard, descriptors necessarily include an outline of the cardiac silhouette taken as a single closed curve, with the top and bottom of the heart approximated by straight lines arbitrarily imposed to close the curve (Cook et al. 1979). A chain code provides for the shape of the heart which is reflected in Fourier descriptors.

Six experiments were performed:

(a) The radiologist interpreted posterior/anterior (PA) and lateral radiographs of a data set of 387 cases, 58 of which were normal.

(b) An automated computer model traced the cardiac outlines on the PA projections of a data set of 388 cases, 59 of which were normal (one case added to the data set above). Using a combination of Fourier pairs and measurements as predictive features which were automatically extracted, all cases were classified as to diagnosis (Tsiang et al. 1974).

(c) The cardiac outlines of 387 cases were hand-drawn on the PA views. The automated computer model extracted selected features including Fourier pairs and classified all cases as to diagnosis (Brooks et al. 1974).

(d) For each of 908 cases, 555 of which were normal, the cardiac silhouette was interactively divided into the five segments that normally are used by humans as diagnostic features. The segments were described by Fourier pairs, and using these features and related measurements, as well as a pairwise decision rule, the computer model classified all cases as to diagnosis (Choy 1976).

(e) The same 908 cases were analyzed by the computer model, and all cases were classified with a stepwise decision rule (Choy 1976).

(f) The same 908 cases were analyzed by the computer model, substituting Walsh descriptors for Fourier descriptors, and seven were calculated for each segment. Thirty-eight descriptors were calculated for each study. A pairwise decision rule was used (Cook et al. 1979).

The classification results of the six experiments are shown in Table 2.

With success in the automated diagnosis of rheumatic heart disease, it seemed that measurements alone could substitute for whatever logic humans were using in making this diagnosis. However, with congenital heart disease, a breakthrough in automated diagnostic accuracy was achieved only through interactive marking of the five segments of the silhouette which normally are evaluated by humans in the diagnostic process. Note that through combining the best of automation and expert medical knowledge, a level of accuracy is achieved which is 26–36% better than either automated or human diagnosis alone.

Table 2

A comparison of six experiments performed to diagnose five classes of congenital heart disease from a data set including normal chest films

Cases	Method	N	ASD	VSD	PS	PDA	TET	Overall[a]
387	Radiologist	58	43	29	20	16	54	41
388	Computer	67	19	36	12	6	20	40
387	Hand-drawn	78	55	29	21	26	63	51
908	Interactive	87	39	55	41	50	63	67
908	Stepwise	83	75	59	60	92	81	77
908	Walsh	93	53	21	51	21	26	70

N normal ASD atrial septal defect VSD ventricular septal defect
PS pulmonic stenosis PDA patent ductus arteriosis TET tetrology of Fallot and other cardiac anomalies
a All results testing except stepwise which is training

Problem 2: to distinguish between three textures commonly seen in bone: Paget's disease, metastatic carcinoma and normal trabecular bone.

Identification of textural abnormality is a major problem in radiographic film interpretation. In the lung, the textural patterns are normally present as summations of the shadows of ribs, blood vessels, bronchi and supportive structures, and have an orientation radiating from the hilum to the lung periphery. In bone, texture largely represents the spongy network of trabecular bone in the ends of long bones and in flat bones, the orientation of which is highly predictable and related to the stresses applied to the bone. Changes in the orientation of trabecular bone result from changes in stress, but disease states may alter the texture of trabecular bone. The human interpretation of abnormalities of trabecular bone is a sophisticated and difficult process associated with a higher rate of error than most such decision processes.

The distinction between Paget's disease and cancer can be a difficult problem for the radiologist, both diseases involving extensive modification of the texture of bone, but at times appearing very much alike.

Two experiments were performed (Lodwick et al. 1979a,b):

(a) Each of two skilled radiologists interpreted films from a test set of images consisting of 68 showing normal bone, 40 showing bone with Paget's disease and 55 showing bone with metastatic carcinoma. Each film was covered by a mask with a window which showed only the selected area of abnormal trabecular bone. No cortical bone could be visualized. No clinical data were provided.

(b) An automated texture identification model, using the spatial gray-level dependence method, scanned, preprocessed and classified each case under identical viewing circumstances to the human experiment.

The results are given in Table 3.

Table 3
Experimental results of human and automated diagnosis of
three bone conditions from radiological images

Method	Correct diagnosis (%)			
	Normal	Paget's disease	Carcinoma	Overall
Radiologist 1	75.0	97.5	87.3	86.6
Radiologist 2	79.4	92.5	76.9	82.9
Computer training	94.1	77.5	94.6	88.7
Computer testing	91.2	70.0	87.2	82.3

While this study is promising for the automated identification of texture, there are clearly problems which need to be resolved such as increasing the size of the training sample. A major problem for the clinician is that he cannot extract a meaning from mathematically defined texture features which is consistent with his own interpretation. In short, he can see and appreciate the statistical results of the texture program, but cannot visualize in a feedback way what the computer is "seeing." On the other hand, the texture investigators have very little theory to guide the development of texture-discriminating systems, and have limited knowledge of the kind and nature of the clinical information that can be obtained from the radiographic images. It would seem that at least a partial resolution might be forthcoming if the extracted data could be used to reconstruct the texture patterns as they are "seen" by the computer.

Problem 3: to examine the effectiveness of a computer-aided diagnostic (CAD) system to properly classify a test library of 245 cases representing 20 decision classes.

Four radiologists, entering data into a CAD model by means of a CRT, and after committing themselves to their own diagnosis, entered their selected signs and measurements for each case into the computer memory. Double-read signs and measurements, and the pathologist's diagnosis from tissue samples, were used as the standards for accuracy. The challenge of this human diagnostic problem is reflected by an average of 55.4% human accuracy for the four experimental subjects. Three out of four had improved scores with CAD (Lodwick et al. 1981, unpublished results). The ceiling of diagnostic accuracy for the computer was 79.6%. Detailed error analysis led to consideration of how humans use diagnostic information. An average of 22.8 signs employed for diagnosing each case is inconsistent with experimental evidence that human short-term memory can accept only 2–3 bits of information (Miller 1956), leading to a hypothesis that humans combine certain signs with qualities of dependence into minipatterns which appear synonymous with the "chunks" of chunking theory (Simon 1974, Sterling 1982). Further study indicates that human recognition of a class of such minipatterns which reflect rate of growth is significantly correlated with the accuracy of the diagnosis selected.

If it is true that humans perceive an image not as a collection of signs or features but rather as from three to eight minipatterns of features, then perhaps our approach to automated analysis can benefit from application of this knowledge towards the improvement of machine accuracy in classifying disease.

7. Conclusion

Any program aimed at automating diagnosis must be truly interdisciplinary; the involvement of clinicians must be dedicated and not tangential. The development of a research team must evolve through mutual interest of basic scientists and clinical scientists in attaining the goal of better diagnosis through automation.

See also: Image Analysis: Extraction of Quantitative Diagnostic Information; Quality Assurance in Diagnostic Radiology; Computerized Axial Tomography; Digital Fluorography

Bibliography

Ballard D, Sklansky J 1973 Tumor detection in radiographs. *Comput. Biomed. Res.* 6: 299–321

Becker H C, Nettleton W J, Meyers P H 1964 Digital computer determinations of a medical diagnostic index directly from a chest x-ray image. *IEEE Trans. Biomed. Eng.* 11: 67–76

Brooks R C, Dwyer S J, Lodwick G S 1974 Computer diagnosis of congenital heart disease. *Proc. IEEE Int. Conf. Systems, Man and Cybernetics.* Institute of Electrical and Electronics Engineers, New York

Chow C K, Kaneko T 1973 Boundary detection and volume determination of the left ventricle from a cineangiogram. *Comput. Biol. Med.* 3: 12–26

Choy A C-K 1976 Decision rules and decision tree approach in pattern recognition theory. Ph.D. dissertation, University of Missouri, Columbia

Conners R W, Harlow C A 1980 A theoretical comparison of texture algorithms. *IEEE Trans. Pattern Anal. Mach. Intell.* 2: 204–22

Cook L T, Harlow C A, Tully R J, Lodwick G S 1979 Fourier–Walsh shape descriptors and applications to radiographic image processing. *Proc. 6th Conf. Computer Applications in Radiology and Computer-Aided Analysis of Radiological Images.* Institute of Electrical and Electronics Engineers, New York, pp. 111–14

Hall D L, Lodwick G S, Kruger R P, Dwyer S J, Townes J R 1971 Direct computer diagnosis of rheumatic heart disease. *Radiology* 101: 497–509

Harlow C A, Conners R W 1979 Image analysis segmentation methods. *Proc. Dahlem Conf. Biomedical Pattern Recognition and Image Processing.* Dahlem Konferenzen, Berlin

Henschke C I, Hessel S J, McNeil B J 1979 Automated diagnosis in radiology. *Invest. Radiol.* 14: 195–201

Hessel S J, Herman P G, Swensson R G 1978 Improving performance by multiple interpretations of chest radiographs: Effectiveness and cost. *Radiology* 127: 589–94

Huang H K 1981 Biomedical image processing. *Crit. Rev. Bioeng.* 5: 185–271

Jagoe J R, Patton K A 1975 Reading chest radiographs for pneumoconiosis by computer. *Br. J. Ind. Med.* 32: 267–72

Kundel H L, Revesz G 1976 Lesion conspicuity, structured noise, and film reader error. *Am. J. Roentgenol.* 126: 1233–38

Lehr J L, Lodwick G S, Farrell C, Braaten O, Virtama P, Koivisto E L 1976 Direct measurement of the effect of film miniaturization on diagnostic accuracy. *Radiology* 118: 257–63

Lehr J L, Parkey R W, Harlow C A, Garrotto L J, Lodwick G S 1970 Computer algorithms for the detection of brain scintigram abnormalities. *Radiology* 97: 269

Lodwick G S, Conners R W, Harlow C A 1979a The diagnosis of bone disease using automatic texture analysis methods. *Proc. 6th Conf. Computer Applications in Radiology and Computer-Aided Analysis of Radiological Images.* Institute of Electrical and Electronics Engineers, New York, pp. 122–28

Lodwick G S, Conners R W, Yu J, Flandreau H 1979b Preliminary report, the importance of feedback in computer-aided medical decision making. *Proc. 4th Illinois Conf. Medical Information Systems, Urbana, Illinois, May 11–13.*

Lodwick G S, Haun C L, Smith W E, Keller R F, Robertson E D 1963 Computer diagnosis of primary bone tumors: A preliminary report. *Radiology* 80: 273–75

McCullough E C 1977 Factors affecting the use of quantitative information from a CT scanner. *Radiology* 124: 99–107

Miller G A 1956 The magical number seven, plus or minus two: Some limits on our capacity for processing information. *Psychol. Rev.* 63: 81–97

Simon H A 1974 How big is a chunk? *Science* 183: 482–88

Steiner L, Bergvall U, Zwetnow N 1975 Quantitative estimation of intracerebral and intraventricular hematoma by computer tomography. *Acta Radiol., Suppl.* 346: 143–54

Sterling J J 1982 The utilization of sign information by radiologists diagnosing bone tumors. Ph.D. Thesis, University of Missouri, Columbia

Tsiang P, Harlow C A, Lodwick G S 1974 The computer analysis of chest radiographs. *Proc. Association for Computing Machinery Conf., San Diego, California, November.*

Ziskin M C, Negin M, Piner C, Lapayowker M S 1975 Computer diagnosis of breast thermograms. *Radiology* 115: 341–47

G. S. Lodwick

Computers in Cardiology

Heart disease is an emotive topic and thus it is perhaps not surprising that new, technologically based investigative techniques are continuously being developed in cardiology. Computers, for example, now have an advanced diagnostic role in cardiac investigation. Indeed, their application in cardiology is perhaps one of the most important uses of computers in medicine.

1. Electrocardiography

1.1 Routine Electrocardiogram Interpretation

Computer interpretation of electrocardiograms (ECGs) was first attempted in the late 1950s. The introduction of microprocessor technology has contributed to the dissemination of the technique, particularly in North America and Japan but it is still true to say that less than 10% of ECGs recorded in Europe are currently interpreted by computer.

The principal reason for investigating the use of computers in the interpretation of ECGs was to increase the accuracy of interpretation by removing observer bias and variation. Secondary to this was the hope that cardiologists might be relieved of the chore of ECG interpretation. In addition, it was suggested that computer assistance would be welcomed in many hospitals in which ECGs were often unreported.

The ECG normally consists of a presentation of twelve different recordings of cardiac electrical activity (see *Electrocardiography*). By placing electrodes on the right and left arms and left leg, three "leads" measuring the potential difference between two limbs can be recorded. If E_R, E_L and E_F denote the potential at the respective limbs, then the three bipolar leads denoted I, II, III are given by

$$I = E_L - E_R$$
$$II = E_F - E_R$$
$$III = E_F - E_L$$

and hence at any instant in the cardiac cycle, $I + III = II$. By using appropriate circuitry it is possible to record the potential variation at each limb, denoted aVF, aVR, aVL. For example, the augmented unipolar limb lead aVF has the form

$$aVF = E_F - \tfrac{1}{2}(E_L + E_R)$$

From symmetry it follows that $aVR + aVL + aVF = 0$.

The potentials at the three limbs can be combined to form a central terminal with respect to which potentials elsewhere on the body can be measured. The form of the equation is

$$Vn = E_n - \tfrac{1}{3}(E_R + E_L + E_F)$$

Conventionally, six points on the precordium are chosen and denoted V1–V6. Thus the twelve-lead ECG consists of three bipolar leads I, II, III, three augmented unipolar limb leads aVR, aVL, aVF and six unipolar chest leads V1–V6. Because of the relationship between the limb leads, only two need be recorded if a computer is used since the other four can be calculated therefrom. For example, if only I and II are recorded, then

$$aVF = II - \tfrac{1}{2}I$$

Where computer techniques are not involved, all twelve leads would be recorded on paper.

A typical computer-based system functions as follows. The conventional twelve-lead ECG is recorded on a

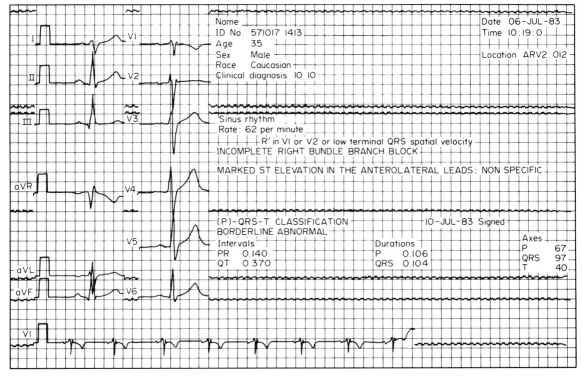

Figure 1
An example of a computer-generated ECG interpretation and ECG printout (courtesy of Siemens Elema)

microprocessor-based electrocardiograph. In view of the inherent redundancy within the twelve-lead ECG, only eight independent leads I, II, V1–V6 are recorded truly simultaneously and converted to digital form within the electrocardiograph. Details of the patient's name, age, clinical classification, etc., can be input to the system using a small alphanumeric keyboard. Such a micro-processor-based system has facilities for testing its own circuitry and can advise the operator of any faults (e.g., poor lead connection). If the hospital has a local area network with wiring from each bedside to the central computer the ECG data can be sent in digital form to the computer for interpretation. The ECG can be inter-preted within minutes after receipt of data, with the report (including interpretation) being available at any number of output stations situated at different locations in the network (see Fig. 1). Clerical staff in the cardiol-ogy office keep track of ECG requests and are able to advise whether or not an ECG has been recorded and interpreted. A system of this type also has facilities for physicians to check the ECG reports before they are printed if it is wished to adopt such a procedure. It also has facilities for storing ECGs and reports for several years.

The analysis which on a large system would take between 30 and 60 s consists of some signal pre-processing to remove noise and the effects of baseline

wander (if present), arrhythmia analysis, a wave classification procedure to group together chosen beats of similar morphology so that an averaged cardiac cycle can be used for wave detection and measurement, and ultimately the diagnostic interpretation including a com-parison with previous ECGs from the same patient.

Alternative approaches do exist. One alternative is to transmit the ECG in groups of three leads in analog form to a central system with results being returned in digital form to a reporting station. A second alternative is to have the ECG reported at the bedside if the microprocessor-based electrocardiograph has adequate storage facilities for both an interpretative program and the ECG data. This technique is becoming increasingly popular in view of its lower costs. In Japan, several thousand such systems are currently in use. One draw-back of this method in its present stage of development is the inability to undertake comparison of serial tracings from the same patient in order to check for sequential changes. In addition, it does not allow a central office to keep track of the ECG recordings being undertaken throughout the hospital. The bedside interpretation system can be linked to a central system for storage of results and serial comparison but this is more expensive. A third approach is for the ECG to be recorded on a digital cassette and then taken to a central computer for interpretation. This has the advantage of avoiding the need to install a data transmission system.

Computers have been employed for population studies. Programs now exist for automatic ECG classification based on a standard (Minnesota) code. (This approach has been shown to be at least as accurate as a team of technicians coding the ECG, but is still prone to coding error.)

1.2 Arrhythmia Monitoring

The use of digital computer-monitoring facilities, particularly in the coronary care unit, is increasing (see *Monitoring Equipment in Coronary and Intensive Care*).

There are two basic approaches to the technique. The first, the distributed approach, is to have a microcomputer within each bedside monitoring unit checking for rhythms which are thought to merit the attention of nursing staff. These may trigger simple high or low heart rate alarms or may set off a warning when ventricular extrasystoles occur above a preselected rate. Trends of heart rate (over several hours) can also be displayed at the bedside. Such units may be interconnected to a central display.

An alternative approach is to have a centralized computer facility capable of monitoring up to 16 beds simultaneously. Such systems are now very sophisticated and able to produce trends of heart rate over the previous 12 or 24 h together with histograms of the frequency of ectopic beats. The ECG may also be displayed, together with heart rate for example, on a video screen on the central monitoring console, where a rhythm interpretation is provided.

The major problem with all such systems is the reliability and accuracy of interpretation. Where possible, major alarms should be based on QRS waveform analysis or the absence of the waveform as the case may be. A detailed analysis of the P wave may often be unreliable because of artifact due to patient movement.

A recent development has been to monitor a patient's ECG by telemetry (see *Biotelemetry*) during the recovery phase following myocardial infarction when the patient can be mobilized at an early stage but still be under nursing surveillance.

1.3 Ambulatory Monitoring

Arrhythmia monitoring is important for the diagnosis of many transient arrhythmias, for the investigation of the therapeutic value of many antiarrhythmic drugs, and also for the elucidation of other abnormalities such as unexplained fainting or dizziness, particularly in the elderly. In ambulatory monitoring for multiples of 24 h periods, small portable analog tape recorders are used which are replayed for analysis from 60 up to 480 times the recording speed (see *Ambulatory Monitoring*). So-called 24 h analyzers are available that permit the rapid scan of the recorded tracings in either 6 or 24 min but these are subject to limitations, the main one being the vast amount of information to be assimilated by

technical staff, perhaps under pressure to analyze a certain number of recordings per session, while constantly observing an oscilloscope trace. To facilitate the analysis of such recordings and to permit a greater amount of useful data to be obtained, computer-assisted techniques have been introduced. In summary, these permit analysis of the 24 h recording to provide data on heart rate on a minute-by-minute basis, together with the incidence for example of aberrant beats if required. Some computer systems automatically capture any arrhythmias and store them on disk for subsequent print-out, thereby avoiding the need to monitor the replayed ECG continuously. Various forms of summarized data printout can be provided to simplify the reviewing of the recording.

Two more recent developments set the pattern for the future. The first is the introduction of a system that transfers the ECG in digital form on to a computer disk at $480 \times$ recording speed. The complete 24 h record is then scanned by software which categorizes each cardiac cycle. A technician can then decide which categories are normal before a report is printed detailing the frequency of occurrence of abnormalities. Any section of recording can be retrieved from disk and viewed.

The second development is the real-time analysis of the ambulatory ECG within a small pocket-sized battery-powered microprocessor-based recorder. Each cycle is classified as above and at the end of 24 h of use, a complete summary is instantly available via an appropriate replay unit which can also write out the full 24 h of a two-channel ECG recording.

Blood pressure can also be recorded over a 24 h period and analyzed by computer to provide a trend analysis as for the ECG.

1.4 Exercise Electrocardiography

During exercise, the ECG may show excessive artefact due to respiration or movement and thus the use of a computer for averaging cardiac cycles is of considerable help. For example, one exercise testing system uses a small array processor to provide a continuous display of an incrementally averaged cycle from all twelve leads of the ECG during exercise. Measurement of important components of the ECG waveform is also made by the system on the averaged signal although interpretation is deliberately avoided. Arrhythmias occurring on exercise can be captured by the system and stored on disk for later printout.

In exercise electrocardiography various protocols are used for exercise testing (e.g., bicycle ergometer or treadmill with exercise to exhaustion or to onset of symptoms) (see *Cardiac Function: Noninvasive Assessment*). A computer can be used to implement the chosen protocol by controlling such parameters as treadmill speed and gradient.

Another use of the computer in exercise and in resting electrocardiography is the production of so-called iso-potential maps. If, for example, a 4×4 array of electrodes is evenly positioned on the chest, the averaged

ECG signals from each can be used to produce the maps at any instant in the cardiac cycle. The method used is as follows. The value of potential at each electrode at one instant is noted. For one row of four electrodes, interpolation techniques can be used to estimate values of potential at intervening points. The procedure is repeated for all rows and then interpolation of the data by columns can proceed. Thereafter all points having a selected value of potential can be linked to produce a series of equipotential lines similar to weather-map isobars. For some forms of display the area between the lines can be filled in with some character (Fig. 2), or if

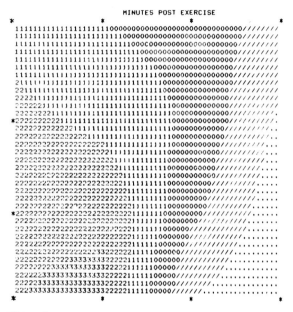

Figure 2
A map of potential variation over the left precordium following exercise. The different characters represent 1/10 mV ranges of positive potential (numbers) and negative potentials (other symbols); for example, 0 represents potentials in the 0 to 0.1 mV range, / represents values in the 0 to −0.1 mV range, etc.

a color TV display is available, color maps can be produced, even on-line during recording. The interpretation of maps is based on a knowledge of normal patterns of positive and negative zone movement throughout the cardiac cycle.

2. Cardiac Catheterization

Computer facilities are now considered to be an integral part of equipment for physiological monitoring at cardiac catheterization. Computer-based systems can undertake on-line calculation of peak blood pressures and these can be printed if required, along with pressure waveforms on the recording paper. This is the most important use of computer assistance in on-line work at cardiac catheterization, although many other measurements can be facilitated with the use of a computer. For example, pressure gradients across valves can be calculated from pressure waveforms obtained during catheter withdrawal from one chamber to another. The catheterization laboratory computer can also assist in the calculation of blood shunts from different parts of the heart when various oxygen values are input to the system.

In many teaching hospitals a single computer system is shared between several adjacent cardiac catheterization laboratories, each being able to obtain results on-line. Data are also stored on the system disk thereby permitting automated production of the investigative report.

The catheterization laboratory computer system can also be utilized for analysis of left ventricular angiograms. Conventionally the angiogram is projected on to a writing table on which the physician may trace round the outline of the left ventricle. The computer then calculates the ventricular volume making corrections for the movement of the heart during systole. Semi-automated methods have also been introduced whereby the physician traces round an initial frame of the angiogram and the equipment is able thereafter to calculate ventricular volumes at different times throughout the cardiac cycle leading to automatic calculation of ejection fraction. Other more mathematical data such as wall stress can also be calculated from a knowledge of ventricular wall measurements.

3. Echocardiogram Analysis

A number of microprocessor-based systems for analysis of M-mode echocardiograms are now available. These all require the physician to input various reference points on the echocardiogram such as the posterior and anterior wall reflections. The posterior left ventricular wall thickness, left ventricular internal diameter, and intraventricular septal thickness are commonly measured by computer-assisted techniques leading to the estimation of left ventricular volume and the total left ventricular weight.

Computers are also used for signal processing of the two-dimensional echocardiogram to give an improved image of the internal structure of the heart (see *Echocardiography*). A standard technique, in many ways similar to that used for radionuclide studies, can be adapted to enhance the image and provide a color display.

4. Radionuclide Studies

Computer assistance is also well established in cardiac radionuclide studies (see *Dynamic Cardiac Studies*). Indeed, dynamic studies would be impossible without either the computer or some equivalent technique. Perhaps the most commonly used method is that of nuclear ventriculography. In this method ^{99m}Tc is injected into an arm vein and thereafter two approaches may be adopted. The first is simply to measure the activity as it passes through the heart for the first and second time and thereby visualize each of the heart chambers in turn. The rate of washout of the isotope can be used to calculate ejection fraction. A more common approach is to wait until the isotope has reached equilibrium, that is, until it is dispersed throughout the blood and then to take a succession of "pictures," each triggered in synchronism with the ECG to look at usually 16 successive phases in the cardiac cycle. Since the activity which is to be measured is lower with the latter method, a few hundred cardiac cycles are used for averaging and ultimately, with further computer processing, a set of 16 images displaying the ventricular wall motion can be obtained (cf. a cine film) and hence areas of abnormality can be detected. The procedure may also be repeated following an exercise test.

A second technique used is to inject ^{201}Tl into an arm vein and wait until it has been taken up by the myocardium. A gamma camera linked to a computer will then provide a display of the radioactivity distribution in the heart muscle from which areas of maximum and minimum uptake can be identified. It is possible to outline a region of interest on the computer display and then have the computer program quantitate the radioactivity within that region. Areas of low activity are often related to poor myocardial perfusion. This test can also be undertaken following exercise when areas of abnormality are more often detected in patients with diseased coronary arteries.

5. Modelling

Computers have also figured in various models of the electrical activity of the heart. They have been used in solutions of the forward problem—the calculation of body surface potentials given the distribution of an electrical source within the model. Conversely, the inverse problem can be solved with computer assistance where a definite internal configuration of equivalent cardiac generator is assumed. Basically the computer is used to solve a number of simultaneous equations—calculations which would be prohibitive without such assistance.

See also: Cardiac Pacemakers: Computerized Data Handling

Bibliography

Macfarlane P W (ed.) 1985 *Computer Techniques in Clinical Medicine*. Butterworth, London

P. W. Macfarlane

Computers in Clinical Biochemistry

One of the earliest applications of computers in clinical biochemistry was in handling the vast amount of data generated by automated analytical instruments. Subsequently it became evident that the computer had a significant contribution to make in many other aspects of laboratory work, such as the processing of test requisitions, production of daily logs and printing patients' reports (Whitehead et al. 1968). Laboratory computer systems were also able to improve quality control, perform very sophisticated calculations, and provide for long-term storage and retrieval of patient files. More recently, with the introduction of microprocessor technology, both instrument operation and data acquisition and processing have been considerably simplified.

The system may be operated in two different modes. With on-line operation there is direct linkage of the analytical instruments to the computer by means of suitable interfaces which, depending on design, may carry out some preprocessing of the data. Alternatively, with off-line operation, data are transcribed onto punched cards, paper tape or other machine-readable media which are batch processed at a later stage. The advantages and disadvantages of both methods have been reviewed by Toren and Eggert (1978).

1. Computers in Laboratory Management

It is important to define clearly the roles of the computer in the laboratory. Three major divisions have been made.

1.1 Specimen Reception and Data Handling

Computer involvement depends to some extent on the types of laboratory instrumentation, the mode in which the system is operated and the form of specimen identification in use.

(a) *Specimen and patient identification.* A unique number must be assigned to a specimen sample and its aliquots, with the corresponding data stored on the computer. The method employed is dependent on whether the bulk of the laboratory instrumentation can recognize a unique code attached to a patient sample matching the output results. If not, patient identification data (PID) are supplied to the computer along with all the relevant test codes. Minimum information should consist of surname, forename, sex, hospital, ward, consultant, hospital number, laboratory number if appropriate, and the tests required. Worksheet production depends on which of these identification methods is employed. The possibility of using the National Insurance number as a "universal" hospital number to facilitate linkage of records in the UK remains to be explored.

(b) *Data processing and file management.* Most of the data will have been preprocessed or evaluated by the analytical staff. Residual processing constitutes such tasks as derivation of the indices of organ function or correction of results because of, for example, changes in instrument sensitivity or protein-binding of some plasma constituents. Verification of quality-control (QC) data may be performed by the system, but apart from storing this information on file, this is probably best performed by the analytical staff since immediate feedback is necessary. Data processing, as opposed to data checking (see Sect. 1.2), is relatively minimal at this stage.

The computer's main function is thus file management, collating results with the correct PID. Since results arrive at different intervals, files need to be updated and so facility in handling complex files is just as important as calculative ability.

File management should ensure that data files which have been reported are removed from the daily result file and copied into the cumulative data file for that patient. It is usual and sufficient to keep on disk the latest five sets of results; the report form is flagged to indicate the existence of earlier data which can be retrieved from the archive listing if necessary.

File management should also supply information on patients in the daily file whose results are not yet complete, and provide adequate information on them to minimize staff involvement in unproductive tasks.

(c) *Report forms.* These should be supplied on continuous stationery and the format should be carefully considered from the respective points of view of the biochemist and clinician. It is usual for the duty biochemist to screen outgoing reports for abnormalities of any kind (e.g., transcription errors, results outside normal ranges or incompatible with diagnosis, rapid swings in results from test to test), and this can be facilitated by providing space for cumulative reports so that comparisons can be made on trends.

(d) *Archiving.* Archiving is an essential and valuable task which the computer can perform. Several types of archive are needed for maximum flexibility and can be outlined as follows:

(i)   QC archives—a daily printout of QC data and storage of back-up copies on magnetic tape and disk for trend analysis on a weekly and monthly basis to monitor overall equipment performance.

(ii) Instrument archives—a daily printout of all tests performed which would be of value in following up incomplete reports and for derivation of cumulative work-load statistics.

(iii) Patient report archives—a daily alphabetical printout of all reports issued and storage of a copy on magnetic tape. This means that previous results are always available to the wards.

1.2 Data Validation Checks

Given suitable algorithms, many procedures can be transferred to the computer, releasing the biochemist for more productive work in following up the biochemically difficult or unusual patient and making his expertise available to the clinical staff for discussion and advice.

The types of data validation checks which can be performed are outlined below.

(a) *Alert checks.* Laboratories have normal or reference ranges for test values, and a value outside the expected range is classed as abnormal, requiring the attention of the biochemist and clinician, in which case the appropriate alert message (HIGH or LOW) is printed alongside the result.

Although the clinical limits of normality are wider than those of the biochemistry laboratory, the results should be flagged on the basis of the biochemical ranges, leaving the decision to intervene to the clinician.

(b) *Delta checks.* To operate these checks it is necessary to implement cumulative reporting. For a given patient, delta-check methods compare the differences (deltas) between the present results and corresponding previous test values with a table of given thresholds (Sheiner et al. 1979). If the delta exceeds its threshold, then the present values are suspected of being erroneous. The error may be only apparent, resulting from large swings in the patient's biochemistry but, importantly, it may be a real error. Two important types of error can be detected by this means, specimen mislabeling, and a transcription or other error in reporting the test results. These types of error are not detectable by any other means; for example, by checking control specimens against quality-control limits. The computer succeeds here because of its high-speed calculative abilities. Such checks are not really feasible for the unaided biochemist. Problems may arise with false positives and suitable algorithms and discriminant functions must be constructed to local needs (Sheiner et al. 1979).

(c) *Discordance checks.* These are often used for laboratory instrumentation checks. For example, enzyme activity may be determined either by end-point analysis or by kinetic analysis, and both may be available in case of instrument failure. Different reference ranges apply to each method (differential analytical techniques) but there is a mathematical relationship between them which can be held on a data file. To verify correct operation of the instruments, samples may be analyzed on both and the discordance checks are then used to verify the results. During instrument failure, the same principles may be applied to correct the results from the back-up instrument. The basic requirements for quality control of test results have been summarized by Walters (1973).

1.3 General Laboratory Management and Policy

To facilitate laboratory management it is necessary to have data on patterns of tests requested, on the number and types of tests requested by wards, the average sample turnround and other similar factors. Such information can be obtained from the archives and the computer can provide such information, together with

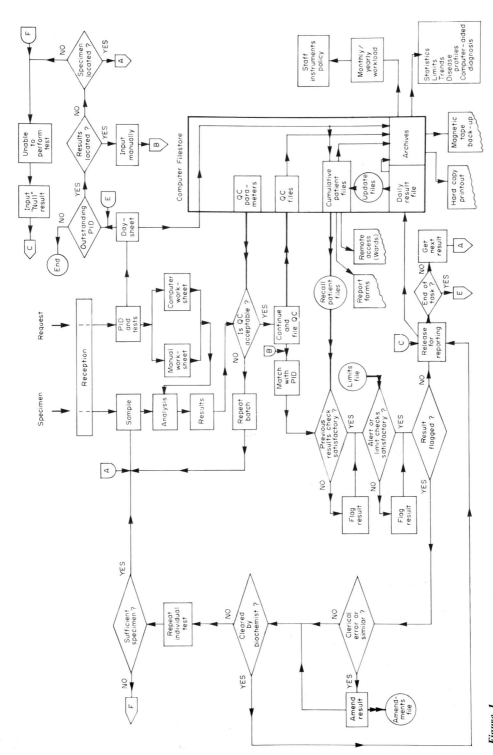

Figure 1
Flow chart of the minimal facilities required for a laboratory computer system

231

statistics and trend analysis, in a short interval of time. Changes in laboratory policy, e.g., staffing levels and types of instrumentation, can also be influenced by such analysis, enabling decisions to be made with greater confidence, as well as supplying evidence to justify new investment.

2. Computer Systems

When purchasing a computer system, a detailed systems analysis on the existing manual system is required so that unnecessary procedures are not transferred. As already shown, the requirement for a main computer system is predominantly one of data management, though this may depend on the degree of sophistication required. Certainly, powerful and flexible file structures are necessary, and a great deal of thought should be given to the programming language.

The systems analyst should also deal with questions of data input/output and number of visual display units (VDUs) and line-printers necessary, allowing for possible future expansion. The type of backing store for the files needs also to be considered. This will normally be demountable hard disks for the day-to-day operation of the filing system, as these have the relatively short access times (50 ms), rapid data transfer (500 kilobytes per second) and the large capacity necessary ($\geqslant 20$ megabytes). Archiving of the material should be done daily as a back-up in case of system failure or disk head-crash and the system should have magnetic tape drives available for this as a relatively economic means of mass storage.

The system should be multiuser (several operators performing different jobs but in the same language) and capable of assigning highest priority to the routine work.

The most important factor in any system is software which is extremely expensive to create from scratch. Efforts should be made therefore to see if an existing manufacturer's package is suitable, or if another laboratory has "core" software that could easily be modified to local requirements. Many institutions are attempting to set up common hardware and software and this trend is to be encouraged; software which is proven in one environment can then be used in other laboratories without expensive and time-consuming reprogramming. Consultation between laboratories and main hospital computing systems is also helpful, so that formats are compatible when laboratory computers come on-line to the main patient data base and interfacing problems are minimized.

Data held within the system and in archives are confidential and unauthorized access should be prevented by user identification codes and passwords.

Figure 1 indicates the minimum facilities necessary for a clinical laboratory computer system.

3. Other Developments

A logical extension of the ability of the computer to generate and store an organized and well-formatted data base is in automated screening and computer-aided diagnosis. Much of the pioneering work and subsequent exploitation has been carried out in the USA (Wolf 1975, Toren and Eggert 1978). Toren and Eggert have shown how interpretation might be achieved by Bayesian logic, decision trees using Boolean operators or by the development of unique software. They also stressed the need for better communications between hospital computer systems for the earlier recognition of drug interactions. They indicated that further developments would include methods for improving real-time quality control and possibly a mobile clinical chemistry laboratory as one solution to the problem of positive-sample identification.

See also: Clinical Biochemistry: Automation; Microcomputers: An Application in the Clinical Laboratory

Bibliography

Sheiner L B, Wheeler L A, Moore J K 1979 The performance of delta check methods. *Clin. Chem.* 25: 2034–37
Siemaszko F (ed.) 1978 *Computing in Clinical Laboratories.* Pitman, London
Toren E C Jr, Eggert A A 1978 *Computers in the Clinical Laboratory: An Introduction.* Marcel Dekker, New York
Walters A R 1973 What to look for in a computerised laboratory information system, Pt. 2. *Lab. Med.* 4: 32–38
Whitehead T P, Becker J F, Peters M 1968 Data processing in a clinical chemistry laboratory. In: McLachlan G, Shegog R A (eds.) 1968 *Computers in the Service of Medicine, 1: Essays on Current Research Applications,* Vol. 1. Oxford University Press, London, pp. 113–33
Wolf P L 1975 Utilization of computers in biochemical profiling. In: Enlander D (ed.) 1975 *Computers in Laboratory Medicine.* Academic Press, New York, pp. 81–101

B. Clark and R. B. Singh

Computers in Neurology

Computers now play a major role in patient diagnosis in neurology and neurosurgery. Electromyography equipment used in the diagnosis of neuropathies and myopathies has a microcomputer as an integral part, and the diagnostic imaging techniques of CT scanning and NMR imaging would not be possible without powerful dedicated computers. In studies of the management of head injury, computers have been used to amass and analyze a large multinational databank of patients in order to predict outcome and to enable different management methods to be compared.

There are few items of equipment used in neurological sciences departments which do not have a built-in computer or microcomputer, and there are few offices and laboratories which do not have a general purpose microcomputer. The examples of computer use given below highlight areas where the procedure is inherently dependent on a powerful computer.

1. Electromyography (EMG)

General electromyography involves measuring the electrical potentials produced in muscle in response to a stimulus which can be visual, auditory or electrical (see *Electromyography*). Highly sensitive amplifiers with common mode rejection ratios of better than 100 000:1 at 50 Hz are required to amplify the detected signals before digitization at rates of up to 1 µs per point. EMG analyzers are typically designed around a standard 16 bit microcomputer. The various tests are controlled by firmware, and data analysis display and printout are handled by the computer. Downloading facilities are available for the transfer of data to another computer for archiving of results if required.

A number of specialized EMG tests have been developed such as the motor unit counting method of Ballantyne and Hansen (1974). The number of motor units (or groups of fibers) in a muscle can be estimated by studying the electrical discharge produced by the contracting fibers. Initially, a small electrical stimulus is applied to the nerve supplying the muscle. This is gradually increased until a signal is picked up by an electrode attached to the surface of the skin overlying the muscle. The area under the discharge curve is computed and the results displayed on a VDU. As the stimulus is gradually increased, the investigator can observe whether the response obtained is similar to the one previously detected or whether an additional unit has been stimulated as indicated by a significant increase in area. This process is continued until six or seven units have been recruited. A supramaximal stimulus is then applied and the size of the discharge obtained from all the units is measured. The number of units can be calculated by simple proportion.

2. CT Scanning

The original EMI scanners (Ambrose 1973, Hounsfield 1973) were built around a Data General Eclipse S200 computer which was used to control the mechanical operations, perform the mathematical computations of tissue attenuation, display the image and store it on disk or magnetic tape for future reference or remote display. In subsequent generations of CT scanner the principles remain the same but the computing required is more powerful (see *Computerized Axial Tomography*). The third or fourth generation CT scanner would typically have a computer with 128 K of 16 bit memory, two 5 megabyte system disks, an 80 megabyte data disk with capacity for 600 images, and perhaps a 120 megabyte magnetic tape unit with 1 megabyte floppy disk drive. Up to twenty different viewing functions are normally included to enable images to be manipulated in a variety of ways and to make quantitative measurements from regions of interest. A dedicated fast processing unit can be used for data collection and processing using 32 bit floating point arithmetic.

3. Nuclear Magnetic Resonance Imaging

As with CT scanning, the computer processing consists of machine control, data collection, image reconstruction and image analysis (see *Nuclear Magnetic Resonance Imaging*). The X, Y and Z magnetic field gradients have to be switched on and off in a pattern determined by the imaging plane selected (transverse, sagittal or coronal) and rf excitation and measurement have to be timed according to the pulse sequence selected (e.g., inversion recovery, spin-echo or multiple echo). The data collected is in the form of phase and frequency information, and has to be transformed to the spatial domain by a fast Fourier transform (FFT) first in the frequency direction and then, on completion of data acquisition, in the phase direction. Before this transformation some frequency shifting, averaging and filtering may be required to reduce the noise in the final image. The FFT computation requires typically in excess of four million floating point addition, multiplication, division and square root operations for an image matrix of 256×256 and eighteen million for a 512×512 image. A dedicated array processor would be used to enable this computation to be done in a few seconds (Tracy 1984). After the image has been produced the data are managed in much the same way as with a CT scanner, with archiving on disk and magnetic tape, and image analysis on a dedicated imaging console. Typically, contrast variation, quantitation within regions of interest, profiling and image enlargement facilities are available.

The images produced by an inversion recovery sequence are normally weighted towards T_1 dependence with a slight T_2 component, and those produced by a spin-echo sequence are more heavily weighted towards T_2 than T_1. Pure T_1 and T_2 images can be produced by solving the relaxation equations which govern the inversion recovery and spin-echo sequences (Crooks et al. 1983).

4. Databank Analysis of Prognosis of Patients with Severe Head Injury

The effectiveness of various drugs or treatment regimes can often best be assessed by carefully controlled studies on large patient populations (Jennett et al. 1979). Originally designed to quantify the prognosis of patients with severe head injury, this project has developed the means for comparison and assessment of treatment methods. Considerable computing power is required to perform the discriminate analysis procedures used to analyze the data and to store the large databank of information on patients from Scotland, the Netherlands and the USA.

See also: Microcomputers; Neurosurgery: Physiological Monitoring; Computerized Image Analysis in Radiology

Bibliography

Ambrose J 1973 Computerized transverse axial scanning (tomography)—II. Clinical application. *Br. J. Radiol.* 46: 1023–47

Ballantyne J P, Hansen S 1974 A new method for the estimation of the number of motor units in a muscle. *J. Neurol. Neurosurg. Psychiat.* 37: 907–15

Crooks L E, Ortendahl D A, Kaufman L et al. 1983 Clinical efficiency of nuclear magnetic resonance imaging. *Radiology* 146: 123–28

Hounsfield G N 1973 Computerized transverse axial scanning (tomography)—I. Description of system. *Br. J. Radiol.* 46: 1016–22

Jennett B, Teasdale G, Braakman R, Minderhoud J, Heiden J, Kurze T 1979 Prognosis of patients with severe head injury. *Neurosurgery* 4: 283–89

Tracy R W 1984 The expanding field of medical electronics: NMR image reconstruction. *New Electron.* (April): 43–45

D. J. Wyper

Cryobiology

Cryobiology was transformed from an art into a science when, in 1949, Polge, Smith and Parkes found, quite by accident, that they could freeze-store spermatozoa successfully, using glycerol to protect the cells from injury. Since then, cryobiology has separated into three distinct but interrelated disciplines: cryopreservation, cryosurgery, and ultrastructural cryofixation, the latter discipline being a technique for preserving the structural details of cells by freezing them rapidly.

1. Frozen–Thawed Cells: Survival Curves

Biological cells contain at least 80% water. Thus the physics that underlies the freezing of biological cells is complex because water—whether it be pure water or water with substances dissolved in it—forms a greater variety of solid-phase structures when it freezes than any other substance. As the temperature of water is lowered, it remains liquid beyond its normal freezing point (that is, it supercools) prior to the nucleation of randomly displaced, slow-moving molecules. The amount of supercooling is greater if homogeneous nucleation in bulk liquid is taking place rather than heterogeneous nucleation around solid particles or at surfaces. The amount of supercooling influences considerably the structural form of the ice crystals and their growth rate after nucleation.

Ice-crystal growth rate at a given amount of supercooling reaches a maximum value with increasing solute concentrations, decreasing with yet higher concen-trations to a rate less than that for pure water. During the liquid–solid phase an electric potential develops, which also reaches a maximum value with increasing solute concentration. Ice-crystal formation is thus a complex process. When one considers that each individual biological cell is surrounded by a membrane with a temperature-dependent permeability which affects crystal growth, the difficulty of finding a quantitative explanation of the phenomenon becomes readily apparent.

Setting aside the problems which still await satisfactory solution, the physical events which occur when individual cells or groups of cells are lowered in temperature until they freeze can be described qualitatively with reference to Fig. 1. As cooling proceeds down to

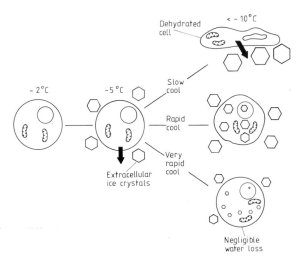

Figure 1
Schematic representation of the physical events which occur in cells during freezing

approximately −5 °C, the cells and surrounding medium remain unfrozen due both to supercooling and to the depression of the aqueous freezing point by protective solutes. Between about −5 °C and −15 °C, ice crystals form in the external medium either homogeneously or heterogeneously. During this period the interior of the cell remains unfrozen and supercooled because the plasma membrane prevents the penetrative growth of ice crystals into the cell. The supercooled water within the cells has a higher chemical potential than the partly frozen extracellular solution and this potential difference causes water to flow from within the cell across the membrane; once outside the cell the water freezes. This tendency towards cell dehydration is influenced principally by the rate of cooling. If the cooling rate is sufficiently slow, water transport across the cell membrane remains in equilibrium with the chemical potential. The cell thus becomes dehydrated, and excessive and possibly lethal solute concentrations build up within it. When the cooling rate is very rapid there is insufficient time for water to pass across the membrane before the cell is destroyed as a result of intracellular freezing.

A particular type of cell has a so-called survival signature with an optimum cooling rate for maximum cell viability and an otherwise considerably increased probability of cell death on either side of this optimum rate. Figure 2 shows some typical survival signatures for mouse marrow stem cells, yeast cells, and human erythrocytes. In relation to the survival curve, the different research areas in cryobiology can be characterized as follows:

(a) cryopreservation—the freezing of cells or organs within the narrow band of cooling rates on either side of the optimum rate;

(b) cryosurgery—in which cooling rates exceed the "optimum" rate as much as possible, in order to destroy infected tissue; and

(c) ultrastructural cyrofixation—the principal objective of which is the retention of the morphology within a cell by freezing it as rapidly as possible; cooling rates at the extreme upper end of the survival curve are required.

Other factors additional to cooling rate, however, can affect the survival signature significantly, including, for example, the minimum temperature achieved during the freeze–thaw cycle, the thawing rate, the concentration distribution of cryoprotective agents, and blood flow in animal tissue.

2. The Fundamental Field Equations

Realistic solutions for the temperature field equations are a necessary starting point for a satisfactory quantitative explanation of the physical problems involved in cryobiology. Irrespective of the geometry or shape of the boundary of any cryobiological specimen that is undergoing freezing and thawing, the temperature field can be represented by Fourier's heat conduction equation, because natural convection currents are mostly suppressed by the cellular structure of biological materials and any blood flow in the microcapillaries of animal tissue can conveniently be included in the internal heat generation term of the equation. Consequently, for the solid frozen region s:

$$\alpha_s \nabla^2 T_s = \frac{\partial T_s}{\partial t} \tag{1}$$

and for the liquid region l:

$$\alpha_l \nabla^2 T_l + \dot{q}/\rho c = \frac{\partial T_l}{\partial t} \tag{2}$$

where T is the temperature in the region at any instant of time t, α (equal to $k/\rho c$) is the thermal diffusivity, ρ the density, k the thermal conductivity and c the specific heat of the medium which may or may not have internal metabolic heat generation $\dot{q}$.

Equations (1) and (2) are coupled together by the heat-flow condition at the freezing front interface given by

$$\rho_s L \frac{d n_i}{d t} = k_s \frac{\partial T_s}{\partial n}\bigg|_{n=n_i} - k_l \frac{\partial T_l}{\partial n}\bigg|_{n=n_i} \tag{3}$$

where L is the latent heat of fusion and $\boldsymbol{n}$ is the vector normal to the moving interface.

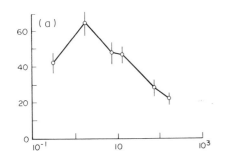
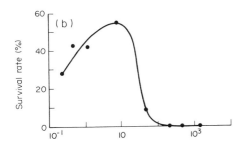
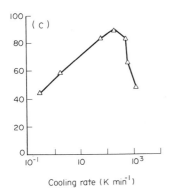

Figure 2
Survival of frozen–thawed cells as a function of cooling rate controlled to $-75\,°C$, transferred to liquid nitrogen at $-196\,°C$ and thawed rapidly at $500–1000\,K\,min^{-1}$: (a) marrow stem cells (after Leibo et al. 1970); (b) yeast cells (after Mazur and Schmidt 1968); (c) human red blood cells (after Miller and Mazur 1976)

Equations (1–3) can be solved numerically provided the boundary conditions are specified (Bald and Crowley 1979). The initial boundary condition for the temperature throughout the region and the temperature at the moving interface are pertinent factors common to all freezing problems in cryobiology. However, the boundary condition at the surface of the region which defines the behavior of the complete freeze–thaw cycle is seldom if ever defined or specified. This is one of the major shortcomings in experimental cryobiology to date.

In addition to the foregoing equations which describe the temperature field, equations completely analogous to Eqn. (1) can be used to describe, for example, the concentration distributions of cryoprotectants throughout the region. Because the concentration at any specific location in the region is a function of temperature, the temperature and concentration field equations are coupled. Fortunately, there are numerical solutions to these coupled equations.

3. Cryopreservation

As already indicated, Polge et al. (1949) revolutionized the field of cryobiology by successfully freeze-storing spermatozoa using glycerol as a cryoprotective agent. Since 1949 the viability of certain types of cell and tissue for long-term storage at low temperature using cryoprotectant agents such as glycerol and dimethyl sulfoxide (DMSO) has been very successful, particularly so in the field of animal breeding, using the technique of artificial insemination with semen that has been cryopreserved. However, the survival rates obtained with other types of biological cells, tissues and organs have not been satisfactory. A contributory cause for these disappointing survival rates is well illustrated by considering a typical storage experiment, in which an ampoule of about 1 cm in diameter is loaded into a cooling bath for freeze storage. The usual procedure is to measure and record the temperature of the sample during its freezing and thawing, by means of a thermocouple that is inserted inside the sample holder. As long ago as 1960, however, Rinfret in the USA demonstrated quite clearly that the percentage hemolysis (red-cell destruction) that occurred within a specimen sample depended quite dramatically on the position of the cells relative to the ampoule wall. The results of his experiment are summarized in Fig. 3. Thus, for successful experimentation in the preservation of biological specimens at low temperature, it is necessary to be able to predict the temperature fields within the specimens. (This is equivalent to solving Eqns. (1–3) using appropriate boundary conditons.) Measuring the temperature at some specified location within the sample makes it exceedingly difficult to predict the temperature at other locations as the new boundary conditions created are not easily determined. Future cryopreservation experiments must include a means of measuring the surface boundary condition such as the outside temperature of the ampoule wall as

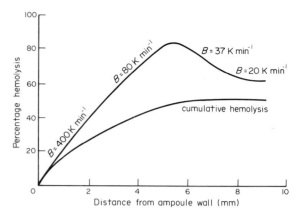

Figure 3

Percentage hemolysis as a function of distance from the container wall for a blood–lactose preparation cooled by immersion in a liquid bath at $-77\,°C$ and then warmed by immersion in a water bath at $40\,°C$. Upper curve indicates percentage loss of cells with cooling rates just prior to freezing shown on curve. Lower curve gives total hemolysis with distance (after Linde Co. 1960 Report to US Office of Naval Research Contract NONR 3003(60) (NR 105.28). Department of the Navy, Washington, DC)

a function of time, allowing numerical solutions of Eqns. (1–3) to be obtained. This approach could then be used to improve the statistical count of cell viability and consequently lead to a better understanding of optimum cooling and warming rates for a given cell type.

The success of cryopreservation experiments greatly depends on the selection of the best cryoprotective agent. This becomes increasingly important as progress is made from the preservation of small groups of cells or small tissues to the preservation of whole organs. The concentration of cryoprotectants and salts throughout the organ must be ascertained by equations analogous to Eqn. (1) and by coupling the thermal and mass diffusion equations included in the assessment of the experimental results.

The mechanism by which a cryoprotective agent enables cells to withstand injury as a result of freezing is not completely understood. However, it is known that the presence of agents of low molecular weight slows the increase in solute concentration within the cells as their temperature is lowered. Cells remain viable which at lower values of cooling rates referenced in Figs. 1 and 2 would otherwise be damaged. This may be at least a partial explanation of the preservation mechanism.

4. Cryosurgery

The object of any cryosurgical procedure, in complete contrast to cryopreservation, is the destruction of an infected tissue region by enclosing it within a frozen ice ball, while ensuring that damage to the surrounding tissue is minimal. Since 1961, when Cooper in the USA described the application of a liquid-nitrogen-cooled probe for use in neurosurgery (Cooper and Lee 1961), cryosurgery has expanded to encompass almost all branches of surgery.

The previous discussion on the survival curve for a specific cell type showed that cell destruction was most likely if the cooling rate was much greater than an optimum value. *In vitro* studies of cell viability and destruction have shown, additionally, that to ensure destruction the infected region should be cooled to below about −40 °C and that the frozen region should be rewarmed at the slowest rate possible. Slow rewarming encourages the recrystallization of intracellular ice, causing the cell membrane to rupture.

In addition to the above conditions for *in vitro* cell destruction, clinical experience would suggest that a complementary process of superimposed ischemic infarction (loss of blood supply) is also involved in *in vivo* cryosurgery. This blood deficiency resulting from obliteration of the microcirculation within the frozen tissue can often destroy any tissue which might survive the corresponding *in vitro* freeze–thaw cycle.

Clinical evidence also suggests that the *in vivo* cryolesion is unique in that, despite the differing cooling rates within the ice ball, the pattern of cellular death is uniform, a sharp demarcation occurring between the damaged and undamaged areas. According to experimental observations on rat liver (Smith and Fraser 1974), the demarcation appears to occur at the −15 °C isotherm within the ice ball and is attributable to ischemic infarction. Thus the minimum temperature required in the infected region during cryosurgery lies between the −15 °C value suggested by *in vivo* experience and the −40 °C suggested by *in vitro* experiments. Numerous studies of *in vivo* cell destruction have demonstrated also that better control over any possible regrowth of a tumor following cryosurgery can be achieved by using repetitive freeze–thaw cycles rather than a single freeze–thaw cycle. The improvement thus obtained is due probably to a combination of better thermal contact between probe and tissue and small changes in the thermal properties of tissue after each successive freeze. The success or failure of a particular cryosurgical procedure is closely linked to the solution of the temperature field equations (1–3), subject to the appropriate boundary conditions. The surface boundary condition which is applicable in this case is the temperature–time function at the cryoprobe tip, a function which is set by the design and capabilities of the cryosurgical equipment used.

Modern cryosurgical equipment extracts the enthalpy from the region of tissue to be treated using one of the following techniques:

(a) flowing a boiling cryogenic liquid such as liquid nitrogen to the cryoprobe tip, reducing the tip temperature to near the boiling point of the nitrogen (−196 °C),

(b) applying the Joule–Thomson expansion principle in expanding a high-pressure gas (nitrous oxide or carbon dioxide) at the probe tip to produce minimum probe tip temperatures in the region of −60 °C, or

(c) direct spraying of liquid nitrogen on to the surface to be treated.

For the treatment of common skin infections such as warts, a cotton swab or copper probe dipped in liquid nitrogen is still commonly used.

Until the recent availability of computer software for solving Eqns. (1–3) using finite element numerical methods (Comini and Del Giudice 1976, Bald 1981), both the approximate analytical and numerical solutions to these equations assumed an instantaneous initial constant temperature for the surface boundary condition. It has therefore been difficult until now to relate the ice-ball size and shape achieved in successful cryosurgical operations to the actual temperature transients experienced by the tissue. Blood deficiency and metabolic heat generation within the infected region during cryosurgery can be included in the solution to the temperature field equations by suitable selection of the quantity $\dot{q}$ appearing in Eqn. (2).

The absence of pain following cryosurgery suggested that nerve tissue is destroyed or inactivated during the procedure. This led Lloyd et al. (1976) to introduce a new clinical technique which they called cryoanalgesia. Using nitrous oxide gas in a probe operating on the Joule–Thomson principle, predetermined nerve centers were frozen to relieve intractable pain. The method has since been extended to the relief of postoperative pain.

The physical problem in cryoanalgesia is not the destruction of some known infected tissue region but the encompassment of the appropriate nerve center within the ice ball to reduce its temperature to below about −20 °C. Thus, in this case, the appropriate placement of the probe tip relative to the nerve center is a necessary precondition for solving the temperature field problem. Cryoprobes therefore incorporate an electrical stimulator so that the relevant nerve center can be located prior to freezing.

5. Ultrastructural Cryofixation

Ultrastructural cryofixation is a new technique for freezing cells or tissue as rapidly as possible to preserve their *in vivo* morphological structure. The frozen specimens can then be prepared for detailed examination by electron microscopy or electron microprobe analysis (x-ray microanalysis). The variable sizes and shapes of the specimens which are quick frozen for ultrastructural studies govern the cooling rates that can be obtained for given cell types.

The morphology of the cell structure is retained best by minimizing the sizes of the ice crystals formed. The experimental observation of Van Venrooij et al. (1975)

in Holland show that the cooling rates should be greater than 10 000 K s^{-1} to minimize crystal growth. The results of their work on crystal size—obtained from experiments on 20% glycerol solution in a 1 mm diameter cylinder—are shown in Figs. 4 and 5.

To obtain freezing rates in excess of 10 000 K s^{-1}, three different cooling methods have been used. These are listed below in descending order of cooling rate achievable:

(a) the so-called slamming method—a thin slice of tissue (~100 μm) is rapidly slammed against a precooled metal block;

(b) rapidly plunging specimens of various geometries into a subcooled bath of cryogenic liquid, which is maintained near its melting point; and

(c) spray freezing the surface of small biological samples using a pressurized spray of cryogenic fluid such as propane.

Although the slamming method is potentially the fastest method of freezing specimens the technique is limited in its applications. Thus rapid plunging into different cryogenic fluids is the most common technique employed. Considerable disagreement exists, however, as to the best cryogenic fluid to use to produce the quickest freeze and therefore the best retention of ultrastructure.

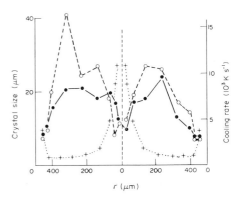

Figure 5
Crystal sizes and cooling rates as a function of the location within the replica with a diameter of 1 mm: ○, length of the crystals in the radial direction; ●, length in the direction perpendicular to the radial direction (the left vertical axis indicates the value of these dimensions); +, cooling rates calculated from the results of the numerical scheme applied to the freezing sphere (diameter 1.145 mm) on corresponding sites (the right vertical axis indicates the cooling rates). Note that the crystal size is inversely proportional to the cooling rate (after Van Venrooij et al. 1975)

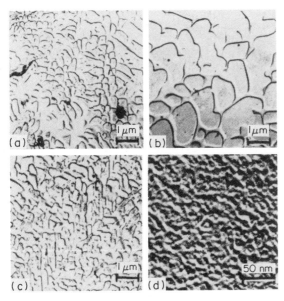

Figure 4
Photomicrographs of ice crystals formed in a 1 mm diameter cylinder: (a) crystals from the center of the cylinder; (b) crystals from an area halfway between the center and the border; (c) crystals near the border; and (d) crystals in a spray-frozen preparation (after Van Venrooij et al. 1975)

When a specimen at room temperature T_R is rapidly plunged into a subcooled liquid maintained near its melting point, T_M, four distinct, successive stages of cooling occur:

(a) Stage 1—the biological sample travelling at a uniform velocity V loses heat to the fluid as a result of forced convective boiling. A stable vapor film cannot be maintained around the sample as in the case of metallic objects.

(b) Stage 2—transient conduction into the surrounding liquid layers occurs, over a short period, as soon as the sample comes to rest.

(c) Stage 3—transient conduction rapidly diminishes and very turbulent natural convection takes over.

(d) Stage 4—the sample temperature now governed by laminar natural convection follows the normal Newton cooling curve. In ultrastructural cryofixation this final stage is of negligible importance.

The best ultrastructure is realized in the outer surface layers of the sample during the initial plunge (Stage 1). If vapor formation is kept to a minimum during this stage then the mean surface heat transfer coefficient $\bar{h}$ for a spherical sample of radius a is given by:

$$\bar{h} = \frac{K_L}{2a} [2 + (0.4 Re^{0.5} + 0.06 Re^{0.67}) Pr^{0.4} (\mu_0/\mu_\infty)^{0.25}] \quad (4)$$

where Re is the Reynolds number ($Re = 2a\rho_L V/\mu_L$); Pr is the Prandtl number ($Pr = \mu_L C_L/K_L$); ρ_L, C_L, K_L and μ_L are the density, specific heat, thermal conductivity and viscosity of the liquid bath, respectively; the viscosity ratio μ_0/μ_∞ is near unity.

The most rapid freeze during the plunge stage is produced by the fluid with the highest value for the product $\bar{h}(\bar{T}_0 - T_M)$, where

$$\bar{T}_0 = \tfrac{1}{2}(T_R + T_0)$$

and is the mean temperature of the sample surface during plunging.

Theoretical calculations show that nitrogen near its melting point of 63.1 K is the fluid with the best potential for morphological preservation provided excessive vapor formation can be prevented (e.g., by hyperbaric freezing, at a pressure of at least 33.5 atm, while maintaining the liquid in a nonequilibrium state at 63.1 K (Bald and Robards 1978)).

6. Future Developments

An exciting long-term aspect of cryobiology is the possible cryopreservation of whole human organs and eventually the cryopreservation of humans themselves. Humans who are deceased and currently preserved in liquid nitrogen have a finite, limited shelf life of some 32 000 years, after which time ionizing radiation will have destroyed all their cells.

Because of the difficulties of controlling the coupled temperature and mass diffusion fields during the freezing and thawing of samples of increasing physical size, successful whole-organ preservation is exceedingly difficult. A solution can only come from better controlled experiments on cell and tissue preservation that can produce more accurate survival signatures, which may then be extended to apply for physically larger samples. The development or discovery of a cryopreservative suitable for all cell types within a given organ is a necessary prerequisite for successful organ preservation.

Cryosurgery has proved to be a very successful technique for the treatment of certain skin cancers (well-circumscribed rodent ulcers and well-defined squamous cell epitheliomas on sun-exposed skin). When equipment of more sophisticated design and control becomes available, more clinicians will be encouraged to adopt the technique as an alternative to conventional surgery, and for pain relief. The development of multiple-tipped probes will make it possible to destroy massive tumors that cannot be managed with the cryosurgical equipment that is presently available. With respect to the ultrastructural cryofixation of small samples of biological material, it is probable that the best possible results have already been achieved. The main future advance in ultrastructural cryofixation will be the development of techniques for preserving the structure of larger speci-

mens at the same freezing rates as are currently obtainable in the case of small samples.

A better understanding of the physical and chemical changes which take place in the structure of molecules during freezing and thawing is essential for future progress in all aspects of cryobiology.

See also: Electron Microscopy: Freeze-Fracture Replication

Bibliography

Arnott J 1851 *On the Treatment of Cancer, by the Regulated Application of an Anaesthetic Temperature.* Churchill, London

Ashwood-Smith M J, Farrant J 1980 *Low Temperature Preservation in Medicine and Biology.* Pitman, Tunbridge Wells

Bald W B 1981 New technique for determining thermal history and concentration gradients in biological specimens during cryopreservation, cryosurgery and ultrastructural studies. *Cryo-Lett.* 2: 201–6

Bald W B 1982 Optimizing the cooling block for the quick freeze method. *J. Microsc. (Oxford)* 131: 11–23

Bald W B, Crowley A B 1979 On defining the thermal history of cells during the freezing of biological materials. *J. Microsc. (Oxford)* 117: 395–409

Bald W B, Fraser J 1982 Cryogenic surgery. *Rep. Prog. Phys.* 45: 1381–434

Bald W B, Robards A W 1978 A device for the rapid freezing of biological specimens under precisely controlled and reproducible conditions. *J. Microsc. (Oxford)* 112: 3–18

Comini G, Del Giudice S 1976 Thermal aspects of cryosurgery. *J. Heat Transfer* 98: 543–49

Cooper I S, Lee A St J, 1961 Cryostatic congelation: A system for producing a limited controlled region of cooling or freezing of biological tissues. *J. Nerv. Ment. Dis.* 133: 259–63

Gerhardt T C 1885 Lupus-Behandlung durch Kälte. *Dtsch. Med. Wochenschr.* 11: 699

Gill W, Fraser J, Carter D C 1968 Repeated freeze–thaw cycles in cryosurgery. *Nature (London)* 219: 410–13

Hebra F 1866 *On Diseases of the Skin including the Exanthemata*, Vol. 1. The New Sydenham Society, London, pp. 355–56

Hobbs P V 1974 *Ice Physics.* Oxford University Press, Oxford

Leibo S P, Farrant J, Mazur P, Hanna M G, Smith L 1970 Effects of freezing on marrow stem cell suspensions: Interactions of cooling and warming rates in the presence of PVP, sucrose, or glycerol. *Cryobiology* 6: 315–32

Lloyd J W, Barnard J D W, Glynn C J 1976 Cryoanalgesia: A new approach to pain relief. *Lancet* ii: 932–34

Mazur P, Schmidt J 1968 Interactions of cooling velocity, temperature and warming velocity on the survival of frozen and thawed yeast. *Cryobiology* 5: 1–17

Meryman H T 1966 *Cryobiology.* Academic Press, London

Miller R H, Mazur P 1976 Survival of frozen–thawed human red cells as a function of cooling and warming velocities. *Cryobiology* 13: 404–14

Morris G J, Clark A 1981 *Effects of Low Temperatures on Biological Membranes.* Academic Press, London

Openchowski P 1883 Sur l'action localisée du froid, appliqué a la surface de la région cortical du cerveau. *C.R. Séances Soc. Biol.* 4: 38

Polge C, Smith A U, Parkes A S 1949 Revival of spermatozoa after vitrification and dehydration at low temperatures. *Nature (London)* 164: 666

Rinfret A P 1963 Cryobiology. In: Vance R W (ed.) 1963 *Cryogenic Technology.* Wiley, New York, Chap. 16, pp. 528–77

Smith A U 1970 *Current Trends in Cryobiology.* Plenum, London

Smith J J, Fraser J 1974 An estimation of tissue damage and thermal history in the cryolesion. *Cryobiology* 11: 139–47

Van Venrooij G E P M, Aertsen A M H J, Hax W M A, Ververgaert P H J T, Verhoeven J J, Van der Vorst H A 1975 Freeze etching: Freezing velocity and crystal size at different locations in samples. *Cryobiology* 12: 46–61

Von Leden H, Cahan W G 1971 *Cryogenics in Surgery.* Lewis, London

W. B. Bald

Cybernetics, Biological

Technology has always played a crucial role in attempts to understand the human mind and body: the study of the steam-engine contributed concepts for the study of metabolism, and electricity has long been part of the study of the brain. In 1748, La Mettrie published *L'Homme Machine* and suggested that such automata as the mechanical duck and flute player of de Vaucanson indicated the possibility of one day building a mechanical man that could talk. In the following century machines were built which could automatically adapt to changing circumstances—perhaps the best known example being Watt's governor for his steam engine. This led to Maxwell's 1868 paper *On Governors* which laid the basis for the theory of negative feedback, as well as the study of system stability. At the same time, Bernard observed that physiological processes often form circular chains of cause and effect which could serve to counteract disturbances in such variables as body temperature, blood pressure, and glucose level in the blood (homeostasis).

1943 was the key year for bringing together the notions of control mechanism and intelligent automata. Craik's *The Nature of Explanation* viewed the nervous system "as a calculating machine capable of modelling or paralleling external events." In the same year, Rosenblueth, Wiener and Bigelow published *Behavior, Purpose and Teleology.* Engineers had noted that if feedback used in controlling the rudder of a ship is too brusque, the rudder overshoots and compensatory feedback yields a larger overshoot in the opposite direction, and so again and again as the system goes into wild oscillations. Wiener and Bigelow had asked Rosenblueth if there were any corresponding pathological condition in humans, and Rosenblueth offered the example of intention tremor associated with injury to the cerebellum. This led the three scientists to suggest that feedback systems could provide the basis for the analysis of the brain as a purposive system explicable only in terms of circular processes, from nervous system to muscles to the external world returning thence by receptors.

1943 also saw the publication of *A Logical Calculus of the Ideas Immanent in Nervous Activity* by McCulloch and Pitts in which they offered their formal model of the neuron as a threshold logic unit, building on the neuron doctrine of Ramon y Cajal, and the excitatory and inhibitory synapses of Sherrington. A major stimulus for them was the work of Turing who in 1936 had made plausible the claim that any effectively definable computation could be carried out by a machine equipped with a suitable program. McCulloch and Pitts demonstrated that each such program could be implemented using a finite network, with loops, of their formal neurons. Thus, as electronic computers came to be built toward the end of World War II, it was understood that whatever they could do could be done by a network of neurons. A number of McCulloch's contributions are collected in McCulloch (1965).

Four subdisciplines of cybernetics have crystallized from the earlier concern with the integrated study of mind, brain and machine:

(a) *Biological control theory.* The techniques of control theory, especially the use of linear approximations, feedback, and stability analysis, came to be widely applied to the analysis of diverse physiological systems such as the stretch reflex, thermoregulation, and the control of the pupil (Milsum 1966).

(b) *Neural modelling.* The Hodgkin–Huxley analysis of the action potential, Rall's models of dendritic function, analysis of lateral inhibition in the retina, and the analysis of rhythm-generating networks are examples of successful mathematical studies of single neurons, or of small or highly regular networks of neurons, which have developed in conjunction with microelectrode studies (Gel'fand 1971, MacGregor and Lewis 1977, Szentágothai and Arbib 1975).

(c) *Artificial intelligence.* This is a branch of computer science devoted to enabling computers to exhibit aspects of intelligent behavior, such as playing checkers, solving logical puzzles or understanding (suitably restricted portions of) a natural language such as English (Nilsson 1980). While many practitioners of artificial intelligence look solely for contributions to technology, there are many who see their field as intimately related with cognitive psychology.

(d) *Cognitive psychology.* The concepts of cybernetics also gave rise to a new form of cognitive psychology which sought to explain human perception and problem-solving not in neural terms but rather in some intermediate level of information-processing constructs (Lindsay and Norman 1977).

In conclusion, biological cybernetics is not limited to the study of any particular concept such as feedback, or the study of "internal models of the world." For example, one important new area is the study of "cooperative computation," based on the study of cooperative phenomena in physics, which sheds light on embryology, neural interactions, and also ecological systems (Arbib 1986, Haken 1978, Levin 1978).

See also: Biofeedback; Biological Control Theory

Bibliography

Arbib M A 1986 A view of brain theory. In: Yates F E (ed.) 1986 *Selforganizing Systems: The Emergence of Order*. Plenum, New York
Gel'fand I M (ed.) 1971 *Models of the Structural–Functional Organization of Certain Biological Systems*. MIT Press, Cambridge, Massachusetts
Haken H 1978 *Synergetics, an Introduction: Nonequilibrium Phase Transitions and Self-Organization in Physics, Chemistry and Biology*, 2nd edn. Springer, Berlin
Levin S A (ed.) 1978 *Studies in Mathematical Biology, Pt. 1: Cellular Behavior and the Development of Pattern; Pt. 2: Populations and Communities*. Mathematical Association of America, Washington, DC
Lindsay P H, Norman D A 1977 *Human Information Processing: An Introduction to Psychology*, 2nd edn. Academic Press, New York
McCulloch W S 1965 *Embodiments of Mind*. MIT Press, Cambridge, Massachusetts
MacGregor R J, Lewis E R 1977 *Neural Modelling: Electrical Signal Processing in the Nervous System*. Plenum, New York
Milsum J H 1966 *Biological Control Systems Analysis*. McGraw-Hill, New York
Nilsson N J 1980 *Principles of Artificial Intelligence*. Tioga, Palo Alto, California
Szentágothai J, Arbib M A 1975 *Conceptual Models of Neural Organization*. MIT Press, Cambridge, Massachusetts
Wiener N 1961 *Cybernetics: Or Control and Communication in the Animal and the Machine*, 2nd edn. MIT Press, Cambridge, Massachusetts

M. A. Arbib

Cyclotrons

Since its invention by E O Lawrence in 1931, the cyclotron has been the nuclear particle accelerator of choice for the production of radioisotopes by charged-particle-induced nuclear reactions. Nuclear reactors, which produce neutron-rich radioisotopes by neutron bombardment, supplanted cyclotrons in the 1950s and 1960s as the major producers of radioisotopes, but the latter are still the dominant producers of neutron-deficient radioisotopes. As a result of significant improvements in cyclotron technology, the number of cyclotrons in dedicated use in biomedical applications has increased rapidly, from one machine in 1964, 25 in 1975, to some 50 in 1985. Another 30–40 cyclotrons are

currently used part-time in biomedical research programs. This trend to increased utilization of cyclotrons has been reinforced by the recent development of positron emission tomography (see *Positron Emission Tomography*), a procedure which enables complex physiological and biochemical processes to be studied *in vivo* (Wolf 1981). The neutron-deficient (and hence positron emitting) radioisotopes ^{11}C, ^{13}N, ^{15}O and ^{18}F have been especially useful in these studies of function and metabolism. As the longest lived of these isotopes, ^{18}F, has a half-life of only 110 min, hospitals and research institutions that undertake positron emission tomography programs need a cyclotron nearby.

1. The Cyclotron Principle

If q and m are, respectively, the charge and mass of a particle moving with velocity v in an arc of radius r, in a plane normal to a uniform magnetic field B, the equation of motion of the particle, by Newton's second law, is

$$qvB = mv^2/r \qquad (1)$$

An expression for the rotational frequency f of the particle follows directly from Eqn. (1):

$$f = qB/2\pi m \qquad (2)$$

and the kinetic energy E of the particle is

$$E = B^2r^2q^2/2m \qquad (3)$$

It is clear from Eqn. (2) that the frequency of rotation of the particle is independent of its speed. This fact underlies the principle of the cyclotron: the same electrodes can be used over and over again to give a series of impulses that are regularly spaced in time and which thus accelerate particles to energies many times that represented by the maximum potential at the electrodes.

2. Nonrelativistic Cyclotrons

In its simplest configuration, a cyclotron (see Fig. 1) consists of an electromagnet; a set of electrodes coupled to a radiofrequency power source; an ion source to

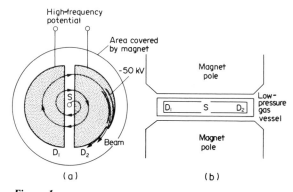

Figure 1
(a) Plan and (b) vertical section through the Lawrence cyclotron showing the position of the dees

introduce ions such as protons, deuterons, ^{3}He or ^{4}He; and a vacuum chamber that encloses the magnet pole faces, the electrodes and the region within which the ions are accelerated. The most elementary electrodes are formed of a shallow, flat "pill-box," cut along a diameter into two D-shaped sections (dees), and placed midway between the magnet poles, in a plane perpendicular to the field *B*.

The cut in the pill-box provides a gap across which the radiofrequency alternating voltage connected to the dees acts upon the ions as they rotate in the field. The ions are introduced into the gap between and near the center of the dees. Some of the ions, which are continuously produced by the source, enter the gap at a time in the alternating voltage cycle such that they receive an impulse which both accelerates and directs them toward the median plane both of the dees and the magnetic field. The ions then follow a circular path, around the center of the dees, crossing the gap twice per revolution. The path of the ions is a series of semicircles of increasing radii as energy (and thus speed) is added to the particles by the alternating field at each crossing of the gap. Ultimately, only those ions which cross the gap just after the voltage has peaked continue the spiral path of connected semicircles outward along the median plane (Richardson 1965). These ions remain in phase with the alternating voltage, their energies being increased twice per revolution.

The focusing effect of the electric field is proportional to the time the ions spend within the gap. Thus, as the speed of the particles increases, the focusing becomes less effective. With the classical (nonrelativistic) cyclotron, the focusing required to hold the beam on the median plane after the first few turns is provided by the magnetic field itself. Although the magnetic field lines are parallel at the center, the field does decrease slightly at larger radii and thus the field lines "bulge out," that is, are concave outward and symmetric around the center. The radially inward directed force on the ions moving through this field thereby develops an axial component which is directed toward the median plane and which has zero value on that plane. The loss of synchronism with the alternating voltage introduced by this decrease of *B* in Eqn. (2) is accommodated by allowing the ion angular velocity to lag or lead the cyclotron frequency *f* at various parts of the acceleration cycle and by maximizing the dee voltage to minimize the number of turns required to reach the desired energy (Richardson 1965).

A greater violation of Eqn. (2) occurs as the relativistic mass *m* of the accelerated ions increases. Sector focusing provides the solution for this problem, and is a feature of present generation, compact, high-current cyclotrons, which have been found so useful in biomedical work.

3. Sector Focusing or Azimuthally Varying Field (AVF) Cyclotrons

The fundamental conflict between the need for a decreasing field for axial focusing and an increasing one to counterbalance the increasing relativistic mass of the accelerated ions was elegantly resolved with the development of sector focusing (Richardson 1965). The magnetic field is shaped by configuring the pole faces into a set of ridges and valleys extending radially or spiralling outward from the central region. The ions as they spiral outward pass through a field which alternates in magnitude—sectors of high magnetic field over the ridges and low magnetic field over the valleys. Thus, the field varies azimuthally. Axial focusing is the net effect of such a field, the restoring force toward the median plane increasing as the "flutter" or difference between hill and valley fields increases. An AVF cyclotron is designed so that the average field traversed by the ions in their rotation can be increased as the ion orbits expand, the flutter being maintained relatively constant by the choice of the relative widths of hills and valleys, and by small additions to the height of the hills or the valleys. The magnetic field of an AVF cyclotron can be profiled in this way to match precisely the relativistic requirement of Eqn. (2), that is, made "isochronous" over the full range of particle energies of interest in biomedical applications. Although AVF cyclotrons have been built that produce 90 MeV protons the most useful maximum energy for a general-purpose biomedical research and production cyclotron lies within the 40–45 MeV range (Wolf and Jones 1983).

The characteristics of sector focusing have made several important improvements in cyclotrons possible, in addition to the higher energy capability. The axial focusing strength of the flutter field is much greater and consequently the amplitude of axial oscillation much less than in the classical case. In addition, the isochronous operating mode eases the requirement for maximal dee voltages and the consequent limitations on spacing.

At present, there are several companies worldwide that offer AVF cyclotrons in the proton energy range 8–45 MeV. These machines are relatively compact as AVF operation allows a reduction in magnetic pole size and spacing; also the demand on power (both for the magnet and for the radiofrequency field) is less.

The productivity of cyclotrons—with respect to radioisotope production—is now essentially target limited, as the strong axial focusing of AVF operation provides particle beams of ample magnitude.

4. External Beams from Cyclotrons

Only in a few special cases can the accelerated beam, as it reaches the maximum energy orbit, be used within the cyclotron. For general purpose applications, the beam must be extracted and directed through a beam line system to targets positioned at some distance from the cyclotron.

The efficient extraction of the circulating beam of positive ion cyclotron appears in practical terms to be dependent on the use of an electrostatic deflector (Richardson 1965). The deflector consists of a pair of closely spaced, parallel electrodes, across which a dc potential is applied to deflect the beam out of its normal orbit and direct it outward through the fringing field of the magnet. The inner electrode, called the septum, is made usually of thin tungsten strip so as to intercept as little of the beam as possible. It is held at ground potential. The outer electrode, called the deflector, is cooled to provide some heat dissipation for the two electrodes. Both are contoured to the desired beam trajectory and extend for some distance along the trajectory. As the circulating beam expands with increasing energy, a fraction of it, usually about 50%, misses the septum on one turn and enters the space between the septum and deflector on the next. The $v \times \mathscr{E}$ force on the moving ions in the electric field $\mathscr{E}$ has the effect of reducing the magnetic field within the channel. Thus, the ion beam curvature is reduced as it is introduced into the extraction trajectory. Magnetic channels are also provided in some machines to aid the focusing of the outgoing beam which is radially dispersed as it travels through the fringing field of the magnet.

There is one important alternative to deflector extraction (Richardson 1965) for the acceleration and extraction of high currents of protons and deuterons—the acceleration of negative ions of the hydrogen atoms present in the positive ion source. The H^- ions (essentially one proton plus two electrons) can be injected into the cyclotron and accelerated in the same fashion as positive ions, since the additional mass of the two electrons produces only a small change (0.1%) in the isochrony condition. However, the binding energy of the second electron is small (0.755 eV) and thus the acceleration lifetime is reduced by scattering. An AVF cyclotron's acceleration lifetime is also affected by the electric field induced ($v \times B$) in the vicinity of the ions as they move through the magnetic field. Apparently these problems are not limiting up to an energy of 42 MeV because several negative ion cyclotrons that deliver 200 μA of protons and 50 μA of deuterons at this energy have been manufactured and installed.

In comparison with electrostatic deflection, beam extraction from a negative ion cyclotron is elegantly simple. A thin foil (approximately 30 μg cm^{-2}) is placed in the beam path at an orbital position corresponding to the desired energy and azimuthally situated so that the particles exit from the vacuum tank at the desired point. The foil strips off both electrons with high efficiency. The now positive ions, with reverse curvature, emerge rapidly through the fringing field. With care they can be brought out with little loss of the beam quality as occurs with electrostatic deflector extraction. A useful additional feature of negative ion extraction is the possibility of providing two extracted beams simultaneously. This could be very useful in production operations where two radioisotopes are needed concurrently or where the full current capability of the cyclotron exceeds that of a single target.

Although the deflector-extracted external beam from positive ion cyclotrons can be used directly, methods used to focus the beam internally are not sufficiently satisfactory for the beam to be utilized optimally. At the exit port the beam usually has an elliptical shape with a major axis diameter of 1–2 cm and a divergence angle of about 20 mrad. Use of a beam transport system employing a magnetic quadrupole or doublet or triplet system allows the beam to be focused at some distance from the cyclotron, to a spot size variable in diameter from 5 mm up to the width of the beam transport tube (i.e., 5–10 cm); transmission efficiency is 90% or better (Wolf and Jones 1983). Generally the quadrupole is followed by a switching magnet which directs the beam to one of several transport tubes (as many as seven are used at some installations) leading to target assemblies that are dedicated to specific applications.

5. Cyclotrons for Biomedical Application

Although a full range of analytical techniques is available to cyclotron users (Chaudhri 1975)—charged-particle-induced and neutron-induced activation, and charged-particle-induced fluorescence—cyclotrons are used mainly for the production of radioisotopes. The other major application is the production of neutrons for neutron therapy, to which purpose about half of the biomedically dedicated cyclotrons are devoted at least part-time (see *Neutron Sources*).

A useful classification of cyclotrons, based primarily on their capability for radioisotope production, has been introduced by Wolf and Jones (1983). They categorize the machines into four "levels," differentiated by the maximum energy of protons produced. The maximum energies of deuterons and the helium ions can be estimated from Eqn. (3). Actually, the heavier ions can have somewhat higher energies than predicted by Eqn. (3) because the precise isochronous conditions vary for the different ion types. The field for each ion type is tuned by making significant changes in the average fields and/or by placing circular trim coils at strategic radii.

The Wolf categories are adopted in the following considerations of commercially available cyclotrons for biomedical applications.

5.1 Level I ($E_{p,d} \leqslant 10\,MeV$)

These are really single-particle machines, deuterons being included because of the historical importance of several 8 MeV deuteron machines that have been used in biomedical research. The burgeoning interest over the past decade in positron tomography has stimulated intensive interest in the techniques for producing the radioisotopes ^{11}C, ^{13}N, ^{15}O and ^{18}F (CNOF). It is now clear that these nuclides can all be produced by proton-only cyclotrons with isotopically enriched targets (Wolf and Jones 1983). The producing reactions are ^{14}N(p, α)^{11}C or ^{11}B(p, n)^{11}C; ^{13}C(p, n)^{13}N; ^{15}N(p, n)^{15}O; ^{18}O(p, n)^{18}F. Only one cyclotron of this class is available at present—from Orbit Incorporated, Oak Ridge, USA. The fixed proton energy, nominally 8 MeV, and output current, 50 µA, have been chosen to provide adequate production of CNOF to support a clinical research program in positron tomography. The cyclotron has been designed to minimize the shielded volume required for the machine and the CNOF target systems (Highfill 1983). The plane of acceleration is vertical and the extracted beam is directed downward so that the target assemblies can be set into the ground at the side of the cyclotron. Target changing is accomplished by remotely controlled translation of the cyclotron. Because of the close spacing of accelerator and targets, and the recessing of the targets into the ground, the minimum area required for the cyclotron/target vault is 100 ft² (9.3 m²). An additional 150 ft² (14 m²) of unshielded space is adequate for power supplies and control console. The accelerator and target systems are fully automated to limit the number of support staff needed for routine radioisotope production to one professional (nuclear chemist or radiopharmacist); an electronics technician with experience of computers is responsible for routine maintenance.

5.2 Level II ($E_p \leqslant 20\,MeV$)

These are generally multiple-particle machines, that is, they produce beams of protons and deuterons; some provide ³He and ⁴He also. A multiplicity of cyclotrons of this class is commercially available (Wolf and Jones 1983). In general, the proton and deuteron output currents are specified at 50 µA. All the major cyclotron manufacturers have at least one model in the Level II energy range. The suppliers are Scanditronix, Uppsala, Sweden; Japan Steel Works, Muroran, Japan; and Sumitomo-CGR, Tokyo, Japan.

Although the primary application of Level II cyclotrons is CNOF production, all of these machines produce proton beams in the upper part of the energy range (16 or 17 MeV). The additional reactions available to these cyclotrons are $^{10}B(d, n)^{11}C$; $^{16}O(p, \alpha)^{13}N$; $^{12}C(d, n)^{13}N$; $^{14}N(d, n)^{15}O$ and $^{20}Ne(d, \alpha)^{18}F$. These machines are all manifestly capable of supporting a full-scale, clinical positron tomography program (Wolf 1981).

The minimum shielded vault area (including the shielded wall area) is 300–400 ft² (28–37 m²) for all these machines (Wolf 1981) because (a) the external beam transport system and target assemblies need significant floor area, and (b) shielding must be provided for the neutrons produced in the reactions. An additional 200–300 ft² (19–28 m²) of unshielded space is ordinarily specified for the power supplies and control console.

The level of automation of cyclotron and targetry operation of these machines varies from machine to machine, and with time, as the manufacturers improve their products. A machine which has not been fully automated would seem to require for its support a full-time cyclotron operator/engineer (who would also be qualified in electronics maintenance) as well as a nuclear chemist/pharmacist.

5.3 Level III ($E_p \leqslant 45\,MeV$)

With the exception of the 42 MeV negative ion machines (which are no longer available), the cyclotrons in the Level III class are multiple ion machines providing all four particles (protons, deuterons, ³He and ⁴He). Several models are available from each of the suppliers listed under Level II and also from CGR MeV, Courbeville, France.

The potential applications cover the full range of biomedical research and production possibilities. Copious quantities of CNOF can be produced as well as other important radionuclides such as ^{67}Ga, ^{111}In, ^{201}Tl and ^{123}I. This class of machine is also being used in an extensive program of cancer therapy evaluation in the USA.

Space and shielding requirements are major considerations, particularly for cyclotrons at the high end of the energy range. A typical installation with an isocentric focusing system for cancer therapy requires 2400 ft² (220 m²) of heavily shielded space, much of it taken up with concrete walls 6 ft (2 m) thick.

The lower energy machines in this class have more modest space and shielding requirements when they are not being used for cancer therapy. Still, the siting of such a machine is a major consideration; significant additions to an existing building are usually required.

The staff support requirements are also significantly greater than for Level II machines, a minimum of several professionals and technicians being required.

5.4 Level IV ($E_p \leqslant 200\,MeV$)

Machines have been produced in the proton energy range 50–60 MeV. The applications considerations are similar to those machines at the upper end of Level III.

See also: Radionuclide Generators

Bibliography

Chandhri M A 1975 The potential and applications of cyclotrons in biomedical fields. In: Joho W (ed.) 1975 *Proc. 7th Int. Conf. Cyclotrons and their Applications.* Birkhäuser, Basel, pp. 447–56

Highfill R R 1983 Design and status of a specialized cyclotron for positron tomography. *IEEE Trans. Nucl. Sci.* 30: 1765–67

Richardson J R 1965 Sector focusing cyclotrons. In: Farley F J M (ed.) 1965 *Progress in Nuclear Techniques and Instrumentation*, Vol. 1. North Holland, Amsterdam, Chap. 1, pp. 1–101

Wolf A P 1977 *Medical Cyclotrons*, Vol. I: *Medical Radionuclide Imaging*. International Atomic Energy Agency, Vienna, pp. 343–54

Wolf A P 1981 Special characteristics and potential for radiopharmaceuticals for positron emission tomography. *Semin. Nucl. Med.* 11: 2

Wolf A P, Jones W B 1983 Cyclotrons for biomedical radioisotope production. *Radiochim. Acta* 34: 1–8

J. Neiler

D

Decision Theory

Four important reasons why physicians today should be more systematic than formerly in their approach to patients have been identified by Fineberg (1981). These are the explosive growth of medical knowledge and information, the increasing technological complexity of medicine, the tightening constraints on available economic resources, and the increasing wish of patients to be better informed about their illnesses. Indeed, much can be done to aid the caring physician in arriving at a decision which will best meet the needs of an individual patient, although no formalism can ever replace a physician's sensitivity and concern for a patient.

1. How Doctors Make Decisions

Much research, known as process tracing, has been done over the last few years to determine how doctors make decisions. It transpires that there simply is no one monolithic pathway to decision making. The same doctor will use quite different methods in differing circumstances (Leaper et al. 1973) and this has given rise to the criticism in some quarters that doctors are rather empirical in what they do. Criticisms about the mystique of medicine usually cite this lack of coherence by doctors as evidence of fallibility on their part. This is not strictly fair. Doctors are adaptive rather than empirical: they adapt interviews to the circumstances. It is only when they come to analyze patients' features in difficult cases that their approach tends to become empirical and thus amenable possibly to assistance from formal theory.

2. Formalizing Medical Decisions

Consider the following case: a 14 year old boy is admitted to hospital with suspected appendicitis. The doctor interviews him and examines him thoroughly, but nevertheless remains quite uncertain as to whether the boy actually does have an inflamed appendix. A decision must be made whether or not to operate.

In decision-theory jargon this is a 2 × 2 decision. The doctor really has only two choices (Table 1), each of which can be right or wrong. However, in formalizing the decision, the consequences of both correct and incorrect choices must be taken into account. This is also illustrated in Table 1. If the doctor opts for surgery and the patient does not have appendicitis, an unnecessary operation will have been performed. If an operation is not carried out and the appendix is inflamed, then it may rupture and a disastrous peritonitis ensue.

There are thus two possible choices and two possible outcomes for each choice (hence the term 2 × 2 table). This does not *per se* give formal guidance on a solution to the problem. To offer formal guidance, a new concept has to be introduced—one which has given rise to considerable discussion and debate—the concept of "utility."

Table 1

2 × 2 Table showing possible decisions and outcomes for the case of a patient with suspected appendicitis

	Decision	
	Operate	Do not operate
Appendicitis	correct	possible perforation
No appendicitis	unnecessary operation	correct

3. The Concept of Utility

Whatever the doctor does, and whatever the outcome, that outcome will have a certain utility for the patient. This is usually expressed on a scale extending from zero to one, one being the best possible outcome, e.g., painless recovery to full health, and zero the worst outcome, usually the patient's death.

For the case under consideration, outcomes can be assigned values (Fig. 1). Clearly the best outcome would be No. 4, i.e., the patient does not have appendicitis, the doctor does not operate, and the patient goes home next day quite recovered. An arbitrary value of 1.0 is allocated to this outcome. Possibly the next best outcome is No. 1. The patient does have appendicitis, but the doctor

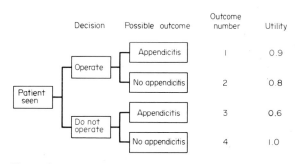

Figure 1

Decisions, outcomes and utilities for the case of a patient with possible appendicitis spots it, operates, and cures him. The patient is cured, but nevertheless has the discomfort of an operation. Hence a value of 0.9 is allocated to this outcome, less than 1 but not much less.

Clearly two other outcomes, No. 2 and No. 3, are also possible. In the former, not only has the patient undergone the discomfort of an operation, but the operation itself was an unnecessary operation. So, if a value of 1 is assigned to the best outcome (No. 4) and 0.9 to the next best outcome (No. 1) then the value of outcome No. 2, an unnecessary operation, would be lower (say 0.8).

Outcome No. 3 is the only other possibility. Its value must be even lower; for if the patient does have appendicitis, and the doctor does not operate, then the patient's chances of dying (due to a ruptured appendix) are known to increase significantly. Even allowing for the miracles of modern antibiotics, a much lower value must be assigned to this outcome, say a value of 0.6, as in Fig. 1.

4. Some Simple Mathematical Concepts

Returning to Fig. 1, only one further problem needs to be considered to be able to make a formal decision analysis of the situation. It has been shown that four outcomes are possible (each with its own consequence and value to the patient). The problem is that the doctor does not know whether the patient has appendicitis or not, but can only assign a probability P to this and by implication therefore a probability $(1 - P)$ that the patient does not have appendicitis.

The doctor has two alternatives (to operate or not) and should choose whichever has the highest utility to the patient. The utility of each choice amounts to the sum of the utilities of the two outcomes (multiplied in each case by the probability of that outcome occurring). Thus the utility of operating U_{op} can be written:

$$U_{op} = P_a U_1 + (1 - P_a)U_2 \qquad (1)$$

where P_a is the probability of the patient having appendicitis, and U_1 is the utility of outcome 1, etc. Similarly, the utility of not operating U_{nop} can be written:

$$U_{nop} = P_a U_3 + (1 - P_a)U_4 \qquad (2)$$

These general equations can be used to look at the choice the doctor faces. Suppose first that the doctor is sure the patient has appendicitis. Then P_a is equal to 1, and thus $(1 - P_a)$ is zero. If these values are substituted into the equations, the overall value of operating is $(1 \times 0.9) + (0 \times 0.8) = 0.9$, whereas the value of not operating is $(1 \times 0.6) + (0 \times 1) = 0.6$. Clearly the doctor should operate here.

These extreme conditions are not relevant here: at the outset it was noted that the doctor was "quite unsure" as to whether the patient had appendicitis. In other words both P_a and $(1 - P_a)$ were 0.5. What should the doctor do in this instance? Substitution into the equations indicates that in these circumstances the doctor

should operate. The value of doing so amounts to $(0.5 \times 0.9) + (0.5 \times 0.8) = 0.85$. Likewise, by similar substitution, the value of not operating amounts to $(0.5 \times 0.6) + (0.5 \times 1) = 0.80$. Therefore, formal analysis of the problem indicates that if absolutely unsure the doctor should operate because the overall utility to the patient (bearing in mind that the doctor may be wrong) is higher.

In fact, this can be illustrated rather well graphically (Fig. 2). This figure traces (for each level of probability)

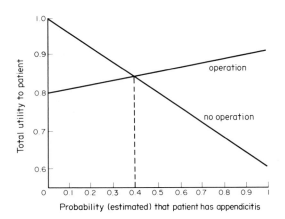

Figure 2
Relative utilities of operative and nonoperative treatment for estimated probabilities of appendicitis, constructed using Eqns. (1) and (2). Note that the crossover point is at a probability of 0.4

the relative values of operative and nonoperative treatment. On the basis of this analysis, formal decision theory clearly indicates that if the doctor is more than about 40% sure the patient has appendicitis, then an operation should be carried out. Such a diagram is quite crucial to an understanding of decision theory·in medicine. For although here it has been applied to a relatively simple problem, it can be applied (at least in theory) to any medical problem, however complex.

5. Some Practical Problems

When these concepts of formal decision making are put into practice, problems emerge which have presented an insuperable obstacle to the use of formal decision theory in medicine for some physicians.

Many doctors argue that although the analysis presented above is valid in principle it fails to take into account other factors about the patient which the doctor automatically assesses. Referring back to the example in Fig. 2, the analysis offered by formal decision theory may be invalidated by a number of circumstances. For example, if the patient in question suffers from hemophilia (and is therefore likely to be at increased risk from bleeding), then the risks of operation, and hence the U

values in the equation, would be greatly changed. Thus many doctors argue that formal decision analysis is not "holistic," even though they concede that for many patients the analysis is valuable.

There are further problems. Perhaps the most serious concerns the allocation of numbers to the values attached to P and U. As regards the former, the doctor's own estimates of the probability were used. Studies have shown, however, that doctors are highly fallible in this respect (Leaper et al. 1972). The profound effect this might have is obvious.

Suppose, for example, a hundred cases are followed up and it is found that when a particular doctor assesses the probability of appendicitis as 25%, actually some 50% of the patients have appendicitis. Should 25% or 50% be specified for the value of P? Clearly this would make all the difference between deciding to operate or not to operate.

Unfortunately, this inability to specify accurate "guesstimates" of probability has been demonstrated, and has led to suggestions that computer estimates of probability based on large series of real-life patients should be used for this purpose (de Dombal 1979). Whatever the outcome of this debate, the inability of doctors to specify risks in numerical terms remains a major obstacle to the real-life introduction of formal analysis.

The most formidable obstacle remains the assigning of values to the utilities which are attached to the outcome of each mode of treatment. In the example given, quite a rational value can be allocated where operation is concerned and also where the diagnosis is missed and the appendix ruptures. For example, it is known from experience that rupture of the appendix increases the mortality risk by a factor of four or five; this can be allowed for in our equation. But what of the value of an "unnecessary" operation? To a healthy adult male this might not matter very much. The risk is not great and an appendix scar is no great handicap. Nonetheless others may feel differently. A young girl concerned about her appearance, or an old person in whom the risks of surgery are much greater, might well have totally different opinions on how "unnecessary" the operation is. Indeed a number of studies have shown that even in the same circumstances the patient and the doctor may sometimes have totally different viewpoints and would attach totally different values to U_2 in Eqn. (1) (de Dombal and Hall 1979). (The interested reader may care to explore this point further by altering the values of the various utilities in the equation and the figure. In doing so the value of formal decision analysis will truly become apparent, for it emerges that such decision theory becomes an extremely useful tool with which to explore medical decision making.)

To illustrate this point further, it becomes obvious when the values of P and U are altered in Eqn. (1) (and Fig. 2) that some aspects never change whatever their values. For example, if it is certain the patient has appendicitis it is always better to operate.

By contrast, however, it also becomes clear that there is a region (between 20% and 80%) where a small alteration of P and U values can sway the decision. This region has been dubbed the close call zone (Lau et al. 1982) and is currently the subject of much study and speculation.

Nevertheless, the point should not be lost in the present instance that formal decision theory has, with all its limitations, delineated three zones: (a) a zone in which it is always best to operate, (b) a zone in which it is never wise to operate, and (c) a close call zone in which the decision to operate depends upon such factors as the doctor's estimate of probability, the reliability of the estimate and the patient's U values (or, put another way, the doctor's estimate of the patient's wishes). In this way, decision theory attempts to mirror the real-life situation and, as in the present simple instance, it can be argued that it does so in a manner both useful and valid.

6. Current Research

As might be anticipated the prospects which this type of analysis have opened up have provoked a good deal of comment, controversy and research over the last few years (Ziporyn 1983). The most encouraging occurrence is the emergence of groups of workers (such as the Society for Medical Decision Making in the USA and the Royal College of Physicians Workshop in the UK) who are beginning to study the problem on a multidisciplinary basis.

Particular research effort and speculation have been devoted to two topics. First, attempts have been made in a variety of medical spheres to define circumstances which lie outside the close call zones, i.e., where the decision is clear-cut whatever (within reason) the allotted P and U values. Here clear guidance can be and has been offered to the practising doctor in a number of areas. The Harvard group, for example, have shown most elegantly and clearly via decision analysis that the treatment of certain types of cancer should be modified (McNeil et al. 1978). Secondly, much effort has been devoted to the study of utility and the problem of obtaining from the patient some idea of the value he or she would attach to a particular outcome of treatment. A variety of methods have been employed here, ranging from complicated "gaming" methods (asking a patient what betting odds they would accept in order to obtain relief of symptoms) to the more simple expedient of asking the patient what he or she is prepared (or can afford) to pay.

Such a diversity of studies merely reflects an inability to formalize and put into mathematical terms the patient's feelings and wishes, and this remains a major obstacle to the introduction of formal decision theory in medicine.

Nonetheless, in delineating areas where the decision is clear-cut, in delineating close call zones and above all in focusing attention upon the problems of medical decision making, decision theory has already proved valuable. It seems unlikely at the present time that decision theory will achieve more than this. Specifically, it seems unlikely that in the close call zone formal decision theory will ever replace the concerned, caring doctor. But if it does no more than formalize problems, point to areas of doubt and concern for further study, offer clear guidance in other areas, and above all focus attention upon the medical decision itself, then the advent of decision theory in medicine will have been of value.

Bibliography

de Dombal F T 1979 Acute abdominal pain. An OMGE survey. *Scand. J. Gastroenterol.* 14 (Suppl. 56): 29–43
de Dombal F T, Hall R 1979 The evaluation of medical care from the clinician's point of view: What should we measure, and can we trust our own measurements? In: Alperovitch A, de Dombal F T, Gremy F (eds.) 1979 *Proc. IFIP Working Conf.* North Holland, Amsterdam
Fineberg H V 1981 Medical decision making and the future of medical practice: Editorial. *Med. Decis. Making* 1: 4–6
Lau J, Levey A S, Kassirer J P, Pauker S G 1982 Idiopathic nephrotic syndrome in a 53-year-old woman: Is a kidney biopsy necessary? *Med. Decis. Making* 2: 497–519
Leaper D J, Gill P W, Staniland J R, Horrocks J C, de Dombal F T 1973 Clinical diagnostic process: An analysis. *Br. Med. J.* 3: 569–74
Leaper D J, Horrocks J C, Staniland J R, de Dombal F T 1972 Computer-assisted diagnosis of abdominal pain using "estimates" provided by clinicians. *Br. Med. J.* 4: 350–54
McNeil B J, Weichselbaum R, Pauker S G 1978 Fallacy of the five-year survival in lung cancer. *New Engl. J. Med.* 299: 1397–401
Ziporyn T 1983 Medical decision making: Analyzing options in the face of uncertainty. *J. Am. Med. Assoc.* 249: 2133–41

F. T. de Dombal

Defibrillators

Under normal circumstances the musculature of the heart contracts in a precisely controlled manner such that the four chambers of the heart empty and fill with blood in a given sequence and at a rate which varies, but within a limited range. Occasions do arise, however, when the desired synchrony of operation is either partially or completely lost—conditions known as atrial fibrillation and ventricular fibrillation, depending on the heart chambers affected. It then becomes necessary to intervene to restore the heart to its normal pattern of operation. Often successful in effecting this restoration is the application of a high-energy electrical shock to the heart, either directly to the heart itself if the chest is open or, more usually, indirectly across the closed chest wall. The technique is known as defibrillation (or sometimes

countershock) and the devices used for the application of the carefully controlled shock are known as defibrillators. The function of a defibrillator is to provide (during the period of the shock) sufficient current to stimulate the entire musculature of the heart so that, upon cessation of the current, all muscle fibers enter their refractory periods together, with a consequent resumption of normal operation.

1. Types of Defibrillator

Early defibrillators employed alternating current and were accordingly known as ac defibrillators. An alternating current (50 Hz) of some 6 A was passed for 0.25–1 s; repeated shocks were often necessary to obtain defibrillation and attempts to effect atrial defibrillation often resulted in the induction of the more serious condition of ventricular fibrillation. As a consequence, the ac defibrillator has been superseded by the direct-current defibrillator.

Direct-current defibrillators are invariably of the capacitor discharge type in which a capacitor is charged up to a high voltage over a short period of time (generally less than 10 s) and the resultant stored energy then discharged via electrodes across the patient. Differences between types of dc defibrillator arise from the various electrical networks used to shape the discharge waveform. Simplistically, the dc defibrillator consists of a capacitor discharging across the patient (Fig. 1). The discharge energy W is given by

$$W = \tfrac{1}{2}CV_0^2$$

where V_0 is the initial voltage of the fully charged capacitor and C is the capacitor value. This simple circuit is not acceptable in practice since, to achieve a given energy discharge, a high initial voltage is required. Since the decay of capacitor voltage V is exponential [$V = V_0 \exp(-t/RC)$, where R is the patient resistance $\sim 50\,\Omega$], the high voltage only affects the patient for a few microseconds (curve A of Fig. 2). Nevertheless, the high current associated with such a high voltage is capable of damaging ventricular musculature and hence it is more common to limit the current by inserting an inductive component in the output circuit (Fig. 3).

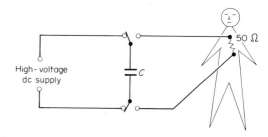

Figure 1
Basic *RC* network

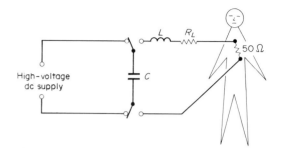

Figure 3
RLC circuit which, if underdamped, produces the Lown waveform

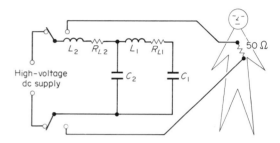

Figure 4
Delay-line or dual-peak monophasic discharge circuit

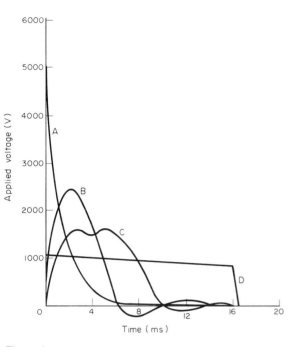

Figure 2
Four defibrillator waveforms: curve A, the waveform of a basic *RC* circuit; curve B, *RLC* (Lown) waveform; curve C, dual-peak monophasic waveform; and curve D, truncated discharge waveform

The dc defibrillator-incorporating inductance is known as an *RLC*-type (resistance–inductance–capacitance) defibrillator. Since an inductor has a dc resistance (about 40 Ω for a 0.1 H inductor) an energy loss occurs between transmission from the storage capacitor and delivery to the patient. This loss is some 30% of the stored energy. Thus the energy delivered to the patient is only 70% of that available at the capacitor. An appreciation of the difference between stored energy and delivered energy is essential when using dc defibrillators and it is particularly important to know whether the energy meter of a particular device is calibrated to indicate stored energy or delivered energy.

2. Output Waveforms

With an *RLC* circuit it is possible (by manipulating values of *R*, *L* and *C*) to produce output waveforms that are either overdamped, underdamped or critically damped. Most dc defibrillators in common use produce an underdamped waveform known as the Lown waveform (Lown et al. 1962), shown as curve B in Fig. 2. The peak voltage reached using an *RLC* circuit is somewhat less than half of that seen using the basic *RC* circuit, while the effective portion of the discharge is extended to about 5 ms compared with less than 2 ms for the *RC* circuit. A basic circuit is illustrated in Fig. 3.

There is some risk of damage to the myocardium, even at the reduced peak voltage of the Lown-type waveform. In order that the voltage may be further reduced, a variation of the *RLC* circuit has been developed to produce the delay-line or dual-peak monophasic defibrillator discharge waveform (Balagot et al. 1964), shown in Fig. 4. The dual *LC* network is designed to lower the peak voltage while increasing the duration of the pulse so that total delivered energy is the same as for the single *RLC* device. In the case of the dual *LC* network device the total energy is given by

$$W = \tfrac{1}{2}(C_1 + C_2)V_0^2$$

where C_1 and C_2 are the values of the two storage capacitors. The output waveform is shown as curve C in Fig. 2.

A further development of the dc defibrillator waveform aimed at extending the philosophy of reduced-peak voltage and extended pulse is known as the truncated discharge waveform (curve D in Fig. 2). A much larger main storage capacitor (about 1000 μF) than in the defibrillator circuits already described is used. A basic circuit for the device is given in Fig. 5. Because of the large value of the capacitor it is possible to store the same amount of energy at a much lower voltage and as a consequence there is no need for an inductor to limit the current during discharge. The time constant of the circuit is also much reduced and although the discharge curve is still exponential, it has a very flat slope. The

249

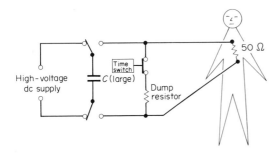

Figure 5
Truncated discharge circuit

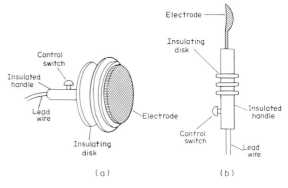

(a)   (b)

Figure 6
Electrodes used in cardiac defibrillation: (a) one of a
pair of paddle-type electrodes that is applied against
the chest wall, and (b) a spoon-shaped internal
electrode that is applied directly to the heart

amount of energy discharged is therefore almost linearly
related to time, and the energy delivered may be determined by controlling the period during which the defibrillator output is connected to the patient. Assuming
that a 1000 μF capacitor is charged to 1000 V the stored
energy W is 500 J. If the patient presents a 50 Ω load, a
discharge of 250 J will be effected in approximately
20 ms. After this period the defibrillator ouput to the
subject is terminated and any remaining charge on the
capacitor dumped across a dummy load within the
defibrillator. In terms of electrical efficiency charge
dumping puts the truncated waveform device at something of a disadvantage although this is more than offset
by the clinical advantages associated with the reduced
voltage levels involved.

3. Operation

Although the most important use of a defibrillator is
ventricular defibrillation—the restoration of a fibril-
lating heart (no effective rhythm present) to a normal
pattern of operation—it is also used to terminate heart
rhythms which are abnormal and potentially dangerous.
The procedure involving a defibrillator in this latter
instance is termed cardioversion. It is important to
ensure that the discharge of a defibrillator used to
correct an abnormal rhythm does not precipitate ventricular fibrillation. Avoidance of ventricular fibrillation is
achieved by synchronizing the discharge with the electrocardiogram (ECG) so that discharge occurs during (or
immediately after) the downward slope of the ECG
R-wave when the heart is in its absolute refractory
period. Thus the possibility of a discharge during the T
wave when the heart is vulnerable to ventricular
fibrillation is avoided. The majority of defibrillators are
equipped with circuitry which allows for the triggering
of the discharge by sensing the R wave from the ECG.
Thus when the operator closes the firing switch on the
defibrillator electrodes, discharge is delayed until the
arrival of an R wave.

The electrodes used to deliver the large current discharge to the patient are called paddles. The paddles
consist of metal disks 8–10 cm in diameter (pediatric

paddles being somewhat smaller) embedded in insulated
handles which protect the operator from shock (Fig. 6a).
In each handle there is a thumb-operated button switch
which discharges the defibrillator once the electrodes
have been placed in position on the patient's chest. The
paddles are normally connected to the defibrillator by
self-coiling high-duty PVC cables. Due to the continuous
extension and flexion of these cables, their failure is not
uncommon and it is good practice to x ray them from
time to time to check their integrity. Good electrical
contact between the electrodes and the patient's chest is
obtained by using either saline-soaked gauze pads or an
electrically conductive gel, thus avoiding skin burns. It
is important to ensure that the paddle handles are free
of gel to preclude the possibility of the operator receiving
a shock. It is also important to ensure that the gel does
not form a conductive path between the two electrodes
on the patient's chest, short-circuiting the current away
from its target—the heart.

Manufacturers usually also provide electrodes that are
designed for internal use (Fig. 6b), that is, for use when
the chest is open, the electrodes being placed directly on
the heart wall. The defibrillator's electronic circuitry
ensures that when internal electrodes are used (or external electrodes in cardioversion) the energy discharge is
limited to a maximum value of 50 J.

Some defibrillators charge automatically, at switch-
on, to a preset level and continue to recharge to that level
immediately after each discharge. Other defibrillators
are charged on demand by depressing a button on the
control panel of the defibrillator and, more recently, by
depressing a selector on the paddle handle itself. Most
defibrillators are fitted with a capacitor charge dump
facility; when the defibrillator is switched off, the capacitor is discharged automatically and not left in a charged
condition liable to shock an unsuspecting operator. The
dump facility operates usefully, too, when the defibrillator is charged but not thereafter discharged. Since

many defibrillators have integral cardioscopes, simply switching the unit off to avoid having a charged but idle device is not acceptable.

4. Energy Measurement

Most defibrillators have stored-energy measuring circuits which indicate (via an analog or digital meter) the amount of energy available at the capacitor terminals. The circuit is simple and consists of a device which measures the voltage across the capacitor (usually through an attenuator) and multiplies it by a constant related to the value of the capacitor. For the truncated waveform defibrillator the stored energy closely approximates to the delivered energy but for those circuits containing inductance, energy losses will occur due to the resistive components of the inductors. It is therefore necessary to calibrate the output relative to the meter reading. The circuit used for calibration is also used for periodic checking of the output.

Two basic calibration circuits are in use: thermal, and electronic types. The thermal energy-measuring circuit consists of a $50\,\Omega$ carbon resistor in which a thermocouple is buried. Discharge of the defibrillator into the resistor causes heating which is measured as a temperature rise by the thermocouple. The rise in temperature may be equated to energy discharge into the resistor. This method takes an unduly long time since the resistor must be allowed to cool between tests.

The electronic energy-measuring circuit depends on the relationship:

$$W = \frac{1}{R} \int_{t_0}^{t} V^2 \, dt$$

The manner in which electronic hardware may be used to measure the discharge energy is shown in Fig. 7. The voltage at the dummy load terminals is monitored throughout the period of the discharge pulse (usually through an attenuator to avoid subjecting the electronics to high voltages) and is squared by a multiplier circuit, the output of which is integrated and multiplied by a constant related to the attenuator and the load resistor value. The resultant voltage is measured by the energy output meter. Since the truncated-discharge-waveform defibrillator may have a pulse duration in excess of 20 ms, the integrator is capable of integrating over at least this period of time, with a hold circuit such that the energy reading remains after the pulse has finished.

5. Maintenance

Since defibrillators are used in emergencies to save lives it is essential that they are carefully maintained and regularly checked to ensure their continued readiness. The following aspects of their operation merit attention:

(a) the condition of main discharge relays,

(b) the integrity of patient leads and electrodes,

(c) the calibration of the energy meter,

(d) the integrity of R wave synchronization circuits, and

(e) the operation of the integral cardioscope, if fitted.

In addition, the earthing of the equipment chassis should be regularly tested so that the safety of operating personnel as well as that of the patient is ensured.

See also: Medical Electrical Equipment: Safety Aspects

Bibliography

Balagot R C, Druz W S, Ramadan M, Lopez-Belio M, Jobgen E, Tomita N, Sadove M S 1964 A monopulse dc current defibrillator for ventricular defibrillation. *J. Thorac. Cardiovasc. Surg.* 47: 487–504

Cromwell L, Arditti M, Weibell F J, Pfeiffer E A, Steele B, Labok J A 1976 *Medical Instrumentation for Health Care.* Prentice Hall, Englewood Cliffs, New Jersey

Finlay J B 1974 Proposals for the selection and testing of dc defibrillators. *Biomed. Eng.* 9: 517–21

Lown B 1968 Intensive heart care. *Sci. Am.* 219(1): 19–27

Lown B, Neuman J, Amarasingham R, Berkovits B V 1962 Comparison of alternating current with direct current electroshock across the closed chest. *Am. J. Cardiol.* 10: 223–33

A. C. Selman

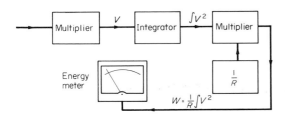

Figure 7
Defibrillator energy-metering circuit

Dental Diagnosis

Dentistry is centrally concerned with the state of health of the teeth and supporting structures as well as any related anatomical or physiological features. Dental diagnosis is usually called for in the context of a specific problem such as pain or injury, but is also carried out during regular dental examinations. In both cases, the dental examination must be considered in the context of the general health of the patient.

The actual dental examination consists, in the first instance, of a visual inspection and a manual palpation of the gingivae and teeth. A range of tools and methodologies are used to facilitate this inspection. To assess those parts of the teeth and oral structures not available to visual survey, x rays, ultrasound and electrical procedures are used.

1. Optical Aids

The main requirements for visual inspection of the dental tissues are ready access and good illumination. Access may be aided by a tongue depressor. Illumination may be effected with an extraoral lamp which is usually mounted on a counter-balanced arm and which may be directed as desired into the mouth of the sitting or reclining patient. Small mirrors of 2–3 cm diameter, mounted on the end of a long, slim handle, may be used to light up and view the inaccessible backs of teeth. Plane mirrors produce the most accurate image of the tissues, but in some instances, the focusing and image magnification obtainable with spherical mirrors make these more useful.

In recent years, the use of fiber-optic bundles for intraoral illumination has grown. In such instruments, light from a 150 W quartz or tungsten halogen projector lamp is conducted by a bundle of some 10^2–10^6 glass fibers (each of about 10^{-5} m in diameter) to the area of the mouth being examined. Since the bundle is flexible, the tip may be directed as desired. The tip may be flat, angled or equipped with a mirror and it may be cold sterilized. The light intensity is varied by controlling the filament current, either continuously or in discrete steps.

The ability to direct a controlled light intensity to the back of a tooth with fiber-optic devices has facilitated the emergence of tooth transillumination examinations. Such examinations can reveal cracks and opaque, carious lesions in the tooth.

2. Mechanical Aids

The two simplest hand tools used for dental diagnosis are explorers and probes. Explorers are very sharply pointed, pencil-shaped instruments, with many different designs of tine (the narrow part close to the tip) and tine flexibilities. Since carious lesions greatly reduce the hardness of dental enamel, the sharp point of an explorer will "catch" when gently dragged over the carious surface. The many shapes of tine allow access to remote aspects of the teeth. Explorers are also useful for assessing the marginal fit of restorations.

Probes are similar to explorers except that the tips are blunt and rounded. They are mostly used for measuring the depth of the gingival crevice in periodontal disease; millimeter graduations along the tine allow the depth to be measured quantitatively.

In the diagnosis of periodontal problems, the rigidity of the anchorage of the tooth is another useful measure. The most common method of qualitatively assessing this parameter is by use of the manual tap test. A finger tip is held behind the tooth and the outer surface of the tooth is gently tapped with the hard end of an instrument. The more rigidly the tooth is anchored, the less is the transmission of impulsive force to the finger.

Some efforts have been made to quantitatively measure the lateral mobility of teeth as a function of frequency. Resonances of about 1 kHz have been observed for incisors, but these methods have not yet found wide acceptance in clinical diagnosis.

3. Tooth Vitality Tests

The vitality of the nerve in a particular tooth is an important factor in determining the treatment for a carious lesion. The application of heat to the tooth crown by means of a heated ball of gutta-percha causes the tooth temperature to rise. If the nerve is vital, the elevation of temperature is readily felt. Blowing cold air on the crown of the tooth may likewise be used to assess the sensation of cold by the tooth nerve. Another cold test is to place a wad of cotton wool saturated with ethyl chloride on the tooth crown. Evaporation of ethyl chloride cools the tooth and a vital nerve senses the cold.

Electrical stimulation pulp testers are widely used to test nerve vitality. These devices produce pulsed dc voltages up to a maximum of some 800 V at repetition rates of 10–60 kHz. An electric current is passed from the generator via an electrode placed on the tooth crown and contacting the crown through a lump of toothpaste (effectively an electrolytic solution). The current then passes through the crown and the pulp to return via an indifferent electrode on the nearby gingivae. The minimum voltage level to cause nerve stimulation is determined. The tooth must be kept dry during this procedure and the active electrode must not touch the gingivae, as the patient could receive an uncomfortable shock. Tooth resistance values have been reported in the range 1–55 MΩ; observed values of capacitance fall in the range 100–25 pF. The threshold voltages can be as much as 200 V in healthy teeth, and the current flowing at threshold can be anything from 1 to 20 μA, depending on the shape and structure of the tooth.

A similar electrical method has been used to measure the length of the root canal before and during endodontic procedures. Here, the resistance of the path between an electrode placed on the crown end of the root canal and a second electrode attached to adjacent gingivae is measured. The main component of this resistance is the long narrow root canal filled with soft tissue and surrounded by less conducting dentine. If this canal is assumed to be of uniform cross section, the resistance is directly proportional to the length. This method has not yet been widely used and further study is needed to establish its potential.

4. Dental Radiography

X radiography is the diagnostic technique of record in most areas of dentistry. The radiopacity of the hard tissues, and the relative radiolucency of carious enamel, soft tissue and gas, produce contrast in the radiographic images of these different features and tissue types.

Radiographic images are produced as shadows cast by the structures interposed between the x-ray source and the x-ray-sensitive film. For clear, dimensionally accurate images, a number of ideal conditions should be met: (a) the source of the x rays should be as small as possible to avoid double imaging effects around the periphery of a shadow; (b) the distance from source to object should be as long as possible to ensure that the x rays are close to parallel; (c) the distance from object to film should be short so that the dimensions in the image are not over-expanded by the spreading of the beam; (d) the object and film should be parallel so that image dimensions are in proportion to those of the object; and (e) the x rays should impinge on both object and film at right angles.

Three main types of dedicated x-ray equipment are in common use in dentistry. In the most common type, the film is fixed inside the mouth and the x-ray generator is fixed extraorally. A panoramic system has the narrowly collimated source and the film positioned extraorally on opposite sides of the head, and they are both rotated together about the patient's head to produce a panoramic projection of the oral/facial region. A third type yields larger scale views of the jaws and of the temporomandibular joint by placing a stationary x-ray source and a stationary film on opposite sides of the patient's head.

4.1 Dental X-Ray Generators

The x rays are generated by bombarding a tungsten target with high energy electrons in an evacuated vessel (see *X-Ray Production*). The x rays produced are polychromatic (i.e., they have a range of wavelengths). The longest wavelengths (softest x rays) are readily absorbed in tissues and produce a poor image. Shorter wavelengths (harder x rays) give good images.

The energy distribution in the primary beam of x rays is set by the voltage between the anode and cathode in the generator tube. This is usually an ac voltage that produces radiation only during half of the "on" time of the tube. The voltage used in these instruments is specified in terms of the peak kilovoltage (kVp), and values from 50 to 110 kVp are typical for dental devices.

The intensity of the x rays produced is controlled by the tube current—the flux of electrons flowing within the tube. Values between 5 and 20 mA are used. Before emerging from the tube housing, the diverging beam of x rays is usually filtered with the equivalent of 1.5 or 2 mm of aluminum, to block the softest rays and optimize the beam for imaging.

The instrument is equipped with a timer to control the duration of x-ray generation and so effect adequate exposure of the film for imaging. In order to minimize the duration of the "on" time and thus the dose of radiation to the patient, the film chosen should have as fast a response time as possible; collimation of the beam also helps in this respect. The beam cross section is restricted to a shape and area as close as possible to that of the receiver film. This can be achieved with an aperture in a lead screen at the outlet port of the tube housing; the aperture has the required beam dimensions. The cylindrical distance-setting pipe fitted on this port may also be impregnated with lead. In some cases the collimator is a lead cylinder with the inner cross section of the approximate dimensions of the film. This cylinder serves to set the minimum distance from source to the structures to be imaged. Different lengths of cylinder are available and may be readily interchanged. The cylinder also helps the operator to direct the beam to the region of interest in the mouth.

4.2 X-Ray Films

The size, shape and distribution of the silver halide crystals in the emulsion on the films for dental purposes have been formulated to make them maximally sensitive to x rays and to minimize the exposure time needed. Film speed is conventionally measured in units of reciprocal röntgens and dental films range in speed from 1.5 to 24 (the fastest) of these units. Films for extraoral use come in two forms, namely nonscreen- and screen-film types. The nonscreen film is strongly sensitive to x-ray wavelengths but not to visible light. It usually has a layer of emulsion on both sides of the acetate sheet, to maximize the sensitivity. The screen film, on the other hand, is mainly sensitive to blue light. It is mounted in a cassette, and sandwiched between two screens coated with fluorescent material. When x rays strike these screens, a blue fluorescence impinges on the film.

Rapid development of the exposed film is desirable so that assessment and diagnosis may proceed without excessive delay. In the new automatic processors, the complete cycle of development, fixing and drying can be finished in about 90 s.

4.3 Positioning of Patient and Film

For intraoral applications, two approaches to film placement are used to achieve the best accuracy: the paralleling technique and the bisecting-angle technique. In the former the film is positioned parallel to the central axis of the tooth and the source is placed up to 40 cm away. The image is then only slightly larger than the actual tooth due to the small divergence of the beam between tooth and film. In the bisecting-angle technique, the central or axial ray of the beam is set perpendicular to a plane midway between the central plane of the tooth and the plane of the film. The length of the shadow cast is close to the length of the tooth, especially if the source is far from the tooth.

Film holders have been developed to facilitate these film–tooth–source positioning arrangements. These are usually made of cold-sterilizable polymeric materials and consist of a rod gripped between the teeth. On one end of the rod the film is set in a holder at a variable but known angle and the other end of the rod provides a reference against which the cylinder from the generator tube housing may be positioned and oriented.

Another aid to x-ray imaging in the oral cavity is a contrast medium, which may be a barium sulfate- or an iodine-containing solution. Such a medium can be painted over the soft tissues and allows the x-ray visualization of the mucosa, and therefore the measurement of the thickness of those tissues.

Radiation protection is an important aspect of dental radiography and every effort must be made to minimize the dose to the patient (especially the patient's gonads), the dentist, and the dental assistant. The patient should be provided with a leaded protective apron. Leaded collars or other such protection should be fitted on the thyroidal region. The dental operator should also wear a protective apron and stand behind the patient at some 90–135° to the direction of travel of the beam. Since x-radiation intensity falls off inversely as the square of the distance from the source, it is also advisable to stand at least 2 m from the source. If there is a radiation protection barrier available, the operator should stand behind it.

5. Other Techniques

Ultraviolet light is used in conjunction with a visually clear but fluorescing liquid which attaches to dental plaque, in order to view the location and extent of plaque. This ultraviolet light should be soft, i.e., of wavelengths between 280 and 380 nm—the so-called UVA and UVB. Care should be taken to minimize the beaming of this light into the patient's and operator's eyes (see *Ultraviolet Radiation: Potential Hazards*).

Ultrasound has been used in a pulse–echo mode to measure the thickness of the soft tissues in the oral cavity. Such measurement is based on measuring the time taken for a short pulse of ultrasound to travel through the tissue layer and, upon reflection, to travel back to the point of origin. Since the speed of propagation of ultrasound in soft tissues is approximately 1540 m s^{-1}, the thickness of the tissue may be calculated.

Efforts have also been made to measure the thicknesses of the tooth layers and to investigate whether the ultrasound propagation properties of carious enamel differ from those of normal enamel. However, none of these techniques has found its way into general diagnostic practice at the present time.

See also: Preventive Dentistry; Teeth: Physical Properties

Bibliography

American Dental Association 1981 *Dentist's Desk Reference: Materials, Instruments and Equipment*, 1st edn. American Dental Association, Chicago
Barr J H, Stephens R G 1980 *Dental Radiology: Pertinent Basic Concepts and their Applications in Clinical Practice*. Saunders, Philadelphia
Bomba J L 1971 Fiber optic lighting systems: Their role in dentistry. *Dent. Clin. North Am.* 15: 197–218
Lynch M A (ed.) 1977 *Burket's Oral Medicine: Diagnosis and Treatment*, 7th edn. Lippincott, Philadelphia
Noyes D H, Solt C W 1973 Measurement of mechanical mobility of human incisors with sinusoidal forces. *J. Biomech.* 6: 439–42
Wuehrmann A H, Manson-Hing L R 1981 *Dental Radiology*, 5th edn. Mosby, St Louis

M. Hussey

Dental Enamel: Crystallography

Dental enamel consists of about 96 wt% of a microcrystalline, oriented, inorganic material which is closely related to hydroxyapatite, $Ca_5(PO_4)_3OH$. The remainder is made up of a heterogeneous proteinaceous material ($\sim 0.1\%$) and water (3.9%). The chemical composition of the inorganic component (in wt%) is approximately Ca, 33.6–39.4; P, 16.1–18.0; CO_2 (present as carbonate), 1.95–3.66; Na, 0.25–0.90; Mg, 0.25–0.56; Cl, 0.19–0.30; and K, 0.05–0.30 (Brudevold and Söremark 1967). There are from 50 to 2000 µg per gram of fluoride ion (often much more in surface enamel) and, usually, variable amounts in the range 0–400 µg per gram of Fe, Zn, Sr, Cu, Mn, Ag and other elements.

The chemical composition of the mineral component of the other calcified tissues (bone, dentine and cementum) is rather similar to enamel, but their crystal sizes are much smaller. As a result, the investigation of their crystal structures is very much more difficult, but there is no reason to assume that they are significantly different from the inorganic component of enamel. However, because the crystals are smaller, there is a greater possibility that surface adsorbed species on the crystals will affect the chemical composition. Although it had previously been thought that there was a substantial amount of amorphous calcium phosphate in bone (but not dental enamel), recent radial distribution function analysis of x-ray powder diffraction measurements has shown that no significant amount of amorphous calcium phosphate can be detected in chick bone mineral of any age or stage of development (Glimcher et al. 1981).

Hydroxyapatite usually has a hexagonal structure with space group $P6_3/m$, lattice constants $a = 9.422$ Å and $c = 6.883$ Å, and two formula units per unit cell. The essential feature of the structure is the approximately hexagonal-close-packed arrangement of the phosphate tetrahedra (Elliott et al. 1973). Channels parallel to the hexagonal (*c* axis) arise naturally from this arrangement. One-third of these channels are filled with columns of hydroxyl ions arranged in a head-to-tail fashion. The other two-thirds of the channels are filled with columns of calcium ions which account for two-fifths of the calcium content of the unit cell. The remaining calcium ions are located in triangles around each oxygen atom of a hydroxyl ion.

If stoichiometric hydroxyapatite is prepared, the *b* axis is doubled giving a monoclinic (pseudohexagonal) structure with space group P2$_1$/*b* (Elliott et al. 1973). This doubling comes from an ordered arrangement of the direction of the hydroxyl ion columns which is easily lost if there is nonstoichiometry (or substitutions) in the structure. Hence mineral hydroxyapatite, in which about 10% of the OH$^-$ ions are replaced by F$^-$ ions, is hexagonal. Many other substitutions are possible, but the important ones in enamel are: OH$^-$ replaced by Cl$^-$ and $\frac{1}{2}$CO$_3^{2-}$; Ca^{2+} replaced by Sr^{2+}; and PO$_4^{3-}$ by CO$_3^{2-}$, although in this case the charge-compensating mechanism is uncertain.

Dental enamel has an x-ray powder diffraction pattern very similar to hexagonal hydroxyapatite, but the *a* axis is slightly enlarged so that the lattice constants *a* and *c* are approximately 9.44 Å and 6.88 Å, respectively. On pyrolysis at 400 °C, there is a contraction in the *a* axis of about 0.02 Å, which is mainly ascribed to the loss of lattice water—H$_2$O for OH$^-$ and/or HPO$_4^{2-}$ for PO$_4^{3-}$ substitutions (LeGeros et al. 1978).

The structure of dental enamel has been investigated directly by Young and Mackie (1980) using the Rietveld method of whole-pattern fitting for x-ray and neutron powder diffraction data. They confirmed the general similarity with hydroxyapatite, but also provided results consistent with some H$_2$O in orientational disorder on the *c* axis (the axis on which the OH$^-$ ions lie in hydroxyapatite) and a small amount of Cl$^-$ replacing OH$^-$ ions.

The position of the carbonate in dental enamel has been a topic of great controversy in the past, and at present not all aspects of this are universally agreed. Infrared absorption spectroscopy is a particularly useful technique for investigating this problem especially if polarized radiation is used. This is possible because the apatite crystals have a preferred orientation with their *c* axes approximately parallel to the enamel prisms (see *Teeth: Electron Microscopy Studies*). The evidence from infrared spectroscopy indicates that the carbonate ions are in two different environments both oriented with respect to the apatite lattice. By analogy with synthetic compounds, one of these environments (which represents about 10–20% of the total) appears to be carbonate ions replacing hydroxyl ions in which the plane of the carbonate ion is approximately parallel to the *c* axis. The remainder appears to correspond to the replacement of phosphate ions. In this latter substitution, the plane of the ion is approximately perpendicular to the *c* axis. It also seems possible that some of the

carbonate is surface bound, particularly in bone with its smaller crystals, and also that some of the carbonate might be in the form of HCO$_3^-$ ions.

X-ray powder diffraction, neutron powder diffraction and infrared spectroscopy do not give any direct information about the structural basis for the variable Ca:P ratio of dental enamel. This problem is usually considered in relation to the problem of the variable Ca:P ratio of synthetic precipitated apatites. The most generally accepted explanation for this in the synthetic apatites supposes coupled lattice substitutions in which it is assumed that there are Ca^{2+} and OH$^-$ ion deficiencies and sometimes lattice water. Most of the experimental evidence for these is indirect as it comes from chemical analyses (including the amount of pyrophosphate formed on heating), thermogravimetric analyses and lattice constant measurements. There is, however, direct evidence for some hydroxyl ion deficiency from neutron diffraction and quantitative infrared and laser-Raman spectroscopy. Two other suggestions have been made to explain the variable Ca:P ratio which could contribute in some circumstances. These are the formation of interlayered structures of octacalcium phosphate (Ca$_8$H$_2$(PO$_4$)$_6$.5H$_2$O) and hydroxyapatite, and the adsorption of variable amounts of HPO$_4^{2-}$ on the surface of the crystals.

The apatite crystals in developing enamel are usually lath- or tapelike with a width of 200–400 Å, approximately 10–30 Å in thickness and at least 1 or 2 μm long. Electron diffraction has shown that the extended faces have indices (100) and ($\bar{1}$00) and that the crystal is extended in the *c* axis direction. This habit is inconsistent with a hexagonal space group and it has been suggested that the reason for this is that the apatite crystallites form by an in situ hydrolysis reaction from octacalcium phosphate which is triclinic and has a bladelike habit. It has alternatively been suggested that these first-formed crystals are monoclinic.

X-ray diffraction line-broadening methods have been used to investigate crystal size and inhomogeneous strain in hippopotamus enamel. Approximate mean dimensions for rodlike particles of 410 × 1600 Å were found after inhomogeneous strain was taken into account (this had a negligible effect on the measured width, but was very significant for the length). In another study in which no corrections for inhomogeneous strain were made it was shown that when the fluoride content increased from 70 to 670 μg g^{-1} in human enamel the apparent uncorrected crystal length increased from 1420 ± 100 Å to 2740 ± 240 Å.

Lattice defects probably play an important part in determining the reactivity of biological apatite, and edge dislocations have been observed by direct observation of the apatite lattice under the electron microscope. There is evidence for a screw dislocation running down the center of the enamel crystals parallel to the *c* axis, as this central region dissolves first in carious enamel.

See also: Dental Fluoridation; Teeth: Physical Properties

Bibliography

Brès E F, Barry J C, Hutchison J L 1984 A structural basis for carious dissolution of the apatite crystals of human tooth enamel. *Ultramicroscopy* 12: 367–72

Brudevold F, Söremark R 1967 Chemistry of the mineral phase of enamel. In: Miles A E W (ed.) 1967 *Structural and Chemical Organization of Teeth*, Vol. 2. Academic Press, New York, pp. 247–77

Centre National de la Recherche Scientifique 1975 *Physico-chimie et Crystallographie des Apatites d'Intérêt Biologique*, Colloques International No. 230. Centre National de la Recherche Scientifique, Paris

Driessens F C M 1982 *Mineral Aspects of Dentistry*. Karger, Basel

Elliott J C, Mackie P E, Young R A 1973 Monoclinic hydroxyapatite. *Science* 180: 1055–57

Fearnhead R W, Suga S (eds.) 1984 *Tooth Enamel IV*. Elsevier, Amsterdam

Glimcher M J, Bonar L C, Grympas M D, Landis W J, Roufosse A H 1981 Recent studies of bone mineral: Is the amorphous calcium phosphate theory valid? *J. Cryst. Growth* 53: 100–19

LeGeros R Z, Bonel G, Legros R 1978 Types of "H_2O" in human enamel and in precipitated apatites. *Calcif. Tissue Res.* 26: 111–18

Posner A S, Betts F, Blumenthal N C 1979 Bone mineral composition and structure. In: Simmons D J (ed.) 1979 *Skeletal Research: An Experimental Approach*. Academic Press, New York, pp. 167–92

Williams R A D, Elliott J C 1979 *Basic and Applied Dental Biochemistry*. Churchill Livingstone, Edinburgh

Young R A, Brown W E 1982 Structures of biological minerals. In: Nancollas G H (ed.) 1982 *Biological Mineralization and Demineralization*. Springer, Heidelberg, pp. 101–41

Young R A, Mackie P E 1980 Crystallography of human tooth enamel: Initial structure refinement. *Mater. Res. Bull.* 15: 17–29

J. C. Elliott

Dental Fluoridation

The beneficial effects of fluoride were first recognized in the late 1930s, with the observation of reduced dental decay (caries) where fluoride occurred naturally in drinking water. There is now overwhelming evidence of a considerable reduction in caries when drinking water is fluoridated at the optimum level of 1 ppm; claims that this constitutes a health hazard have not been substantiated (Royal College of Physicians 1976). Topical fluoride (small amounts at a higher concentration applied directly to the tooth surface) is also of value: thus almost all toothpastes sold in the UK and the USA now contain about 0.1% fluoride, and evidence for their beneficial effect is rapidly accumulating.

Dental decay is a dynamic process which can be partially reversed under favorable conditions. It starts as demineralization just below the enamel surface, and is produced by organic acids formed from sugar in the dense microbial plaque which rapidly collects on all teeth. Fluoride may have some effect on tooth decay by modifying enzyme activity in this plaque, or by changing its aggregation and adhesion properties, but the main mechanism probably involves the enamel itself, reducing the effect of acid attack and favoring mineral deposition.

1. Tooth Structure

Tooth enamel is over 95 wt% mineralized, consisting of bundles of needle-shaped crystals interspersed with a little water and organic material. These crystals belong to an isomorphous series of calcium phosphates known as apatites, with the stoichiometric formula: $(Ca)_{10}(PO_4)_6(X)_2$ in which X can be OH, F, or another halide, located in columns along the screw hexagonal axis at half unit-cell intervals. In mineralized tissues the X ion is mainly OH, although the outer surface of tooth enamel contains 0.2–0.4% fluoride. Apatite can accommodate any mixture of OH and F as X, with hydroxyapatite (OH only) more soluble than fluorapatite. Recent findings (Arends et al. 1984, ten Cate 1984) stress the importance of fluoride in the solution.

2. Mechanisms of Fluoride Action

Suggested mechanisms of fluoride action range from those based on details of the crystal structure, through pure considerations of thermodynamic solubility, to diffusion kinetics and exchange within the inter-crystallite pore structure.

(a) Using neutron diffraction, NMR and infrared spectroscopy, Young et al. (1969) showed that minority fluoride ions in mineral hydroxyapatite are randomly distributed down the X-ion columns, and are sometimes hydrogen bonded to two neighboring OH ions, reversing their relative orientation. They proposed that small quantities of fluoride might thus block OH diffusion in enamel crystallites and reduce the rate of acid dissolution. Recent theoretical evidence makes this elegant theory less likely, and attention has shifted from the crystal interior towards the influence of fluoride in the fluid phase between the crystallites. Nevertheless, increased entropy due to fluoride-promoted disordering of OH in these columns combined with the effects of OH–F hydrogen bonding should bring the solubility of the solid solution below that of either end member of the series; this property has been observed (Moreno et al. 1977) for some synthetic apatites.

(b) Uptake at specific dissolution sites may allow small amounts of fluoride to reduce the solubility of the whole apatite crystal. Hydroxyapatite takes on the solubility characteristics of fluorapatite in the presence of dissolving CaF_2 (Brown et al. 1977) which could result from deposition of a thin coating of fluorapatite, either over the whole crystal, or else at specific sites such as screw or edge dislocations (but see also (c) and (d) below). Fluoride also tends to promote formation of larger crystallites, giving a reduced surface area for attack.

(c) Using thermodynamic arguments Brown et al. (1977) suggested that increased fluoride activity within the enamel pores lowers the $Ca(OH)_2$ activity, reducing outward diffusion of calcium, and shifting the dynamic balance away from dissolution and towards precipitation.

(d) Arends et al. (1984) proposed that preferential sorption of fluoride ions, hydrogen bonded to fixed protons in the electrical double layer of the enamel crystallites, would reduce enamel solubility by ejecting surface phosphate groups, and blocking proton (acid) attack.

(e) A problem in understanding the operation of topical fluoride, including fluoridated toothpastes which seem to work much better than originally thought, has concerned the transient nature of the application. However, fluoride is known to be preferentially deposited in early carious lesions (Robinson et al. 1982), so that during subsequent acid challenges this can return to solution and raise the local F^- activity. Furthermore it is known (Arends et al. 1984) that many topical fluorides (including some fluoridated toothpastes) can deposit CaF_2 directly onto the tooth surface, providing an additional reservoir of fluoride ions.

Our understanding of the mechanisms whereby fluoride reduces tooth decay is increasing rapidly. Water fluoridation is still the best and most cost-effective method for dealing with one of the most expensive diseases afflicting man, but it is now clearer how topical fluoride may operate as a useful additional factor.

See also: Dental Enamel: Crystallography; Teeth: Physical Properties; Preventive Dentistry

Bibliography

Arends J, Nelson D G A, Dijkman A G, Jongebloed W L 1984 Effect of various fluorides on enamel structure and chemistry. In: Guggenheim B (ed.) 1984 *Cariology Today*. Karger, Basel, pp. 245–58
Brown W E, Gregory T M, Chow L C 1977 Effect of fluoride on enamel solubility and cariostasis. *Caries Res.* Vol. 11 (Suppl. 1): 118–36
Moreno E C, Kresak M, Zahradnik R T 1977 Physiochemical aspects of the fluoride–apatite system relevant to the study of dental caries. *Caries Res.* 11: 142–60
Robinson C, Weatherell J A, Hallsworth A S 1982 Alterations in composition of permanent human enamel during carious attack. In: Leach S A, Edgar W M (eds.) 1983 *Demineralisation and Remineralisation of Teeth*. IRL Press, Eynsham, UK, pp. 209–23
Royal College of Physicians 1976 *Fluoride, Teeth and Health: Summary of a Report on Fluoride and its Effect on Teeth and Health from the Royal College of Physicians of London*. Pitmans, London
ten Cate U 1984 Effect of fluoride on enamel de- and remineralisation *in vitro* and *in vivo*. In: Guggenheim B (ed.) 1984 *Cariology Today*. Karger, Basel, pp. 231–36
Young R A, Lugt W van der, Elliott J C 1969 Mechanism for fluorine inhibition of diffusion in hydroxyapatite. *Nature (London)* 223: 729–30

G. H. Dibdin

Dental Force Analysis

Perhaps the most important function of the dentition is mastication (or chewing), during which the maximum dental force is exerted on the masticatory apparatus including the teeth, the peridontal ligament (membrane), and the supporting alveolar bone of the maxilla and mandible. This force must also be exerted on and by all dental restorations, such as fillings, fixed bridges, crowns, partial or removable dentures in which some of the teeth are replaced, and complete dental prosthesis in which all the teeth are replaced. Therefore, dental force analysis becomes a key aspect of the theory and practice of dentistry.

In this article we shall first present a theoretical analysis of the masticatory force and then describe some selected applications to restorative dentistry. Of particular interest is the force analysis relating to the shape of the natural human molar, on which the maximum force of mastication is applied. Heretofore in the science of dentistry, the role that the tooth surface cuspal shape and tooth axis orientation plays in the growth and function of the teeth has been generally neglected. The analysis presented clearly demonstrates why and how the natural dentition (e.g., the molars) comprise the anatomical shape and have the axis orientation that is actually found in humans.

Also, a method is outlined for the setting of the occlusal surface of the teeth of artificial dentures that will best tend to stabilize the dentures during mastication. Until now, many other techniques have been employed by dentists to stabilize the denture, which in most instances did not take into account this maximum force exerted during mastication. However, control of these forces through the proper design of the occlusal (chewing) surface of the artificial teeth can significantly improve the functioning of the restoration.

1. The Masticatory Force

The maximum force on the dentition is exerted during mastication, or chewing, with the posterior teeth. This force was measured by early researchers and is approximately 5–10 kg.

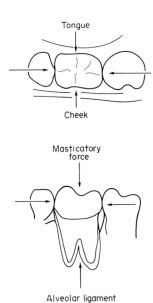

Figure 1
The forces (indicated by arrows) applied to the lower
first molar, namely the force of the cheek, the tongue,
the adjacent teeth, the alveolar ligament and the food
force

Focusing our attention on the lower first molar (see
Fig. 1), we can distinguish the force of the cheek against
the tooth, of the tongue against the tooth, the forces due
to contact of the adjacent teeth and, finally, the force
exerted on the tooth by the bony structures in reaction
to the force exerted by the food on the occlusal, or
chewing, surface of the tooth. Of these forces, the food
or masticatory force is the greatest and will be of
primary concern in this article.

Of course, if the tooth does not move, then considerations of elementary physics dictate that the sum of the
forces applied to the tooth must be zero; if the sum of
the forces is not zero, the tooth will be displaced. When
the food force is applied, it is balanced by the bony
structures which apply a reactive force through the
peridontal ligament, or membrane. This membrane distributes the reactive force along the surface of the root
of the tooth.

The maximum force occurs when chewing tough
foods, such as steak, rather than when chewing brittle or
shaped foods, such as carrots or grapes. This food force
is applied over the approximately 0.1 in.^{2} (6.45 mm^2)
surface of the tooth, resulting in a pressure of approximately 100–200 lb in.$^{-2}$ (0.7–1.4 MPa) on the tooth
surface.

From a physiological point of view, the integral of
the force over time is the important factor, particularly
when concerned with physiological bone reaction to the
applied force. Thus the biodental situation is slightly
different from that of ordinary physics.

2. Mastication

Since the maximum force is exerted during mastication,
it is interesting to discuss some facts about this process.
First, people chew on one side at a time; second, during
chewing the teeth do not contact because the food bolus
remains between the occlusal surfaces of the upper and
lower teeth. A classic experiment to illustrate this point
was carried out by Jankelsen et al. (1953) who connected
opposing metallic inlays on upper and lower molars
through wires to a voltage source so that a current would
flow and be recorded when actual upper- and lower-
tooth surfaces met. Jankelsen then observed that during
mastication the teeth did not make contact until the
bolus was swallowed, at which point the teeth did
actually meet. Thus tooth contact during mastication
occurs only during swallowing (see Fig. 2). The origin of

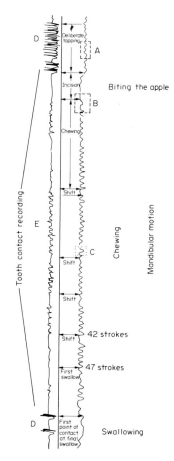

Figure 2
Composite graph of vertical motion and tooth
contact during eating of crisp apple: A, strokes
during rapid tapping of teeth; B, incising movement;
C, masticatory cycles; D, tooth contact deflections; E,
current leakage deflections (after Jankelson et al.
1953)

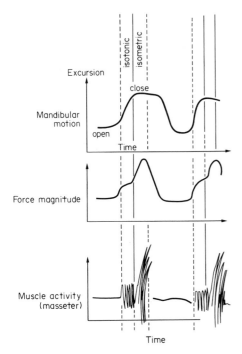

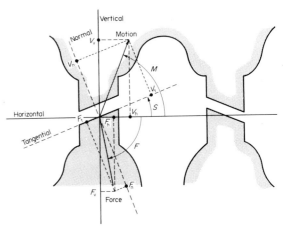

Figure 3
The mandibular motion, force magnitude and muscle activity as a function of time during mastication illustrating both the isotonic (approximately constant force) and isometric (approximately constant closure) phases of the chewing cycle

Figure 4
Illustration of the angles M, S, F, and the horizontal and vertical and tangential and normal components of both the force and final closing chewing displacement

the masticatory force lies primarily with the masseter and temporal muscles. Figure 3 illustrates aspects of the chewing cycle showing mandibular motion, force magnitude and muscle activity. After the initial opening motion, there is an isotonic phase followed by an isometric phase, and it is during this latter isometric phase that the maximum chewing force is developed.

3. The Direction of the Masticatory Force

In this section an analysis of the chewing force or functional force of mastication is given, and in the following sections some applications of the theory are discussed.

First, it should be noted that it is the direction of the maximum force exerted by the food on the teeth that is of greatest clinical and physiological importance. In particular, it can be shown that this direction (in the frontal plane) is given by the following formula:

$$\tan F = \frac{K \tan S + \tan (M - S)}{K - \tan (M - S) \tan S}$$

where (see Fig. 4) the angle F is the direction of the force with respect to the horizontal, the angle S is the slope of

the cuspal peak plane with respect to the horizontal, and the angle M is the so-called "angle of chew," that is, the direction with respect to the horizontal of the final closing motion during mastication (see Fig. 5). It should be pointed out that the horizontal could be any line just as long as all of the angles F, S and M are referred to this same line. The constant K is a function of the physical properties of the food being chewed (e.g., the

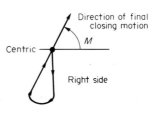

Figure 5
Illustration of the angle of chew. If a card is attached to the upper jaw (maxilla) and the lower jaw (mandible), then a chewing tracing will be obtained from which the angle of chew can be measured for the patient

shear modulus A and the compression modulus B) and of the cuspal slope e, and is given by the following formula (Ledley 1954):

$$K = A/B + \tan^2 e \ (1 - A^2/B^2) - \ldots$$

The motion of the lower jaw with respect to the upper may be roughly described as an opening movement with a downward path, followed by a lateral movement toward the side on which the chewing is taking place, and the final movement upward from the lateral position toward the so-called "centric" position. The mechanics of this chewing motion have been demonstrated by many investigators, including Kurth (1949), Boswell (1951) and Hildebrand (1931), using extraoral devices and x-ray motion pictures. The angle of chew is a habitual characteristic of each individual person and has been shown to be independent of the type of dentition that the person may have. For instance, Kurth made several different types of artificial dentures for the same patient and found that the angle of chew was the same regardless of the dentures being worn.

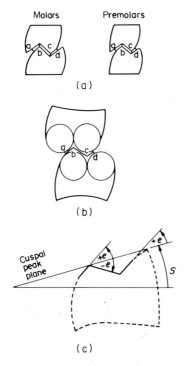

(a)

(b)

(c)

Figure 6
Functional surfaces of anatomic teeth and cuspal peak plane. (a) For molar and bicuspid teeth, area of slope ab + area of slope cd = area of slope bc. (b) Cusps represented schematically by inscribed circles. Geometrically, ab + cd = bc. (c) Occlusal surface slope S for anatomic teeth; angle S is between cuspal peak plane and horizontal. Cusps make angles $+e$ and $-e$ with respect to the cuspal peak plane

The slope of the cuspal peak plane S is determined by first placing a plane on the tops of the cusps of the teeth and measuring the angle between this plane and the horizontal (see Fig. 6). The cuspal slope e is the angle of the cusps with respect to the cuspal peak plane. We also assume that the chewing surface of the lingual (tongue) side of the buccal (cheek) cusp is equal to the sum of the chewing surfaces of the buccal on both cusps of the lower molar; that is, in Fig. 6, bc is equal to ab + cd. The physical properties of the food being chewed can be measured by the shear modulus A divided by the compression modulus B of the food substance, which for tough steak turns out to be approximately

$$A/B = 0.14$$

This constant value A/B was measured at the National Bureau of Standards utilizing standard "tough" meat as determined by the US Department of Agriculture (Ledley 1954). The cuspal slope is defined as the angle between the cuspal peak plane and the actual slope of the cusp.

There is also an angle of chew in the sagittal plane (see Fig. 7). The formula is analogous to that given above. Note that in this case the final closing motion of the lower jaw is slightly forward. Thus the situation can be visualized in three dimensions. However, in many applications it is the frontal component of the force vector that is most important.

Tables can be computed from the formulas, giving for various cuspal slope ranges the force direction for combinations of angle of chew and cuspal peak plane slope. Table 1 shows values for cuspal slopes of 18–21°; for a comprehensive set of tables the reader is referred

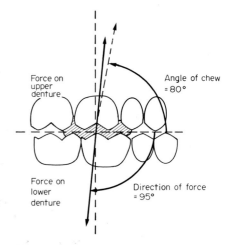

Figure 7
Calculated force direction in sagittal plane on lower denture and equal and opposite force on upper denture when dentures have teeth with a cuspal slope of 30°, patient an anteroposterior angle of chew 80°, and food chewed is tough meat

Table 1
Force direction for combinations of angle of chew and cuspal peak plane slope (cuspal slope = 18–22°)

Cuspal peak plane slope S / Angle of chew M

S \ M	90	88	86	84	82	80	78	76	74	72	70	68	66	64	62	60	58	56	54	52	50
40	63	64	64	65	67	68	69	70	72	73	75	77	79	81	84	86	89	93	97		
39	63	64	65	66	67	68	69	71	72	73	75	77	79	81	83	86	89	92	96		
38	64	65	66	66	67	69	70	71	72	74	75	77	79	81	83	86	88	91	95	99	
37	64	65	66	67	68	69	70	71	73	74	75	77	77	79	81	83	85	88	91	94	98
36	65	66	67	68	68	69	71	72	73	74	76	77	77	79	81	83	85	88	90	93	97
35	66	66	67	68	69	70	71	72	73	75	75	76	77	79	81	83	85	87	[90]	93	96
34	66	67	68	69	70	70	71	73	74	75	76	78	79	81	83	85	87	[90]	92	95	99
33	67	68	68	69	70	71	72	73	74	75	77	78	79	81	83	85	87	89	92	95	98
32	67	68	69	70	71	72	72	73	75	76	77	78	80	81	83	85	87	89	92	94	97
31	68	69	70	70	71	72	73	74	75	76	77	79	80	81	83	85	87	89	91	94	97
30	69	69	70	71	72	73	74	74	75	77	78	79	80	82	83	85	87	89	91	94	96
29	69	70	71	72	72	73	74	75	76	77	78	79	81	82	83	85	87	89	91	93	96
28	70	71	71	72	73	74	75	76	76	77	79	80	81	82	84	85	87	89	91	93	96
27	71	71	72	73	74	74	75	76	77	78	79	80	81	83	84	85	87	89	91	93	95
26	71	72	73	73	74	75	76	77	78	78	79	81	82	83	84	86	87	89	91	93	95
25	72	73	73	74	75	76	76	77	78	79	80	81	82	83	85	86	87	89	91	93	95
24	73	73	74	75	75	76	77	78	79	80	80	81	83	84	85	86	88	89	91	93	95
23	73	74	75	75	76	77	78	79	80	81	82	83	84	85	87	88	89	91	93	95	
22	74	75	75	76	77	77	78	79	80	81	82	82	83	85	86	87	88	[90]	91	93	95
21	75	75	76	77	77	78	79	80	80	81	82	83	84	85	86	87	89	90	91	93	95
20	76	76	77	77	78	79	79	80	81	82	83	84	84	85	87	88	89	90	92	93	95
19	76	77	77	78	79	79	80	81	82	82	83	84	85	86	87	88	89	91	92	93	95
18	77	78	78	79	79	80	81	81	82	83	84	85	86	86	87	89	[90]	91	92	94	95
17	78	78	79	79	80	81	81	82	83	84	84	85	86	87	88	89	90	91	93	94	95
16	78	79	80	80	81	81	82	83	83	84	85	86	87	88	88	89	91	92	93	94	96
15	79	80	80	81	81	82	83	83	84	85	86	86	87	88	89	[90]	91	92	93	95	96
14	80	80	81	82	82	83	83	84	85	85	86	87	88	89	[90]	90	91	93	94	95	96
13	81	81	82	82	83	83	84	85	85	86	87	88	88	89	90	91	92	93	94	95	97
12	81	82	82	83	84	84	85	85	86	87	87	88	89	[90]	91	92	92	93	95	96	97
11	82	83	83	84	84	85	85	86	87	87	88	89	[90]	90	91	92	93	94	95	96	97
10	83	83	84	84	85	86	86	87	87	88	89	89	90	91	92	93	94	94	95	97	98
09	83	84	85	85	86	86	87	87	88	89	89	[90]	91	92	92	93	94	95	96	97	98
08	84	85	85	86	86	87	88	88	89	89	[90]	91	91	92	93	94	94	95	96	96	99
07	85	85	86	87	87	88	88	89	89	[90]	91	91	92	93	94	94	95	96	97	98	99
06	86	86	87	87	88	88	89	[90]	[90]	91	91	92	93	93	94	95	96	97	98	98	99
05	86	87	87	88	89	89	[90]	90	91	91	92	93	93	94	95	96	96	97	98	99	
04	87	88	88	89	89	[90]	90	91	92	92	93	93	94	95	95	96	97	98	99		
03	88	88	89	89	[90]	91	91	92	92	93	93	94	95	95	96	97	98	98	99		
02	89	89	[90]	[90]	91	91	92	92	93	94	94	95	95	96	97	97	98	99			
01	89	[90]	90	91	91	92	93	93	94	94	95	95	96	97	97	98	99				
00	[90]	91	91	92	92	93	93	94	94	95	96	96	97	97	98	99	99				

Direction of the force

to Ledley (1954). As will be shown below, for application to complete denture prosthesis, a force of direction of 90° is particularly important; therefore, a box is placed around this value in Table 1.

4. Applications

There are many applications of this force analysis. It can be used to explain the shape of the teeth and why teeth have cusps, to devise techniques for optimizing the stability of complete dentures, and for the analysis of the forces in orthodontics and so forth. In the following, we shall consider a few of these applications.

The lingual cusp of the human molar is "higher" than the buccal cusp (see Fig. 8). That is to say, if a line that bisects the tooth is drawn from the root through the central fossa, then with respect to a perpendicular, the cuspal peak plane will have a slope that rises in the lingual direction. The reason for this can be explained as follows. In the normal dentition, the resultant force of mastication on the tooth must pass through the apex of

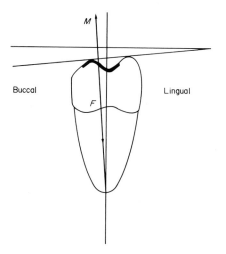

Figure 8
Explanation of tooth anatomy. The lingual cusp of the lower first molar is higher than the buccal cusp as illustrated by aligning the tooth along an access from the root apex through the cuspal fossa. However, for natural dentition, the teeth are so inclined in the mandible that a line through the center of force and the root apex is colinear with the final closing displacement motion (i.e., angle of chew). To obtain such a force direction with respect to the cuspal peak plane, this line must be perpendicular to the plane, giving rise to the necessity for the lingual cusp to be in effect "higher" than the buccal cusp. The thick line represents the chewing surface of the tooth

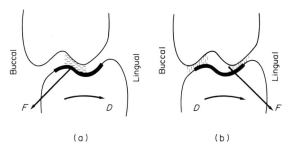

Figure 9
Illustration of how the cusps tend to keep the natural dentition properly aligned. (a) A lingual displacement results in a buccally directed force tending to restabilize the tooth. (b) A buccal displacement results in a lingually directed force against stabilizing the tooth. The thick line represents the chewing surface of the tooth

the root in order that there be no displacing moment. By symmetry, this force must be applied at the center of the lingual surface of the buccal cusp, as shown in the figure. From our formula, it is easy to see that this can only occur when the tooth is so positioned in the lower jaw that the angle of chew is along the line of the force direction, and hence the force direction must be perpendicular to the cuspal peak plane. This can only occur when the lingual cusp is "higher" than the buccal cusp, as shown.

Using the force analysis method it is easy to explain why, in the course of evolution, human teeth have developed cusps. The function of the cusps is primarily to stabilize the natural tooth so that it will not move out of line, buccally or lingually. Figure 9 illustrates this point. It is worth noting that, if the lower tooth were displaced lingually, then the food force on the lingual surface of the buccal cusp would be greater than the force on the buccal surfaces of the tooth, because there would be greater compression of the food at this lingual surface than at the buccal surfaces. This would tend to push the tooth in the opposite (buccal) direction. On the other hand, if the lower tooth were displaced buccally, then the food force on the buccal surfaces of the cusps would become greater than that on the lingual surface

of the buccal cusp, tending again to displace the tooth in the opposite (lingual) direction.

5. Application to Complete Denture Prosthesis

One of the most important applications of the force analysis is in the design of complete dentures to optimize their stability during function. In Fig. 10 we illustrate a frontal section through the lower jaw showing stable and unstable force directions for complete dentures. If the

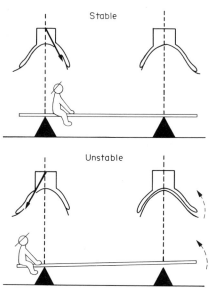

Figure 10
Stabilizing and unstabilizing forces for artificial dentures: force inside lower arch, tending to stabilize denture; force outside arch, tending to dislodge denture

force direction lies inside the lower ridge arch, the lower denture will tend to be stabilized, whereas if the force direction lies outside the ridge arch then the denture will tend to be unstable, or rise on the side opposite the chewing. The man on the see-saw illustrates this point.

For our force direction formula, we observe that the angle of chew of each patient is a habit of the patient, and therefore the only factors under the control of the dentist are the tooth surface slope and cuspal inclination. Hence, to stabilize the lower denture, the dentist can measure the patient's angle of chew, choose a force direction that will lie inside the arch, and construct the dentures that will result in this force direction by setting the cuspal peak plane of the teeth at the angle given by the appropriate chart. For example, if the angle of chew of the patient were 80°, then for teeth of cuspal slopes between 18 and 22°, and a desired force angle of 89°, the appropriate cuspal peak plane slope S should be 5° (see Table 1).

A number of factors are involved in choosing the appropriate force direction. A force that may stabilize the lower denture will have an equal and opposite reactive force applied to the upper denture, which may tend to make the upper denture unstable. Of course, in general the upper denture, having a larger surface area on the mucous membranes, tends to have better retention than the lower denture. In any event, the relative width of the upper and lower arches is an important factor to be considered. Another important factor is the placement of the teeth relative to the ridges, which itself may be predetermined by the size of the patient's tongue,

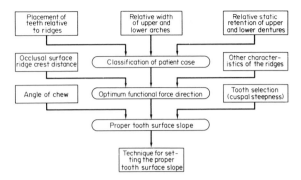

Figure 11
Factors to be considered in the functional force control technique

or cheeks, and other clinical considerations. The so-called "vertical dimension," or normal jaw closure that the dentures are to maintain, will dictate the distance from occlusal surface to ridge crest distance, which is another factor to be considered. Finally, there is the tooth selection, which embodies the cuspal steepness. Figure 11 is a flow chart that could be followed to optimize denture stability when constructing complete dentures.

Figure 12 illustrates ranges of force directions that will tend to stabilize dentures, depending upon the patient's classification with respect to relative widths of the upper and lower arches. Three classes are defined, depending

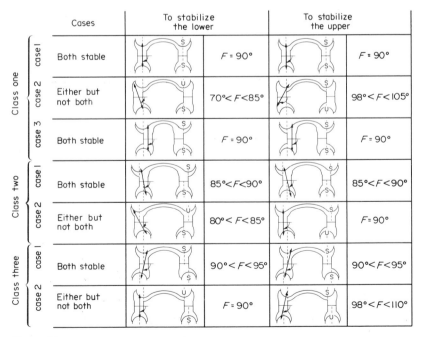

Figure 12
Classification of cases

upon whether the upper and lower arches are approximately equal in width, or the upper arch is wider than the lower arch, or vice versa. All of these classes occur in the population, with no one of them being predominant. Within each class several cases can occur, depending upon whether the artificial teeth will be placed directly over the ridge, buccal to the ridge, or lingual to the ridge, as shown. For each such class and case that exists, the figure gives the force direction range which will tend to stabilize the lower denture or to stabilize the upper denture, or both, when this can be accomplished.

6. Tooth Shape to Eliminate Horizontal Force Component

In complete denture design, it is frequently desirable to prevent the dentures from having a tendency to slip

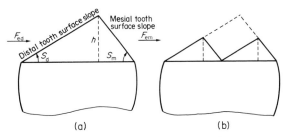

Figure 13
(a) Mesiodistal tooth shape with nomenclature used. S_d = distal tooth surface slope; S_m = mesial tooth surface slope; h = height of peak; F_{ed} = effective distal force vector; F_{em} = effective mesial force vector. (b) Equivalent tooth shape actually used in tooth forms

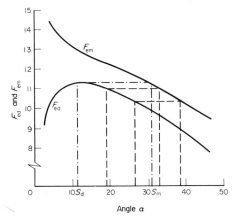

Figure 14
Graph of the functions F_{ed} and F_{em} versus the angle α. F_{ed} = effective distal force magnitude; F_{em} = effective mesial force magnitude; $\alpha = S_d$ for the lower curve and S_m for the upper curve; S_d = distal tooth surface slope; S_m = mesial tooth surface slope. Here M equals $86°$

forward or backward. This can be done by eliminating the horizontal mesiodistal force component. This can be accomplished by appropriately designing the artificial tooth surface so that the mesial slope bears a certain relationship to the distal slope of the cusps. Although a sagittal section of a molar appears roughly as in Fig. 13a, we can for our purposes imagine it to be as shown in Fig. 13b.

In Fig. 14 we plot a graph of the effective horizontal distal force vector F_{ed} as a function of the angle α and, similarly, for the effective mesial force vector F_{em}. If the mesiodistal force is to be zero, then F_{ed} must equal F_{em}.

The question is, then, what are the combinations of the mesial two-surface slope S_m and the distal two-surface slope S_d that will satisfy our desired condition? Three such results are shown in Fig. 14. In general, we wish the angles S_d and S_m to be as small as possible, which is the case where we choose the angle S_d corresponding to the maximum of the curve F_{ed}. Thus, the dashed dotted line in Fig. 14 gives the optimum pair of angles S_d and S_m. The curves of Fig. 14 were developed from the force direction formula as adapted for the geometric situation presented in Fig. 13 and, in this case, for a sagittal angle of chew of 86°. Table 2 gives

Table 2
Tooth shapes to eliminate horizontal component of force in the sagittal plane

Sagittal angle of chew M	Distal slope S_d	Mesial slope S_m
88°	9.6°	22.0°
86°	11.7°	31.2°
84°	13.0°	38.5°
82°	14.0°	44.5°

combinations of minimum mesial and distal slope angles to eliminate the horizontal force direction for various angles of chew.

See also: Teeth: Physical Properties; Preventive Dentistry; Dental Diagnosis; Dental Materials

Bibliography

Boswell J V 1951 Practical occlusion in relation to complete dentures. *J. Prosthet. Dent.* 1: 307–21
Hildebrand G Y 1931 *Studies in the Masticatory Movements of the Human Lower Jaw.* Walter de Gruyter, Berlin
Jankelsen B, Hoffman G M, Henderson J A Jr 1953 The physiology of the stomatognathic system. *J. Am. Dent. Assoc.* 46: 375–86
Kurth L E 1949 Mandibular movement and articular occlusion. *J. Am. Dent. Assoc.* 39: 37–46
Ledley R S 1954 The relation of occlusal surfaces to the stability of artificial dentures. *J. Am. Dent. Assoc.* 48(4): 508–26
Ledley R S 1955a Mastication and denture stability. *J. Am. Dent. Assoc.* 50(2): 241–42

Ledley R S 1955b A new method of determining the functional forces applied to prosthetic appliances and their supporting tissues. *J. Prosthet. Dent.* 5(4): 546–62

Ledley R S 1965a Tooth shape to eliminate horizontal force component. *J. Dent. Res.* 44(2): 402–4

Ledley R S 1965b An important property of the masticatory force. *J. Dent. Res.* 44(2): 405–7

Ledley R S 1968 Theoretical analysis of displacement and force distribution for tissue-bearing surface of dentures. *J. Dent. Res.* 47(2): 318–22

Ledley R S 1969a Numerical experiments with linear force-displacement tooth model. *J. Dent. Res.* 48(1): 32–37

Ledley R S 1969b *Engineering Analysis of Dental Forces: Theory and Application.* National Biomedical Research Foundation, Silver Spring, Maryland

Ledley R S 1971 Dental forces and mastication. *J. Texture Stud.* 2(1): 3–17

Ledley R S, Huang H K 1968 Linear model of tooth displacement by applied forces. *J. Dent. Res.* 47(3): 427–32

Ledley R S, Huang H K 1969 *Theoretical Analysis of Force Distribution on Lower Partial Dentures with Symmetric Saddles*, ASME Publication 69-BHF-2. American Society of Mechanical Engineers, New York

Ledley R S, Huang H K, Pence R G 1971 Quantitative study of normal growth and eruption of teeth. *Comput. Biol. Med.* 1(3): 231–41

R. S. Ledley

Dental Materials

The successful practice of dentistry is critically dependent on a wise choice of material and on its correct manipulation. Advances in dental materials science offer the practitioner not only products of improved properties, but also new techniques for improved prevention of tooth decay.

1. Requirements

The requirements governing the suitability of materials for use in the oral cavity are strict. The materials used should have no adverse effects on their environment; thus, substances which are toxic, irritant, allergenic, mutagenic or carcinogenic are unacceptable. The materials must be able to withstand the conditions prevailing in the mouth, and therefore resistance to corrosion, low solubility, and freedom from chemical attack by saliva and dietary constitutents are important considerations. Adequate mechanical properties (e.g., strength, modulus, hardness, impact strength and fatigue strength) are necessary to resist the forces of mastication. Other factors which should be considered include physical properties (e.g., thermal expansion, thermal diffusivity and density), and surface and adhesive properties. Aesthetic considerations—color, translucency and refractive index—must also be taken into account.

Considerable research is being devoted to the development of materials and techniques for bonding to the dental tissues. It is often important to know the rheological properties of materials, many of which are supplied as powders to be mixed with liquids, or as pastes, which subsequently harden. Rheological measurements are important to understand the manipulation of the materials both as supplied and as mixed, and also in determining the rate of change of viscosity of the material with time, so that working times and setting times are known. Viscoelastic properties of set materials are also relevant, and are assessed by measurement of creep and stress relaxation, and by the application of oscillatory techniques.

2. Metallic Materials

Pure gold can be used as a filling material. Gold foil, the surface of which is free of adsorbed gases, is packed into a prepared tooth cavity. Application of sufficient force causes welding and work hardening of the metal. However, this technique is time consuming and is therefore not in widespread use.

Gold is more frequently used as a constituent of casting alloys for inlays, crowns, bridges and occasionally for the framework of a partial denture. Gold is the principal component, owing to its good corrosion resistance. Copper, silver, platinum and palladium are usually present, and zinc may be included as a scavenger for oxygen. Some of these complex alloys are designed to be heat treated to give optimum strength and hardness.

Base-metal casting alloys, particularly those containing chromium (e.g., Co–Cr alloys), are now widely used for cast partial dentures. The presence of chromium renders the alloy resistant to corrosive attack. Alternative alloys such as Ag–Pd and Al–bronze have also been suggested.

Austenitic stainless steel is used occasionally as a denture base, and frequently in wires from which orthodontic appliances are constructed.

The most widely used filling materials are metallic. Dental amalgam is prepared by mixing a powdered alloy with mercury; the amalgamation reactions cause setting. A typical conventional alloy composition might be 69%Ag–27%Sn–3%Cu–1%Zn. These constituents are fused together by the manufacturer to form an ingot, which is cut into filings. Alternatively, spherical particle alloys are produced by an atomization process. The principal component of these alloys is Ag_3Sn (the γ phase). On mixing with mercury, some of the γ phase dissolves in the mercury, giving two reaction products, plus a core of unreacted Ag_3Sn:

$$Ag_3Sn + Hg \rightarrow Ag_2Hg_3 + Sn_{7-8}Hg + Ag_3Sn \text{ (unreacted)}$$

$$\gamma + Hg \rightarrow \underbrace{\gamma_1 + \gamma_2}_{\text{matrix}} + \underbrace{\gamma}_{\text{core}} \quad (1)$$

More recent amalgam alloys have produced materials which are essentially free of the γ_2 phase. These are copper-enriched alloys (often called high-copper alloys) and may be either blended alloys or single-composition materials.

The blended alloys consist of two parts (by weight) conventional alloy plus one part of Ag–Cu eutectic spheres. Two reactions are of importance: firstly, the reaction described in Eqn. (1) occurs with the subsequent reaction of the γ_2 phase with Ag–Cu:

$$Sn_{7-8}Hg + Ag\text{--}Cu \rightarrow Cu_6Sn_5 + Ag_2Hg_3$$
$$\gamma_2 + Ag\text{--}Cu \rightarrow Cu_6Sn_5 + \gamma_1 \qquad (2)$$

The Cu_6Sn_5 appears as a halo surrounding the Ag–Cu particles.

Single-composition alloys are usually ternary or quaternary formulations, in spherical or spheroidal form, of composition 60%Ag–25%Sn–15%Cu, 40%Ag–30%Sn–30%Cu or 59%Ag–24%Sn–13%Cu–4%In.

Copper-enriched alloys show a number of advantages over conventional materials. Lack of the γ_2 phase imparts improved corrosion resistance. If present this phase will produce products such as SnO_2 and $Sn(OH)_6Cl$ as well as free mercury which can react with some of the remaining γ phase. It has been suggested that this causes mercuroscopic expansion which may lead to breakdown of a restoration at the margin between amalgam and tooth structure; that is, the expanded material protrudes away from the supporting tooth structure and fractures. Amalgam without the γ_2 phase also shows less creep and develops strength more rapidly, with less likelihood of early fracture.

3. Polymeric Materials

Synthetic polymers have found widespread use in dentistry. Rigid polymers are used for denture-base construction, in denture repair, for making artificial teeth, as constituents of composite restorative materials (see Sect. 5), and as sealants bonded to tooth enamel to prevent decay. Flexible polymeric materials are applied as resilient linings for dentures, and in certain types of impression material (see Sect. 7). Water-soluble polymers are constituents of adhesive dental cements (see Sect. 6).

Many types of polymer have been tried with varying degrees of success for the construction of dentures, notably vulcanite, bakelite, cellulose nitrate, nylon, epoxy resins and poly(vinyl chloride).

The polymer of choice today for denture construction is poly(methyl methacrylate). This is supplied as a powder–liquid formulation. On mixing the powder (polymer plus benzoyl peroxide initiator) with methyl methacrylate monomer (a liquid containing a stabilizer such as hydroquinone), a dough is formed which is packed into a mold, and heated (typically 72°C for 16 h) to polymerize the monomer. These materials are particularly good in terms of aesthetic appeal. However, two limitations are evident. Firstly, the lack of radiopacity can present serious diagnostic problems if a denture, or part of a denture, is accidentally inhaled or ingested. Of the many experimental attempts to solve this problem, none has proved successful. Secondly, it would be desirable to have polymers with greater impact and fatigue strength. Dentures can be made from high-impact polycarbonates, but the cost and inconvenience of injection molding has limited their use.

Poly(methyl methacrylate) is also used for repairing, relining and rebasing dentures. In this instance, a room-temperature polymerizing material is used. A tertiary amine such as dimethyl-*p*-toluidine is incorporated into the monomer liquid to activate the peroxide initiator. These self-curing resins are not recommended for denture-base construction, as the resultant polymer is weaker and is more likely to suffer permanent deformation in service, as evidenced by higher creep values.

The polymerization of methyl methacrylate is a free-radical addition reaction; benzoyl peroxide decomposes to yield two free radicals:

$$(C_6H_5CO_2)_2 \rightarrow 2(C_6H_5CO_2^{\cdot})$$
$$C_6H_5CO_2^{\cdot} \rightarrow C_6H_5^{\cdot} + CO_2 \qquad (3)$$

Propagation then takes place:

$$R^{\cdot} + M \rightarrow R\text{--}M^{\cdot}$$
$$R\text{--}M^{\cdot} + M \rightarrow R\text{--}M\text{--}M^{\cdot}, \text{ etc.} \qquad (4)$$

where $R^{\cdot}$ represents the free radical from benzoyl peroxide and M represents a monomer molecule. Termination of the reaction occurs by the combination of two free radicals, giving a stable molecule:

$$R\text{--}M\text{--}M \ldots M^{\cdot} + R^{\cdot} \rightarrow R\text{--}M\text{--}M \ldots M\text{--}R \qquad (5)$$

or by a transfer process:

$$R\text{--}M\text{--}M \ldots M^{\cdot} + RH$$
$$\rightarrow R\text{--}M\text{--}M \ldots MH + R^{\cdot} \qquad (6)$$

Poly(methyl methacrylate), or acrylic resin, is also used for the fabrication of the teeth of dentures (see Sect. 4).

In addition to the methacrylate polymers, dimethacrylate resins are important, particularly for sealing pits and fissures of teeth prone to decay; these give cross-linked materials and also polymerize by a free radical mechanism. Pretreatment of the enamel is necessary to secure bonding. This is achieved by etching the surface with phosphoric acid (usually 30–40%). Meticulous attention to practical detail is important to secure durable bonding to enamel. These techniques can also be used to bond orthodontic appliances directly to teeth.

The dimethacrylate resins currently available offer a choice of techniques for activating the polymerization reactions. Some products are chemically activated, and others require the application of ultraviolet or visible radiation of the appropriate wavelength.

Table 1
Principal components of acid–base reaction cements

Type of cement	Cement powder	Cement liquid	Related materials
Zinc oxide–eugenol	zinc oxide	eugenol	some cements contain *o*-ethoxybenzoic acid (EBA cements)
Zinc phosphate	zinc oxide	phosphoric acid	occasionally, cements contain copper or silver salts in the cement powder
Zinc polycarboxylate	zinc oxide	aqueous solution of polyacrylic acid	polyacrylic acid may be in solid form as a powder component
Silicate cements	fluorine-containing aluminosilicate glass	phosphoric acid	silicophosphate cements have a powder which is essentially a mixture of zinc oxide plus aluminosilicate glass
Glass–ionomer cements	fluorine-containing aluminosilicate glass	aqueous solution of polyacrylic acid	

Resilient polymers are used as linings for dentures, for example, where there is atrophy of the denture-bearing area. Silicone polymers can be used, but their bonding to acrylic denture-base material is poor. Plasticized methacrylates bond well to poly(methyl methacrylate), but loss of plasticizer occurs in service, with a resultant loss of resilience.

In some instances temporary linings in dentures are used as tissue conditioners where there is inflammation of the oral mucosa. Similar products are also used for recording impressions of the oral tissues under functional stress. These materials are viscoelastic, and are usually composed of a polymer such as poly(ethyl methacrylate) which is mixed with an alcohol–ester mixture (e.g., 5–20% ethyl alcohol in butyl phthalylbutyl glycolate) to form a gel.

4. Ceramics

Porcelain usually contains a relatively high percentage of kaolin (China clay), plus quartz, and a flux to fuse at a suitable temperature to bind the material together. However, dental porcelain, which is used for the construction of jacket crowns, contains little or no kaolin; it is principally composed of quartz, feldspar (as a flux), and other lower fusing fluxes. As supplied for dental use, the ingredients have been fused together (to permit pyrochemical reactions to occur), then the resultant mass is quenched and ground. In the dental laboratory, this powder is mixed with water and molded to the required shape on a die, prior to firing at the appropriate temperature.

Aluminous porcelain has become widely used in the last few years; this contains a dispersion of 40–50% fine alumina which strengthens the material by hindering crack propagation. Care is taken to ensure that there is a close match between the coefficients of thermal expansion of the porcelain and alumina, so that the interface between the two is comparatively stress free. Pure alumina can also be used as a reinforcing insert in a ceramic restoration.

Techniques have been developed for bonding porcelain to metallic materials, to take advantage of the excellent esthetic characteristics of the former and the good mechanical properties of the latter. The alloys of choice may be gold or silver with platinum and palladium, or Ni–Cr. An alternative is to bond the porcelain to platinum foil.

Other applications of ceramics in dentistry include their use in cement powders (see Sect. 6), as constituents of some laboratory materials (see Sect. 8), for making artificial teeth, and as components of polymer–ceramic composite filling materials (see Sect. 5).

5. Composites

Polymer–ceramic composite materials are widely used for anterior restorations, having largely replaced silicate cements for this purpose (Table 1). As supplied, these consist of:

(a) monomers capable of polymerizing to form a set material,

(b) inorganic fillers to give stronger and more rigid materials with lower coefficients of thermal expansion,

(c) silane coupling agents with which the inorganic components are treated in attempts to secure good bonding with the polymeric phase,

(d) polymerization inhibitors to give stability on storage,

(e) uv stabilizers to reduce discoloration of the set material, and

(f) components to act as activators and/or initiators of the polymerization reaction.

The principal monomer is usually a viscous dimethacrylate, such as the reaction product (Fig. 1a) between

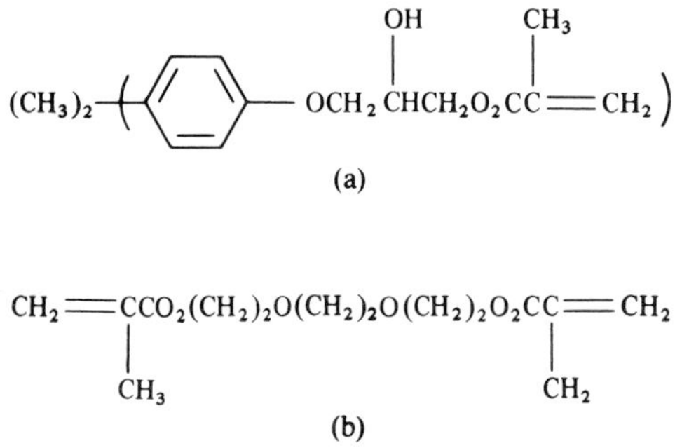

$$(CH_3)_2 \left(\underset{\text{(a)}}{\bigcirc} - OCH_2 \overset{OH}{\underset{|}{C}}HCH_2O_2C \overset{CH_3}{\underset{|}{C}} = CH_2 \right)$$

$$CH_2 = \overset{}{\underset{CH_3}{C}}CO_2(CH_2)_2O(CH_2)_2O(CH_2)_2O_2C \overset{}{\underset{CH_2}{C}} = CH_2$$

(b)

Figure 1
Structure of: (a) Bowen's resin, (b) triethylene glycol dimethacrylate

bisphenol A and glycidyl methacrylate (known as bis-GMA or Bowen's resin). Urethane dimethacrylates are alternatives. Usually a lower molecular weight monomer of lower viscosity is also included (e.g., triethylene glycol dimethacrylate, Fig. 1b).

Inorganic fillers are usually present up to about 78 wt%, and may be crystalline quartz or glasses (including barium containing glass to confer radiopacity). The particles of these fillers are usually 20–50 μm. Recently microfine silica, particle size of about 0.05 μm, has been employed; the fine particles are prepared in a matrix of polymer which is then pulverized and used as the filler. The inorganic content of such products is from 30 to 60 wt%. In general, the microfine materials are not as strong and rigid as the more conventional products, but it is easier to obtain a restoration with an acceptably smooth surface from them.

The dental practitioner is faced with a choice of curing systems for composites. Commonly, amine–peroxide systems have been employed. These, however, have a limited working time, since there is a continuous increase in viscosity from the start of mixing. Command-setting systems are available where activation is achieved by the application of either ultraviolet or visible light. Visible-light-activated materials are beginning to be widely used.

Two recent developments are of significance: (a) new composites or resin-bonded ceramics with up to 87 wt% filler, intended as a replacement for dental amalgam; and (b) chemical bonding agents to secure adhesion of composites to dentine.

6. Cements

Cements are a special class of material and find widespread dental application. Such substances are used:

(a) as linings for cavities, to protect the pulp of the tooth from chemical damage by constituents of filling materials, or to give thermal insulation under large metallic restorations;

(b) as temporary restorations;

(c) for filling deciduous teeth;

(d) for luting cast metal and porcelain restorations;

(e) for application to exposed pulps of teeth; and

(f) as root-canal filling materials.

Most of the available materials are acid–base reaction cements. A basic or amphoteric powder (proton acceptor) is mixed with an acidic liquid (or proton donor). Some of the powder reacts with the liquid to form a gel salt, which acts as a matrix to bind together a core of remaining unreacted powder.

The principal powders are zinc oxide and ion-leachable fluorine containing aluminosilicate glasses. The liquids are aqueous phosphoric acid, eugenol and an aqueous solution of polyacrylic acid. The available acid–base reaction cements are summarized in Table 1. The cement containing polyacrylic acid is adhesive to tooth enamel, if care is taken in the technique of application of the material.

Miscellaneous materials include calcium hydroxide cements and varnishes which contain a resin dissolved in a volatile solvent.

Table 2
Impression materials

Classification	Type of material	Setting reaction
Nonelastic	plaster of paris	hydration of calcium sulfate hemihydrate to form the dihydrate; contains additives to control setting expansion and time
Nonelastic	impression paste	reaction between zinc oxide and eugenol to give zinc eugenate
Nonelastic	impression compound or composition	cooling of a thermoplastic material
Elastic	hydrocolloid: agar	gel formation on cooling
Elastic	hydrocolloid: alginate	cross-linking of alginate molecules by Ca^{2+}
Elastic	elastomer: polysulfide ⎫	
Elastic	elastomer: polyether ⎬ set by polymerization and cross-linking	
Elastic	elastomer: silicone ⎭	

7. Impression Materials

Where a restoration or appliance is to be fabricated outside the mouth, it is first necessary to obtain an impression, from which a model, cast or die can be prepared. The ideal impression material should be capable of plastic deformation on insertion into the mouth (and sufficiently fluid to record fine detail), and should deform elastically on displacement from the oral tissues.

A wide range of materials is available, some of which are nonelastic when set and so cannot be used when undercuts are present. The available types of products are listed in Table 2.

8. Laboratory Materials

In the dental laboratory, plaster of paris type materials are used for casts and dies. Waxes are used as pattern materials, e.g., in the preparation of two-part molds for denture construction and as patterns for gold restoration. Refractory investment materials are required as molds for cast restoration. These contain silica and a choice of binding agents—either gypsum (for lower fusing alloys such as those of gold), phosphate, or silica binding systems (e.g., for cobalt-base alloys which melt above 125°C). Laboratory materials also include solders, fluxes, abrasives and polishing agents.

See also: Teeth: Physical Properties; Preventive Dentistry; Dental Force Analysis

Bibliography

American Dental Association 1981 *Dentist's Desk Reference: Materials, Instruments and Equipment.* American Dental Association, Chicago

Combe E C 1981 *Notes on Dental Materials*, 4th edn. Churchill Livingstone, Edinburgh

Horn H R 1981 Symposium on composite resins in dentistry. *Dent. Clin. North Am.* 25(2)

Phillips R W 1982 *Skinner's Science of Dental Materials*, 8th edn. Saunders, Philadelphia

Smith D C, Williams D F (eds.) 1982 *Biocompatibility of Dental Materials*, Vols. 1–4. CRC Press, Boca Raton

E. C. Combe

Digital Fluorography

Medical imaging is one of the most widely used diagnostic modalities in modern medical practice. In the UK alone, over 22 million examinations are undertaken annually (Kendall et al. 1980). The imaging techniques involved are conventional radiography (for over 75% of the examinations), fluoroscopy (10–15%), as well as computerized tomography (CT), ultrasound, and radio-

nuclide imaging. While images have been produced in digital format for some time in the case of CT, ultrasound and radionuclide examinations, the application of digital techniques to fluoroscopy, or photoelectric imaging, means that the major imaging market is now being addressed by this technology. Further, the possibility of employing photoelectronic systems as general radiographic transducers is already being considered (Capp et al. 1981). This would lead to departments of medical imaging using digital techniques exclusively, a development which would have considerable impact on medical imaging and on health care in general.

Digital fluorographic systems (see Fig. 1) were developed initially for angiographic examinations (Mistretta

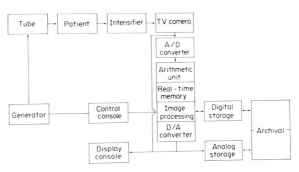

Figure 1
Basic components of a digital fluorographic system

et al. 1981). They offered the clinician the immediate potential of visualizing vascular structures more clearly, as well as the possibility of introducing contrast media by intravenous injections for particular examinations. Medical imaging systems are designed to perform particular clinical examinations. In this article the design factors which affect the physical performance of digital fluorographic systems are discussed: special attention is given to the design aspects of photoelectronic systems which are being considered as radiographic transducers.

The physical performance of digital fluorographic systems is usefully considered in two main areas: (a) image quality measured in terms of resolution, noise, signal-to-noise ratio and threshold contrast; and (b) radiographic capability resulting from framing rates, the ability to undertake gated studies, and the need for and capability of image-processing packages.

1. Spatial Resolution

For radiology, image resolution should match that of the eye because radiographic images are interpreted by trained observers. It is worth noting—in the interest of cost effectiveness—that the overall resolution of an imaging system can be no better than that of the component with the poorest resolution. Hence, improvements in any single component may produce little improvement generally, because of limitations elsewhere in the system.

It is useful to consider resolution in relation to four aspects of digital fluorographic systems:

(*a*) *Display mapping.* This relates the input image field size to the display matrix size. This is obviously governed by many interrelated factors but does provide a first indication of the smallest resolvable element. The largest possible display matrix for given field size is desirable for optimum displayed resolution and there has been a steady increase from 256×256 pixels to 512×512 and more recently to 1024×1024 formats. Increased display matrix size requires fundamental improvements in system design which can usually be obtained only at increased cost. It is pointless to increase matrix sizes if other more basic factors limit the resolution.

(*b*) *Geometric unsharpness.* This is governed by focal-spot size of the x-ray tube and the distances from the focus to patient and from the patient to the imaging device. Digital fluorography employs high repetition, high x-ray output exposures, thus tube loading is important. Hence focal spot size—an important aspect of tube loading and therefore geometric unsharpness—needs to be considered.

(*c*) *Movement unsharpness.* This is due to involuntary movement of the patient or individual organs within. The amount of movement unsharpness in any single x-ray image or frame is dependent on the speed of the anatomical motion and the exposure time per frame. A wide variation of anatomical speeds are encountered clinically including motions with speeds up to $400 \, \text{mm s}^{-1}$ encountered within the heart. Time-interval studies can employ several seconds of imaging data, thus movement unsharpness between single frames may be relevant.

(*d*) *The combination of unsharpness from individual components in the total system.* In digital fluorography these components are the image intensifier, the optical coupling, the TV camera, the data collection, the data display system and the observer's visual process. System unsharpness affects all the factors controlling the display mapping as outlined in (*a*).

It is worthwhile to consider in more detail the physical basis of the interrelated factors which control display mapping. One of the more important aspects is associated with the nature of the image produced and displayed by video or television. In fluoroscopic systems the image is read off the image intensifier by a TV camera which employs a line scan reading process. The image is displayed on a TV monitor similarly by means of a line scan process. Intuitively one associates a single TV line with the fundamental element of resolution. There are, however, other factors which determine the element of resolution in line scan systems, namely the psycho-physical requirements that a certain number of TV lines must occupy the smallest dimension of an object if it

is to be: (a) detected, (b) recognized and (c) identified (Rossell and Willson 1973). For detection approximately 2 lines are required per minimum object dimension, for recognition approximately 8 lines, and for identification 13 lines. Hence, if a radiologist is to predict with a degree of confidence the state—healthy or otherwise—of a contrast-medium-filled vessel lying in the direction of the scan lines, approximately 4 TV lines per vessel thickness are required. The minimum diagnostically recognizable object—where diagnostically recognizable implies something more than pure detection—depends, therefore, upon the display mapping, that is, the number of TV lines which are employed to image the total input field size. Ideally, for maximum resolution the input field size should be kept as small as possible, and the largest possible number of scan lines and/or matrix size used. One cannot continue to increase the number of TV scan lines or the display matrix size without encountering important design constraints.

Line-scan imaging systems do not actually give one TV line's worth of resolution for every scan line. The vertical resolution V_R (number of effective TV lines produced) is given by

$$V_R = K N_s \tag{1}$$

where N_s is the number of scan lines and K is the Kell factor (Kell et al. 1934). A Kell factor of approximately 0.7 is usually employed, thus 100 scan lines are required for 70 TV lines' worth of resolution. For a 525-line system, V_R is approximately 350, and is 618 for a 945-line system.

In any imaging system it is desirable to have equal vertical and horizontal resolution. The number of TV lines determines the vertical resolution. The horizontal resolution H_R, corresponding to the number of effective resolution elements along a single line, is given by

$$H_R = \frac{b_f}{\text{frames per second} \times \text{lines per frame}} \tag{2}$$

where b_f is the frequency bandwidth of horizontal and vertical time bases. For a 525-line system with 343 effective TV lines, a bandwidth of 3.3 MHz is required, whereas for a 945-line system, 10.5 MHz is necessary. Attempts to increase both horizontal and vertical resolution require large increases in the bandwidth of the TV system if the number of frames per second remains constant.

The number of scan lines employed in the TV camera and display monitor is fundamental in relation to the resolving capability of digital fluorographic systems and it is important to consider how this aspect of resolution compares with that of other components in the system. Figure 2 represents modulation transfer functions (MTFs) of the important imaging components. At a focus-to-film distance (FFD) of 100 cm and an object-to-film distance (OFD) of 10 cm, a 1 mm x-ray focal spot will resolve eight line pairs per millimeter ($8 \, \text{lp mm}^{-1}$) which is a significantly greater resolution capability than the other components possess. The effect of larger focal

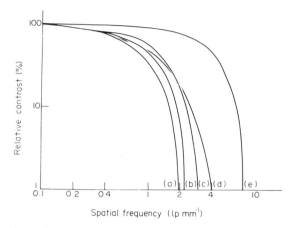

Figure 2
Modulation transfer functions of (a) an intensifier/TV system (22.5 cm image intensifier/1000 TV lines), (b) movement unsharpness (2.0 cm s^{-1}, 20 ms exposure), (c) movement unsharpness (0.5 cm s^{-1}, 20 ms exposure), (d) a modern image intensifier, and (e) a 1 mm focal spot

spots or poorer geometry would be noticeable, however. Modern image intensifiers have resolving capabilities of some 4 lp mm^{-1}. Since a line pair is equivalent to 2 TV lines, 22.5 cm intensifiers can accommodate 1800 TV lines and 35 cm intensifiers nearly 3000 lines.

The resolving capability of image intensifier/TV systems is mainly dictated by the TV camera. Conventional vidicons with a 2.5 cm input face employ up to 800–1000 scan lines. Employed with a 22.5 cm intensifier, this is equivalent to approximately 2 lp mm^{-1} resolution. In order to utilize the full resolving capability of image intensifiers, TV cameras with many more scan lines must be employed. Vidicons with over 2000 scan lines are available (Roehrig et al. 1977) but these have larger input faces. Employing TV cameras with larger input faces will probably require different optical couplings between the intensifier and TV camera. Multiple intensification stages may be necessary to ensure adequate light intensity at the camera face for each resolution element in the scan process.

Figure 2 shows the MTFs for movement unsharpness arising from both 0.5 and 2 cm s^{-1} velocities with a 20 ms exposure time and standard radiological geometry. These curves highlight the critical role played by movement in defining the upper limit of resolution. In order to minimize movement unsharpness in each stored image frame some digital fluorographic systems employ repetitive x-ray exposures of short duration (pulsed mode). Movement unsharpness between frames may, however, lead to misregistration errors when performing subtraction studies.

Pulsed-mode operation involves relatively high x-ray exposures for each image frame in order to reduce the quantum noise level (see Sect. 2). The noise may be

reduced by employing continuous x-ray exposure of lower intensity and integrating the requisite number of video frames. Under these conditions each integrated image will contain integrated motion blur.

2. Noise

Whereas spatial resolution determines the smallest size of object which can be resolved, noise sets the limit on the smallest contrast difference which can be detected. Every stage in a multistage process introduces some noise or statistical uncertainty into the data transfer process. In fact, the accuracy or quality of data transfer in imaging systems can be measured in terms of the comparative noise level or detective quantum efficiency DQE (Dainty and Shaw 1974), where

$$DQE = \frac{(\text{noise in})^2}{(\text{noise out})^2} \qquad (3)$$

The main types of noise in digital fluorographic imaging systems are:

(a) quantum noise due to the finite number of x-ray photons employed to produce an image,

(b) beam-shot or electron-beam reading noise due to the finite number of electrons employed in reading the image from the TV camera input face or target, and

(c) general electronic noise due to thermal effects in circuit components.

In designing any radiological imaging system it is particularly important to ensure that the main source of noise is that due to the quantum nature of the x-ray input. The system is then x-ray quantum limited and ensures that optimum use is being made of the level of x-ray exposure employed, at least with respect to noise. In fact, most of the early development in fluoroscopic systems was concerned with producing an image which was bright enough for photopic vision and, therefore, for maximum visual acuity (Ter-Pogossian 1967), while maintaining an x-ray quantum noise limit. Digital fluorographic systems use high x-ray exposures in order to reduce quantum noise. The magnitude of system component noise relative to the quantum noise level is thus an important factor.

Figure 3 outlines the quantum transfer of information through an image intensifier for both fluoroscopic and fluorographic input exposure rates, some 100 μR s^{-1} and 1 mR s^{-1}, respectively. At the intensifier input there is a decrease in the number of x-ray quanta due to their fractional absorption in the input phosphor. In the fluoroscopic process, the image integration period of the eye must also be considered since an image is only available for approximately 0.1 s. A large quantum gain results from the conversion of x-rays to light, followed by a decrease as the light is converted to electrons. Finally, there is a large quantum increase at the output

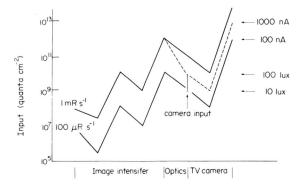

Figure 3
The quantum transfer of information through an image intensifier/TV system for fluoroscopic and fluorographic dose rates

of the intensifier due to the intensifier's acceleration of electrons. Only a relatively small fraction of the light emitted from the back face of the intensifier can be coupled to the TV camera, and there is also some loss at the input face of the TV camera. These losses are more than overcome by the large electronic gain provided by the photoconductive process occurring at the intensifier/TV camera coupling stage. TV cameras operate over a given range of light-input intensities. The lower limit of intensity must be sufficient to overcome any dark current in the device; the upper limit is set by saturation constraints. The upper and lower values indicated in Fig. 3 are those applicable to a plumbicon camera (Haan et al. 1964) and give rise to the indicated operating range for TV camera currents. The lower limit of TV camera current is dependent also upon temporal modulation effects (Franken and Scheren 1972), the value of 100 nA shown in Fig. 3 being merely an indication of its magnitude. As the input x-ray exposure rate is increased, the overall illumination of the camera face increases also. To avoid saturation and the possiblity of damaging the camera, the optical coupling between the intensifier and the camera is decreased by means of an iris diaphragm. To deduce the upper and lower limits of x-ray exposure the camera illumination E has to be considered. It can be expressed as

$$E = \pi \dot{D} G V / 4f^2 \qquad (4)$$

where $\dot{D}$ is the intensifier x-ray exposure rate (mR s^{-1}), G is the conversion factor (Cd m^{-2} per mR s^{-1}), V is the optical coupling transmittance and f is the lens number (Kuhl 1969). With these units E is given in lux. The light intensity incident upon the TV camera face gives rise to a signal current i_s (in amps), where

$$i_s \approx 10^{-7} \dot{D} G V / f^2 \qquad (5)$$

The upper limit of this current is set by the overload conditions and the lower limit by the system noise.

Assuming $i_s < 5 \times 10^{-7}$ A (to prevent overload) and $i_s > 10^{-8}$ A (to exceed system noise) then

$$10^{-1} < \dot{D} G V / f^2 < 5 \qquad (6)$$

These limitations are only approximate but they serve to indicate the design constraints imposed at this stage of the system. As the input dose rate is increased, f must be altered to maintain satisfactory operating conditions since G and V are normally fixed. Changes in dose rates by a factor of 10–20 can probably be accommodated in individual intensifier/TV systems. This is, however, an important consideration if one attempts to employ a single photoelectronic system to undertake both fluoroscopy, with dose rates at the imaging device as low as 10–50 μR s^{-1}, and radiography, with dose rates as high as 1–10 mR s^{-1}.

The total system noise per pixel σ_s, including both beam shot noise and general electronic-component noise, is given by

$$\sigma_s = [\Delta f (K_1 i_s + K_2 + K_3 \Delta f^2)]^{1/2} \qquad (7)$$

where K_1, K_2 and K_3 are constants, i_s is the TV camera signal current and Δf, the electronic bandwidth which is inversely related to the time taken to read one picture element or pixel (Roehrig et al. 1977). In an area containing N pixels the noise will be reduced by a factor of $N^{1/2}$.

Frequency bandwidth plays a major role in defining system noise. The next section shows that it is possible to increase the signal-to-noise capability of the TV camera by reducing the bandwidth.

3. Contrast and Signal-to-Noise Ratio

The primary x-ray contrast C_p in the transmitted x-ray beam is given by

$$C_p = \frac{N_1 - N_2}{\langle N_1 \rangle} = \frac{N_1 - N_2}{N_a} \qquad (8)$$

where N_1 and N_2 are numbers of photons, $\langle N_1 \rangle$ is the average value of N_1, and N_a signifies average number of photons (Moores 1982). The signal-to-noise ratio (S/N) at the input to the intensifier is given by

$$(S/N)_{in} = C_p (N_a A)^{1/2} \qquad (9)$$

where A is the area of a region with contrast C_p relative to the background. The smallest detectable (threshold) contrast C_{pt} is therefore given by

$$C_{pt} = K / (N_a A)^{1/2} = K / (S/N)_{max} \qquad (10)$$

where K is the threshold signal-to-noise ratio, normally in the region of 3–5 (Rose 1948), and $(S/N)_{max}$ is the maximum achievable signal-to-noise ratio. For quantum limited systems, $(S/N)_{max}$ is determined by the number of photons employed, where $N_1 \gg N_2$ and

$$(S/N)_{max} = N A / (N_a A)^{1/2}$$

High exposure rates can be employed in digital fluorography to reduce quantum noise and hence minimize threshold contrast. Under these circumstances system noise and maximum achievable system signal-to-noise ratio are important. For linear subtraction it can be shown that

$$(S/N)_{\text{lin}} = \left[\frac{1}{(S/N)_p^2} + \frac{1}{(S/N)_{\text{TV}}^2} \right]^{1/2} \quad (11)$$

(Cohen et al. 1982), where $(S/N)_p$ is the photon signal-to-noise ratio and $(S/N)_{\text{TV}}$ that of the TV system. To ensure an x-ray quantum-limited digital image, $(S/N)_{\text{TV}}$ should be as high as possible. As indicated in Sect. 2, electronic frequency bandwidth plays an important role in defining system noise. By using amplifiers of much narrower bandwidth it is possible to increase the signal-to-noise ratio of the TV camera beyond that achieved at normal frame rates (Roehrig et al. 1977). This approach to noise reduction is termed slow scan and is employed to read short, relatively high-intensity x-ray exposures with minimum quantum noise. If high signal-to-noise ratios are to be maintained it is important to ensure that the analog-to-digital (A/D) converter following the camera has sufficient discrimination. Most units now employ 10 bit A/D conversion which ensures that 1000:1 signal-to-noise ratios can be accommodated.

Another approach to the reduction of noise involves frame integration. In this approach several video frames are integrated. This has the effect of reducing both quantum and system noise and places less stringent requirements on TV camera signal-to-noise ratios. It does mean, however, that the final image is an average in respect of unsharpness caused by movement.

The signal-to-noise ratio given in Eqn. (11) may be written in terms of x-ray photon number (Cohen et al. 1982), such that

$$(S/N)_{\text{lin}} = \left[\frac{1}{N_a A} + \frac{T}{(N_a A)^2} \right]^{-1/2} \quad (12)$$

where T is a constant, involving intensifier and TV camera factors. For a quantum-limited exposure regime, $(S/N)_{\text{lin}}$ is proportional to $(N_a A)^{1/2}$, whereas for a system noise limit, $(S/N)_{\text{lin}}$ is proportional to $N_a A$.

Measurements undertaken on one particular unit (Cohen et al. 1982) indicate that the transition from quantum-noise-dominant to video-noise-dominant regimes occurs at an exposure per frame of approximately 340 µR.

4. Contrast Detail

An extremely useful measure of imaging-system performance is provided by a contrast-detail diagram in which threshold contrast is plotted against the diameter of a circular target area (Hay and Chesters 1976). Contrast detail measurements have been undertaken on a digital fluorographic system for different exposures and different field sizes (Macintyre et al. 1981). Their results showed clearly that, for small detail or resolution limited imaging, the design characteristics of the digital system, that is, input field size and numbers of scan lines, play a major role in defining the behavior of the system in relation to contrast detail. For larger detail (>2 mm diameter), where quantum efficiency is important, the digital system performed significantly better than a film subtraction technique in which the exposures employed were higher. One reason for the potential improvement in the contrast-detail behavior of digital fluorographic units over film/screen systems even at lower exposure lies in their ability to display the image optimally. Subtracted images contain a limited range of data since relevant information is superimposed on a uniform background. The relevant information can be displayed at maximum gain or gamma. The effect of display gain or gamma on TV images has been investigated (Oosterkamp 1974). It has a marked effect on threshold contrast, particularly for gammas less than unity.

Threshold contrast behavior in clinical digital fluorographic practice is affected by certain biological factors. Not least of these is the actual concentration of contrast agent which can be made to flow in vessels of particular diameters. High-contrast boluses of 150–300 mg of iodine per milliliter normally exist only in close proximity to injection sites. With a concentration of 150 mg iodine ml^{-1} the contrast of a 1 mm vessel exceeds 10%—well above threshold.

Dilution to concentrations of 15–30 mg iodine ml^{-1} can be expected in the arterial vasculature if the injectate has been introduced intravenously. A 1 mm vessel with an iodine concentration of 15 mg ml^{-1} will have a contrast of some 2.5% at diagnostic x-ray energies. This level of contrast is close to threshold level at digital fluorographic x-ray exposure rates.

Threshold contrast is also affected by patient movement and by scattered radiation.

5. Radiographic Performance and Image Processing

Image quality is important but so too are the radiographic and image processing facilities provided by digital fluorographic units. As indicated previously, framing rates are in part governed by noise requirements in each frame. Normal video frame rates are in the region of 30 full images per second and A/D converters with 10 bit wordlength and 10 MHz frequency are employed to digitize them. Tube loading will not permit the use of extremely high exposures at this frequency. Consequently, either frame integration or pulse exposures with slow scan TV operation can be employed to reduce noise. Whichever approach is employed a key element in the process is the real-time memory shown in Fig. 1. The size of this memory dictates both the image matrix size and frame frequency which can be handled in real time. For example, time-interval-difference studies involving subtraction of sequential frames can be undertaken in real time at a rate dictated by the size of the memory. One major area of development has been the increase in this memory size. A memory size of 3 megabytes is now standard and larger sizes are being introduced.

A generator which can be triggered from an external control is required for gated studies. The central processing unit provides the necessary pulse sequence. Alternatively, this pulse sequence may be derived from physiological monitoring equipment.

A wide variety of image-processing software packages are now available. Straightforward subtraction with the facility to "remask" in order to overcome misregistration is standard. Packages are obtainable for quantitative flow studies and for heart volume analyses. In fact, digital fluorography now unites cardioangiography and video densitometry in a single facility.

Movement blurring is an extremely important phenomenon and it cannot be eliminated in every examination. Software packages enable part of an individual frame to be reregistered by shifting an appropriate number of pixels horizontally or vertically or even rotating them. Frequency filtering of an image in order to eliminate undesirable structures, for example, the heart in coronary angiography, is also under development.

These facilities expand the capability for interrogating the original data. It must not be forgotten, however, that a trained observer is an extremely powerful datahandling entity. Nonetheless, given the current trends in both hard and software developments in information technology, more sophisticated automatic image analysis packages are likely to appear in the not-too-distant future. It is important that the radiological community assesses their performance adequately to ensure their cost-effective implementation.

See also: Image Analysis: Extraction of Quantitative Diagnostic Information

Bibliography

Capp M P, Nudelman S, Fisher D, Ovitt T W, Pond G D, Frost M M, Roehrig H, Seeger J, Oimette D 1981 Photoelectronic radiology department. *Proc. Soc. Photo-Opt. Instrum. Eng.* 314: 2–8

Cohen G, Wagner L K, Rauschkolb E N 1982 Evaluation of a digital subtraction angiography unit. *Radiology* 144: 613–17

Dainty J C, Shaw R 1974 *Image Science: Principles, Analysis and Evaluation of Photographic-Type Imaging Processes.* Academic Press, London

Franken A A J, Scheren W J L 1972 The influence of the camera tube on the temporal modulation transfer function in diagnostic X-ray television. *Medicamundi* 17: 121–23

Haan E F, van der Drift A, Schampers P P M 1964 The "plumbicon", a new television camera tube. *Philips Tech. Rev.* 25: 133–80

Hay G A, Chesters M S 1976 Threshold mechanisms in the presence of visible noise. In: Hay G A (ed.) 1976 *Medical Images: Formation, Perception and Measurement, 7th L H Gray Conf.* Wiley, Chichester, pp. 208–20

Hillman B J, Ovitt T W, Nudelman S, Fisher H S, Frost M M, Capp M P, Roehrig H, Seeley G 1981 Digital video subtraction angiography of renal vascular abnormalities. *Radiology* 139: 277–80

Kell R D, Bedford A V, Trainer M A 1934 An experimental television system. *Proc. Inst. Radio Eng.* 22: 1246–52

Kendall G M, Darby S C, Harries S V, Rae S 1980 *A Frequency Survey of Radiological Examinations carried out in National Health Service Hospitals in Great Britain in 1977 for Diagnostic Purposes,* NRPB Report 104. National Radiological Protection Board, Didcot

Kuhl W 1969 X-ray image intensifiers today and tomorrow. *Medicamundi* 14: 57–61

Macintyre W J, Pavlicek W, Gallagher J H, Meaney T F, Buonocore E, Weinstein M A 1981 Imaging capability of an experimental digital subtraction angiography unit. *Radiology* 139: 307–13

Mistretta C A, Crummy A B, Strother C M 1981 Digital angiography: A perspective. *Radiology* 139: 273–76

Moores B M 1983 Screen/film combinations for radiography. In: Sharpe R S (ed.) 1983 *Research Techniques in Nondestructive Testing,* Vol. 6. Academic Press, London, pp. 289–308

Oosterkamp W J 1974 New concepts and progress in instrumentation for cine and video radiology. *Medicamundi* 19: 79–84

Roehrig H, Nudelman S, Capp M P, Frost M M 1977 X-ray image intensifier video system for diagnostic radiology, Pt. 1: Design characteristics. *Proc. Soc. Photo-Opt. Instrum. Eng.* 127: 216–25

Rose A 1948 The sensitivity performance of the human eye on an absolute scale. *J. Opt. Soc. Am.* 38: 196–208

Rossell F A, Willson R H 1973 Recent psychophysical experiments and the display signal-to-noise ratio concept. In: Biberman M L (ed.) 1973 *Perception of Displayed Information.* Plenum, New York, pp. 167–232

Ter-Pogossian M M 1967 *The Physical Aspects of Diagnostic Radiology.* Harper and Row, New York

B. M. Moores

Doppler Blood Flow Measurement

The ultrasonic measurement of blood flow by the reflection technique depends on the Doppler shift in the frequency of ultrasound backscattered from the moving ensemble of blood cells. The method gives directional information about velocity and is safe, noninvasive and convenient to use. Although the calculation of blood flow volume rate is subject to several errors, the method appears favorable in comparison with alternative techniques. The technique can also be used during invasive procedures.

1. Basic Principles

If two coplanar transducers are arranged so that one receives reflected ultrasonic energy transmitted by the other, the received signal is shifted in frequency by the Doppler effect if the reflector (or ensemble of scatterers such as blood) is moving with respect to the transducers. The difference f_D between the transmitted and received frequencies is given by

$$f_D = \frac{2vf}{c} \cos \gamma \qquad (1)$$

where v is the speed of the reflector towards the source of ultrasound, f is the frequency of the transmitted ultrasound, c is the speed of ultrasound and γ is the angle between the directions of the ultrasonic beam and the movement of the reflector (Wells 1977).

The Doppler measurement of blood flow depends on the detection of a backscattered ultrasonic signal of adequate amplitude. Optimal conditions are obtained by an appropriate compromise between attenuation in the tissue between the transducer and the flowing blood (which increases roughly in proportion to the frequency) and the backscattering from the blood (which increases roughly in proportion to the fourth power of the frequency). The optimal frequency f_{opt} (in MHz) for the Doppler detection of normal hematocrit blood is given by $f_{opt} \simeq 90/d$ where d is the soft-tissue distance (in mm) between the transducer and the blood (Reid and Baker 1971). For example, in the case of a blood vessel situated about 10 mm below the surface of the skin (such as the carotid artery in the neck), the optimal ultrasonic frequency is about 9 MHz, and the Doppler shift frequency for a flow velocity of 100 mm s^{-1} and an angle γ equal to 45° (from Eqn. (1)) is 850 Hz. Similarly, detection of flow in the ascending aorta by means of a transducer situated in the suprasternal notch, where the range is about 60 mm, is optimal at a frequency of about 1.5 MHz; but considerations of beam divergence, which increases with decreasing frequency for a given size of transducer, might lead to a choice of 2–3 MHz as the best compromise in this situation.

2. Instrumentation

2.1 Ultrasonic Doppler Flowmeters

The simplest type of ultrasonic Doppler blood-flow detector uses continuous waves of ultrasound (Atkinson and Woodcock 1982). The transmitter drives a transducer at constant amplitude and frequency. The probe usually has separate transducers for transmitting and receiving (to minimize the direct transfer of energy from the transmitter to the receiver, which might otherwise overload the receiver), and the receiving transducer detects ultrasound backscattered by structures lying along the ultrasonic beam within the patient. The backscattered ultrasound consists of a mixture of signals, some of frequency equal to that of the transmitter (due to reflections from stationary structures) and some

shifted in frequency by the Doppler effect (due to reflections from moving structures and backscattering from flowing blood). These signals are mixed in the demodulator of the receiver, the output from which contains the difference frequencies between the transmitted wave and the Doppler-shifted received waves. These difference frequencies, which generally fall in the audible range, are filtered out and amplified. The resultant output signal can be listened to, or it may be fed to an analyzer (see Sect. 2.2).

The resolution in azimuth and elevation of a continuous-wave Doppler system depends on the cross-sectional dimensions of the ultrasonic beam at the position of the moving target under investigation (Wells 1970). Such a system does not have any range discrimination, however, and all moving targets lying along the ultrasonic beam contribute to the Doppler frequency spectrum provided that the signals to which they give rise are of sufficient amplitude to be detected.

Range discrimination can be given to a Doppler system by pulsing the transmitted ultrasound and gating the received signal (Wells 1969). The duration of the transmitted pulse (to which the receiver gate is made equal) determines the length of the resolution cell (defined as the sensitive volume of the system), and the position of the resolution cell along the ultrasonic beam is determined by the delay between the transmission of the pulse and the triggering of the receiver gate. Thus, both the position and length of the resolution cell can be controlled. Each transmitted pulse provides a new sample of Doppler information, and this is used to update a sample-and-hold circuit, the output from which is indistinguishable (from the point of view of the operator) from that of a continuous-wave system.

In a pulsed Doppler system of this type, the maximum pulse-repetition rate is limited by the round-trip transit time between the transducer and the resolution cell, and by the reverberation decay time. It is well known that if a signal waveform has an upper frequency of f_{max} in its spectrum, it is possible to convey all the information in the spectrum provided that it is sampled at a frequency of at least $2f_{max}$. This sets an upper limit to the maximum unambiguously measurable vector velocity at any given penetration for a given ultrasonic frequency.

Simple Doppler systems are only capable of measuring the magnitude of the Doppler-shift frequency, and not the direction of flow. The directional information can be extracted from the ultrasonic Doppler signal by determining whether the received frequency is higher (corresponding to flow towards the transducer) or lower (flow away) than that of the transmitter (Coghlan and Taylor 1976). Conceptually the simplest way of doing this is to use upper and lower sideband ultrasonic frequency filters. More commonly, however, the received ultrasonic signal is separately mixed with two signals, both at the transmitted frequency but differing in phase by 90°. The resultant two-channel signals can be processed in one of three different ways. Firstly, using time-domain processing, a logic circuit can determine which signal leads the other in phase, and can switch the Doppler signal to appear either at a "forward flow" or at a "reverse flow" output point. The advantage of the method is that it is relatively inexpensive to incorporate into an instrument. The main disadvantage, however, is that the technique cannot cope with simultaneous forward and reverse flow. The second method avoids this important disadvantage by using phase-domain processing by shifting the signals in each channel by 90° and adding them to the signals in the opposite channel. The resultant signals appear either at a "forward flow" or at a "reverse flow" output point. The main difficulty in implementing the method is that the 90° phase-shifting networks have to operate over the whole of the Doppler-shift frequency spectrum. The third method of directional separation employs frequency domain processing. The signals in each channel are mixed with two signals at the same pilot frequency but separated by 90°. The resultant signals are added to produce an output signal in which zero flow corresponds to the pilot frequency (in the audible range) with forward and reverse flow signals, respectively, in the upper and lower sidebands.

2.2 Instruments for Analyzing Doppler Signals

Many useful clinical applications of the Doppler method depend on the user simply listening to the signals. The next stage in complexity is to use a ratemeter to measure such quantities as the fetal heart rate from cyclic variations of the Doppler signal.

The analysis of Doppler blood flow signals requires more sophistication. A common (but not altogether satisfactory) approach is to measure the root-mean-square frequency by means of a zero-crossing counter (Lunt 1975). The output may be displayed as a time-interval histogram. It is generally much better, however, to subject the signal to frequency-spectrum analysis. Techniques include the sound spectrograph, the time compression analyzer, and multiple filter banks, but the fast-Fourier-transform method is preferable because of its rapid response and high resolution.

Display of the frequency spectrum is the most complete way of presenting the data. It has to be remembered, however, that the energy density of the spectrum is not necessarily an accurate representation of the blood flow. The two main reasons for this are that the ultrasonic beam may not uniformly insonate all the flowing blood, and that the Doppler receiver may have a high-pass filter to remove the lower-frequency signals which include not only those arising from the walls of the blood vessel but also from the slower-moving blood. Strictly speaking, the only really reproducible measurement which can be made is that of the maximum Doppler shift frequency; this can be extracted by means of a maximum frequency follower (Skidmore and Follett 1978). Despite the potential errors, however, the average Doppler shift frequency is also very informative; this can be extracted by means of an average velocity detector (Arts and Roevros 1972).

3. Analysis of Doppler Blood Flow Waveforms

3.1 Peripheral and Cerebrovascular Flow Studies

Arterial disease may modify the character of the blood-flow pulse detected by the ultrasonic Doppler technique.

The blood pressure pulse at the ankle can be measured noninvasively using the ECG R-wave as a timing reference and measuring the delays in the arrivals of different parts of the pressure wave at a Doppler probe positioned over the posterior tibial artery, as the pressure in sphygmomanometer cuff is changed (Johnston and Kakkar 1974).

Another approach to arterial characterization depends on measurements of the transit time of the arterial pulse through the suspect segment of artery, and the pulsatility index of the Doppler waveforms at the input and output sites of the segment, the ratio of which is called the damping factor of the segment (Woodcock 1970). Since the transit time can be either long or short, and the damping factor can be either high or low, arterial disease of the lower limb can be separated into four distinct classes. Lower-limb disease can be more selectively described in terms of the Laplace transform of the blood-flow velocity–time waveform which provides indices related to arterial stiffness, proximal lumen size and distal peripheral impedance (Bird et al. 1980).

Blood flow in the carotid arteries is complicated by reflection from the descending aorta during diastole. In the presence of occlusive disease, one of the most sensitive tests is to look for signs of spectral broadening in the Doppler frequency spectrum during systole. Quite accurate results can be obtained, at least in more advanced disease, by measuring the ratio of the peak velocities in systole and in diastole (Atkinson and Woodcock 1982).

3.2 Cardiac Studies

By placing a Doppler probe in the suprasternal notch, it is possible to obtain information about flow in the ascending or transverse aorta. This reflects left-heart function (Light 1977). Doppler shifted signals obtained from blood flowing in the jugular vein depend on right-heart function (Kalmanson et al. 1974).

Potentially valuable information can be obtained by measuring blood flow within the heart. Since the blood-flow velocities may be high, it is sometimes advantageous to use a continuous-wave Doppler system to avoid the problem of ambiguity in a pulsed Doppler system (see Sect. 2.1). When combined with real-time imaging, however, range discrimination can be very helpful in interpreting the Doppler signals. In addition to the applications in the study of murmurs, valvular diseases, left ventricular outflow tract obstruction, atrial and ventricular septal defects and coarctation of the aorta, the ratio of the flow velocities on each side of a valve seems to be quantitatively related to the pressure drop (Hatle et al. 1978).

4. Measurement of Blood-Flow Volume Rate

Essentially there are two approaches to measuring the volume rate of blood flow through a vessel.

4.1 Diameter and Average Velocity Approach

The conceptually simpler approach depends on the measurement of the blood-vessel diameter (from which the cross-sectional area can be calculated) and the average velocity of blood flow through the vessel (averaged over the period for which it is required to measure the flow volume rate). Figures 1a and 1b illustrate two ways in which this approach can be implemented.

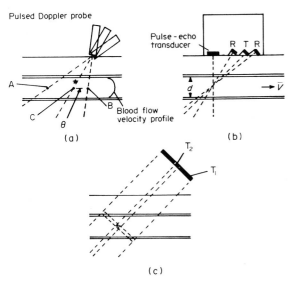

Figure 1

Approaches to blood-flow volume measurement:
(a) pulsed Doppler image—flow profile method,
(b) pulse-echo diameter—average velocity method, and
(c) spectral analysis—attenuation compensation method

The method shown in Fig. 1a uses a pulsed Doppler imaging system. The orientation of the blood vessel is determined from images made in planes A and B, assuming that the vessel is not curved and that the position of maximum flow velocity lies at the geometrical center of the vessel, which is itself assumed to be cylindrical. This allows the angle θ between the directions of the flow and of the ultrasonic beam to be estimated in plane C. In plane C, the pulsed Doppler system is used to measure the blood-flow velocity profile from many (typically ten) closely-spaced sample positions along the ultrasonic beam, over a sufficiently long time to allow the average blood flow velocity to be calculated, assuming that the blood-flow profile has circular symmetry.

An alternative method is shown in Fig. 1b. In this case, the diameter d of the blood vessel is measured by the pulse-echo transducer orientated so that the echoes from the blood vessel walls are maximized (in which situation the ultrasonic beam is normal to the vessel axis). The mean velocity $\bar{V}$ of blood flow is measured by a continuous-wave Doppler system operating with the transmitting transducer T and two receiving transducers R.

Both these methods of implementing this approach to blood-flow volume measurement are subject to related sources of error. Errors in the measurement of vessel diameter (either in the form of the flow profile or the direct measurement) are squared when the cross-sectional area of the vessel is calculated. Thus, an error of 10% in diameter measurement leads to an error of more than 20% in the flow-volume estimate. If the vessel is not cylindrical, or if it pulsates during blood flow, additional errors may be introduced. Secondly, because the cosine of the angle between the directions of the ultrasound beam and the blood flow is used in the calculation of blood-flow speed, errors in the measurement of this angle lead to errors in the flow volume estimate which depend on the magnitude of the angle. For example, if the angle is 30°, an error of 5° leads to an error of about 5% in the flow volume; but if it is 70°, a 5° error in the angle leads to an error of about 25% in the flow volume. The estimates of the average velocity are also subject to errors. The first method depends on the flow velocity profile having circular symmetry. The second method depends on uniform insonation of the flowing blood; this is certainly not achieved using an ultrasonic beam designed for optimal imaging performance, since it is as narrow as possible to give the best resolution. Uniform insonation of all but the smallest blood vessels (in which accurate flow measurement may be impossible because of the difficulty of determining flow velocity profile or vessel diameter) depends on having a defocused beam or, better, a specially formed beam (Fu and Gerzberg 1983).

4.2 Spectral Analysis Approach

The frequency spectrum of the Doppler-shifted signal detected from a uniformly insonated section of blood flowing through a vessel (which can be of arbitrary shape) has a power distribution (first moment) which reflects the contribution towards the flow volume made by each component of the blood according to the flow velocity distribution. In order to obtain the absolute value of the flow volume rate, it is only necessary to compensate for the attenuation of ultrasound in the tissue lying between the transducer and the blood vessel. As shown in Fig. 1c, this can be done by measuring the backscattered ultrasonic power from a sample volume known to be positioned entirely within blood inside the insonated section of the vessel. The transducer T_1 provides the uniform insonation, while T_2 provides the small sample volume. This approach, known as the attenuation-compensated first-moment blood-flow-volume ratemeter, has the advantage of being independent of both the angle between the directions of the ultrasonic beam and the blood flow (except that this angle must be chosen to give an adequate Doppler-shift frequency spectrum) and the diameter of the blood vessel (Hottinger and Meindl 1979).

Possibly because this approach is conceptually quite difficult, there has been little interest in its implementation. The main technical problems are those of obtaining the necessary uniform insonation for spectral measurement and the necessary small sample volume for attenuation compensation. In practice, this can best be done by means of an annular transducer array with appropriate conditions of amplitude and phase grading (Fu and Gerzberg 1983).

See also: Blood Flow: Invasive and Noninvasive Measurement; Cerebral Blood Flow: Regional Measurement; Echocardiography; Ultrasound in Medicine

Bibliography

Arts M G J, Roevros J M J G 1972 On the instantaneous measurement of blood-flow by ultrasonic means. *Med. Biol. Eng.* 10: 23–34
Atkinson P, Woodcock J P 1982 *Doppler Ultrasound and its Use in Clinical Measurement*. Academic Press, London, pp. 27–31, 91–113, 134–39
Bird D R, Skidmore R, Woodcock J P 1980 The value of Doppler transfer function analysis in the diagnosis of aorto-iliac occlusive arterial disease. In: Baird R N, Woodcock J P (eds.) 1980 *Diagnosis and Monitoring in Arterial Surgery*. Wright, Bristol, pp. 121–26
Coghlan B A, Taylor M G 1976 Directional Doppler techniques for the detection of blood velocities. *Ultrasound Med. Biol.* 2: 181–88
Fish P J 1981 A method of transcutaneous blood flow measurement: Accuracy considerations. In: Kurjak A, Kratochwil A (eds.) 1981 *Recent Advances in Ultrasound Diagnosis*. Excerpta Medica, Amsterdam, pp. 110–15
Fu C-C, Gerzberg L 1983 Annular arrays for quantitative pulsed Doppler ultrasonic flowmeters. *Ultrason. Imaging* 5: 1–16
Hatle L, Brubakk A, Tromsdal A, Angelsen B 1978 Non-invasive assessment of pressure drop in mitral stenosis by Doppler ultrasound. *Br. Heart J.* 40: 131–40
Hottinger C F, Meindl J D 1979 Blood flow measurement using the attenuation-compensated volume flowmeter. *Ultrason. Imaging* 1: 1–15
Johnston K W, Kakkar V V 1974 Noninvasive measurement of systolic pressure slope: A reliable index of the presence of peripheral arterial occlusive disease. *Arch. Surg.* 108: 52–56
Kalmanson D, Veyrat C, Chiche P, Wichitz S 1974 Non-invasive diagnosis of right heart diseases and left-to-right shunts using directional Doppler ultrasound. In: Reneman R S (ed.) 1974 *Cardiovascular Applications of Ultrasound*. North Holland, Amsterdam, pp. 361–70
Light L H 1977 Aortic blood velocity measurement by transcutaneous aortovelography and its clinical applications. In: Bom N (ed.) 1977 *Echocardiology: Doppler*

Applications and Real-Time Imaging. Nijhoff, The Hague, pp. 233–43
Lunt M J 1975 Accuracy and limitations of the ultrasonic Doppler blood velocimeter and zero crossing detector. *Ultrasound Med. Biol.* 2: 1–10
Reid J N M, Baker D W 1971 Physics and electronics of the ultrasonic Doppler method. In: Bock J, Ossoinig K (eds.) 1971 *Ultrasonographia Medica*, Vol 1. Wien Med. Akad., Vienna, pp. 109–20
Skidmore R, Follett D H 1978 Maximum frequency follower for the processing of ultrasonic Doppler shift signals. *Ultrasound Med. Biol.* 4: 145–47
Uematsu S 1981 Determination of volume of arterial blood flow by an ultrasonic device. *J. Clin. Ultrasound* 9: 209–16
Wells P N T 1969 A range-gated ultrasonic Doppler system. *Med. Biol. Eng.* 7: 641–52
Wells P N T 1970 The directivities of some ultrasonic Doppler probes. *Med. Biol. Eng.* 8: 241–56
Wells P N T 1977 *Biomedical Ultrasonics*. Academic Press, London, pp. 355–57
Woodcock J P 1970 The significance of changes in the time/velocity waveform in occlusive arterial disease in the leg. In: Filipczynski L (ed.) *Ultrasonics in Medicine and Biology*. Polish Scientific, Warsaw, pp. 243–50

P. N. T. Wells

Dosimetry of Internally Administered Radioactive Substances

Internal radiation dosimetry is important for the evaluation of the potential risk to patients receiving radioactive substances for diagnostic purposes, and for assessing the effectiveness of radioactive substances which are administered for therapeutic purposes. It is also important to be able to estimate absorbed dose when radioactive material has been ingested accidentally.

The dose absorbed by the tissues from internally administered radioactive material is determined by several factors. These include: (a) the distribution of the radioactive material within the body which is governed by the chemical form of the material and the route of its administration; (b) the biological turnover of the material which, along with the radioactive decay of the radionuclide, determines the total number of nuclear disintegrations which affect the tissues of interest; and (c) the energy imparted to these tissues and to neighboring tissues per nuclear disintegration. Any additional radionuclides, present as contaminants, must also be taken into account.

The formalism used widely for expressing the factors which determine the absorbed dose is that developed by the Medical Internal Radiation Dose Committee (MIRD) of the Society of Nuclear Medicine.

1. Estimation of Absorbed Dose

There are three components in the estimation of absorbed dose: the total number of nuclear disintegrations

Table 1

Radioactive decay of ^{125}I (half-life 60.2 days). The fractions of the emitted energy absorbed in the thyroid gland and in the total body are given

Principal emissions	Mean number per disintegration	Mean energy (MeV)	Δ_i (g mGy MBq^{-1} h^{-1})	ϕ_i (thyroid)[a]	ϕ_i (total body)[a]
γ ray 1	0.068	0.0355	1.379	0.149[b]	0.756[b]
Kα x ray	1.116	0.0274	17.597	0.149[b]	0.756[b]
Kβ x ray	0.240	0.0312	4.298	0.149[b]	0.756[b]
K internal conversion electron	0.746	0.0037	1.595	1	1
L internal conversion electron	0.107	0.0309	1.892	1	1
M internal conversion electron	0.080	0.0347	1.595	1	1
KLL Auger electron	0.137	0.0227	1.784	1	1
KLX Auger electron	0.058	0.0264	0.892	1	1
LMM Auger electron	1.490	0.0029	2.487	1	1
MXY Auger electron	3.590	0.0008	1.649	1	1

a For a source in the thyroid gland. b Absorbed fraction for photon energy of 0.030 MeV (MIRD Pamphlet 5), which is sufficiently accurate for present purposes

affecting the tissue; the energy emitted per nuclear disintegration for each type of emission; and the fraction of energy absorbed for each type of emission.

The total number of nuclear disintegrations affecting the tissues of interest is the cumulated activity A which is the integral to infinity of the activity time curve in the tissues containing the source. This is conveniently expressed in units of megabecquerel-hours (MBq h). Much of the uncertainty in the estimation of absorbed dose arises from the inaccuracy in determining cumulated activity using empirical data obtained from studies in animal or human subjects (Gillespie and Orr 1969). However, in certain instances, radioactive uptake in an organ is very rapid, with biological release of the radioactive material occurring in a single exponential manner. Thus, if A (MBq) is the initial radioactivity in the organ, λ_B(h^{-1}) is the biological decay constant and λ_P(h^{-1}) is the physical decay constant, the cumulated activity $\tilde{A}$, by integration, is given by

$$\tilde{A} \text{ (MBq h)} = A/(\lambda_B + \lambda_P)$$

The energy emitted per nuclear disintegration for each type of emission i is related to the equilibrium dose constant or the mean energy emitted per unit cumulated activity Δ_i, which is expressed in the units g mGy MBq^{-1} h^{-1}. Values of Δ_i for the radionuclides ^{125}I

and ^{99m}Tc are presented in Tables 1 and 2, respectively. Data for other radionuclides are given in MIRD Pamphlets 4 and 6 (Dillman 1969, 1970).

The fraction of energy absorbed ϕ_i for each type of emission is the proportion of the emitted energy per unit activity Δ_i that is deposited in the tissue in which the absorbed dose is being estimated. As a rule, $\phi = 1$ for particulate emissions and for low-energy (< 10 keV) photon emissions within the source region, but for more penetrating photon emissions, $\phi < 1$. Since penetrating radiation arising from within the source region deposits energy within that region and in surrounding target regions, absorbed fractions for both source and target regions must be known. Such data have been computed from simulated photon histories (the Monte Carlo method) for selected photon energies in the range 0.01–4 MeV for various source and target organs within a heterogeneous mathematical phantom of the human body (Snyder et al. 1969). Values of ϕ_i for the photon emissions from ^{125}I and ^{99m}Tc are given as examples in Tables 1 and 2. More recent tables contain values of the absorbed dose per unit cumulated activity (S factors) for various radionuclides where the values of Δ_i and ϕ_i have been combined for each type of emission (MIRD Pamphlet 11, Feller et al. 1977).

The three components of the estimation of absorbed

Table 2

Radioactive decay of ^{99m}Tc (half-life 6 h). The fractions of the emitted energy absorbed by the stomach and its contents, for a source in that region, and by the total body, for a source distributed throughout the body, are given

Principal emissions	Mean number per disintegration	Mean energy (MeV)	Δ_i (g mGy MBq^{-1} h^{-1})	ϕ_i (stomach + contents)	ϕ_i (total body)
γ ray 2	0.883	0.141	71.44	0.101[a]	0.354[a]
K internal conversion electron	0.088	0.120	6.08	1	1
L internal conversion electron	0.011	0.138	0.86	1	1
M internal conversion electron	0.004	0.140	0.30	1	1
K internal conversion electron	0.010	0.122	0.68	1	1

a By interpolation from published absorbed fractions for photon energies of 0.1 and 0.2 MeV (MIRD Pamphlet 5) dose may be combined to provide an estimate of absorbed dose D within a region of mass m (g):

$$D \text{ (mGy)} = \tilde{A} \, \Sigma \, \phi_i \Delta_i / m$$

Alternatively, if the appropriate S factor is known, the absorbed dose is the product of the cumulated activity and the S factor.

2. Calculations of Absorbed Dose

Example 1

The dosimetry of the radioiodines has been studied in detail (MIRD Dose Estimate Report No. 5). A simplified account of the calculations for ^{125}I is given here as an example.

The organ most affected by an intake of radioiodine is the thyroid gland which utilizes iodine in the synthesis of thyroid hormone (see *Radioiodine: Clinical Uses*). If it is assumed that 25% of an intravenous or oral dose of radioiodine is taken up by the thyroid gland within 24 h of administration and that the biological half-life of iodine in the human gland is 65 d, the cumulated activity $\tilde{A}$ of ^{125}I in the gland for an administered dose of 37 MBq is given by

$$\tilde{A} = 0.25 \times 37 \times 24 / \lambda_B + \lambda_P = 1.00 \times 10^4 \text{ MBq h}$$

where

$$\lambda_B = 0.693 / 65 \text{ d}^{-1} \quad \text{and} \quad \lambda_P = 0.693 / 60.2 \text{ d}^{-1}$$

The remainder of the administered radioactivity is distributed uniformly throughout the body and excreted rapidly. It does not contribute substantially to the overall absorbed dose and will be neglected here. From the data in Table 1 where the values of Δ_i and corresponding values of ϕ_i for ^{125}I are listed it may be calculated that the mean energy deposited in the thyroid gland per unit cumulated activity $(\Sigma \phi_i \Delta_i)$ is $15.36 \text{ g mGy MBq}^{-1} \text{ h}^{-1}$. Thus the absorbed dose to a thyroid gland of mass 20 g is given by

$$D \text{ (thyroid)} = 1.00 \times 10^4 \times 15.36 / 20 = 7680 \text{ mGy}$$

The appropriate ϕ_i values for the total body for a source of ^{125}I in the thyroid gland are also given in Table 1. Thus in the case of the total body the mean energy deposited per unit cumulated activity in the thyroid is $29.48 \text{ g mGy MBq}^{-1} \text{ h}^{-1}$. The absorbed dose averaged throughout the total body (mass 70 kg) is therefore

$$D \text{ (total body)} = 1.00 \times 10^4 \times 29.48 / 70 \times 10^3$$
$$= 4.21 \text{ mGy}$$

Example 2

^{99m}Tc is widely used in medicine and, as an example, the dosimetry of ^{99m}Tc given intravenously as sodium pertechnetate will be considered here. Experimental data in humans show that, in resting subjects, the critical organ

likely to receive the greatest absorbed dose is the stomach wall, due to accumulation of radioactivity in the contents of the stomach. For present purposes, only an account of the absorbed dose calculation for the stomach wall and total body is given; complete details of the dosimetry can be found in the MIRD Dose Estimate Report No. 8.

The biological time course of pertechnetate in the stomach can be described as a sum of exponential terms. Thus after an intravenous administration of sodium pertechnetate, the fraction of the dose in the stomach as a function of time (ignoring physical decay of ^{99m}Tc for the time being) is given by

Dose fraction in stomach

$$= 0.035 \exp(-7.92t) - 0.477 \exp(-0.63t)$$
$$+ 0.424 \exp(-0.07t) + 0.019 \exp(-0.004t)$$

where t is in hours. Similarly, the fraction of the dose in the total body is given by

Dose fraction in total body

$$= 0.066 \exp(-0.63t) + 0.259 \exp(-0.07t)$$
$$+ 0.675 \exp(-0.004t)$$

Applying the physical decay constant of ^{99m}Tc $(0.693/6 = 0.116 \text{ h}^{-1})$ to each exponential term, as to the one term in Example 1, the cumulated activity in the stomach for an administration of 37 MBq is given by

$\tilde{A}$ (stomach)

$$= 37 [0.035/(7.92 + 0.116) - 0.477/(0.63 + 0.116)$$
$$+ 0.424/(0.07 + 0.116) + 0.019/(0.004/0.116)]$$
$$= 67.0 \text{ MBq h}$$

Likewise the cumulated activity in the total body is given by

$$\tilde{A} \text{ (total body)} = 263.0 \text{ MBq h}$$

From the data in Table 2 where the values of Δ_i and the corresponding values of ϕ_i for ^{99m}Tc are given, it may be calculated that the mean energy deposited in the stomach wall and contents per unit cumulated activity in the stomach $(\Sigma \phi_i \Delta_i)$ is $15.14 \text{ g mGy MBq}^{-1} \text{ h}^{-1}$. Thus the absorbed dose to the stomach wall and contents (mass 400 g) is given by

$$D \text{ (stomach wall and contents)} = 67.0 \times 15.14 / 400$$
$$= 2.54 \text{ mGy}$$

An alternative approach to the calculation is to employ the S factor for a ^{99m}Tc source within the contents of the stomach irradiating the wall of the stomach. The appropriate S factor is $3.51 \times 10^{-2} \text{ mGy MBq}^{-1} \text{ h}^{-1}$, thus the absorbed dose to the stomach wall is given by

$$D \text{ (stomach wall)} = 67.0 \times 3.51 \times 10^{-2} = 2.35 \text{ mGy}$$

which is similar to the previous estimate.

Table 3

Estimates of absorbed dose from some common radiopharmaceuticals

Radiopharmaceutical	Absorbed dose (mGy MBq^{-1} administered)	
	Total body	Critical organ
[99m]Tc macroaggregated albumin	0.002	0.072 (lung) 0.007 (liver)
[99m]Tc diethylenetriamine-pentaacetic acid (DTPA)	0.004	0.011 (kidney) 0.095–0.149 (bladder)
[99m]Tc dimercaptosuccinate (DMS)	0.004	0.192 (kidney) 0.022–0.076 (bladder)
[99m]Tc pertechnetate	0.004	0.068 (stomach) 0.035 (thyroid)
[99m]Tc sulphur colloid	0.005	0.092 (liver) 0.057 (spleen)
[99m]Tc iminodiacetates (HIDA)	0.004	0.245 (gall bladder) 0.082 (upper large intestine)
[99m]Tc glucoheptonate	0.002	0.046 (kidney) 0.076–0.216 (bladder)
[99m]Tc methylene diphosphonate (MDP)	0.002	0.010 (skeleton) 0.008 (kidney) 0.119 (bladder)
[99m]Tc human serum albumin	0.004	0.022 (kidney) 0.007 (liver)
[123]I iodide	0.008	3.510 (thyroid)
[123]I hippuran	0.002	0.006 (kidney) 0.257 (bladder)
[123]I hexadecanoic acid	0.008	0.011 (heart)
[201]Tl chloride	0.057	0.316 (kidney) 0.278 (thyroid)
[67]Ga citrate	0.070	0.243 (lower large intestine) 0.157 (bone marrow)
[133]Xe gas (spirometer volume 5×10^{-3} m^3)	0.001	0.003 (lung)
[111]In-labelled leukocytes	0.143	1.323 (liver) 5.508 (spleen)
[75]Se-selenomethyl cholesterol	1.512	6.588 (adrenal) 3.780 (liver)
[75]Se-selenomethionine	2.430	6.750 (liver) 6.210 (kidney)
[131]I-iodocholesterol	0.254	13.230 (adrenal) 1.917 (liver)

The absorbed dose to the total body may be found from the relevant Δ_i and ϕ_i in Table 2. In this case the mean energy deposited in the total body per unit cumulated activity in the body in 33.21 g mGy MBq^{-1} h^{-1}. Thus the absorbed dose to the total body is

$$D \text{ (total body)} = 264.0 \times 33.21/70 \times 10^3$$

$$= 0.125 \text{ mGy}$$

Details of the absorbed doses from several radiopharmaceuticals incorporating ^{99m}Tc and other radionuclides are given in Table 3.

See also: Occupancy Principle; Ionizing Radiation: Absorption in Body Tissues; Radiopharmaceuticals: Preparation and Quality Assurance; Radiation Protection: Internal Exposure

Bibliography

Dillman L T 1969 Radionuclide decay schemes and nuclear parameters for use in radiation dose estimation, MIRD Pamphlet 4. *J. Nucl. Med.* 10: Suppl. 2

Dillman L T 1970 Radionuclide decay schemes and nuclear parameters for use in radiation dose estimation, MIRD Pamphlet 6. *J. Nucl. Med.* 11: Suppl. 4

Feller P A, Sodd V J, Kereiakes J G 1977 Using the S tables of MIRD Pamphlet 11. *J. Nucl Med.* 18: 747

Gillespie F C, Orr J S 1969 The prediction of dose due to an internal radioisotope by application of the occupancy principle. *Phys. Med. Biol.* 14: 639–44

Loevinger R, Berman M 1968 A schema for absorbed-dose calculations for biologically-distributed radionuclides, MIRD Pamphlet 1. *J. Nucl. Med.* 9: Suppl. 1

Medical Internal Radiation Dose Committee 1975 "*S*" *Absorbed Dose Per Unit Cumulated Activity for Selected Radionuclides and Organs*, MIRD Pamphlet 11. Society of Nuclear Medicine, New York.

Medical Internal Radiation Dose Committee 1975 Summary of current radiation dose estimates to humans from ^{123}I, [124]I, ^{125}I, [126]I, [130]I, ^{131}I and [132]I as sodium iodide, MIRD dose estimation Report No. 5. *J. Nucl. Med.* 16: 857–60

Medical Internal Radiation Dose Committee 1976 Summary of current radiation dose estimates to normal humans from ^{99m}Tc as sodium pertechnetate, MIRD dose estimation Report No. 8. *J. Nucl. Med.* 17: 74–77

Snyder W S, Ford M R, Warner G G, Fisher H L 1969 Estimates of absorbed fractions for monoenergetic photon sources uniformly distributed in various organs of a heterogeneous phantom, MIRD Pamphlet 5. *J. Nucl. Med.* 10: Suppl. 3

T. E. Hilditch

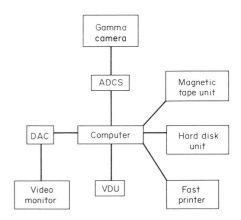

Figure 1
Block diagram of a typical gamma camera/computer system. ADCS and DAC represent analog-to-digital and digital-to-analog converters, respectively

Dynamic Cardiac Studies

The use of radionuclides to examine the cardiovascular system dates from 1927. In recent years, the development of gamma cameras, their interfacing to small computers and the availability of suitable man-made radioisotopes as tracers has enabled the noninvasive investigation of cardiac function in many clinical centers. Images of the heart can be acquired for the assessment of cardiac function and the detection of shunts between the left and right side of the heart, as a bolus of radiotracer makes its initial passage through the heart (first-pass studies). Alternatively, equilibrated tracer within the cardiac blood pool can be imaged, many sequential images being acquired throughout the cardiac cycle, each built up over a number of cycles using the R wave of the electrocardiogram as a cycle phase reference (see *Electrocardiography*). This latter technique enables visual and quantitative assessment of total or regional ventricular function from a cine display of the successive images forming the cardiac cycle.

The success of both these methods of dynamic cardiac study has led to a tremendous increase in the use of radionuclides in the diagnosis of heart disease and in monitoring the patient's response to treatment.

1. Instrumentation

A gamma camera interfaced to a small computer is required for acquiring, processing and storing dynamic cardiac studies (Fig. 1). The camera consists essentially of a large-diameter (typically 40 cm) sodium iodide scintillation crystal (behind a lead collimator), and of associated photomultipliers, with electronics to detect and map the source positions of the γ rays emanating from the patient and incident upon the crystal (see *Radionuclide Imaging*). An image of the distribution of the radiopharmaceutical within an organ is built up from the γ-ray events. Image resolution is 1–2 cm. Photon count rates of 50 000 per second are possible in dynamic cardiac studies. The images are stored in matrix form within the computer; 64×64, 128×128 or 256×256 element matrix arrays being used to record the number of γ-ray events within each pixel relating to a particular part of the organ under study (see *Image Analysis: Extraction of Quantitative Diagnostic Information*). The larger the matrix used, the greater are the memory and storage requirements (e.g., a single, 64×64 image requires 4 kilobytes of memory storage whereas a 256×256 image requires 64 kilobytes of storage). Hard disks, floppy disks or magnetic tape are used to store the data acquired (Fig. 1). A video monitor driven from a digital-to-analog converter presents a visual image of the data acquired by the computer. At least 64 kilobytes of computer memory is needed for storing programs and images. Figure 2 depicts a typical clinical scene, in which a 45° left anterior oblique image is being acquired.

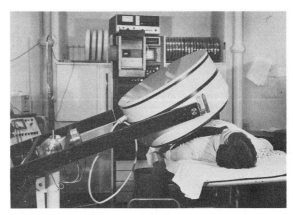

Figure 2
Patient being imaged supine under a gamma camera

2. First-Pass Studies

2.1 Image Acquisition

The passage through the heart of a radioactive bolus introduced intravenously can be followed in a rapid sequence of gamma-camera images. ^{99m}Tc is the radioisotope used, since it has a single γ-ray energy at 140 keV and a half-life of 6 h. Typically an injection of 750 MBq of pertechnetate or technetium-labelled human serum albumin in 1 ml volume is administered. The patient is normally imaged from the 45° left anterior oblique position. This gives good separation of the chambers of the heart. The gamma camera and associated computer then record each scintillation event either in preset time images, or in list mode for 30 s (see Sect. 3.1) for subsequent reconstruction of the images, each typically of 1 s duration. Figure 3 shows a sequence of four 1 s images, as a bolus passes through the heart.

2.2 Qualitative Analysis

It is possible to derive useful information from visual analysis of the sequential images produced by the computer on a television monitor. The bolus can be traced as it arrives at the superior vena cava and passes through the right heart, lungs and left heart. The technique can be helpful in assessing congenital heart disease (e.g., in right-to-left shunts) where there is early visualization of tracer in the left ventricle. Left-to-right shunts can also be detected visually from the sequential images.

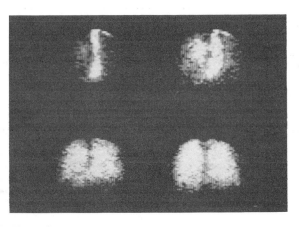

Figure 3
A sequence of four images each of 1 s acquired as part of a first-pass study. Right ventricle and then lungs are visualized

2.3 Quantitative Analysis

By using the computer and a marker on the monitor, regions of interest can be selected on the images. Curves of activity against time can be produced for different parts of the heart to assess the passage of the radioactive tracer in more detail. The temporal width of the injected bolus can be measured from a region-of-interest over the superior vena cava, and shunt size can be assessed by detailed analysis of right ventricle, left ventricle and pulmonary activity–time curves.

The changes of activity in the left-ventricle region-of-interest appear as shown in Fig. 4, if temporal resolution is adequate. The high-frequency change in activity is due

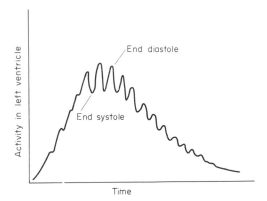

Figure 4
Typical shape of an activity–time curve from a region-of-interest over the left ventricle during a first-pass study

to the beating of the left ventricle, the activity being highest when the chamber is fully expanded (end diastole) and lowest when it is fully contracted (end systole). From this high-frequency component it is possible to measure the ejection fraction of the ventricle, which is perhaps the most important parameter in determining the state of ventricular function. Ejection fraction EF is defined as

$$EF = \frac{EDV - ESV}{EDV} = \frac{SV}{EDV} \qquad (1)$$

where EDV is the end-diastolic volume, ESV is the end-systolic volume and SV is the stroke volume of the left ventricle. Since the activity within the blood is proportional to volume, ejection fraction can be measured as

$$EF = \frac{EDC - ESC}{EDC - Bgd} \qquad (2)$$

where EDC is the end-diastolic counts, ESC the end-systolic counts, and Bgd is the necessary background correction. As the technique is noninvasive, monitoring the effects of drug treatment or of surgery by measuring ejection fraction is simpler than using the alternative highly invasive (and expensive) method of contrast angiography.

3. Gated Cardiac Studies

3.1 Acquisition of Studies

In gated heart studies, many sequential images (typically 16 to 64) are acquired per cardiac cycle. In certain phases of the cycle, however, the counts are so low that they cannot provide a good quality image. Thus some technique is needed for adding the counts obtained over a number of cardiac cycles and in relation to their position in the cycle. The R wave on the electrocardiogram signal indicates depolarization and occurs when the heart is fully expanded. It thus provides a suitable reference point on the cycle for initiating and synchronizing data acquisition. For the first sixteenth of a cycle the computer can begin building up the first image in a given memory location and then switch to the next image, and so on. At the next R wave the computer switches back to the first image and stores the events recorded during the first sixteenth of the second cardiac cycle in the same memory location. Thus each of 16 successive phases of the cardiac cycle are imaged and stored within the computer's memory over several hundred beats, as shown in Fig. 5 (taken from a typical study). Image 1 shows the ventricle fully expanded (end diastole) and image 7 shows the ventricle fully contracted (end systole).

Data are acquired after some 5 min following intravenous injection as in the case of first-pass studies. The delay allows the radiopharmaceutical to equilibrate within the vascular system. Imaging in the 30°–45° left

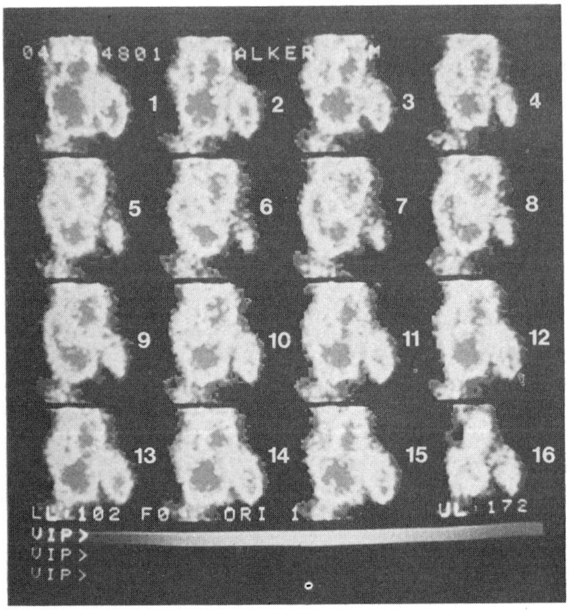

Figure 5
Sixteen images throughout the cardiac cycle from a gated equilibrium study. Image 1 is end diastole and image 7 is end systole

anterior oblique position gives good separation of left and right ventricles for measurement of ejection fractions, but other projections are also useful (e.g., anterior), which gives an excellent view of the apex of the left ventricle.

Using the technique described above, difficulty may be experienced in coping with varying or irregular heart rates, since the length of each cycle is unknown until the occurrence of the subsequent R wave. The counts from each cycle may not be added up correctly with phase, smearing out the acquired images. Various methods are used to tackle this problem, such as buffering in the computer memory (not storing the images until the length of a beat is known), varying the length of aquisition time intervals as the heart rate varies, or acquiring data in list-mode. In list-mode, all data from the gamma camera are stored directly on a hard disk (this is necessary due to its data storage and transfer rate capacity) as a series of addresses, along with timing signals and the R wave. When the data are processed into images, the length of each beat is known. Thus it is possible to divide each beat accurately; short ectopic beats can be excluded. List-mode acquisition is obviously more time consuming and makes a greater demand on memory-storage capacity, but the reconstructed images are more accurate.

3.2 Qualitative Analysis

Global and regional left ventricular function can be assessed by sequentially displaying each of the images in a cine loop. Poorly functioning areas can be readily identified; aneurysms, where a region of myocardium bulges out as the rest contracts, can be visualized. Other procedures such as displaying cine contours or alternating between the end-diastolic and end-systolic image are available to assess cardiac wall motion.

3.3 Quantitative Analysis

As in first-pass studies, a region-of-interest around the left ventricle and the counts within this region can be plotted for each image in the cardiac cycle. Figure 6 shows a typical curve obtained (after subtraction of suitable background) from which the ejection fraction can be calculated using Eqn. (2). There are two problems to be considered in this technique: defining the edge of the ventricle, and defining a suitable background.

Various methods, including manual drawing, second derivative definition of the edge and isocount contour techniques are used to define the left ventricle region. Background is normally taken as a narrow region, drawn (either manually or automatically) around the left ventricle. Semiautomated methods give better interobserver reproducibility, and the results when compared with ejection fractions obtained from contrast angiography give correlation coefficients typically of 0.85–0.95.

Other useful parameters for assessing left ventricular function can be obtained from the activity–time curve,

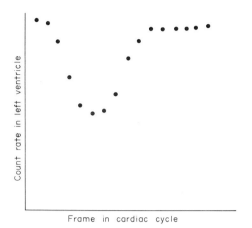

Figure 6
A typical activity–time curve achieved from a region-of-interest over the left ventricle in a gated equilibrium study. The ejection fraction was found to be 0.45

e.g., peak and mean injection rates, filling rates, and systolic time intervals. Changes in any of these parameters during various stress tests indicate the state of the left ventricle and the disease state of patients.

4. Applications

Due to the noninvasive nature of the studies, there are many possible applications both for the diagnosis of coronary disease and for monitoring patient response to therapy or surgery. These include:

(a) detection of patients with coronary artery disease,

(b) assessment of patients with acute infarcts,

(c) detection of aneurysms,

(d) assessment of patients for coronary arteriography,

(e) effects of drug therapy,

(f) investigation of congenital heart defects,

(g) right ventricular function studies,

(h) evaluation of ventricular wall motion, and

(i) post-operative assessment in coronary artery bypass surgery.

The influence of various physiological stresses (e.g., supine bicycle exercise, isometric stress or cold stress) on cardiac function can be investigated to try to detect abnormal responses in groups of patients under investigation. These are areas of current active research. Some of the major applications are discussed below.

4.1 Detection of Ischemic Heart Disease

Sequential studies on patients at rest and during stress may show abnormality in left ventricular response in patients with coronary artery disease. Normal patients here show an increase in global ejection fraction, whereas a drop in global ejection fraction with associated regional wall motion abnormalities occurs where there is coronary disease. This type of test is sensitive, specific and of great potential. Along with cold and isometric stress tests it is being carefully evaluated in many centers—reportedly with varying degrees of success.

4.2 Effects of Drug Therapy on Left Ventricular Function

Studies, pre- and post-therapy, allow assessment of individuals' response to treatment, and may be useful in drug-dose optimization. The effects of nitroglycerine in patients with coronary artery disease have been studied, the results showing both improved ejection fraction and wall motion. The administration of digoxin to patients in heart failure has also been assessed, patients showing a significant increase in ejection fraction when on digoxin. The cardiotoxicity of adriamycin, a commonly used chemotherapy agent, has also been assessed.

4.3 Assessment of Left Ventricular Function

Dynamic cardiac studies of global function and regional wall motion function can be used to determine the suitability of patients for coronary bypass surgery—obviating the need, to some extent, for highly invasive catheterization procedures. Repeat studies can monitor post surgical improvement in function.

5. Concluding Remarks

The use of radionuclides provides an important routine and research role for the assessment of cardiac disease. The clinical demand for this type of study is increasing. During the next few years, greater availability of new radiopharmaceuticals, such as labelled fatty acids for myocardial imaging and the short-lived radionuclide ^{195m}Au, as well as advances in imaging techniques and processing systems, will allow even better assessment of cardiac function.

See also: Cardiac Function: Noninvasive Assessment; Computers in Cardiology; Radionuclides: Clinical Uses; Echocardiography

Bibliography

Maisey M 1980 *Nuclear Medicine: A Clinical Introduction.* Update Books, London
Parker R P, Smith P H S, Taylor D M 1978 *Basic Science of Nuclear Medicine.* Churchill Livingstone, London
Strauss H W, Pitt B, James A E (eds.) 1974 *Cardiovascular Nuclear Medicine.* Mosby, St Louis
Wagner M N (ed.) 1969 *Principles of Nuclear Medicine.* Saunders, Philadelphia

W. Martin

E

Ear Anatomy and Physiology

The ear is connected to the brain via the auditory nerve, and the process of hearing is served by more than 10^8 nerve cells. Although the conversion of sound into nerve impulses occurs in the inner ear, mechanisms which modify the characteristics of the incoming sound begin to operate in the external part of the ear. In this article, advances in understanding the anatomy and physiology of the peripheral organ, particularly with respect to transduction processes in the inner ear and their relation to the auditory function of the whole organism, are discussed.

1. Outer and Middle Ear

The external ear (pinna) and the outer ear canal (see Fig. 1) constitute a very broadly tuned (Q factor ≈ 1) "ear trumpet" with a peak gain of about 14 dB at 3 kHz, falling to 2 dB at about 250 kHz and 6 kHz. Even the rather flat human pinna, because of directionally sensitive diffracting properties, helps in discriminating a sound's direction, particularly in distinguishing up/down and front/back.

In the middle ear, the geometry of the ossicular chain, which connects the ear drum with the oval window of the cochlea, matches the acoustic impedance of the external air to that of the inner-ear fluids, thus optimizing the transfer of sound energy. The factors above determine the range of hearing, from about 20 Hz to 20 kHz, and also the shape of the normal human "audiogram" which represents the threshold detectable sound energy as a function of frequency for an untraumatized auditory system.

There are two middle-ear muscles: tensor tympani, which acts on the eardrum, and stapedius, which acts on the stirrup bone (stapes). When contracted they attenuate low-frequency and all-frequency sound inputs, respectively. During speech they attenuate sound input to the cochlea from the speaker's own larynx, being activated before vocalization starts. They also contract some 50 ms after a loud external sound arrives.

2. Inner Ear

The transduction of sound into nerve signals occurs in the inner ear in the snail-shaped cochlea. Hair cells

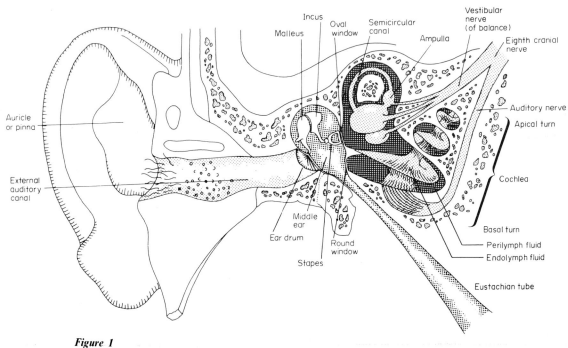

Figure 1
Ear anatomy

situated between the basilar membrane and the tectorial membrane have distortions imposed upon their "hairs" (stereocilia) when travelling waves are generated in the basilar membrane by perturbations in the cochlea fluids. These occur when sound pressure variations incident upon the ear drum are transmitted to the oval window of the inner ear by the final bone in the ossicular chain, the stapes.

3. Mechanoelectrical and Nervous Processing Within the Cochlea

3.1 Stimuli

Most natural sounds are not as brief as 1 ms "clicks" nor as long lasting as "pure tones." Nevertheless, the click and the tone are well suited to probe the timing and tuning, respectively, of auditory mechanisms, particularly in peripheral parts of the nervous system including the cochlea. Special mechanisms develop more centrally, both pre- and post-natally, to analyze the variations in amplitude and frequency encountered in environmental sounds, including speech. To investigate the auditory central nervous system therefore, well-defined sets of frequency- and amplitude-modulated tones are to be preferred as stimuli (Kay 1982).

3.2 Frequency Selectivity of the Cochlea

For sounds $\geqslant 30$ dB suprathreshold and frequencies 0.5–2 kHz, where hearing is most sensitive, people can detect changes of the order of 1% in energy and of the order of 0.1% in frequency. This high acuity for frequency change (contrast the 6% musical semitone) can usefully guide a description of cochlear mechanisms. (Signalling of loudness is essentially by a convergent integration between many cochlear outputs. Each of these has a limited frequency passband and intensity range.)

Cochlear outputs must contain all the information needed to underpin frequency acuity, yet cochlear mechanisms are themselves rather broadly tuned; this is necessary to allow good temporal resolution of transients. However, the tuning of cochlear output into each acoustic nerve fiber is sharper (Q value ≈ 10) than the tuning of the outer and middle ear. The mechanisms determining this tuning have recently been much clarified.

Around 1863, Helmholtz, with only strings and pipes as apparatus, considered two mechanisms for audio-frequency analysis:

(a) A sharp mechanical tuning of the basilar membrane, regarding it as a closely packed but independently vibrating set of parallel, taut and resonant transverse strings so that one string vibrates most for a given sound frequency. High frequencies are represented near the cochlear base, where the width of the basilar membrane is least and the "strings" are short: low frequencies generate vibrations near the cochlear apex where the membrane is widest. To determine frequency, the brain inspects the size of signals from individual hair cells in the narrow strip of sensory receptors that stretches along the organ of Corti, and identifies the most excited cell.

(b) A very broadly tuned periphery, placing the whole burden of frequency discrimination upon the central nervous system, not excluding the impressive neuronal machinery that lies centrally to the receptors but is within the cochlea itself (Spoendlin 1973).

Helmholtz favored mechanism (a). By observing the audibility of musical trills, Helmholtz realized that the tuning of peripheral mechanical vibrations must be to some extent damped to allow the trills to be perceived, but he could not assess damping accurately enough by testing temporal resolution, having no electronics to generate brief clicks that could be approximated closely in time. A mechanical oscillator tuned as sharply as the whole human response (bandwidth (Δf) ≈ 1 Hz at 1 kHz) struck by a transient would oscillate with a time constant $\tau \approx 300$ ms ($\tau = 1/(\pi \Delta f)$), implying temporal discrimination of that order of magnitude. However, modern experiment shows that people can distinguish two brief, vastly suprathreshold clicks as being separate, when they are sounded at time separations of about 3 ms. This in contrast implies rather broad tuning bandwidths ($\Delta f = 1/(\pi \tau)$) of order 100 Hz, possibly wider bands than Helmholtz expected.

At about the same time, Mach found that a sound must last a certain minimum time for a definite pitch to be perceived, and in this century it has been shown that sounds lasting 100 ms or more are needed for the best discrimination of changes in frequency or amplitude. A sound also needs to be loud enough for best discrimination ($\geqslant 30$ dB suprathreshold). These minimal sound exposures for optimal performance show that passive mechanical tuning alone is not an adequate explanation for the acuity of frequency discrimination, but it could be a first mechanism.

Helmholtz's idea of an orderly spatial segregation along the cochlear partition for the vibrations resulting from different sound frequencies has been confirmed by later work, and in the direction predicted. His analogy between piano strings and the cochlea was to some extent apt: short strings for high frequencies, long for low frequencies.

However, von Békésy (1960) cut a longitudinal slit in the cochlear partition. It did not gape, so there is little transverse tension in the putative taut strings, and thus the piano analogy is not exact. Also, by direct ultramicroscopic observation of vibrations of the (recently dead) human cochlear partition, he showed that a spatial sorting of vibration does occur as expected, but the mechanical response is widely tuned. There is an optimal frequency for vibration at any one place, but frequencies differing from the optimal by several hundred hertz are still effective.

An objection to von Békésy's work is that dead material was used. Also, even working ultramicroscopically, the vibrations needed for visibility were above the pain threshold (130 dB) rather than at physiologically normal hearing levels, perhaps overloading the cochlea and distorting its vibration.

Experiments from the 1960s onwards on living material improved the sensitivity in the 70–80 dB region (in which sound is loud but not unusually so) by exploiting the Mössbauer effect. These experiments measured the Doppler shift in the γ-ray frequency of an emitter sitting on the cochlear partition. The measurements showed that in monkeys and guinea pigs as in people, the mechanical tuning bandwidth is of the order of 100 Hz (i.e., it cannot discriminate between frequencies closer than 100 Hz). In contrast, the whole organism can detect frequency changes of the order of a few hertz.

There exists an intracochlear electrical potential, the "cochlear microphonic," which follows the amplitude and waveform of incident sinusoidal sound. It increases monotonically with sound energy from below the threshold detectability of sound to near the pain threshold. It has no threshold itself and is a graded response over a very wide range; thus it is certainly not derived from nerve impulses but is rather the extracellular sign of intracellular receptor potentials that follow the local vibrations of the basilar membrane. Tasaki et al. (1952, 1954), by inserting fine electrodes through different points in the cochlear wall, discovered at very moderate sound levels (20 dB) a spatial sorting and tuning of the cochlear microphonic very like that of the mechanical vibrations.

As a final confirmation of broad peripheral tuning, the response width of individual acoustic nerve fibers measured in animals and people is also found to be of order 100 Hz (Evans 1982). Thus the spatial sorting according to frequency ("tonotopicity") and the tuning bandwidth of cochlear outputs in individual acoustic nerve fibers are closely correlated with the spatial sorting and tuning of the cochlear vibrations.

3.3 The "Second Filter"

Measurements such as those described (in Sect. 3.2) did seem to indicate that the response of a nerve fiber might be a little more sharply tuned than the mechanical vibration. In particular, the high-frequency cutoff side of the nervous response looked steeper than for the vibration itself. For this reason, the idea of a "second filter" was proposed. This would act somewhere between the vibratory input and the nervous output. It would be within the cochlea, since nerve tuning exists before there is time for any extracochlear feedback mechanism to act and it is present even if the cochlear nerve is cut. There is plenty of nervous machinery within the cochlea to undertake this special task if need be. However, the need for such a second filter has recently been severely challenged.

Work on amphibians and reptiles first suggested that a major revision of views on cochlear tuning might be necessary. The usual spatial sorting of frequency in nervous outputs is found in frogs and toads yet there is no moving basilar membrane. The receptor cells sit on an immobile part of the inner ear. It was thought, therefore, that the tectorial membrane may be specially significant.

In the alligator lizard, orderly tonotopic sorting and tuning of auditory nerve outputs are also found. The basilar membrane does not vibrate with a frequency-sensitive travelling wave, but as a whole. There is a tectorial membrane in the basal regions of the lizard's cochlea, yet there is none in the apical regions where nerve fibers nevertheless show clear tonotopic organization and tuning (Holton and Weiss 1983). Thus, it seems that neither a basilar membrane travelling wave nor even a tectorial membrane is essential in all species for finding tonotopicity of tuned afferent outputs.

In the turtle, Crawford and Fettiplace (1981) measured intracellular ac hair cell potentials, related in size to the instantaneous sound pressure, saturating at some tens of millivolts amplitude at about 100 dB sound pressure level. There is a tonotopic arrangement; cells with best frequency response near 30 Hz are near the cochlear apex, while basal cells represent frequencies near 700 Hz. (The basilar membrane in this animal is very short (~ 1 mm), with its frequency range correspondingly restricted.) It is notable that the tuning of the intracellular potentials from the turtle hair cell is as sharp as the output of its acoustic nerve fiber. Thus, any second filter would have to act between the vibration and the receptor cell's response, not later. An electromechanical resonance reported for these hair cell responses was suggested as the mechanism for this filter.

Extrapolation of experimental results from one animal to another may be unjustified, particularly in species so different as turtles and humans. However, some of the mechanisms that generate the tuned tonotopic arrangements found in these disparate animals might also be found subserving tonotopicity and tuning in mammals, where the basilar membrane does vibrate and a tectorial membrane does exist. This raises the question: if a second filter following the hair cells is unnecessary in the turtle, do we need to look for one in humans?

Armchair speculation has been curbed by the successful recording of intracellular responses from mammalian (guinea pig) hair cells by Russell and Sellick (1980, 1983) and Russell (1983). Their experimental animal is small but, like humans, is a mammal living on land with most sound of interest to it being airborne. It is found that the guinea pig hair cell has an intracellular dc resting potential, in the absence of sound, of about -45 mV. In the presence of sound, a positive-going depolarization is superimposed upon this, increasing in size with increasing sound energy by as much as 20 mV. The time course of this dc receptor potential follows the envelope of any amplitude change in the incident sound. There are also ac receptor potentials that follow the instantaneous waveform of the sound, certainly at low frequencies (and maybe at high frequencies, but there the self-capacitance of the recording electrode might limit their observation). In spite of some interpretive difficulties, it seems certain that both ac and dc intracellularly recorded potentials from guinea pig inner hair cells have tuning bandwidths identical with those of their afferent output nerve fibers. Here again (but now in a mammal) it proves unnecessary to invoke a second filter between the receptor cell and the nerve output to sharpen the tuning of the latter. The electromechanical tuning in the guinea pig hair cell, unlike that in the turtle's, is apparently free from resonance.

Khanna and Leonard (1982), remembering that sharpening by the putative second filter had been found to be vulnerable to acoustic trauma, performed very careful measurements of the mechanical vibration in the feline cochlea using laser interferometry of light reflected from a gold mirror weighing 10^{-8} g sitting on the moving membrane. They rejected all results that were not stable with time or showed other signs of possible cochlear trauma. They found that the mechanical vibration in the untraumatized cochlea was as sharply tuned as the final nervous response. Sellick et al. (1982) confirmed this conclusion for guinea pig untraumatized cochlea using the Mössbauer technique to measure the vibration. They found this to be as sharply tuned as the intracellular receptor potentials and as the nervous output. Having now been found unnecessary in two smaller mammals, the second-filter concept may also be unnecessary in humans, and the human cochlea mechanical vibrations may be tuned as sharply as the eighth (auditory) nerve response. This should not be too surprising since von Békésy's essential and pioneering work, showing broader mechanical tuning in the human, was on the traumatized, dead and overloaded cochlea. What has been found vulnerable to trauma is the sharpness of tuning of the vibratory pattern itself (i.e., the first filter).

The nervous output in these mammals and in humans is nevertheless tuned no more sharply than $Q \approx 10$. There is therefore a lot of nervous processing, which is left to the more central auditory nervous system, to achieve the even sharper effective tuning of the whole human response to frequency change. This is about two orders of magnitude sharper than the tuning at the acoustic nerve. There are about 10^7 neurons between the cochlea and the output from the primary auditory cortex available to perform this and other tasks. It is now quite clear that whatever the purpose of the elaborate intracochlear nervous machinery between the hair cells and the auditory nerve, it is not for subserving a proposed second filter between the cochlea vibrations and the afferent eighth nerve output from the peripheral organ.

4. Continuing Research

Most output fibers in the acoustic nerve come from the inner hair cells: the function of the outer hair cells is uncertain. The tuning of the intracellular potentials in the inner hair cells fits the tuning of the velocity of the

sound vibration better than that of the amplitude. This and histological evidence suggest that the hairs are free to move in the cochlear fluids and are distorted by viscous drag from the moving fluid. The outer hair cell responses seem better correlated with vibration amplitude: their stereocilia are embedded in the tectorial membrane (Dallos 1982).

With all these results the whole field of cochlear physiology is in a turmoil of research effort. Among the more interesting arguments is whether the newly found sharpness of mechanical tuning needs active processes to support it or not. The discovery of actin filaments (and other proteins associated with muscular contraction) in the stereocilia of the hair cells and in the cuticular plate from which they grow (Flock et al. 1981, 1983, 1984) has led to proposals that these determine the stiffness of the hairs and hence their vibratory response. It is even suggested that the hairs of the outer hair cells, being embedded in the tectorial membrane, may influence the coupling between the basilar and tectorial membranes and thus affect the mechanical tuning of the cochlear partition (Sellick et al. 1982). Therefore, electrical influences on the outer hair cells transmitted from the central nervous system through efferent nerve connections of the olivocochlear bundle might influence mechanical tuning (Art et al. 1982, Brown and Nutall 1984). The possibilities abound and with such enthusiastic speculation there are certainly real advances to be made.

See also: Audiometers; Artificial Mastoid; Artificial Ear; Hearing Aids; Sound: Biological Effects

Bibliography

Art J J, Crawford A C, Fettiplace R, Fuchs D A 1982 Efferent regulation of hair cells in the turtle cochlea. *Proc. R. Soc. London, Ser. B* 216: 377–84

Barlow H B, Mollon J D (eds.) 1982 *The Senses.* Cambridge University Press, Cambridge

Brown M C, Nutall A L 1984 Efferent control of cochlear inner hair cell responses in the guinea-pig. *J. Physiol. (London)* 354: 625–46

Crawford A C, Fettiplace R 1981 An electrical tuning mechanism in turtle cochlear hair cells. *J. Physiol. (London)* 312: 377–412

Dallos P 1982 Cochlear physiology. *Annu. Rev. Psychol.* 32: 153–90

Evans E F 1982 Functional anatomy of the auditory system. In: Barlow H B, Mollon J D 1982, pp. 275–89

Flock A, Cheung H C, Flock B, Utter G 1981 Three sets of actin filaments in sensory cells of the inner ear. Indentification and functional orientation determined by gel electrophoresis, immunofluorescence and electron microscopy. *J. Neurocytol.* 10: 133–47

Flock A, Orman S 1983 Active control of sensory hair mechanics implied by susceptibility to media that induce contraction in muscle. *Hearing Res.* 11: 261–66

Flock A, Stelioff D 1984 Graded and nonlinear mechanical

properties of sensory hairs in mammalian hearing organ. *Nature* (*London*) 310: 597–99

Holton T, Weiss T F 1983 Frequency selectivity of hair cells and nerve fibres in the alligator lizard cochlea. *J. Physiol.* (*London*) 345: 241–60

Kay R H 1982 Hearing of modulation in sounds. *Physiol. Rev.* 62: 894–975

Khanna S M, Leonard D G B 1982 Basilar membrane tuning in the cat cochlea. *Science* 215: 305–6

Russell I J 1983 Origin of the receptor potential in inner hair cells of the mammalian cochlea: Evidence for Davis' theory. *Nature* (*London*) 301: 334–36

Russell I J, Sellick P M 1980 Responses of hair cells to low frequency tones and their relationship to the extracellular receptor potentials and sound pressure level in the guinea pig cochlea. In: Syka J, Aitkin L (eds.) 1981 *Neuronal Mechanisms of Hearing.* Plenum, New York, pp. 3–15

Russell I J, Sellick P M 1983 Low frequency characteristics of intracellularly recorded potentials in guinea pig cochlear hair cells. *J. Physiol. (London)* 338: 179–206

Sellick P M, Pattuzi R, Johnstone B M 1982 Measurement of basilar membrane motion in the guinea pig using the Mössbauer technique. *J. Acoust. Soc. Am.* 72: 131–41

Spoendlin H 1973 The innervation of the cochlear receptor. In: Møller A R, Boston P (eds.) 1973 *Basic Mechanisms in Hearing.* Academic Press, New York, pp. 185–230

Tasaki I, Davis H, Eldridge D H 1954 Exploration of cochlear potentials in guinea pig with a microelectrode. *J. Acoust. Soc. Am.* 26: 765–77

Tasaki I, Davis H, Legouix J-P 1952 The space–time pattern of the cochlear microphonics (guinea pig) as recorded by differential electrodes. *J. Acoust. Soc. Am.* 24: 502–19

von Békésy G 1960 *Experiments in Hearing.* Wiley, New York

R. H. Kay

Echocardiography

The use of ultrasound in cardiac diagnosis was pioneered by Inge Edler in the mid-1950s. In recent years, echocardiography has become an established investigative method. It is an important technique because it is non-invasive and also because it provides information which cannot be obtained using other techniques.

1. Methods and Outline of Diagnostic Applications

The frequency of the ultrasonic transducers usually used in adult cardiology is 2.25 MHz; higher-frequency transducers (3–7 MHz) are used in pediatric cardiology.

1.1 M-Mode Echocardiography

In M-mode echocardiography a single static ultrasound beam is directed into the heart (Fig. 1). The M (motion)-mode echocardiogram plots the position of the echoes set up at tissue interfaces against time. Echo

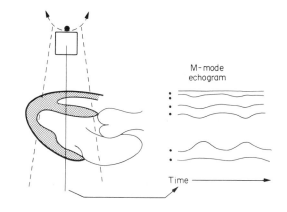

Figure 1
Derivation of the M-mode and (broken lines) two-dimensional echocardiogram (Reproduced with permission from *Br. Med. Bull.* 36: 261–66 (1980))

position, on the vertical axis, indicating the depth of the reflecting structure from the transducer (near echoes at the top) and its motion are displayed on a chart recorder. M-mode echocardiography provides detailed information about valve and wall motion but has important limitations: some cardiac structures are inaccessible to M-mode examination, and because at any instant echoes are received only from the reflecting surfaces in the path of the ultrasound beam, it is a poor method with which to examine the spatial relationships within the heart.

1.2 Two-Dimensional Echocardiography

With this method, electronic or mechanical means are used to sweep the ultrasound beam rapidly through an arc (Fig. 1). Echoes are received essentially simultaneously from the reflecting structures along its path and the result is a moving real-time echo image of a sector of the heart. The transducer can be positioned to image the heart either in its long axis or in a series of short axes. Long-axis sector scans are presented with the aortic root to the right and the chest wall to the top (as in the M-mode presentation) and short-axis scans are presented as viewed from below with the subject recumbent (Fig. 2). Two-dimensional echocardiography provides less detailed information about motion than the M-mode does, but all cardiac structures are accessible and the two-dimensional technique is of particular value in the investigation of spatial relationships.

M-mode and two-dimensional echocardiography are complementary techniques which in combination provide comprehensive information about cardiac anatomy and function. With an electronic sector scanner it is possible to record simultaneous M-mode and two-dimensional echocardiograms, the line for M-mode examination being selected with reference to the two-dimensional display. This coupling of the two techniques allows both to be used to the best advantage.

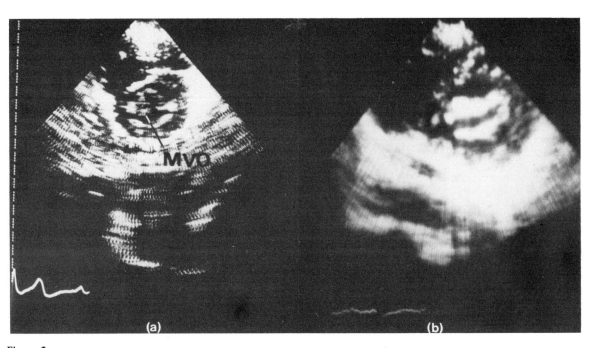

Figure 2
Short-axis two-dimensional echocardiograms (diastole) of (a) a normal and (b) a rheumatic (stenotic) mitral valve. MVO: mitral-valve orifice (after Barnett and Morley 1983. *Clinical Diagnostic Ultrasound.* © Blackwell Scientific Publications, Oxford. Reproduced with permission)

1.3 Contrast Echocardiography

This is an extension of the routine examination. Gaseous microbubbles are intense reflectors of ultrasound and act as an acoustic contrast agent. When, during echocardiographic examination, a bolus of a biologically compatible fluid (usually 5% dextrose) is injected into a peripheral vein, the microbubbles within the bolus are seen to opacify the right heart; they are filtered out in the pulmonary circuit and, in the absence of a right-to-left shunt, do not enter the left heart. Opacification of the left heart is thus diagnostic of a right-to-left shunt and the site of opacification indicates the level of the shunt. It is likely that acoustic contrast agents which can traverse the lungs and which are nontoxic will be developed. Contrast echocardiographic studies of the left heart following a peripheral venous injection will then be feasible and it should be possible to detect left-to-right shunts echocardiographically.

1.4 Doppler Echocardiography

This important development provides information about the characteristics of intracardiac blood flow (see *Doppler Blood Flow Measurement*). Ultrasound pulses of known frequency (1–3 MHz) are directed into the heart; backscatter from the moving blood cells produces a variety of frequency shifts. Spectral analysis (by fast Fourier transform) of the composite Doppler shift signal produces a display of the direction and velocity of blood flow against time. A combination of the pulsed-wave and continuous-wave Doppler techniques is used. With pulsed-wave Doppler, flow is examined within a sample volume which is at a known depth and which, with simultaneous two-dimensional imaging, can be positioned to record velocities at any selected site within the heart. The maximum velocity which can be detected in the pulsed-wave mode is limited by the relatively low pulse-repetition frequency required to operate at depth, and continuous-wave Doppler is necessary for accurate measurement of high velocity flows within the heart. Cardiac Doppler permits noninvasive quantification of valvular stenoses, detection of valvular regurgitation, and detection and quantification of intracardiac shunts.

Echocardiography may be used in the investigation of all forms of heart disease and, because examination can be repeated as required, it can be of particular value in their serial assessment. Echocardiographic investigation can be definitive, e.g., in the assessment of a hemodynamically insignificant cardiac murmur, or it can be used to rationalize subsequent invasive studies.

2. Valvular Heart Disease

2.1 Mitral Valve

During diastole, the normal mitral valve opens, partially closes (as the left ventricle fills) and then reopens (as the

291

left atrium contracts); during systole, the closed valve is gradually displaced forward (see *Heart Valves*). These movements are reflected in the configuration of the normal M-mode echocardiogram: in diastole, the mitral anterior and posterior leaflets are M and W shaped, respectively, and both are directed anteriorly throughout systole (Fig. 3a). Measurement of the mitral diastolic closure slope provides a rough index of the left ventricular filling rate. Long-axis, two-dimensional echocardiograms display both mitral leaflets, their subtending chordae tendineae and the posterior papillary muscle. The mitral valve orifice can be imaged on appropriate short-axis scans (Fig. 2).

Both M-mode and two-dimensional echocardiography will not only reliably differentiate a normal from an abnormal mitral valve but will also distinguish between mild and severe disease. In rheumatic mitral valve disease, the M-mode echocardiogram typically shows thickening of the leaflet echoes, a reduced diastolic closure rate and anterior diastolic motion of the posterior leaflet (Fig. 3b); multiple echoes persisting at low gain settings reflect calcification or severe fibrosis. The thickening and the restricted motion of the leaflets are similarly obvious on two-dimensional echocardiograms; the two-dimensional technique has the advantage that it will also detect disorganization of the subvalvar apparatus. With two-dimensional echocardiography, accurate quantification of mitral stenosis is possible: measurements of the mitral-valve orifice made from short-axis two-dimensional scans agree closely with the mitral valve area measured at operation. M-mode quantification is less satisfactory—while there

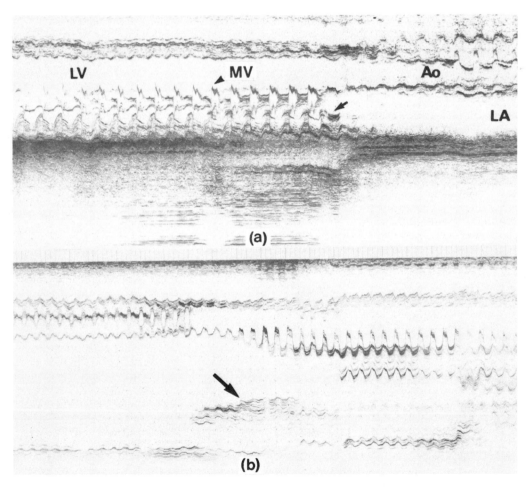

Figure 3
(a) Slow M-mode sweep from left ventricle to aorta—the mitral and aortic valves are normal, there is calcification of the mitral ring (arrow), and the left ventricle is dilated: LV, left ventricle; LA, left atrium; MV, mitral valve; Ao, aorta (after Barnett and Morley 1983. *Clinical Diagnostic Ultrasound*. © Blackwell Scientific Publications, Oxford. Reproduced with permission). (b) Slow M-mode sweep from aorta to left ventricle from patient with rheumatic mitral valve disease—the left atrium is grossly dilated and contains a calcified thrombus (arrow)

is a useful correlation between diastolic separation of the leaflets and the mitral-valve area, the mitral diastolic closure rate is an unreliable index of the severity of mitral stenosis.

Whereas calcific degeneration of the mitral ring (an important cause of mitral regurgitation in the elderly) is characterized on M-mode examination by a band of dense echoes behind a normal leaflet echogram (Fig. 3a), myxomatous degeneration of the mitral valve (floppy mitral valve) is characterized by slight thickening of the leaflet echoes with increased diastolic excursion and, when there is significant leaflet prolapse, by major systolic sagging of the echogram. Two-dimensional echocardiography can provide additional information about the direction and extent of mitral prolapse but otherwise adds little to the M-mode assessment of these two disorders.

Regurgitant flow across the mitral valve can be identified by Doppler ultrasound but not by M-mode and two-dimensional echocardiography. However, it should be appreciated, firstly, that given an appropriate murmur and evidence of a disorganized rheumatic valve or of a classically floppy valve, the presence of mitral regurgitation can clearly be inferred from the M-mode examination, and secondly, that disordered systolic apposition of the leaflets of a rheumatic valve may be defined on short-axis two-dimensional scans allowing even minor degrees of regurgitation to be predicted.

2.2 Aortic Valve

On M-mode examination of the aortic valve, echoes are usually recorded from only two of its three cusps, i.e., from the right and noncoronary cusps. Echoes from the cusps of the normal aortic valve are thin; in systole, the right and noncoronary cusps move towards the anterior and posterior walls of the aorta respectively, and in diastole, the closed cusps are recorded as a single line in the center of the aortic root (Fig. 3). The same two cusps are imaged on long-axis two-dimensional echocardiograms but all three cusps and the valve orifice can be imaged on short-axis scans through the aortic root. The characteristic systolic doming of a congenitally stenotic but still mobile aortic valve can then be displayed on long-axis two-dimensional scans but this configuration makes it impossible to detect such valves by M-mode echocardiography. However, with this exception, and provided that care is taken with the technical aspects, M-mode echocardiography will differentiate a normal from an abnormal aortic valve. The bicuspid aortic valve is characterized by an eccentric diastolic closure line. Rheumatic disease (depending on its severity) produces varying degrees of thickening and restricted separation of the cusps; with gross disorganization and calcification, cusp separation can no longer be defined and an immobile echo complex is recorded within the aortic root. Calcification of a congenitally abnormal or of a previously normal aortic valve produces M-mode features indistinguishable from those of the calcified rheumatic valve. Two-dimensional

short-axis echocardiography can add usefully to the assessment by demonstrating the extent to which its cusp is involved by disease and by displaying the valve orifice. Two-dimensional echocardiographic estimates of the aortic orifice are reported to agree reasonably well with the measurements made at operation.

Regurgitant aortic flow can be identified by Doppler ultrasound but not by M-mode or two-dimensional echocardiography. However, aortic regurgitation can be reliably inferred in the presence of diastolic oscillation of the mitral valve (produced by the regurgitant stream) on the M-mode echogram and it can be predicted with reasonable accuracy when failure of diastolic apposition of the cusps is defined on short-axis two-dimensional scans through the aortic root. Finally, M-mode and two-dimensional echocardiographic estimates of the left ventricular cavity size, wall thickness and wall motion can provide indirect evidence of the severity of stenotic or regurgitant lesions of the aortic valve.

2.3 Tricuspid Valve

M-mode echograms can be recorded from the anterior leaflet of a normal tricuspid valve and occasionally from its septal leaflet. The echogram from the tricuspid anterior leaflet resembles the mitral anterior leaflet echogram and the two are similarly altered by disease. All three leaflets of the tricuspid valve can be imaged with two-dimensional echocardiography (Fig. 4a) and, as with the mitral valve, thickening, reduced mobility, disordered apposition and prolapse of the leaflets can be detected.

2.4 Pulmonary Valve

The pulmonary valve M-mode echogram can be difficult to record; it reflects motion of the posterior cusp of the pulmonary valve and is thus broadly similar to the echogram recorded from the noncoronary cusp of the aortic valve. Access to the pulmonary valve is easier with two-dimensional echocardiography—the posterior cusp of the normal pulmonary valve is usually visible on short-axis scans at aortic valve level (Fig. 4a) and one of the anterior cusps may also be seen. The echocardiographic diagnosis of disorders of the pulmonary valve is unsatisfactory. Thus, though systolic doming of a stenotic valve may be defined by two-dimensional echocardiography, the configuration of the valve makes M-mode diagnosis impossible. Furthermore, the echocardiographic techniques cannot reliably diagnose pulmonary atresia or pulmonary infundibular stenosis.

2.5 Prosthetic Valves

Echocardiography has a role in the assessment of prosthetic valves. In the case of mechanical prostheses this role is limited: although disk and ball excursion can be recorded on the echocardiogram, the function of modern radiopaque prostheses can be studied more accurately by fluoroscopy. In contrast, the function of

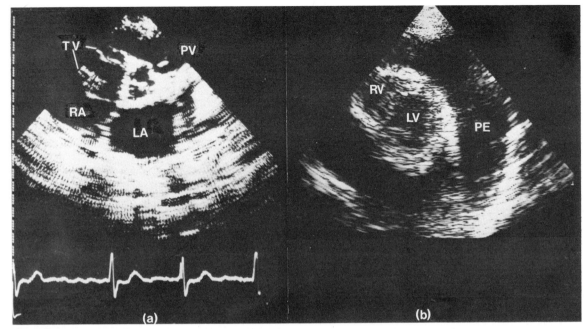

Figure 4
(a) Short-axis two-dimensional echocardiogram through the aortic root from patient with a floppy mitral valve and probable myxomatous degeneration of the aortic and tricuspid valves (TV, tricuspid valve; PV, pulmonary valve; RA, right atrium). (b) Short-axis two-dimensional echocardiogram at ventricular level. The echo-free space surrounding the heart indicates a pericardial effusion (PE)

biological prostheses cannot be examined by non-invasive radiological techniques and here echo-cardiography is the investigative method of choice—the structure and motion of the leaflets can be studied on M-mode echograms and unseating of the prosthesis can be detected on the real-time display.

3. Infective Endocarditis

Natural or prosthetic valves may become infected by bacteria or fungi; the result can be the potentially fatal destruction of the valve and its supporting apparatus. Vegetations are the hallmark of infective endocarditis; they can be detected by echocardiography and by no other investigative technique. On the M-mode echo-cardiogram, vegetations are recorded as dense, shaggy echoes and on two-dimensional images, as rounded or elongated projections from the infected valve. It must, however, be emphasized that only relatively large vegetations can be detected and that failure to demonstrate them does not exclude the diagnosis. Quite apart from the detection of vegetations, echocardiography can demonstrate disruption of the valve, and serial echocardiographic studies are thus of great importance in assessing the progress of patients with infective endocarditis.

4. Pericardial Effusion

The echocardiographic recognition of pericardial effusion depends on the demonstration of an echo-free space separating the epicardial and pericardial surfaces of either the left or right ventricle. With both M-mode and two-dimensional echocardiography, a posterior pericardial effusion is detected as an echo-free space behind the left ventricular posterior wall (pericardial fluid rarely collects behind the left atrium) and an anterior pericardial effusion is detected as an echo-free space separating the epicardial surface of the right ventricle from the chest wall. Two-dimensional echo-cardiography (Fig. 4b) provides additional information about the distribution of the pericardial fluid and can simplify the differential diagnosis from a left pleural effusion and other potential sources of diagnostic confusion. Small or large pericardial effusions can usually be differentiated by M-mode echocardiography but more precise quantification is possible with the two-dimensional technique.

5. Left Ventricular Function and the Investigation of Ischemic Heart Disease

With M-mode echocardiography only the ventricular septum and the posterior wall of the left ventricle can be studied. Nonetheless, a slow M-mode scan of the ventricular long axis (Fig. 3) can provide useful information about cavity size and wall motion. With two-dimensional echocardiography, a comprehensive assessment of cavity shape and wall motion is feasible. Thus, the ventricular septum and posterior wall of the left ventricle can be imaged on long-axis scans from the parasternal position, the anterolateral and inferior walls can be imaged on appropriately orientated long-axis scans from the apical position, and the entire ventricular circumference can be imaged on short-axis two-dimensional echocardiograms (Figs. 2 and 4a).

M-mode echocardiography is used in the investigation of left ventricular function but it has important limitations in this context. Thus, measurements of the dimension of the left ventricular cavity (just below the mitral valve) at end-diastole and end-systole can be used to calculate fractional shortening (of the internal dimension). This reflects circumferential fiber shortening and with the important provisos that cavity shape and the ventricular contraction pattern are normal, fractional shortening correlates with the ejection fraction measured by other techniques. The peak velocity of circumferential fiber shortening (V_{cf}) can also be measured from the M-mode echocardiogram; as there are fewer potential sources of measurement error it is probably a more reliable noninvasive index of left ventricular performance than is fractional shortening. However, the same provisos about cavity shape and contraction pattern stand. The M-mode echocardiogram can be digitized and the rate of change of the left ventricular dimension can be investigated. With reference to a simultaneous apex cardiogram or to a simultaneous phonocardiogram it is then possible to detect incoordinate contraction, relaxation and filling of the left ventricle.

On the assumption that the left ventricular cavity approximates an ellipsoid, ventricular volumes and hence the ejection fraction can be calculated by cubing the echocardiographic dimensions. However, as departures from this ideal geometry are common and particularly because measurement errors are magnified by the cubing process, these calculations have been abandoned in favor of fractional shortening and peak V_{cf}.

Early attempts to measure left ventricular ejection fraction by cross-sectional echocardiography were unsuccessful, since the systems used did not have the resolution necessary to outline the cavity with precision, and because of the inaccuracies involved in the single-plane estimation of ejection fraction in the presence of asynergy. With biplane imaging and improved resolution, more accurate measurement is possible but problems remain: adequate two-dimensional images cannot always be obtained and the analysis is time consuming. Contrast echocardiography of the left heart may assist the delineation of the left ventricular cavity and may thus further improve the accuracy of two-dimensional echocardiographic measurement of ejection fraction.

Furthermore, it is theoretically possible to derive and digitize multiple M-mode echocardiograms from the two-dimensional echocardiogram and, though this is not yet a practical proposition, an accurate echocardiographic assessment of the overall contraction pattern and thus of the overall function of the left ventricle is feasible. Meanwhile, it should be noted that the radionuclide ejection fraction, which is independent of ventricular geometry, is clearly superior to the comparable echo-derived indices of left ventricular function.

Wall motion in the ischemic heart is often asynergic. Much is made of the inaccuracies of the M-mode assessment of function in the presence of asynergy but it should be remembered that the echocardiographic dimension remains a rough index of ventricular size and that this information alone may be of considerable practical value. Similarly, whether or not two-dimensional echocardiographic estimates of ejection fraction prove to be accurate there is no doubt that a globally hypokinetic and a normally or near normally functioning left ventricle can be easily and usefully differentiated on the real-time display.

Since only the ventricular septum and the posterior wall can be examined, M-mode echocardiography can provide only limited information about wall motion. These limitations do not apply to the two-dimensional technique and where it is possible to detect wall-motion abnormalities developing in the course of acute myocardial infarction and during exercise stress testing. Similarly, while only posterior myocardial aneurysms can be detected by M-mode echocardiography, both posterior and anterior anteroapical aneurysms can be imaged with the two-dimensional technique. With careful attention to technical details, intracavitary thrombi can be imaged by two-dimensional echocardiography; there are important therapeutic implications and this particular aspect of the echocardiographic assessment of the ischemic heart is likely to be widely applied.

6. Congenital Heart Disease

In congenital heart disease, abnormalities may be single but are often multiple and anatomically complex. Two-dimensional echocardiography therefore has advantages over the M-mode in the investigation of these disorders.

The presence of an atrial septal defect may be reliably inferred when right ventricular dilatation, reversed motion of the ventricular septum, and mitral and tricuspid prolapse are recorded on the M-mode echocardiogram. Atrial septal defects may be imaged directly on two-dimensional echocardiograms but false positives can occur due to echo dropout. Peripheral venous injection of an acoustic contrast agent during the two-dimensional examination provides a definitive diagnosis: all atrial septal defects have some degree of right-to-left shunting and a trail of contrast echoes can therefore be detected crossing through the defect. By displaying volume overload of the left ventricle, M-mode echocardiography can provide indirect evidence of a left-to-right shunt through a ventricular septal defect or a patent ductus arteriosus. However, only very large defects of the membranous septum can be directly demonstrated by this technique and direct evidence of a patent ductus cannot be obtained by M-mode echocardiography. With two-dimensional echocardiography, imaging of a patent ductus arteriosus, though difficult, is feasible but again only the large ventricular septal defects associated with, for example, Fallot's tetralogy or atrioventricular canal abnormalities can be detected by this technique; imaging of small defects of the muscular septum is at best unreliable. Right-to-left shunts at ventricular level can be detected by contrast echocardiography of the right heart.

Inversion of the atria and ventricles can be diagnosed by two-dimensional echocardiography: imaging of the inferior vena cava (which always drains into the right atrium) defines atrial situs and ventricular situs can be defined by a number of anatomical features, e.g., the position of the atrioventricular valves and the degree of trabeculation within the ventricle. Transposition of the great arteries can be reliably detected by two-dimensional echocardiography and the differential diagnosis between Fallot's tetralogy, truncus arteriosus and pulmonary atresia, which is difficult with M-mode echocardiography, is greatly simplified by two-dimensional imaging.

See also: Blood Flow: Invasive and Noninvasive Measurement; Cardiac Function: Noninvasive Assessment; Ultrasonic Image Analysis

Bibliography

Edler I 1961 Atrioventricular valve motility in the living human heart recorded by ultrasound. *Acta Med. Scand.* 170, Suppl. 370: 85–113

Feigenbaum H 1980 Echocardiography. In: Braunwald E (ed.) 1980 *Heart Disease: A Textbook of Cardiovascular Medicine.* Saunders, Philadelphia, pp. 96–146

Goldberg S J, Allen H D, Sahn D J 1980 *Pediatric and Adolescent Echocardiography: A Handbook*, 2nd edn. Year Book Medical, Chicago

Hatle L, Angelsen B 1985 *Doppler Ultrasound in Cardiology: Physical Principles and Clinical Applications*, 2nd edn. Lea and Febiger, Philadelphia

Henry W L, DeMaria A, Gramiak R, King D L, Kisslo J A, Popp R L, Sahn D J, Schiller N B, Tajik A, Teicholz L E, Weyman A E 1980 Report of the American Society of Echocardiography committee on nomenclature and standards in two-dimensional echocardiography. *Circulation* 62: 212–17

Sahn D J, DeMaria A N, Kisslo J, Weyman A 1978 Recommendations regarding quantitation in M-mode echocardiography: Results of a survey of echocardiographic measurements. *Circulation* 58: 1072–83

Upton M T, Gibson D G 1978 The study of left ventricular function from digitized echocardiograms. *Prog. Cardiovasc. Dis.* 20: 359–84

J. C. Rodger

Electric and Magnetic Fields: Biological Effects

The biological effects of high-energy electric and magnetic fields have long been studied because of their possibly hazardous effects on the health of workers in or around power installations where electric field strengths are some $5 \, kV \, m^{-1}$ and upward, and magnetic field strengths are typically in excess of $5 \, mT$, and often two or three orders of magnitude higher. These studies have assumed a greater significance with the development of some newer technologies, which have put a greater number at risk (Tenforde 1979). Personnel involved with superconducting-magnet energy-storage devices (with $\sim 4 \, T$ fields in their interiors) can be exposed to fields of $5–8 \times 10^{-2} \, T$. With nuclear fusion reactors, magnetic-field operation levels are of the order $1–3 \times 10^{-2} \, T$, reduced by shielding to $10^{-4} \, T$. Processes associated with particle accelerators, e.g., bubble-chamber operation, can expose operators to magnetic field strengths of $0.6–1.5 \, T$ in magnitude for a few minutes several times during a working day.

The biological effects of low-energy fields are of particular interest because of their current use in promoting the healing of injured tissues. Two distinctly different methods of treatment are in vogue. In one method, a direct electric current ($10–20 \, \mu A$) is passed through the injured tissue. In the other method, the tissue is exposed to a pulsed magnetic field of $1–2 \times 10^{-4} \, T$. The first method is more fully supported by experiments, both *in vivo* and *in vitro*, but both methods are apparently beneficial clinically.

1. Strong Fields

Electric fields below high-voltage ($400 \, kV$) overhead power transmission lines are of the order $5–10 \, kV \, m^{-1}$; in substations, field strengths are even higher. Studies conducted in the UK (Male and Morris 1982) suggest that such fields have no adverse effect on health. However, a more recent epidemiological study (Perry 1983) has revealed an association between exposure to environmental, power-frequency magnetic fields and suicide; the proportion of suicides in the group studied was 40% greater than in the control group.

Humans exposed to strong magnetic fields greater than $10^{-2} \, T$ experience excitation of the visual pathway which manifests itself as flickering or shimmering light patterns; this may be due to the generation of phosphenes in the retina. In lower mammals, various physiological and histopathological changes are reported but little progress has been made in their study because of the lack of standardization in experiments conducted. The effects of electromagnetic fields—on EEG patterns as a measure of brain-cell response, and on memory stimulation—have been used to monitor interaction between external fields and the human central nervous system. The desired induced voltage gradients are obtained using an amplitude-modulated radiofrequency carrier—147 MHz and 450 MHz modulated over the EEG frequency range 1–35 Hz (Bawin 1973). Some of these studies suggest that calcium ion efflux from brain cells may be changed by electromagnetic fields. A 15–20% decrease of calcium ion efflux was observed in the frequency range 6–16 Hz at electric field gradients of 10–56 V m^{-1}. This is of relevance in relation to energy transfer across biological membranes; transfer may be controlled by rates of movement of ions, in particular divalent calcium ions. An intensity "window" effect is observed in the sense that the biological calcium efflux effects only occur at certain ranges of frequency and amplitude; interesting to note in view of similar windows reported in studies of low-intensity pulsed magnetic fields (Pilla 1979). The whole subject of biological controls or switches whereby a change in flux of one ion can amplify a series of reactions is of great interest (see Williams referred to in Hastings 1983). Information flow—from environments external to individual cells causing them to change function or activity—involves the movement of small molecules. Biological control processes also depend on small molecule or ion movements. Some control processes work, in the context of the cell's normal metabolism, to maintain a steady state but other controls (switches) act rapidly. If electrical fields affect ionic flow they will have a consequent effect on cell switching. Certain ions, and processes with definite activation energies, are involved; this may be a partial explanation for the observed "windows" within which electrical fields produce effects.

Detailed physiological studies (involving measurement of ECG, EEG, blood pressure and body temperature) are being conducted on human subjects by Silny at the Helmholtz Institute in Aachen, Germany. No physiological effects have been reported other than visual evoked potential effects with corresponding shimmering visual pattern and flickering of the ciliary muscle. Studies of the effects of magnetic fields on the mutagenicity of the plant *Tradescantia* have also been reported. No effect—using fields up to 3.7 T—has been observed on the plant's mutagenicity. It is believed that an 8 h (over 24 h) exposure to a 4 T field would result in a similar—or even a lower—mutation frequency to that produced by a 0.25 mSv x-ray exposure.

Orientational effects on retinal rods (1 T fields) and on pollen tube growth (14 T fields) have been reported (Sperber et al. 1983).

1.1 Standards of Permitted Exposure

Various standards of permitted exposure to high-energy fields are presently under consideration. A typical proposal could be that, in respect of dc fields and extended exposure (up to 1 h), the whole body or head should not be exposed to fields greater than 2×10^{-1} T, the limiting field strength being extended to 20 T for the exposure of hands and arms. These limits would be increased tenfold for short-duration exposures (up to 15 min).

2. Low-Energy Fields

Considerable interest has developed around the medical use of low-energy electric and magnetic fields to promote healing, particularly the healing of ununited bone fractures. This therapeutic technique has developed from observations on electrical phenomena in bone. Biopotentials measured from skin electrodes sited just above a long bone become more negative by about 20 mV above a fracture, and it has been speculated that this might be the stimulus for the bone to begin healing. Bone is also known to generate strain-related potentials of a piezoelectric nature and it has been proposed that these potentials play a part in the well-known process whereby external mechanical forces affect bone growth. The biopotentials may have their origin in the periosteum (the membrane ensheathing bone) but the strain-related effects are clearly associated with the osteons, the lamellar units within cortical bone. No clear physiological effect arising from these phenomena has yet been proved (Hastings 1983).

It is still a controversial area, but two basic treatment techniques are used clinically. The first uses stainless-steel electrodes (cathodes) implanted into the nonunion site in the bone, delivering 10–20 µA direct current. In present practice, four cathodes are used, each delivering 20 µA continuously for twelve weeks, during which time the limb is immobilized in plaster. An 83% healing rate is claimed based on results from 324 nonunions (Brighton et al. 1981). This same group has analyzed the factors that adversely affect healing by this technique. It would appear that a synovial pseudoarthrosis (false joint) and a large interfragmentary gap inhibit the healing effect completely. The humerus is the least successful bone to treat.

The upper limit of current density for osteogenesis is some 3–5 µA mm^{-2}, destructive processes being observed above this level. True current density *in vivo* may be greater than anticipated due to protein deposition causing a reduction in effective electrode area. In the region of the electrode, oxygen depletion is observed with concomitant increase in pH. Low oxygen levels are known to favor the processes involved in osteogenesis.

The second method of treatment employs a parallel pair of Helmholtz aiding coils arranged on opposite sides of the fracture which is completely immobilized in a plaster cast (Bassett et al. 1978). The net result of such an arrangement of air-cored coils is an electric field component along the bone surface, in opposite directions on opposite sides of the bone. A pulsed magnetic field (2 G) is used, producing a voltage gradient in the bone—from the fast rise time of the pulse—of 1.0–1.5 mV cm^{-1}.

In the original coil-pair system, the waveform was a 300 µs pulse at a repetition rate of 75 Hz. The treatment lasted for 12–16 h per day, for up to 6 months. This waveform is now used for congenital nonunions. For acquired nonunions a pulse train is used—5 ms bursts repeated at a rate of 10–15 Hz. The individual pulses are

200 µs wide, repeated at 230 µs intervals. A specific "window" effect is claimed for an induced electric field between 1 and 1.5 mV per centimeter of bone. A strict regime of non-weightbearing is followed until signs of new bone formation are present, at which point controlled weightbearing is permitted. Success rates vary according to site and type of nonunion. However, an overall healing rate of 83% has been obtained for acquired nonunions of the tibia. Other air-cored coil systems have been used (Corfield and Jones 1981) with similar success rates.

A different system uses an iron-cored coil with a pulse repetition rate of 1 Hz and a voltage gradient of approximately one order of magnitude less than that referred to above. Watson and Downes (1979) report a success rate of 64% for treatment of 50 patients.

A clinical trial (Barker 1984), although for a small series, shows little difference between the success rate of fractures treated with pulsed electromagnetic fields and a control group. In fact, the rate was slightly better for the latter. A careful study of systems used clinically (Anderson 1984) has pointed out the importance of careful geometrical alignment of the coils to ensure uniformity of the magnetic field and shows that when a metal implant is already present the field will be considerably distorted. This work set out to study the possible effect of the fields on collagen orientation using synthetic macromolecules as models but results were negative for clinical field strengths. Higher fields showed some effect (Anderson and Hastings in press).

Stimulation of bone growth associated with direct current passed through implanted electrodes is reasonably well established from animal trials. Work in this field has been reviewed by Watson (1981). It is difficult to produce a nonunion in animals to study the effect experimentally and, of course, the treatment of a fresh animal or human fracture is a very different matter from the treatment of a long-standing human nonunion, the biological condition of the tissues being different. Hence the direct relevance of these results with animals may be questioned. Although there is sufficient evidence to be assured that bone will grow when a cathodic electrode is inserted, accurate definition of the experimental model is still required, as is further study of the actual mechanism of growth. It is claimed that the work on animals provides sufficient basis to support clinical work, although no double-blind controlled clinical trials have been performed.

Supportive evidence from experiments with animals is less clear cut with regard to external pulsed electromagnetic fields because of the lack of a control group. Although changes have been seen in beagle dogs, an acceleration of fracture repair being reported, the significance level was 7% (Bassett 1974). Other workers using the same equipment as Bassett obtained completely negative results and similarly so with a different system (Enzler et al. 1980). In contrast, Christel et al. (1981) reported a significant increase in the tensile strength of healing rat osteotomies in comparison with separate control animals. They noted the importance of a frequency range (11–15 Hz) and reported an optimum dose amplitude effect. This so-called "window" effect is also postulated by Pilla following his experiments on blood cells and toad bladder membranes. However, in spite of this uncertainty, evidence from clinical work continues to accumulate (Goldberg 1982) and the system has been extended to the treatment of femoral head avascular necrosis (Bassett 1981).

The need for good orthopedic treatment is emphasized by exponents of both systems and this necessity apart from any effect of the electrical systems in the attainment of results should be evaluated. The chance that a purely placebo effect is operating must, of course, be eliminated. Whether it is a magnetic field per se that is significant, or the induced electric field, or direct current or electrode reactions, requires more investigation. Controlled clinical trails run by impartial bodies are also required. Precise definition of technical data is a further requirement, together with controlled, objective methods of measuring and assessing the physiological effects. A study of blood flow and tissue oxygen concentration during magnetic-field exposure using a machine designed to treat soft-tissue injury showed, in a double-blind trial, no significant effects attributable to the field (Railton and Newman 1983).

2.1 Related In Vivo Studies

The purpose of wide-ranging *in vivo* studies additional to those carried out on bone has been to elucidate the biochemical and structural changes occurring in soft tissues subjected to various forms of electrical and magnetic stimulation, in the hope that the methods used for bone healing can be applied to the repair of nerves and other soft tissues. Increased vascularization and blood flow, and beneficial effects on repair processes in skin and ligaments, have been variously reported following stimulation. Studies of wound repair and peripheral nerve regeneration are of obvious importance. Using direct current stimulation, the growth of cancellous bone into porous implants has been enhanced to the extent that the interfacial shear strength of porous intramedullary implants is increased when direct current stimulation is used. This could be of value in prosthesis stabilization either in the immediate post-operative period or in the later treatment of a prosthesis that has loosened (Biological Repair and Growth Society 1981–82). Limb regeneration in amphibians has also been studied and it is observed that the presence of electric fields inhibits regeneration. In fact, the process is not unlike stump formation in nonregenerating vertebrates.

Disuse osteoporosis in immobilized rats (Bassett et al. 1979) and in denervated rats (Brighton et al. 1982) is significantly inhibited when the rats are subjected to stimulation. This observation could be of significance in relation to the treatment of human female osteoporosis although, in fact, this condition is biochemical in origin.

2.2 In Vitro Experiments

In vitro studies have been widely used (Brighton et al. 1979) to elucidate the mechanisms underlying the biological results of stimulation and to study the processes per se that operate in a controlled system. Both organ culture using, for example, chick embryo rudiments, and cell culture with well-characterized cell populations have been employed. Studies have also been conducted to examine the effects of fields on growth and orientation in microorganisms. Changes in calcium ion flux and in cAMP (cyclic adenosine monophosphate) metabolism are observed in tissues subjected to pulsed magnetic fields. Studies on chick embryos revealed changes in values of cAMP when the field was switched off after a prolonged period of culture (186 h) and again when it was switched on, the response observed from the shafts of the embryonic bones being different from that of the cartilage. The shaft cAMP values showed initial significant increases at both the commencement and cessation of each period of stimulation, whereas cartilage cAMP values decreased markedly. Both shaft and cartilage cAMP values then returned towards control values but with wide fluctuations (Jones 1982).

As referred to earlier, certain waveforms may be more clinically effective than others. It is now emerging (from *in vitro* studies) that some waveforms greatly enhance cellular DNA production, whereas others have a minimal effect on its production (Bassett 1974). Studies on cultures of chick embryos (Fitton-Jackson et al. 1981, Pilla 1979) show, for example, that a simple pulse of the form used clinically in the treatment of ununited fractures has its main effect on cellular calcium uptake at a repetition rate of 15 Hz with a second peak effect at 133 Hz, at which rate cell proliferation is inhibited. In these *in vitro* studies, the temperature at the location of the cells must be strictly controlled. Should the temperature be allowed to rise the results obtained could be very misleading.

Typical of cell culture work is the use of cloned rat osteosarcoma cells to study electrical perturbation effects on cell proliferation. Alkaline phosphatase production and DNA synthesis are commonly measured indicators of cell activity. A proportionality between cell activity and frequency (2–25 Hz range) has been noted, the inhibition of alkaline phosphatase production being greater in the early stages of growth when the cells were growing vigorously. A 5.2 ms pulse train repeating at 15 Hz is used, as in bone-fracture work. Other waveforms give similar results. Mixtures of chondroblasts and fibroblasts stimulated by a 15 Hz, 5 ms pulse train show increased glycosaminoglycan production. The work of Pilla (1979) and Keyser and Gutsman (1980) on red blood cells of frogs has been important in establishing that a frequency window exists between 40 and 70 Hz outside which dedifferentiation of the cells is inhibited. Similar effects are reported from Pilla's studies with toad bladder membranes. Microorganism cultures studied include *E. coli*, yeasts and slime molds. Results show dependence of cell activity on frequency, field strength and aeration. Using cultures of *E. coli* grown under controlled conditions a threshold magnetic field strength is observed, below which no effects are apparent. Above the threshold the results are periodic. The mean generation time of the cells is unaffected by an applied 50 Hz field up to a peak field strength of 480 T followed by a decrease ($\sim 4\%$) above 480 T and a further rise at ~ 800 T. Based on the size of the cells, it has been shown that the field strength which is effective corresponds to a flux linkage of one flux quantum per dividing cell (Aarholt and Smith 1982).

The advantage of such *in vitro* studies, as opposed to *in vivo* studies, is that the experimental conditions can be controlled more precisely, allowing a more significant analysis of the effects. However, the physics of the system actually used is not always defined and as yet the relative importance of magnetic and electrical fields is not yet known.

Ferromagnetic orientation has been observed in certain bacteria obtained from swamps (Blakemore 1978). Marine fish are sensitive to weak electric fields (voltage gradients as low as 0.01 V cm^{-2}) in their normal environment. Sharks, rays and skates are particularly responsive. Similar stimuli may affect birds and insects.

3. Conclusions

A knowledge of the electrical properties of biological materials is important for an understanding of physical and biological growth and repair process mechanisms. The effects of external fields on the growth and repair processes of living systems have been used to elucidate the characteristics of the fields that are determinant in producing these effects. Lack of standardization in experimental conditions makes assessment of the validity of some observations difficult. The fact that patients are being successfully treated makes it more necessary to be able to control the effects produced by low-strength fields to suit a range of clinical conditions. With reference to high-energy fields the increasing likelihood of hazardous exposure of personnel underlines the necessity for studies on possibly linked psychiatric and brain cell effects. The increasing number of people with cardiac pacemakers adds to the need for studies to define safety limits for exposure to electrical and magnetic fields in industrial and domestic environments.

Bibliography

Aarholt E, Smith C W 1982 Effects of low-frequency magnetic fields on bacterial growth rate: Weak field effects on cellular metabolism. *IEE Conf. Some Biological Effects on Electric and Magnetic Fields*. Institution of Electrical Engineers, London

Anderson A M 1984 The effect of pulsing electromagnetic fields on polymerising systems. Ph.D. Thesis, North Staffordshire Polytechnic

Anderson A M, Hastings G W (in press) The effect of pulsing magnetic fields on polymerisation of acrylamide and the reconstruction of collagen. *Biomaterials*

Barker A T, Dixon R A, Sharrard W J W, Sutcliffe M L 1984 Pulsed magnetic field therapy for tibial non-union. *Lancet* i: 994–95

Bassett C A L, Mitchell S N, Norton L, Pilla A A 1978 Repair of non-unions by pulsing electromagnetic fields. *Acta Orthop. Belg.* 44: 706–24

Bassett C A L, Pawluk R J, Pilla A A 1974 Augmentation of bone repair by inductively coupled electromagnetic fields. A surgically non-invasive method. *Science* 184: 575

Bassett C A L, Shink M, Mitchell S N 1981 Pulsing electromagnetic fields effects in avascular necrosis a preliminary clinical report. *Trans. Biol. Repair Growth Soc.* 1: 38

Bassett L S, Tzitzikatakis G, Pawluk R J, Bassett C A L 1979 Prevention of disuse osteoporosis in the rat by means of pulsing electromagnetic fields. In: Brighton et al. 1979, pp. 311–31

Bawin S M, Garalas R J, Adly W R 1973 Effects of modulated VHF fields on the clinical nervous system. *Brain Res.* 58: 365–84

Biological Repair and Growth Society 1981–82 *Trans. Biological Repair and Growth Society*. Society of Medical Education, University of Pennsylvania, Philadelphia, Pennsylvania

Blakemore R 1978 Ferromagnetic orientation in bacteria. In: Tenforde T S (ed.) 1978 *Magnetic Fields in Biological Systems*. Plenum, New York, pp. 13–15

Brighton C T, Black J, Pollack S R (eds.) 1979 *Electrical Properties of Bone and Cartilage*. Grune and Stratton, New York

Brighton C T, Friedenberg Z B, Black J, Heppenstall R B, Esterhai J L Jr 1981 Treatment of non union with constant direct current. *Trans. Biol. Repair Growth Soc.* 1: 16

Brighton C T, Katz M J, Pollack S R 1982 Prevention of denervation/disease osteoporosis in the rat with a capacitating coupled electrical field. *Trans. Biol. Repair Growth Soc.* 2: 51

Christel P, Pilla A A 1981 Pulsating electromagnetically induced current (PEMIC) modulation of bone repair: Effect of waveform configuration on rat radial osteotomies. *Trans. Biol. Repair Growth Soc.* 1: 35

Corfield J R, Jones E R L 1981 An externally applied pulsed magnetic field treatment for delayed union of bone fractures. *Trans. Biol. Repair Growth Soc.* 1: 40

Enzler M A, Waelchi-Suter C, Perren S M 1980 Prophylaxe der Pseudarthrose durch magnetische Stimulation? Experimentelle Überprüfung der Methode nach Bassett an Beagle Hunden. *Unfallheilkunde* 83: 188–94

Fitton-Jackson S, Jones D B, Murray J C, Farndale R W 1981 The response of connective and skeletal tissues to pulsed magnetic fields. *Trans. Biol. Repair Growth Soc.* 1: 85

Goldberg A A J 1982 Analysis of computerised data-base of more than 11,000 patients with ununited fractures submitted for noninvasive electrical treatment. *Proc. Symp. Factors in Fracture Union*. Biological Engineering Society, Royal College of Surgeons, London

Hastings G W 1984 Structural and mechanistic considerations in the strain-related electrical behaviour of bone. In: Hastings G W, Ducheyne P (eds.) 1984 *Structure Property Relationships in Biomaterials*, Vol. III, *Natural and Living Biomaterials*. CRC Press, Boca Raton, Florida, pp. 151–60

Jones D B 1982 The effect of pulsed magnetic fields on cAMP metabolism in chick embryo tibiae. *Trans. Biol. Repair Growth Soc.* 2: 29

Keyser H, Gutsman F (eds.) 1980 *Biochemistry*. Plenum, New York

Male J C, Norris W T 1982 The nature of exposure to low frequency electric fields near electric power transmission plant. *IEE Conf. Some Biological Effects of Electric and Magnetic Fields*. Institution of Electrical Engineers, London

Perry E S 1982 Environmental power frequency magnetic fields and suicide. *IEE Conf. Some Biological Effects of Electric and Magnetic Fields*. Institute of Electrical Engineers, London

Pilla A A 1979 Electrochemical information transfer and its possible role in the control of cell function. In: Brighton C T et al. 1979, pp. 455–89

Railton R, Newman P 1983 Magnetic field therapy—Does it affect soft tissue? *J. Orthop. Sports Phys. Ther.* 4: 241–46

Sperber D, Dransfeld K, Maret G, Weisenseel M H 1982 Oriented growth of pollen tubes in strong magnetic fields. *Naturwissenschaften* 68: 40–41

Tenforde T S (ed.) 1979 *Magnetic Field Effects on Biological Systems*. Plenum, New York

Watson J 1981 Electricity and bone healing. *Proc. IEE* 128: 329–35

Watson J, Downes E M 1979 Clinical aspects of the stimulation of bone healing using electrical phenomena. *Med. Biol. Eng. Comput.* 17: 161–69

G. W. Hastings

Electric Response Audiometry

The technique of electric response audiometry (ERA) is used to evaluate the function of the auditory system objectively by measurement of electrophysiological activity evoked by a transient acoustic stimulus. An individual response is rarely of sufficient magnitude to be detectable in the concurrent spontaneous electrical activity, and time-domain averaging is used to extract the response. The main objectives of ERA are to infer the behavioral threshold and the site, type and severity of dysfunction in the auditory system. Information about the different levels of the system is obtained from subtechniques, such as electrocochleography, brain-stem electric response audiometry and electroencephalographic audiometry.

See also: Audiometers; Objective Audiometry; Electrocochleography; Brain-Stem Electric Response Audiometry; Electroencephalic Audiometry

Bibliography

Gibson W P R 1978 *Essentials of Clinical Electric Response Audiometry*. Churchill Livingstone, London

S. Gatehouse

Electrocardiography

Electrocardiography is the recording on a voltage–time graph of potentials existing on the skin as a result of the electrical activity of the heart. The standard electrocardiogram (ECG) consists of twelve recordings of voltages between various pairs of electrodes placed on the skin of the limbs and chest. The basic methods of electrocardiography have remained unchanged for over 40 years, but advances in technology have greatly simplified the procedure so that the ECG can be recorded quickly in the patient's home or in a health center, as well as in hospitals. The analysis of ECGs remains largely an empirical exercise although increasing understanding of the genesis of the ECG has allowed a scientific approach to be adopted in some areas of the subject. Electrocardiography is one of the commonest medical investigations. It can give valuable information to the clinician in widely varying pathological conditions, such as disturbances of blood electrolytes, endocrine diseases, and pulmonary disease, as well as in all aspects of heart disease.

Von Kölliker and Muller investigated the electrical activity of isolated frogs' hearts in 1856, and in 1887, A D Waller discovered that one can detect this electrical activity on the surface of the body in man, using bowls of saline in which the right hand and left foot were immersed. Einthoven measured the activity in 1912, using a string galvanometer in conjunction with an optical recording device. He described three "lead" connections, between right arm and left arm (standard lead I), between right arm and left leg (standard lead II), and between left arm and left leg (standard lead III). He postulated that the heart was in the center of the triangle

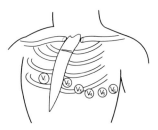

Figure 2
The positions on the chest for the unipolar chest leads

formed by these "leads" and that conduction to the surface of the body was uniform in all directions (see Fig. 1). Wilson (1934) added further lead connections termed unipolar or V leads. He used an exploring terminal placed at one of six positions on the chest over the heart and a central terminal achieved by connecting the leads from the limbs together, having inserted a 5000 Ω resistance into each (see Fig. 2). Three unipolar limb leads introduced by Goldberger complete the 12-lead ECG. These are obtained by connecting the exploring electrode to the right arm (AVR), left arm (AVL) and left leg (AVF), and measuring the potential difference between the exploring electrode and the central terminal.

The electrical activity of the heart results from a process of electrical depolarization and repolarization of the surface membrane of the cardiac muscle cells. In the resting state, the outside of the membrane is positively charged while the inside is negatively charged—the potential difference being 90 mV across the membrane. This potential is due to the relative concentration of potassium, sodium and chloride ions inside and outside the cell. Movement of these ions across the membrane neutralizes the potential difference (depolarization). The process of depolarization starts in the right atrium and spreads throughout the atria to the ventricular septum and thence to the walls of the left and right ventricles. This process precedes, and is a prerequisite for, mechanical contraction of the heart.

These events produce on the ECG a series of deflections from the baseline. The first deflection is a low amplitude, broad deflection due to atrial depolarization (P wave). After an interval of 0.1–0.2 s from the onset of this wave there is a bigger deflection due to ventricular depolarization (QRS complex) followed by a lower broader deflection (T wave) due to ventricular repolarization (see Fig. 3a).

Analysis of the ECG is invaluable in the diagnosis of disturbances of cardiac rhythm and conduction (passage of electrical stimuli), enlargement of the cardiac chambers, ischemia or necrosis of heart muscle, disease of the myocardium owing to infections, toxic chemicals or drugs, and disease of the pericardium. The ECG also shows characteristic changes in the presence of excess or deficiency of potassium, calcium and magnesium. Hypothyroidism produces a characteristic ECG waveform.

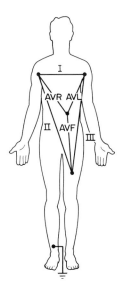

Figure 1
The standard and unipolar limb leads and Einthoven's triangle

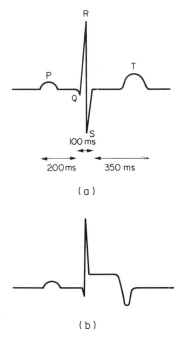

(a)

(b)

Figure 3
(a) Appearances and time relationships of the ECG.
(b) The ECG in acute myocardial infarction showing elevation of the ST segment and inversion of the T wave

Perhaps the most important use of the ECG is in the diagnosis of coronary artery disease. Within a short time of a coronary artery occluding and producing myocardial infarction, the ST segments of the ECG become elevated (up to 10 mm above the baseline). Thereafter the Q waves deepen and the T waves become inverted (Fig. 3b). The cause of these changes is a current of injury passing between the damaged heart muscle and the surrounding undamaged muscle giving rise to ST segment elevation. The damaged area transmits the intracavity ECG giving rise to the Q wave. The pattern of repolarization is interrupted by the damaged tissue resulting in the inverted T wave. The leads in which these changes appear indicate the area of the heart that is damaged. Although these abnormalities subside to some extent, the ECG usually retains some signs of damage indefinitely.

Myocardial infarction is frequently complicated by disturbances of cardiac rhythm and conduction which may be fatal if not detected and treated immediately. It is important therefore to monitor one lead of the ECG continuously by means of an oscilloscope when a patient is at risk.

In angina pectoris, ischemia of the myocardium occurs when oxygen requirements exceed the supply, for example on exertion. Although the ECG at rest may be normal, recordings made during and after exercise reveal

the diseased condition. Most cardiac departments are equipped with bicycle ergometers or treadmills for this purpose. A further application of electrocardiography is in the diagnosis of intermittent disorders of heart rhythm by means of 24 h ECG recordings on a small portable tape recorder, carried by the patient. The tape can be replayed on a scanner at 60 times the recording speed. Any disturbances of heart rhythm are readily detectable. This type of ECG monitoring can be undertaken throughout the patient's normal day (see *Ambulatory Monitoring*).

ECG telemetry is a similar but more restricted type of monitoring. An ambulatory patient's ECG is transmitted to a receiving station where it is displayed on an oscilloscope or recorder.

See also: Physiological Measurement; Monitoring Equipment in Coronary and Intensive Care; Vectorcardiography; Computers in Cardiology

Bibliography

Shamroth N 1985 *Introduction to Electrocardiography*, 6th edn. Blackwell Scientific, Oxford

J. F. Robinson

Electrocochleography

Electrocochleography (ECoG) is a form of electric response audiometry (see *Electric Response Audiometry*) which measures cochlear and auditory nerve potentials. The active recording electrode can be surgically placed on the bony promontory of the middle ear, the tympanic membrane, or the external auditory meatus. The main components of the response are the cochlear microphonic potential, the cochlear summating potential and the compound auditory nerve action potential, and these occur within 10 ms of the onset of the acoustic stimulus. ECoG has been used primarily to estimate the behavioral audiometric thresholds but may be used to identify some of the sites and types of dysfunction in the auditory system. The surgically invasive nature of the technique is its main drawback and, in recent years, the development of brain-stem electric response audiometry (see *Brain-Stem Electric Response Audiometry*) has considerably reduced the need for ECoG.

See also: Audiometers; Objective Audiometry

Bibliography

Ruben R J, Elberling C, Salomon G 1976 *Electrocochleography*. University Park Press, Baltimore, Maryland

S. Gatehouse

Electroconvulsive Therapy

The fact that the clinical conditions of schizophrenia and epilepsy are antagonistic has led to the belief that the induction of convulsions similar to those associated with epilepsy may be beneficial in the treatment of certain psychiatric disorders, particularly in cases of depressive psychosis, mania and schizophrenia. The technique of electroconvulsive therapy (ECT) as a method of inducing convulsions was first used on humans in 1938 although experimentation on animals had been in progress at the laboratories of Professor Ugo Cerletti since 1935. The introduction of ECT was precipitated largely by the shortcomings inherent in drug-induced convulsions, notably with drugs such as pentylenetetrazole, in which conditions of confusion, amnesia and nausea were frequent side effects of the drug therapy.

The manner in which ECT works as a treatment remains far from clear although a number of hypotheses exist. One theory is that the convulsion affects cerebral pathology directly; another is that the ECT effect is essentially psychological, particularly in terms of the more beneficial aspects of amnesia, a frequent by-product of the treatment. A third theory proposes that the effect is biochemical and another considers the effect to be electrophysiological. Each of the theories is substantiated to at least some extent by experimental data although at present there is little clinical agreement as to which process dominates.

The waveform originally employed in the administration of the convulsing current was a 50 Hz sinewave having an adjustable voltage in the range 50–150 V. This was applied to the patient for 0.5–0.7 s at 0.3–0.6 A. The sinewave was found in practice to have similar undesirable side effects as the drug therapy, i.e., amnesia, confusion and nausea. Attempts to reduce the intensity of the side effects have been made by using a rectified sinewave. In this technique the current was lowered to levels of approximately 40 mA for 1.0–2.0 s. More recent ECT machines have a modified sinewave such that the first 40% of each cycle is chopped off. With this waveform the voltage used is 400 V peak-to-peak for 1.7 s in the case of the rectified waveform. Some ECT machines have a facility for a technique called "crescendo offset" in which the current is increased in steps from zero to the operating current over a period of about 0.5 s rather than subject the patient immediately to the full current.

A number of machines are able to produce an output waveform having a square or rectangular shape. A typical unit has a square-wave output with a pulse width of 300 µs, a pulse amplitude of approximately 400 V and a patient treatment period of 5 s. Perfectly acceptable results can be achieved using square waveforms which significantly reduce the amount of electrical energy needed to effect a treatment. There is evidence to suggest that ECT equipment will in future have square waveform outputs exclusively. Various other waveforms have been tried at various times with different frequencies, voltages, currents and treatment durations all determined in a relatively subjective manner.

The diversity of opinion concerning waveforms is reflected in the various designs of equipment. Nonetheless, despite the variations, the underlying technical design remains simple and not greatly removed from the original equipment. Basically an ECT machine consists of a device or devices to produce the waveform, a circuit to permit adjustment of voltage and current, a timer device and a method of conducting the current to the patient (Fig. 1).

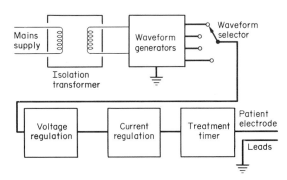

Figure 1
Block diagram of electroconvulsive therapy apparatus

The electrodes originally used by Cerletti in 1938 were flat metal disks applied to the frontoparietal regions. Electrode positioning patterns are divided basically into unilateral (in which the electrodes, and hence current, only apply to one side of the brain) and bilateral (in which both hemispheres of the brain are affected). Various sites for the active electrode have been used. Normally electrode gel is used which helps to avoid burns. As with other aspects of ECT there is no concensus concerning the best position for the active electrode. The outputs of many ECT units are derived from the discharge of a capacitor, and exact magnitudes and durations are outside the control of the operator, being dependent on the load impedance presented by the patient's head and the patient electrodes. To remove this uncertainty, most new ECT apparatus is of constant current output design.

In terms of electrical safety it is necessary to consider the interests of both the patient and the operator (see *Medical Electrical Equipment: Safety Aspects*). To protect the patient the inherent design should be such that a circuit failure within the machine does not in itself render it dangerous. Even if the machine is safe in design, certain features need regular servicing to ensure continued safe operation. In particular, the following merit careful examination:

(a) patient isolation circuitry,
(b) current and voltage control circuitry,
(c) timers and associate circuitry, and
(d) electrodes and patient leads.

In addition, the earthing of the equipment should be regularly checked for continuity. In the UK it has been stipulated that post 1981 ECT apparatus should be designed and manufactured in accordance with the British Standard BS5724 (1979).

Bibliography

British Standards Institution 1979 *The Safety of Medical and Electrical Equipment* Pt. 1: *General Requirements*, BS 5724. British Standards Institution, London

Fink M 1979 *Convulsive Therapy: Theory and Practice*. Raven Press, New York

A. C. Selman

Electrodermal Audiometry

Electrodermal audiometry (EDA) is a form of electric response audiometry (see *Electric Response Audiometry*) which measures a change in skin properties following perception of a sound stimulus. The technique was developed in the 1950s and has been found to be inherently unreliable. EDA has been totally superseded by electroencephalic audiometry, electrocochleography and brain-stem electric response audiometry and has no routine clinical applications at present.

See also: Audiometers; Electroencephalic Audiometry; Electrocochleography; Brain-Stem Electric Response Audiometry; Objective Audiometry

Bibliography

Bordley J E, Hardy W G, Richer C P 1948 Audiometry with the use of galvanic skin-resistance response. *Bull. Johns Hopkins Hosp.* 82: 569

Burk K W 1958 Traditional and psychogalvanic skin-response audiometry. *J. Speech Hear. Res.* 1: 275–78

Chaiklin J B, Ventry I M, Barrett L S 1961 Reliability of conditioned GSR pure-tone audiometry with adult males. *J. Speech Hear. Res.* 4: 269–80

Doerfler L G, McClure C T 1954 The measurement of hearing loss in adults by measurement of galvanic skin response. *J. Speech Hear. Disord.* 19: 184–89

Hind J E, Aronson A E, Irwin J V 1958 GSR auditory threshold mechanisms: instrumentation, spontaneous response and threshold definition. *J. Speech Hear. Res.* 1: 220–26

McCleary R A 1950 The nature of the galvanic skin response. *Psychol. Bull.* 47: 97–117

S. Gatehouse

Electroencephalic Audiometry

Electroencephalic audiometry is a form of electric response audiometry (see *Electric Response Audiometry*) which, at present, is mainly concerned with the change of electrical activity at the vertex (the top most point of the skull, on the scalp) evoked by a sound stimulus. Known as the slow vertex electric response (SVER), it is recorded using gross surface electrodes and is widely distributed over the central and frontal areas of the scalp, with its components occurring as much as 300 ms after the onset of the stimulus. The principal application of this technique is as a method of behavioral threshold estimation for patients who are unable or unwilling to cooperate in conventional audiometric procedures (examples are hyperactive children and the multiply handicapped). The SVER is used routinely for medicolegal assessments where its objectivity, namely a lack of dependence on patient motivation, is a considerable advantage.

See also: Audiometers; Brain-Stem Electric Response Audiometry; Objective Audiometry

Bibliography

Gibson W P R 1978 *Essentials of Clinical Electric Response Audiometry*. Churchill Livingstone, London

S. Gatehouse

Electroencephalography

The intact brain may be studied by recording its electrical activity. The activity of the individual neurons is recorded using microelectrodes placed within a few micrometers of the cell body by advancing the electrode through the cortex. The mass action of a large number of cells can be recorded using electrodes that are relatively large compared with the individual cells. These electrodes can be placed within the tissue (see *Intracerebral Electrodes*), on the surface of the cortex, or on the scalp. The recording of the scalp activity is known as an electroencephalogram (EEG); such activity is quasirhythmic and in the frequency band of about 0–100 Hz. The amplitude is of the order of tens of microvolts and the activity varies from place to place on the head. The EEG changes with the mental state of the subject and shows characteristic patterns of activity during sleep, coma, epileptic seizures and other disturbances of cerebral origin. The activity can be changed by external stimuli and characteristic waveform patterns are evoked which reflect both the physical characteristics and the significance or meaning of the stimulus (see *Evoked Potentials*).

The EEG is obtained by amplifying the electrical activity derived from the electrodes placed on the scalp. The electrodes are usually disks of chlorided silver held in place by an adhesive or elastic harness. Good contact with the tissue is obtained by cleaning or abrading the skin beneath the electrode. Twenty or more electrodes are placed at preset sites; many channels of activity are recorded simultaneously. Because of the small amplitude of the signal and the presence of external electrical and magnetic fields in the frequency band of interest (usually 50 or 60 Hz mains interference) high-gain differential amplifiers with a high common-mode rejection ratio are used. The data are usually displayed on a multichannel pen recorder.

One of the main objectives of conventional recordings is to measure the distribution of electrical potential over the scalp and to infer the location of the source or sources giving rise to it. Since the electrical activities are widespread over the scalp there is no "neutral point" on the head against which the potentials can be measured. The presence of the electrical activity of the heart (millivolts compared with microvolts) restricts the use of a noncephalic electrode that might be assumed to be undisturbed by brain activity. This electrocardiogram may be reduced by subtraction techniques.

There are three basic methods of deriving electrical signals from such an electrode array. These are commonly described as bipolar, common reference and average reference. In the bipolar derivation, each channel is connected between two (usually adjacent) electrodes, both of which are likely to be affected by appreciable brain potentials. When using the common reference derivation, one electrode, usually chosen to minimize the possibility of picking up potentials from the brain, is common to all channels. In the average reference system the average potential of all (or most) of the electrodes in the array is generated by connecting all electrodes to a "star point" through high resistors. This star point is then used as a common reference for all channels. In a modification of the average reference system (called source derivation), the activity of a particular electrode in the array is referred to the average of those electrodes immediately surrounding it.

Although the data arising from a set of electrodes are obviously the same whatever the method of derivation, the various techniques can be used to emphasize particular features. For example, the bipolar technique, which measures local gradient of the potential field, gives a different result from the common reference system that measures the potentials against the common (often incorrectly presumed inactive) point. The combination, in the input circuit of the amplifier, of time-displaced rhythmic activity from two electrodes can give rise to ambiguities that make interpretation in terms of sources very difficult.

The EEG is often contaminated by artefacts from extracerebral sources; for example, the presence of relatively large potentials from eye movements and blinks (because of the change of the corneal–retinal potential field), and from the electrical activity generated by the musculature of the scalp. Fortunately this latter effect is of relatively high frequency (50 Hz or more) and can be reduced by restricting the upper limit of the amplifier bandwidth to 30 Hz. The measurement of the steady potentials between parts of the brain is difficult because

of differences in electrode potentials and steady potentials generated in the skin. Thus, for routine recording, the bandwidth of the amplifiers is usually restricted at the lower end to approximately 1 Hz by high-pass filters.

The outer layer of cortex beneath the scalp electrodes is the main source of the electrical potentials recorded. If the activity in the deep structures is evoked by, for example, auditory stimuli it can be detected on the scalp by signal enhancement techniques such as the averaging of signals.

See also: Electroconvulsive Therapy

Bibliography

Binnie C D, Rowan A J, Gutter T 1982 *A Manual of Electroencephalographic Technology*. Cambridge University Press, Cambridge, UK
Cooper R, Osselton J W, Shaw J C 1980 *EEG Technology*, 3rd edn. Butterworth, London
Geddes L A 1972 *Electrodes and the Measurement of Bioelectric Events*. Wiley, New York
Klass D W, Daly D D (eds.) 1979 *Current Practice of Clinical Electroencephalography*. Raven, New York
Thompson R F, Patterson M M (eds.) 1973 *Bioelectric Recording Techniques*, Pt. A: *Cellular Processes and Brain Potentials*. Academic Press, New York
Thompson R F, Patterson M M (eds.) 1974 *Bioelectric Recording Techniques*, Pt. B: *Electroencephalography and Human Brain Potentials*. Academic Press, New York

R. Cooper

Electromyography

Electromyography is the technique of recording the electrical responses (action potentials) generated by muscle as a result of either voluntary muscular contraction or spontaneous muscle-fiber activity, or in response to stimulation of the nerve supplying the muscle. The principal application of electromyography is in clinical neurophysiology but it is also employed in kinesiology (the study of limb movement) and in the provision of control signals for prosthetic devices (see *Prostheses, Myoelectrically Controlled*).

1. Instrumentation

A recording system or electromyograph (see Fig. 1) consists of electrodes applied to the patient, high-impedance amplifiers, an oscilloscope display, a loudspeaker and, in most cases, a chart-recording device, which produces an electromyogram (EMG). A voltage or current pulse generator is also provided to permit external stimulation of nerves.

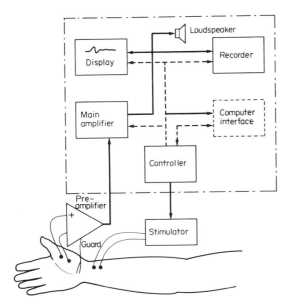

Figure 1
Schematic representation of an electromyograph

1.1 Electrodes

Two main types of electrodes are used in clinical electromyography: needle electrodes, which are inserted into the muscle through the skin, and surface electrodes, which are placed on the skin (Fig. 2). The amplitude and shape of the action potentials detected (Fig. 3) depend on the configuration of the electrode, the number of active muscle fibers, and the distance between the electrode and the muscle fibers. Fine single-fiber needle electrodes (Fig. 2a) record action potentials from only a

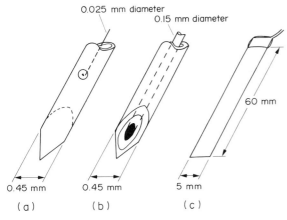

Figure 2
Main types of electrodes used in electromyography: (a) single-fiber needle electrode, (b) concentric needle electrode used in clinical electromyography, and (c) surface electrode (silver foil, 0.1 mm thick)

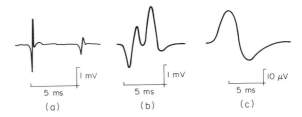

Figure 3
Example of action potentials recorded with: (a) single-fiber needle electrode, (b) concentric needle electrode, and (c) surface electrode

few muscle fibers. The larger concentric needle electrodes (Fig. 2b), more commonly used in clinical electromyography, can record from some 50 to 200 muscle fibers. A surface electrode (Fig. 2c) is capable of recording from the whole muscle or large parts thereof.

1.2 Electromyograph

Recording of the action potentials requires a differential amplifier with high input impedance, high common-mode rejection ratio and low noise level. The frequency bandwidth of the amplifier should be variable in the range 2 Hz to 20 kHz. Some electromyographs include preamplifiers in which the patient electrodes are isolated electrically from the rest of the mains-powered equipment, further reducing unwanted interference signals (e.g., from the mains supply) and increasing patient safety. The amplified signal is displayed on an oscilloscope and drives a loudspeaker. In clinical practice the audio signal can be as useful as the displayed signal. The nerve stimulator included in the electromyograph generates voltage or current pulses of variable amplitude (0–400 V or 0–100 mA, respectively) and variable width (0.05–1 ms). The stimulator output is synchronized with the oscilloscope sweep and is transformer coupled in order to reduce stimulus artefact resulting from conduction of the stimulus pulse through the tissue fluids and to provide greater safety for the patient.

Permanent recordings are produced on ultraviolet or electrostatic recorders. For many applications short-term storage (as provided by a storage oscilloscope) is sufficient. The displayed action potentials may then be measured or a Polaroid photograph taken from the oscilloscope display.

Most electromyographs now incorporate digital recording and measuring facilities ranging from a signal averager to a microprocessor-based system that can store and analyze action potentials.

2. Clinical Applications

Muscle fibers are organized in functional units called motor units, each motor unit consisting of a neuron, its axon and the muscle fibers it supplies (Fig. 4). Electromyographic measurements of motor unit activity and nerve conduction velocity are used clinically in the investigation of neuromuscular disease.

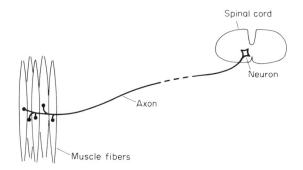

Figure 4
Schematic diagram of a motor unit

At rest, no motor-unit potentials (MUPs) are generated by normal, healthy muscle, but low-amplitude, short-duration action potentials (fibrillations) may be detected in diseased muscle if the muscle fibers have lost their nerve supply. The presence of abnormally large or small MUPs or a high incidence of MUPs with more than four phases all indicate neuromuscular disease when generated during weak voluntary muscular contraction. The abundance of motor-unit activity at maximal voluntary contraction is used as an index of the motor-unit complement in the muscle.

Measurement of nerve-conduction velocity is used to detect slowing in the propagation of nerve pulses. The nerve is stimulated at two separate sites and the times between the stimuli and the onset of the evoked muscle responses—latencies—are measured. The latency difference and the distance between the two sites determine the nerve-conduction velocity. Repetitive nerve stimulation at frequencies of 3 or 10 Hz, or sometimes up to 50 Hz is used to test neuromuscular transmission.

3. Quantitative Electromyography

Quantification of the qualitative changes noted in conventional electromyography has been attempted. Amplitude and duration of MUPs recorded with concentric needle electrodes have been measured (Buchthal and Rosenfalck 1963).

The single-fiber electrode (Fig. 2a) can record from two muscle fibers belonging to the same motor unit (Stålberg and Trontelj 1979). The variability in the time interval—jitter—between the action potentials from the two muscle fibers is measured. Increased jitter is caused by defects in neuromuscular transmission or nerve conduction. Single-fiber electromyography can also give a measure of muscle density in the motor unit.

The number of functioning motor units in a muscle can be estimated by an on-line computerized method (Ballantyne and Hansen 1974a). A small sample of motor units are recruited sequentially by increasing the intensity of a series of electrical stimuli to the nerve supplying the muscle. The computer records the summated MUPs and by comparison with the supramaximally evoked muscle action potential (all motor units responding together) the number of motor units is estimated. An extension of this technique allows the MUPs from the sample to be analyzed in detail. Some neuromuscular diseases (e.g., the neuropathies) show a decrease in the number of motor units and exhibit changes in the MUP parameters.

Over the years, a number of automated methods for diagnosis of neuromuscular diseases have been introduced. One example is frequency analysis of the EMG, whereby the signal is passed through a series of electronic filters and the result plotted as a histogram of frequency distribution. Another method provides a smoothed spectrum using a digital computer (Scott 1967). Some disease processes in the muscle have been shown to alter the frequency spectrum of the EMG. Another technique measures the mean amplitude and the number of changes in the sign of the slope of the EMG recorded at constant muscle tension (Willison 1964). Few of these automated techniques have proved to be of value in clinical electromyography.

See also: Muscle; Gait Analysis

Bibliography

Aminoff M J 1978 *Electromyography in Clinical Practice.* Addison-Wesley, London
Ballantyne J P, Hansen S 1974a A new method for the estimation of the number of motor units in a muscle. *J. Neurol., Neurosurg. Psychiatry* 37: 907–15
Ballantyne J P, Hansen S 1974b Computer method for the analysis of evoked motor unit potentials, *J. Neurol., Neurosurg. Psychiatry* 37: 1187–94
Buchthal F, Rosenfalck P 1963 Electrophysiological aspects of myopathy with reference to progressive muscular dystrophy. In: Bourne G H, Golarz M N (eds.) 1963 *Muscular Dystrophy in Man and Animals.* Hafner, New York, pp. 194–262
Goodgold J, Eberstein A 1972 *Electrodiagnosis of Neuromuscular Diseases,* Williams and Wilkins, Baltimore, Maryland
Lenman J A R, Ritchie A E 1977 *Clinical Electromyography,* 2nd edn. Pitman Medical, Tunbridge Wells
Scott R N 1967 Myo-electric energy spectra. *Med. Biol. Eng.* 5: 303–5
Stålberg E, Trontelj J V 1979 *Single Fibre Electromyography.* Mirvalle, Old Woking
Willison R G 1964 Analysis of electrical activity in healthy and dystrophic muscle in man. *J. Neurol., Neurosurg. Psychiatry* 27: 386–94

S. Hansen, D. L. Sutton and J. P. Ballantyne

Electron Linear Accelerators

The linear accelerator (linac) is a device for the indirect acceleration of electrons, that is, acceleration without the use of voltages equivalent to the energy gain of the electron.

The simplest method of achieving linear acceleration is by the use of a cylindrical cavity so dimensioned that when supplied with radio-frequency (rf) power at a chosen frequency, it supports an electromagnetic field pattern with a component of electric field along the axis of the cylinder. A cavity operating in the transverse magnetic (TM_{01}) mode is shown in Fig. 1. Reentrant sections may be used to decrease the transit time of the electron in the field and allow it to gain energy nearly equal to that of the peak field across the cavity.

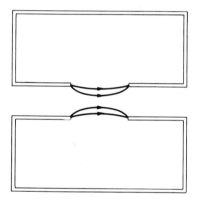

Figure 1
Reentrant standing-wave cavity showing electric field

In principle such a device could be built to operate at low frequency but the dimensions of the cavity required would be too large to make this practicable. The construction of resonant cavity accelerators had to await the development of high-power microwave sources in World War II.

Several accelerators using resonant cavities were constructed in the post-war years. Either a single or a small number of cavities were used, fed by a single magnetron or triode. Energies in the region of 1 MeV were realized. Energy values were limited by cavity breakdown due to the high field gradients. The notable exception to this was an accelerator built at the Massachusetts Institute of Technology. It used 21 magnetrons locked in phase to drive 21 6-element cavities realizing a total energy of 18 MeV (Demos 1951).

In parallel with resonant cavity work, Fry in the UK (Fry et al. 1948) and Hansen in the USA (Ginzton et al. 1948) were developing the travelling-wave (TW) accelerator concept in which wave and electron travel with approximately the same velocity along the acceleration path. When rf power is fed into a suitably dimensioned, smooth-walled cylindrical waveguide, a travelling wave is propagated unidirectionally along the tube. The phase velocity of the wave is greater than the velocity of light, negating the possibility of wave and electron travelling together. However, Cutler (1944) showed that the phase velocity of the wave could be reduced to values less than the velocity of light by the introduction of circular-holed irises in the center of the waveguide. His analysis was carried out for an infinite series of infinitely thin irises. This was extended by Walkinshaw (1948) to the use of "thick" irises having a practical spacing of 3 to 10 irises per guide wavelength. This work led to the development of the first two practical travelling-wave linacs in the UK and USA.

1. Operation of the Travelling-Wave Linac

The field configuration for a TM_{01} wave in an iris-loaded waveguide is shown in Fig. 2a, and Fig. 2b illustrates in

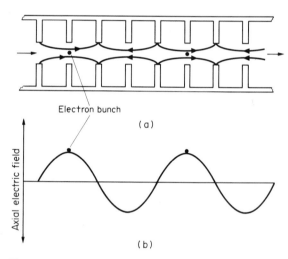

Figure 2
(a) Travelling-wave accelerator tube ($\pi/2$ mode, 4 irises per wavelength), and (b) associated axial electric field

more pictorial form the accelerating field experienced by the electron. It should be noted that these structures yield a low group velocity, the velocity at which power may be propagated. In practice, group velocities vary between $c/30$ and $c/200$ (where c is the velocity of light), depending on the detailed design. A finite time (generally less than 1 µs) is required to establish the final values and simultaneously store energy in the waveguide. This time is known as the filling time. The electrons are grouped or bunched either externally to the waveguide or at the early stages of acceleration. These bunches (between 1° and 30° wide) are then positioned at or near the peaks of the accelerating field for acceleration to the final energy, the wave velocity then being effectively equal to the velocity of light.

For small beam currents, the energy gain V (in MeV) in a length of waveguide is given by

$$V = \frac{(qW)^{1/2}[1 - \exp(-\alpha L)]}{\alpha L} L$$

where L is the acceleration path length (in m), W is the power fed to the waveguide (in MW), and α is the attenuation coefficient (in nepers m^{-1}). The attenuation coefficient is dependent on the conductivity of the surface of the waveguide, the quality of the machining and geometry. The series impedance of the waveguide structure, q, is expressed in MΩ m^{-2}. The value of q is determined by the geometry of the iris structure and is proportional to $(\lambda/a)^4$, where a is the radius of the iris hole and λ is the wavelength when the phase velocity is equal to the velocity of light. Typical practical values of q lie in the range 10–50 MΩ m^{-2}. A useful parameter which is a good measure of the quality of the structure is the shunt impedance Z, where

$$Z = q/2\alpha \, (\text{M}\Omega \, \text{m}^{-1})$$

Z is almost independent of the value of q but is largely dependent on the smoothness of the waveguide interior walls and the method of waveguide assembly. A typical good-quality travelling waveguide in the 3 GHz region has a shunt impedance of 50 MΩ m^{-1}.

Having chosen the value of q required to give the required field gradient, α is determined. The internal diameter of the cylindrical part of the waveguide is chosen to give the required local phase velocity.

Almost all linacs which were built between the late 1940s and the late 1960s used the travelling-wave principle. Development carried out at Los Alamos, New Mexico on a periodic standing-wave accelerator provided a structure of higher shunt impedance than that of conventional travelling-wave structures, with the same energy gain for the same rf power being obtained in a shorter length. This facility is convenient in many applications and the standing-wave structure has, during the 1970s, largely superseded the travelling-wave structure. A modern example of the standing-wave structure is shown in Fig. 3.

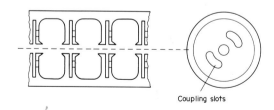

Coupling slots

Figure 3
Biperiodic standing-wave accelerator tube

2. Choice of Frequency and Power Sources

The shunt impedance of an accelerating structure, whether travelling or standing wave, varies with the square root of the frequency. Thus, from this viewpoint, a high operating frequency is indicated. However, as a general rule the power output of available rf sources, whether magnetrons or klystrons, decreases with increasing frequency and thus the choice of operating frequency is not solely determined by shunt impedance. In addition, while the relative machining accuracy in the manufacture of the structure remains constant whatever the frequency, the absolute machining tolerance becomes very small at high frequencies. While linacs have been built at various frequencies between 1 and 10 GHz, the vast majority operate in the 3 GHz region.

A short list of microwave sources which have been used to power linacs is given in Table 1. At energies up

Table 1
Microwave sources used to power linacs

Tube type	Frequency (MHz)	Peak power (MW)	Average power (kW)
Magnetron	9250	1	1
Magnetron	3000	2	3
Klystron	3000	2	60
Klystron	2856, 3000	25	25
Klystron	1200	20	40

to 10 MeV and beam powers up to 1 kW, the majority of linacs use the 2 MW, 3 kW magnetron. Above 10 MeV, klystrons are normally used. For very high energies requiring a considerable number of power sources, klystrons provide the only practical solution. A series of klystrons may be driven from a single, highly stable, low power source, since typical gains are between 40 and 60 dB.

3. Components of a Linac

A linac consists of an evacuated accelerating tube (usually provided with a cylindrical focusing coil, at least in the early stages of acceleration) which is fed from the microwave source through a suitable mode transformer via a rectangular waveguide containing an insulating gas (Fig. 4).

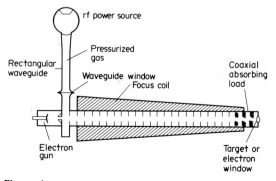

Figure 4
Layout of accelerator system

An electron gun operating at an injection voltage in the range 10–200 kV (depending on the detailed design) injects electrons axially into the buncher, the early part of the accelerating tube. Roughly 50% of the electrons injected into the buncher are trapped and form into a bunch 2–30° wide, again depending on the detailed design. The remainder of the electrons lose energy and ultimately collide with the walls of the accelerating tube. Sometimes a prebuncher consisting of an axial resonant cavity is interposed between the gun and buncher. This can serve to: (a) form a crude bunch by velocity modulation, (b) decrease the bunch width produced in the buncher, and (c) materially decrease those electrons which are rejected and which strike the tube walls. It is sometimes required to leave some of the possible bunch stations in the accelerating wave vacant. This may be easily effected by operating the prebuncher at a subharmonic frequency.

The microwave source normally operates in pulse mode, being fed by a modulator supplying pulses of between 30 and 250 kV, depending on the source type. Typical magnetrons operating at 3 GHz with powers of 2–3 MW require a modulating voltage of 40–60 kV, whereas klystrons of the same frequency yielding powers of 15–30 MW require voltages of 150–250 kV.

Modulators usually use a pulse-forming network which is discharged by a hydrogen thyratron into the primary coil of a pulse transformer; the secondary coil, feeding the microwave tube, is frequently bifilar so that the filament of the tube may be fed without the use of an additional high-voltage insulating transformer.

Figure 5 shows a modulator of this type providing a duty cycle, the product of pulse-repetition frequency and pulse duration (a few microseconds) in the 10^{-3} region. For higher-duty cycle pulsed linacs (10^{-2}), the size and cost of the pulse-forming network and pulse transformer make it desirable to use a high-vacuum-switch tube which, unlike a thyratron, can be turned on and off. The tube applies voltage pulses, of duration determined by the signal on the grid of the switch tube, directly to the power tube from a dc power supply of the required final voltage.

The electron gun is generally pulsed by the same modulator that pulses the power tube, though not necessarily at the same voltage. The gun pulse may be derived from a tapping on the pulse transformer secondary.

Guns may be either diodes or triodes. With the diode, the injected electron pulse has the same duration as the rf pulse unless a deflection system is used between gun and accelerating tube, thus allowing a shorter pulse to be injected. The pulse duration from a triode gun is determined by a signal applied to its grid through a suitable isolating transformer or video link. The triode may be pulsed on its cathode or, alternatively, a dc supply may be used. At very short pulse widths (a few nanoseconds) of interest in both physics and chemistry, the triode gun is preferred.

In the continuous-wave (cw) linac, the modulator is no longer required, thereby simplifying matters considerably. A high voltage power supply feeds either one klystron or several klystrons in parallel.

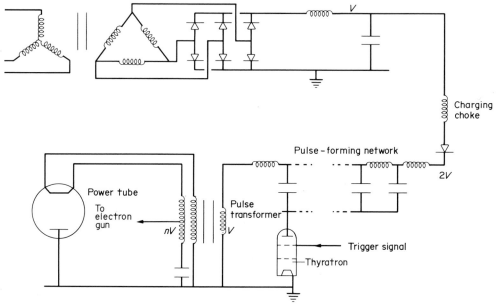

Figure 5
Modulator layout

4. Types of Linac

4.1 Radiotherapy

About 90% of all linacs constructed since 1953 have been in the 4–10 MeV range and have been powered by a 2 MW pulsed magnetron operating at or near 3 GHz; the electrons being stopped in a heavy-metal target (usually tungsten) provide a powerful source of x rays for radiotherapy. The x-ray dose rate obtained is normally 2–4 Gy min^{-1} at a distance of 1 m from the target, the x-ray field being flattened over a 30–50 cm diameter circle (at 1 m) by means of a suitably shaped metal absorbing cone. Most of these accelerators are mounted on a rotating gantry to simplify patient treatment (see *Radiotherapy: Linear Accelerators*).

Some of these linacs provide an electron output also. Due to the small depth of tissue penetrated by 4–10 MeV electrons, the usefulness of the electron output in treating patients is very limited.

There are also appreciable numbers of linacs of variable energy—in the range 10–35 MeV. These provide, in addition to x rays, electron beams of sufficient penetration to be of practical interest for radiotherapy. These machines are normally klystron powered, and, because of their greater size and complexity, are almost invariably of fixed construction. The electron beam is transported to the treatment head by a system of magnets mounted on a rotating gantry giving the same geometry as for a gantry-mounted accelerator.

4.2 Industrial Uses

Linacs are used as x-ray sources in industrial radiography, and the majority are basically similar to the smaller medical accelerators. To a lesser extent linacs have found application elsewhere in industry for the sterilization of pharmaceutical products and the cross-linking of long-chain polymers such as polyethylene. This is usually achieved by direct electron irradiation. To achieve sufficient penetration a high energy is required. On the other hand, the energy must not be so high that it generates radioactive products by the (γ, n) reaction. The energy is thus limited to around 10 MeV. Linacs with electron beam powers of some tens of kW are in operation or envisaged.

4.3 Research Applications

While the total number of research linacs that have been built is small in comparison with the number of radiotherapy machines, the investment in research types is far greater because of their size and complexity.

Diversity is great among research linacs. They range from nanosecond pulse machines of a few MeV to the giant 2-mile long GeV machine built at Stanford University in California. Many are in the 50–200 MeV range and are used for nuclear structure research; a few in the 10–50 MeV range are used for photon activation analysis by using the (γ, n) and other photon induced reactions to produce daughter products which are radio-active and which are identifiable by the energy of the x rays emitted and by their half-lives (see *Photon Activation Analysis*). This technique is complementary to neutron activation analysis (see *Neutron Activation Analysis*).

Finally, there are the cw accelerators, most of which are used as the accelerating structures in cw microtrons. These are of two types. One type uses a superconducting accelerating structure where the rf power required to maintain the accelerating field is minimal, the power supplied being only that required to accelerate the electron beam. The other type is used at room temperature, and some 70% of the power is used to maintain the accelerating field. At first sight the superconducting system seems preferable but the technology of the construction of superconducting cavities is very difficult. Thus, many of the systems under construction have room-temperature structures.

Bibliography

Cutler C C 1944 *Electromagnetic Waves Guided by Corrugated Conducting Surfaces*, Report No. CRB45/593. Bell Telephone Laboratory, Murray Hill, New Jersey
Demos P T, Kip A F, Slater J C 1952 The MIT linear electron accelerator. *J. Appl. Phys.* 23: 53–65
Fry D W, R-S-Harvie R B, Mullett L B, Walkinshaw W 1948 A travelling-wave linear accelerator for 4-MeV electrons. *Nature (London)* 162: 859–61
Ginzton E L, Hansen W W, Kennedy W R 1948 A linear electron accelerator. *Rev. Sci. Instrum.* 19: 89–108
Graffunder W 1949 Die Entwicklung des linearen Accelerators. *Helv. Phys. Acta* 22: 233–60
Mullett L B, Loach B G 1948 Experimental work on corrugated waveguides and associated components for electron linear accelerators. *Proc. Phys. Soc.* 61: 271–84
R-S-Harvie, R B 1948 Travelling wave linear accelerators. *Proc. Phys. Soc.* 61: 255–70
Walkinshaw W 1948 Theoretical design of linear accelerator for electrons. *Proc. Phys. Soc.* 61: 246

M. Kelliher

Electron Microprobe Analysis

Electron microprobe analysis, otherwise known as electron-probe x-ray-emission microanalysis or x-ray microanalysis, is a technique for determining the local concentration of chemical elements in a specimen. The technique involves the spectroscopic analysis of the x rays generated within the microscopic regions of the specimen by a finely focused beam of electrons. The detection of a characteristic line usually provides positive identification of the presence of an element, the x-ray spectra of the elements being much simpler than the optical spectra. Also, the method detects elements rather than compounds because the chemical binding has a negligible effect on the x-ray spectrum. Data are obtained in the form of local point analyses, line scans or images of the distribution of an element with a spatial resolution in the range 0.1–1 μm. The method is quantitative and may be used for all elements for which x-ray diffractors and detectors are available, that is, elements with atomic number $Z > 5$. For elements with $Z > 8$ in biological tissues, the minimum measurable amount is commonly 10^{-16} g, although the use of an energy-dispersive x-ray detector for elements with $Z > 15$ may reduce this to 10^{-20} g. The minimum measurable weight fraction is of the order of 0.01%. However, this limit is the concentration in the locally analyzed area of the specimen as it exists under the electron probe, and would include, for example, the embedding medium for an embedded specimen.

1. Instrumentation

Figure 1 shows the most important components of a conventional microprobe analyzer. These are an electron gun and magnetic lenses which produce a finely focused

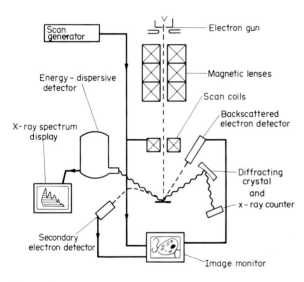

Figure 1
The electron microprobe

beam of electrons; scanning coils or plates to perform scans or to position the electron beam on the specimen (placed within the vacuum of the electron optical column); electron detectors to form images using back-scattered or secondary electrons as in the scanning electron microscope (SEM); and one or more x-ray detectors with which to perform the analyses. Data can be extracted by setting the x-ray detector to record the x-ray line which is characteristic of one element, then using the x-ray intensity variations detected as the electron beam is scanned over the specimen to modulate

the brightness of the display tube. Thus a map is formed, showing the distribution of the chosen element in the specimen. When accurate quantitation is required, analyses may be performed by identifying the region of interest on the scanning image and then recording the x-ray signals for many elements with the static electron probe localized on the area of interest.

In transmission electron microscopes (TEM) used for microanalysis, the conventional transmission image rather than a scanning image is usually used to localize the region of interest before focusing the electron beam onto the region to perform an analysis. Some TEMs have scanning attachments, and the electron signal transmitted through the thin specimen while the beam is scanned is used to form an image (scanning transmission electron microscopy). The advantage of this method is that the finest electron probe (and hence the highest analytical spatial resolution) is utilized.

In general two types of x-ray detectors are used, each having advantages for microanalysis. The diffracting crystal spectrometer, which is usually an integral part of the conventional electron microprobe, has a higher x-ray energy resolution (~ 5 eV) and thus can separate interfering lines more readily, resulting in lower measurable concentrations. However, it can only measure one element at a time, has a low x-ray collection efficiency ($\sim 10^{-4}$) and is mechanically complex. The solid-state energy dispersive (ED) detector which, because of its mechanical simplicity, is often fitted retrospectively to SEMs or TEMs, can measure the lines from all elements with $Z > 11$ simultaneously and thus is ideal for qualitative analysis. It can have a high collection efficiency of up to 0.1, and can be used to measure x-ray signals from elemental masses which would produce no characteristic signal detectable above background on a diffracting spectrometer. It has a relatively poor x-ray energy resolution (~ 150 eV) and thus complex computer manipulation of the recorded data to extract quantitative information is often required.

2. Quantitation

The spatial resolution of the analysis is determined by the sideways scattering of the electron beam in the specimen, since this determines the volume of x-ray generation. This may be between 1 μm and 10 μm in thick specimens, but the scattering may be minimized by preparing samples as thin (~ 0.1 μm) sections. For thick specimens, the theory of quantitation is complex but sound, provided the electrons are confined to a homogeneous region (e.g., bone). However, quantitative results are dubious when this is not so, as is often the case in soft tissue. For thin specimens the theory is simpler and not affected by specimen nonuniformity. The accuracy of absolute quantitation is about 10% although relative concentrations of a single element in different regions of the specimen can be measured to within a few percent, limited only by the statistical accuracy of x-ray counting.

See also: Scanning Electron Microscopy; Transmission Electron Microscopy; Clinical Chemistry: Physics and Instrumentation

Bibliography

Hall T A 1971 The microprobe assay of chemical elements. In: Oster G (ed.) 1971 *Physical Techniques in Biological Research*, Vol. 1A, *Optical Techniques*. Academic Press, New York, pp. 158–275

Hall T A, Echlin P, Kaufmann R 1974 *Microprobe Analysis as Applied to Cells and Tissues*. Academic Press, London

Hutchinson T E, Somlyo A P 1981 *Microprobe Analysis of Biological Systems*. Academic Press, New York

Lechene C P, Warner R R (eds.) 1979 *Microbeam Analysis in Biology*. Academic Press, London

Revel J P, Barnard T, Haggis G M 1984 *The Science of Biological Preparation*. O'Hare, Chicago

P. Nicholson

Electron Microscopy: Freeze-Fracture Replication

Freeze fracture (also known as freeze cleave and freeze etching) involves low-temperature (173 K) cleavage of rapidly frozen biological specimens, resulting in fracture planes which offer novel views of cellular structures. Electron microscopic examination of specimens is done indirectly via a stable replica of the cleaved specimen, made in vacuo at about 173 K. The technique, first successfully applied to biological specimens by Steere (1957), involves four preparation steps.

1. Freezing of the Specimen

Unlike other specimen preparation methods for electron microscopy of biological materials, freeze fracture employs physical fixation as a means of preserving cellular ultrastructure. In theory, this avoids the exposure of cells to the inimical chemical agents used as fixatives (e.g., osmium tetroxide, glutaraldehyde) or as stains (e.g., salts of lead or uranium) in the more traditional methods of specimen preparation.

During freezing of the specimen, the aim of which is to preserve intracellular anatomy in a life-like state, the most disruptive and potentially lethal effects are caused by ice crystal growth with accompanying dehydration. Both these processes can be avoided or at least reduced to an acceptable level if the cooling rate between the freezing point of cytoplasm and the recrystallization temperature (the so-called critical interval) is sufficiently rapid. Alternatively, the magnitude of the critical interval can be reduced by prior impregnation of the specimen with cryofixatives (cryoprotectants, antifreezes) such as glycerol or dimethyl sulfoxide, or by freezing specimens under high pressure (2×10^8 Pa). Of these alternatives, very high cooling rates are preferable to the use of cryofixatives, which can induce reversible alterations in membrane distribution and structure. Relatively high cooling rates can be achieved by quenching small specimens [1 mm^3 pieces of tissue or small drops (10 μl) of cell suspension] supported on a copper or aluminum disk, into a suitable liquid coolant at a temperature just above its melting point. Liquid propane (mp = 85 K) or Freon-22 (mp = 113 K) cooled in liquid nitrogen, or nitrogen at its melting point (63 K) is frequently used. After initial freezing, specimens are maintained at temperatures below 173 K during all subsequent procedures.

2. Fracturing of Specimens

Specimens can be split under liquid nitrogen using a cooled blade and suitably designed hinged holders prior to transfer to the vacuum chamber of the freeze-fracture apparatus. Alternatively, they can be cleaved with a cooled microtome device after transfer to the precooled (77 K) specimen table in the vacuum chamber. Once cleaved, the exposed surfaces of the specimen are susceptible to condensation of water vapor and hydrocarbons from the vacuum plant and should therefore be replicated without delay. However, at this stage, where the specimen surface consists of areas of ice and fractured or partially exposed whole cells, further elements of the specimen may be exposed by partial freeze-drying (etching). During etching, ice sublimes off the fracture faces at a controlled rate depending on specimen temperature and chamber pressure, thus exposing further elements of the specimen.

3. Replication

The aim of the replication process is to produce an exact copy of the specimen surface; the copy must be stable and recoverable after the specimen is brought back to atmospheric pressure and ambient temperature. Replication normally follows fracturing (or etching) without delay. First, the topography of the exposed faces is accentuated by shadowing from an angle with a mixture of platinum and carbon. The mixture is evaporated in vacuo using resistance heating electrodes or preferably an electron-beam evaporation source. Second, the differentially deposited areas of platinum are joined together by evaporating a continuous layer of relatively electron-transparent carbon over the fractured surface. This produces an intact replica which can be recovered after the specimen melts, and later freed from contaminating fragments of specimens before examination by electron microscopy.

4. Interpretation

It is widely agreed that fractures through frozen biological material preferentially follow the planes of biomembranes, the split occurring along the interior hydrophobic domain of the lipid bilayer. Figure 1a shows a

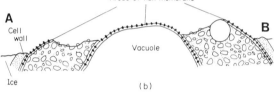

Figure 1
(a) Electron micrograph from a replica of a freeze-fractured yeast cell; the arrow shows the direction of platinum/carbon shadowing.
(b) Diagrammatic representation of the profile (along the line AB) of the fracture face shown in (a). The fractured cell is embedded in ice and shows cross fractures of the cell wall and cytoplasm. The biomembranes in the fracture face show areas of half membrane, the outer layer of the bimolecular lipid leaflet having been cleaved off with other parts of the cell during the fracturing process

replica of a yeast cell with a fracture following the plasma membrane, then dipping through the cytoplasm before splitting part of a cytoplasmic vacuolar membrane. Numerous particles (intramembrane particles) can be seen on the fractured membrane faces. Figure 1b shows diagrammatically the path of the fracture through the specimen. Using suitable holders, both halves of the cleaved specimen can be replicated and their fracture faces matched after examination of the replicas by electron microscopy. The agreed nomenclature for membrane fracture faces is E for the exterior-half (lipid monolayer) and P for the interior-half membrane. Each monolayer has sides consisting of an original membrane surface (PS or ES) and a newly revealed fracture surface (PF or EF).

See also: Scanning Electron Microscopy; Transmission Electron Microscopy; Cryobiology

Bibliography

Hudson S C, Rash J E, Graham W F 1979 Introduction to sample preparation for freeze-fracture. In: Rash J E, Hudson S C (eds.) 1979 *Freeze Fracture: Methods, Artifacts and Interpretations.* Raven, New York, pp. 1–10
Steere R L 1957 Electron microscopy of structural detail in frozen biological specimens. *J. Biochem. Biophys. Cytol.* 3: 45–60

J. H. Freer

Electrooculography

The eye is an electric dipole, the cornea being positive with respect to the retina. This standing potential of approximately 6 mV between the front and back of the eye was utilized until the late 1950s to register and measure eye movements, and electrooculography was the term used to describe this application and the investigation of nystagmus. However, in more recent years it has been discovered that changes in this standing potential that occur as the eyes are taken from a dark-adapted to a light-adapted state give useful clinical information in relation to some functional aspects of the retina. The term electrooculography is now used in this context to describe the clinical use of a test which is essentially a comparison of the standing potentials in light-adapted and dark-adapted eyes.

The presence of a potential difference between the front and back of the eye was first demonstrated by Du Bois-Reymond (1849). It was later shown by Meyers (1929) that eye movements produce electrical activity which can be measured by surface electrodes. However, at this time it was thought that these electrical changes were related to muscle action potentials, and it was not until later that their behavior was attributed to the existence of a standing potential (Mowrer et al. 1936). The earliest accounts of electrooculography as a test of retinal function are those of François et al. (1955). The procedure which is now most widely used in the clinical situation was described by Arden et al. (1962), and the consequent Arden index is now used as a measure of retinal function.

1. Clinical Test

The standing potential between the anterior and posterior poles of the eye has to be measured indirectly. Electrodes can be placed at either side of both eyes, or lateral to each eye, and referenced to a central forehead electrode. Since the eye is an electric dipole, signals are generated by reciprocating lateral eye movements, and the size of the signal is proportional to the size of the corneoretinal potential. Red lights or light-emitting diodes are normally used to stimulate the horizontal eye movements since they produce no change in the corneoretinal potential. In the simplest form of test, two fixation points are set to determine the extent of the lateral movements and the patient is simply asked to look briskly between them. A more suitable procedure is to ask the patient to follow a strip of lights switched on in succession. This produces a sinusoidal movement of the eyes. The sinusoidal signals thus generated (frequency typically 0.5 Hz) can be easily filtered to exclude artefacts caused by blinks and slow baseline shifts. Signals picked up are amplified by an isolated amplifier. The patient is asked to make the appropriate eye movements for 10–20 s of each minute throughout a 30 min test, to provide a series of 30 measurements for each eye. The amplitudes of the signals detected are proportional to the size of the corneoretinal potential (Fig. 1). The

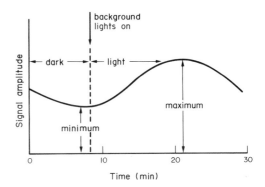

Figure 2
Normal response curve for one eye. The maximum and minimum potential values are used to calculate the Arden index

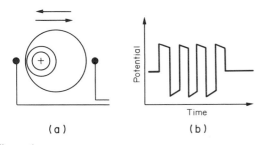

Figure 1
(a) Repeated lateral movements of eye through a fixed arc, and (b) the signal production, the amplitude of which is proportional to the standing potential of the eye

signals may be recorded simply on a chart recorder, although more elaborate types of equipment have been used (Jackson 1980).

At the beginning of the test, the subject is placed in the dark. As the eyes become dark-adapted the corneoretinal potential falls to reach a minimum value after some 10 min. If a bright stimulus is then applied, the sharp change in illuminance excites a damped oscillation of the ocular electric dipole with a period of some 25 min which decays completely within 3–4 periods (Taümer et al. 1974). The Arden index is simply the ratio of the maximum potential in the light-adapted eye to the minimum potential in the dark-adapted eye, expressed as a percentage (Fig. 2). Normal values of the Arden index range from 150 to 340 (Adams 1973).

Any change in illuminance of the retina will evoke a change in the corneoretinal potential. Hence when dark-adapting the patient initially, it is necessary to reduce the ambient lighting slowly to minimize the effect of a sharp decrease in illuminance. The amount of change in illuminance when the light stimulus is applied is also important, and it is recommended that at least 300 lx be used to produce a maximal response with a minimum of patient discomfort (Jackson 1979). If the media of the

eye are opaque, a greater (but difficult to estimate) illuminance is needed.

2. Clinical Uses

There are a variety of pathological conditions which give rise to an abnormal electrooculogram; the usual departure from normality being a low response to light stimulus and a subsequent low Arden index. Probably the most dramatic results are obtained in cases of retinitis pigmentosa, where virtually no response occurs giving, commonly, Arden indices of around 100.

When the media of the eyes are opaque (e.g., after ocular trauma) and it is not possible to evaluate any damage to the retina by direct observations, it is feasible to use electrooculography to get an indication of retinal function. Another important use of electrooculography is in monitoring the retinal condition of patients using the drug chloroquine which produces a reduction in retinal function. Here, the periodic evaluation of retinal function is helpful in patient management. Electrooculography can also be used to evaluate vascular lesions of the eye.

3. Concluding Remarks

At the present time, the exact nature of the process that produces the electrooculographic response is unclear, but the time scale over which the response occurs tends to exclude a direct relationship with neural activity. There is evidence that the potential itself originates in the pigment epithelium, and also that the metabolism of visual purple (rhodopsin), the light-sensitive chemical of the rods, is associated with standing potential variations (Arden and Kelsey 1962). Consequently, a normal electrooculograph implies functioning rods, functioning pigment epithelium, functional contact between neural and pigment layers, and a sufficient choroidal blood supply. The test itself is a general one and therefore can give only an overall measure of retinal function. However, it is relatively easily performed and interpreted. Together with other electrophysiological and ophthalmological tests it is a useful clinical tool.

See also: Electroretinography; Vision

Bibliography

Adams A 1973 The normal electro-oculogram (EOG). *Acta Ophthalmol.* 51: 551–61

Arden G B, Barrada A, Kelsey J H 1962 New clinical test of retinal function based upon the standing potential of the eye. *Br. J. Ophthalmol.* 46: 449–67

Arden G B, Kelsey J H 1962 Changes produced by light in the standing potential of the human eye. *J. Physiol. (London)* 161: 189–204

Du Bois-Reymond E 1849 *Untersuchungen über Thierische Elektricität.* Reimer, Berlin, pp. 256–57

François J, Verriest G, De Rouck A 1955 Modification of the amplitude of the human electro-oculogram by light and dark adaptation. *Br. J. Ophthalmol.* 39: 398–408

Jackson S A 1979 The optimum illuminance level for clinical electro-oculography. *Acta Ophthalmol.* 57: 665–67

Jackson S A 1980 Automated electro-oculography: A microprocessor application example. *J. Med. Eng. Technol.* 4: 285–89

Meyers I L 1929 Electronystagmography: A graphic study of the active currents in nystagmus. *Arch. Neurol. Psychiatry* 21: 901–18

Mowrer O H, Ruch T C, Miller N E 1936 The corneo-retinal potential difference as the basis of the galvanometric method of recording eye movements. *Am. J. Physiol.* 114: 423–28

Taümer R, Hennig J, Pernice D 1974 The ocular dipole: A damped oscillator stimulated by the speed of change in illumination. *Vision Res.* 14: 637–45

S. A. Jackson

Electroretinography

An electroretinogram (ERG) is a recording of the pattern of transient electrical signals that develop across the retina of the eye when it is stimulated by light. Although the first ERG was made more than a century ago and an extensive body of information regarding the electrical activity of the retina now exists, the phenomenon is still not completely understood.

1. Electrical Signals

In vertebrates, the electrical activity of the retina is characterized by three prominent deflections, termed the a-, b- and c-waves (Fig. 1a). An upward displacement indicates a positive shift in the electrical potential of the front surface of the retina with respect to the back. The

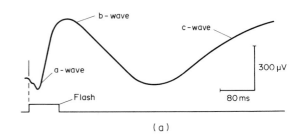

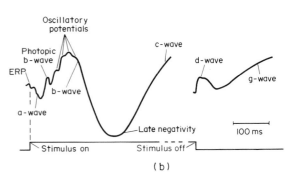

Figure 1
(a) Principal and (b) additional components of the electroretinogram

ERG is, however, more complicated than this simple nomenclature suggests, since under particular circumstances, additional (either more or less prominent) peaks appear. In Fig. 1b, the appearance of some of these peaks relative to the position of the a- and b-waves is shown. Included are the early receptor potential (ERP), the photopic b-wave, and the oscillatory potentials, as well as peaks termed the d-wave, the g-wave, the late negative or after potential, and others. Each peak stands out only in response to an appropriate stimulus. The peaks are not all seen simultaneously. The e-wave, which has been most extensively studied in the frog, may be essentially the same signal as the human g-wave (Skoog et al. 1977).

Various analyses have been undertaken to determine whether these individual waves can be considered as separate components that are related to specific retinal processes. It has been found that the early receptor potential arises from the outer segments of the retinal receptors, in response, and with virtually no delay, to stimulation with very intense light (Cone 1964). The initial part (negative going) of the a-wave is believed to come from the inner part of the retinal receptors (Murakami and Kaneko 1966). Because of the time delay between the appearance of this negative-going part of the a-wave and the stimulus, it is known as the late receptor potential. A release of potassium ions, taking place near the inner plexiform layer of the retina (Fig. 2) and also at a more distal site, activates cells (Müller cells) that extend through the retina and which generate

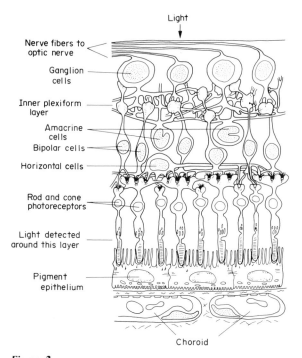

Figure 2
The structure of the retina

the b-wave (Newman 1979). The c-wave is produced by the pigment epithelial cells at the back of the retina as a result of a local decrease in potassium concentration (Oakley and Green 1976). The individual wavelets that make up the oscillatory potential apparently have their origins at different depths within the layers of the retinal bipolar cells (Wachtmeister and Dowling 1978). Despite the tentativeness and incompleteness of current knowledge, it is clear that these waves are indicative of the transmission of visual information through the retina.

The components of the ERG may also be separated with respect to retinal function, most usefully in terms of photopic and scotopic action (Armington 1974). Under some conditions, the b-wave breaks up into two smaller waves, one initiated by the action of the cone system of the retina and the other by the rod system (see *Vision*). The a-wave can also be split into photopic and scotopic segments. Two of the components listed above, the d- and e-waves, are produced by the termination of a stimulus, whereas the others are produced by its onset. The e-wave is believed to be a scotopic off-effect and the d-wave a mixed photopic–scotopic off-effect (Tomita et al. 1979).

2. Recording Methods

The ERG is readily made from a human subject using standard electrophysiological equipment. A recording electrode cannot be placed directly on the retina, but the ERG signals can be detected satisfactorily via an electrode built into a contact lens worn by the subject (Fig. 3). The electrode is referenced electrically to the cheek or forehead. Under many recording conditions the retinal electrical activity is quite small (see below) and not easily distinguished from a background of electrical interference. Here, the technique of signal averaging is useful for isolating the ERG signals from background noise.

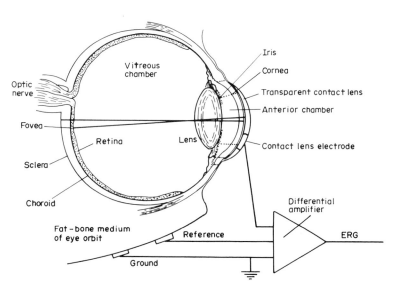

Figure 3
Transparent contact lens containing one electrode, shown on horizontal section of right eye. Reference electrode is placed on right temple

The character of an elicited response may be entirely dependent on the form of stimulus that is used. Two classes of stimulus used are flashing light stimuli and pattern reversal. A flash stimulus, even though imaged on a small local area of the retina, is distributed broadly as stray light scattered and reflected within the eye. Flashes are used when a simple technique of stimulation is needed for the investigation of the retina as a whole. Responses elicited by single flashes preferentially elicit scotopic activity; flashes flickering at a rate of 20 Hz or more elicit photopic activity. Pattern-reversal stimuli are used for the investigation of local retinal areas. The subject views a recurrent pattern of stripes or checks (Fig. 4). The light and dark areas are interchanged

Figure 4
A patterned stimulus

periodically, and this light and dark reversal produces stimulation within the image area. Because the stray light produced by an alternating pattern is steady, there is no stimulation of the retina outside the area of the image. Only a local response is obtained. Responses to patterned stimuli are photopic.

The magnitude of the ERG signals is dependent on the recording conditions. A single flash presented to the dark-adapted eye may elicit a response of the order of 500 µV or more. With light adaptation the same stimulus may produce a response too small to be detected without signal averaging. The amplitude of the ERG signals increases with stimulus luminance until saturation for the particular testing condition is achieved. The amplitude of the response increases also in proportion to the size of retinal area stimulated. Patterned stimuli subtend small visual angles at the retina in proportion to the total visual field. Thus, since only a small fraction of the retina is stimulated, the responses generated are small ($\sim 10\,\mu V$). With signal averaging, responses can be obtained with fields that subtend angles at the retina of as little as 2°. Even with signal averaging, however, responses of amplitude less than 1 µV are difficult to investigate.

3. Applications of Electroretinography

Electroretinography has both practical and research applications. It has been used as a tool to investigate the physiological mechanisms that underlie various psychophysical processes, and knowledge obtained from such basic studies has been useful for evaluating the results of various clinical investigations. The principal psycho-

physical findings have been brought together elsewhere (Armington 1974).

The human ERG is a useful tool for following visual development (Armington 1974). Electrical activity is barely detectable at birth, but develops rapidly soon thereafter and virtually full amplitude is reached at twelve months. The ERG finds its chief clinical applications, however, in the diagnosis of visual disorders. By taking into account the different origins of the electroretinographic waves and the specific physical characteristics of the stimulus, it is often possible to reach conclusions on these disorders that are attainable in no other way.

Stimulus flashes, covering the full retina, are useful in distinguishing various forms of retinal degeneration (Berson 1981, Carr and Siegel 1982). In progressive degeneration (retinitis pigmentosa) the a- and b-waves are severely reduced in amplitude, the response often being too small to be recorded at all (extinguished) even in the early stages of the disease. Thus, the ERG is clearly abnormal. Since retinitis pigmentosa is hereditary in origin, electroretinography (which can detect retinal impairment at an earlier age than any other method of detection) is of particular value for testing children who may suffer from the disease although its symptoms have not yet developed. In other conditions specific components of the ERG may be absent. For example, the scotopic components of the ERG are absent in subjects afflicted with congenital night blindness. In color blindness, the photopic components of the ERG are affected. Indeed, the principal forms of color blindness can be distinguished from one another by using colored stimuli and appropriate recording conditions (see *Color Blindness*). In the early phases of developing retinopathy in diabetics, a reduction in the oscillatory potentials can be observed before other components of the ERG are affected.

Patterned stimulation, although a recent development, is already finding clinical uses. Since sharpness of focus on the retina determines the effectiveness of a pattern, electroretinography can be used to test visual acuity. It will test retinal acuity over more widespread areas of the retina than can conventional methods of testing which deal only with the central visual field (Millodot and Riggs 1970). In amblyopia, the late negative potential and the b-wave are reduced in amplitude in response to patterned stimulation (Sokol and Nadler 1979). There is also some evidence to indicate that in optic-nerve disease the ERG derived in response to patterned stimulation is attenuated or absent, whereas the response to flashes of light is unchanged (Fiorentini et al. 1981).

See also: Electrooculography; Color Vision; Visual Fields and Thresholds

Bibliography

Armington J C 1974 *The Electroretinogram*. Academic Press, New York

Berson E L 1981 Electrical phenomena in the retina. In: Moses R A (ed.) 1981 *Adler's Physiology of the Eye: Clinical Application*. Mosby, London, pp. 466–529

Carr R E, Siegel I M 1982 *Visual Electrodiagnostic Testing: A Practical Guide for the Clinician*. Williams and Wilkins, Baltimore, Maryland

Cone R A 1964 Early receptor potential of the vertebrate retina. *Nature (London)* 204: 736–39

Fiorentini A, Maffei L, Pirchio M, Spinelli D, Porciatti V 1981 The ERG in response to alternating gratings in patients with diseases of the peripheral visual pathway. *Invest. Ophthalmol. Vis. Sci.* 21: 490–93

Millodot M, Riggs L A 1970 Refraction determined electrophysiologically: Responses to alternation of visual contours. *Arch. Ophthalmol.* 84: 272–78

Murakami M, Kaneko A 1966 Differentiation of PIII subcomponents in cold-blood vertebrate retinas. *Vision Res.* 6: 627–36

Newman E A 1979 B-wave currents in the frog retina. *Vision Res.* 19: 227–34

Oakley B II, Green D G 1976 Correlation of light-induced changes in retinal extracellular potassium concentration with the c-wave of the electroretinogram. *J. Neurophysiol.* 39: 1117–33

Skoog K-O, Welinder E, Nilsson S E G 1977 Off-responses in the human d.c. registered electroretinogram. *Vision Res.* 17: 409–15

Sokol S, Nadler D 1979 Simultaneous electroretinograms and visually evoked potentials from adult amblyopes in response to a pattern stimulus. *Invest. Ophthalmol. Vis. Sci.* 18: 848–55

Tomita T, Matsuura T, Fujimoto M, Miller W H 1979 The electroretinographic c- and e-waves with special reference to the receptor potential. *Jpn. J. Ophthalmol.* Suppl. 23: 15–25

Wachtmeister L, Dowling J E 1978 The oscillatory potentials of the mudpuppy retina. *Invest. Ophthalmol. Vis. Sci.* 17: 1176–88

J. C. Armington

Electrosurgery

Electrosurgery (also known as surgical diathermy) is a technique which uses high-frequency currents to cut tissue and coagulate blood. While low-frequency currents can be hazardous depending on their magnitude, the stimulating effects of electric current progressively decrease at increasing frequencies above 1 kHz and are negligible at frequencies above 100 kHz. At frequencies between 100 kHz and 10 MHz, heating is the only significant effect produced in tissue. Tissue is almost purely resistive at these frequencies.

An electrosurgery unit (Fig. 1) is thus essentially a high-frequency current generator. The patient is connected across the output of the generator, forming its load. One pole of the generator is connected to a fleshy part of the patient via a large surface area electrode—the neutral or patient plate. The other pole is attached to a pointed or very small surface electrode—the active electrode. This is held in a pen-like holder by the surgeon and is used to apply current to the selected site of operation. The surgeon is able to switch the generator on and off, usually by means of a foot switch. The current passing can be controlled by a front-panel control on the instrument.

Current is thus applied to the tissue through a very small surface area, and is returned to the generator via the bulk of the body and a large surface-area neutral plate. All but a very small fraction of the power developed in the patient is dissipated at the site of application of the active electrode. The heat generated there is sufficient to cut tissue and coagulate blood.

The work of d'Arsonval on the frequency-dependent effects of current on the human body laid the foundation for the use of high-frequency current in medical treatment and in surgery. Early devices produced high-voltage, low-current waveforms and were used to treat warts and skin lesions. The development of lower-voltage, higher-current generators using vacuum tubes

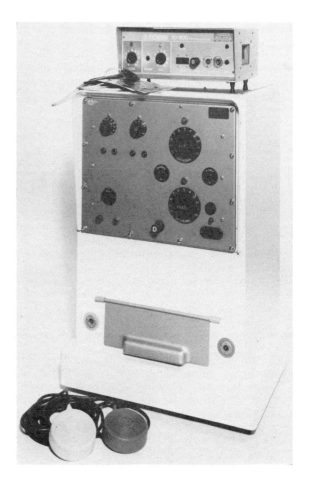

Figure 1
A typical valve/spark-gap electrosurgery unit and a comparable solid-state unit (on top)

led to instruments with good cutting properties. However, as these vacuum tubes were neither readily available nor cheap, the definitive development of electrosurgery and its commercialization were carried out by W T Bovie in the USA using spark-gap generators. Valve generators did not, however, disappear and these two types of generators, often combined into one instrument, formed the basis of all electrosurgery units until the 1970s when high-voltage power transistors became available. This led to the development of modern solid-state electrosurgery apparatus.

1. High-Frequency Generators

All modern electrosurgery units use frequencies in the range 200 kHz to 10 MHz. This avoids both the low-frequency problems of unwanted stimulation and the problems of the increasingly nonohmic heating effects of very high frequency current. Most units operate at frequencies around 1 MHz.

The precise effect of current depends on the power used, the shape of the active electrode and the output waveform. This last factor has been the subject of much empirical development over the years, although little quantitative work appears to have been done.

A fine-intensity arc from a pointed electrode produces a very rapid heating of tissue over a very small area. This causes microexplosive boiling of intracellular fluid, the energy developed being absorbed by the creation of steam. The tissue is thus cut without dissipation of heat in surrounding tissue. Sinusoidal current waveforms appear to have good cutting properties.

Coagulation of blood by heat to reduce the need for individual ligatures of many small blood vessels is a very long-established procedure. Two coagulation techniques using high-frequency current are employed and have come to be known as fulguration and desiccation. Fulguration is the destructive charring of tissue induced by sparking. Here, a fine high-intensity arc is undesirable, so flat spatula-shaped or ball electrodes are employed. In consequence high peak voltages are necessary to establish an arc. These are of relatively low power to avoid microexplosive cutting of the tissue. The ratio of peak voltage to rms voltage in a waveform is known as the crest factor. Waveforms with a high crest factor produce good fulguration.

The other method of coagulation, desiccation, is the straightforward ohmic heating of tissue resulting from contact between the active electrode and tissue. Here, the active electrode is often a pair of forceps clamping a blood vessel through which high-frequency current is passed. In these cases, while the power used must be adequate, the current waveform is not significant.

Many general-purpose electrosurgery units provide a blended or "cutting with hemostasis" output which combines cutting properties with a fixed or variable amount of coagulating effect.

These largely empirical findings have led to the use of valve oscillators to produce cutting waveforms. For technical simplicity self-rectifying circuits are used, producing discontinuous waveforms modulated 100% at supply mains frequency. The power available is typically 300–400 W maximum, delivered into a matching load of typically 400 Ω but with a fairly broad output characteristic for loads of between about 100 Ω and 2 kΩ.

The more difficult requirement for good coagulation (fulguration) has been met by circuits which include a spark gap in an LC resonator, again powered from an ac supply at mains frequency, which produce a burst-fired, damped sinusoidal waveform, rich in harmonics and with a high crest factor. The output characteristics are again broad with respect to load, but usually peak at a lower resistance value. The maximum power available is also usually lower.

In the recent development of solid-state units, difficulties have been met in matching some aspects of the performance of the older valve/spark-gap instruments. Most solid-state units are designed around a master oscillator running at 500 kHz which drives a power stage loaded by the primary coil of a step-up, high-frequency output transformer. The master oscillator produces reasonable quality sine waves with little modulation, giving excellent, smooth results when used for cutting. The production of high crest-factor waveforms, effective for fulguration, is difficult. The typical solution is to gate the master oscillator at, say, 10 μs on, 30 μs off, thus retaining a high peak voltage but reducing total power.

With the valve/spark-gap equipment the blended output is obtained by activating both high-frequency generators simultaneously and mixing the outputs from both at the output terminal. With the solid-state equipment, this is less easy since there is only the one master oscillator. The blended effect is usually achieved by gating the master oscillator with a different mark space ratio from that used for coagulation alone; typically 50 μs on, 50 μs off might be used. This retains a coagulating effect while increasing the cutting effect.

It has been found empirically that the coagulating performance of spark-gap generated waveforms is very good. Early attempts to match this performance using solid-state technology were not very successful. However, further sophistication of designs has led to very much improved equipment.

Figure 1 shows a typical valve/spark-gap instrument with a comparable solid-state unit on top. One set of foot pedals and one neutral plate and lead are shown. These are essentially the same for both models.

2. Arrangement of Output Circuits

The arrangement of the output circuits is important in relation to patient safety. Three conceivable electrical hazards exist.

First, any flow of direct current above a few microamps through the patient would result in blistering and ulceration at the site of the negative pole. This is easily prevented by coupling the output of the high-frequency generator through a capacitor to the output terminal. Thus, there is no dc circuit through the patient. All electrosurgery equipment has this design feature built in.

Second, current at mains supply frequency poses an electrical shock hazard to the patient. Some leakage current at mains frequency is unavoidable but modern safety standards strictly limit the maximum permitted levels. The patient attached to a diathermy is connected via a large surface-area conductive electrode. If this electrode is earthed, the hazard to the patient resulting from the catastrophic failure of associated mains powered equipment would be considerable. Modern practice is never to earth the neutral plate directly. It may be referenced to earth using capacitors whose values are chosen to present a sufficiently high impedance at mains frequency to limit fault currents to acceptable values. However, the neutral plate may be isolated from earth at high frequencies also. The implications of these two alternative arrangements are discussed below.

The third potential hazard, that of unwanted burns, results from the possible flow of high-frequency current from the patient via a contact with conductive areas other than the neutral plate. These may be monitoring electrodes or an inadvertent contact with the metal parts of the operating table. Such sites are usually of small contact area. If the neutral plate is poorly attached to the patient, or poorly sited, or inadequately connected to the unit, it is possible that the sites of contact with other conductors may form lower-impedance return paths for earth-referenced current. The small surface areas of these alternative pathways may lead to high current density and consequent burns at the alternative contacts. This has proved a very real risk at the site of monitoring electrodes, especially where needle electrodes have been used. Monitoring instruments may well be isolated at supply mains frequency but present a relatively low impedance to earth for high frequencies. Modern practice is to install limiting resistors at the patient end of monitoring electrode leads. These limit the flow of high-frequency current to earth via the monitoring equipment. The alternative approach is to isolate the neutral pole of the high-frequency generator so that in normal use no significant current can flow to earth via these alternative sites since the high-frequency circuit is not earth referred.

These considerations—the risk of low-frequency shock and the risk of high-frequency burns—have resulted in two possible arrangements for the output circuits of electrosurgery apparatus; namely, the capacitively earth-referred system (Fig. 2) and the isolated system (Fig. 3). While it might appear initially that the high-frequency isolated system is clearly superior, the matter merits discussion since both systems are in common use.

In practice, the earth-referred system is the only one readily achievable for equipment working at 1 MHz and higher. At these frequencies, stray capacitance makes it

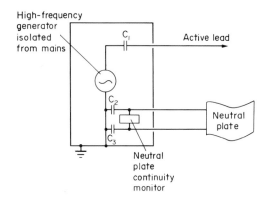

Figure 2
Earth-referred system of output circuit of an electrosurgery unit. Here, C_1 is the dc blocking capacitor, typically about 1 nF, and C_2 and C_3 are earth-referencing capacitors, typically about 0.01 μF

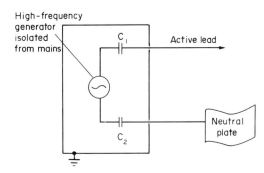

Figure 3
High-frequency isolated system of output circuit of an electrosurgery unit. Here, C_1 is typically 1 nF and C_2 is optional

impossible to achieve an adequately isolated output. Thus, the earth-referred system is used on the more traditional valve/spark-gap equipment. The major disadvantage of this system is the risk of burns at monitoring electrodes or sites of contact with earthed metal. At these frequencies, the operating table, though not directly earthed, is close to earth potential owing to its physical size and consequent stray capacitive coupling to earth.

The risk of burns arises if the return path via the neutral plate ceases to be the return path of significantly the lowest impedance. This can occur if the attachment of plate to the diathermy unit is poor or if attachment of plate to patient is poor. Various design features are incorporated to attempt to monitor these conditions. It is now universal practice to monitor continuity between equipment and neutral plate using a two-wire lead and a low-frequency monitoring current from an isolated low-voltage source. In the event of continuity failure the diathermy output is inhibited and an alarm sounded. To monitor the attachment of the neutral plate to the patient is more difficult. Two principles have been tried. The first uses an additional monitoring circuit to pass a small test current from the active electrode through the patient to neutral electrode and back to the electrosurgery unit. However, this requires the surgeon to touch the patient with the active electrode before starting the procedure. This is not the technique used in most circumstances where an arc is required (the surgeon draws an arc by placing the active electrode close to the patient). The second approach is to use a high-frequency current-difference monitor. This compares the current in the active output with the current in the neutral plate. Any significant difference must be due to current return via paths other than the neutral plate. The problem with this system is the choice of an appropriate response time for the monitor to inhibit the output of the unit: too fast a response time results in much nuisance tripping, whereas too slow a response will permit application of electrosurgery in short bursts (a common technique) even in fault situations.

In practice, most earth-referred units have only the simple plate/lead continuity monitor and rely on good nursing practice to position and apply the neutral plate properly.

The advantage of the earth-referred system is that the properly applied plate fixes the potential of the patient. Thus, open-circuit operation of the unit presents no hazard; nor does inadvertent earthing of the active output.

The fully isolated system, if properly designed, overcomes most of the disadvantages of the earth-referred system. Poor continuity between patient and neutral plate or plate and equipment is seen as a high impedance in an isolated circuit, and the current available from the unit is limited. Thus, no additional continuity-monitoring circuits are required, though in practice most manufacturers incorporate the simple neutral lead and plate monitor. However, a considerable disadvantage arises if, owing to stray capacitance in the output circuits, the isolation of the output is less than adequate. In these circumstances the patient is neither isolated nor firmly held at earth potential. Open-circuit activation of the electrosurgery unit may cause the patient to float up to some undetermined high-frequency potential with the risk of inadvertent current flow from the neutral plate via other sites to the unit. Indeed, the raising of the patient's potential during open-circuit activation of the diathermy is a problem with even a well-isolated unit. A number of manufacturers have overcome this by incorporating a neutral-plate voltage monitor in their design, that monitors the potential with respect to earth on the plate and alarms if it rises above a preset voltage, usually about 100 V.

Overall, the isolated system has more advantages than the earth-referred system but must be well designed to achieve adequate isolation. This is unlikely to be possible in equipment working above about 500 kHz.

3. Neutral Electrodes

The neutral electrode used in electrosurgery is known by a variety of other names: neutral plate, patient plate, electrosurgery grounding pad and earth plate are common alternatives. It plays a vital role in ensuring patient safety whatever type of output circuit is chosen. Various materials and designs have been used for neutral electrodes. The most common in the early days was a soft lead plate which could be bent to follow the curve of the patient at the site of application. In order to increase contact with the patient, the plate was usually wrapped in a saline-soaked bandage. If this dried out during the course of a long operation, all contact between patient and plate was lost. In use with the older earthed or earth-referred equipment, this has been a significant cause of burns at alternative contact sites and at monitoring electrodes. Lead neutral plates have almost completely disappeared in current practice.

There are three main types of neutral plate in use today. The first type is a large surface area (say 600 cm^2) semirigid stainless-steel plate on which the patient is placed. Good contact is ensured by the patient's weight, though contact area is significantly less than the total plate area. The plate cannot be molded to match patient contour, so sites of application are limited.

The second type are semidisposable flexible foil electrodes of around 250 cm^2 area. These are usually made of aluminum foil, in some cases plastic backed, and are held in an appropriate neutral-plate holder. They are used dry and where possible the patient is laid on them so that contact is made with a fleshy part. However, they can be strapped to, say, the upper leg because of their malleability. Approximately ten sessions per plate is a reasonable maximum usage, but if a plate has become significantly kinked it should be replaced sooner. These "dry plate" electrodes have proved most effective in use (Whelpton 1980).

Disposable electrodes are the third type of neutral plate. These are strictly for single use only and are usually smaller in area (~ 180 cm^2) than the foil type. Many are pregelled with a conductive electrode gel and all have some type of adhesive which allows them to be stuck to any suitable site on the patient. Their cost per patient is between 50 and 100 times that of the multiuse dry plates.

Some manufacturers offer so-called pediatric-sized disposable neutral electrodes. There is no evidence, however, that surgeons use less power when operating on infants. Indeed, there is no theoretical reason for assuming lower powers to be adequate in comparable procedures. Thus, the size of the neutral electrode required for an infant is no smaller than for an adult, and small-size electrodes should not be used. A very effective technique is to attach the neutral plate to a large piece of paper-backed aluminum foil on which the infant is positioned. At the end of the procedure the foil can be wrapped around the infant to help conserve body heat (Dobbie 1969).

4. Low-Power and Bipolar Equipment

It is current practice to design general-purpose electrosurgery units with a maximum output power no greater than 400 W. However, very many procedures require much less power; thus units with a maximum power output of around 170 W are becoming more common. These are totally adequate for most procedures except those involving the fluid-filled bladder. Units operating at this power level are particularly useful for electrosurgery down the biopsy channel of a flexible endoscope. In these particular circumstances considerable care must be taken with the neutral-plate arrangements to ensure adequate patient and operator safety (Hanwell 1979).

Many procedures in, for example, dentistry, ophthalmic surgery, chiropody and dermatology require powers of less than 50 W. In using these much lower powered units the inherent risks of diathermy are significantly reduced. Indeed, some units operating at this level use no neutral plate as such. They rely on the extraneous coupling of the patient to earth, either capacitive coupling or via ill-defined ohmic contacts, e.g., a hand on the arm of a dental chair, to provide a return path to the earth-referenced unit.

Another type of electrosurgery equipment to be considered is the bipolar unit. These units also generate output powers of up to 50 W maximum. Bipolar units have no neutral electrode. Instead, bipolar forceps are used in which the two halves are insulated from one another except at the tips, and active and neutral leads are brought to either half. The current flows through the tissue held between the tips of the forceps. The current path is thus very strictly defined and the coagulating effect is restricted to a very small volume of tissue. This technique is particularly useful in plastic surgery, in burn surgery and in neurosurgery. Many general-purpose electrosurgery units now incorporate a bipolar unit as an alternative instrument within the same cabinet.

5. Future Developments

Engineering developments will center around the use of state-of-the-art electronic components. Already units are being produced which incorporate a microcomputer to carry out various control functions. VMOS transistors are being used in the output stages of the electronics. Some units have a display of output scaled in watts. These have led to great confusion since surgeons are used to working with an output control arbitrarily scaled 0–10. Constant-current as opposed to constant-voltage units have been suggested.

In clinical terms the trend is towards lower-power units and to units designed for more specialized use. Endoscopy is a good example. Units that have additional monitoring circuits to improve safety when used with a flexible endoscope are also available. The use of bipolar electrosurgery will increase. The development of a bipolar instrument for use down a flexible endoscope

would be a very useful advance. The development of other specialized instruments is likely to have an impact (Hashmi and McCarthy 1979).

See also: Medical Electrical Equipment: Safety Aspects

Bibliography

Brown B H, Johnson S G, Betts R P, Henry L 1977 Burns thresholds to radiofrequency leakage currents from surgical diathermy equipment. *J. Med. Eng. Technol.* 1: 277–81

Dobbie A K 1969 The electrical aspects of surgical diathermy. *Biomed. Eng.* 4: 206–16

Hanwell A E 1979 Electrical hazards in fibre optic endoscopy. *Proc. HPA Conf. on Fibre Optics and Endoscopy.* Hospital Physicists' Association, London

Hashmi M S, McCarthy J P 1979 Cauterizing iris scissors. *Br. J. Ophthalmol.* 63: 754–57

Health Equipment Information 1980–85 *Evaluation of Surgical Diathermy Units*, Report Nos. 84, 99, 122, 139. Department of Health and Social Security, London

Lacourse J R, Miller W T 1985 Effect of high frequency current on nerve and muscle tissue. *IEEE Trans. Biomed. Eng.* 32: 82–86

Mitchell J P, Lumb G N, Dobbie A K 1978 *A Handbook of Surgical Diathermy*, 2nd edn. Wright, Bristol

Pearce J 1985 Current electrosurgical practice: Hazards. *J. Med. Eng. Technol.* 9: 107–11

Taunton J C 1981 Surgical diathermy: A review. *J. Med. Eng. Technol.* 5: 175–83

Whelpton D 1980 Performance evaluation of surgical diathermy indifferent electrodes. *Clin. Phys. Physiol. Meas.* 1: 59–70

J. McCarthy

Endoscopes

Endoscopes, which may be rigid or flexible, are devices which facilitate the examination of body cavities inaccessible to the human eye. They require a means of illumination and an image transmission system (Fig. 1). Modern versions incorporate almost universally a bundle of glass fibers to conduct the illumination from an external light source to the tip of the instrument. Most

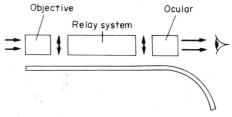

Figure 1
Schematic diagram of an endoscope

rigid instruments utilize a train of lenses as an image transmission system but the fully flexible fiberoptic endoscope uses a specialized image-transmission glass-fiber bundle.

1. Optics of Fiber Illumination

The illumination fiber bundle of a modern endoscope consists of 10–100000 individual glass fibers, each of diameter 30–70 μm. Each fiber comprises a glass core surrounded by a thin outer coating of a glass of a lower refractive index, which insulates the core from the effects of surface contamination, defects, and contact with adjacent fibers.

The fiber transmits light by multiple total internal reflection at the interface between the core and the cladding (Fig. 2). Total internal reflection occurs when the angle of incidence exceeds the "critical angle." This results in a limited cone of acceptance with a semiangle α given by $\sin \alpha = (n^2 - n_1^2)^{1/2}$, where n and n_1 are the refractive indices of the core and cladding, respectively. Although each reflection is total, transmission losses occur due to absorption by the glass and amount to approximately 30% per meter.

Glass fibers are extremely flexible by virtue of their fine diameters. The maximum bending which can be tolerated by the fiber without fracture is given by $R_F = Ed/2\sigma$, where R_F is the minimum radius of curvature, E is the modulus of elasticity, σ the ultimate failure stress, and d is the diameter of the fiber. Since glass is

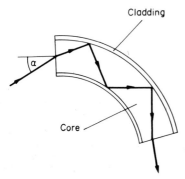

Figure 2
Multiple total internal reflection in a coated fiber

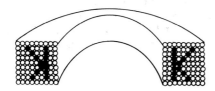

Figure 3
A coherent fiber bundle

a brittle material, σ is critically dependent on the presence of small cracks on the surface. For bulk glass, $E/2\sigma$ is typically 1000–1500 but the relative absence of surface flaws in drawn fiber gives values in the region of 10–30, so that a single fiber of 30 μm diameter may be bent to a minimum bend radius of <1 mm.

2. Flexible Endoscopes

The modern fully flexible endoscope incorporates two types of fiber bundle. Illumination is provided from an external light source via a bundle of randomly assembled fibers (incoherent bundle), whereas a second type of fiber bundle (coherent bundle) is used to transmit the image formed by the objective lens to the eyepiece (Fig. 3). In a coherent bundle the individual fibers are carefully ordered, each having the same relative position on the front face of the bundle as on the back face. It is this ordering of the fibers which allows the image to be transmitted, since individual fibers only carry an average of the incident light and do not transmit image information. The outer cladding of the fibers and the spaces in between them do not transmit light, so the image is composed of spots of light, each the diameter of the inner core of the fibers. The image quality depends on the size of the fibers and the accuracy with which they are ordered.

The calculation of the resolution of a fiber-optic system is complex; the size of the smallest resolvable object is dependent upon the working distance. Modern endoscopes when working at their shortest range will usually resolve finer detail than that resolvable by the naked eye.

3. Rigid Endoscopes

There are three different types of relay systems used to transmit the image from the objective to the ocular in modern rigid endoscopes.

(*a*) *Thin lens relay system* (Fig. 4a). In this system, doublet lenses corrected for chromatic aberrations are employed. The diameter of the individual lenses is greater than the lens thickness and the lens separations are usually 10–20 times that of the lens thickness. The image formed by each relay stage becomes the object for the next adjacent stage and so on for as many stages as are required by the length of the endoscope.

(*b*) *Rod lens relay system* (Fig. 4b). Initially developed by Hopkins, the lenses work in much the same way as a conventional thin-lens relay, but the thickness of each relay lens is typically ten times that of its diameter and the air gaps between the individual lenses are correspondingly smaller. This construction gives certain optical advantages, particularly in giving a brighter overall image.

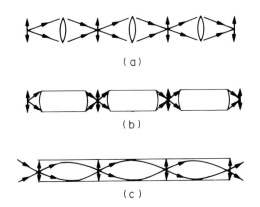

Figure 4
Optical systems of rigid endoscopes: (a) thin-lens relay, (b) rod-lens relay, and (c) Selfoc relay

(*c*) *Selfoc relay system.* A solid glass rod is used which has been treated so that the refractive index at the center of the rod is greater than that at the edge. This structure continuously refracts the light as shown in Fig. 4c. Generally, Selfoc systems have a poor image quality in comparison to other relay systems, but they have advantages for small diameter systems (<2 mm diameter).

See also: Fiber Endoscopy

Bibliography

Hopkins H H 1976 The physics of the fiber-optic endoscope. In: Schiller K F R, Salmon P R (eds.) 1976 *Modern Topics in Gastrointestinal Endoscopy*. Heinemann, London, pp. 15–62

A. E. Hanwell and J. G. Richards

Evoked Potentials

Spontaneous electrical activity from the brain of living animals can be modified in a consistent and detectable way by sensory stimulation, voluntary movement, or the performance of certain cognitive tasks. The study of such evoked changes in the ongoing activity of the brain has developed into a major area of research, loosely called the study of evoked potentials or, more properly, event-related brain potentials. Because of the small size of these potential changes, which, when recorded through the scalp, are usually within the microvolt range, their study involves enhancement of the signal-to-noise ratio by the summation of a large number of such responses (signal averaging), a technique developed for this purpose in the late 1940s. Although this was initially carried out with small, purpose-built analog computers, it is now more common to employ digital averagers.

Potentials evoked by visual, auditory or electrical stimulation may show characteristic alterations in their waveform and latency in certain pathological states. Their recording has found many applications in clinical diagnosis and in testing hearing in early infancy (see *Electroencephalic Audiometry*).

1. Historical Development

Richard Caton first discovered the spontaneous electrical activity recordable from the surface of the brain of animals (rabbits and monkeys) in 1875. He noted a negative swing in the potentials in those areas of the gray matter which appeared to be functionally active at any one time. He observed that such "negative variations" of the potential could be produced by visual stimulation in the area of the cortex associated with eye movements and that potential changes were similarly associated with mastication and head rotation, being detected in the areas of the brain which produced these movements on electrical stimulation.

Some fifty years later Hans Berger discovered that the spontaneous electrical activity of the human brain could be recorded, at a much lower voltage level, from electrodes placed on the surface of the scalp. His study confirmed that the spontaneous activity of the human brain could be modified by sensory stimulation or voluntary movement and also by cognitive tasks such as mental arithmetic. Any of these activities could be shown to block temporarily the ongoing α rhythm, which had a characteristic frequency of 8–14 Hz. Berger regarded the α rhythm as a whole brain response, but Adrian and Matthews (1934) subsequently showed that it was particularly influenced by visual attention or inattention and originated largely in the occipital lobes near the primary visual receiving areas in the striate cortex.

2. Superimposition and Averaging Techniques

In 1947, Dawson introduced a new technique which enabled very much smaller evoked potential changes to be detected in scalp electrode recordings. Since these changes follow closely and systematically on the arrival of the afferent sensory volley at the cortex, evoked potentials show a fixed temporal relationship with the evoking stimulus. In an individual trial these responses may disappear in the spontaneous background activity of the normal electroencephalogram (see *Electroencephalography*), but this background activity is unrelated to the stimulus. Dawson was able to improve the signal-to-noise ratio by superimposing photographically the responses from a large number of trials. The common features of the consistent responses reinforce each other and become apparent, while the random background activity becomes less and less obtrusive as more trials are accumulated. In practice the method was effected by photographing the responses from successive sweeps on an oscilloscope display, the stimulus being used to trigger each sweep. This method enabled Dawson to record the much smaller evoked potentials, induced in the primary receiving area of the cortex for somatosensory stimulation, following the electrical stimulation of a peripheral nerve in the upper or lower limb. The method was also used in the same year to record the pathologically enhanced responses in a patient with myoclonic epilepsy.

Four years later Dawson introduced a major improvement in the recording technique by arranging for the evoked potentials from a larger number of trials to be summed in an analog computer, rather than superimposed photographically. His original equipment incorporated a bank of capacitors, connected sequentially by a rotating switch, to sample the input voltage at fixed intervals following the delivery of the stimulus. The summation or averaging method had the advantage of giving a relatively "clean" record in which the improvement in the signal-to-noise ratio was proportional to the square root of the number of input trials. Both superimposition and averaging techniques demand the stimulus-locked recording of a large number of responses. They have the disadvantage of requiring a sharp-fronted, accurately timeable stimulus. Hence the evoked response technique cannot easily be used for many natural forms of stimulation, which are not, by their very nature, sudden in onset.

Dawson's original averager was a small purpose-built analog computer, as were a number of the other averaging devices which followed. Since the early 1960s, however, it has been more usual to use a small digital computer. This has the advantage of greater flexibility in sampling rate and can be interfaced with a larger computer to provide facilities for more sophisticated forms of signal analysis; for example, analysis of variance, principal-component analysis or discriminant-function analysis. These methods have proved particularly useful in the study of the cognitive event-related potentials. A short account of the historical development of evoked response recording, with references to the original papers, may be found in Barber (1980).

3. Near-Field and Far-Field Recording

The ability to record evoked potentials at a distance from the brain surface through the intervening layers of scalp and brain meninges depends on volume conduction of the electrical field through the interposed media. Potentials generated in the cerebral cortex near the surface of the brain are particularly favorably placed for recording in this way because the generator neurons are all oriented perpendicular to the cortical surface, and when an area of cortex is synchronously activated by a volley of afferent impulses there is a coherent current flow between the terminal dendrites just under the cortical surface and the cell bodies situated at somewhat deeper levels. The potentials at the cortical surface

become electrically positive or negative in a succession of waves or components, the latency and polarity of which tend to be fairly stereotyped, at least for the early components of short latency, following a particular stimulus.

In addition, it has been demonstrated that potentials evoked at certain subcortical levels by the passage of the afferent volley through various structures on its way to the cortex can be recorded. These potentials are known as far-field potentials, since they are recorded at very small amplitudes at greater distances from the generator site. A discussion of the physical principles underlying the propagation of near-field and far-field potentials can be found in Callaway et al. (1978).

4. Auditory, Visual and Somatosensory Evoked Potentials

Since all nerves conduct their messages in essentially the same manner, as a frequency-modulated train of all-or-nothing impulses, evoked potentials can be elicited in response to many different modalities of sensation, including auditory, tactile, electrical and visual stimulation. Because of the requirement for a sharp-fronted stimulus and a synchronous afferent volley, the types of stimulus found to be most effective have been clicks or photic flashes or (in the case of somatic sensation) the electrical stimulation of a sensory nerve trunk through the intact skin with a pulse of short duration (typically 10 or 100 μs). The spatial topography and latency of the resulting evoked potentials are, in the first instance, related to the site of the generator along the afferent pathway and the time taken by the afferent volley to reach it. Thus a cortical response to stimulation of the median nerve at the wrist can be detected at about 18–20 ms, the exact interval depending on the arm length and height of the individual concerned; whereas earlier components arising in the brachial plexus, cervical spinal cord and brain-stem can be recorded at 8–10, 11–13 and 14–16 ms, respectively. Similar cortical responses from stimulation of nerves in the leg have a latency of more than 30 ms.

In the auditory and visual modalities the response undergoes much later elaboration after reaching the cortex. The most prominent potentials occur 100 ms or so after the stimulus, notwithstanding that intracortical recordings have confirmed a minimum latency of 14 ms for the arrival of the first afferent impulses at the primary auditory receiving area; the earliest latency for a bright flash being about 28 ms.

5. Clinical Applications of Evoked-Potential Recording

Evoked-response audiometry has established itself as a useful objective technique for the investigation of suspected deafness in infants or in older people unable or unwilling to cooperate in conventional audiometric testing. The auditory evoked potential may also provide useful diagnostic information to the neurologist or ear, nose and throat surgeon in the investigation of demyelinating disease or vascular and compressive lesions involving the brain-stem (Halliday 1982).

The visual evoked potentials arising from a reversing black-and-white checkerboard stimulus have proved particularly sensitive to the presence of demyelinating lesions of the optic nerves or posterior visual pathways, and marked delays of the cortical potentials may be seen in these cases—frequently in the absence of other clinical signs (see *Electroretinography*). For this reason the method has been widely adopted as a diagnostic test in hospitals dealing with disorders of the eye or nervous system. Abnormalities of the visual evoked potentials may be seen in many other disease states, including tumors, vascular lesions and retinopathies.

The somatosensory evoked potential has similarly proved of value in the investigation of lesions affecting various parts of the sensory nerve pathways, including peripheral neuropathies, cervical spondylosis, and lesions of the spinal cord and brain-stem. In some forms of epilepsy, particularly those associated with myoclonic jerking, the cortical response itself may be very greatly exaggerated and this too has proved diagnostically useful. In recent years, the responses have been employed to provide continuous monitoring of the integrity of the spinal cord during corrective surgery for scoliosis and other spinal deformities.

See also: Bioelectricity

Bibliography

Adrian E D, Matthews B H C 1934 The Berger rhythm: Potential changes from the occipital lobes. *Brain* 57: 355–85

Barber C (ed.) 1980 *Evoked Potentials*. MTP Press, Lancaster, UK

Bodis-Wollner I (ed.) 1982 Evoked potentials. *Ann. N.Y. Acad. Sci.* 388

Callaway E, Tueting P, Koslow S H (eds.) 1978 *Event-Related Brain Potentials in Man*. Academic Press, New York

Chiarenza G A, Papakostopoulos D (eds.) 1982 *Clinical Applications of Cerebral Evoked Potentials in Pediatric Medicine*. Excerpta Medica, Amsterdam

Cobb W, Morocutti C (eds.) 1967 The evoked potentials. *Electroencephalogr. Clin. Neurophysiol.* Suppl. 26: 1–218

Courjon J, Mauguière F, Revol M (eds.) 1982 *Clinical Applications of Evoked Potentials in Neurology*. Raven, New York

Dawson G D 1947a Cerebral responses to electrical stimulation of peripheral nerve in man. *J. Neurol. Neurosurg. Psychiat.* 10: 134–40

Dawson G D 1947b Investigations on a patient subject to myoclonic seizures after sensory stimulation. *J. Neurol. Neurosurg. Psychiat.* 10: 141–62

Dawson G D 1954 A summation technique for the detection of small evoked potentials. *Electroencephalogr. Clin. Neurophysiol.* 6: 65–84

Desmedt J E (ed.) 1977 *Visual Evoked Potentials in Man: New Developments*. Clarendon Press, Oxford

Halliday A M (ed.) 1982 *Evoked Potentials in Clinical Testing*. Churchill Livingstone, Edinburgh

Katzman R (ed.) 1964 Sensory evoked response in man. *Ann. N.Y. Acad. Sci.* 112

Lehmann D, Callaway E (eds.) 1978 *Human Evoked Potentials: Applications and Problems*. Plenum, New York

Storm Van Leeuwen W, Lopes Da Silva F H, Kamp A (eds.) 1975 *Evoked Responses*. Elsevier, Amsterdam

A. M. Halliday

F

Fast Neutron Therapy

Neutrons having energies above a few MeV have sufficient penetration in tissue to treat human cancers in a similar way to conventional radiotherapy with photons (x or γ rays). Both neutrons and photons are indirectly ionizing radiations but there are fundamental differences in the way in which energy is released in tissue. Photons interact with atomic electrons whereas neutrons interact with nuclei. In both cases it is the resulting secondary charged particles which ionize the medium. However, the density of ionization along the tracks of secondary nuclear particles (protons, α particles and recoil nuclei) is much greater than that along the tracks of recoil electrons. It is this factor which is responsible for the different biological effects.

1. Rationale for Fast Neutron Therapy

At least half of all cancer patients receive radiotherapy with photons at some stage in their disease. The treatment is, however, by no means always successful, even for tumors of limited extent treated with the most modern x-ray equipment. What is needed is some way of altering the biological response, to increase the effect on the tumor relative to that on the surrounding normal tissues, often referred to as the therapeutic ratio.

The first trial of fast-neutron therapy was conducted in California from 1938 to 1943 using some of the first cyclotrons developed by E O Lawrence. The rationale was empirical, in the hope that the therapeutic ratio would be improved. Although some surprising tumor regressions were seen, the general conclusion was unfavorable to the use of neutrons, mainly owing to severe late reactions in normal tissue.

With hindsight, it is now clear that this trial was undertaken too early, before there was any understanding of the biological effects of neutrons and in the absence of reliable equipment so that a safe and satisfactory schedule of treatments could not be established.

During the 1950s and 60s some of the reasons for radioresistance in tumors were elucidated. One of these is the presence of hypoxic but viable cells in tumors and the radioprotective effect of anoxia. Anoxic cells need three times the dose of photons given to aerobic cells for a given level of sterilization, whereas with fast neutrons the factor is about 1.7. As nearly all cells in normal tissues are aerobic this is a strong argument for using neutrons. Tumor cells may be radioresistant for other reasons, such as the repair of sublethal and potentially lethal radiation damage and the presence of a moiety of cells in a resistant part of the cell cycle. The effects of more densely ionizing radiation are in general less dependent on such factors. In addition, by the mid-1960s cyclotrons had become much more reliable. As a result it became reasonable to have another look at fast neutron therapy, and in 1966 studies were initiated at the Hammersmith Hospital, London, using a medical cyclotron. Today trials of neutron therapy are in progress in several countries.

2. Neutron Production

The energy of fission neutrons from reactors or ^{252}Cf is too low for beam therapy, although ^{252}Cf in small tubes has been used for interstitial and intracavitary treatments (see *Brachytherapy*).

The most economical method for generating an intense beam of fast neutrons is by the reaction ^{3}H(^{2}H,n)^{4}He. The peak of the excitation function occurs at 100 keV and neutrons of 14–15 MeV energy are given off isotropically. The main difficulty with this method is the loss of tritium from the target due to heating and sputtering during operation at high power. Attempts have been made to overcome this by accelerating a mixture of deuterons and tritons which would continuously replenish the target but in practice the life of the tubes has been short and erratic.

The alternative is to use a high-energy accelerator such as a cyclotron (see *Cyclotrons*). Most centers attempting neutron therapy now use cyclotrons, usually accelerating deuterons on to a target of beryllium. At Hammersmith Hospital 16 MeV deuterons are used but the penetration of the resulting neutron beam (mean energy around 7 MeV) is too low for treatment of deep-seated tumors.

Modern isochronous cyclotrons can accelerate protons to nearly double the energy which can be given to deuterons, with a corresponding increase in neutron energy. All the newer installations will use protons in the range 35–70 MeV and beryllium targets. The neutron spectrum from protons contains a sizeable proportion of low-energy neutrons but this can be reduced by using relatively thin targets or a hydrogenous filter.

Figure 1 shows the neutron intensity in the forward direction from various reactions. A typical dose of neutrons is 1 Gy which should be given in less than 5 min. If a factor of 2 is allowed for filtration or use of a thin target, the dose rate from a thick target needs to be at least 0.4 Gy min^{-1}. With 50 μA on the target and a treatment distance of 150 cm, Fig. 1 indicates minimum energies of 20 MeV for deuterons and 40 MeV for protons.

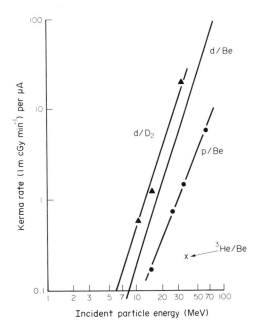

Figure 1
Neutron intensity in the forward direction from thick targets of ^{2}H and Be bombarded by deuterons and Be bombarded by protons (after Catterall and Bewley 1979. © Academic Press, London. Reproduced with permission)

3. Interactions of Fast Neutrons with Tissues

Interactions consist of scattering, which can be elastic or inelastic, and nuclear reactions. Elastic scattering by hydrogen is the most effective method whereby neutrons give up their energy, because a neutron can give all its kinetic energy to a recoil proton and because hydrogen is the commonest atom in most tissues. Nuclear reactions such as ^{16}O(n, α)^{13}C remove all the energy of the neutron but the cross sections are much smaller than for scattering and there is often a threshold energy below which the reaction does not occur at all.

Neutrons which have been reduced to thermal energies by repeated scatterings may escape from the body or may be removed by the reactions ^{1}H(n, γ) ^{2}H and ^{14}N(n, p) ^{14}C.

The energy released in matter by neutron interactions is called the kerma (kinetic energy released per unit mass) (see *Neutron Kerma Values*). The importance of elastic scattering by hydrogen causes kerma in different tissues to be roughly proportional to the concentration of hydrogen. As a result subcutaneous fatty tissue receives about 15% more energy from fast neutrons than do most other soft tissues, whereas bone receives 25–50% less. These differences are in the opposite sense to those occurring under photon irradiation (see Table 1).

Table 1
Kerma factors (relative to kerma in muscle) for neutrons and photons in various biological media

Radiation	Muscle	Fatty tissue	Solid bone	Small cavity in bone
Fast neutrons	100	114	60–70	65–75
^{60}Co γ rays	100	101	96	104–110
250 kV x rays	100	95	150–170	150–170

Source: Catterall and Bewley 1979. © Academic Press, London. Reproduced with permission

The reduced absorption of energy in bone is demonstrated vividly in neutrograms (see *Neutron Radiography*).

4. Dose Distribution in Tissue

In conventional photon radiotherapy it is usual to measure the distribution of dosage in water as a model for soft tissue. With neutrons the exact chemical composition of the medium is more important. Special liquid mixtures have been made to simulate tissues but their densities are usually too great. Water is a reasonable representation of most tissues and has the great advantage of being universally available. Figure 2 shows some neutron depth doses in comparison with high-energy photon beams. It is clear that protons of about 60 MeV are needed to provide a beam comparable with beams from modern x-ray machines.

Dose distributions in tissue from fast neutrons can be modified in similar ways to dose distributions from

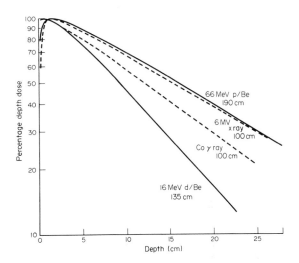

Figure 2
Percentage depth–dose curves for two energies of neutrons compared with x and γ rays. Field size 10×10 cm in each case

photons. Thus percentage depth doses increase with neutron energy, filtration, target–skin distance and field size. One factor which is different is that neutron beams are always accompanied by γ radiation which originates in the target, the collimation system and the patient. Gamma rays have a lower biological effectiveness than fast neutrons, usually by a factor between 2 and 5, so it is best to measure the two components separately. This becomes more difficult to do as the neutron energy rises. The partial dose of γ radiation varies with the irradiation conditions but is usually in the range 5–20% of total dose.

Close to the surface there is a transition zone, in which secondary charged-particle equilibrium is established (see Fig. 2) which results in sparing of the skin. This is also analogous to the transition zone with photons but is governed by the range of the recoil protons. Close to the surface, where there is a deficiency of recoil protons, most of the absorbed dose is derived from heavier ions such as α particles and recoil nuclei of short range. These ions give more densely ionized tracks than protons and have greater biological effect per unit dose. Consequently, the reduced dosage close to the surface may not be reflected in a comparably reduced biological effect but nevertheless a significant degree of skin sparing can be obtained.

5. Collimation and Shielding

Shielding against fast neutrons is more difficult than against photons. There is no regular increase in absorption coefficient with atomic number as there is with photons. Indeed, hydrogen is the most efficient material per unit mass. However, per unit thickness, dense materials such as iron and tungsten are more effective than hydrogenous materials. For bulk shields iron is the most practical material, but since this is not very opaque to neutrons of around 1 MeV it is best to surround the iron with a layer of hydrogenous material such as polyethylene. Boron in the polyethylene reduces the production of γ radiation and an outer layer of lead can be added to reduce the γ radiation even more.

Activation of the shield gives rise to residual γ radiation after the neutrons are switched off. Care must be taken to avoid elements with large cross sections for induction of radioactivity in which energetic γ rays are emitted. Aluminum is particularly bad in this respect, as is manganese which is a common component of steel.

At present most neutron-therapy installations use interchangeable inserts to define the field size. This can be a serious impediment to using the best treatment plan for each patient. Adjustable collimators are, however, feasible and are now coming into use.

Neutron beams fixed in a single direction are also very inconvenient. Deuterium–tritium generators can be rotated around the patient like a cobalt unit. To obtain an angled beam from a cyclotron is more difficult and entails bending the charged-particle beam up and round

in the form of a hook before it strikes the target, the whole assembly being able to rotate about an axis along the line of the beam emerging from the cyclotron. This is another improvement which is included in the more recent installations.

6. Miscellaneous Techniques

The techniques of treatment and dosimetry are broadly similar to those used with photons but with important differences. Absorbed dose is usually measured with ionization chambers (see *Neutron Dosimetry*). For measurements on the patient, or *in vivo* dosimetry, nuclear activation offers a very convenient method. The response is to the neutron component alone but this is not a serious disadvantage as the γ-ray component has little biological effect.

The neutron spectrum may change with depth in the patient and particularly outside the main beam. To allow for this, the ideal material would have a cross section whose variation with neutron energy matches the variation in kerma per unit fluence. Different elements approximate to this ideal to varying degrees over different regions of neutron energy. For the neutron beam at Hammersmith Hospital (16 MeV d on Be) the reaction ^{115}In(n, n')^{115m}In is satisfactory in this respect, judging by a comparison between lateral profiles measured in a tissue-equivalent phantom with indium and a measurement of absorbed dose of neutrons. ^{115m}In decays with a half-life of 4.5 h and emits a γ ray of 336 keV. The longish half-life is convenient for measuring the dose given from several fields during a single treatment session.

Other possible techniques of *in vivo* dosimetry are track registration in plastic foils and the change in resistance induced in silicon diodes. At present, however, these methods cannot match the precision and convenience of the activation method. Thermoluminescent dosimetry (TLD) suffers from the disadvantage that its sensitivity to γ rays is greater than its sensitivity to fast neutrons.

If bolus is used to help in treatment planning, it should have the same scattering and attenuation properties as soft tissue. This requires about 10% of hydrogen. Recoil protons from the bolus will eliminate any skin sparing which may be obtainable with the neutron beam but they may be removed by placing a thin layer of non-hydrogenous material against the skin. Lead is the most convenient material as it can be molded easily to fit awkward shapes.

Wedge filters can be used as in conventional radiotherapy. Polyethylene is a convenient material as it does not become activated except from neutrons above the threshold for the reaction ^{12}C(n, 2n)^{11}C which is at 20 MeV. On the other hand, the wedges can be smaller if made of iron.

The positioning of the treatment field can be checked by taking a neutron radiograph during treatment.

7. Clinical Results

The response of normal tissues and tumors has not been qualitatively different after neutrons than after protons. There have, however, been quantitative differences depending on the end point observed. This is partly a reflection of radiobiological findings which have shown substantial differences in the relative biological effectiveness (RBE) of neutrons for different tissues. Relative to γ rays, the RBE of fast neutrons at the dose levels used in normal radiotherapy have been found to vary between one (intermitotic death of white blood cells and physiological effects on rats' stomachs) to five (damage to brain and spinal cord in rats).

Another reason for differences is the wide variation of neutron kerma values in different tissues (Table 1). The increased dosage to fatty tissues is related to the increased incidence of subcutaneous fibrosis after neutron therapy, whereas the reduced dosage to bone is reflected in the lack of bone damage.

The most striking difference between the response of patients to neutrons and photons has been the reaction of brain tissue. Attempts to treat brain tumors with fast neutrons have resulted in severe demyelination and degeneration in normal brain tissues to an extent not normally seen after treatment with photons. This is likely to be connected with the high RBE of neutrons for brain tissues measured in rats. There seems to be no therapeutic margin between sterilization of brain tumors (gliomas) and unacceptable damage to normal brain tissues.

Most existing neutron-therapy installations produce beams fixed in a horizontal direction and of rather poor penetration into the body. As a result the most extensive trials of fast-neutron therapy have been for cancer in the mouth and throat. These cancers are not too deep and their response can be assessed relatively easily.

The first trial in this area was conducted at Hammersmith Hospital from 1971 to 1975, with 161 patients. The proportion showing complete regression of the tumor (on clinical judgement) was about 60% in the neutron series and about 25% in the photon series. There were virtually no recurrences in the neutron series. On the other hand, survival was not significantly different in the either series.

Several factors have a bearing on these results. The patients mostly had very advanced disease and many were elderly (there was no age limit). Mortality was 50% by 9–12 months, limiting the possibility of late recurrences to become manifest. The dose of radiation was more variable on the photon side and often rather low. This arose because it was a multicenter trial, in which patients entered at other hospitals and allocated to photons were mostly treated at the hospital of origin. Each hospital was allowed to follow its own treatment policy. Another consequence was that follow-up in the photon series was usually less complete than in the neutron series. Nevertheless, the photon series treated at Hammersmith Hospital also showed a significant difference from the neutron series. The best sites for neutrons were floor of mouth, antrum and salivary glands. The worst site was larynx and pharynx where the full neutron dose needed to control the tumor produced a high level of morbidity.

Several more recent trials in these sites at other centers are still under way. Preliminary results do not show the same advantage of neutrons, particularly the almost complete lack of recurrences. It is not yet clear why the conclusions of the trial at Hammersmith should be so different; one factor may be the selection of patients who had, on the whole, less-advanced disease in the later trials; another may be the dose levels used and the degree of morbidity accepted.

Centers in the USA have mostly used mixed schedules in which two or three treatments per week were given with neutrons and the remainder with photons. It is likely that this would dilute any possible benefit of neutrons and would tend to conceal differences in response.

Trials in other sites are at an even earlier stage, e.g., for cancers of the bladder and rectum, melanomas and soft-tissue sarcomas.

All of these trials have compared the best modern radiotherapy using ^{60}Co γ rays or megavoltage x rays from linear accelerations with neutron beams which are inadequate from a technical point of view. The neutron beams are fixed in direction, usually horizontal. Only a few field sizes are usually available and often the collimators have to be changed by slow and inconvenient mechanical devices. Often wedge filters and field-shaping blocks cannot be used simultaneously. Depth doses and skin sparing are usually inadequate.

A new generation of cyclotrons is now under construction which should overcome most, if not all, of these problems. Higher depth doses and better skin sparing will be combined with moving beams and adjustable diaphragms. Only when neutron therapy is technically equivalent to conventional treatment can a proper comparison be made.

See also: Neutron-Capture Therapy; Neutron Sources; Radiobiology: Charged and Uncharged Particles; Radiotherapy: Treatment Planning; Radiobiology: Kinetic Basis of Normal Tissue Response to Radiation

Bibliography

Bewley D K, Meulders J-P, Page B C 1984 New neutron sources for radiotherapy. *Phys. Med. Biol.* 29: 341

British Journal of Radiology, Suppl. 17, 1983 Central axis depth-dose data for use in radiotherapy

Broerse J J 1976 *Monograph on Basic Physical Data for Neutron Dosimetry*, EUR 5629e. European Economic Community, Brussels

Catterall M, Bewley D K 1979 *Fast Neutrons in the Treatment of Cancer.* Academic Press, London

Fowler J F 1981 *Nuclear Particles in Cancer Treatment.* Hilger, Bristol

Hall E J 1978 *Radiobiology for the Radiologist*, 2nd edn. Harper and Row, London

Hall E J, Graves R G, Phillips T L, Suit H D (eds.) 1982 Particle accelerators in radiation therapy. *Int. J. Radiat. Oncol., Biol. Phys.* 8(12)

Raju M R 1980 *Heavy Particle Therapy*. Academic Press, New York

D. K. Bewley

Fetal Monitoring

In the intrapartum period of childbirth, that is, between the onset of labor and delivery, the fetus may suffer hypoxia or asphyxia if the blood flow to the placenta is restricted or if the umbilical cord is compressed. In the antepartum period (conception to onset of labor) the fetus may be poorly nourished as a result of insufficient placental blood flow, or due to reduced placental surface area. This may lead to impaired respiratory function, hypoxia and neurological damage, or even stillbirth.

To detect fetal distress, or better still, to predict it and thereby reduce fetal mortality and morbidity, it is common practice to monitor the condition of the fetus *in utero*. Indeed, stethoscopic auscultation of the fetal heart has been practised for many years. However, using such a method it is impossible to measure the heart rate continuously, the values obtained being only averages over a period of 15 or 30 s. Furthermore, the rate cannot usually be measured during periods of uterine contraction. In recent years, however, electronic fetal monitoring equipment has been developed which does provide the clinician with a means of detecting fetal heart action over long periods, including periods of uterine contraction.

Changes in fetal heart rate give an indication of the well-being of the fetus and can be correlated with changes in uterine activity. For this reason fetal monitors record the fetal heart rate (FHR) and the uterine activity simultaneously, presenting them side by side on a single strip chart. The recording is then examined for baseline FHR changes, and for the magnitude and duration of any accelerations or decelerations of FHR and their temporal relationship to uterine contractions and fetal movements. Another aspect of the recording is the variation of the FHR. Normally the FHR varies continually and the magnitude and frequency of these changes is referred to as the variability. Changes in this pattern of variability can provide important indications.

1. Detection of Fetal Heart Action

A number of techniques are available for detecting fetal heart action. The fetal electrocardiogram (FECG) can be obtained by the direct connection of an electrode to the fetus or, alternatively, to the maternal abdomen. Fetal heart valve or heart wall motion can be detected using various Doppler ultrasound methods. A phono-transducer can also be used to pick up the audible sounds produced when the heart valves close.

1.1 Fetal Electrocardiography

During labor, the most reliable signal is obtained via the direct connection of an electrode to the fetal scalp after the rupture of the amniotic membranes. There are various designs of scalp electrode. Three in current use are shown in Fig. 1, the spiral electrode types being the

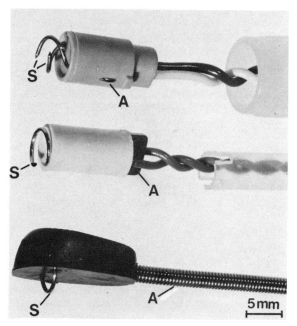

Figure 1

Three typical fetal scalp ECG electrodes: each type provides two connections, the one marked S to the fetal scalp and the other marked A to the fetal abdomen

most widely used. The spiral electrode is inserted into the fetal scalp and from there one electric path connection to the heart is provided via the head, the neck and upper abdomen of the fetus, while the second connection is through a small plate electrode which is in contact with vaginal fluids, via the uterus, the placenta and umbilical cord, to the lower abdomen of the fetus. The magnitude of the signal is about 1 mV peak; good quality recordings can be obtained even during contractions. Alternatively, the FECG can be picked up via electrodes placed on the maternal abdominal surface but here the signal strength is much less (50–200 µV), and during labor the signal is often partially obscured by the maternal ECG and by signals from uterine and other muscular action. This method is successful only during the antepartum period, while the mother is lying in a relaxed state. The signals are particularly weak at around 30 weeks of gestation, probably due to the presence of vernix caseosa.

To ensure good quality FECG signals it is necessary to overcome interference (at 50 or 60 Hz) from the local mains supply wiring as well as the effect of the variable high source impedance of the signal. The latter effect can be reduced by using an amplifier with an input impedance which is very high in comparison to the source impedance. The amplifier should also have a high common-mode rejection ratio to reduce the effects of unwanted signals which are common to both inputs. The signal quality may be further improved by feeding back a small inverted version of the interference signal to a reference connection on the mother—usually the under-surface of the connector which joins the scalp electrode leads and the screened cable to the monitor (Fig. 2).

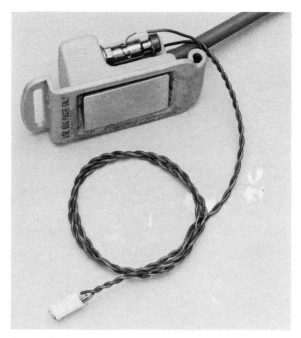

Figure 2
The "leg plate" provides a connection between the scalp electrode leads and the screened cable to the monitor. The metal plate on the underside is used to provide an active reference connection

When derived from an abdominal connection, the FECG has to be distinguished from the larger and usually slower maternal ECG. Maternal QRS complexes recognizable by their greater amplitude can be suppressed before they enter the FECG detector. When fetal and maternal signals coincide, fetal beats are sometimes lost. Further circuitry is therefore added to replace missing beats on the basis of the preceding last few cardiac cycles—over three cycles at most to avoid the production of spurious recordings. A solitary maternal QRS complex may, alternatively, be stored and subtracted from subsequent larger maternal complexes, leaving only the fetal component.

1.2 Ultrasonic Detection of Fetal Heart Rate

Ultrasonic detection of fetal heart-rate action is an attractive technique because it can be used prior to the rupture of the amniotic membranes which surround the fetus. Here, a beam of ultrasound of frequency 1 or 2 MHz is directed towards the fetal heart. Owing to the Doppler effect the frequency of the ultrasound energy reflected from moving structures, such as the fetal heart valves and heart wall, differs from that of the incident beam. The four main valvular movements (mitral closure, aortic opening and closure, and mitral opening) produce a Doppler frequency shift of approximately 1 kHz. Movements towards the transducer produce an increase in frequency, movements in the opposite direction a decrease. In fetal heart detectors, which are used simply to detect the fetal heartbeat, this directional information is ignored. Similarly, early versions of fetal monitors were also insensitive to the direction of valve movement. The tracings produced tended to display a variability significantly greater than that obtained using the direct FECG method. Recordings can be improved by selecting the signals corresponding to one direction of movement only or by using a signal autocorrelation technique. The variability observed with this latter method is claimed to be very similar to that from direct FECG.

The fetal heart must remain within the sound field during the period of the recording. To ensure this, while allowing normal movement of the fetus, a large volume must be insonated using either a single, wide-angle transducer, or an array of transducer elements. Nonetheless it may still be necessary to adjust the transducer position as labor progresses.

1.3 Phonocardiography

In fetal phonocardiography the sounds produced by the heart are picked up by a sensitive microphone placed on the maternal abdomen. Until recently the difficulties caused by fetal and maternal movements or ambient noise have meant that this technique was of little value compared with ultrasound. However, a redesigned transducer (Talbert et al. 1984) has produced greatly improved results. Since the method has inherent advantages over ultrasound (it is passive, i.e., no energy is transmitted into the patient, and it is less directional), it may well become much more widely used.

2. Fetal Heart-Rate Monitoring

In FHR determinations the heart beat must be detected reliably at a consistent point in each cardiac cycle, for example, the rapid upswing of the R wave of the QRS complex, or the QRS peak. An electronic pulse is generated at the chosen point in the cycle and passed to the tachometer (rate) circuit which provides an output to the recorder proportional to the instantaneous FHR. The quality of the recorded tracing depends to a great extent on the ability of the signal processing circuits to extract and isolate the FHR signal under all conditions. Given a constant amplitude, noise-free signal and a slowly changing rate, this is not difficult, but conditions are rarely so ideal. Filtering is thus required to remove noise, automatic gain control to compensate for variations in signal amplitude, and logic circuitry to reject artefacts. Such artefact rejection can be based, for example, on the likelihood of the FHR not exceeding 240 beats per minute (bpm) and changing by no more than ± 5 bpm from one beat to the next.

As has been mentioned, the FHR is usually in a state of continual change. Normal tracings exhibit two forms of variation: beat-to-beat changes (referred to as short-term variability), and cyclic changes, with a frequency of approximately 5–8 cycles per minute (referred to as long-term variability). Visual estimates of the magnitude of variability usually relate to cyclic changes. A typical classification is as follows: changes of 6–25 bpm are regarded as normal, <5 bpm as minimal or signifying absence of variability, and >25 bpm as increased variability. The variability parameter is important because low variability can be a warning of fetal demise. However, low variability is also associated with fetal prematurity or fetal sleep; it may also arise from maternal disease or from the effect of drugs. Increased variability is usually due to a poor signal-to-noise ratio although occasionally it may indicate cardiac compensation following a reduction in blood flow.

Methods of quantifying long- and short-term variability have been suggested, such as the differential index (DI) for short-term variability and the interval index (II) for long-term variability (Yeh et al. 1973). The DI is the coefficient of variation of the duration of the FECG cycles over a 30 s period. The interval index is the standard deviation of the coefficient of variation of the differences between successive FECG cycles. Neither these nor, indeed, many of other similar indices have gained widespread acceptance. Only one currently available commercial monitor provides a measurement of variability—recording the average of the coefficient of variation of adjacent cycles over the time of 256 cycles, excluding periods of uterine contraction.

3. Uterine Activity Monitoring

Uterine contractions, prior to the rupture of the amniotic membranes or when the cervix is closed by the fetus, cause an increase in intrauterine pressure. This indication of uterine activity is most reliably measured by placing a miniature pressure transducer mounted on the tip of a catheter inside the uterus (Fig. 3a). Somewhat less ideal is an external transducer which can be coupled to the amniotic fluid via a saline-filled catheter (Fig. 3b, Fig. 4). The reliability of this latter technique

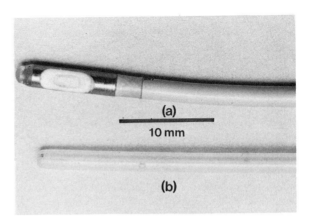

Figure 3
(a) Miniature transducer mounted at tip of catheter for insertion into uterus, and (b) tip of catheter used with pressure transducer in Fig. 4. (Note holes in side of catheter in addition to opening at end)

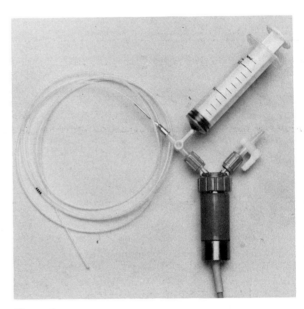

Figure 4
Typical pressure transducer with catheter, taps and syringe for filling and flushing the system with sterile saline

is poor, due to the possibility of the holes at the catheter tip becoming blocked or the catheter kinking; these effects obviously attenuate the pressure sensed by the tranducer. During the antepartum period (or during the intrapartum period if the membranes have not ruptured), uterine activity is detected externally using a tocotransducer (Fig. 5). Unfortunately, this device responds not only to uterine muscle tension but also to the movement of the uterus relative to the abdominal wall

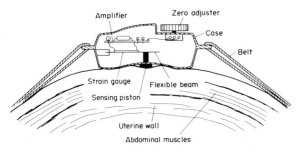

Figure 5
Tocotransducer for detection of uterine activity

and to the tightness of the abdominal belt retaining the transducer. However, the method gives a usable indication of the occurrence and duration of contractions, and being noninvasive it is widely used, even during labor.

The detection of myoelectric signals from the contracting uterine muscle is being investigated as an alternative to pressure measurement but to date a suitably reliable system for routine use has not been reported.

On the standard charts used in fetal monitoring, pressure can be recorded over the range 0–100 mm Hg (13.3 kPa). The basal tone within the abdomen is approximately 15 mm Hg. The recorder is thus usually set approximately at this value to avoid the loss of information during a period of reduced pressure, which could result from a change in the height of the transducer relative to the tip of the fluid-filled catheter. Some instruments incorporate automatic pressure zeroing circuits to keep the recorded pressure at a constant level between contractions. This obviates the chore of continually checking the baseline value of pressure over a period which may last some hours.

Intrauterine pressure transducers, catheter-coupled pressure transducers and tocotransducers all employ strain gauges arranged in the arms of Wheatstone bridge circuits. For pressure measurement these strain gauges are mounted on a flexible diaphragm which is responsive to uterine pressure. In the tocotransducer, the gauges are mounted on a beam which is deflected by movements of a piston in relation to the case (Fig. 5).

To reduce the risk of infection, intrauterine transducers are normally stored in a column of suitable sterilizing fluid (e.g., gluteraldehyde). External pressure transducers are coupled to the uterine cavity via a sterile-saline-filled disposable sterile catheter. It has been suggested that a slow continuous flow of the saline through the catheter towards the uterus helps to prevent the catheter becoming blocked. External tocotransducers do not need to be sterilized but do need careful cleaning to avoid the build-up of deposits which could restrict free movement of the sensing piston.

A number of indices were proposed to summarize uterine activity, such as Alexandria or Montevideo units.

These gave only a poor indication of the total uterine activity and have been replaced by measurement of the area under the pressure–time curve. In the Corometrics 112 monitor the total area above zero pressure is integrated over each ten-minute period. Since changes in baseline pressure can easily occur, particularly in the case of the fluid-filled catheter, and persist for long periods, the resulting readings may be of doubtful value. In the Sonicaid FMR3 the active area (the area above the baseline pressure) is measured and summated for a period of 15 min (Fig. 6). To do this, a reliable method

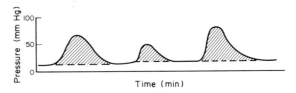

Figure 6
Active area measurement for quantification of uterine activity. The active area (shaded) is summated for a period of 15 min

of detecting the start and finish of each contraction is required. In spite of this and the added difficulty of dealing with hypertonic contractions this procedure may provide the best index of uterine activity. Its use for controlling oxytocin infusion for inducing labor has been advocated (Steer 1977).

4. Telemetry

There has been some criticism of intrapartum monitoring on the grounds that it keeps the mother in bed for long periods of time. Telemetry of fetal heart and uterine activity signals within the area of a normal ward could be used to overcome this problem. Commercially available equipment uses two frequency multiplexed channels on a radio-frequency carrier (150 MHz, <10 mW). A typical transmitter carried by the mother measures about 100 × 50 × 20 mm and weighs approximately 500 g. Usually the signals transmitted are the FECG from a scalp electrode and the uterine pressure obtained via an intrauterine catheter.

The recent improvements in Doppler systems may make it possible to use these with telemetry. Then, if the wearing of two retaining belts, for an ultrasound transducer and for a tocotransducer, were acceptable it would thus be possible to use noninvasive methods of detecting both uterine activity and FHR while allowing the mother complete freedom of movement.

5. Patient Safety

In any technique where electrical connection or contact is made to a patient, safety is a factor of paramount

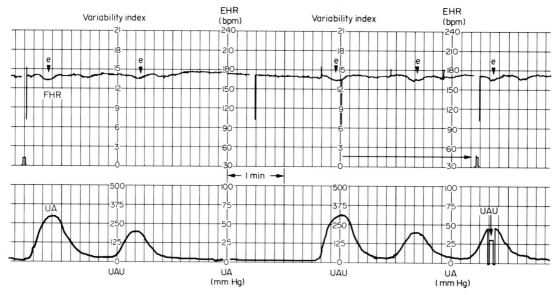

Figure 7
Example of early decelerations of fetal heart rate. This recording from a fetal monitor shows FHR in the upper tracing and uterine pressure or activity (UA) on the lower tracing. The decelerations (marked e) in FHR are of the early type. Note, these coincide with and resemble an inverted version of the uterine contractions which are indicated by the peaks on the UA tracing. This tracing also contains a few small decelerations of the late type. Also shown here is automatic quantitation of FHR variability and of uterine activity in uterine activity units (UAU) which is the area above the baseline integrated over 10 min (courtesy of Dr M J Whittle)

importance. To ensure safety during fetal monitoring, both the heart rate and uterine activity monitoring channels should be electrically isolated. Electrical isolation techniques are well established (see *Medical Electrical Equipment: Safety Aspects*). Transformer circuits operating at a few kilohertz are commonly used to power the transducer and to couple the transducer-developed signals to the the main signal processing system. Electrooptical isolation is used in some equipment. Both methods can ensure that under normal circumstances less than 10 µA RMS flows through the patient to ground, and in any single fault condition, no more than 50 µA.

6. Intrapartum Monitoring

During labor, the baseline FHR (the value of the FHR between contractions) is normally in the range 120–160 bpm. Higher rates (tachycardia) can be caused by fetal hypoxia, by maternal fever, or by the effect of drugs such as atropine; their cause must be diagnosed and corrected. A rate lower than normal (bradycardia) may indicate a dying fetus, but only if preceded by the signs of fetal stress caused by hypoxia, for example baseline tachycardia or decreased short-term variation (variability) of some types of deceleration. Bradycardia

is also observed in cases of heart block and during the second stage of labor, but if a normal level of variability is observed, bradycardia is not regarded as ominous.

Decelerations and accelerations are transient changes in FHR; accelerations are usually benign, but decelerations are potentially ominous. Decelerations can be divided into three types: early, late and variable. In early decelerations the change in FHR "mirrors" and is coincident with the uterine contractions (Fig. 7). This type of change is attributed to the parasympathetic system being affected by compression of the fetal head and is not associated with danger to the fetus. Late decelerations are those which begin after the onset of a contraction (perhaps even 15 s later than the contraction peak) and which continue after the contraction is over (Fig. 8). They indicate reduced blood flow into the uteroplacental system due to increased intrauterine pressure. Sufficient oxygen is available to the fetus from the blood already in the placenta during the period of increasing pressure but if the flow to the placenta is reduced during this phase, the oxygen available towards the end of the phase, and shortly after, is insufficient. Fetal hypoxemia results, causing the fetal heart rate to fall. Unlike decelerations of the early type, late ones cannot be disregarded. Even if small they are always a sign of hypoxic stress and of danger for the fetus. Stronger contractions are very likely to produce greater decelerations (Fig. 8).

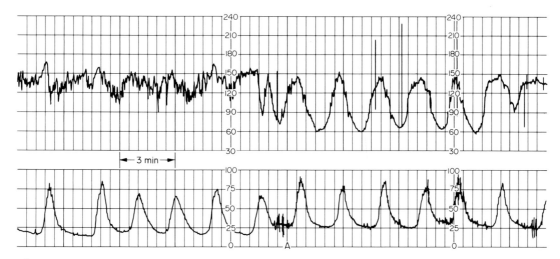

Figure 8
Example of late decelerations of FHR. Note the increase in depth of decelerations with increase in peak uterine pressure after point A. Also apparent is a decrease in FHR variability from a normal level at start of tracing (courtesy of Dr M J Whittle)

The third type, the variable deceleration (Fig. 9), is the most common. Not only are these decelerations exceedingly variable in shape but decreases from, and return to, the baseline rate are always rapid. An acceleration frequently marks the start and finish of the decelerations. Variable decelerations are associated not only with contractions but also with fetal movements, and are caused by compression of the umbilical cord. Most variable decelerations cause no concern. Only when they are severe, dropping to less than 70 bpm for more than 60 s, is the fetus likely to be suffering from hypoxic stress which, if unrelieved, will lead to asphyxia.

If signs of fetal distress are noted alleviating action must be taken. Nonoperative methods of rectifying the condition are first tried and only if these fail is assisted or cesarean delivery considered. Nonoperative methods can be as simple as changing the maternal position, in the case of late decelerations, to correct maternal hypotension. Increased perfusion of the placenta is then very likely. In the case of variable decelerations a change of maternal position may remove pressure from the umbilical cord and allow the blood to flow normally to the fetus. The intravenous infusion of fluids can also help reduce maternal hypotension. Short period adminis-

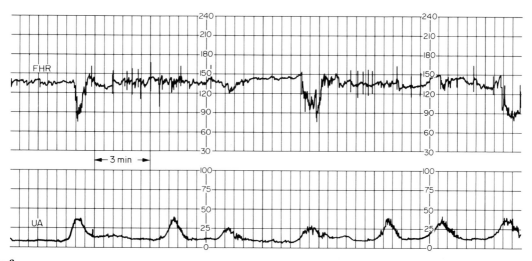

Figure 9
Example of variable deceleration in FHR. Note the rapid fall and rise of deceleration. It is typical that these do not occur with every contraction indicated on uterine activity (UA) recording (courtesy of Dr M J Whittle)

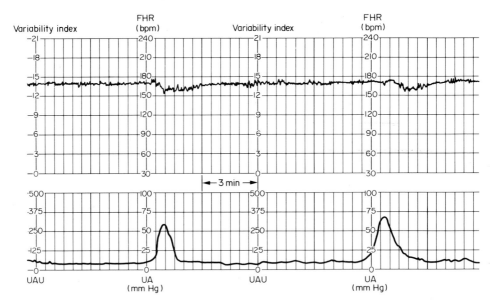

Figure 10
Example of antepartum stress test. Decelerations in FHR in upper tracing commence with onset of contraction (lower trace) and persist for about 90 s after the end of contraction. These decelerations are an ominous sign (courtesy of Dr M J Whittle)

tration of 100% oxygen to the mother increases oxygen transfer to the fetus. Reduction of uterine activity is also worthy of attention, particularly if hypertonic contractions are occurring following the use of oxytocin or other contraction stimulants.

7. Antepartum Monitoring

From about 28 weeks of gestation onwards, the fetus can be monitored with the same noninvasive techniques as used in the intrapartum period. There are two types of test, stress and nonstress, for monitoring the state of the fetus. Both are aimed at distinguishing fetuses that are actually at risk of suffering hypoxic stress or damage from the larger group which on the basis of other clinical indications may possibly be at risk.

In the stress test, also known as the oxytocin challenge test, the form of any decelerations in heart rate which result from a small reduction in placental perfusion when oxytocin is administered are observed. In practice it may be that spontaneous contractions (which occur naturally from time to time, well before term) are seen before administering the oxytocin. The contractions and the fetal heart rate are recorded over a period of 10–20 min. As in intrapartum monitoring, late decelerations (Fig. 10) are a danger sign and their occurrence appears to be a sensitive method of detecting potentially dangerous uteroplacental insufficiency. This method cannot, of course, predict other potentially

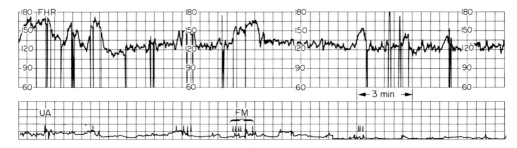

Figure 11
Example of nonstress test. Upper trace indicates FHR detected by ultrasound, and lower trace shows uterine activity and fetal movement. The vertical marks (FM in lower trace) were produced by the mother when she felt fetal movements. Note the accelerations in FHR coinciding with fetal movement—the normal response of a healthy fetus (courtesy of Dr M J Whittle)

dangerous conditions such as occlusion of the cord or abruption of the placenta.

Nonstress testing relies on recording reactions in the form of accelerations of the FHR in response to uterine activity, fetal movements or external stimulation. Such accelerations represent the normal response of the sympathetic division of the autonomic nervous system and are generally regarded as reassuring. In this test the mother is asked to operate a push button whenever she feels a fetal movement and a distinctive mark is recorded on the uterine activity trace. The occurrence of two or more accelerations in response to movements, contractions, or other stimuli in a 20 min period is regarded as an indication that the fetus is in good condition (Fig. 11). It must be remembered, however, that fetal sleep and drugs may prevent the fetus reacting. Thus the test may need to be repeated.

8. Computer Processing

The interpretation of recordings from fetal monitors, or cardiotocographs, depends on visual recognition and assessment of the changes in heart rate and their relationship to uterine activity and other events. Ideally, continuous and consistent attention to the cardiotocograph (CTG) recording is required. For this reason much effort is being expended in developing suitable computer systems to carry out this function. Notable in this field are the working groups within the EEC Project on Perinatal Monitoring (van Geign 1984). Their objectives are, firstly to produce a set of sample recordings with interpretations agreed by a panel of experts, and secondly to write computer software which can reliably reproduce these concensus interpretations. If these goals are achieved, expert interpretative ability will become more widely available.

See also: Ultrasound in Obstetrics; Ultrasound in Medicine; Echocardiography

Bibliography

Cibils L A 1981 *Electronic Feto-Maternal Monitoring, Antepartum/Intrapartum.* Nijhoff, The Hague
Department of Health and Social Security 1984 *Evaluation of Cardiotocographs,* Health Equipment Information No. 133. DHSS, London
Freeman R K, Garite T J 1981 *Fetal Heart Rate Monitoring.* Williams and Wilkins, Baltimore
Journal of Perinatal Medicine 1984 12: 221–84. Symposium on New Methods in Perinatology, May 7–8
Klavan M, Laver A T, Boscola M A 1977 *Clinical Concepts of Fetal Heart Rate Monitoring.* Hewlett–Packard, Waltham, Massachusetts
Pearce J M, Willson K 1983 Fetal heart rate monitors— Some clinical and technical aspects. *Br. J. Hosp. Med.* 30: 123–31
Steer P 1977 The measurement and control of uterine contractions. In: *The Current Status of Fetal Heart Rate Monitoring and Ultrasound in Obstetrics.* Royal College of Obstetricians and Gynaecologists, London
Talbert D G, Dewhurst J, Southall D P 1984 New transducer for detecting fetal heart sounds: Use of compliance matching for maximum sound transfer. *Lancet* i: 426–27
van Geign H P (ed.) 1984 *Perinatal Monitoring,* Report No. PM 84-10-1. Commission of the European Communities, Amsterdam
Yeh S-Y, Forsythe A, Hon E H 1973 Quantification of fetal heart beat-to-beat interval differences. *Obstet. Gynecol.* 41: 355–63

J. E. E. Fleming

Fiber Endoscopy

Endoscopic instruments are devices which utilize light to illuminate and provide useful information about internal organs otherwise inaccessible to direct examination. Fiber endoscopes entered the arena in the late 1950s, prior to which endoscopy was pursued by the use of rigid or semiflexible instruments based on classical lens systems. In gastroenterology the limitations which these older instruments imposed earned endoscopy the somewhat ironic description of "this narrow peeping specialty." Rigid systems still have their place, and under certain circumstances their advantages. This is particularly true, for example, of laparoscopy and cystoscopy. In some of these instruments the rod-lens system has replaced the classical lens system. With improvements in the design of endoscopic instruments it has now become possible to incorporate accessory aids for both diagnosis and therapy.

The biggest single advantage of the fiber endoscope over its rigid or semirigid predecessor is its flexibility. There are few passages in human anatomy that are naturally straight and that can therefore readily and safely accommodate a straight instrument. Fiber endoscopy has therefore considerably increased the diagnostic and therapeutic impact of clinical endoscopy.

Endoscopy has many applications in clinical medicine. It is used for cystoscopy (inspection of the bladder) and associated procedures; in gynecology for the laparoscopic examination of the pelvic organs, and to an increasing extent for laparoscopic methods of treatment; in hepatology and related disciplines where laparoscopic examination of the liver and other abdominal organs is indispensable; in otorhinolaryngology (e.g., for the inspection of palatal and laryngeal movements); in rheumatology and orthopedic surgery for arthroscopy (the inspection of joints), which may be therapeutic as well as diagnostic; and in other smaller clinical fields. Many of these investigations require rigid rather than flexible endoscopes.

There are, however, two areas where the flexible fiber endoscope has become indispensable. The first of these is bronchoscopy: while it has not displaced the rigid orthodox bronchoscope it has enabled the bronchoscopist to enter more deeply into the bronchial tree, to pass an instrument where for technical reasons a rigid bronchoscope was inappropriate or impassable, and to do all this in the conscious cooperative patient. It is, however, in the field of gastroenterology that there has been the greatest growth in fiber endoscopic procedures, and it is with the clinical applications of fiber endoscopy in gastroenterology that this article will be mainly concerned.

Using a fiber endoscope it is relatively easy to visualize directly the organ in question, be it normal or abnormal, and to observe its appearance, shape, size, contents and movements. The requisite samples may then be taken under direct vision for histopathological, cytological and other forms of examination. In addition, accurate still photographic, cinefilm and video-tape records of the examination may be made in full color. These are the diagnostic applications of fiber endoscopy. Increasingly, endoscopy is also used in therapy, e.g., the the removal of ingested foreign bodies, the dilatation of esophageal strictures, the electrosurgical removal of polyps, the treatment of bleeding lesions, and the enlargement of the orifice of the ampulla of Vater (sphincterotomy) for the nonsurgical removal of gallstones. Endoscopy also finds major application in research, and in screening for malignancy.

There is no one all-purpose gastroenterological fiber endoscope, as each type is individually designed to be used in a specific anatomical area, e.g., an esophago-gastro-duodenoscope, an enteroscope (for the small bowel) and a colonoscope. Instruments also vary in diameter to allow for the age and size of the patient, and in basic design, e.g., single channel for routine purposes, and double channel for "operating" endoscopes. Leading makers include the Fuji Photo Optical Company, the Olympus Optical Company, the American Cystoscope Makers, and many others, e.g., the Machida Endoscope Company, and the Asahi Optical Company.

Endoscopes are usually forward viewing though there are indications for fore-oblique viewing (e.g., injection of esophageal varices) and side-viewing (e.g., endoscopic retrograde cholangiopancreatography, ERCP). All instruments share the following characteristics: they have a coherent viewing bundle and an incoherent cold-light light-guide bundle; extreme flexibility; controlled distal-tip maneuverability in four directions; push button air insufflation and suction; a lens-cleaning facility; a channel for biopsy, cytology, measuring and operative devices; the possibility to attach teaching side arms; and facilities for still and cine color photography, video recording and closed-circuit television (see Fig. 1). Unfortunately they also share the characteristics of high capital and maintenance costs. In addition, a brief mention should be made of certain special characteristics, such as two operating channels rather than one in an endoscope, variable sized channels, the presence of a distal bridge or forceps raiser, and constructing the

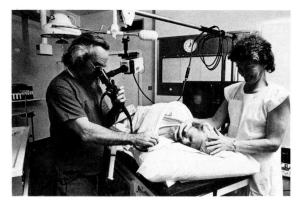

Figure 1
Upper gastroenterological fiber endoscopy. A sedated patient is seen undergoing esophago-gastro-duodenoscopy. Note that the endoscope is linked by a special attachment to a video camera and hence to a visual display unit (VDU) which in turn is linked to a video recorder (out of view). The procedure is being undertaken in a specially designed endoscopy room

endoscope so that it may be safely used for electro-surgical maneuvers.

There are three major anatomical areas covered by gastroenterological fiber endoscopy: the esophagus, stomach and upper duodenum (esophago-gastro-duodenoscopy or OGD), the pancreatic and biliary field (endoscopic retrograde cholangiopancreatography or ERCP), and the colon and rectum (colonoscopy and fiber sigmoidoscopy). The small bowel below the upper duodenum and above the terminal few centimeters of the ileum does not lend itself readily to fiber endoscopy, though enteroscopy is technically possible. As laparoscopy, also known as peritoneoscopy, is performed using the rigid instrument it is outside the scope of this article.

The commonest disorder for which endoscopy is employed diagnostically is dyspepsia. This is usually due to benign gastric or duodenal ulceration, but may be associated with malignancy (e.g., cancer of the stomach) or minor mucosal abnormalities such as superficial erosions or duodenitis. A negative endoscopy performed by a competent experienced endoscopist will rule out the presence of such conditions. However, an abnormality once found can, as already stated, not merely be visualized by endoscopy but samples can be taken, using cytology brushes and biopsy forceps. Such specimens are then referred to the requisite specialist, e.g., a cytologist or a histopathologist, for fuller assessment. Commoner abnormalities in the esophagus will include esophagitis (inflammation of the lining of the esophagus), carcinoma, benign or malignant strictures (areas of pathological narrowing) and motility disorders.

There are many indications for therapeutic rather than purely diagnostic endoscopy in the upper gastrointestinal tract. It is possible, for example, to dilate esophageal strictures using various devices passed through the operating channel of the endoscope, the actual procedure of dilatation being monitored radiologically. Similarly, after dilating a malignant stricture, a plastic tube can be positioned in the length of esophagus involved in an inoperable carcinoma. Varices (varicose veins) of the esophagus can now be injected with sclerosing fluids under endoscopic control in an attempt to prevent their bleeding or rebleeding.

Upper gastrointestinal bleeding is a common medical emergency. The cause of such bleeding must be positively identified before rational treatment—over and above emergency resuscitation—can be instituted. Endoscopy is now used routinely for this purpose. In addition there have been many attempts at therapeutic endoscopy in such bleeders: everything from the endoscopic application of clips and glue to electrocautery and laser beams have been tried. All such methods remain experimental and of uncertain value.

There is another endoscopic procedure which is well established and noncontroversial; namely, the removal of ingested foreign bodies. Smaller foreign bodies can easily be removed from the upper gastrointestinal tract using special graspers and other attachments in association with fiberoptic endoscopes. For larger foreign bodies, such as false teeth or sausages impacted in the esophagus, it is still customary to employ a rigid esophagoscope.

Enteroscopy, i.e., viewing the small bowel below the level of the upper duodenum, is not a routine pursuit. Enteroscopes cannot easily be advanced along the small bowel, and various special maneuvers such as passing them along a previously swallowed line, have been tried. The examination is in the experimental stage.

The bile duct and pancreatic duct join just proximal to their common opening in the upper duodenum, the papilla of Vater, and the investigation of these two systems via the papilla by endoscopic means is now well established. Here, however, the endoscopist relies not so much on what he can see as what the radiologist can visualize. The endoscopist introduces a fine cannula into the papilla, a radiopaque substance is injected via this cannula and, selectively, the biliary and/or pancreatic duct systems are visualized radiographically. As regards the pancreas, there is no other nonoperative way in which the pancreatic duct can be opacified, though there are many other ways in which pancreatic structure and function can be assessed. This method of examination (endoscopic retrograde pancreatography or ERP) is useful in the diagnosis of cancer of the pancreas, pancreatitis (inflammation of the pancreas) and in complications of the latter, e.g., pseudocyst formations. The place of endoscopic retrograde cholangiography (ERC) is less certain now that it is competing with computerized axial tomography, ultrasonography and percutaneous transhepatic cholangiography as well as the time-honored radiological investigations of cholecystography (visualization of the gallbladder) and cholangiography (visualization of the biliary duct system). The important diagnoses here are gallstones and intrinsic or extrinsic carcinoma. It should be added that ERCP (ERC and ERP) is technically more difficult than OGD.

A relatively recent development of ERCP is endoscopic sphincterotomy (papillotomy) where the orifice of the papilla of Vater is electrosurgically incised and enlarged, using diathermy, to enable an impacted gallstone to pass. This procedure is now used extensively in older subjects with recurrent stones when surgery seems contra-indicated. It is also possible endoscopically to place small drainage tubes in a bile duct blocked, for example, by a malignant growth.

The lower bowel or colon is another fertile field for the endoscopist. Though the lowest 25 cm or so may be viewed with a rigid sigmoidoscope, there are certain advantages in using a flexible fiber sigmoidoscope. However, a fully flexible fiber endoscope is inevitably required for the higher reaches of the colon. Total colonoscopy, i.e., viewing the whole colon to the ileocecal valve and indeed including the lowest few centimeters of the ileum, is now commonplace. The indications for diagnostic colonoscopy include the search for a carcinoma and polyps, the assessment of inflammatory bowel disease (e.g., ulcerative colitis and Crohn's disease), and the investigation of lower gastrointestinal bleeding. The chief indication for operative colonoscopy is the snaring and electrosurgical removal of colonic polyps, such polyps or rather certain subgroups of polyps representing a premalignant condition.

In deciding to proceed to endoscopy the gastroenterologist must be certain that this approach is superior to other available diagnostic or therapeutic alternatives. In general it is correct to say that diagnostic endoscopy in both the upper and the lower gastrointestinal tract is more accurate than its major complementary procedure, contrast radiology. Nevertheless, other factors must be taken into consideration: these must include discomfort, risk and cost. Fiber endoscopy is, broadly speaking, more uncomfortable, riskier and more expensive than radiology. The risks are multifarious, the chief of these being perforation. Nevertheless it is on balance increasingly the diagnostic procedure of choice. The same judgement must be made when it comes to therapeutic endoscopy where this must be weighed against other methods such as surgery. Again, endoscopy has become established in certain fields, e.g., polypectomy, where it is easier, cheaper and safer.

Reference has earlier been made to endoscopy as a research tool and its use in screening procedures. It is clear that fiber endoscopy of the gastrointestinal tract, including ERCP, has contributed greatly to basic and clinical research in the field of gastroenterology over the last decade or more. For example, no clinical trial involving the assessment of a substance which it is claimed may have ulcer-healing properties could now be conducted in the absence of regular endoscopic assessments of the patients included in the trial. As to screening, endoscopy is playing an increasingly important role in the routine screening of high risk patients with regard to the development of gastrointestinal cancer. One is reminded, for example, of the use of endoscopy in Japan where cancer of the stomach is unduly common, and the screening of patients with long-standing ulcerative colitis who are at greater risk regarding the development of cancer of the colon.

In conclusion, fiber endoscopy has become an indispensable part of both diagnostic and therapeutic gastroenterology, and to a lesser extent has also achieved a place in other clinical disciplines such as thoracic medicine. It is certain that its importance can only grow as instrument design becomes ever more sophisticated and new possibilities are exploited.

See also: Endoscopes

Bibliography

Bennett J R (ed.) 1981 *Therapeutic Endoscopy and Radiology of the Gut*. Chapman and Hall, London

Cotton P B, Williams C B 1982 *Practical Gastrointestinal Endoscopy* (2nd edn.). Blackwell Scientific, Oxford

Schiller K F R (ed.) 1978 *Clinics in Gastroenterology*, Vol. 7, No. 3: *Endoscopy*. Saunders, London

Schiller K F R, Salmon P R (eds.) 1976 *Modern Topics in Gastrointestinal Endoscopy*. Heinemann Medical, London

K. F. R. Schiller

Fluorimetry

When light interacts with matter it may pass through with little absorption or it may be partly or wholly absorbed, causing an energy transfer and consequent elevation of atoms and molecules in the absorbing material to an excited state. The return to ground state occurs after some 10^{-8} s, leading to a release of energy. This phenomenon is termed fluorescence if the energy is emitted in the form of light. When the absorber is a dilute solution, the amount of light absorbed is very small and the following equation holds:

$$F = [I_0(2.3\epsilon cd)][\phi]$$

where F is the fluorescence intensity, I_0 is the intensity of the exciting light in quanta per second, c is the concentration of the solution, d is the optical depth of the solution, ϵ is the molecular extinction coefficient, and ϕ is the quantum efficiency of fluorescence. Therefore, at low concentration, if the exciting wavelength and intensity are constant, F is proportional to the concentration of the fluorescing compound.

The application of fluorimetric methods of analysis has gradually developed since the publication of a paper on fluorescence by Stokes in 1852, although the fluorescent properties of various plant materials had been noticed before. Most molecules either fluoresce directly or can be made to do so after some form of chemical reaction. This offers great possibilities for recognizing and analyzing compounds of biological origin.

The use of fluorimetry as a method of analysis in biology and medicine has expanded in parallel with the availability of low-cost reliable fluorimeters. Basically, a fluorimeter consists of a light source, a sample cuvette and a detector, usually a photocell or photomultiplier linked to a meter and/or recorder. The excitation source must supply energy in the form of light of appropriate wavelength. Initially light from an arc lamp or other device was used, but today high-pressure gaseous-discharge lamps are generally employed, and the use of lasers is increasing. Mercury arc discharge lamps have been widely used as excitation sources because they are stable, inexpensive and have a long life. In some instances a lamp with a high emission in a specific spectral region is required. For example, a zinc lamp with emission wavelengths in the range 468–481 nm is used for steroid analysis. A filter or monochromator is used between the source and sample to select and isolate light of the desired wavelength.

Sample preparation is important and some purification may be required to remove other fluorescent material. Care must be taken to use solvents which do not themselves fluoresce and do not absorb light or quench fluorescence under the conditions used. With filter fluorimeters, silica test tubes or cells are used as sample holders, provided they are not scratched and consequently do not scatter light. For excitation wavelengths below 320 nm, quartz sample cells are used.

Also important in fluorimetry is the consideration of temperature effects, sources of contamination, standardization of samples, and effects involving interference with light intensity, e.g., from absorption of light in the sample, from incident light, or the fluorescence quenching of emergent light.

Fluorimetry may be used in medicine and biology to locate or identify a particular compound or it may be used to estimate the concentration of a compound in a particular medium.

Quantitative fluorimetric analysis generally involves manual extraction of the compound being studied and its dilution in a suitable solvent to yield a suitable sample in the sample cuvette. The limits of detection vary between instruments but usually lie between 0.1 and 0.001 µg ml^{-1}. Instruments vary from inexpensive single-beam fluorimeters to expensive sensitive instruments which can analyze at low fluorescence levels using very small samples.

Using ultraviolet light as a sample excitation source, it is possible to identify or locate certain compounds in tissues and in solution by microscopic or macroscopic examination. At a microscopic level, elastic tissue or necrotic tissue can be observed under a fluorescent microscope, and vitamin A, riboflavin and porphyrins have been studied this way. When the compound of interest does not fluoresce naturally it is possible, in some cases, to use fluorescent dyes with which the compound binds. The resulting complex may be located fluorimetrically.

The fluorescence of certain dyes has been used to trace blood vessels by injecting acidic fluorescent dye into the blood stream. This could be useful in coronary angiography and in the visualizing of the blood vessels supplying blood to the limbs and other organs. The technique may also be used in oncology, to locate a tumor and indicate its size by tracing the vasculature supplying it with blood.

Fluorimetry is widely used in biology, particularly in the quantitation of compounds in plant and animal tissues and fluids. Nearly all methods of analysis require some form of initial sample purification. Because of the extreme sensitivity of fluorimetry, the purity of the solvents used and the cleanliness of the glassware are critical; even laboratory dust may cause problems. Solvent extraction, paper or thin-layer chromatography, or ion-exchange resins may be used for sample preparation but sometimes no initial purification is needed because of the dilutions used.

The range of compounds analyzed by fluorimetry in clinical laboratories has varied over the years but has included inorganic ions, proteins and peptides, amino acids, catecholamines, porphyrins, steroids, vitamins, drugs and enzymes.

The analytical systems used in biology and medicine in the detection and analysis of these compounds have evolved over the years from the visual location of fluorescent spots on chromatograms viewed under ultraviolet light (see *Chromatography*) to completely automated systems for fluorimetric analysis. There are fluorescent scanners for paper or thin-layer chromatograms and fluorescence detectors for gas–liquid chromatograms, and a new approach combines laser excitation with high-performance liquid chromatography and the use of samples in droplet form to remove problems associated with the use of cuvettes. A recently developed system for separating and analyzing cells on the basis of their fluorescence and light-scattering properties is shown in Fig. 1. Cells that exhibit native fluorescence or that have been tagged by suitable fluorescent probes (biological fluorochromes) are introduced in suspension to the center of a liquid stream and then passed one at a time through a focused laser beam. Each cell is individually characterized by the intensity, color or polarization of the fluorescent light it emits while it is in the beam. When biological fluorochromes are used, excitation by the laser occurs at a selected

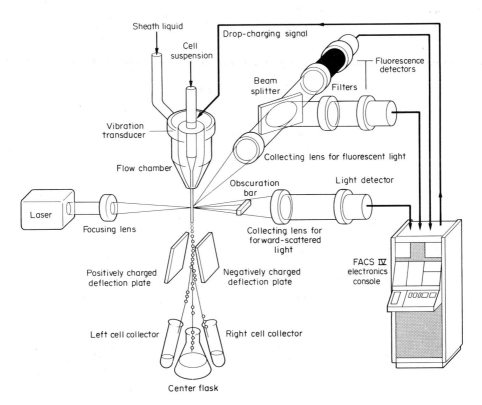

Figure 1
Cell sorting on the basis of fluorescence and light-scattering properties (courtesy of Becton Dickinson Ltd.)

wavelength and fluorescence occurs at longer wavelengths. This spectral shift allows the use of selective optical filters to eliminate unwanted signals, permitting the cell to be identified solely on the basis of its affinity for the fluorescent material used. The method has research applications in many disciplines including cancer immunology, hematology, cell-cycle analysis and quantitative cytology. One particular example of its use is the quantitation of the relative cellular RNA content of control and RNAase-treated HeLa and CHo cells—from the measurement of the red fluorescence from the cells stained with acridine orange.

Another important analytical technique is fluoroimmunoassay, which has some advantages over radioimmunoassay in that there is no radiation hazard involved and therefore no regulations restricting use and handling. Also, the labelled antigen is reasonably stable. The assay is based on the competition between antigen and labelled antigen for antibody combining sites:

$$Ag + Ag^* + Ab = AgAb + Ag^* \, Ab$$

where Ag = antigen, Ag* = labelled antigen and Ab = antibody to Ag. In spite of the drawbacks of fluoroimmunoassay (i.e., susceptibility to quenching, photolability and lack of suitable instrumentation) the technique is developing rapidly.

Analytical methods based on luminescence offer considerable advantages in the clinical laboratory. Fluorimetric, phosphorimetric, chemiluminescent and bioluminescent methods are very sensitive, often highly specific, and are rapid and easily automated. They require only small samples, which is an advantage in some situations. Interference from other compounds is a potential disadvantage.

See also: Clinical Chemistry: Physics and Instrumentation

Bibliography

Bodansky O, Stewart C P 1976 Fluorimetry and phosphometry in clinical chemistry. *Adv. Clin. Chem.* 13: 161–269
O'Donnell C M, Solie T N 1978 Fluorometric and phosphorometric analysis. *Anal. Chem.* 50: 189–205
Stokes G C 1852 On the change of refrangibility of light *Philos. Trans. R. Soc. London, Ser. A* 142: 463–562
Udenfriend S 1969 *Fluorescence Assay in Biology and Medicine*, Vol. 2. Academic Press, New York

J. I. Blair

Forensic Applications of Chromatography

Chromatography is a technique of separation first described by the Russian botanist M S Tswett in 1903. He separated flower pigments into bands of different colors: hence the term "chromatography." Today, the simple procedure has been developed into a wide range of analytical techniques which can be used to resolve complex mixtures of materials with similar properties. The apparatus required ranges from the simple (for paper chromatography) to the very complex (for high-performance liquid chromatography or gas chromatography).

This article describes the various types of chromatography, and their application in forensic science. The use of chromatography in clinical chemistry is described elsewhere (see *Chromatography*).

1. Paper Chromatography

Paper chromatography is the simplest and the cheapest of the separation techniques. A sheet of absorbent paper (blotting paper is adequate) approximately 10 cm × 10 cm is used. A solution of the mixture to be separated is spotted on to the paper by means of a fine glass capillary tube, about 1 cm from one edge. The spot size should be kept as small as possible. Reference standards can then be spotted similarly along a line parallel to the edge of the paper and through the original spot. This edge is then dipped to a depth of a few millimeters into a trough containing the solvent which will effect the separation. The solvent rises up the chromatography paper by capillary action and reaches the top in a few hours. As the solvent passes, the components of the mixture partition themselves between the eluting solvent (the mobile phase) and the water held by hydrogen bonding to the cellulose of the paper (the stationary phase). The rate of advance of any particular component depends on its partition coefficient (i.e., its relative affinity for the two phases) in the system being used. As a result, components move along with the solvent system at different rates thus effecting their separation. When the solvent reaches a predetermined height (e.g., the top edge), the paper is removed from the supply of solvent and dried. The components of the mixture may be visible if they are dyestuffs but more usually they are revealed by spraying the paper with a color-forming reagent. The color of the individual component may be sufficient to give identification but more usually the position on the paper is required. This measurement of position is known as the R_f value and is a measure of the distance travelled by the component divided by the distance travelled by the solvent. This is illustrated in Fig. 1.

The R_f of compound X in the mixture is the same as that of compound X run as a standard; this is also the case for compounds Y and Z. More usually compound X is unknown. By reference to a table of R_f values of compounds obtained by using that particular chromatography system, a number of possible identifications for X is found, say compounds H, I, J, K, L and M. By changing to a solvent system of different polarity and repeating the chromatography, another reference table for the new solvent system gives a number of possible identifications for X, say E, F, G, J, N and O. J has the same R_f value as X in both systems, that is, compound

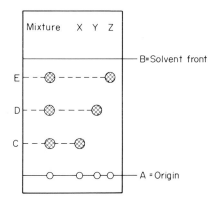

Figure 1
Schematic separation of a three-component mixture by paper chromatography. R_f of component $X = AC/AB$; R_f of component $Y = AD/AB$; and R_f of component $Z = AE/AB$

X may be tentatively identified as J. This is confirmed by chromatographing compounds J and X separately and as a mixture in a third system of different polarity. If the materials do not separate then X may be identified as J.

The simple example described illustrates the use of measurements in paper chromatography to identify an unknown. The technique is one of comparison and its scope is limited by the size of the reference tables. In forensic toxicology the colors produced by spray reagents with drugs give additional information which helps with identification. In the past, paper chromatography was a major technique in the forensic toxicology laboratory but it has now been superseded by thin-layer and other forms of chromatography. Its major disadvantages are slow development times (several hours) and its unsuitability for use with corrosive reagents such as concentrated sulfuric acid which destroys the paper. A full description of the theory and practice of paper chromatography together with lists of solvent systems and tables of R_f values suitable for many assays is available (Smith 1969).

2. Thin-Layer Chromatography

Thin-layer chromatography (TLC) is the technique most widely used in drug analysis. As recently as the early 1960s it was a research technique, largely because there was no suitable kit available to prepare thin-layer plates reproducibly and rapidly. Today, simple spreading equipment and precoated plates are readily available.

Thin-layer plates are prepared by coating a sheet of glass, 20 cm × 20 cm, with a slurry of absorbent in a suitable volume of water to a depth of 0.25 mm. The plate is allowed to dry at room temperature for about 15 min and is then activated by heating to 110 °C for 30 min. The plates are allowed to cool in a desiccator where they can be stored until required. Thereafter the method of use is similar to that of paper chromatography.

Thin-layer chromatography is essentially adsorption chromatography. The sample distributes between a layer of solid adsorbent (usually silica gel), the stationary phase, and the moving liquid phase. Separations are achieved by the movement of the liquid phase over the stationary phase. Normally the solvent is left to run up the plate a distance of about 15 cm. This takes between 20 and 40 min. Thereafter the plates are removed from the solvent and the drug is visualized by the application of color reagents. Highly corrosive reagents based on concentrated sulfuric acid can be used, as well as the type of reagents used in paper chromatography.

In forensic toxicology, use is made of the colors produced with sulfuric-acid-based reagents. Sulfuric acid produces pink colors with phenothiazine-type tranquilizers. Marquis's reagent, 1 part formaldehyde to 100 parts concentrated sulfuric acid, produces purple colors with opiate drugs. Mandelin's reagent, 1 g ammonium vanadate in 100 g of concentrated sulfuric acid, produces a play of colors from blue to purple to brick red to orange with strychnine. Full tabulations of these color reactions together with R_f values for drugs in two solvent systems are available in the literature (Rentoul and Smith 1973).

3. Column Chromatography

Historically, column chromatography is the earliest form of chromatography. Here, the absorbent is poured into a glass tube fitted with a plug of glass wool at the bottom to form a column. The mixture or crude extract is placed at the top of the column and the eluting solvent is carefully added. Now if we picture the thin-layer chromatogram upside down, the components that run furthest on the layer for the same solvent system elute first from the column. For colored materials, the colored bands are easy to collect. For drugs, fractions are collected and the eluate from the column is divided into samples as it comes through. These fractions are evaporated down and the residue weighed to find the components of the mixture.

Column chromatography is not really applicable in modern toxicology where one is dealing with microgram levels of drugs. It does find application, however, as a purification step in the identification of anabolic steroids in athletes' urine samples. A special application employs XAD2—an ion-exchange resin. Short columns packed with beads of this resin are available commercially and are used for the direct extraction of drugs from urine samples. A buffered urine sample is poured directly through a column of the resin and the drugs eluted with an organic solvent. The eluate is evaporated and concentrated extract analyzed by thin-layer chromatography.

4. High-Pressure Liquid Chromatography

High-pressure liquid chromatography or high-performance liquid chromatography is a modern form of column chromatography. A short column is packed under pressure with very fine particle size absorbent. A high-pressure pump forces the eluting solvent through the column at a constant flow rate of the order of 1–5 ml min^{-1}. The sample, placed on the top of the column by injection or by using a sample loop, is eluted through the column by the moving solvent. As it passes through the column, it is separated into its individual components which emerge sequentially from the end of the columns. As they emerge, they are "seen" by a sensitive detector, usually an ultraviolet spectrometer which is set to monitor a suitable wavelength. The signal from the spectrometer is amplified and displayed on a strip chart recorder. A typical tracing is illustrated in Fig. 2.

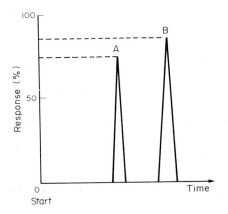

Figure 2
Schematic separation of a two-component mixture by high-performance liquid chromatography

The time after injection for individual components to emerge is measured. This time or a time relative to a chromatographed standard is compared to a list of retention times for drugs chromatographed on the same system. This information is used to identify the drug as before. In addition, the peak height, or more accurately, the peak area is in direct proportion to the amount of drug in the specimen. The quantity of drug in the sample is measured by comparing the response with standards of known concentration.

Only within the last decade has high-performance liquid chromatography found its place in forensic laboratories. Although, as a technique, it still receives a great deal of attention from the research worker, it does have some drawbacks. The equipment and prepacked columns are relatively expensive. It lacks, too, the sensitivity of gas liquid chromatography. Sample preparation, however, is kept to an absolute minimum. To measure a drug, one may only need to precipitate proteins in a serum sample and inject aliquots of the

supernatant. In contrast with gas liquid chromatography (see below), many compounds can be chromatographed directly without time-consuming derivative formation. The detection system is nondestructive. As a result, the effluent from the detector can be collected and further tests such as radioimmunoassay or mass spectrometry can be carried out on the purified extract obtained.

5. Gas Chromatography and Gas–Liquid Chromatography

Gas chromatography, although postulated as early as 1941, was first used in analytical chemistry in 1952. It has since developed and is, today, the most useful and versatile analytical technique used in forensic science. Analyses are rapid, sensitive and easily made quantitative with surprisingly small amounts of sample.

As with other forms of chromatography, a mixture is transported over a stationary phase, usually a wax or grease which has been coated on to a fine-mesh, inert granular powder. This liquid coating, which has low volatility at the operating temperature, is chosen to suit the separation required. The coated powder is packed into a glass tube, 1–2 m long, coiled to suit the geometry of the oven of the instrument in use. The moving phase, a gas, usually nitrogen or helium, is continually passed through the column and out into a detection device. The oven of the instrument containing the column is maintained at a temperature between room temperature and 300 °C depending on the analysis required. For some, more complex, analyses, various settings of temperature may be programmed.

The sample is injected on to the start of the column, and because of the high temperature it is vaporized. The carrier gas tries to force it through the column; the liquid phase tries to hold it back. The rate of progress of a particular component depends a great deal on its partition coefficient between the phases, the selective retardation effecting the separation. The components emerge sequentially from the column and are swept into a detection (essentially ionization detection) device. The passage of a component through the device produces a change in an electrical signal which is amplified and displayed on a strip chart recorder. A typical separation is similar to that shown for HPLC (Fig. 2). The measurements OA and OB represent the time that the individual components have been retained on the column. This retention time can be used, with the aid of tables of collected reference data for the particular column at the settings used, to identify tentatively the unknown. A column with different polarity will produce another retention time which can be used with a similar set of tables to confirm the identity. In forensic science, this type of application of gas chromatography permits the identification of drugs obtained in street seizures, drugs in body fluids, solvents in blood samples taken from glue sniffers, paraffin and other accelerants in residues from arson cases, and the identification of paint samples.

Quantitation of a component in a mixture is achieved by measuring the peak size or area. This is directly related to the amount of material "seen" by the detector. The area of the peak can be measured and compared with peak areas resulting from the chromatographing of known amounts of the compound. In practice, the response of the sample of unknown amount is measured electronically against a known amount of a reference compound added to the mixture. This is then compared with the relative response of a known amount of sample which has been chromatographed with the same amount of added reference compound. This procedure is in daily use in forensic laboratories to detect, identify and measure components of mixtures. It is used to identify and measure alcohol in samples of blood and urine taken in relation to the Road Traffic Act. The blood or urine sample taken from the driver is carefully packaged and transported to the laboratory. There, the packaging and seals on the containers are carefully examined to ensure that the specimen has not been tampered with. The details of the specimen are documented and only then is it analyzed. A measured amount of the specimen is diluted with a known amount of *n*-propanol dissolved in water. An aliquot of this mixture is injected into the gas chromatograph. This produces two peaks, one for ethanol, and the other for *n*-propanol. Both are measured electronically and the ethanol content is calculated by comparing the relative response for the ethanol to the *n*-propanol in the driver's specimen to the relative response for ethanol to *n*-propanol in an identically treated solution of a weighed amount of ethanol in water. The dilution of the driver's specimen is then repeated and an aliquot is injected into a second gas chromatograph equipped with a different column. Again the relative response is compared with that of an identically treated standard solution of ethanol in water. The retention times from both systems confirm the identity of the ethanol in the specimen. The measurements of the alcohol in the specimen should agree within 2%.

6. Concluding Remarks

Without the techniques of chromatography, the forensic scientist would be severely handicapped in routine daily tasks. The separation of closely related drugs and metabolites would be almost impossible without it. The major drawback of the techniques, when used for identification purposes, is that they are techniques of comparison. All chromatography systems are as good as the reference tables of R_f values or retention times. Identification is achieved by careful measurement and observation, using two or more systems of different characteristics. For paper or thin-layer chromatography, identical R_f values and identical color reactions are essential; slight variations indicate that further investigation is necessary.

Some fundamental information can be obtained to assist with the identification of the complete unknown. Specific chemical reactions aimed at functional groups on molecules will produce characteristic colors or form derivatives which will alter the R_f values or retention times. For the identification of the complete unknown, chromatography is best used as a purification device. The purified substance obtained from thin-layer or paper chromatography, or from the effluent of the high-pressure chromatograph can be subjected to other techniques available to the analytical chemist such as infrared spectrometry, nuclear magnetic resonance spectrometry or mass spectrometry. Work on the interfacing of high-performance liquid chromatographs with mass spectrometers is progressing but the combined technique is still very much a research tool. The mass spectrometer coupled directly to a gas chromatograph, replacing the detector of the gas chromatograph, is perhaps the most powerful and sensitive combination of analytical instruments available to the forensic scientist. The unknown substance as it emerges from the gas chromatograph can be analyzed almost instantaneously by the mass spectrometer. When coupled to a computer the cracking pattern obtained can be compared in a few seconds with a large library of material. If the material is completely unknown then its mass spectrum can be used to identify fragments of the molecule which can be built up to identify the unknown.

See also: Chromatography; Forensic Applications of Spectroscopy

Bibliography

Cunliffe F, Piazza P B 1980 *Criminalistics and Scientific Investigation.* Prentice-Hall, Englewood Cliffs, New Jersey
Davies G (ed.) 1975 *Forensic Science.* American Chemical Society, Washington, DC
Kind S, Overman M 1972 *Science Against Crime.* Aldus Books, London
Rentoul E, Smith H 1973 *Glaister's Medical Jurisprudence and Toxicology*, 13th edn. Churchill Livingstone, Edinburgh
Smith I (ed.) 1969 *Chromatographic and Electrophoretic Techniques*, Vol. 1: *Chromatography.* Heinemann, London
Walls H J 1974 *Forensic Science: An Introduction to Scientific Crime Detection*, 2nd edn. Sweet and Maxwell, London

J. S. Oliver

Forensic Applications of Spectroscopy

The terms spectroscopy or spectrometry cover a wide range of procedures used in forensic toxicology. The toxicologist employs colorimetry, ultraviolet and visible spectroscopy, mass spectrometry and occasionally infrared spectroscopy to search for drugs and common poisons in body materials. Use is also made of atomic absorption, a spectrophotometric technique for the detection and measurement of metallic poisons. γ-Energy spectroscopy in the context of neutron activation analysis is employed for the detection, identification and measurement of some elements present in trace amounts. The technique of radioimmunoassay which is used to detect and measure low levels of drugs in small biological samples involves a form of spectroscopy, the radiation detector and associated pulse height analyzer being analogous to a colorimeter. Less frequently used in the field of toxicology than in forensic science is the technique of x-ray fluorescence spectroscopy. Using an x-ray fluorescence spectrometer coupled to a minicomputer one can, for example, identify counterfeit coins in seconds.

1. Absorption Spectroscopy

Spectroscopy is a technique that makes use of the dispersal of a band of radiation into its separate energies or wavelengths. (Consider for example, the simple dispersal of visible light by raindrops to form the familiar rainbow.) Colored objects absorb some of the component wavelengths of an incident beam of light and transmit or reflect the others. A dye which appears red absorbs the yellow and blue colors of light; a dye which appears yellow absorbs red and blue. By holding a dissolved dye up to light its color can be seen. It can also be seen instrumentally by sequentially shining the individual colors or wavelengths of visible light through the dye solution and instrumentally measuring the intensity or strength of the transmitted light, thus giving a measurement of the amount of light absorbed. A calculation of absorbance plotted against wavelength produces an absorption spectrum for the dyestuff. If a red dye is examined, there will be peaks of absorbance in the blue and yellow regions of the spectrum. The addition of more dye will not change the color, only its intensity; that is, the absorption spectrum will have the same overall shape although the peaks of absorbance will be more pronounced. The overall shape of the spectrum can be used to identify the dystuff; the intensity of the absorption can be used to measure the quantity present. This absorption of light is governed by the Beer–Lambert law, expressed mathematically as

$$\log_{10}(I_0/I) = Kcd$$

where I_0 is the intensity of the initial light beam, I is the measured intensity having passed through the solution, K is the molar extinction coefficient, c is the concentration of the solution used in moles per liter, d is the thickness in centimeters of the solution traversed by the light beam, and $\log_{10}(I_0/I)$ is the absorbance or optical density.

In practice, since K is a constant for any one compound, and fixed path length cells are used, a linear calibration curve may be obtained for a plot of absorbance against concentration. Absorbance need not now

be calculated. Modern instruments sample incident and transmitted light automatically and give either a reading or a plot of absorbance, with absorbance on the ordinate scale and wavelength on the abscissa.

2. Ultraviolet Spectroscopy

Ultraviolet spectroscopy is perhaps the most important form of optical analysis used in forensic toxicology. Instruments which scan over the wavelength range 190 nm to at least 350 nm are employed in daily routine. Quantum energies in the ultraviolet and visible part of the spectrum are sufficient to raise electrons—in suitable chromophoric groups of the molecule of interest—from the ground state to excited or energized states. Particular atoms or groups of atoms forming the molecule determine the energy necessary for those transitions and hence the wavelength of the light absorbed. Ultraviolet spectroscopy is used following the initial extraction of a drug from biological material into a suitable organic solvent. The drug can then be extracted into dilute acid or alkali and placed in the measuring cell of the instrument. With, for example, barbiturates, a change in the pH of the solution from 14 to 10.5 causes a characteristic shift in the absorption curve and helps to identify the drug. For drugs extractable from alkaline solutions, a scan taken in acid can assist identification. Careful note is made of the wavelength of maximum absorbance (λ_{max}), also shoulders and inflexions on the absorbance curve, and the position of minimum absorbance. This derived information may, with care, be used in conjunction with tables of data published in suitable reference books (Rentoul and Smith 1973, Clarke 1969, 1975) to attempt to identify the unknown substance. Such identification is not absolute since closely related drugs and metabolites may not be differentiated thereby. However, the procedure can be used both as a guide and to estimate drug concentration. The technique is nondestructive so the drug solution may be processed for further analysis by high-performance liquid chromatography, the λ_{max} indicating a suitable monitoring wavelength, or by gas chromatography or gas chromatography–mass spectrometry, where the estimate of the drug concentration may be used to determine dilution factors to avoid overloading the equipment and hence wasting time and sample in needless repeats of assays.

3. Infrared Spectroscopy

Instruments used for infrared spectroscopy are similar to those used for ultraviolet and visible spectroscopy. Longer wavelength energy sources are employed which cause an increase in the vibration, stretching and bending of bonds between atoms in a molecule. The spectrum obtained is much more detailed than that obtained in ultraviolet spectroscopy. While ultraviolet spectroscopy will enable one to distinguish an absorbing species in the presence of coextractable but weakly absorbing materials, a high sample purity is required for infrared spectroscopy. Infrared spectroscopy is of particular use in the examination of fibers, oils, paints, or drugs seized in the street, where the absorbance spectrum provides either absolute identification or a fingerprint for comparison with a reference source found, say, at the scene of a crime or in the possession of a suspect. The forensic toxicologist, however, is obliged to purify perhaps a microgram of drug in the presence of large amounts of coextractable material. Whilst such purification is possible, best use of the available sample is made if analysis is performed by mass spectrometry.

4. Mass Spectrometry

Mass spectrometry is not a new technique. However, the coupling of mass spectrometers to gas chromatographs has led to its more widespread use in forensic science laboratories. This combination of instruments can be used for the rapid identification of any compound that can be chromatographed, particularly if a library of compounds is available for comparison—via a minicomputer coupled with the instruments.

The mass spectrometer has a source (a means of ionizing a sample), an analyzer and a detector. The sample has access to the source through a direct insertion probe for solid and liquid samples, a leak inlet from a reservoir for gases and calibrants. One or two capillary tube inlets carry the effluent from the gas chromatography column. Once in the source the sample receives an electrical charge, either positive or negative, by being exposed, say, to an electron beam. (Although negative ion mass spectrometry is becoming more common, most instruments are equipped for positive-ion operation.) Molecules gain energy in the process which leads to their reproducible fragmentation into positively charged particles. Since the source is held at a positive potential, these ions are repelled into the analyzer.

The analyzer separates the ions according to their mass-to-charge ratios (in practice according to their mass) by the action of a magnet or a quadrupole mass filter. A charged particle entering a magnetic field is deflected. Heavier particles are deflected less than light ones with the same charge. As a result a stream of particles entering the field will fan out according to their masses, the lighter particles being deflected more than the heavier ones. A detector positioned to capture the lightest ion can be made to "see" the other particles if they can be deflected into it. This is achieved by increasing the strength of the magnet by increasing the power supply to it. As an ion impinges on the detector a generated electrical signal is amplified and recorded on the light sensitive paper of an oscillographic recorder. A series of peaks corresponding to the masses and intensities of the fragments of the unknown constitute the mass spectrum of the sample. Alternatively, the detector signal can be recorded for processing and display using a dedicated computer-based data system.

The quadrupole mass filter operates by causing the ions that leave the source to oscillate in a radiofrequency field. By electronically sweeping through a range of frequencies, ions of different mass to charge ratios can be passed sequentially into the detector and a mass spectrum obtained as before.

The mass spectrum reveals the cracking or breakdown pattern of the compound of interest. By consulting library files, either manually or, more rapidly, by using a computer, an identification of the unknown is achieved. If no identification results, then detailed examination of the spectrum allows possible identification of molecular weight and fragmentation pattern from which one can tentatively construct a possible molecular structure for the unknown. By obtaining a sample with this molecular structure and determining its mass spectrum for comparison, the identity of the unknown can be refuted or confirmed.

The mass spectrometer is used to identify plastics, fibers, paints, drugs obtained in street seizures, and less easily identified drugs and poisons in biological specimens. A specialized role is the monitoring of a characteristic fragment of a drug, the mass spectrometer being used as an extremely sensitive and selective detector for gas chromatography. This technique, mass fragmentography, can be used to detect and measure drugs at very low concentrations in very small samples. The detection limit is typically a few picograms.

An important forensic use of the mass spectrometer is the elucidation of reaction pathways used in illicit laboratories to manufacture drugs of abuse. As well as identifying intermediate compounds in the synthesis, the identification of trace amounts of these in seized drugs can verify the synthetic route chosen and possibly identify the manufacturing laboratory.

The inorganic mass spectrometer can be used to identify trace elements in small amounts of sample for either identification or comparative purposes. It can assist in identifying fragments of glass.

5. Atomic Absorption Spectroscopy

Atomic absorption spectroscopy makes use of the ultraviolet and visible parts of the spectrum. It is used in the rapid determination of minute amounts of metals in biological specimens. In the vapor phase, atoms can have a ground or low-energy state or an excited or high-energy state. If exposed to radiation of appropriate energy, ground state atoms may make a transition to the excited state. Radiation of the correct wavelength passing through vaporized metal causes excitations in proportion to the number of atoms present and hence is absorbed in proportion to the number of atoms present, giving a basis for measurement. Detection and identification rely on the discrete absorbance wavelengths of each metal.

In practical terms, the elements must be removed from the biological matrix. This can be accomplished by digestion techniques using hot concentrated acids or alkalis. The residues from these procedures can then be aspirated directly into the flame of an atomic absorption spectrometer. The flame has a temperature of approximately 2500 °C, sufficient to produce free atoms in the gaseous state. A hollow cathode lamp is used to emit the discrete wavelength required to excite the element sought for, and the absorbance of the light passed through the flame is measured. The use of standard solutions of known concentrations of the element allows the plotting of a calibration curve. This may be done automatically by modern instruments. The absorbance for the unknown is converted into a concentration reading either automatically or by manual comparison with the calibration curve. A small electric furnace may be used instead of a flame to vaporize the elements in the light beam.

Atomic absorption spectroscopy is rapid, specific and very sensitive. Its major use in forensic toxicology is the detection, identification and measurement of metallic poisons. In forensic science in general, it is used to determine barium and antimony in gunshot residues on hands and clothing, trace element impurities in street drugs with a view to tracing a common source, and for comparison of soil samples from the scene of a crime to those found on a suspect's shoe. Although multielement lamps are indeed available, the main disadvantage of atomic absorption spectroscopy is the need to replace lamps between analyses of different elements. This changing of lamps does consume both time and sample, although sample requirements are small.

6. Colorimetry and Absorptiometry

Colorimetry makes the most practical use of the visible region of the spectrum (see *Colorimetry*). If spot tests that produce colors for poisons are available, these same tests can generally be adapted to produce colored solutions that can be used to measure the quantity of poison present. Perhaps the best example is the blue color produced by an alkaline solution of sodium dithionite in the presence of paraquat ions. This spot test for use in urine and stomach contents can be used directly on deproteinized plasma or homogenized tissue samples. The absorbance of the blue color produced can be compared directly with the absorbance of similarly treated paraquat solutions of known concentrations to determine the concentration in the biological specimen. Similar tests exist for salicylates, paracetamol, phenols, some metals and toxic anions.

Absorption spectrometry in the visible region has long been used by biochemists to determine the rates of reaction of enzymes. By careful temperature control of the sample cell and by monitoring optical density change at timed intervals, the reaction kinetics can be determined. This technique has been adapted and is the basis

of the enzyme multiplied immunoassay technique (EMIT) for the detection and measurement of drugs. An antibody which will bind to the drug is produced by injecting a drug protein complex into a host animal. (The animal's immune system reacts to this foreign protein and forms antibodies which bind to it.) After a suitable interval, say six months, a serum sample taken from the host animal contains these antibodies. They bind to both the foreign protein and the drug. To use this as the basis of an assay, it is necessary to label a suitable enzyme by attaching to it the drug of interest in such a way that the enzyme can function normally, but is inhibited if the antibody subsequently attaches to the drug. The addition of free drug, drug enzyme complex and drug antibody to a solution gives rise to competition within the solution. Both the free drug and the drug bound to the enzyme compete for the antibody. The more drug present, the fewer enzyme drug molecules that are deactivated by the antibody, that is more enzyme is available for the reaction with a suitable substrate. In the system chosen, lysozyme acting on bacterial cell wall suspension (*M. luteus*), the optical density of the substrate is decreased by the action of the enzyme and at a rate proportional to the amount of the free or active enzyme. This rate of change in optical density can be measured and related to the concentration of drug present. Kits are available which will permit the screening of urine samples for the common drugs of abuse within a matter of minutes. The technique is rapid and in the outlined form consumes only 50 µl of urine. Similar assays using a different enzyme system are available for the measurement of many other drugs in serum samples.

Since cross reactions can and do occur, the technique is of use to the forensic toxicologist as a presumptive test and must be followed up with rigorous analysis before evidence can be given in court.

7. Neutron Activation Analysis

Neutron activation analysis is a very sensitive technique for the detection and measurement of an element within a sample (see *Neutron Activation Analysis*). The technique is so sensitive that it is possible to determine the arsenic content of 1 mm length of a single hair accurately. This can be likened to finding a needle in a haystack since arsenic typically amounts to less than 0.5 ppm of a hair.

The technique is based upon the ability of the nucleus of an atom to pick up a neutron when bombarded by such particles in a nuclear reactor. The reaction of interest is

$$\text{nucleus} + \text{neutron} \rightarrow \text{excited nucleus}$$

The excited nucleus is deexcited by emitting a particle, a proton or neutron, or emitting γ rays. If the resulting nucleus is radioactive it can be identified from the decay pattern of the radiation which it emits. This is characteristic of a particular element and forms the basis of qualitative analysis. Quantitative analysis is achieved by comparing the number of disintegrations per second from the sample with the number of disintegrations per second from a known quantity of the pure element (called the standard) irradiated in the reactor alongside the sample. Before the rate of disintegration can be measured for the element contained within the sample, it is necessary to separate the radionuclide of interest from all other radioactive nuclei which could otherwise interfere with the analysis. The pure radionuclide can be isolated from the sample in one of two ways. It can either be chemically removed from the sample or, if it is a γ-ray emitter, it can be sorted according to the energy of its emission instrumentally using a γ-ray spectrometer.

In the first case, conventional chemical methods are used to extract the element from the sample after dissolving it in a suitable solvent enriched with a known amount of the inactive element of interest added to the radioactive mixture to increase its bulk. The recovery of the added material may be used to correct any losses during the analysis. Microchemical techniques are thus generally avoided and reagents need not be specially purified. A clean-up procedure then follows to eliminate all unwanted radioactive species which may still be clinging to the radionuclide of interest. The rate of disintegration of the pure radionuclide is then determined using simple equipment such as a Geiger–Müller tube for β emitters and a sodium iodide detector for γ emitters.

In the second case, the sample is taken from the reactor to a solid state detector coupled to an electronic sorting device which separates the γ rays from the different radioactive species according to their energy—a γ-ray spectrometer. This energy is characteristic of the particular radionuclide, and once separate, the γ rays of interest can be counted. This latter procedure is of particular importance in forensic science. It is nondestructive and allows the performance of a multi-element analysis.

Neutron activation analysis has similar uses to atomic absorption spectroscopy in forensic science. Its major advantage over the latter is the potential for the preservation of a sample for subsequent production in court. The major disadvantages are the possible radioactive hazard and the need to transport specimens to and from the reactor.

8. Radioimmunoassay

Analogous to the visible spectrometer as a detector for the enzyme multiplied immunoassay technique, the γ spectrometer has a place in both the clinical and forensic laboratory as a detector for the radioimmunoassay procedure for the detection and measurement of drugs in serum, plasma and urine. As with EMIT, antibodies to the drug in question must be raised. Instead of labelling an enzyme with the drug of interest, the drug

itself is labelled with a suitable radioactive isotope. Most frequently used isotopes are tritium, a β emitter, and ^{125}I, a γ emitter. Superficially, the assay is similar to the EMIT assay. To a sample containing an unknown amount of drug is added a known amount of labelled drug. Antibody added will combine with both labelled and unlabelled drugs in proportion to the relative amount of each present. Addition of a suitable adsorbent will bind the free drug, both radioactive and nonactive, that has not been bound to the antibody. Following centrifuging to remove the adsorbent with the free drug, the radioactivity of the remaining solution is measured. The greater the concentration of drug in the sample, the less the radioactive drug bound to the antibody. Hence a lower level of radioactivity is measured. By comparing the activity with similarly treated standard solutions of drug of known concentrations, a measurement of the drug level can be obtained.

Although this technique is rapid and has exceptionally high sensitivity, it is not specific for a particular drug. Cross reactions or interference can occur with similar drugs, especially within the same group of drugs. This is acceptable clinically where the technique is used to monitor known drugs that have been given to a patient, but is useful only as a presumptive test to the forensic scientist. Further confirmatory analysis must be performed. Where no other suitably sensitive techniques exist as, for example, in the detection of LSD in plasma samples, the technique can be used as a detector for high-performance liquid chromatography, where it can be used on aliquots of the column effluent. The results obtained can be used to construct a chromatogram from which both measurement and identification can be obtained.

See also: Clinical Chemistry: Physics and Instrumentation; Radioimmunoassay; Photon Activation Analysis; Forensic Applications of Chromatography

Bibliography

Clarke E G C 1969 *Isolation and Identification of Drugs*, Vol. 1. Pharmaceutical Press, London
Clarke E G C 1975 *Isolation and Identification of Drugs*, Vol. 2. Pharmaceutical Press, London
Cuncliffe F, Piazza P B 1980 *Criminalistics and Scientific Investigation*. Prentice-Hall, Englewood Cliffs, New Jersey
Davies G 1975 *Forensic Science, Symposium of the 168th Meeting of the American Chemical Society*. American Chemical Society, Washington, DC
Kind S, Overman M 1972 *Science Against Crime*. Aldus Books, London
Rentoul E, Smith H 1973 *Glaister's Medical Jurisprudence and Toxicology*, 13th edn. Churchill Livingstone, Edinburgh
Walls H J 1974 *Forensic Science: An Introduction to Scientific Crime Detection*, 2nd edn. Sweet and Maxwell, London

J. S. Oliver

G

Gait Analysis

Scientists and clinicians have long had an interest in the cyclical movements of the body during walking. The researcher may need information relating to the design of assistive devices or replacements for the skeletal structure; the clinician may use gait and its variation as a measure of the treatment applied to a patient; physiotherapists, prosthetists and orthotists may assess the gait of a patient with a view to deciding an appropriate treatment modality. For all of these individuals, gait analysis means significantly different things and various techniques may be used to assess the phenomenon. Gait analysis is usually performed in a laboratory or clinical situation in which the patient walks in a straight line on a level surface at more or less uniform speed. Thus, the measurements and assessments may not correspond closely to the behavior of the patient at large.

Progression occurs by controlled alteration in the angulation between segments of the limbs. Although the fundamental movement is forwards, there are important related vertical, mediolateral and rotational movements which need to be taken into account.

In the normal individual, passive angular movements at the joints can be measured moderately easily on an examination table. Typical values are quoted in Table 1

Table 1
Approximate ranges of angular movement in the healthy normal human

Angular movement	Hip	Knee	Ankle
Flexion	113°	134°	48° (plantar flexion)
Extension	28°	10° (hyperextension)	30° (dorsiflexion)
Abduction	48°		18° (eversion)
Adduction	31°		33° (inversion)
Internal rotation	45°		
External rotation	45°		

for the hip, knee and ankle. In level walking, the range of movement involved is considerably less than the free motion so measured; Fig. 1 shows classical data from the work of the research team at the University of California in the late 1940s and early 1950s. These data relate to a normal healthy individual. Even with considerable restriction of these ranges of movement the disabled person may be able to move with moderate freedom. It

should be noted that the angles are measured as the projections of the limb segments onto appropriate orthogonal planes and do not correspond to "Euler" angles (that is, to parameters that specify the orientation of a rigid body relative to a fixed, three-dimensional axis system). This latter system of representation would be necessary if complex analytical analysis were to be undertaken to calculate the true angular velocities and accelerations. With regard to accelerations, it is of interest to note that in fast walking the horizontal component of linear acceleration of the ankle in the direction of progression may be of the order of 4*g*.

Maintenance of the mean body position during activity while the legs adopt their various configurations requires the development of force in appropriate muscular or ligamentous structures. If the masses of all body

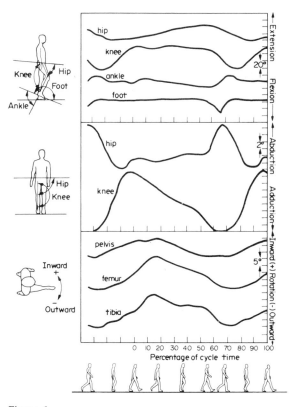

Figure 1
Variation of the angulation of the segments of the leg during the walking cycle. Time is expressed as a percentage of the time for one complete cycle commencing at zero with heel contact (after Eberhart et al. 1947)

segments are known, and their accelerations in specific instants in time can be determined, one may infer the loading transmitted between segments of the body during the statically determinate phases corresponding to support on one foot or on the other. During normal walking, however, there are change-over phases when both feet are in contact with the ground. The duration of these phases amounts approximately to a total of 25% of the time of each walking cycle at $1.5 \, \mathrm{m \, s^{-1}}$ mean walking speed. Similarly, at these change-over phases the loads in the leg segments tend to be near their maximum values, and the inverse dynamics problem of determining the loads becomes impossible. Experimental measurement of the separate loads transmitted to each foot is frequently undertaken. For each foot, there are six unknown quantities corresponding to the three components of force and the moments about a set of reference axes.

The components of force developed between ground and foot during a single ground contact are generally reported in terms of the direction of progression, mediolateral direction and vertical direction. Figure 2 shows the typical variation of these components expressed as a percentage of the time for one complete cycle. One

might expect the vertical component to have a constant value of 100% of body weight, except when the load is changing over from the left to the right foot and vice versa. The strong cyclical effects superimposed are due to the vertical acceleration of the trunk and limbs during the walking cycle. Similarly, during walking on a level surface with negligible wind resistance, the anterior–posterior component of force between ground and foot would be expected to be zero. In fact the body undertakes cyclical accelerations and decelerations on top of the mean velocity of progress which correspond to some extent to the shape of ground–foot force recorded for this anterior–posterior component. There is, however, a definite overlap between the curves for the two legs when forces on the left foot and right foot are being exerted in opposite directions. The advantage of this situation is shown in Fig. 3; note that in early and

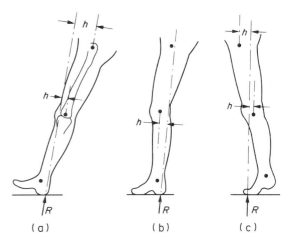

Figure 3
Three views of a leg from the side in walking, showing the appropriate direction of the resultant ground-to-foot force R relative to the anatomical structures. The lever arm of the force relative to the knee and hip joints is indicated by h

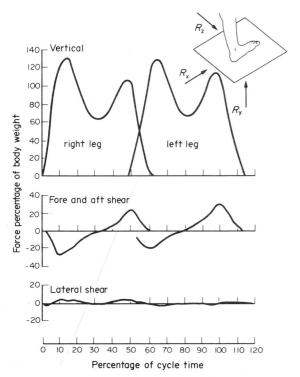

Figure 2
Variation with time of the forces developed between ground and foot in the vertical, fore and aft, and mediolateral directions. Positive directions for each component of force acting by the ground on the leg are shown in the sketch

late stance the ground–foot force has a line of action closer to the axes of the knee and hip joint than would be the case if its direction were vertical. Similarly, the mediolateral component of force might be expected to be zero for straight-line, forwards walking, but the side-to-side displacements involve accelerations reflected in this component force. By themselves, these lines of force are not of very great significance but taken in conjunction with the limb configuration they allow the resulting loading transmitted between limb segments to be determined immediately. The anterior–posterior view of the leg during gait is shown in Fig. 4, together with the direction of the vector for ground–foot force in the vertical plane. In most analyses for the joints of the leg during walking, the forces and couples due to gravity

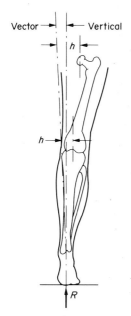

Figure 4
View of a leg from the rear during the stance phase
of walking, with the ground reaction force *R* shown
by position and line of action. The offset of the line
of action from the center of the hip and knee joint is
shown by the dimension *h*

and acceleration can be taken to contribute less than
10% to the overall values. The resultant forces transmitted between segments can be taken to be the ground
forces. The moments are therefore the ground force
multiplied by the appropriate lever arm exemplified by
h in the figure. In Fig. 3, the corresponding information
is shown with regard to the plane of progression at three
instants of time during the support phase for one leg.

Information with regard to the moments transmitted
between leg segments is usually reported relative to the
ground reference axes on the assumption that the test
subject is walking along one of the reference axes and
that his ankle and knee joints are parallel to the ground
system. This is an approximation as can be seen from
Fig. 1, which shows the rotations about the vertical axis
at the various positions. The variation of these moments
with time during the cycle of level walking is shown in
Fig. 5 for normal level walking. In each case the leg is
divided into three parts so that the "free body diagrams"
may be used to define positive moment actions. The load
actions transmitted between body segments generally
correspond to the development of tension in ligamentous
or muscle structures. If the ankle is considered in the
situation shown in Fig. 3a and Fig. 4, there will obviously be very little tension in the muscles controlling
inversion or eversion. However, in the early part of the
stance phase, the position of the line of action resultant
force posterior to the axis of ankle joint will require

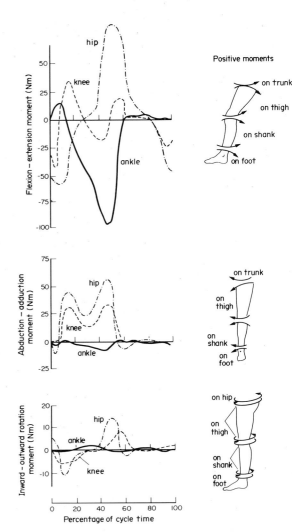

Figure 5
Variation with time of the moments transmitted
between the segments of the leg during the walking
cycle

tension in the anterior tibial muscles. A corresponding
situation is shown in Fig. 3b. After the heel has risen
from the ground (Fig. 3c), the force vector must move
forward under the foot. In this situation, the large lever
arm of the external force relative to the ankle joint must
produce considerable tension in the achilles tendon and,
consequently, in the gastrocnemius and soleus muscles.

For the knee joint, Fig. 4 shows that the line of action
of the resultant force in the frontal view passes medial
to the joint center. Across the knee joint this combination of longitudinal load and bending couple may be
transmitted as shown in Fig. 6a as a varying pressure
across the medial and lateral surfaces of the joint. For
greater eccentricity of the external load action, however,
the situation as shown in Fig. 6b may arise, whereby all

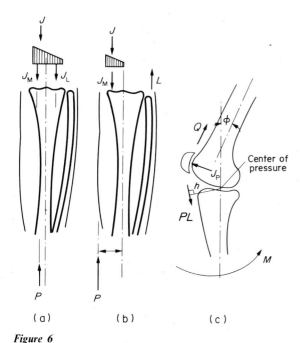

force tends to flex the knee against the resistance of the patella tendon and against the tension in the quadriceps muscles (Fig. 6c). The knee joint axis in flexion–extension is not fixed and the effective instantaneous center of rotation varies with the angle of flexion. However, if a normal knee is viewed using x rays at a series of flexion angles, the point of closest approach betwen femur and tibia is seen to remain effectively stationary on a line drawn along the internal margin of the posterior cortex of the tibia. It is a reasonable assumption that the line of action of the joint force passes through an axis defined in this way; the calculation of the force in the patellar ligament is straightforward.

The force transmitted at the hip joint is calculated from the analysis of the abductor muscle force in one-legged standing (Fig. 7). The joint force may have values between 1.9 and 2.5 times body weight depending on the angle of tilt of the pelvis. In walking, the abduction moments at the hip are greater than in standing; flexion and extension load actions are transmitted as shown in the Fig. 3 and Fig. 5. The hip joint

Figure 6
(a) Posterior view of the leg showing the distribution of pressure at the knee joint, where the resultant load from below P lies near the center line of the joint. It is then reasonable to assume that the joint pressure J varies in some way and, as indicated, is greater in the medial compartment than in the lateral compartment for the obliquity shown. (b) Loading at the knee joint corresponding to offset of the resultant force P so that its line of action is outside the confines of the joint capsule. The tension L is the ligamental force. (c) Conventional lateral view of the flexed knee, showing the component forces transmitted by the patellar ligament (PL) the quadriceps muscle (Q) and the patello-femoral joint (J_p)

of the joint loading is concentrated in the medial compartment. Additionally, tension in structures lateral to the knee joint may be required. These structures may be either the lateral collateral ligament or the fascia lata. Considering loading in the plane of progression as shown in Fig. 3a and Fig. 3c (early and late in stance) the knee moment is positive indicating that the external loading is tending to hyperextend the shank of the thigh: thus tension is required in structures at the back of the knee joint. If the leg is in full extension then the load may be taken by the posterior fibers of the joint capsule, but if, as is more usual, the leg is slightly flexed, the loading will be developed either by the gastrocnemius muscle or by the muscles inserted onto the hamstrings. The gastrocnemius muscle acting at the ankle carries the extending moment at the knee during late stance. Similarly the hamstring muscles, because of their extending action at the hip, carry the knee extending moment in early stance. At the mid-stance phase, exemplified by Fig. 3b, ground

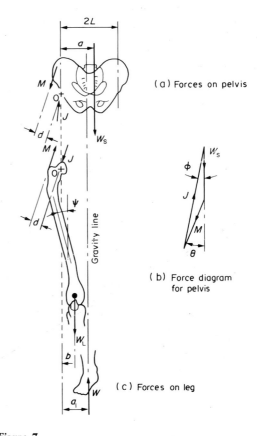

Figure 7
Analysis of the loading of the hip joint during one-legged standing (after McLeish and Charnley 1970)

is basically spherical and the angular position of the femur may be maintained by a minimum of six muscles presuming that throughout its free range of movement the position is controlled by muscles which can exert tension only. In fact, 21 muscles act at the hip joint. Dynamic equilibrium equations, therefore, do not suffice for the calculation of the different loads in each muscle and hence the joint loads. Various methods have been suggested for calculating the forces taken by individual muscles. The parameters utilized include the following:

(a) muscle force equal to a uniform stress × physiological cross-sectional area;

(b) minimization of physiological energy cost;

(c) minimization of muscular fatigue;

(d) simplification of the system by treating muscles in groups corresponding to their anatomical position and phasic EMG, calibration of EMG signals obtained in controlled isometric situations; and

(e) calibration of EMG signals obtained in controlled isometric situations.

Generally, for the major muscles, methods based on these different parameters, give similar values for load transmission, although at times they predict unreasonable values for forces in small muscles. The values for hip joint force and knee joint force thus calculated when patients are walking in a straight line on a level surface in laboratory conditions are shown in Fig. 8.

In mechanical terms the direction of the force transmitted to a structure is of as much significance as its magnitude and this is equally true with respect to the stressing of body structures and the stressing of implants which may be introduced to replace or support body structures. The directions of hip joint forces relative to the long axis of the femur at their maximum values (Fig. 8) are shown in Fig. 9 both in frontal view and in the plane of progression.

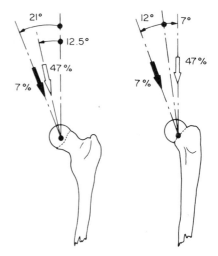

Figure 9
Frontal and lateral views of the head of femur, showing the directions of the resultant joint force relative to the line joining the center of the head to the center of the femoral condyles, at the instants of maximum force shown in Fig. 8

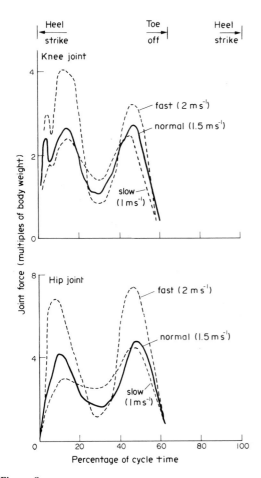

Figure 8
Variation of knee and hip joint forces with time during the walking cycle at various speeds

The data available for the biomechanics of the leg correspond for the most part to the activities of standing and locomotion. These activities develop the load conditions that are most relevant to muscular loading, to the design of orthopedic implants, and the usage of artificial limbs. For the upper extremity there is no comparable group of activities so frequently undertaken and corresponding to such well-defined load values for which the biomechanics can be evaluated. With regard to the wrist, hand and fingers, there are a few activities (typing, piano playing, mixing dough, operating levers such as on the handlebars of wheeled vehicles) in which load is developed and movement occurs simultaneously, corresponding to work output or input. In most everyday activities, however, the hand and fingers are utilized to grasp an article so that it may be maneuvered, using the muscles of the upper arm and shoulder. For the upper extremity as for the lower extremity, if any analysis of loading is to be undertaken, it is important that the exact conditions at the interface between the body and the external component be known. For instance, there have been many analyses of the loading in the joints of the finger based on the assumption that external loading is in the form of a normal pressure over a specific area of the finger pad. Investigations have shown, however, that substantial frictional forces and couples may be developed in almost any activity. Such frictional loadings have significant effects on the resultant load actions at the segmental interfaces.

The standardized configuration for the analyses of finger joint loading is the two-finger pinch. Here the loads at the proximal interphalangeal and metacarpophalangeal joints are 54 N and 102 N, respectively, in maximal isometric activity. The corresponding values of loading at these joints in turning a faucet having four radial arms has been calculated to be 52 N and 170 N, respectively.

There is little comprehensive data on the loading of the wrist, elbow and shoulder joints during athletic activities. In clinical biomechanics the maximal loadings are likely to be found in patients with diseases such as rheumatoid arthritis, who, because of disability in a leg, use their arms to assist major translations of the body, as in turning in bed or rising from a chair. Figure 10 shows the variation with time of the three component moments at the wrist, the elbow, and the shoulder when a normal individual undertakes a movement, simulating rising from a chair, in which the hips and knees are maintained in extension while body weight is lifted from the arms of the chair by the combined action of both arms. For the elbow, these moments imply a force on the joint of 1500 N, and on the humero-scapular joint a maximum force of 1800 N. Both of these values refer to tests involving a subject of 60 kg body mass. There has as yet been no analysis of wrist joint loading which can offer comparable information.

There is considerable interest in the mechanics of the lumbar spine due to the widespread incidence of low back pain, but the analysis of loading of the spine is complicated by the number of structures able to carry load and by uncertainty with regard to the actual loads to be transmitted. There is little information on the mass properties of the portion of the trunk above the sections of interest, which may be the major loading element in forward leaning. With regard to the anatomical structures, it is obvious that compression is developed at the lumbar spinal joints corresponding to the applied load and to the tension in the posterior spinal musculature. In forward bending, the external load and the load of the supported trunk exert a bending moment, which may also be partially transmitted by compression in the abdominal cavity. Weightlifters certainly increase this pressure by means of a tight belt and by means of

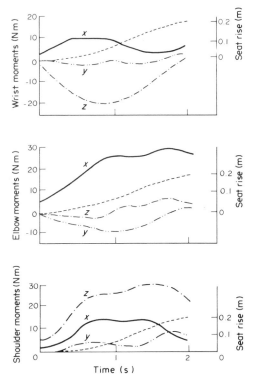

Figure 10
Variation with time of the moments transmitted at the wrist, elbow and shoulder during a maneuver in which the test subject rose from a seated position with the use of the arms pressing against arm rests. The dotted curve shows the displacement of the trunk during the procedure. In each case the y direction is along the axis of the forearm or upper arm, the z direction corresponds to an axis parallel to the elbow flexion axis and the x direction is perpendicular to the other two

controlled respiration. Reported analyses of the force developed during weightlifting suggest an approximate figure of 400 N for compression at the lumbar spine. Pressure values as high as 3 MPa have been measured in the spine during certain lifting activities.

In their normal function, the joints are loaded in a dynamic fashion while undertaking movement of one joint surface relative to another. In these circumstances, the calculation of the pressures developed must take account of the complicated pressure–viscosity relationships for synovial fluid, and the local compliance of the cartilagenous surfaces of the bones.

To date, there has been no exact measurement or calculation of the pressure gradients to be expected in normal activities, nor of the maximum pressures developed. It is possible, however, to calculate nominal average pressures taking into account the calculated loads and the known anatomical areas of the articulations. While this may be a reasonably accurate procedure for the hip joint, which is spherical with close congruence between the two parts, in the case of the knee, which is bicompartmental with no congruence between the femur and the tibia, such calculations could be less than accurate. Reported values for pressure at some of the major joints are: 1.4 MPa at the hip, 1.9 MPa at the knee, 3 MPa at the elbow, 4.3 MPa at the interphalangeal joint of the index finger and 2.9 MPa at the metacarpophalangeal joint of the index finger, the test subject being either in maximum voluntary loading or in rapid movement. It is interesting to note that these pressure values lie within a comparatively small range. This is not unexpected if it is supposed that the material at each joint has the same basic load carrying capacity.

See also: Biomechanics; Artificial Limbs and Locomotor Aids: Evaluation; Artificial Joints, Implanted; Bone: Mechanical Properties

Bibliography

Eberhart H D, Inman V T, Saunders J B de C 1947 *Fundamental Studies of Human Locomotion and Other Information Relating to Design of Artificial Limbs.* University of California Press, Berkeley, California
Grieve D W, Miller D, Mitchelson D, Paul J P, Smith A J 1975 *Techniques for the Analysis of Human Movement.* Lepus, London
McLeish R D, Charnley J 1970 Abduction forces in the one-legged stance. *J. Biomech.* 3: 191–200
Steindler A 1935 *Mechanics of Normal and Pathological Locomotion in Man.* Thomas, Springfield, Illinois
Winter D 1979 *Biomechanics of Human Movement.* Wiley, New York

J. P. Paul

Gamma-Ray Detectors

Emitters of γ radiation are used for many medical and nonmedical purposes. One important use is in medical diagnosis, in which the activity of a radionuclide administered to a patient is measured. γ Emitters are also used in *in vitro* studies, particularly in radioimmunoassay (see *Radioimmunoassay*). Although very low activities are often used, large sources may be employed, for example, in the treatment of cancer or for the radiography of welded joints.

Whatever the purpose, it is necessary to detect and measure the radiation emitted, whether for safety reasons or for quantitation of activity. This article is mainly concerned with the measurement of radioactivity from γ emitters for medical tests and for research. Even within this limitation, the activities to be measured cover an extremely wide range. The activity administered for organ imaging in a patient may be 500 MBq or more,

whereas for a blood-volume measurement it may be 0.2 MBq and in the blood sample taken for the latter measurement, the activity may be 10^3 times smaller.

No one instrument can be used to measure this wide range of activities. For the highest activity samples, ionization chambers are often employed, whereas for the lowest activities, scintillation counters are the norm. For intermediate levels, Geiger–Müller counters are suitable but are used largely for radiation protection purposes. Semiconductor detection systems play a smaller specialized role.

1. Ionization Chambers

A special form of ionization chamber known as a reentrant chamber is commonly employed to determine activity by measuring the ionization produced in air or other gas by the γ rays emitted by the radioactive material. A cross section of a typical instrument is shown in Fig. 1. A potential of 200–300 V is applied between

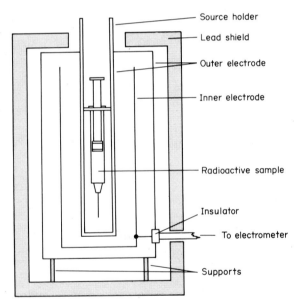

Figure 1
Reentrant ionization chamber

the inner and outer electrodes and the resulting ionization current is fed to a high-impedance electrometer. The first ionization chambers contained air in equilibrium with the atmosphere, and for accurate work, corrections were necessary to take into account variations in temperature and pressure. Modern instruments are often operated at a few atmospheres of pressure with a resultant increase in sensitivity and avoiding the need for the correction factor. A digital read-out display is often provided. The range of activity measured depends on the isotope, and may extend from 1 kBq to over 100 GBq. Although provision is made for adjusting the read-out according to the radionuclide decay scheme and energies, considerable inaccuracies may arise with low-energy emitters such as ^{125}I ($\sim 30 \, keV$) because of sample self absorption and consequent geometry-dependent effects. However, the reentrant volume is, in general, made large so that different volumes and shapes may be accommodated. The ionization chamber thus provides a means of measuring the radioactivity of γ emitters which is both reliable and relatively insensitive to geometric factors.

2. Geiger–Müller Counters

Geiger–Müller counters operate on a similar principle to that of the ionization chamber but are filled with an inert gas. A higher voltage is applied across the electrodes which usually take the form of a hollow cylinder and a central wire. When a single ionizing event occurs, an ionization avalanche is generated in the tube with considerable amplification of charge (around 10^6) allowing the event to be detected. The magnitude of the output pulse is independent of the photon energy of the γ radiation and hence no energy spectral analysis of the radiation is possible. Because of the relatively long time that the positive ions take to migrate to the negative electrode, the pulse-pair resolving time may be as great as a few hundred microseconds. This sets a limit on the maximum count rate of the instrument. Geiger–Müller tubes are inherently insensitive to γ radiation but this is partially overcome by having the wall or outer electrode made of lead or steel. Ion-producing interactions occur in the wall of the tube rather than in the gas, and the tube is triggered by the secondary electrons thus set in motion. This type of counter is used almost entirely for radiation safety checks, e.g., for monitoring benches for contamination or for checking radiation intensities around γ sources.

3. Scintillation Detectors

Scintillation counting, which involves measuring the light generated in scintillator material when it absorbs radiation, is the most widely used method of measuring the activities of γ-emitting radionuclides. A solid scintillator is the most convenient form, with the γ emitter at some distance from it—unlike the system for measuring β emitters in which the isotope and scintillator form an intimate mixture (see *Beta-Particle Detection*). The activity of a β–γ emitter is assessed by measuring its γ component. Only if the γ ray is of high energy or emitted in a small percentage of disintegrations is the increased sensitivity afforded by the β emission likely to be utilized.

Several types of solid scintillator have been used and each has advantages and disadvantages. An ideal scintillator should be an efficient absorber of γ rays, and thus its atomic number and density should be high in order to convert the γ-ray energy efficiently to light. The material must of course be transparent to light. Its physical and mechanical properties should enable the material to be easily worked and used in the intended environment. It should be possible to produce the scintillator in large sizes at reasonable cost.

Pioneer workers measured α-particle emission by counting the number of scintillations produced in zinc sulfide but the method, besides being extremely tedious, proved to be unsuitable for γ-ray measurements. In 1947 it was discovered that naphthalene could be used as detector material in conjunction with photomultipliers, which produce an electric pulse proportional to the light emitted by the material. Crystalline anthracene also proved to be an effective scintillator. Inorganic scintillators such as sodium iodide and organic plastic scintillators as well as liquid scintillators have since been discovered.

Plastic scintillators are currently available with widely varying properties and in extremely large volumes, for example, in relatively inexpensive 1 m diameter cylinders or 3.5 m long sheets. This has led to their use where large detectors are necessary, for instance in whole body counters (see *Radionuclides: Whole-Body Monitors*). Their light output is low, however (around 60% of that from anthracene). Their relative density (about 1) and effective atomic number are also low thus they absorb γ rays poorly—mainly by Compton interactions. This predominance of Compton absorption results in a very poor energy resolution (typically 23% half width at half maximum height for ^{137}Cs) making identification between different radionuclides virtually impossible. Plastic scintillators have the advantage of being easily machined and having a very short scintillation pulse width and decay time, of the order of 1–2 ns.

Thallium-activated sodium iodide is the most widely employed scintillator material although more expensive and not available in such large sizes as plastic scintillators. Relative density (about 3.7) is markedly greater. Furthermore, effective atomic number is much higher and thus radiation absorption is due to the photoelectric absorption process to a much greater extent. The energy resolution is considerably better (typically 8% full width at half maximum height for ^{137}Cs), permitting the separation and identification of the various radiations from many radionuclides. Light output is relatively high (230% of that of anthracene) and this is particularly important in reducing statistical variation in the measurement of low-energy radiation.

The decay of the light pulses takes longer in sodium iodide than in plastic (around 200 ns) but this is not a major disadvantage. Sodium iodide is hygroscopic, however, and must therefore be encapsulated, usually in aluminum, although for low energies beryllium may be used. The scintillations exit through a sealed transparent window. Sodium iodide may be machined into many shapes and sizes despite its affinity for water. Crystals may be as large as 50 cm in diameter by 1 cm thick for use in gamma cameras (see *Radionuclide Imaging*). In detectors, the crystals are less disk-like, the largest (around 15 cm diameter and 10 cm thick) usually being encountered in whole-body counters. Cylindrical crystals, 5 × 5 cm, are often employed for diagnostic *in vivo* investigations. Although sensitivity increases with the area of detector exposed to the radiation, considerations of practicalities and expense rule out the use of larger detectors. Sensitivity also varies with thickness of detector, and a thickness of 5 cm is chosen as a compromise. Although many commonly used γ rays have energies in the range 100–150 keV and are virtually totally absorbed by a crystal only 1 cm thick, high energies may require a crystal thickness of more than 5 cm for total absorption. To limit the radiation detected to that from the area of interest, a lead collimator is attached to the lead shielding which surrounds the scintillator, reducing the background radiation level. The cross section of such a detector with its associated photomultiplier is shown in Fig. 2.

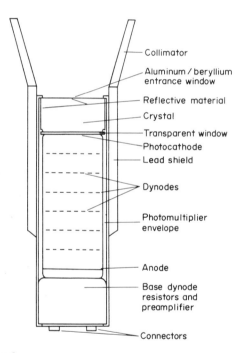

Figure 2
Scintillation crystal with photomultiplier

For measuring sample activity, a "well" detector is employed in which a hole (for housing the sample) is machined, usually axially, to about three-quarters of the total depth of the scintillator crystal. By this means, almost 4π geometry is obtained and the limit of detectable activity is markedly lowered. One of the chief applications of the well detector is in radioimmunoassay in which large numbers of samples are counted. Automatic sample changers are manufactured for this type of

application which deal with up to a thousand samples. A more recent approach has been the introduction of multicrystal counters which typically consist of a dozen well detectors which measure twelve samples simultaneously. This permits a rapid counting of samples but the practical difficulty of shielding against interference from neighboring samples usually limits the technique to the counting of low energy emitters.

Whether the detector is for *in vivo* or *in vitro* use, the same electronic circuitry is used. The light from the detector crystal, which is coated with a reflecting material, impinges on the photocathode of a photomultiplier tube, releasing electrons which are accelerated to the first dynode. Each electron then releases several electrons at this dynode, which are accelerated to the second dynode. This process is repeated at each dynode, giving a total amplification of current of 10^8 or more. The electrical pulse thus generated for each scintillation is then further amplified with convention electronic amplifiers. The magnitude of each pulse is proportional to the energy deposited in the crystal during the original γ-ray interaction and it is therefore possible to perform spectral analysis (pulse-height analysis) at this stage. Pulse-height analysis at its simplest is used to maximize the sample count in relation to the background count or, for *in vivo* systems, to eliminate low energy pulses arising from scattering outside the detector. For some purposes, three or four channels of analysis are used simultaneously. For others, multichannel systems (typically 1000 channels) are employed, particularly for the identification of radionuclides.

After pulse-height analysis the pulses may be fed to a ratemeter and chart recorder, to a scaler controlled by a timer, or to a digital ratemeter and printer or paper-tape punch. In the latest equipment a microcomputer may be incorporated in the apparatus to analyze the data in almost any desired format.

4. Semiconductor Detectors

Scintillation counting exhibits two disadvantages: the bulkiness of the detection equipment due principally to the associated photomultiplier tubes, and the poor energy resolution obtained.

The semiconductor detector behaves much like an ionization chamber, but because solids are about 10^3 times denser than gases, its volume can be considerably smaller. Furthermore, because the ionization produced for a given incident energy is greater, the electrical pulses are large enough to be detected individually with good statistical accuracy. Semiconductor materials commonly used are silicon and germanium but, because both elements have relatively low atomic numbers, the absorption of γ radiation is not as high as is desirable. Owing to difficulties in producing high-purity materials it is usually necessary to introduce another element, lithium, by a special technique known as drifting. This introduction has the effect of cancelling out the impurities and reducing the leakage current measured when no radiation is present. The two commonly used materials are hence designated Si(Li) and Ge(Li).

More recently the production of high-purity germanium has become possible. This may prove particularly useful as, of the two elements, it has the higher atomic number and density.

Semiconductor-based systems have considerably better energy resolution than scintillation counters (Fig. 3)

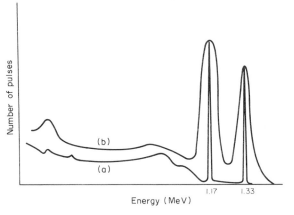

Figure 3
Resolving power of (a) a semiconductor detector (8 mm depletion layer germanium) and (b) a scintillation detector (10×15 cm NaI(Tl)) for ^{60}Co spectra

and are exclusively used to identify or measure particular γ emissions; for example, in neutron activation studies, and x-ray fluorescent analysis or imaging. The detector can be quite small, but for optimal performance it must be cooled to liquid nitrogen temperatures. The associated cryostat thus nullifies the advantage of size. In addition, accidental warming of the system can severely damage the semiconductor. In comparison to scintillation crystals, semiconductor detectors of moderate size are very expensive and are not available at all in the largest sizes.

Cadmium telluride shows promise as a detector for use at ambient temperatures. However, its energy resolution is little better than scintillation systems and at present it is obtainable only in small volumes.

See also: Radioactivity Measurement; Radiation Quantities: Measurement; Radiation Protection and Personnel Monitoring; Radiation Dosimetry

Bibliography

Birks J B 1964 *The Theory and Practice of Scintillation Counting*. Pergamon, Oxford

Debertin K, Mann W B (eds.) 1981 *Gamma and X-ray Spectrometry Techniques and Applications*. Pergamon, Oxford

Dyer A 1974 *An Introduction to Liquid Scintillation Counting*. Heyden, London

Hine G J 1967 *Instrumentation in Nuclear Medicine*. Academic Press, New York

Hoffer P B, Beck R N, Gottschalk A (eds.) 1971 *The Role of Semiconductor Detectors in the Future of Nuclear Medicine*. Society of Nuclear Medicine, New York

Hospital Physicists' Association 1980 *An Introduction to Automatic Radioactive Sample Counters*. Hospital Physicists' Association, London

Mann W B, Ayres R L, Garfinkel S B 1980 *Radioactivity and its Measurement*, 2nd edn. Pergamon, Oxford

Mann W B, Lowenthal G C, Taylor J G V (eds.) 1983 *Applied Radionuclide Metrology*. Pergamon, Oxford

Nicholson P W 1974 *Nuclear Electronics*. Wiley, London

K. G. Leach

Genetic Code

All forms of life are primarily dependent on proteins, both enzymic and structural, for the control of life-related processes. An understanding of the mechanism by which organisms store and express the information necessary for the production of this diverse range of proteins has been the most important advance in biology during the past few decades. It has been shown that proteins, each of which consists of a string of amino acids, are represented in the hereditary material (nucleic acid) by a series of nucleotide bases. Each amino acid is coded for by a specific combination of three of these bases, and together these nucleotide triplets form what is known as the genetic code.

The work of Mendel in the last century initiated the concept that inheritance is carried by particulate factors (genes), and in higher organisms these genes were later identified as being closely associated with the chromosomes. The latter are composed of proteins and nucleic acid. At first it was thought that only the proteins could exhibit sufficient variability to satisfy the requirement of being genetic material, but subsequent studies with microorganisms demonstrated that it was the nucleic acid that carried the hereditary information. For example, it was shown that nucleic acid from dead bacteria of a virulent *Pneumococcus* strain could transform a harmless mutant into a virulent form capable of killing mice. Furthermore, investigations of the transfer of radiolabelled components of T2 bacteriophage showed that the nucleic acid, not the protein component, entered the host bacterium prior to viral replication.

It is now known that most organisms use deoxyribonucleic acid (DNA) as their hereditary material, with a few small viruses utilizing ribonucleic acid (RNA). DNA consists of two polynucleotide chains wound around one another in the form of a double helix. Each chain is composed of a backbone of covalently linked deoxyribose–phosphate groups with nucleotide bases pointing inwards. DNA usually contains four types of base, guanine (G), cytosine (C), adenine (A) and thymine (T), and it has been shown that a guanine on one chain of the double helix is always paired (primarily by hydrogen bonding) to a cytosine on the other chain; similarly adenine and thymine are always paired together.

The other nucleic acid, RNA, is based on a ribose–phosphate backbone bearing the nucleotide bases guanine, cytosine, adenine and, instead of thymine, uracil (U). Unlike DNA, which is primarily located in the nucleus, these molecules are found throughout the cells of higher organisms. RNA is also usually single stranded, and it is found as one of three main types, namely messenger RNA (mRNA), transfer RNA (tRNA) and ribosomal RNA (rRNA).

1. Expression of the Gene as Protein

Full details of the mechanism by which the information contained in a gene is expressed in the production of a protein can be found in Lewin (1974). An outline of the main events is shown in Fig. 1. The section of the DNA

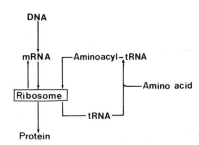

Figure 1
Sequence of events in the production of a protein from information stored in the DNA

comprising the gene in question first unwinds and allows an enzyme, RNA polymerase, to synthesize a complementary copy (mRNA) of one of the DNA strands. This mRNA molecule is formed by every DNA guanine being transcribed as a mRNA cytosine, and so on, so that a DNA sequence of GCAT is transcribed into mRNA as CGUA. (Note that adenine is transcribed as uracil and not thymine.) The mRNA molecule diffuses out of the nucleus to the cytoplasm where it attaches to a ribosome and acts as a template for protein synthesis. The ribosome consists of two subunits which are formed from rRNA and a small group of ribosomal-specific proteins and its function is to allow a series of tRNA molecules (each of which carries a specific amino acid) to bind to the mRNA. Each triplet of bases (codon) of the mRNA is recognized by a complementary triplet of bases (anticodon) on one of the tRNA species. For

example, the mRNA triplet UUU will bind a tRNA molecule with the AAA anticodon (which carries the amino acid phenylalanine). The amino acids are removed one by one from the tRNAs and bound to the growing amino acid chain. As the ribosome moves along the mRNA, more amino acids are attached until the ribosome reaches the end of the mRNA and the peptide is complete.

2. Elucidation of the Genetic Code

It is known that proteins are composed of approximately 20 different types of amino acid, whereas DNA contains only four types of base. Therefore, there cannot be a simple 1:1 relationship between the bases and amino acids, just as it is not feasible for combinations of two bases (i.e., 16 combinations) to code for all of the amino acids. From theoretical considerations, therefore, it appears that each amino acid is coded for by a sequence of at least three bases (i.e., 64 combinations).

Two main approaches have been used to assign each nucleotide triplet to its respective amino acid. The first was to use a cell-free protein synthesis system consisting of ribosomes, aminoacyl-tRNAs and an energy source. Addition of a synthetic polynucleotide, which acts in the same way as a mRNA molecule does in the cell, stimulates the production of a peptide. The amino acids incorporated into this peptide were determined by using an incubation mixture in which one of the 20 amino acids was labelled with [14]C. Using this method demonstrated that poly-U promotes phenylalanine incorporation, poly-A promotes lysine incorporation and poly-C promotes proline incorporation. An extension of this technique employed the use of polynucleotides containing two alternating bases. For example, the use of poly-UGUG produced an incorporation of cysteine and valine, thus indicating that UGU codes for cysteine and GUG codes for valine. This result also strongly supports the concept that each amino acid is coded for by a triplet of nucleotides. By using more complicated polynucleotides a wide range of the nucleotide triplets were assigned to amino acids.

A second approach involved the use of a trinucleotide of known sequence which when added to the cell-free protein synthesis mix causes specific molecules of aminoacyl-tRNA to bind to ribosomes. This mix can then be filtered so that the ribosome–tRNA complex is retained but the unbound tRNA molecules are washed away. Thus by labelling one aminoacyl-tRNA species out of the 20 present it was possible to determine which trinucleotide sequence codes for which amino acid. These two approaches have produced data which are in good agreement.

Using the methods described above, 61 of the 64 codons were assigned to specific amino acids. The remaining three codons, UAA, UAG and UGA did not appear to code for any amino acids and were later found to be signals for peptide chain termination. This explains

Table 1

Assignment of amino acids to each nucleotide codon

First (5′) nucleotide	Middle nucleotide				Third (3′) nucleotide
	Uracil	Cytosine	Adenine	Guanine	
Uracil	phenylalanine	serine	tyrosine	cysteine	uracil
	phenylalanine	serine	tyrosine	cysteine	cytosine
	leucine	serine	STOPª	STOP	adenine
	leucine	serine	STOP	tryptophan	guanine
Cytosine	leucine	proline	histidine	arginine	uracil
	leucine	proline	histidine	arginine	cytosine
	leucine	proline	glutamine	arginine	adenine
	leucine	proline	glutamine	arginine	guanine
Adenine	isoleucine	threonine	asparagine	serine	uracil
	isoleucine	threonine	asparagine	serine	cytosine
	isoleucine	threonine	lysine	arginine	adenine
	methionine	threonine	lysine	arginine	guanine
Guanine	valine	alanine	aspartic acid	glycine	uracil
	valine	alanine	aspartic acid	glycine	cytosine
	valine	alanine	glutamic acid	glycine	adenine
	valine	alanine	glutamic acid	glycine	guanine

a STOP represents codons which terminate peptide synthesis

why neither poly-GAUA nor poly-GUAA could be translated into a peptide in cell-free studies. A complete listing of the assignments of all combinations of the nucleotide triplets is given in Table 1.

It was initially thought that no specific codon served to signal the start of a peptide chain because all 64 triplets had been accounted for and because no common triplet was required for synthetic polynucleotides to be translated *in vitro*. However, the discoveries in *Escherichia coli* that methionyl-tRNA could be formylated in the α-amino position of the methionine residue (which meant that it could only become part of a peptide if it were the first amino acid), and that half of the *E. coli* proteins contained a methionine group at their amino-terminal ends, indicated that bacterial proteins might all be initiated by formylmethionine. Direct evidence supporting this idea was obtained from cell-free studies of bacteriophage mRNAs which when translated *in vivo* might produce a peptide starting Ala–Ser . . . , but *in vitro* produced a peptide starting fMet–Ala–Ser . . . , thus indicating that the proteins are normally initiated by formylmethionine which is then often removed *in vivo* by enzymes not present in the cell-free mix. In a ribosome-binding assay fMet–tRNA will recognize AUG, GUG and UUG but it appears that AUG is by far the most common naturally occurring initiation codon.

It is interesting that the triplets coding for the same amino acid are not distributed at random but are grouped together so that they generally share the same 5′ and middle base. This has the consequence that mutations producing a change in the base at the 3′ position of the codon often have no effect on the amino acid specified. Furthermore, the different amino acids are segregated to a considerable extent on the basis of chemical similarity (hydrophobicity, hydrophilicity, acidity and basicity). Thus a mutation in the 5′ base of any of the six leucine codons would produce a codon specifying another hydrophobic amino acid. It is quite possible that this arrangement of codons has evolved to reduce the potentially harmful effect of possible mutations.

Although the genetic code was elucidated using *E. coli*, it appears to be universal for all living systems. The coding responses of tRNAs from different species have demonstrated that the codon assignments are invariable. Furthermore, mRNAs from one species have been found to be correctly translated *in vitro* and *in vivo* by the protein synthetic apparatus of other species.

3. Overlapping Codes

As has already been stated, the nucleotide sequence of a gene is read in triplets, and produces one peptide from one gene. However, recent studies of the genome of bacteriophage ϕX174 have provided evidence that in this organism the nucleotide sequence can also be read out of phase (see Fig. 2), to allow the production of two proteins from one DNA sequence. The genome of this virus codes for nine proteins and analysis of the nucleotide sequence has demonstrated that not only is the gene for protein B totally contained within the gene for protein A (but in a different reading frame), but also gene E is similarly contained within gene D. This

Protein A	— Glu — Try — Asn —Asn — Ser — Leu —

| Nucleotide sequence | C T T A C C T T G T T G A G T G A T T T
G A A T G G A A C A A C T C A C T A A A |

| Protein B | Met — Glu — Gln — Leu —Thr — Lys — |

Figure 2
Part of the nucleotide sequence of bacteriophage
φX174 DNA coding for the start of protein B and a
section of protein A

remarkable phenomenon, which has allowed the bacteriophage to store more information in its DNA than would by convention have been deemed possible, has provided yet another intriguing insight into the functioning of the genetic code.

See also: Genetic Engineering

Bibliography

Adams R L P, Burdon R H, Campbell A M, Leader D P, Smellie R M S 1981 *The Biochemistry of the Nucleic Acids*, 9th edn. Chapman and Hall, London
Lewin B 1974 *Gene Expression*, Vol. 1, *Bacterial Genomes*. Wiley, New York
Lewin B 1980 *Gene Expression*, Vol. 2, *Eucaryotic Chromosomes*, 2nd edn. Wiley, New York
Sanger F, Air G M, Barrell B G, Brown N L, Coulson A R, Fiddes J C, Hutchison C A, Slocombe P M, Smith M 1977 Nucleotide sequence of bacteriophage φX174 DNA. *Nature (London)* 265: 687–95

M. H. Goyns

Genetic Engineering

Genetic engineering essentially involves the manipulation of DNA sequences and is defined as the formation of new combinations of heritable material by the insertion of nucleic acid molecules, produced by whatever means outside the cell, into any virus, bacterial plasmid or other vector system so as to allow their incorporation into a host organism in which they do not naturally occur but in which they are capable of continued propagation. Technically, the manipulation of DNA molecules in this way is known as recombinant DNA technology. In this article, the application of genetic engineering to the genes of animal (primarily mammalian) cells is considered.

1. Central Role of DNA in Living Cells

A central theory of molecular biology states that the flow of genetic information is from DNA to RNA to protein. DNA, the macromolecule that carries genetic information within cells, comprises two complementary strands, each strand being made of an ordered sequence of four different nucleotide bases: adenine (A), guanine (G), cytosine (C) and thymine (T). The ordering of these four bases in the DNA of living cells serves as a code to store all the information necessary to specify living organisms. The chemical structure of these bases is such that specific weak bonding forces can be made between given pairs of bases. Adenine always pairs with thymine and guanine always pairs with cytosine. This forms the basis of the double-stranded DNA molecule by base complementarity and provides a mechanism by which the molecule can reproduce itself exactly—an important requirement for the lineal perpetuity of genetic information. Thus a sequence on strand 1 (as shown in Fig. 1) would have the sequence in strand 2 as its complement. The thousands of different proteins used in

```
A G G C T T A G C T A T 1
| | | | | | | | | | | |
T C C G A A T C G A T A 2
```

Figure 1
DNA duplex

living cells are encoded by countless variations of the order of these four bases in combinations of different lengths. Each sequence coding for a protein is known as a gene.

To extract this information from DNA and translate it into protein, the second informational macromolecule of cells, ribonucleic acid (RNA), plays a key role. By the same base complementarity rules the DNA can be "read" into a single-stranded messenger RNA (mRNA) molecule, each mRNA having the information to encode a single protein molecule. This is known as transcription of DNA. This process occurs in the nucleus of living cells, and from there the mRNA molecules travel to the cytoplasm to be "read" in turn by ribosomes. The ribosomes "decode" mRNA and string together the correct order of amino acids to form the particular protein specified by a given mRNA molecule; the whole process is known as translation. Figure 2 summarizes the events of transcription and translation which give rise to the protein molecules encoded in the DNA of living cells and which are used in cell growth and function.

In multicellular organisms with different tissues and hence cell types, all the cells of the body contain exactly the same complement of DNA sequences. The different spectrum of proteins found in different cell types is the consequence of differential expression of the available complement of gene sequences. In addition to containing the base sequences encoding protein molecules, DNA

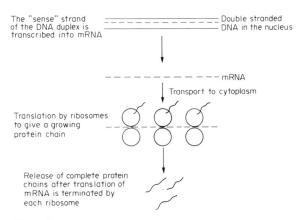

The "sense" strand of the DNA duplex is transcribed into mRNA

Double stranded DNA in the nucleus

mRNA

Transport to cytoplasm

Translation by ribosomes to give a growing protein chain

Release of complete protein chains after translation of mRNA is terminated by each ribosome

Figure 2
Sequence of events from DNA to synthesized protein molecules

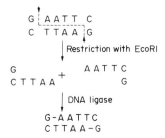

Restriction with EcoRI

DNA ligase

Figure 3
Cutting of a DNA sequence using the restriction endonuclease EcoR1 followed by the rejoining using DNA ligase

also contains specific sequences which function as regulatory elements. These control not only the tissue-specific pattern of gene expression, but also temporal variations in gene sequences selected for use, as is the case during embryonic development of an organism.

2. Recombinant DNA Techniques

2.1 Elemental Components of Gene Cloning

The mechanics of genetic engineering revolve around ways of specifically cutting and joining desired DNA sequences and placing them within a cloning vector or vehicle to permit propagation.

Enzymes, which are protein molecules that perform chemical reactions within cells, are used for cutting and joining DNA sequences. The elemental components of recombinant DNA techniques are (a) restriction endonucleases—enzymes which cut DNA molecules, (b) DNA ligases—enzymes which can join together DNA molecules, and (c) cloning vectors capable of autonomous replication to carry and amplify a purified DNA sequence.

Restriction endonucleases cut DNA molecules in a sequence-specific manner, that is, they recognize a particular DNA sequence in terms of DNA bases and effect a cut only at such sequences. There are many hundreds of different restriction enzymes all purified from different bacterial strains, each having a specific sequence dictating the point of DNA strand scission. For example, the restriction endonuclease EcoR1 from the bacterium *Escherichia coli* will cut DNA at a six-base site (Fig. 3). The cut is between the G and the A in both DNA strands resulting in a staggered cut. This gives rise to complementary single-stranded sequences known as "sticky ends." Restriction enzymes may also cut DNA to give flush or blunt ends in which there are no staggers.

Sticky or blunt ends can be rejoined by the action of DNA ligase, thus affording the means by which different combinations of DNA molecules can be obtained (Fig. 3).

The third essential ingredient for genetic engineering is the cloning vector or vehicle—a DNA molecule that is capable of autonomous replication within a host cell. Such a molecule can act as a carrier for a purified DNA sequence. The most commonly used cloning vectors are bacterial plasmids and suitably modified versions of the bacteriophage λ. Plasmids are small circular DNA molecules found in bacteria in addition to the bacterial chromosome. They replicate independently of the bacterial chromosome and carry genetic information. For example, they carry genes which specify drug resistance. They can pass from one bacterium to another and in this way transfer resistance to antibiotics. The bacteriophage λ is a circular DNA virus (which can also exist in a linear form) that infects bacteria and like plasmids can replicate independently of the host cell's DNA.

By selecting a restriction enzyme that cuts a cloning vector once, the molecule can be opened at a unique position into which a DNA sequence can be inserted or cloned. The various manipulations of recombinant DNA technology involve different ways of bringing about this insertion.

2.2 Strategies for Cloning a Specific Gene Sequence

It is clearly a formidable task to isolate a single desired gene sequence from the entire complement of DNA gene sequences (genome) of animal cells. For example, the genome of the mouse contains sufficient DNA to encode 10^7 gene sequences of 10^3 bases in length. The complexity of this problem can be simplified by adopting two different strategies of gene cloning.

(a) *cDNA cloning.* The first approach uses plasmid vectors in a technique known as cDNA cloning. The starting point in this case is not genomic DNA, but the mRNA population of a tissue expressing the gene sequence of interest at an elevated level. This means that

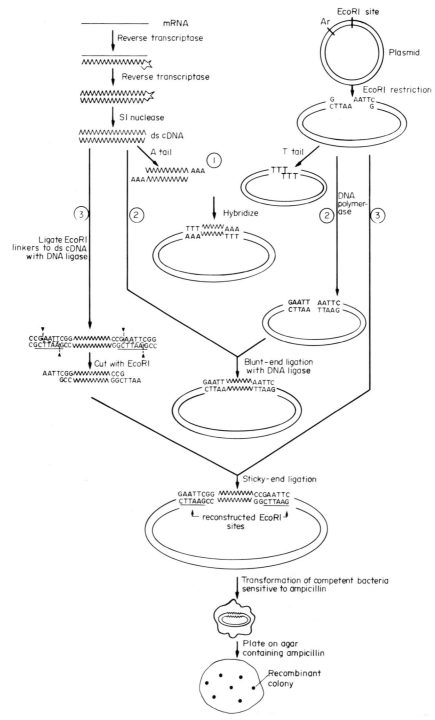

Figure 4
Alternative means of introducing a double stranded cDNA molecule into a plasmid vector: 1, by tailing; 2, by blunt-end ligation; and 3, by linker sequences

the point of departure is a very much less complex mixture of sequences.

The mRNA population can be copied into a single-stranded DNA copy (cDNA) by the enzyme reverse transcriptase (Fig. 4). By further copying with reverse transcriptase the single-stranded cDNA can be converted to the double-stranded form (ds cDNA) by "looping back" on itself. In this way the DNA gene coding sequences of the tissue can be reconstructed from the mRNA population. After removal of the "hairpin loop" by S1 nuclease, which degrades single-stranded DNA regions, the double-stranded DNA molecules can be inserted into a unique restriction enzyme site of a plasmid vector by a variety of methods. However, before this can be done the purified plasmid vector has to be opened at a unique site by the action of a restriction endonuclease such as EcoR1.

The vector is now ready to accept a DNA sequence. The first way of engineering this is by a method called "tailing." This involves extending a "tail" (about 20 bases long) of a single DNA base (using the enzyme terminal transferase) from one end of each DNA strand of the vector and cDNA molecules (Fig. 4). It is arranged such that the base used with the cDNA is complementary to that used on the vector and hence the two molecules can associate or hybridize (via these tails) by base complementarity. For example, the cDNA can be tailed with A and the plasmid vector with T (route 1 in Fig. 4).

Alternatively, the cDNA molecules may be introduced into the vector by blunt-end ligation. The staggered ends of the restricted vector are first repaired or filled in with DNA polymerase to give the necessary flush or blunt ends. The cDNA molecules are then linked to the vector DNA by the action of DNA ligase (route 2 in Fig. 4).

Finally, a third and more elegant means, which is an elaboration of blunt-end ligation, may be used. In this case, the cDNA molecules are first flanked by small stretches of DNA called "linkers," added on by blunt-end ligation. Linkers are synthetic stretches of DNA which can be tailored to contain any desired restriction endonuclease site. In the case of route 3 in Fig. 4, EcoR1 linkers are illustrated. The sites within the linkers are then cut with EcoR1 endonuclease to reveal sticky ends which are complementary to those on the nuclease restricted EcoR1 site of the plasmid vector. The cDNA molecule is then introduced by sticky-end ligation. The great advantage of this method is that it results in reconstructed EcoR1 sites immediately on either side of the cloned DNA sequence. This affords a means of precisely excising the cloned sequence after propagation of the plasmid within a host cell.

For all these methods, conditions are used that permit the introduction of only one ds cDNA molecule per plasmid molecule. Furthermore, recircularization of the plasmid will only occur if a cDNA molecule is inserted. Having created what is known as a cDNA library by these means, the next step is to clone-out individual plasmids containing a single gene sequence. This is done by infecting or transforming bacteria (made competent to accept plasmid DNA through their cell wall) with the recombinant plasmid DNA. Each cell will accept only one plasmid and hence receive only one cloned sequence. The transformed bacteria are then plated onto nutrient agar at a density permitting the isolation of individual colonies. If a plasmid which carries resistance to an antibiotic (e.g., ampicillin, Ar) is used, then the transformation of ampicillin-sensitive bacteria with the recombinant plasmid DNA and subsequent plating on agar containing the antibiotic, means that only those bacteria which have received a plasmid will grow. The normally ampicillin-sensitive bacteria become resistant upon receipt of a plasmid. This allows a direct selection of recombinant colonies containing a single cloned gene sequence. Such colonies can be picked and screened by a variety of methods to identify the nature of the cloned sequence.

(b) *Cloning of genomic DNA.* Using a modified bacteriophage λ as a cloning vector, genomic DNA can be cloned directly. The construction of a genomic DNA library will result in the cloning not only of protein coding sequences, but also associated DNA sequences which may be involved in regulating gene expression. This is important where gene cloning is used as a tool to analyze the mechanisms regulating the differential use of gene sequences in development and tissue differentiation.

The DNA within each chromosome of animal cells exists as a single continuous double-stranded molecule many millions of DNA bases long. Before attempting the cloning of genomic DNA, it is therefore important to break the DNA into smaller, less complex fragments—usually about 2×10^4 bases long. This can be done, conveniently, with a partial digestion of the DNA by a restriction enzyme such as EcoR1.

The linear form of phage λ is used as a vector essentially by excising a central piece of the viral DNA (with the enzyme EcoR1) and replacing it with a genomic fragment. The advantage of λ over plasmid vectors is its ability to accept larger pieces of inserted DNA. Figure 5 shows that the genomic fragments, generated by EcoR1 digestion, are linked to the left and right arms of the engineered λ vector by sticky-end ligation to create a genomic DNA library.

As with plasmid cloning, a positive selection for λ recombinant DNA molecules can be made. The mature λ phage particle has the DNA molecule encapsulated within a protein coat encoded by genes carried on λ DNA. The process of introducing the DNA into the protein coat is known as packaging and greatly improves the efficiency with which the mature virus particle infects host cell. The λ arms alone ligated together will not package efficiently, whereas λ arms carrying a genomic DNA fragment will. In this way recombinant molecules are selected preferentially by carrying out a packaging step *in vitro*. Subsequent infection of a "lawn" of bacterial host cells on an agar plate give rise to amplification of the recombinant virus and eventual lysis

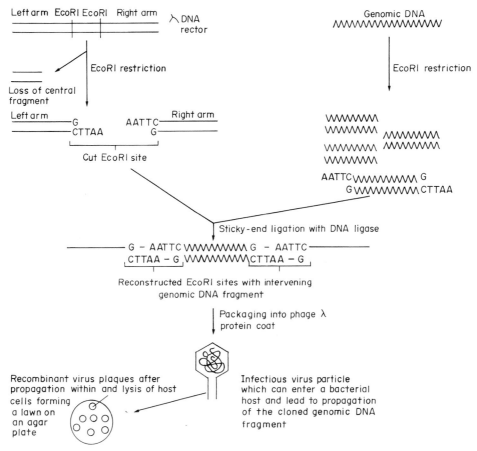

Figure 5
Cloning of EcoR1 cut genomic DNA fragments by sticky-end ligation into a phage λ vector

of the infected cells to release recombinant DNA in the form of a plaque (Fig. 5). The virus in these plaques can be picked and further plaque purified to yield a purified recombinant virus carrying a single genomic DNA fragment isolated from the complete library.

3. Application of Recombinant DNA Technology

3.1 Unravelling Gene Structure and Expression

The advances made in the understanding of gene structure and function resulting from the application of genetic engineering techniques can only be described as meteoric. Four main areas of molecular biology epitomize the startling advances that have been made.

Unlike bacteria, the structure of many genes found in higher organisms does not display an uninterrupted continuity of protein coding sequences. Detailed dissection of many cloned genes from animal cells has shown that the protein coding sequences (exons) within

the DNA are interrupted by other DNA sequences (introns) which do not contribute in any way to the amino acid sequence found in the corresponding protein. For example, the mouse β-globin gene contains two introns of 116 and 612 bases which interrupt the protein coding sequences between amino acids 30 and 31 and 104 and 105 of the β-globin chain (Fig. 6). After transcription, these introns are removed from the transcript with subsequent splicing together of the protein coding sequences found in the functional mRNA. The biological significance of introns is unclear, but a rigid conception of the higher organism genome with simple linear arrays of protein coding sequences has necessarily had to cede to a more flexible and dynamic view of its operation.

This is no more eloquently portrayed than in the fascinating schemes unravelled by recombinant DNA techniques to explain the immune response at the molecular level. By cloning the gene sequences encoding the constant and variable regions of antibody molecules, it has been shown that these represent different gene

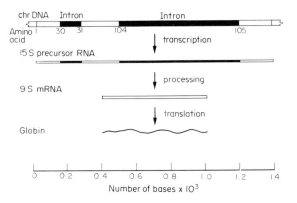

Figure 6

Typical structure of a eukaryotic gene showing the interruption of the mouse β-globin protein coding sequences (exons) with noncoding sequences (introns). After transcription into a 15 S (1500 nucleotide) RNA species, the introns are removed by a splicing mechanism to yield the mature 9 S (650 nucleotide) mRNA which is subsequently translated

entities encoding their appropriate amino acid sequences. The covalent and colinear association of the same constant-region amino acid sequence with different variable-region amino acid sequences in different antibody molecules is due to rearrangements of genomic DNA in antibody producing cells. The constant-region gene is juxtaposed to the appropriate variable-region sequence (of which there are several hundred within the genome), hence creating an mRNA transcript that can give rise to the desired antibody. Further antibody diversity is conferred by somatic mutation in variable-region genes. This dynamic response of genomic DNA to regulating gene expression may be a more general phenomenon in genome regulation, though perhaps not as dramatic as in the case of immunoglobulin genes.

The role of noncoding DNA in regulating the expression of protein coding sequences is being attacked directly by genetic manipulation. This is especially so for those sequences flanking coding genes. The search is being directed primarily to elucidating the nature of sequences controlling the transcription of genes. By precisely dissecting-out defined DNA sequences, and placing them within appropriate vector molecules in front of protein coding sequences in various combinations, their effect upon expression (in terms of transcription) of these sequences can be assayed. Well characterized sequences necessary for gene transcription have been identified by such means.

Finally, genomic cloning has permitted important advances in our understanding of cancer at the molecular level. The cellular equivalents of genes carried on tumor-causing viruses that are known to bring about cancerous transformation have been isolated. These are called oncogenes and a whole range of such oncogenes specific to and active within different types of tumors is

emerging. Some of these genes have been shown to have kinase activity, that is, they add phosphate groups to certain amino acid residues on particular proteins. Therefore, cancer-causing genes are carried as a normal integral part of a cell's genetic constitution. Much work is now being directed to elucidating why in some cells the genes are expressed and how their activation brings about malignancy.

3.2 Medically Important Proteins

The tantalizing prospect of using genetic engineering to produce medically important drugs is beginning to be realized. The ability to introduce a protein coding sequence into a vector that can be propagated in a bacterium raises the possibility of using the bacterium as a protein factory for the foreign gene. This depends upon inducing the bacterial machinery not only to transcribe the cloned DNA but also to translate it into protein. Precisely this has been achieved with a number of important proteins, primarily hormones of therapeutic value.

To facilitate efficient expression of the cloned sequence within the bacterium, it is desirable to place the inserted gene under the control of bacterial regulatory elements. In most of the examples to date, this has been achieved by the fusion of the inserted gene to a bacterial gene which is readily expressed, for example, the penicillinase gene which is carried on many plasmids, or the β-galactosidase gene which is required for the metabolism of lactose.

The hormone somatostatin inhibits the secretion of several hormones, notably growth hormone, insulin and glucagon. Consequently, its possible large-scale production within expressing bacteria may provide a source of the hormone suitable for the treatment of acromegaly, acute pancreatitis and insulin-dependent diabetes. The gene coding for somatostatin (a peptide hormone of 14 amino acids) has been chemically synthesized by stringing together the appropriate DNA bases to code for the known amino acid sequence. By building-in suitable restriction sites at either end of the coding sequence it is possible to introduce the gene using sticky-end ligation into a plasmid vector carrying the β-galactosidase gene such that the two are fused. In this way the somatostatin sequence was placed under the control of β-galactosidase expression. Bacteria transformed with this recombinant have been shown to express active somatostatin which can be purified.

Human growth hormone (HGH) has been cloned (by ligation) from cDNA primed by pituitary mRNA. The recombinants containing HGH were identified by nucleic hybridization. A radioactive-labelled probe very close in its complementarity to HGH DNA (i.e., human chorionic somatomammotropin cDNA) is allowed to associate by base complementarity to DNA (fixed on special paper) isolated from recombinant colonies. Only those recombinants of pituitary cDNA which contained HGH sequences hybridized to the probe and hence were revealed by exposure to x-ray film. By inserting the sequence into the plasmid vector in such a way as to put its expression under bacterial regulatory elements (those which control the β-galactosidase gene), expression of HGH protein was obtained at a high level as detected by radioimmunoassay.

Similarly, the gene coding for insulin, an important hormone for the treatment of diabetes, has been cloned from cDNA prepared against pancreatic islet cells—the cells responsible for producing the hormone in the pancreas. In this example, the pancreatic cDNA molecules were introduced by the tailing method into a site found within the penicillinase gene of the plasmid vector. Recombinants carrying the insulin coding sequences were identified by a process called hybridization-arrested translation. It is possible to translate purified mRNA *in vitro* and identify the encoded proteins by separating the products on a gel. This translation can be arrested by hybridization of the mRNA to a complementary DNA sequence. Hence, by hybridizing pancreatic mRNA to plasmid DNA from recombinant colonies, it is possible to see which recombinant inhibits specifically the *in vitro* translation of insulin mRNA. In this way, recombinants containing the insulin gene were identified and by immunological assays it was shown that insulin could be produced within the bacterial host cell as a hybrid penicillinase–insulin fused protein chain.

As a final example of a gene product whose cloning and expression has created much excitement, interferon cannot escape a mention. Interferon, a protein produced by cells in response to viral infection, acts as an antiviral agent. With this property in mind, much work has been done in developing it as a drug to treat cancers produced by tumor-promoting viruses. cDNA, prepared from the mRNA population of virus-infected human white blood cells, was used for cloning by the tailing method. The cDNA molecules were, once again, inserted into the penicillinase gene of the plasmid vector to take advantage of the regulatory elements controlling the production of penicillinase, and hence bring about the expression of interferon. Recombinant molecules containing the interferon coding cDNA were identified by a method called hybridization-selected translation. For this, recombinant plasmid DNA from purified colonies was fixed to a solid paper support and used to select, by hybridization, its complementary mRNA from the mRNA population of white blood cells. The purified mRNA was then recovered and translated into protein and the products analyzed for interferon activity. Extracts of bacteria identified as carrying the interferon gene also showed the presence of biologically active interferon, thus demonstrating that it had been expressed within the bacterium.

The level of expression within bacteria of some of those cloned genes is still very low. New vectors carrying these genes are being engineered in an attempt to augment the amount of protein made.

3.3 Antenatal Diagnosis

Genetic engineering offers a means of generating probes for genes involved in hereditary lesions of clinical importance.

Changes in genomic DNA, brought about by mutation, can have serious consequences when these involve genes of importance to the normal physiology of the body. Such mutation can involve deletions and inversions of key stretches of DNA, point mutation (changes where one DNA base is replaced by another) and translocations where different chromosomes are involved in a rearrangement of genetic material.

If such changes associated with a particular genetic disease affect a site for a restriction enzyme, then restriction enzyme analysis of DNA can be used to diagnose the condition in a suspected patient. For example, deletion may remove a site, inversion and translocation alter its position and point mutation either abolish an existing site, change an existing site or create a site for a particular enzyme that previously was not present. In all these cases, the normal pattern of DNA fragments obtained after restriction enzyme analysis would be altered in the mutant genome. By having available cloned gene probes for the affected gene, such alterations in restriction enzyme fragment patterns can be revealed using molecular hybridization assays with radioactively labelled probes.

Such an approach need not necessarily be confined to DNA changes within the relevant gene. Nondeleterious variation or polymorphism is common within the genome and often involves changes which create or alter restriction enzyme sites. Such polymorphisms when linked and inherited with the deleterious mutation of interest, can be used as a diagnosis for the mutant gene. The normal restriction fragment pattern is altered by the polymorphism and this is used as a criterion for the inheritance of the deleterious mutation. Any probe that will detect the polymorphism can be used.

An elegant example employing both these approaches is illustrated by the molecular genetic diagnosis of sickle-cell anemia. Sickle-cell hemoglobin is caused by a point mutation in the codon for the sixth amino acid in the β-globin gene. The change A $\rightarrow$ T in this position results in the substitution of valine for glutamic acid. At the same time, this point mutation eliminates the recognition sequence for the restriction enzyme DdeI; a change from CTGAG to CTGTG. Hence, restriction of the sickle-cell β-globin DNA with this enzyme will yield a different pattern of DNA fragments when compared to digests of the normal allele. A polymorphism lying outside the coding region of the globin gene, but inherited with the sickle-cell allele, can also be used to diagnose the condition. Digestion of normal DNA with the restriction enzyme HpaI yields a DNA fragment 7.6×10^3 bases long containing the β-globin gene. Similar digestion of DNA from individuals with the sickle-cell trait, yields a fragment of 13×10^3 bases containing the β-globin gene. This allele linked polymorphism can hence be used to detect the sickle-cell trait.

Amniocentesis or trophoblast biopsy can provide fetal cells from which DNA can be prepared and used in restriction enzyme digest analysis in conjunction with the appropriate gene probe. In this way an early antenatal diagnosis can be performed where a serious genetic condition is suspected. Early identification of a condition provides a means for more effective counselling and, if desired, the option of therapeutic abortion at a stage entailing a reduced risk to the mother.

3.4 Prospect of Gene Therapy

Up to this point two ways in which genes can be cloned and manipulated within bacterial cells have been considered. Recent work has probed the possibility of using genes cloned in this way to engineer mammalian cells as a first step in using recombinant DNA techniques in the treatment of heritable gene defects such as thalassemia.

Cloned genes can be introduced into mammalian cells in three ways. The first is to incorporate the cloned sequence into an animal virus vector, such as SV40, and then infect the host cell with the recombinant virus, hence using the virus ability to infect the cell and gain entrance for the cloned gene. The rabbit β-globin gene has been incorporated into the monkey virus SV40 and placed into host monkey cells. Both rabbit β-globin mRNA and the protein chain have been shown to be synthesized within the host cell.

The second method is known as DNA-mediated gene transfer. In this approach the purified gene sequence (usually within a vector molecule such as a plasmid) in the form of DNA is precipitated with calcium and phosphate onto the host cell. Under these conditions the cell can ingest the DNA. To select those cells which have taken up (or become transformed with) the cloned sequence, usually a second gene acting as a selective marker is used in the transformation mixture. A commonly used marker is the thymidine kinase gene (TK) which codes for an enzyme necessary for the utilization of the DNA base thymidine. By using host cells with a mutation in the TK gene and hence lacking a functional enzyme (TK$^-$), only those cells which are transformed with a functional TK gene will grow when placed in a special selective medium. The TK gene from the herpes simplex virus has been cloned and is used in DNA-mediated gene transfer for selective purposes. This TK gene can be incorporated into the same plasmid carrying the cloned gene sequence of interest or cotransformed as a separate piece of DNA. Thus when TK$^-$ host cells in selective medium are transformed with DNA containing the TK gene and the cloned sequence of interest, only those cells which receive the TK gene and express it will grow. In this way it is possible to directly select for those cells which have received the cloned gene. A number of studies have successfully exploited this method and obtained TK$^+$ transformants, hence showing that it is possible to obtain expression of an introduced gene.

The third means of placing cloned gene sequences within mammalian cells is by microinjection. This involves injecting DNA directly into the cell nucleus with the use of an extremely fine glass needle. Using a TK$^-$ mouse cell line as host, separate recombinant plasmids carrying the herpes TK gene and the human β-globin gene have been injected directly into the mouse nucleus. Correct expression of the injected TK gene was obtained since the TK$^-$ defect of the host was corrected. β-Globin mRNA was detectable but not the protein. Furthermore, the level of expression (in terms of mRNA) of the human gene within the mouse cell was very low compared to the level found in transferred genes.

These preliminary findings indicate that it is possible to effect genetic engineering within mammalian cells. The major problems at present are those of obtaining expression of the introduced gene under the normal control elements found within its cell type of origin and at an equally high level. This may depend upon integrating the transferred gene at the correct position within the host genome. It would seem, therefore important to develop means by which this integration can be directed. Such efforts are currently in progress.

Successful transfer, stable integration and expression of mammalian genes would permit treatment of a number of hereditary diseases. For example, the thalassemias are conditions which effect the oxygen carrying red cells of the blood and are due to defects in the genes coding for the various protein chains of hemoglobin. Where bone marrow transplants to replace the blood-producing cells of the bone marrow are not possible (possibly because of the lack of a compatible donor), then the introduction of functional globin genes to correct the defect in the patient's own blood-producing cells may be a means of effecting a cure. Such strategies for treating hereditary diseases are, as yet, in their infancy.

See also: Genetic Code

Bibliography

Anderson W F, Diocumakos E G 1981 Genetic engineering in mammalian cells. *Sci. Am.* 245: 60–93

Berg P 1981 Dissections and reconstructions of genes and chromosomes. *Science* 213: 286–303

Gilbert W, Villa-Komaroff L 1980 Useful proteins from recombinant bacteria. *Sci. Am.* 242: 68–82

Goeddel D V, Heyneker H L, Hozumi T, Arentzen R, Itakura K, Yansura D G, Ross M J, Miozzari G, Crea R, Seeburg P 1979 Direct expression in *Escherichia coli* of a DNA sequence coding for human growth hormone. *Nature* (*London*) 281: 544–48

Itakura K, Hirose T, Crea R, Riggs A, Heyneker H, Bolivar F, Boyer H W 1977 Expression in *Escherichia coli* of a chemically synthesized gene for the hormone somatostatin. *Science* 198: 1056–63

Lewin R 1981 Biggest challenge since the double helix. *Science* 212: 28–32

Logan J, Cairns J 1982 The secrets of cancer. *Nature* (*London*) 300: 104–5

Marx J L 1981 More progress on gene transfer. *Science* 213: 996–97

Marx J L 1982 The case of the misplaced gene. *Science* 218: 983–85

Old R W, Primrose S B 1980 *Principles of Gene Manipulation: An Introduction to Genetic Engineering.* Blackwell Scientific, Oxford

Rigby P W J 1982 The oncogenic circle closes. *Nature (London)* 297: 451–53

Robertson M 1983 Clues to the genetic basis of cancer. *New Sci.* 9: 688–91

Shows T B, Sakaguchi A Y, Naylor S 1982 Mapping the human genome, cloned genes, DNA polymorphisms and inherited disease. *Adv. Hum. Genet.* 12: 341–452

Villa-Komaroff L, Efstratiadis A, Broome S, Lomedico P, Tizard R, Naber S, Chick W, Gilbert W 1978 A bacterial clone synthesizing proinsulin. *Proc. Nat. Acad. Sci. U.S.A.* 75: 3727–31

Williamson R 1981 The cloning revolution meets human genetics. *Nature (London)* 293: 10–11

Williamson R 1983 Cloned genes and the study of human inherited disease. *Hosp. Update* 9: 25–32

N. Affara

H

Hearing Aids

About one in ten people in the UK suffer from a hearing loss, the effects of which may be minimized by the use of a hearing aid. In general, hearing aids operate by presenting to the ear sounds of greater intensity than would have been there in the absence of the aid. The amplification of the original sound is now usually performed electrically by battery-powered aids. Nevertheless, circumstances may exist which make the old-fashioned nonelectric acoustic aids preferable. A relatively new technique, the cochlear implant, can help certain people who have no measurable hearing and who derive no benefit from conventional amplifying aids.

1. Acoustic Aids

Acoustic aids rely entirely on the physical properties of sound to achieve amplification. Simply cupping a hand behind the ear, for example, reflects extra sound into the ear and gives an increase in sound pressure level (SPL) of some 8 dB at a frequency of around 1.5 kHz. An ear trumpet, which operates on the same principle (though resonance may play a part also) may give a gain of over 20 dB in the frequency range 500–3000 Hz. The speaking tube is also available but operates somewhat differently. It conveys high intensity sound (with little loss) from the mouth of the speaker to the ear of the listener. During conversational speech sound pressure levels are about 95 dB at the mouth. Very much higher pressure levels are produced if the mouth is close-coupled to a speaking tube. Thus the speaking tube, if properly used, can enable even people who are severely deaf to hear better.

Acoustic aids are of limited efficacy and social acceptability and thus they are likely to be used only by those who cannot cope with aids of more modern design. Elderly or arthritic people who find the small controls on a modern electric hearing aid difficult to use may find the management of the larger acoustic aids simpler. Acoustic aids also find application in countries where electrically powered aids are unobtainable.

2. Electrically Operated Hearing Aids

The essential components of the hearing aids in common use which amplify sound by electrical means are: a microphone, an amplifying device or circuit, and an earphone or receiver to present the amplified sound to the ear. The location of the hearing aid on the person is a convenient first means of describing the aid. They may be worn behind the ear (post-aural), in the ear, in a spectacle frame, on a headband or simply "body worn" on another part of the body—often in the breast pocket. Figure 1 illustrates some of these aids. The three component parts mentioned above, plus the battery, are contained within the one housing, with the exception of the body-worn aid, the earphone of which is linked to the aid with flexible wire.

The maximum acoustic output of hearing aids provides a basis for classifying them acoustically (see Table 1). However, a more complete description requires a statement also of (a) the frequency range over which the aid usefully operates, and (b) the maximum amplification, or gain, of the aid. The International Electrotechnical Commission (IEC) has produced a set of standards, IEC 118, *Hearing Aids*, Parts 1–12, which establishes measuring methods for these and other parameters. To facilitate the interpretation of measured values of these parameters, it is essential to adhere to the procedures for measurement laid down in these standards.

In Fig. 2 these basic acoustic parameters are presented graphically. Each curve shows the output of the aid as a function of frequency for a given sound pressure level input. The gain control of the aid is at its maximum setting for these measurements and thus the maximum amplification of the aid can be deduced. For example, for an input level of 50 dB (lowest curve) and at a frequency of 1 kHz, the output of the aid is 110 dB. The (maximum) gain of the aid at 1 kHz is thus 60 dB. The maximum overall gain—some 68 dB—occurs at 2.5 kHz

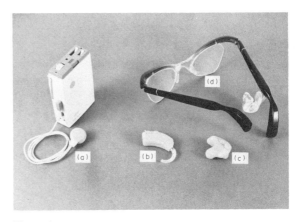

Figure 1
Some types of common electrically operated hearing aids: (a) body-worn aid, (b) post-aural aid (hook links into hollow plastic tube of earmold),
(c) in-the-ear aid, and (d) post-aural aid mounted in spectacle frame with attached earmold

Table 1

Hearing aids issued free on the British National Health Service. All hearing aids listed have a pick-up coil facility for use with an induction loop system or telephone coupler

Type of aid	Series	Code	Type of battery	Forward-facing microphone	Tone control	Output control	Max. gain (dB)	Max. output (dB SPL)
Behind-the-ear	BE10	BE11	CP1(675)	No	Yes	No	46	122
		BE12	CP1(675)	No	No	No	45	120
		BE13	CP1(675)	Yes	No	No	50	121
		BE14	CP1(675)	No	Yes	No	45	120
		BE15	CP1(675)	No	Yes	No	46	125
		BE16	CP1(675)	Yes	Yes	No	47	119
		BE17	CP1(675)	Yes	Yes	No		
Behind-the-ear	BE30	BE31	CP1(675)	No	Yes	Yes	53	130
		BE32	CP1(675)	Yes	Yes	Yes	60	133
		BE33	CP1(675)	Yes	Yes	Yes	57	134
Behind-the-ear	BE50	BE51	CP1(675)	Yes	Yes	Yes	64	138
		BE52	CP1(675)	Yes	Yes	Yes	63	136
Bodyworn	BW60	BW61	CP6(Mn 1500)		Yes	Yes	60–76	129–146
Bodyworn	BW80	BW81	CP6(Mn 1500)		Yes	Yes	88–93	146–148

approximately. At inputs of 60 dB and 70 dB, the output at all frequencies can be seen to have increased essentially in direct proportion to input level, but at higher input levels this is not the case. Over the useful part of the frequency range, the output for an input of 90 dB is no greater than for an input of 80 dB. That is, the output of the aid has a maximum value, referred to as the saturated sound pressure level (SSPL) of the aid.

The shape of these curves which illustrate the frequency response of the aid is determined largely by resonances etc. in the earphone. Usually, however, user operated and preset simple bass and treble cut tone controls are provided so that frequency response can be altered somewhat.

Some patients who suffer from sensorineural hearing loss may experience discomfort when subjected to the

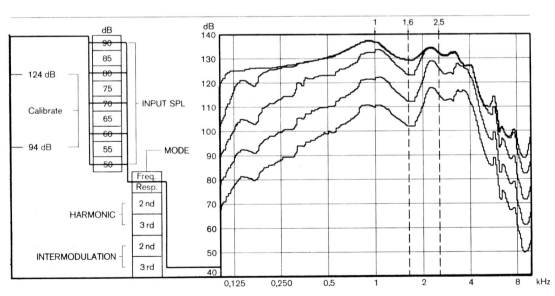

Figure 2

Overall acoustic performance of a typical post aural hearing aid in response to sound pressure level inputs of 50, 60, 70, 80, 90 dB

very high levels of sound that modern high-power aids can produce and thus it is of prime importance to limit their maximum acoustic output. Various means are adopted. Relatively simple "peak clipping" circuits may be used, or more complex automatic gain controls which automatically reduced the gain of the aid at high input levels. A further development is the splitting of the input signal into a number of frequency bands or channels, the signal in each channel being processed separately. These latter aids are known as multichannel compression aids.

The use of another type of aid, a frequency transposing aid, may be appropriate for patients with a profound hearing loss at high frequencies but much less loss at low frequencies. In one make, the signal frequency band 4–8 kHz is shifted to 0–4 kHz, the frequency components of the signal within the band retaining their original relationships to each other. The low frequency component of the original signal is maintained unchanged and added to the transposed signal prior to acoustic output.

3. Bone Conduction

Some people who have conductive deafness can hear better through the bones of the skull than through the ears, the sound being transmitted by vibration. Thus, instead of an earphone, a vibrator is held against the mastoid bone behind the ear by means of a headband. Bone conduction spectacle aids are produced but reinforced spectacle frames can also be used to hold a bone vibrator, used with a body-worn aid, in place.

4. Cochlear Implants

A cochlear implant may help certain people who have no measurable hearing. Tiny wire electrodes are implanted in or on the cochlea of the ear in series with a small induction coil. The electrical output of a hearing aid of specific design worn behind the ear is inductively coupled to the electrodes. Such implants are at present being actively investigated. An important factor in their use is that a long learning period is required for the patient to identify different sounds, the auditory nerve being directly stimulated by electrical signals. For this reason it has been found advantageous in developing the technique to involve patients who have previously been able to hear. For the implant to succeed, the auditory nerve running from the cochlea must, of course, be intact, although the cochlea itself will be nonfunctional. Only the totally deaf should be considered for a cochlear implant as the operation may destroy all normal hearing, depending upon the type of implant.

5. Earmolds

An earmold is an essential and critical part of a conventional hearing aid system, being the interface between the acoustic output of the aid and the ear. Molded usually in plastic—either a hard or soft acrylic—it should fit snugly and comfortably in the ear. Thus each mold is made to fit the individual patient, an impression first being taken of the ear using, for example, cold curing silicone rubber from which the mold is cast—a technique similar to that used in the manufacture of dentures. The acoustic output of the aid may be led into the ear canal either (see Fig. 3) through a hole drilled in

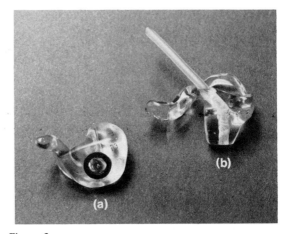

Figure 3
(a) A typical earmold for a button earphone, and (b) an earmold with a tube attachment for use with a post-aural aid

the mold or, if the hearing aid earphone is integral with the aid, through a flexible plastic tube which is part of the earmold. If the earphone is separate from the aid, then a spring clip ring is fitted into the earmold to hold the earphone in place.

The earmold acts, importantly, as an acoustic seal in the ear, preventing amplified sound from reentering the microphone. This, when it occurs, is known as acoustic feedback, and is often evidenced by a sustained whistling sound from the aid. Sometimes a second hole may be drilled through the mold, linking the air enclosed by the mold in the ear with the atmosphere. Venting the mold in this way can improve wearer comfort. It can also modify the overall frequency response of the aid by reducing the low frequency output. However, venting increases the likelihood of acoustic feedback.

6. Nonacoustic Input to Hearing Aids

Hearing aid users have considerable difficulty in understanding speech against a background of other masking and unwanted sounds or noises, thus the hearing aid microphone should be placed as close as possible to the speaker to maximize signal-to-noise ratio. Although in theory it is possible to do this with a body-worn aid (the aid placed close to the speaker and the separated receiver (earphone) connected to it by means of a long lead), in practice the method is frequently inconvenient and cumbersome. However, methods of overcoming the problem have been developed.

6.1 The Inductive Pickup Coil ("Telecoil")

The inductive pickup coil (a small coil made from many turns of fine wire) is fitted to many hearing aids. The coil is installed inside the case of the hearing aid. It generates an electrical signal in response to a magnetic field which can be produced from a loop of wire placed around the room and fed via a microphone/amplifier combination. The loop of wire by which the magnetic field is produced has to be installed in the first instance, but once this has been done the hearing aid wearer can hear whoever is speaking into the microphone as though his own hearing aid was adjacent to the speaker. A switch on the hearing aid allows the user to switch out the hearing aid microphone when the aid is being used with the induction loop, thereby minimizing even further the effect of local distractive sounds. Usually this switch is a three position one, combining the function of an on/off switch. The positions are marked "O" (off), "T" (telecoil) and "M" (microphone). The intensity of magnetic field required of a loop to work in conjuction with hearing aid inductive pick-up coils has been specified in the IEC standard 118, Part 4 and this formal recognition has led to an increased usage of the system.

6.2 Radio and Infrared Transmitter/Receivers

A problem arises when it is desired to operate several independent induction loop systems which are adjacent to one another. Magnetic fields cannot easily be constrained within a room, nor is it possible to select one source out of several, and thus there is interference between the systems. Instead of using a magnetic field to convey the speaker's voice to the hearing aid wearer, radio transmission may be used. The signal from a microphone placed close to the speaker is used to modulate a radio wave. The transmitted wave is received and demodulated by a radio receiver carried by the hearing aid wearer, and the audio signal further amplified to suit his requirements. Radio receivers can of course be tuned to a transmission of a particular frequency, so interference is avoided if neighboring systems use transmitters operating on different frequencies. The Department of Trade and Industry (Radio Regulatory Division) (UK) has reserved certain frequencies around 170 MHz in the VHF band for use by radio microphone hearing-aid systems.

Infrared radiation may be used instead of radio waves. The principles are the same in that an infrared transmitter and receiver are required, but as infrared radiation (unlike radio) cannot pass through the walls of a room, interference between systems cannot normally occur and no special tuning devices are necessary. Such systems do not work well out-of-doors or in sunlit rooms as there is a great deal of infrared radiation in sunlight.

Radio systems in particular have been widely used in schools for deaf children. In many of these systems the hearing aid is incorporated in the radio receiver, but some couple the audio output into the child's own hearing aid either inductively via the telecoil or using a direct electrical input (see below).

6.3 Direct Electrical Input

A limited number of hearing aids are fitted with a miniature socket into which the audio signal in electrical form may be fed directly. This enables such aids to be used conveniently in conjunction with the radio microphone system described, or alternatively they may be fed with the sound from a television, cassette player etc.

7. Group Hearing Aids and Auditory Trainers

The use of radio transmission systems in schools for deaf children has already been mentioned. The requirement that the teacher's voice be presented free of background noise can be met with such systems. However, it is often considered desirable that the pupils hear their own voices, and those of their classmates, again as free from background noise as possible. The radio system does not suit this purpose, and group hearing aids may be used instead. The group hearing aid consists, typically, of up to ten places, each place having an associated microphone, amplifier and earphone set for the pupils. The microphone may be on a boom attached to the earphone, to form a "headset." The teacher is also provided with a microphone. A control panel on the group aid allows the teacher to route the audio signals in various combinations through the aid, so that for example the pupils can talk and listen to one another, or the pupils can talk and listen only to the teacher, or the teacher can talk to an individual pupil, etc. The amplifiers usually have a comprehensive set of output and tone controls, which can be accurately set to enable the output to be adjusted to suit the needs of the pupil using it. As a rule, group hearing aids are used only with profoundly deaf children. Similar to group hearing aids are auditory trainers but they are for teachers working with one or two pupils only. Some auditory trainers can be coupled together to form group hearing aids.

See also: Ear Anatomy and Physiology; Audiometers; Artificial Ear; Artificial Mastoid; Sound: Biological Effects

Bibliography

Beagley H A (ed.) 1981 *Audiology and Audiological Medicine.* Oxford University Press, Oxford
Pollack M C (ed.) 1980 *Amplification for the Hearing Impaired*, 2nd edn. Grune and Stratton, New York
Watts W J (ed.) 1983 *Rehabilitation and Acquired Deafness.* Croom Helm, London

M. C. Martin

Heart–Lung Machines

The interior of the heart was the last frontier to be crossed by the surgeon. In 1812 Le Gallois wrote: "If one could substitute for the heart a kind of injection of arterial blood, either naturally or artificially made, one would succeed easily in maintaining alive indefinitely any part of the body whatsoever" (Belt et al. 1920). This prophecy was fulfilled in 1953 when Gibbon performed the first successful open-heart operation on man, after 20 years of development work in the animal laboratory (Gibbon 1954).

Heart–lung machines are devices which take over the pumping and gas-exchange functions of the heart and lungs temporarily, so that the "bypassed" heart can be operated upon. Although extracorporeal pumps were devised to bypass the heart alone (with the patient's lungs effecting gas exchange) the closeness of the heart and lungs and the availability of efficient artificial oxygenators made it convenient and safer to bypass both heart and lungs with an extracorporeal pump oxygenator.

The basic apparatus (Fig. 1) comprises the cannulae which make connection with the patient's aorta and great veins, tubing conduits, a pump, an oxygenator, a heat-exchanger, and a filter. A subsidiary circuit is used to retrieve blood spilt in the operative field.

The blood-contacting surfaces of the circuit must be nonwettable, inert, nonallergenic, pyrogen-free and sterile, and are made of medical-grade plastics. The entire circuit is used once only and is disposable.

1. Pumps

All heart–lung machines use roller pumps (Fig. 2) similar to that originally developed by De Bakey (1934). The pumps are capable of an infinitely variable output flow within the range 0–$8 \, \mathrm{l \, min^{-1}}$.

Until recently, a relatively nonpulsatile flow pattern has been used, on the assumption that if a physiological flow rate was supplied the wave form was unimportant. Some studies (Taylor 1981), however, have demonstrated clearly that most organs prefer a pulsatile flow with an amplitude of at least 25 mm Hg. This work has led to the application of roller pumps with an intermittent action which produce a pulsatile flow closely mimicking normal heart action (Fig. 3). These pumps have the added benefit that they can be used in a counter-pulsation mode to assist a failing heart. For this purpose the pump is activated phasically by the electrocardiogram so that its stroke volume is delivered during cardiac diastole. Thus the circulation is augmented between each heart beat. This form of circulatory assistance is only applicable, of course, in association with partial heart–lung bypass and requires cannulation of the great vessels under operation conditions.

2. Oxygenators

The gas-exchange function of the lungs (i.e., oxygen delivery and carbon dioxide release) can be provided artificially by three methods. In the first and most widely used method, oxygen is dispersed in the blood in the form of bubbles; in the second, transfer occurs through a membrane; and in the third, blood is filmed over screens or disks in an atmosphere of oxygen.

Disk and screen oxygenators are not now used in clinical practice since they are relatively inefficient and cause too much damage to the formed elements of blood. Bubble oxygenators have found the widest application, since Clark and his coworkers demonstrated the

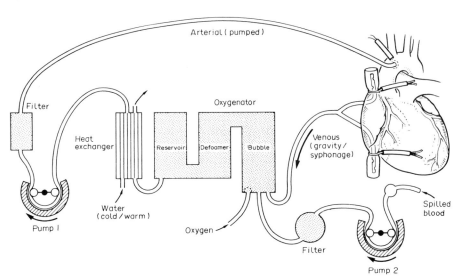

Figure 1
A heart–lung machine and its connections to the patient's circulation

Figure 2
A mechanical roller pump: the operating head of a pump module on the Stockert heart–lung machine

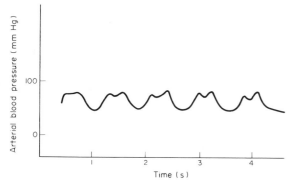

Figure 3
The blood-pressure waveform recorded from a radial artery cannula during pulsatile cardiopulmonary bypass. Mean pump flow = $3.48 \, l \, min^{-1}$

antifoaming qualities of silicone compounds (Clark et al. 1950). Commercially available bubble oxygenators comprise a series of semirigid plastic compartments in which oxygen is bubbled through a column of venous blood, the resultant foam is debubbled through silicone-coated screens and the arterialized blood is collected for onward transmission by the pump to the patient's aorta. These oxygenators have a gas-exchange capacity which amply covers the blood flow rates required in practice (i.e., $0–8 \, l \, min^{-1}$). The direct exposure of blood to gas bubbles causes some minor damage to blood elements and denaturation of blood proteins, and foaming has a defibrinating and platelet-activating action (Clowes et al. 1954). These effects are minimized by reducing the gas-flow to blood-flow ratio to less than 1:1. A gas

mixture of 97.5% O_2 and 2.5% CO_2 is used, so as to maintain a correct arterial carbon dioxide tension of about 40 mm Hg.

In membrane oxygenators, the blood and gas are separated by a microporous membrane (Teflon or silicone rubber) which is supported on an open-weave plastic mesh. The thickness of the blood film can be decreased by inflating a "shim" which compresses the folds of the membrane and increases oxygen transfer without affecting carbon dioxide elimination.

Theoretically, the use of a membrane oxygenator should eliminate the hazards of microbubbles, bubble-induced hemolysis and protein denaturation, and reduce platelet activation. In practice, however, for bypass procedures of less than two hours, these advantages are offset by the increased complexity of the circuit and difficulty in controlling the extracorporeal volume. The resistance to blood flow in a membrane oxygenator is such that the blood must be pumped through it. This requires the use of an extra pump between the venous reservoir and the oxygenator.

Alternatively, if nonpulsatile flow is accepted, the oxygenator must be interposed between the pump and the patient.

Membrane oxygenators are presently used when a long procedure is anticipated or in pediatric practice where damage to the relatively small blood volume presents a greater hazard than in the adult.

3. Filters

Donor blood and autologous blood after contact with extracorporeal surfaces contain abnormal particles which can block the microvasculature of the body during perfusion.

These particles consist mainly of microaggregates of platelets, fibrin strands and the debris of cellular degradation, and their damaging effect has been demonstrated in the capillary network of the lungs, brain and kidney. Consequently, most authorities recommend the use of a filter in extracorporeal circuits, between the pump and the arterial cannula. The design of arterial line filters is still evolving, and the ideal has not yet been reached. If the pore size of the filter is too small, its use is associated with a progressive rise in resistance to flow as the filter screen becomes blocked. Furthermore, there is evidence that the filter itself can damage the blood passed through it. Several types of filter have been evaluated with reference to pore size and distribution, blood-flow pathways, thrombogenicity and the effect of the filter on blood-cell counts. Scanning electron microscopy has been used to study the filtered deposits (Guidoin et al. 1977). A filter of pleated polyester with a pore size of 40 μm is widely used.

4. Heat Exchangers

A device to alter the temperature of the blood during its passage through the extracorporeal circuit is essential to prevent progressive uncontrolled cooling of the patient. The heat exchanger comprises two compartments: one transmits blood, the other takes the form of a water jacket. Variable thermostats regulate the temperature of the circulating water, and blood temperature is monitored. Normally the water-jacket temperature is maintained at 40°C and this maintains a normal blood temperature (37°C).

For some operations the blood temperature is purposely lowered to achieve cooling of the patient (elective hypothermia). This is done to reduce the metabolic requirements of the body and allow a reduction in pump output.

The heat exchanger is commonly an integral part of the oxygenator or it may be a separate module added to the circuit between the oxygenator and the arterial filter.

When the blood temperature is being altered, care is taken to avoid abrupt temperature gradients, since these can cause damage to the blood cells or the release of gas from solution in the plasma.

5. Priming Solutions

The heart–lung machine and its associated circuitry requires a priming volume of about 2 l for adults. The fluid usually used is a modified Ringer's solution which is isotonic and isosomolar with blood. The infusion of this volume of crystalloid solution into the patient at the beginning of heart–lung bypass causes dilution of the blood, so that the "packed cell volume" is reduced from the normal 45% to about 26%. The associated reduction in blood viscosity aids capillary flow during bypass. The excess fluid and electrolytes are excreted in the urine over the subsequent 4–5 h.

See also: Hemodynamics

Bibliography

Belt A E, Smith H P, Whipple G H 1920 Factors concerned in the perfusion of living organs and tissues. Artificial solutions substituted for blood serum and the resulting injury to parenchymal cells. *Am. J. Physiol.* 52: 101–20
Clark L C, Gollan F, Gupta V B 1950 The oxygenation of blood by gas dispersion. *Science* 111: 85–87
Clowes G H A, Neville W E, Hopkins A, Anzola J, Simeone F A 1954 Factors contributing to success or failure in use of pump oxygenator for complete bypass of heart and lung: Experimental and clinical. *Surgery* 36: 557–79
De Bakey M E 1934 Simple continuous-flow blood transfusion instrument. *New Orleans Med. Surg. J.* 87: 386–89
Galletti P M, Brecher G A 1962 *Heart–Lung By-pass: Principles and Techniques of Extra-Corporeal Circulation.* Grune and Stratton, New York
Gibbon J H Jr 1954 Application of a mechanical heart and lung apparatus to cardiac surgery. *Minn. Med.* 37: 171–85
Guidoin R, Taylor K, Bain W H 1977 Blood filter evaluation. *Biomater. Med. Devices, Artif. Organs* 5: 317–36
Melrose D G 1981 Historical review and introduction. In:
Longmore D B (ed.) 1981 *Towards Safer Cardiac Surgery.* MTP, Lancaster, UK, pp. 259–65
Reed C C, Clark D K 1975 *Cardiopulmonary Perfusion.* Texas Medical Press, Houston
Taylor K M 1981 Why pulsatile flow during cardiopulmonary bypass? In: Longmore D B (ed.) 1981 *Towards Safer Cardiac Surgery.* MTP, Lancaster, UK, pp. 483–500

W. H. Bain

Heart Valves

The relatively simple function of the human heart valves allowed recognition of their role in the production of the circulation of the blood long before most other human physiological activities were understood. Clear understanding of the anatomy, physiology and pathology of heart valves is currently of great practical relevance to the fields of investigative cardiology, cardiac surgery and bioengineering, as a result of recent advances in these fields. Accurate assessment of valve anatomy and function is routine clinical practice and replacement of diseased valves by prostheses is well established.

1. Anatomy

The heart contains two types of valves: the atrioventricular valves and the ventriculoarterial valves.

The atrioventricular valves are situated between the thin-walled, low-pressure receiving chambers of the heart (the atria) and the high-pressure, muscular pumping chambers of the heart (the ventricles). The orifice between the atria and ventricles contains a fibrous annulus which gives rise to the leaflets of the atrioventricular valves. These leaflets are composed chiefly of collagen and elastin fibers embedded in a mucopolysaccharide ground substance containing fibroblast cells responsible for growth and repair. They are covered by a single layer of vascular endothelium, and are tethered by chordae tendineae (collagenous extensions of the leaflets), chiefly from their free edges, to papillary muscles which arise from the inner aspect of the ventricular myocardium. The tricuspid valve, located between right atrium and right ventricle, has three leaflets, and the mitral valve, located between left atrium and left ventricle, has two.

The ventriculoarterial valves are situated between the high-pressure muscular pumping chambers or ventricles and the major arteries leaving the heart. Each valve consists of three semilunar-shaped leaflets attached at a scallop-shaped annulus surrounding the orifice of each major artery. The leaflets are collagen and elastin fibers in a polysaccharide ground substance with fibroblast cells, and there is an endothelial lining. The pulmonary valve is located between right ventricle and pulmonary

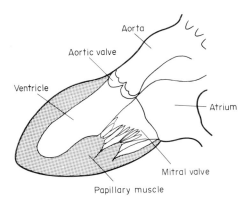

Figure 1
Left-heart anatomy

artery, and the aortic valve is located between left ventricle and aorta.

Figure 1 shows the anatomical features of the mitral and aortic valves.

2. Physiology

The heart valves function as one-way valves, allowing muscular contraction of the ventricles to impart unidirectional flow to blood within the heart. Opening and closing of the valves is largely a passive phenomenon, in response to pressure relationships on either side of the valve. Active contraction of the papillary muscle during ventricular contraction does occur and prevents prolapse of the atrioventricular valves into the atrium during ventricular contraction (systole). Figure 2 shows the pressure relationships between the left atrium, left ventricle and aorta throughout one cardiac cycle and indicates the points of opening and closing of the left-sided

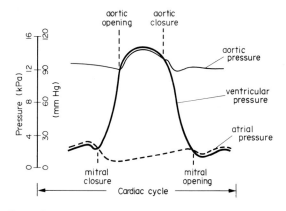

Figure 2
Left-heart pressure relationships throughout one cardiac cycle

heart valves. A pressure drop across the valve during the flow phase and regurgitation through the valve during the closed phase are not detectable with normal heart valves.

3. Pathology

The commonest congenital valve abnormality is valve stenosis or obstruction due to incomplete separation of leaflets during their development. This most commonly affects the pulmonary and the aortic valves.

Rheumatic fever may be followed after many years by a chronic, fibrosing reaction in the heart valves—usually the mitral and aortic valves—resulting in either valve stenosis due to increasing fusion of valve leaflets, or regurgitation due to retraction and shortening of the scarred leaflets, or a combination of hemodynamic effects.

Syphilis is now a rare cause of destruction of the aortic valve annulus and leaflets which results in valve regurgitation.

Infective endocarditis is a consequence of bacterial (most commonly *Streptococcus viridans*), fungal or rickettsial infection of a heart valve which is usually already abnormal. The infection produces rapid valve destruction and valve regurgitation.

4. Pathophysiological Effects of Valve Dysfunction

Stenosis of a valve causes a pressure drop across the valve and the effects seen are usually the result of the increased pressure proximal to the obstructed valve. Thus, with mitral stenosis, increased venous pressure in the pulmonary veins results in lung congestion, stiffening of the lungs (giving shortness of breath) and the tendency to pulmonary edema. Stenosis of an aortic valve causes left-ventricular myocardial hypertrophy, and ultimately left-ventricular failure.

Regurgitation results in pressure effects proximal to the valve. With mitral regurgitation, this causes pulmonary congestion, shortness of breath and the development of pulmonary edema. With aortic regurgitation there is left-ventricular hypertrophy and dilatation of the heart. The rapid loss of arterial pressure during diastole as a result of regurgitation of blood into the left ventricle produces a rapid fall of arterial pressure, readily detectable clinically (Braunwald 1984).

5. Methods of Assessment of Valve Function

(*a*) *Clinical assessment.* The clinician is guided by the patient's history and symptoms in assessing the likely pathology and the severity of valvular dysfunction, and is helped by the detection of cardiac murmurs. Murmurs arise due to turbulence in blood flow through the heart.

Systolic murmurs over the mitral or tricuspid valves indicate regurgitation of blood through these valves, whereas diastolic murmurs indicate obstruction to blood flow across the valves. Over the pulmonary and aortic valves, a systolic murmur indicates obstruction to blood flow and a diastolic murmur indicates regurgitation. The quality, localization and radiation of these murmurs enable a fairly accurate diagnosis to be made in most cases.

(*b*) *Radiology*. Radiology of the heart and lung provides valuable evidence of specific chamber enlargement in the heart, or pulmonary congestion (e.g., left-atrial enlargement with pulmonary congestion suggests mitral valve disease). It is also possible to see calcification in abnormal heart valves in this way.

(*c*) *Echocardiography*. Echocardiography enables movements of heart-valve leaflets to be quantitated and the thickness and quality of the leaflet tissue to be assessed (see *Echocardiography*).

(*d*) *Cardiac catheterization*. Cardiac catheterization gives evidence of valve dysfunction by measuring pressures in the various cardiac chambers (see *Cardiac Catheterization*). Thus the pressure drop across a stenotic valve enables the degree of stenosis to be quantitated. Similarly, the elevation of pressure and regurgitant pressure wave form proximal to an atrioventricular valve will allow assessment of the severity of regurgitation. Cineangiocardiography gives further evidence of valve regurgitation. An injection of a bolus of radiopaque medium into the aorta or the left ventricle will give evidence of regurgitation at the aortic or mitral valves, respectively, if they are regurgitant, and allows a semiquantitative assessment of severity.

6. Heart-Valve Replacement

The first successful removal of a diseased aortic valve and replacement with a prosthetic valve was performed in 1960 by Harken (Harken et al. 1960), and the replacement of a mitral valve by Starr was also undertaken in 1960 (Starr and Edwards 1961). Nowadays replacement surgery is commonly performed and in the UK in 1981 over 4500 patients underwent valve replacement surgery. Replacement still entails an operative mortality somewhere in the region of 5–10% in the best centers, due largely to problems of myocardial ischemic damage during surgery, bleeding, or cardiopulmonary bypass complications.

Early in the development of prosthetic heart valves, the following criteria for a satisfactory valve were recognized:

(a) valve function should be comparable to that of a natural heart valve, and should allow unobstructed forward flow with minimal back flow over the range of heart rate encountered physiologically and pathologically,

(b) the valve should not be subject to the deposition of platelets and blood clot,

(c) the valve should not be subject to overgrowth of host tissue which could impair valve function,

(d) the valve should not damage blood components (i.e., it should not cause hemolysis), and

(e) the valve should be durable, ideally outlasting the patient's life expectancy.

These criteria have still not all been met in one valve, a fact attested to by the wide variety of prosthetic heart valves in clinical use.

7. Types of Prosthetic Heart Valve

Prosthetic heart valves are frequently divided into two categories: mechanical valves, and biological (or tissue) valves. This division arises partly from the nature of their component materials and partly from their clinical behavior.

7.1 Mechanical Valves

Mechanical valves are manufactured from materials such as stainless steel, various metallic alloys, silicone rubber or pyrolite carbon. The hemodynamic function of mechanical valves is less than ideal, but continuous improvements in design have resulted in a large number of clinically acceptable valves currently available. Mechanical valves function with a transvalve pressure loss in the order of 5–20 mm Hg (0.6–2.7 kPa) at heart rates between 60 and 120 beats per minute, and give ventricular filling flow rates between 100 and 300 ml s^{-1}. This effectively gives a mild degree of mitral stenosis for valves used in the mitral position, but in the aortic position, this degree of pressure loss is not a significant problem in clinical practice (Scotten et al. 1981).

Throughout their history, mechanical valves have been troubled by the problems of platelet deposition and thrombus formation on the valve. This may either interfere with valve function, or a thrombus may break free and embolize in the systemic circulation to cause a wide variety of complications depending on the site of impaction, e.g., cerebral complications such as hemiplegia. Indefinite oral administration of systemic anticoagulants such as Warfarin is required to minimize the risk of thromboembolism. This requires regular checks of prothrombin time and carries a risk of hemorrhage, particularly where other complaints such as peptic ulceration are present.

Mechanical valves of early design were often associated with hemolysis, at times severe enough to result in recurrent anemia.

Durability has been a problem with some valves. Lipid absorption by silicone-rubber heart-valve poppets gave rise to ball-valve variance, but this problem has largely been solved, either by improved methods for curing the silicone rubber or by the use of stellite alloy hollow metallic poppets. Disk valves manufactured of delrin have shown wear, but this problem has largely been surmounted by the use of pyrolite carbon disks. Wear has occurred on cloth-covered struts in mechanical valves but design modifications have reduced this problem. Most of the currently available mechanical valves are, for practical purposes, likely to be durable enough to cover the life expectancy of most patients.

In practice, the use of mechanical valves has the disadvantages of long-term anticoagulation and the small risk of thromboembolism, but the reassurance of long-term durability.

A large number of prostheses have been designed and the more commonly available prostheses in present clinical practice are briefly described.

(*a*) *Starr–Edwards caged ball prosthesis* (Fig. 3). This is the best known of all the prosthetic heart valves. It consists basically of a metal ring or orifice surrounded by a fabric sewing ring for insertion, and a ball occluder held in a retaining cage. The valve can be used for replacement of either the atrioventricular or the ventriculoarterial heart valves.

Flow through such a valve is affected by the valve ring dimensions (the primary orifice), the space between the valve ring and the ball in the open position (the secondary orifice), and the space between the ball and the surrounding aorta or left ventricle (the tertiary orifice) which is dependent on the size and shape of the aorta or left ventricle into which the cage projects. Initially the valve had a large cage with large secondary orifice, but obstruction at the tertiary orifice necessitated design modifications with reduction in cage size. Early valves had a silicone-rubber ball and bare metal struts to the cage. Early problems of ball-valve variance resulted in valve failure, but this problem has largely been overcome and the silastic ball valve with the bare struts (Model 6120 mitral prosthesis and Model 1200 aortic prosthesis) is still in frequent use and has given excellent long-term clinical results.

The development of the Starr–Edwards prosthesis is a good illustration of the practical problems encountered in prosthetic heart-valve design. In an attempt to reduce the incidence of thromboembolism arising from the valve, the metal struts were covered with a porous polypropylene cloth—as bare metal surfaces are known to favor the formation of thrombus. This, however, reduced the clearance between the poppet and the cage, and build-up of fibrin on the cage increased the risk of ball entrapment. A further modification, in which the occluder consisted of a hollow metallic poppet of stellite alloy, obviated the problem of ball variance but increased the likelihood of cloth wear on the struts and seat. This resulted in a further modification in which a composite seat of cloth and metallic alloy studs was used, and later a further modification in which metallic tracks were added on the inside of the cage struts.

Because of the problems of accommodating a cage in a small ventricle, attempts to design a "low-profile"

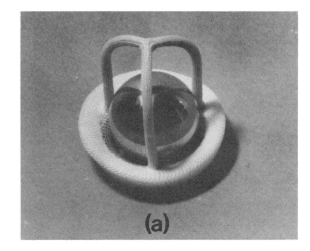

(a)

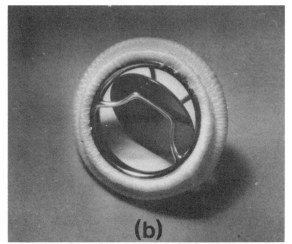

(b)

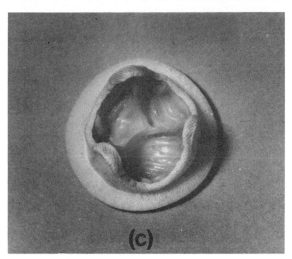

(c)

Figure 3
Prosthetic heart valves: (a) Starr–Edwards, (b) Bjork–Shiley, and (c) Carpentier–Edwards

prosthesis resulted in a number of prostheses employing disk occluders (Scotten et al. 1981, Murphy et al. 1983).

(*b*) *Starr–Edwards disk valve*. In this valve, a plastic disk is restrained within a cage consisting of four metal struts. Flow characteristics are not as good with this valve, but there is less protrusion of the cage into the ventricle when used in the mitral position.

Another type of mechanical valve is the tilting disk prosthesis, in which the disk occluder rotates about an axis, giving a much wider orifice area than the centrally occluding valves.

(*c*) *Bjork–Shiley prosthesis* (Fig. 3b). This is the best known tilting disk prosthesis, a device in which the disk occluder rotates about an axis, giving a much wider orifice area than the centrally occluding valves. A centrally occluding disk, initially made of delrin and, more recently (to reduce wear problems) of pyrolitic carbon, is restrained by two U-shaped struts, obviating the need for a hinge. The valve has a low profile and good hemodynamic function, and has a good clinical record (Murphy et al. 1983).

7.2 Biological or Tissue Valves

Biological or tissue valves are either totally or in-part derived from living tissues. The aortic valve, either of human or animal origin, may be used mounted on a frame to replace either the atrioventricular or ventriculo-arterial valves. Alternatively, some form of collagen sheet such as pericardium, fascia lata, or dura mater can be used and fashioned on a three-pronged stent to form a three-leaflet valve—the biological component of such a valve being the flexible valve leaflets. The reason for using biological or tissue valves is the low incidence of valve thromboembolism and the freedom from the need for long-term anticoagulation. The major problem that has plagued the use of these valves since their introduction has been tissue degeneration with valve failure. The practical implications of using tissue valves in clinical practice today are the minimal incidence of thromboembolism without the need for long-term anticoagulation, but a continued uncertainty about the long-term durability of such valves may imply the need for repeat surgery.

(*a*) *Allograft heart valves*. A normal human aortic heart valve removed shortly after death was first transplanted into the aortic valve position of a patient by Donald Ross in 1962 (Ross 1962). The allograft valve (i.e., tissue transplanted between members of the same species) has given excellent clinical function when used for aortic valve replacement in a direct valve transplant. The valve requires some form of sterilization in its preparation and this was initially done by using betapropiolactone, γ irradiation, ethylene oxide, or exposure to mixtures of antibiotics. The valves function as nonliving grafts and have normal hemodynamic function without the risk of thromboembolism. There are obvious difficulties in acquiring such valves and considerable skill is needed for their accurate insertion. These valves can be mounted onto a three-pronged frame and turned upside down for use in the mitral position. Unfortunately, late degeneration of the valve has marred its clinical success, and although antibiotic sterilization appears to give better results than the earlier methods, the problem of late degeneration has severely curtailed the use of this valve.

(*b*) *Glutaraldehyde-preserved xenograft prostheses*. The aortic valve of the pig has been used for commercial preparation of prosthetic heart valves. The pig's aortic valve is mounted on a three-pronged frame for ease of insertion into either the atrioventricular or ventriculo-arterial position. The valve is sterilized and treated with glutaraldehyde resulting in increased cross-linkage of collagen molecules. This has the effect of reducing antigenicity and considerably improving the tensile strength of the valves. It also stiffens the leaflets with the consequence that hemodynamic function is not as good as in the naturally occurring valve. Pressure loss across these valves is comparable to that across the more recent mechanical valves. The valves are free from the problem of thromboembolism, but the late occurrence of calcification and degeneration in the tissues has given rise to some concern. At present there is uncertainty about the long-term future of these valves beyond the first decade. The two best known commercially available prostheses of this sort are the Hancock prosthesis and the Carpentier–Edwards prosthesis (Fig. 3c).

The glutaraldehyde-preserved pericardial valve is a valve in which bovine pericardium is treated with glutaraldehyde to increase collagen cross-linkage and improve strength. The resultant sheet of strengthened collagen is used to fashion a three-leaflet valve on a three-pronged stent which mimics the naturally occurring aortic valve. This valve can be used to replace the atrioventricular or ventriculoarterial valves. Hemodynamic function is particularly good in this valve. It is commercially available as the Ionescu–Shiley, the Hancock pericardial or the Edwards pericardial valve, which have small pressure losses across the valve, large orifice areas, and freedom from thromboembolism. Uncertainty about the long-term durability remains, as with other tissue valves (Gabbay et al. 1984).

Most currently available prosthetic heart valves provide similar hemodynamic performance in the mitral position. In the aortic position, there is similar hemodynamic performance from all the presently available prosthetic heart valves, with the exception of the smaller sizes of the glutaraldehyde-prepared porcine valves which are moderately stenotic.

7.3 Survival Rates

Five-year survival for patients having successful mitral-valve replacement in recent years has been reported from various implantation centers at between 70 and 90% using both tissue and mechanical valves. Embolus-free survival for five years after mitral-valve replacement has varied between about 80 and 96% for both mechanical valves with anticoagulation and biological valves without anticoagulation.

Five-year survival after aortic-valve replacement has been reported at between 70 and 86% for both tissue and biological valves. Five-year embolus-free survival for aortic valve prostheses has been reported at between 87 and 97% for both biological valves without anticoagulation and mechanical valves with anticoagulation.

Current results of valve replacement give much better clinical prospects than for untreated valvular heart disease. However, problems remain and the ideal heart-valve prosthesis is not yet in existence (Cohn et al. 1984, Brais et al. 1985).

See also: Implanted Prostheses: Tissue Response

Bibliography

Brais M P, Bedard J P, Goldstein W, Koshal A, Keon W J 1985 Ionescu–Shiley pericardial xenografts: Follow-up of up to 6 years. *Ann. Thorac. Surg.* 39: 105–11
Braunwald E 1984 Valvular heart disease. In: Braunwald E (ed.) 1984 *Heart Disease. A Textbook of Cardiovascular Medicine.* Saunders, Philadelphia, pp. 1063–1135
Cohn L H, Allred E N, DiSesa V J, Sawtelle K, Shemin R J, Collins J J 1984 Early and late risk of aortic valve replacement: A 12 year concomitant comparison of the porcine bioprosthesis and tilting disc prosthetic aortic valves. *J. Thorac. Cardiovasc. Surg.* 88: 695–705
Harken D E, Soroff H S, Taylor W J 1960 Partial and complete prostheses in aortic insufficiency. *J. Thorac. Cardiovasc. Surg.* 40: 744–62
Hurst J W 1974 *The Heart, Arteries and Veins.* McGraw-Hill, Tokyo
Macmanus Q, Grunkemeier G L, Lambert L E, Tepley J F, Harlan B J, Starr A 1980 Year of operation as a risk factor in the late results of valve replacement. *J. Thorac. Cardiovasc. Surg.* 80: 834–41
Murphy D A, Levine F H, Buckley M J, Swinski L, Daggett W M, Akins C W, Austen W G 1983 Mechanical valves: A comparative analysis of the Starr–Edwards and Bjork–Shiley prostheses. *J. Thorac. Cardiovasc. Surg.* 86: 746–52
Oyer P E, Miller D C, Stinson E B, Reitz B A, Moreno-Cabral R J, Shumway N E 1980 Clinical durability of the Hancock porcine bioprosthetic valve. *J. Thorac. Cardiovasc. Surg.* 80: 824–33
Ross D N 1962 Homograft replacement of the aortic valve. *Lancet* ii: 487
Scotten L N, Racca R G, Nugent A H, Walker D K, Brownlee R T 1981 New tilting disc cardiac valve prostheses. *In vitro* comparison of their hydrodynamic performance in the mitral position. *J. Thorac. Cardiovasc. Surg.* 82: 136–46
Sebening F, Klovekorn W P, Meisner H, Struck E 1979 *Bioprosthetic Cardiac Valves.* Deutches Herzzentrum, Munich, FRG
Starr A, Edwards M L 1961 Mitral replacement: Clinical experience with a ball valve prosthesis. *Ann. Surg.* 154: 726–40
Walker D K, Scotten L N, Brownlee R T 1984 New

generation tissue valves: Their *in vitro* function in the mitral position. *J. Thorac. Cardiovasc. Surg.* 88: 573–82

D. J. Wheatley

Hemodynamics

Blood, like any other fluid, flows down a pressure gradient in a manner determined by its rheological properties and by the geometry of the channel along which it flows. Hemodynamics is a branch of fluid dynamics in which the motion of the blood and the forces involved in its circulation are considered.

1. Steady Flow

The volumetric flow rate Q of a fluid is related to the pressure gradient P' by the Poiseuille expression (which itself resulted from an attempt to describe blood flow):

$$Q = P' \pi r^4 / 8\eta \tag{1}$$

where η is the fluid viscosity and r is the vessel radius. The dominant effect of the term r^4 has led to the use of pressure–flow relationships to infer changes in vessel caliber. By analogy with Ohm's law, a variable called fluid-resistance is determined, the complex vascular bed being modelled as a series/parallel combination of many resistive segments.

1.1 Newtonian Viscosity

In the larger blood vessels (say $r > 0.25 \, \text{mm}$) the assumption of Newtonian viscosity is probably reasonable, but elsewhere the effective viscosity decreases with vessel size (see *Blood Viscosity Measurement*). In the case of the narrowest vessels (capillaries), which are of the same order of size as red blood cells, it is no longer appropriate to treat blood as a single-phase material. Studies suggest that red-cell deformability is an important variable in this region. Under normal conditions it is safe to neglect these non-Newtonian effects when considering short-term variations in the fluid resistance of a vascular segment. Changes in cell properties may be significant, however, in extremely sick individuals.

1.2 Established Flow

The inlet length of the aorta covers the entire length of the vessel and is generally taken to be of the order $0.8 \times r \times Re$, where Re is Reynolds number (defined as $\bar{V} R v^{-1}$, where $\bar{V}$ is the average flow velocity and v is the kinematic viscosity of the fluid). The inlet length of a tube is defined as the distance a fluid has to travel down the tube to establish a new steady state as evidenced by the attainment of the fully developed parabolic profile of Poiseuille flow. Local accelerations occur within the entrance length which result in extra pressure losses. In narrower vessels (e.g., the veins and the micro-circulation) entrance effects are detected, although laminar flow is probably realized.

1.3 Laminar Flow

For conventional fluids, laminar or streamline Poiseuille flow is expected for $Re < 1000$. Laminar flow apparently exists when blood flows steadily in regular channels. "Steady" turbulence should not, therefore, occur in the circulation. However, when flow is oscillatory and channels are irregular the situation is much more complex (see Sect. 2.).

2. Oscillatory Flow

Nowhere in the circulation is flow steady. The major oscillations in flow are due to the heart beat but respiration and muscular activity are also contributory influences. In linear systems the steady and oscillatory events may be considered separately and then summed. The circulation is reasonably linear, justifying the determination of vascular resistance as the ratio of the mean pressure difference and the mean flow.

The study of oscillatory flow relationships has been a major preoccupaton of hemodynamicists. Because of the rather regular beat of the heart and the linearity (to a first-order approximation) of the vascular system, a treatment based on the analysis of the steady-state in the frequency domain using Fourier methods has proved successful. (The waves, e.g., of pressure or flow, are considered members of a long series of identical waves, which, indeed, is often the case.) Alternative but less tractable treatments operating in the time domain were developed earlier ("Windkessel" models). They may have a place in the analysis of intracardiac events where the steady-state concept is not truly valid.

Steady-state analysis starts with consideration of the flow resulting from an oscillatory pressure gradient, $P' = M \cos(\omega t - \phi)$, where M is the amplitude or modulus of the pressure change (note that the peak-to-peak variation is $2M$), ϕ is its phase, and ω the angular frequency (expressed in radians).

The response of the fluid depends on its longitudinal or axial impedance which, in turn, is dependent on conduit geometry (here a regular cylinder radius r) as well as on fluid viscosity and intertia. (Blood is essentially incompressible but the vessels are distensible and vascular viscoelasticity has an important influence on wave travel and blood flow, see below.) Viscosity and inertia are of paramount importance and their effects were introduced through the dimensionless parameter α which, like Re, characterizes the balance of fluid resistance and inertial forces:

$$\alpha = r(\omega/v)^{1/2} \tag{2}$$

and the resulting expression for flow is of the form (here simplified)

$$Q = MW \cos(\omega t - \phi - \theta) \tag{3}$$

where W and θ are functions of α. When $\alpha \ll 1$ (very small vessels, very low frequency) resistive effects dominate and Eqn. (3) reduces to the Poiseuille Eqn. (1), indicating a flow change of appropriate amplitude in phase with the pressure gradient. Conversely for $\alpha > 10$ inertia predominates and flow diminishes, lagging the gradient by $90°$ ($\pi/2$ rad) exactly, as would a pure mass for which acceleration is proportional to force.

Indeed, in the aorta where the values for α are the highest, a good representation of the oscillatory flow is obtained by integrating the pressure gradient. In the mammalian circulation, values for α vary between these two extremes. The amplitude and phase terms in Eqn. (3) are tabulated functions (e.g., in McDonald 1974). Flow waves computed with them correspond satisfactorily with actual measurements. It is also predicted that the velocity profile varies with α from its parabolic Poiseuille form for $\alpha \ll 1$ to a flat profile with shear concentrated at the wall for $\alpha > 10$. These predictions have been verified experimentally.

2.1 Vascular Tethering

Considerable axial movements of the blood vessels would be expected if they are considered to be free elastic tubes. As such movements have not been observed, a tethering term was introduced in descriptive mathematical expressions of blood flow to take account of any external restraint loading of blood vessels (McDonald 1974). What little is known about the exact form of restraint is described in Patel and Vaishnav (1980), but it is clear that little movement is possible. The expressions for very stiff restraint become conveniently rather simple and are thus generally used.

2.2 Viscosity

Very little is known about the effects of non-Newtonian viscosity in oscillatory flow, but these effects are unlikely to be large and are indeed neglected in large vessel studies.

2.3 Entrance Effects

For oscillatory flow the entrance length is of the order $3.4 v/\omega$, where v is instantaneous core velocity. The predicted oscillatory entrance length is much shorter than that for the steady component in the aorta. This implies a flat velocity profile. In smaller vessels, shorter entrance lengths and lower values of α modify the profile towards the parabolic form. However, bends, branches, and all departures from simple cylindrical geometry have marked effects on the flow profile. Because of a possible association between vessel wall disease and flow patterns, many theoretical and experimental studies have been reported, generally using rather simple geometries and flows. It is clear that the details of vascular structure are of great importance. There is very little knowledge of what actually occurs *in vivo*.

2.4 Turbulence and Disturbed Flow

When instantaneous Reynolds numbers are computed

for flow into the aortic root it is apparent that the critical number is greatly exceeded during peak flow. While the breakdown of steady laminar flow can be detected *in vitro* by an increased pressure loss or by dye mixing, these methods are inapplicable *in vivo*, and miniature flow sensors are needed in the stream to detect turbulence. (Note that the onset of turbulence is not necessarily associated with the generation of sound which is an altogether more complex process.) It appears that flow often becomes turbulent in the aorta for a brief period following peak flow under exercise conditions and possibly at rest, but the disturbance dies away in diastole. Thus only that portion of the aorta filled by one ejection, maybe a third of its length, will be affected by turbulence. In general, it seems that flow is stabilized in oscillatory conditions and that the critical Reynolds number rises with α, being of the order 150α. Turbulence is unlikely to be harmful in itself but, again, altered flow patterns may affect the vessel wall. Departures from simple cylindrical geometries produce complex disturbances with secondary flows as *Re* increases. Abrupt caliber changes (e.g., stiffened narrowed valves) can lead to the generation of high velocity turbulent jets and pronounced downstream disturbances.

2.5 Nonlinearities

The calculation of a flow from a nonsinusoidal repetitive pressure gradient involves decomposition into harmonics of pressure, calculation of the flow at each harmonic frequency together ideally with the mean term, and the final synthesis of the flow wave as the sum of the harmonic terms. This is valid so long as pressure gradient and flow are linearly related. In addition to the problems listed above, it must be appreciated that the radius of a distensible tube is itself a function of pressure. However, the best estimates indicate that even in the relatively distensible pulmonary artery such effects are not large and that calculated oscillatory flows in segments of vessel devoid of major branches or irregularities agree well with measurement. In practice, the steady flow component is the most troublesome parameter since the mean pressure gradient is very small and is measured, necessarily, from two relatively large absolute pressures.

3. Oscillatory Pressure Gradient

With each heart beat a pressure wave is developed at the aortic root. The wave propagates through the arterial system and can be detected beyond the capillaries. If propagation were lossless and all components travelled at the same velocity c (no dispersion), then

$$dP/dx = (-dP/dt)(1/c) \qquad (4)$$

since $c = dx/dt$. Using Eqns. (3) and (4), flow can be computed from the time differential of pressure. Equation (4) also shows that the peak pressure gradient will occur before peak pressure so that peak flow generally precedes peak pressure, a fact which has puzzled workers in the past. However, Eqn. (4) does not reliably predict flow because in practice the wave speed is not constant and attenuation occurs.

4. Wave Propagation in an Infinite Tube

The simplest expression for the propagation of a pressure wave in an elastic tube is known as the Moens–Korteweg equation:

$$c = (Eh/2r\rho)^{1/2} \qquad (5)$$

where E is the Young's modulus of the tube material, h the wall thickness and ρ the fluid density. (Other propagating disturbances exist, e.g., axial and torsional waves.) For the circulation, Eqn. (5) needs modification to take account of three particular problems: arterial elasticity, fluid dynamical factors, and wave reflection.

4.1 Arterial Elasticity

The arterial wall shows non-Hookean properties, i.e., it gets stiffer as it is stretched, is anisotropic and viscoelastic and subject to external tethering (McDonald 1974).

4.2 Fluid Dynamical Factors

For $\alpha > 10$, the expression for c, modified to take account of items listed above, shows an upper asymptote, but for lower values the situation is much less clear and the experimental evidence rather thin. Similarly, the attenuation factors are influenced by α and by wall properties, rendering the real situation even more obscure. Experiments *in vivo* are seriously hampered by reflections.

4.3 Wave Reflection

Arteries are relatively short, e.g., the human aorta is, say, 50 cm long. Thus with a wave speed of $\sim 5 \, m \, s^{-1}$, the aortic length is approximately 0.1 wavelengths when $f = 1 \, Hz$, a typical resting heart rate. There is ample experimental evidence for the existence of reflections which appear to return from a distributed set of closed-end terminations with a few major discrete points, e.g., the aortic termination. The peripheral vascular bed is much "tighter" hydraulically than the aorta and the two are coupled to some extent by a gradual increase in arterial stiffness and wave speed as one travels away from the heart; matching, however, is not perfect and further reflections arise. Incident and reflected waves interfere, affecting the measured propagation velocity, and the "apparent phase velocity" may differ markedly from that predicted by Eqn. (5) for a wave at this frequency. A wave running into a gradually stiffening and tapering tube grows in amplitude so that peripheral pressure pulses are greater than aortic ones ("peaking") and are amplified rather than attenuated.

4.4 Nonlinearities

The simple expressions take no account of the fact that wave speed is itself a function of pressure (by virtue of non-Hookean arterial wall properties), nor of the convection of the wave by the moving blood so that incident and reflected waves travel at $c \pm v$. Normally v is very much less than c, but in the aorta and under conditions of high flow (e.g., exercise or disease) this may not be true.

5. Hydraulic Input Impedance

The influence of all the factors discussed can be seen in the relation between the pressure and flow in any chosen artery. These are measured and analyzed to extract their Fourier harmonics which results in terms at frequencies of zero (mean terms) and 1, 2, 3, ... times heart frequency for a regular beat. These are then related term by term to produce a frequency spectrum showing amplitude ratios and phase differences for as many terms as contain useful information, normally < 10. The impedance spectrum shows the combined effects of fluid and wave propagation factors, the influence of reflections generally being most apparent. The practical significance of the impedance is that it determines the work necessary to pump blood through a bed. A proper understanding of its determinants will lead to better understanding of the work demands on the heart.

See also: Blood Flow: Invasive and Noninvasive Measurement; Blood Pressure: Invasive and Noninvasive Measurement; Cardiac Output Measurement

Bibliography

Bergel D H (ed.) 1972 *Cardiovascular Fluid Dynamics.* Academic Press, London
Caro C G, Pedley T J, Schroter R C, Seed W A 1978 *The Mechanics of the Circulation.* Oxford University Press, Oxford
McDonald D A 1974 *Blood Flow in Arteries*, 2nd edn. Arnold, London
Milner R W 1982 *Hemodynamics.* Williams and Wilkins, Baltimore
Patel D J, Vaishnav R N 1980 *Basic Hemodynamics and its Role in Disease Processes.* University Park Press, Baltimore

D. H. Bergel

Hormesis and Homeostasis

Luckey (1980, p. 49) has defined hormesis as beneficial stimulation by "... subharmful quantities of any agent to any (biological) system." The phenomenon has sometimes been referred to as the Arndt–Schulz "law" (Schulz 1888, Luckey 1980, Stebbing 1979, 1981). In 1919, in studies of effects of x-irradiation on the flour beetle, *Tribolium confusum*, Davey found that while high level exposures were damaging or lethal, a low-level range existed within which survival of experimental insects was prolonged as compared with controls. Cork (1957) confirmed Davey's "beneficial" radiation effects studies and showed that life prolongation in *T. confusum* occurred in response to moderate exposures to γ radiation as well. Cork considered "... the possibility that the slight destruction of tissue incurred by the light (mild) irradiation may be sufficient to stimulate the repair mechanism ..." He also noted that the irradiated organism was able not only to "restore normalcy in spite of continuing damage, ... but also to cope with other events which under normal conditions might have been lethal."

Comparably, Southam and Ehrlich (1943) observed stimulation of fungal growth by low levels of oak bark extracts that prevented growth at higher levels, a phenomenon they termed "hormesis." In recent years, it has been realized that hormetic phenomena are quite general and involve responses to chemicals and to physical factors such as radiation and cold (Luckey 1980, Stebbing 1981). For example, low doses of toxaphene were found to stimulate growth of the house cricket, *Acheta domesticus* (Luckey 1968). Also, quite low levels of certain antibiotics stimulate growth of both conventional and germfree laboratory animals (Luckey 1978) and the vinegar eel, *Turbatrix aceti* (Luckey 1963). Such stimulatory phenomena have also been referred to as "hormoligosis" (Luckey 1968).

Hundreds of examples of hormesis involving various hazardous or toxic agents have been reported over the last 80 or more years (Luckey 1980, 1982, 1984, 1985). Goodman and Gilman (1955) recognized the hormetic phenomenon in the effects of quinine. They observed that since quinine affects so many biological systems, it has been considered a "general protoplasmic poison," and that "like many other poisons ..., it may stimulate in low ... but depress in higher concentrations." "Other poisons" may include cadmium, arsenic, germanium, nicotine and ionizing radiation. An hormetic phenomenon encompassing effects of both radiation and toxic chemicals is likely to be biophysical rather than biochemical.

In studies of the marine hydroid, *Campanularia flexurosa*, Stebbing (1979) found that every inhibitory variable examined stimulated growth within a narrow range of subinhibitory concentrations, and that hormesis was due to transient overcorrections by control mechanisms to inhibitory challenges well within the capacity of the organism to counteract. Such protective phenomena appear to be examples of the functioning of natural, evolved physiological mechanisms including those of homeostasis (Cannon 1932, Richter 1943, Mitchell 1978, Stebbing et al. 1984), or perhaps, since these are dynamic relationships, *homeorhesis* (Stebbing 1981), at least in multicellular organisms. Such phenomena in colonial organisms are, however, unclear.

Radman (1980), in his examination of the question of SOS repair (in the signal-for-help sense) in mammalian cells, notes: "The terms SOS induction, SOS repair, and SOS functions were coined to designate a set of genetically programmed, inducible and diverse metabolic responses to cellular stress (such as damage to DNA or to its replicative machinery), resulting in an increased chance of cell survival . . . Since SOS induction coordinately and temporarily activates otherwise unrelated cellular pathways, . . . its biological rationale at the cellular level resembles that of stress hormones at the organismic level."

Hormesis has also been referred to as a "generalized aspect of environmental stress etiology" (Laughlin et al. 1981). This could be consistent with the second stage of Selye's "general adaptive syndrome" which also leads to enhanced resistance (Selye 1956, Stebbing 1979).

Growth and homeostasis appear intimately connected. Stebbing (1981, 1982) has observed that it is "necessary to disturb the growth process slightly in order to initiate the action of the control mechanism . . . ." Comfort (1979) has viewed aging as deterioration in ability to maintain homeostasis, suggesting that it may occur only in organisms with a stabilized size at which growth is complete. Aging is known to be accelerated by unrelieved stress.

The relationship between growth, immunity and stress is complex. High stress levels depress growth and derange the immune system. But mouse pups that are stressed by handling display increased brain growth and improved maze-running ability. There are also reports (Luckey 1980, 1982) of stimulation of neurological functions and of neural development due to exposure to low-level ionizing radiation as well as growth stimulation. The immune system and neuroendocrine hormones are involved in complex cybernetic interactions. Low levels of stimulation appear to trigger homeostatic or homeorhetic responses, while high levels destabilize systems. Improved immune competence and surveillance may ensue. As Selye has advised, some stress is not only unavoidable but also desirable or necessary for normal functioning.

From the viewpoint of evolutionary biology, based on natural selection in populations, those individuals less able to cope with natural environmental hazards such as ambient ionizing radiation are at elevated risk of death and deletion of their genes from the gene pool. If this removal is subsequent to reproduction, the effect will be diluted but will still occur due to effects on inclusive fitness. Individuals genetically deficient in genetic repair enzymes (Setlow 1978) could be at elevated lethal risk due to effects of ambient radiation in a manner similar to consequences of defective immune systems. This would be especially marked in geographic regions where natural background radiation levels are comparatively high. One would expect enhanced genetic repair and fewer immune deficiencies in such populations.

For example, about 30 000 persons in a Brazilian town, Guarapari, are exposed continuously, without evidence of biological damage, to an ionizing radiation level of about 600 mrem year^{-1} (Penna Franca 1977, 1981). Also, on the Kerala coast of South India, natural radiation levels of about 1500–3000 mrem year^{-1} have been reported (Kochupillai et al. 1976). Individuals able to cope with such elevated radiation levels compared, for example, with a level of about 65 mrem year^{-1} for Dallas, Texas (Oakley 1972), will have survival advantage in Kerala over those less able to cope. Furthermore, individuals benefitting from ambient radiation will have even greater survival advantage—most likely through hormetic effects.

It is noteworthy that rats (*Rattus rattus*) surviving on Enjebi, in the Enewetak Atoll, some years after the nuclear test explosions, were found to be quite strong, healthy, and not deformed (Bastian and Jackson 1975, Canby and Stanfield 1977). Jackson (1977) commented that in over a decade of work at Enewetak, no abnormal incidence of tumors was seen in rats, and limited chromosome studies of cells derived from bone marrow showed no abnormalities.

Stebbing (1981) has noted that although hormesis has been known for over 90 years (Luckey 1980, 1982), it is not well understood. Solution of the problem has not been advanced by a number of largely epidemiological reports on health effects of low level ionizing radiation that have disregarded the hormesis phenomenon (e.g., Gofman and Tamplin 1972, Gofman 1982, Sternglass 1972), in spite of extensive documentation of the occurrence of apparently beneficial effects of low-level ionizing radiation (Luckey 1980, 1982, Cork 1957, Sanders 1978, Hickey et al. 1981, 1983). Upton (1981) has acknowledged the phenomenon.

Hormesis has been overlooked, misunderstood or ignored for over 100 years. It is now appropriate to recognize this phemonenon and to attempt to study it systematically, unencumbered by emotionalism, subjective judgment, and "statistical" extrapolation (Hickey et al. 1982a,b). While biological effects of ionizing radiation are clearly responses to physical stimuli, it is noteworthy that they appear to trigger response mechanisms similar to responses to certain chemical and biochemical (toxic) stimuli. These may well be part of the broad, general immune response, which is widely viewed as being triggered only by biological stimuli (Hickey et al. 1983, 1984).

Deciphering immunological biochemistry is one of the major thrusts of contemporary medical science. It is clear that all existing life forms evolved in the presence of many ecological hazards—radiation, chemicals, pathogenic organisms, viruses, and predators. Genetic heterogeneity in populations and ecological selective pressures lead continuously to selection of individuals suited at the time to the local environment, including ambient ionizing radiation. Just as nutritional deficiencies are known, it appears appropriate to evaluate the prospect, also, of hormetic deficiencies for better understanding of problems of public health. Furthermore, Luckey and Stone (1960) have cautioned that differentiation between growth stimulation by a non-nutrient action, such as hormesis, and by an essential nutrient is difficult. Since life forms evolved and survive in the presence of various hazards and stresses, they are adapted to them and they may function less well, if at all, in a hazard-free environment.

Among the greatest obstacles to understanding and to therapeutic and other beneficial applications of low-level ionizing radiation and the radiation hormesis phenomenon are erroneous published claims (e.g., Radford 1980, 1983, Gofman 1982) that all levels of ionizing radiation, no matter how low, are hazardous to health in direct proportion to dose or exposure. Such claims are based largely on belief and advocacy of linear and monotonic curvilinear extrapolation of high-level exposure dose–response effects data downward toward zero exposure through the ecologically realistic low-level exposure region. Claims have been made that such unmeasured "data" are reasonably reliable. However, extrapolation is speculation and opinion; it *does not* involve observed data, and so it is not science (Hickey 1985).

Observed data for low-level effects are not easily obtained. However, the existing evidence, imperfect as it may be, indicates that the dose–response curve is J-shaped; high-level exposures are damaging and hazardous, while low-level exposure levels are, on average, stimulatory and beneficial (Luckey 1980, 1982, 1984, 1985, Hickey et al. 1983, 1984).

It has been demonstrated, based on human epidemiological data for 43 urban areas in the USA, that, for example, mortality rates for cancer of the respiratory organs and peritoneum were significantly *lower* [$r = -0.514$ ($p < 0.001$)] in the *higher* than in the lower radiation areas (Hickey et al. 1981). Similar results were obtained for total cancer mortality rate ($r = -0.300$, $p = 0.05$). Comparable results were reported from the People's Republic of China (High Background Radiation Research Group, China 1981). Cancer mortality rates were significantly *lower* in the higher level radiation regions of Guandong Province than in the lower level radiation regions of this province ($p < 0.05$). Such observational evidence is incompatible with the claims, based mainly on personal opinion, that *all* levels of ionizing radiation are hazardous (e.g., carcinogenic, leukemogenic) in direct proportion to exposure. Such claims of hazard, widely reported by the popular media, have led, regrettably, to radiation phobias, and they interfere with beneficial or therapeutic applications of low-level ionizing radiation.

Bibliography

Bastian R K, Jackson W A 1975 ^{137}Cs and ^{60}Co in a terrestrial community at Enewetak Atoll. *4th National Symposium on Radioecology: Radioecology and Energy Resources*. Ecological Society of America, Oregon State University, Corvallis

Canby T Y, Stanfield J L 1977 The rat: Lapdog of the Devil. *Nat. Geogr.* 152(1): 69–87

Cannon W B 1932 *The Wisdom of the Body*. Norton, New York

Comfort A 1979 *The Biology of Senescence*, 3rd edn. Elsevier, New York

Cork J M 1957 Gamma-radiation and longevity in the flour beetle. *Radiat. Res.* 7: 551–57

Cullen T L 1977 Review of Brazilian investigations in areas of high natural radioactivity. Part I. Radiometric and dosimetric studies. In: Cullen T L, Penna Franca E (eds.) 1977 *Int. Symp. Areas of High Natural Radioactivity*. Academia Brasileira de Ciências, Rio de Janeiro, Brazil, pp. 49–64

Davey W P 1919 Prolongation of life of *Tribolium confusum* apparently due to small doses of X-rays. *J. Exp. Zool.* 28: 447–58

Gloag D 1980 Risks of low-level radiation: The evidence of epidemiology. *Br. Med. J.* 281: 1479–82

Gofman J W 1982 Critique critiqued. *Nucl. News* 25(13): 20, 22, 29

Gofman J W, Tamplin A R 1972 Epidemiologic studies of carcinogenesis by ionizing radiation. In: Lecam L M, Neyman J, Scott E L (eds.) 1972 *Proc. 6th Berkeley Symposium on Mathematical Statistics and Probability*, Vol. VI, *Effects of Pollution on Health*. University of California Press, Berkeley, California, pp. 235–68

Goodman L S, Gilman A 1955 *The Pharmacological Basis of Therapeutics*, 2nd edn. Macmillan, New York, p. 1191

Hickey R J 1985 Orwellian "official science"? *Am. J. Epidemiol.* (in press)

Hickey R J, Allen I E, Bowers E J 1982a Creative statistics. *Nucl. News* 25(10): 25, 26, 34

Hickey R J, Bowers E J, Allen I E 1982b Statistical fantasy. *Nucl. News* 25(14): 49, 50, 52

Hickey R J, Bowers E J, Clelland R C 1983 Radiation hormesis, public health and public policy. A commentary. *Health Phys.* 44: 207–19

Hickey R J, Bowers E J, Spence D E, Zemel B S, Clelland A B, Clelland R C 1981 Low level ionizing radiation and human mortality: Multiregional epidemiological studies. A preliminary report. *Health Phys.* 40: 625–41

Hickey R J, Clelland R C, Bowers E J 1984 More comments on radiation hormesis, epidemiology and public health. *Health Phys.* 46: 1159–60

High Background Radiation Research Group, China 1981 Aspects of environmental radiation and dosimetry concerning the high background radiation area in China. *J. Radiat. Res.* 22: 88–100

Jackson W B 1977 Personal communication to R J Hickey

Kemeny J G (chairman) 1979 *Report of the President's Commission on the Accident at TMI*. US Government Printing Office, Washington, DC

Kochupillai N, Verma I C, Grewal M S, Rawlingswami V 1976 Down's syndrome and related abnormalities in an area of high background radiation in coastal Kerala. *Nature* 262: 60–61.

Laughlin R B Jr, Ng J, Guard H E 1981 Hormesis: A response to low environmental concentrations of petroleum hydrocarbons. *Science* 211: 705–07

Luckey T D 1963 Stimulation of *Turbatrix aceti* by antibiotics. *Proc. Soc. Exp. Biol. Med.* 113: 121–24

Luckey T D 1968 Insecticide hormologosis. *J. Econ. Entomol.* 61: 7–12.

Luckey T D 1978 Prelude to the industrial use of dietary

antibiotics. *Fed. Proc.* 37: 107–08; Antibiotics stimulate growth of germfree birds. *Fed. Proc.* 37: 2553

Luckey T D 1980 *Hormesis with Ionizing Radiation.* CRC Press, Boca Raton, Florida

Luckey T D 1982 Physiological benefits from low levels of ionizing radiation. *Health Phys.* 43: 771–89

Luckey T D 1984 Hormesis bei Krebsinduktion durch radioaktive Strahlung hoher Ionisationsdichte. *Z. Phys. Med. Baln. Med. Klim.* (Sonderheft 1) 13: 11–16

Luckey T D 1985 Beneficial physiologic effects of ionizing radiation. In: Leppin W, Meissner J, Börner W, Messerschmidt O (eds.) 1985 *Die Hypothesen im Strahlenschutz*, (Strahlenschutz in Forschung und Praxis, Band XXV). Georg Thieme Verlag, Stuttgart, pp. 184–96

Luckey T D, Stone P C 1960 Hormology in nutrition. *Science* 132 1891–92

Mitchell A R 1978 Salt appetite, salt intake, and hypertension: A deviation of perspective. *Perspect. Biol. Med.* 21: 335–47

Oakley D T 1972 *Natural Radiation Exposure in the United States.* US Environmental Protection Agency, Washington, DC

Penna Franca E 1977 Review of Brazilian investigations in areas of high natural radioactivity. Part II. Internal exposure and cytogenic survey. In: Cullen T L, Penna Franca E (eds.) 1977 *Int. Symp. Areas of High Natural Radioactivity.* Academia Brasileira de Ciências, Rio de Janeiro, Brazil, pp. 29–48

Penna Franca E 1981 Discussion, following paper by Upton (1981). *Ann N.Y. Acad. Sci.* 365: 73–74

Radford E P 1980 Human health effects of low doses of ionizing radiation. The BEIR III controversy. *Radiat. Res.* 84: 369–94

Radford E P 1983 Epidemiology of radiation-induced cancer. *Environ. Health Perspect.* 52: 45–50

Radman M 1980 Is there SOS induction in mammalian cells? Yearly review. *Photochem. Photobiol.* 32: 823–30

Richter C P 1943 Total self-regulatory functions in animals and human beings. *Harvey Lect.* 38: 63–103

Sanders B S 1978 Low-level radiation and cancer deaths. *Health Phys.* 34: 521–38

Schulz H 1888 Ueber Hefegifte. *Pflügers Arch. Ges. Physiol. Mensch Tiere* 42: 517–41

Selye H 1956 *The Stress of Life.* McGraw-Hill, New York

Setlow R B 1978 Repair deficient human disorders and cancer. *Nature* 271: 713–17

Southam C M, Ehrlich J 1943 Effects of extracts of western red cedar heartwood on certain wood-decaying fungi in culture. *Phytopathology* 33: 517–24

Stebbing A R D 1979 An experimental approach to the determinants of biological water quality. *Philos. Trans. R. Soc. London, Ser. B* 286: 465–81

Stebbing A R D 1981 The kinetics of growth control in a colonial hydroid. *J. Mar. Biol. Assoc.* 61: 35–63

Stebbing A R D 1982 Hormesis—The stimulation of growth by low levels of inhibitors. *Sci. Total Environ.* 22: 213–34

Stebbing A R D, Norton J P, Brinsley M D 1984 Dynamics of growth control in a marine yeast subjected to perturbation. *J. Gen. Microbiol.* 130: 1799–1808

Sternglass E J 1972 Environmental radiation and human health. In: LeCam L M, Neyman J, Scott E L (eds.) 1972 *Proc. 6th Berkeley Symposium on Mathematical Statistics and Probability*, Vol. VI, *Effects of Pollution on Health.* University of California Press, Berkeley, California, pp. 145–232

Upton A C 1981 Health impact of the Three Mile Island accident. Health effect of ionizing radiation. *Ann. N.Y. Acad. Sci* 365: 63–75

R. J. Hickey and E. J. Bowers

Hyperbaric Medicine

Hyperbaric medicine is concerned with the treatment of patients in atmospheres of increased ambient gas pressures in pressurized chambers. It has become increasingly important recently in relation to the physiology of deep-sea diving owing to the expansion in exploration for oil.

1. Terminology

At sea level, the atmosphere exerts a pressure referred to as 1 atmosphere absolute (1 ATA). This pressure is equivalent to 101.3 kPa. Most instruments used to record pressure are calibrated in such a way that they ignore ambient pressure. Thus, a gauge recording the pressure in a chamber at 3 ATA would record a pressure of 202.6 kPa, or 2 atm. This pressure is sometimes referred to as 2 atmospheres gauge, but in order to avoid confusion, the pressure inside a chamber should be referred to in terms of total pressure and expressed in either atmospheres absolute or in kilopascals.

As a gas mixture is compressed, the fractional concentration of each constituent remains unchanged, whereas its partial pressure increases. Thus, oxygen in air at 1 ATA has a fractional concentration of 0.2093, and a partial pressure of 0.2093×101.3 kPa $= 21.2$ kPa. At 3 ATA, the fractional concentration is still 0.2093, but the partial pressure is 0.2093×303.9 kPa $= 63.6$ kPa. The presence of water vapor in an atmosphere is an important consideration. The saturated vapor pressure of any liquid is independent of pressure, and is related solely to the temperature of the liquid. Thus the partial pressure of water vapor in the alveoli of the lungs at 37 °C is 6.26 kPa, irrespective of environmental pressure. The partial pressure of oxygen in the trachea during inspiration of air at 2 ATA is therefore $0.2093 \times (202.6 - 6.26)$ kPa.

2. Therapeutic Uses of Hyperbaric Chambers

Hyperbaric chambers are used for two main purposes: (a) at relatively low pressures (1–3 ATA) for the administration of a high partial pressure of oxygen for medical therapeutic purposes, and (b) at much higher pressures (up to about 30 ATA) for the slow decompression of personnel subjected to high environmental pressures during occupational activity (e.g., diving, tunnelling).

2.1 Hyperbaric Oxygen

In the lungs, oxygen moves by passive diffusion from the alveoli into blood contained in the pulmonary capillary vessels. Alveolar gas is saturated with water vapor (exerting a partial pressure of 6.26 kPa at 37 °C) and contains carbon dioxide at a partial pressure of approximately 5.3 kPa. Carbon dioxide is produced continuously by cellular metabolism and is carried in venous blood from its site of production to the right side of the heart and thence into the lungs where it diffuses along a partial pressure gradient into the alveoli. There is therefore a fairly constant partial pressure in the alveoli of approximately 11.5 kPa, exerted by water and carbon dioxide molecules. The partial pressure of oxygen (p_{O_2}) in the alveoli is related to its inspired partial pressure (Table 1).

Table 1
Alveolar p_{O_2} attained when breathing 100% oxygen at environmental pressures from 1–5 ATA, assuming a body temperature of 37 °C and alveolar p_{CO_2} of 5.3 kPa

Environmental pressure		Alveolar pressure
(ATA)	(kPa)	(kPa)
1	101.3	90
2	202.6	191
3	303.9	292
4	405.2	393
5	506.5	494

There are two ways in which oxygen is carried in blood: in chemical combination with hemoglobin and in solution in plasma. Hemoglobin combines with oxygen up to a maximum of 1.34 ml oxygen per gram of hemoglobin. Each 1 kPa increase in partial pressure of oxygen produces an increase of 0.022 ml of oxygen dissolved in plasma. The total oxygen content of 100 ml of blood (c_{O_2}) is given by

$$c_{O_2} = (s_{O_2} \times \mathrm{Hb} \times 1.34) + (p_{O_2} \times 0.022) \qquad (1)$$

where s_{O_2} is the degree of saturation of hemoglobin with oxygen expressed as a fraction of unity, Hb is the hemoglobin concentration of blood expressed in grams per 100 ml of blood, and p_{O_2} is the partial pressure (in kPa) of oxygen in the blood. The p_{O_2} of blood in the pulmonary capillaries approximates to that in the alveoli while normal hemoglobin concentration is 14.6 g per 100 ml of blood. The alveolar p_{O_2} in a healthy individual breathing air at sea level is approximately 13.3 kPa. At this level of arterial p_{O_2}, hemoglobin is 97% saturated with oxygen. Thus, inserting these values in Eqn. (1), the normal oxygen content of blood that has passed through the pulmonary capillaries is $(0.97 \times 14.6 \times 1.34) + (13.3 \times 0.022) = 19.3$ ml per 100 ml of blood. An increase in the p_{O_2} of blood produces only a relatively

modest increase in oxygen content, since hemoglobin is already close to saturation. In order to produce any substantial increase in oxygen content, the arterial p_{O_2} must be elevated to very high levels, thereby increasing the content of dissolved oxygen. In a subject breathing 100% oxygen at 1 ATA, the alveolar p_{O_2} is 90 kPa. Hemoglobin is fully saturated with oxygen, and the oxygen content of each 100 ml of blood is $(1 \times 14.6 \times 1.34) + (90 \times 0.022) = 21.5$ ml. Breathing 100% oxygen at 3 ATA produces an increase in oxygen content related directly to the increase in arterial p_{O_2}. This increase is equal to $(292 - 90) \times 0.022 = 4.4$ ml per 100 ml of blood (see Table 1).

The increase in arterial oxygen content produced by a large increase in environmental pressure may theoretically be sufficient to meet the metabolic requirements of the tissues. With an average cardiac output of 5 l of blood per minute and a normal total metabolic oxygen consumption of 250 ml min^{-1}, only 5 ml of oxygen are extracted from every 100 ml of blood passing around the circulation. Thus oxygen breathing at 3 ATA should ensure sufficient oxygen dissolved in plasma to meet the body's metabolic requirements even in the absence of hemoglobin. However, for reasons stated below, this theoretical improvement in oxygenation is never realized. Nonetheless it may be important in patients who have defects resulting in blood from the venous side of the circulation bypassing the lungs (e.g., patients with an intracardiac lesion resulting in blood being transferred directly from the right to the left atrium or ventricle), or in patients who have large portions of lung which are not receiving inspired gas, but which are still being perfused with blood. Under these circumstances, the increased oxygen content from blood which has passed through the well-ventilated areas is used to oxygenate the poorly oxygenated blood which has bypassed the alveoli, resulting in an acceptable degree of oxygenation in arterial blood.

Hyperbaric oxygen has therefore been used where there is impaired oxygenation of some or all of the tissues. Examples are generalized tissue hypoxia resulting from poor arterial oxygenation because of intracardiac or intrapulmonary shunting of venous blood, and localized impairment of tissue oxygenation resulting from poor blood supply (e.g., in patients with blockages of blood vessels to limbs, intestines, brain (stroke), or heart muscle (myocardial infarction)). In addition, hyperbaric oxygen has been used in newborn babies who have difficulty in breathing adequately.

Results of hyperbaric oxygen treatment in all these groups of patients have been disappointing. A high arterial p_{O_2} causes constriction of blood vessels, reducing blood flow to tissues. In addition, high arterial levels of oxygen cause a reduction in cardiac output. Although the oxygen content of the blood may be increased, the volume of blood reaching the tissues is reduced, and the actual volume of oxygen reaching the tissues remains unchanged. There are, however, four areas in which hyperbaric oxygen therapy has proved useful.

(*a*) *Carbon monoxide poisoning*. Carbon monoxide is more readily bound to hemoglobin than is oxygen, and therefore if inhaled in sufficient concentrations will displace oxygen from its binding sites on the hemoglobin molecule, thus reducing the oxygen carried by the blood and rendering the tissues ischemic. Treatment with hyperbaric oxygen, at 2–3 ATA, provides an adequate oxygen supply to the tissues by increasing the dissolved oxygen content in plasma and accelerates the dissociation of carbon monoxide from hemoglobin molecules.

(*b*) *Infection*. Certain strains of bacteria require conditions of low partial pressure of oxygen. Tetanus and gas gangrene are two conditions which may be alleviated if treated with hyperbaric oxygen.

Some aerobic organisms are sensitive to hyperbaric oxygen. Treatment seems to be of particular benefit if there is surface involvement by the organisms, for example infected burns, ulcers or pressure sores.

(*c*) *Cancer therapy*. Hyperbaric oxygenation raises the intravenous blood oxygen tension to at least normal levels, making cancer cells more sensitive to radiation, while only marginally increasing the sensitivity of nearby normal cells. Favorable results have been reported in combining hyperbaric oxygenation and radiotherapy, and also in combining hyperbaric oxygen and cytotoxic drugs.

(*d*) *Decompression therapy*. Decompression sickness (see below) is the result of gas (usually nitrogen) emerging from tissue in gaseous form as bubbles. The rate of elimination of nitrogen from the tissue is increased if a subject breathes oxygen instead of air, because of the greater diffusion gradient from tissue to blood. Hyperbaric oxygen may be used during normal decompression, but is more frequently used when complications of decompression are present.

2.2 Occupational Decompression

This branch of hyperbaric medicine is expanding rapidly at present because of the enormous increase in diving activity related to oil exploration. Dives to a depth of 250–300 m are becoming commonplace, and the pressures involved here are 26–31 ATA. Any gas inhaled under pressure goes into solution in the body at its alveolar partial pressure. The quantity of gas dissolved in the body depends on its solubility coefficient and the time during which it has been inhaled. The time required for different tissues to become saturated varies inversely with the blood supply. When the ambient pressure decreases the gas passes back out of the tissues into blood, thence to the alveoli. However, if decompression occurs too rapidly, the gas comes out of solution in the tissues or in the circulation and forms bubbles. This may result in joint pains (the "bends") or itching of the skin, or, rather more seriously, involvement of the central nervous system (the "staggers") or the pulmonary circulation (the "chokes").

A chronic condition resulting in bone necrosis in the long bones of the limbs may be seen after repeated exposure to hyperbaric conditions. Tunnel workers who are exposed frequently to air pressures in excess of 3 ATA for up to eight hours at a time may manifest this condition after months or even years of exposure to compressed air conditions.

"Decompression sickness" is seen only after exposure to pressures in excess of 2 ATA. The incidence is related to the time spent under pressure, as well as to the degree of pressure itself. It can be prevented completely by slow decompression, and comprehensive tables are available indicating the rate and duration of decompression required to ensure safety. Decompression may be a lengthy process in excess of one week in a decompression chamber.

3. Types of Pressure Chamber

3.1 Hyperbaric Oxygen Chambers

Hyperbaric oxygen can be administered in two ways. A small chamber in which only the patient is pressurized can be used, provided that no medical or nursing care is required during compression. The chamber is pressurized with 100% oxygen and operates in a working range of 2–4 ATA. This is the only type of chamber used in radiotherapy. The oxygen is usually recirculated, with provision being made for the removal of carbon dioxide.

The second type of chamber is large and designed to accommodate the patient and attendants and may even be adequate in size for surgery. The chambers are pressurized with compressed air, while 100% oxygen is administered to the patient through a face mask or an endotracheal tube. Instruments, samples and other items may be passed to and from the chamber through an airlock.

3.2 Decompression Chambers

These chambers are very variable in size, and are made structurally much stronger than hyperbaric oxygen chambers, because of the much higher pressures ($\geqslant 30$ ATA) to which they are subjected. They are pressurized with the appropriate gas mixture (see below) which is normally recirculated.

4. Problems Associated with Hyperbaric Conditions

4.1 Toxicity

(*a*) *Nitrogen and inert gas narcosis*. The inert gases are all capable of producing intoxication, narcosis, and also anesthesia if their partial pressure in the brain reaches sufficiently high values. The degree to which this occurs correlates with lipid solubility. Xenon is the most narcotic, and produces symptoms at 1 ATA. Nitrogen produces a progressive feeling of euphoria and a decrease in efficiency of performance as the partial pressure increases. In subjects breathing air at 6 ATA, there is a 5% loss of manual dexterity and a 15% loss of cognitive skills. At 9 ATA, even simple tasks become extremely difficult, and at 13 ATA there is a gradual loss of consciousness. Helium is the least narcotic of the inert gases, and it is customary to use oxygen/helium mixtures when pressures in excess of 6–10 ATA are anticipated.

(*b*) *Oxygen toxicity*. Inhalation of oxygen at partial pressures in excess of 51 kPa causes inflammation and congestion of the lungs. If exposure is not stopped, fibrosis eventually develops. The time to onset of this type of oxygen toxicity varies inversely with the inspired partial pressure, from 24–36 h at 100 kPa to only a few hours at 200 kPa.

Toxicity to the central nervous system, consisting of muscle spasms and convulsions, may be fatal if the partial pressure is not reduced rapidly. However, it is uncommon at less than 200 kPa in exercizing individuals, but its incidence and rate of development increases rapidly at partial pressures in excess of this value.

Newborn babies may develop changes in the eye (termed retrolental fibroplasia) if exposed to high partial pressures of oxygen. This condition, which is related to the level of arterial p_{O_2}, may cause permanent blindness if exposure is prolonged.

All these effects of gases are related to the partial pressure and not to the concentration. Thus, the composition of the gas mixture in a hyperbaric chamber must be chosen carefully. The inspired p_{O_2} should not normally exceed 40 kPa unless there is a specific indication for high inspired partial pressures. In a chamber pressurized to 26 ATA, a gas mixture of 1.5% oxygen and 98.5% helium would prevent oxygen toxicity and nitrogen narcosis.

4.2 Barotrauma

Any cavity within the body containing gas will change in volume in accordance with Boyle's law during compression or decompression. During rapid decompression, air trapped in areas of the lung expands and may rupture alveoli (leading to pneumothorax) or allow air to enter the bloodstream (an air embolism, which if sufficiently large can cause cardiac arrest). The air contained in the middle ear normally equilibrates with environmental pressure during swallowing, but if the pharyngotympanic tube fails to open, the tympanic membrane is distorted during change in pressure, and may be perforated. Elective myringotomy should be performed on anesthetized patients being subjected to pressure changes. Subjects with blocked openings between the respiratory sinuses in the skull and the nasal passages may suffer severe pain, particularly during compression. Small pockets of gas sometimes exist around the roots of teeth, and may cause toothache. Gas in the intestinal tract may cause abdominal distention and discomfort during decompression.

4.3 Fire

A 100% oxygen atmosphere is particularly dangerous as it forms a potentially explosive combination with any type of fuel. Any source of ignition must be excluded from oxygen-filled hyperbaric chambers. Air-filled chambers at elevated pressures present an increased risk of conflagration.

Mechanical, electrical and electrostatic sparks are the most likely causes of fire. No apparatus which could be a source of sparks may be taken into a hyperbaric chamber.

Certain anesthetic agents present a fire hazard. Ether and cyclopropane must be excluded from hyperbaric chambers. Although many of the other commonly used volatile anesthetic agents are nonflammable at 1 ATA (e.g., halothane), they may be flammable at higher atmospheric pressures, particularly in the presence of nitrous oxide. There is also a risk of fire from chemical reactions between oxygen and other substances, for example oil or grease. Any potential fuel (e.g., wood, flammable clothing and plastics) should be excluded from pressure chambers. A suitably qualified fire prevention officer should be asked for advice in relation to any hyperbaric installation.

4.4 Apparatus

(*a*) *Pressure-dependent apparatus*. The filling pressure of gas cylinders which might be used in a hyperbaric chamber is very high relative to the likely chamber pressure, and the cylinders therefore discharge their contents normally until equilibration with ambient pressure occurs. Reducing valves or pressure regulators are dependent on the pressure gradient across them and are designed to function normally over a wide range of pressure variation. They therefore perform normally in a pressure chamber.

(*b*) *Flow-dependent apparatus*. For any gas there is approximately a linear relationship between the total gas pressure and the gas density. The effect of compression on gas viscosity is very small in the range of pressures under consideration. Bobbin flowmeters depend for their calibration on gas viscosity at low flows, but on gas density at high flows. Thus, at higher settings, a flowmeter calibrated at 1 ATA gives a reading higher than the actual flow rate if used at hyperbaric pressures. At 2 ATA, an approximate correction factor is given by the equation

$$F_1 = F_0 \left(\rho_0 / \rho_1 \right)$$

where F_1 is the actual flow at 2 ATA, F_0 is the flow reading at 1 ATA, ρ_0 is the gas density at 1 ATA, and ρ_1 is the gas density at 2 ATA. Ball-type flowmeters are not affected by pressurization up to 3 ATA.

Respirometers (e.g., the Wright respirometer) become progressively less accurate as ambient pressure increases, but can be calibrated at the working pressure using a wet spirometer.

Resistance to flow in anesthetic or other gas delivery systems increases with pressure. Most anesthetic vaporizers operate on the principle of splitting the carrier gas, and saturating one of the fractions with vapor (see *Anesthesia Physics*). This mixture is then diluted with the remaining fraction in order to provide a known, relatively low (up to 5%) concentration before delivery to the patient. Since the saturated vapor pressure of any liquid is dependent only on temperature, such a vaporizer should produce the same partial pressure of vapor at a given setting at any ambient pressure. However (because of the characteristics of the flow-splitting valve in conditions of turbulent flow) some inaccuracy arises in relation to the proportion of gas passing through the vaporizing chamber since variation in the flow-splitting ratio occurs with changes in gas density.

4.5 Volume Changes

Any apparatus or container which is not open to the atmosphere is subject to pressure changes across its walls if the ambient pressure is altered. Thus glass bottles or ampoules may implode or explode if not opened to atmosphere. Spirometer systems may siphon water into the circuit. Ink-jet recorders may discharge ink and flood, and even metal containers such as anesthetic vaporizers may explode. Cuffs on endotracheal tubes and intravenous or bladder catheters should be filled with liquid and not gas, and any tube draining a body cavity should be opened to the atmosphere.

5. Monitoring and Measuring Apparatus

The most commonly used medical gas analyzers, including the infrared CO_2 analyzer and the paramagnetic oxygen analyzer, measure the number of gas molecules present (i.e., the partial pressure of the gas), although they are usually calibrated in percentage concentration (see *Medical Gases: Measurement and Analysis*). Thus, they may be used inside a pressure chamber in the knowledge that if calibrated in percentage of gas it is with reference to percentage of 1 ATA (dry). Alternatively, a fine gas sampling tube may be led to an analyzer outside the chamber.

Blood-gas and acid–base estimations can be carried out with appropriate instruments inside a chamber; such measurements are known to be accurate up to pressures of 3 ATA. At higher pressures, it may be necessary to pass blood samples for analysis out of the chamber, although decompression causes bubble formation, making blood-gas analysis unreliable. Other biochemical investigations should be unaffected.

Electrocardiographic monitoring apparatus and oscilloscopes should be left outside the chamber (because of the risk of ignition) and connected to the interior via leads. Such apparatus may be viewed through chamber portholes.

See also: Blood Gas Analysis; Blood Gas Tensions: Continuous Measurement

Bibliography

Bennett P B, Elliot D H 1975 *The Physiology and Medicine of Diving and Compressed Air Work*, 2nd edn. Ballière Tindall, London
Chew H E R, Hanson G C, Slack W K 1969 Hyperbaric oxygenation. *Br. J. Dis. Chest* 63: 113–39
Cox J, Robinson D J Anaesthesia at depth. *Br. J. Hosp. Med.* 23: 144–51
Lambertsen G J 1967 *Underwater Physiology*. Williams and Wilkins, Baltimore, Maryland
McDowall D G 1964 Anaesthesia in a pressure chamber. *Anaesthesia* 19: 321–29
Smith G 1980 Oxygen toxicity. In: Gray T C, Nunn J F, Utting J E (eds.) 1980 *General Anaesthesia*, 4th edn., Vol. 1. Butterworth, London, pp. 551–71
Spence A A, Smith G 1974 Problems associated with the use of anesthetic and related equipment in a hyperbaric environment. *Int. Anesthesiol. Clin.* 12: 165–79

G. Smith and A. R. Aitkenhead

Hyperthermia in Cancer Treatment

Increasing the temperature of biological materials by only a few degrees above normal causes considerable disruption in the biochemistry of cells and may lead to their death. The first reports of the phenomenon being applied in the treatment of cancer appeared more than a century ago. However, attempts made then to cause local hyperthermia were not sufficiently successful and interest in hyperthermic treatment faded. A firm rationale has now been developed, however, for using hyperthermia in cancer treatment, and this in turn has stimulated further developments in the technique.

Some reasons for hyperthermia being of possible use in treating cancer are:

(a) tumors may become hotter than normal tissues in a localized treatment field because of their comparatively sluggish blood supply;

(b) some types of tumor cells may be intrinsically more heat-sensitive than normal cells;

(c) tumors contain cells which may be either hypoxic, which seems not to markedly affect their response to hyperthermia (in contrast to conventional treatment with x rays), or acidic and nutrient deficient, both of which cause cells to be more sensitive to heating;

(d) cells in the DNA synthesizing phase are particularly sensitive to heat but relatively resistant to x rays, suggesting that treatment combining hyperthermia and radiotherapy could be an advantage in some circumstances; and

(e) the combination of heat and anticancer drugs may be beneficial by increasing the uptake of drugs or by enhancing sensitivity to drugs.

This evidence is derived from experimental work on cells, normal tissues and experimental tumors.

The effects of heat alone are different from those of heat as a potentiator of other anticancer modalities, and ways have been found to study the effects separately. After radiation most cells die at mitosis, whereas after heat, nondividing cells lyse just as readily as mitotic cells. Skin, for example, shows radiation damage as radiodermatitis. The injury is enhanced by moderate hyperthermia but remains unaltered qualitatively. In contrast, tissue necrosis occurs rapidly after severe hyperthermia.

With heat alone, once the threshold for injury is reached, a small increase in hyperthermal treatment (temperature or time) will cause a dramatic increase in the probability of tissue necrosis. Thus accurate thermometry and very careful control of heat delivery are clinically essential. When heat only is applied, it is found, with a wide range of cells and tissues, that a change in temperature of 1 °C is equivalent to a two-fold change in heating time, giving an activation energy of approximately $600 \, kJ \, mol^{-1}$ and pointing to protein denaturation as a possible primary mode of injury.

With cells and animal tissues there is a considerable recovery potential between heat doses. In addition, heat causes a transient resistance (termed thermotolerance) to subsequent hyperthermia. This is an extremely large effect, and a second treatment may have to be more than doubled to produce a specified effect if given after a previous heating. Thermotolerance manifests itself in different ways, depending on whether heat is given as a high temperature for a short time or as a lower temperature for many hours (as is the case in whole-body hyperthermia). Thermotolerance with respect to heat as a potentiator of radiation damage appears to be much less than for direct heat injury. The phenomenon is clearly of great clinical importance.

If heat enhancement of the effects of radiation or chemotherapy is greater for tumors than for normal tissues, there is possibly a therapeutic advantage to be gained in using a combined treatment. Unfortunately this has been difficult to determine experimentally, because it is difficult to heat tumors (or even normal tissues) uniformly. The interaction between heat and x rays fades when the two modalities are separated in time. It is lost after a time lapse of 4–5 h, as illustrated by many experiments on cells *in vitro* and on normal tissues. It is possible that a therapeutic advantage may be gained by heating for some 4–5 h after irradiation if hyperthermia causes more direct damage to tumors than to the less sensitive normal tissues.

While the primary mechanisms of hyperthermia injury and thermal tolerance are not yet clear, much recent evidence suggests that the membranes of cells are importantly involved.

A range of techniques is now available for achieving local hyperthermia, mostly involving ultrasound or electromagnetic radiation (microwaves or radio-frequency waves). Current methods of measuring temperature (using metallic thermometers) are inadequate. However, thermometers not affected by the application of hyperthermia are being developed as are methods which do not require the insertion of probes into the body.

Despite experimental difficulties, the clinical results obtained and reported in the literature are extremely encouraging. Tumors in man and other large animals appear to become hotter than normal tissues. Very few complications have been reported.

See also: Thermography; X Rays: Biological Effects; Radiosensitizers

Bibliography

Meyn R E, Withers H R (eds.) 1980 *Radiation Biology in Cancer Research*. Raven, New York, pp. 589–644
Streffer C (ed.) 1978 *Cancer Therapy by Hyperthermia and Radiation*. Urban and Schwarzenberg, Baltimore

S. B. Field

Hypothermia

Man possesses the ability to maintain his central body temperature within close range of 37 °C despite wide fluctuations in the temperature of his environment; this ability classifies him as a homeotherm. The body's responses to changing temperature are of two types and act together: these are autonomic (self-governing) and behavioral (under voluntary control).

Hypothermia has been defined as "the clinical state of subnormal body temperature" (Benazon 1974). For man, this is accepted as a central or core temperature of below 35 °C.

1. Body Temperature

Normal body temperature is difficult to define since physiological variations occur on exercise, in diurnal rhythm and in the luteal phase of the menstrual cycle. Variations are also found at different sites in the body where temperature is measured, such as sublingual, rectal, esophageal and tympanic membrane sites. For these reasons it is helpful to adopt the core-and-shell hypothesis. The body's core consists of the thoracic, abdominal and cranial contents and the deep tissues of the limbs; its temperature is not affected directly by fluctuations in environmental temperature, and normally varies little. The shell consists of all the more superficial layers of the body. They provide important insulation and display temperature gradients throughout their layers. In a warm environment these gradients are small, and much of the shell may reach core temperature.

2. Thermoregulation in Hypothermia

The physiological thermostat, or thermoregulatory control organ, is the hypothalamus. It is always striving to hold the central temperature at a constant value. This value is known as the set point. Any deviation away from the set point initiates thermoregulatory responses. The maintenance of a set point requires that the control mechanisms involved achieve thermal balance, i.e., heat loss and heat gain by the body must be equal. However, the set point may be changed by certain diseases and drugs. Deviations in body temperature away from the set point are detected by the hypothalamus from two input sources: (a) a direct sensitivity to the temperature of its own arterial blood; and (b) through nerve pathways from thermal receptors on the skin. These specialized receptors are situated subepithelially, and their response is transmitted by small myelinated fibers ($A\delta$ type) through the lateral spinothalamic tracts. At extreme temperatures, pain receptors are stimulated simultaneously.

During hypothermia, heat loss exceeds heat gain, and the thermoregulatory responses required to correct the imbalance are those stimulating heat gain and those minimizing heat loss by improving the insulating capacity of the shell.

2.1 Mechanisms for Heat Gain

(*a*) *Increased muscle activity*. This is the most efficient form of heat production, and can produce heat up to ten times that of resting muscle. The activity can be behavioral (voluntary) or under autonomic control (shivering). Shivering is a specialized form of involuntary muscular exercise. Thermosensitive neurons in the spinal cord are stimulated by cold; they transmit impulses to the posterior hypothalamus, and the effector pathway is through the mid-brain motor system back to the spinal cord. Shivering increases heat production to up to four times the muscle's resting rate; hard voluntary exercise can produce more heat, but losses by convection are then greater. Both types of muscle activity are equally subject to fatigue.

(*b*) *Brown fat metabolism*. In the newborn, shivering is ineffective; instead, the sympathetic nervous system stimulates heat production from the highly vascular deposits of brown fat. This thermogenic fat is sited between the scapulae, around the great vessels of the neck, thorax and abdomen, and around the kidneys. It is thought to be responsible for at least 50% of a neonate's heat production; the heat is the end-product of cyclic-AMP (adenosine monophosphate) stimulation in the brown fat cells by noradrenaline.

(*c*) *Increased metabolism*. This leads to increased heat production as a result of hormonal stimulation of metabolism; it is a self-governing effect and is under hypothalamic control. Thyrotrophin releasing hormone stimulates thyroxine and tri-iodothyronine production by the thyroid; these hormones enhance heat production by increasing basal metabolic rate. The adrenal axis is also stimulated by the cold, and the extra noradrenaline secreted increases glucose metabolism.

(*d*) *Passive heat gain from the environment*. Heat may be gained passively by behavioral modifications, such as by radiation from warm surroundings or conduction from hot food and drinks and warmed intravenous fluids. Breathing warmed, humidified air is not an effective means of heat gain, but does help to prevent heat loss.

2.2 Mechanisms for Reducing Heat Loss

(*a*) *Behavioral mechanisms*. Adding to the natural insulating shell by donning additional layers of clothing reduces heat loss by evaporation and by radiation; in addition, layers of stagnant air become heated by the body and trapped between it and the environment. Moving away from areas of high air movement reduces losses due to convection, while prevention of contact with cold objects, especially cold water immersion, minimizes conductive loss.

(*b*) *Autonomic mechanisms*. Reduction in sweating limits evaporative heat loss, and occurs once the skin temperature falls below 33 °C, as long as the central temperature has not risen above the normal range (Benzinger 1969).

Vascular control is an important thermoregulatory effector mechanism. Sympathetic nerve fibers alter blood flow from the core to the surface, especially at the extremities, and provide a fine adjustment to body temperature control. The temperature gradients throughout the shell are varied thus, by making use of the high heat capacity of blood. Therefore, in hypothermia, heat loss is minimized by vasoconstriction; in addition, the close association of the deep arteries and venae comitantes forms a heat exchanger to limit loss of heat to the surface.

3. Physiological Responses and Modifications

As discussed already, the immediate effect of cooling the body is a stimulation of those thermoregulatory mechanisms which either reduce heat loss or lead directly to heat gain. With further cooling, such compensation becomes inadequate and physiological derangements gradually appear.

3.1. Metabolism

As the body temperature falls, all the numerous biological processes gradually slow down, but at varying rates depending on their temperature coefficients. Metabolic and circadian processes are affected first, followed by contractile mechanisms, and then diffusion.

In anesthetized man, the metabolic rate and oxygen consumption are found to fall exponentially with the drop in body temperature, provided that cooling is uniform and that shivering has been inhibited by muscle relaxation (Forrester 1969). Once the core temperature has reached 28 °C, the metabolic rate is about 50% of its normal resting value. If only surface cooling is used, large temperature gradients develop between the superficial and deep tissues, cooling is nonuniform, and oxygen requirements are reduced unevenly.

When first entering cold water, man produces a particularly marked rise in metabolic rate and oxygen consumption as he attempts to swim or tread water, before the fall due to deepening hypothermia supervenes.

3.2 Cardiovascular System

At the onset of hypothermia, the autonomic nervous system stimulates the cardiovascular system to produce a tachycardia and a rise in systolic blood pressure which help to meet the increased oxygen consumption of shivering muscles. In addition, vasoconstriction of vessels in the skin and subcutaneous tissues diverts blood flow centrally into deep capacitance vessels. Evidence of these effects is seen as distended pulmonary vessels on chest x rays and as the so-called cold diuresis, due to an increased renal blood flow.

However, once further cooling of the body core occurs, these immediate effects are superseded by a depression of myocardial performance. This is manifest as a fall in systemic blood pressure and cardiac output, due both to reduced contractility, and to a bradycardia (or other conduction defect), as the sino-atrial node becomes depressed. The peripheral vasoconstriction persists, and the resulting high afterload increases myocardial work in the face of its reduced efficiency.

Conduction abnormalities are common below 31 °C, and they become more varied and unpredictable as hypothermia deepens. They are easily detected from an electrocardiogram (see *Electrocardiography*). Commonly the QRS complex widens, while at the same time the PR and QT intervals lengthen. Classically, a J wave also appears—this is a notched downstroke which either precedes or forms part of the QRS complex, and is most easily seen in the electrical signals from the right side of the heart. ST-segment and T-wave changes are variable and inconsistent with the degree of hypothermia. The lengthened QT interval is the most specific indicator of myocardial activity depressed by cold; it may be due to direct tissue hypoxia and often persists for several days after normothermia has been restored (Muir 1955). The electrocardiogram also usually shows a fine irregularity of the baseline; this is due to fine muscular tremor, present even though shivering may not be clinically obvious. Clinically, atrial dysrhythmias are more common in elderly hypothermic patients than ventricular ones, although the picture is often complicated by additional acid–base and electrolyte disturbances, and by the agent causing hypothermia, such as drug overdose. Below a core temperature of 28 °C, the risk of sudden ventricular fibrillation is considerable, and clinical managment is aimed at rapid rewarming to a temperature above this point. Sinus rhythm may be resumed spontaneously at temperatures above 33 °C.

3.3 Respiratory System

With the onset of hypothermia, respiration is initially increased, causing moderate hypocapnia, but soon after it is gradually depressed in spite of an increased oxygen demand due to shivering. Arterial oxygen tension remains low during the hypothermic state, and carbon dioxide tensions tend to rise. These effects may result from a combination of factors, namely:

(a) reduced minute ventilation, due both to a slower respiratory rate and to a smaller respiratory tidal volume,

(b) increased alveolar capillary block,

(c) increased blood viscosity,

(d) reduced hypoxic pulmonary vasoconstriction,

(e) increased dead space, due to bronchodilation,

(f) alterations in the central control of respiration, so that a hypoxic drive may supervene, and

(g) suppressed cough reflex, which with the reduced alveolar ventilation, predisposes the patient to segmental collapse of the lung and infection.

Airway pressure and compliance measurements show no change in hypothermic conditions.

Respiration normally ceases at a core temperature of 24 °C. In elderly patients suffering from accidental hypothermia, significant preexisting chest infections frequently increase these disturbances in respiratory function, and therefore in tissue oxygenation.

During hypothermia, more oxygen and carbon dioxide can be carried in solution in the plasma, since the solubility of gases in liquids varies inversely with temperatures. On the other hand, the oxygen dissociation curve is shifted to the left at the same time, reducing the unloading of oxygen from oxyhemoglobin at the tissues; however, this shift is often reversed by the coexisting acidosis and hypercapnia. If mechanical ventilation is instituted to correct hypoxemia, care must be taken to avoid simultaneous hypocapnia, which would impair both cardiac output and the availability of oxygen for the tissues.

3.4 Acid–Base Disturbances

If cooling occurred evenly in all tissues as hypothermia progressed, the low temperatures reached would protect the cells from the effects of hypoxia and acidosis, their own metabolic requirements being reduced in parallel. In practice, however, cooling is never even. In accidental hypothermia, due to exposure for example, the skin, subcutaneous tissue and muscle cool faster than the core tissues. Thus, when lowered tissue perfusion fails to match even the reduced oxygen demands, an oxygen debt ensues, leading to the onset of anaerobic metabolism and a metabolic acidosis. Hypothermia lowers tissue perfusion by reducing cardiac output, and by increasing blood viscosity and peripheral resistance in the microcirculation; these mechanisms, as well as hypoxemia itself, all tend to increase metabolic acidosis. Furthermore, the waste products of anaerobic metabolism, mainly lactates and pyruvates, are degraded in the liver; during hypothermia, impaired hepatic blood flow and reduced enzyme activities mean that elimination is delayed, and acidosis continues. This becomes more apparent clinically during rewarming of the hypothermic patient, when vasodilation and an improved cardiac output promote the return of these acids into the systemic circulation.

The normal renal buffering mechanism, i.e., increased retention of bicarbonate ions and increased excretion of hydrogen ions, becomes ineffective during hypothermia. Retention of bicarbonate is impaired by the reduced reabsorptive function of the tubular cells, while increased sodium excretion reduces the elimination of hydrogen ions.

3.5 Renal System

Urine production is affected during hypothermia by two separate mechanisms:

(a) Reduced renal blood flow inevitably lowers the glomerular filtration rate. It is said to be halved once the core temperature has been lowered to 30 °C (Morales et al. 1957).

(b) Sluggish tubular function in terms of reduced active transport of ions means that sodium, glucose and water fail to be reabsorbed properly, while hydrogen ions fail to be excreted.

The result of these two effects is the initial production of large volumes of dilute urine, with less sodium ions and glucose, followed eventually by a gradual tailing off of urine flow.

3.6 Fluid and Electrolyte Changes

Fluid intake and intestinal absorption are both reduced during hypothermia, which together with polyuria leads to a reduced plasma volume. This is compounded by fluid shifts from the circulation into both the interstitum and the cells. The estimation of a patient's fluid requirements is therefore difficult, and the usual criteria become less relevant.

During induced hypothermia, plasma sodium and chloride values remain essentially normal, but the effects of concomitant diseases may alter them in accidental cases; myxedema and pituitary insufficiency are two classic examples. Potassium levels may fluctuate either way, and are frequently found to be normal; hypokalemia results from malnutrition or the dilution effect of glucose infusions, while supervening acute renal failure usually leads to hyperkalemia.

Serum calcium and protein levels are rarely changed significantly, but magnesium levels show a consistent rise; this may aid stabilization of cellular membranes and preservation of cell functions at low metabolic rates. Phosphate levels may be normal or high.

3.7 Gastrointestinal System

The effects of hypothermia on the gut have been little studied. However, its function is depressed, due both to reduced motility and decreased blood flow. Acute dilatation of the stomach secondary to atony is commonly encountered clinically, and at endoscopy multiple erosions and submucosal hemorrhages are seen. Liver function is curtailed as metabolism falls, and the detoxification and synthetic functions are equally affected. This becomes apparent clinically as hyperglycemia due to a temporary glucose intolerance; the cells develop both insulin resistance and impaired utilization of glucose. This hyperglycemia is worsened by both high circulating catecholamines and by hemoconcentration. Free-fatty-acid levels rise in the plasma during hypothermia, as lipolysis is stimulated.

Acute pancreatitis is a common clinical finding amongst hypothermia victims. Serum amylase levels are raised but the precise mechanism is unclear.

3.8 Endocrine Glands

Hypothermia stimulates catecholamine secretion from the adrenal medulla. Catecholamines in turn stimulate cardiac output and promote the shunting of blood from peripheral tissues towards the core during the early stages of hypothermia. Once shivering ceases as hypothermia deepens, catecholamines promote thermogenesis by stimulating adenyl cyclase and thereby encouraging lipolysis. In contrast, adenal cortical secretion is suppressed during hypothermia. Cortisol becomes increasingly protein-bound as body temperature falls, thus reducing its utilization too. The purpose of these suppressive effects is not known, but they are readily reversible on rewarming. Thyroxine levels increase due to pituitary stimulation, but this hormone also becomes more heavily protein-bound. The importance of this effect is not understood since the normal half-life of serum thyroxine is eight days.

3.9 Peripheral Blood

The effects on hemoglobin and white-cell levels during hypothermia are greatly dependent on concurrent conditions and infections, but platelets are affected directly—numbers are reduced and clumping is characteristic. Both effects are reversed on rewarming, but surgery under hypothermic conditions may lead to capillary oozing.

Blood flow through the microcirculation is impaired by the following factors found under the hypothermic conditions:

(a) increased blood viscosity,

(b) high packed cell volumes,

(c) plasma shifts out of the circulation, and

(d) circulating cryoglobulins and cryofibrinogen (these precipitins disappear on rewarming).

As discussed in Sect. 3.4, any reduction in blood flow

hastens the onset of anaerobic metabolism and metabolic acidosis.

3.10 Central Nervous System

The important effect of hypothermia is reduced cerebral blood flow; this results from a combination of depressed cardiovascular performance, raised blood viscosity and increased resistance of the cerebral vessels. Cerebral oxygen consumption is reduced simultaneously.

Clinically, the sequelae resemble the onset of general anesthesia. Consciousness becomes clouded below core temperatures of 33 °C, but is often not lost until 28 °C. Speech becomes slower and dysarthric, with an increasing delay before even simple questions can be answered.

The electroencephalogram (EEG) shows diminishing cortical activity, reflecting reduced neuronal metabolism. At a core temperature of 18 °C, the EEG becomes flat—a reversible phenomenon and not indicative of brain death under these circumstances.

3.11 Neuromuscular System

Nerve conduction becomes progressively slower during hypothermia, and is paralleled by the reduction in acetylcholine release from the neuromuscular junction. Voluntary movement decreases; incoordination and an ataxic gait are obvious at core temperatures below 32 °C. Muscle tone increases simultaneously, and is not unlike the rigidity of Parkinson's disease. Muscle paralysis does not occur until the muscle temperature is as low as 4 °C.

4. Causes of Hypothermia

Hypothermia may be deliberately induced, or may be accidental. Induced hypothermia is produced in a controlled fashion in order to perform specialized surgery, usually cardiac or neurosurgery. Cellular functions are preserved for considerable periods at minimal rates of metabolism, so that complicated operations can be carried out. Accidental hypothermia is defined as a spontaneous drop in core temperature below 35 °C, and its main causes are exposure, disease and drugs. The subclinical chronic hypothermia of the elderly is also classed as accidental; it is probably due to deteriorating function in the hypothalamic thermostat.

4.1 Exposure

Exposure hypothermia may affect any age group. The cold stress placed on the subject exceeds the body's maximum heat production by both physiological and behavioral means. Humidity and air movement are important in modifying the cold stress which man can withstand. If the cold stress involves immersion, behavioral responses are limited and hypothermia progresses more rapidly. After rescue, the core temperature may continue to fall up to a futher 3 °C. This is the so-called after-drop and is due to cold blood from the periphery

reentering the central circulation. Consciousness may become depressed again as a result, and muscle rigidity may reappear. Exposure hypothermia may be precipitated by exhaustion in cold environments; it occurs when muscle glycogen stores become depleted, and so should be preventable by maintaining glucose intake.

4.2 Disease

The many varied disorders predisposing to hypothermia can act in three ways: by reducing normal heat production, by increasing heat loss, or by disturbing thermoregulatory control. Most hypothermia victims not suffering from exposure, especially the elderly, are suffering from such an illness. The disorders can be classified generally thus:

(a) metabolic and endocrine disturbances (e.g., myxedema, hypopituitarism, diabetic ketoacidosis, hypoglycemic attacks and severe uremia),

(b) neurological disturbances (e.g., cerebrovascular accidents, transient ischemic attacks and confusional states; epilepsy can lead to periodic hypothermia),

(c) cardiovascular collapse (e.g., due to myocardial infarction and severe infection, such as acute pancreatitis),

(d) starvation,

(e) prolonged inactivity, usually secondary to rheumatological or neurological disorders, and

(f) skin conditions, causing disproportionate heat loss (e.g., burns and erythroderma).

4.3 Drugs

Most of the drugs implicated either depress the central nervous system or produce vasodilatation, or both. Common examples are ethanol, phenothiazines, barbiturates and antidepressants. In elderly patients with a sluggish metabolism, even recommended doses may predispose to hypothermia, but frequently the drugs have been taken in a self-poisoning attempt, and the combination of central depression and vasodilatation produced exceeds both behavioral and autonomic compensation (Linton and Ledingham 1966).

5. Management of Hypothermia

Correct management of hypothermia victims involves both immediate treatment and hospital treatment.

5.1 Immediate Treatment

This commences as soon as the victim is found, and aims to prevent further heat loss and to preserve life during transport to hospital. The basic requirements are shelter, exchanging wet garments for dry, good insulation, gentle handling, and placing the patient in the recovery position. In cases of cardiorespiratory arrest, immediate resuscitation should be carried out at half-normal rates (Hillman 1972).

5.2 Hospital Treatment

In addition to continuing the prevention of further heat loss and dealing with the underlying cause of the hypothermic state, the principles of hospital treatment have been well described (Ledingham and Mone 1978), and have reduced mortality due to hypothermia from as high as 80% down to 27%. The overall mortality rate is still 50%, as many still die of the predisposing disease.

In summary, the requirements for treatment are as follows.

(*a*) *Oxygen.* This is essential in order to correct tissue hypoxia and acidosis. If spontaneous respiration is absent or inadequate, endotracheal intubation and intermittent positive pressure ventilation (IPPV) are required to achieve oxygenation. IPPV also helps to expand collapsed lung segments and to reduce alveolar edema fluid.

(*b*) *Monitoring.* Continuous monitoring provides early detection of changes in physiological parameters, so that management may be optimal.

Monitoring should be carried out on several functions. The electrocardiogram provides information on changes in heart rate and rhythm and their causes, which can then be treated promptly. Arrhythmias have been reported in association with handling during the introduction of monitoring or therapeutic equipment, and may still occur for up to three days after recovery (Maclean and Emslie-Smith 1977). For measuring arterial pressure, an indwelling cannula is desirable so that monitoring is continuous. In addition it provides a route for regular sampling for blood gas analysis, acid–base balance, urea and electrolytes, blood glucose, packed cell volume and coagulation screens. Monitoring the central venous pressure enables the rate of infusion of colloid and crystalloid solutions during rewarming to be carried out more safely, as vascular resistance changes are considerable during this period. To maintain fluid balance, careful input/output records are required. Urine output is monitored hourly by means of a bladder catheter. A nasogastric tube is passed to aspirate and measure gastric secretions—it can be used later for feeding. Cerebral activity is recorded on neurological observation charts. A cerebral-function monitor should be used if available. Temperature measurement is essential to assess the speed of rewarming and to prevent hyperthermia; more than one core site is chosen for measurement. The core-to-peripheral temperature difference is a reliable index of peripheral perfusion, and therefore an indirect reflector of cardiac output. Temperature probes must be placed accurately for reliable results.

(*c*) *Intravenous fluids.* These are warmed to normal body temperature before infusion. They are needed to correct hypotension, counteract vasoconstriction, prevent sludging and thrombosis by reducing blood viscosity, and to correct electrolyte disturbances.

(*d*) *Specific drug therapy.* Prophylactic broad-spectrum antibiotics are advocated by some workers (Maclean and Emslie-Smith 1977), while others reserve them for a proven infection (Ledingham and Mone 1978), or aspiration pneumonia; the latter is one of the lethal complications of hypothermia, when corticosteroids would be added to the drug therapy. Other drugs are used empirically to treat specific associated diseases, or complications, as they arise.

(*e*) *Rewarming.* There has been considerable controversy over whether hypothermic patients should be rewarmed rapidly by active measures, or whether they should be allowed to rewarm passively (spontaneously) after being placed in a warm environment. It now seems clear that, provided there is adequate oxygenation, fluid replacement and close monitoring, active rapid rewarming reduces mortality in moderate and severe cases of hypothermia (Ledingham and Mone 1980). If the core temperature is raised quickly to over 30 °C, the risk of sudden death from ventricular fibrillation is minimized. Rewarming from 33 °C upwards can be carried out more slowly.

Active rewarming should be undertaken when the core temperature falls below 33 °C, and in any subject who is unable to shiver effectively, such as a premature infant. The methods of rewarming can be divided into three broad categories: external, internal and extracorporeal. External methods are generally suitable for patients with mild hypothermia (32–35 °C) of fairly brief duration. A heated mattress or blanket, or a radiant heat source are frequently used. A rate of rise of 1–2 °C per hour is satisfactory. The application of the heat source over the torso of the body should rewarm the heart and other vital organs preferentially, restoring their normal function first, and reducing the likelihood of hypotension due to peripheral vasodilatation (Fernandez et al. 1970). Immersion of the patient in a hot bath at 40–44 °C is also effective, but is often inconvenient; the patient's limbs should not be immersed, to avoid sudden vasodilatation.

Internal rewarming is required in more severe cases, and a variety of routes have been tried. Peritoneal dialysis with fluid at 38 °C is successful, but carries the risks of infection, electrolyte disturbances and diaphragmatic splinting (Soung et al. 1977). The continuous circulation of warmed fluid through a large intragastric balloon avoids these complications, and preferential mediastinal warming improves cardiovascular function early (Ledingham et al. 1980). However, the tube bearing the balloon is unpleasant for all but unconscious patients. Heating and humidifying the inspired air or gases is simple and safe, but is usually insufficient alone, serving mainly to prevent further evaporative heat loss from the respiratory tract (Lloyd 1973). Neither fluids nor inspired gas should be heated above 44 °C, to avoid heat damage to mucous membranes.

Warming the blood extracorporeally, by making use of the heat exchanger incorporated in hemodialysis machines or cardiopulmonary bypass equipment, is certainly the most rapid method of central rewarming, and has proved life-saving in very severe cases. It requires the services of experienced staff in a specialized medical center.

The regular use of an active regime of rewarming with full oxygenation, continuous monitoring and detailed attention to fluid and electrolyte balance has reduced the overall mortality rate of moderate and severe accidental hypothermia in a prospective 15 year study to 27% (Ledingham and Mone 1980), with only 5% of patients dying before rewarming was complete. Previous reported figures from several sources had quoted mortality rates varying from 40–80% (Mills 1973).

Bibliography

Benazon D 1974 Hypothermia. In: Scurr C, Feldman S (eds.) 1974 *Scientific Foundations of Anaesthesia*, 2nd edn. Heinemann, London, pp. 344–57

Benzinger T H 1969 Heat regulation: Homeostasis of central temperature in man. *Physiol. Rev.* 49: 671–759

Fernandez J P, O'Rourke R A, Ewy G A 1970 Rapid active external rewarming in accidental hypothermia. *J. Am. Med. Assoc.* 212: 153–56

Forrester A C 1969 Hypothermia in anaesthesia. *Med., Sci. Law* 9: 233–41

Hillman H 1972 Treatment after exposure to cold. *Lancet* i: 140–41

Ledingham I M, Douglas I D, Routh G S, Macdonald A M 1980 Central rewarming system for treatment of hypothermia. *Lancet* i: 1168–69

Ledingham I M, Mone J G 1978 Accidental hypothermia. *Lancet* i: 391

Ledingham I M, Mone J G 1980 Treatment of accidental hypothermia: A prospective clinical study. *Br. Med. J.* 280: 1102–05

Linton A L, Ledingham I M 1966 Severe hypothermia with barbiturate intoxication. *Lancet* i: 24–26

Lloyd E L 1973 Accidental hypothermia treated by central rewarming through the airway. *Br. J. Anaesth.* 45: 41–48

Maclean D, Emslie-Smith D 1977 *Accidental Hypothermia*. Blackwell, Oxford

Mills G L 1973 Accidental hypothermia in the elderly. *Br. J. Hosp. Med.* 10: 691–99

Morales P, Carbery W, Morello A, Morales G 1957 Alterations in renal function during hypothermia in man. *Ann. Surg.* 145: 488–99

Muir F H 1955 Some changes observed in the electrocardiogram during operations on patients under hypothermia. *Soc. Cardiol. Tech. J.* 2: 167–72

Soung L S, Swank L, Ing T S, Said R A, Goldman J W, Perez J 1977 Treatment of accidental hypothermia with peritoneal dialysis. *Can. Med. Assoc. J.* 117: 1415–16

A. M. Macdonald

I

Image Analysis: Extraction of Quantitative Diagnostic Information

The information in a medical image is observed as a pattern of varying intensity or color. This pattern can be digitized and stored as an array of numbers in a computer. The aim of quantitative analysis of medical images is to extract information from this medical image which can aid in diagnosis or improve understanding of the physiology of the patient. In order to store the image in a computer it is usually first divided into a square grid of small elemental areas called pixels (picture elements). The value of each pixel is the average intensity of the image in the area occupied by the pixel. The maximum pixel size, if negligible degradation of the image is to occur during digitization, is determined by the resolution of the image. The accuracy with which the pixel value can be stored is determined by the size of the memory unit allocated to each pixel. In order to store the image it may be necessary to use a large amount of memory. Expressing the amount of storage required in bits, from 2^{20} to 2^{28} bits may be needed to store an image, yet the clinical information required may be as simple as the presence or absence of disease which could be stored in only one bit. To extract either a diagnostic classification or provide information on the clinical or physiological state of the patient from an image, a severe reduction in the amount of data in the image is required. This can only be done by progressively discarding irrelevant information using both a knowledge of the clinical problem at hand and the physics of the imaging process until the required information is revealed.

Most medical images represent a two-dimensional projection of a three-dimensional object. However, the use of multiple-slice tomographic techniques has meant that in some cases three-dimensional images are available. A different but important type of three-dimensional image is the dynamic study which consists of a temporal sequence of images. This sort of study is used to follow the changing distribution of a tracer or marker substance within a patient. Both these types of image may be subjected to quantitative analysis.

1. Image Analysis

Each image pixel has a numerical value; the simplest form of analysis would be to use this value directly. Unfortunately, the absolute or even relative value of a pixel is only rarely a useful diagnostic measurement. Some examples do exist. Using tracers labelled with positron-emitting isotopes it is possible to produce images of tracer distribution which by careful calibration may be converted to images of the absolute tracer concentration. These concentrations can then often be related by suitable physiological models to various physiological parameters such as blood flow and oxygen and glucose utilization.

Analysis of pixel values from other types of tomographic techniques is less useful because the property being measured is less directly related to physiology. Although different tissues do, for example, have different x-ray attenuation coefficients, these are not in general specific enough to allow clear classification of tissue type. Apart from the simple point-analysis method described above, most image-analysis techniques seek to reduce the amount of irrelevant information in the image before attempting to extract useful information. Most of these methods are either transform methods, especially those of orthogonal factor analysis, or techniques based on image segmentation.

2. Factor Analysis

A critical step in image analysis is the reduction of the volume of information in the image. An image consists of a set of pixels; the value of each pixel may be thought of as a coordinate in an n-dimensional space, where n is the number of pixels in the image. The dimensionality n of a typical medical image may be very large, especially if a three-dimensional image is being considered. However, inspection of any real image shows that there is a considerable amount of redundant information in the image; adjacent pixels have very similar values. A general method of eliminating this redundancy and reducing the dimensionality of the data is that of factor analysis, especially when the factorization is into orthogonal components. An image, represented as an n-dimensional vector I, may be factored into n mutually orthogonal vectors e such that

$$I = \sum_{j=1}^{n} c_j e_j \qquad (1)$$

where the e_j represent a new set of axes in the n space and the coefficients c_j are the coordinates of I in this space. If a population of vectors of the same type are considered, it can be shown that there is a preferred set of vectors p_j called the principal components of that population which, compared to other sets of orthogonal vectors, on average minimize the error when the summation in Eqn. (1) is limited to the first m terms.

In other words, the error

$$E = \left\| \left(I - \sum_{j=1}^{m} c_j p_j \right) \right\| \qquad (2)$$

402

is on average a minimum for the principal components. In practice it has been shown that E is often still small for m very much less than n and therefore by transforming the image vectors into the space defined by the principal components as axes it is possible to achieve a significant reduction in dimensionality. Principal components techniques require substantial computation and to date have only been used on images of coarse resolution which require relatively few pixels. An example of their use in the classification of pseudoimages produced from ultrasonic Doppler measurements of the common carotid artery has been given by Sherriff et al. (1983) and their use in the detection of lesions in liver and brain radionulcide images has also been reported (Barber 1982).

Factor analysis is more widely used in the analysis of dynamic studies. The vector whose elements consist of the values of a single pixel followed through the study may be called a dixel. All the pixels in a study represent a set of vectors for which principal components may be computed. The true dimensionality of this set of vectors largely depends on the complexity of the underlying physiological processes being investigated in the study. In general the dimensionality usually turns out to be quite low, typically less than five, which means each dixel can be defined as a point in a space of low dimensionality. This space is called the study space (Barber and Nijran 1982, Nijran and Barber 1984), and with certain assumptions about the underlying physiological processes being investigated in the study and using appropriate boundary conditions, it is possible in some cases to identify points in the study space which transform to vectors representing the underlying physiological behavior which may then be quantified using appropriate models. Examples have been given of the extraction of vectors representing the uptake of tracer in the liver, the thyroid gland, the kidney and cardiac function.

3. Image Segmentation

Image segmentation reduces the dimensionality of an image by dividing it into a small number of mutually exclusive regions. Segmentation is therefore concerned with identifying pixels in the image which can be linked together by some common property and treated as a single unit. The extracted segments or regions can then be used to isolate a piece of the image for further analysis. Alternatively, the shape of the segment boundary can be measured and used for classification.

The simplest form of segmentation is by thresholding in which the division into regions is purely on the basis of pixel value. Pixels with a value greater than or equal to some threshold T are assigned to one segment or region, whereas pixels with a value less than T are assigned to a second. This approach is usually ineffective for most medical images except in fairly trivial cases, but a modified version which adjusts the threshold value

based on an analysis of local image statistics has been used to define the left ventricle of the heart in a cardioangiogram. It is more common to find segments defined by a human observer, usually using an appropriate electronic marking device. This is at present the common practice with most dynamic studies (both radionuclide and radiographic). The summation of the pixel values in the region for each image in the sequence may be determined and these totalized values may be expressed as a time curve or vector. This vector then can be analyzed using an appropriate model or by empirical methods to provide a patient classification. If the region chosen contains information about, say, two structures A and B of which A is the structure of interest and it is not possible to separate A and B because they overlap, it may be necessary to find a region representing pure B and subtract the vector determined from this region from the original vector to isolate the behavior of A. It is often quite difficult to find a suitable second region and to determine the subtraction weighting.

Manual identification of regions, although it uses the powerful facilities of the human visual system, can be very variable and unreliable. A considerable amount of effort has been put into automatic methods of segmentation. It often happens that the edge of a structure of interest about which a region is to be drawn is characterized by a rapid local change in intensity; in other words the gradient of intensity is high. The edge of a region can therefore be identified with points of local maximum gradient. The simplest way to calculate the intensity gradient in the horizontal direction is to calculate the difference in value between adjacent pixels. A similar calculation can be done in the vertical direction and it is easily shown that the magnitude of the gradient at a point on the image is given by

$$g = (g_h^2 + g_v^2)^{1/2} \qquad (3)$$

where g_h and g_v are the gradients in the horizontal and vertical directions, respectively. The edge of an object may be found by linking together points of maximum local gradient. The edge detector described above is a very simple one and behaves badly in the presence of noise. More sophisticated methods, both linear and nonlinear, have been used successfully to deal with this problem. In practice an edge image is produced (i.e., an image whose pixel values are local gradient magnitudes) and an edge map produced by thresholding this image. This map may need further processing to remove isolated edge points and to thin-out thick edges and generally link pieces of edge together to produce complete continuous boundaries. Assuming the final boundaries are closed then points within each boundary form a distinct region. Boundaries have also been generated using a combination of thresholding, edge detection, and higher derivatives of pixel value. These have been used, for example, to find the edges of the left ventricle in cardiac function studies. Examples have also been given of the use of edge-detection methods for identifying ribs on chest radiographs and defining the outline of the bones in the hip joint. In the latter case the aim was not to directly segment the image but to use positional measurements taken from the extracted edge contour to diagnose abnormalities of the joint.

Other properties may be used to define edges such as color and texture, but examples of these are less commonly found.

4. Region Shape Analysis

The qualitative use of shape is often employed by radiologists to differentiate between lesions. Simple methods of describing the shape of objects as defined by their boundary have been widely used in cytology and these can be used with other imaging techniques. The basis of these techniques is to extract various measurements from the outline of the objects such as object area, boundary length, and the maximum chord (the longest distance between any two points on the boundary). It is possible to derive dimensionless parameters from these measurements, for example the ratio between the square of the boundary length and the area (which, for example, is a minimum for a circle), and use several of these derived parameters to define a vector which can be used for classification of the boundary shape. A useful property of these derived parameters is their invariance to orientation, position and object size at least within the limits of the quantization errors in the image-digitization process. They are measures of shape.

An entirely different approach is to analyze the shape of the boundary in terms of primitive features, such as convex or concave corners and straight segments, and then describe the boundary as an ordered concatenation of these primitives. This forms a description of the boundary which can be analyzed and classified using the techniques of syntactic analysis.

Three-dimensional images require the use of three-dimensional shape descriptors. It can be shown that if the principal axes of the inertia tensor of a three-dimensional object can be found, after suitable normalization of the boundary shape of the object in the space defined by these axes, the intersection of the boundary of the object with the three principal planes of this space is unique. These contours can then be extracted and analyzed by the methods used for two-dimensional contours.

5. Texture Analysis

Radiologists often use such terms as fine, coarse and smooth to visually characterize the texture of an image within a region, although studies have shown that there are marked differences between different observers in the way these terms are used. A variety of methods for the automatic classification of texture have been devised.

A texture has both short-range order (the texture primitive) and long-range order (the pattern of the primitive). Since both these are usually statistical in nature, textures are usually best characterized by statistical methods. Simple statistical measures of pixel values, such as the standard deviation or skew of pixel values within a region, rarely classify anything but the most extreme texture differences since they do not take into account the spatial structure of the texture. If there are v allowed pixel values in a region then a $v \times v$ matrix may be constructed whose ith row and jth column is the probability that a pixel with value i is horizontally adjacent to a pixel with value j. Such a matrix is called the horizontal cooccurrence matrix and similar matrices may be defined for the vertical and two diagonal directions. These matrices, as probability distributions, may then be characterized by various statistical measures and have proved useful in the classification of textures. The number of matrices computed for a given region may be extended by considering other directions and other distances between pixels. Texture methods based on cooccurrence matrices have mainly been used in radiology for the analysis of pulmonary disease. Texture-analysis techniques are also being used for tissue characterization from ultrasonic images (see *Ultrasound: Tissue Characterization*).

An alternative approach which has been suggested is to compute an edge map of all the edges in the texture pattern and analyze the statistics of the microedges.

See also: Computerized Image Analysis in Radiology; Image Analysis: Receiver Operating Characteristic Curves; Image Analysis: Transfer Functions; Computerized Axial Tomography; Quality Assurance in Diagnostic Radiology

Bibliography

Barber D C 1982 Automatic alignment of radionuclide images. *Phys. Med. Biol.* 27: 387–96
Barber D C, Nijran K S 1982 Factor analysis of dynamic radionuclide studies. In: Raynaud C (ed.) 1982 *Nuclear Medicine and Biology Advances*, Vol. 1. Pergamon, Oxford, pp. 31–34
Di Paola R, Kahn E (eds.) 1980 *Information Processing in Medical Imaging*. Inserm, Paris
Glenn W V, Harlow C A, Dwyer S J, Rhodes M L, Parker D L 1972 Image manipulation and pattern recognition. In: Newton T H, Potts D G (eds.) 1972 *Radiology of the Skull and Brain: Technical Aspects of Computed Tomography*. Mosby, St. Louis, Missouri, pp. 4326–54
Gonzalez R C, Wintz P 1977 *Digital Image Processing*. Addison-Wesley, Reading, Massachusetts
Nijran K S, Barber D C 1984 Analysis of dynamic radionuclide studies using factor analysis. In: Beconinck F (ed.) 1984 *Information Processing in Medical Imaging*. Nijhoff, Dordrecht, pp. 30–45
Pratt W K 1978 *Digital Image Processing*. Wiley, New York
Preston K, Onoe M (eds.) 1976 *Digital Processing of Biomedical Images*. Plenum, New York

Sheriff S B, Barber D C, Martin T R P, Lakeman J M 1983 Mathematical feature extraction techniques applied to the Doppler shifted signal obtained from the common carotid artery. In: Taylor D E M, Stevens A L (eds.) *Blood Flow: Theory and Practice.* Academic Press, London, pp. 235–60

D. C. Barber

Image Analysis: Receiver-Operating-Characteristic Curves

Receiver-operating-characteristic (ROC) curve analysis is a technique for evaluating a diagnostic imaging process as a complete system. The problem in diagnostic imaging can often be reduced to the detection or nondetection of an abnormality in the image. In recent years the techniques of signal-detection theory have been applied to analyze the correlation between the image-interpreter's decision and the input signal to the imaging system.

1. Signal-Detection Theory

In its simplest form, the diagnostic task consists of an observer being confronted with an image which may or may not contain an abnormality and being asked to say whether or not he thinks an abnormality is present. In the language of psychophysics where the method origi-nated (Green and Swets 1966) there are two possible stimuli and two possible responses. If an image contains an abnormality the input to the imaging system contains a signal s. Alternatively, the image may not contain an abnormality and, as far as the diagnosis is concerned, the image can be regarded as noise n. When shown the image the observer may respond S (signal) or N (noise) according to whether or not he perceives the diagnostic abnormality. An imaging system can be assessed using a large number of images, some of which are known to contain a diagnostic signal. The experiment consists of showing an observer each image and recording the response S or N (Fig. 1).

With complete transfer of information by the image, the observer responds S every time the image contains the signal s, and N every time the image represents n, i.e.,

$$P(S|s) = 1 \quad P(N|n) = 1$$
$$P(S|n) = 0 \quad P(N|s) = 0$$

$P(S|s)$, which reads "the probability of S given s" is the conditional probability of the response S when the input to the imaging system is s.

When the transfer of information by the imaging system s is not perfect, these conditional probabilities are not unity or zero but are fractional.

2. Stimulus–Response Matrix

Table 1 shows a stimulus–response matrix for an imaging system. Each cell of the matrix contains a conditional

Table 1
A stimulus–response matrix for an imaging system

Stimulus	Response	
	S	N
s	$P(S\|s)$ true positive	$P(N\|s)$ false negative
n	$P(S\|n)$ false positive	$P(N\|n)$ true negative

probability which is often quoted in the evaluation of a diagnostic test. If the imaging system is good at trans-mitting the information the observer will often respond S when the signal input is s, i.e., $P(S|s)$ will be close to unity. The true-positive ratio, $P(S|s)$, is called the sensitivity of the diagnostic imaging test.

If the observer responds N when the image in fact contains a signal, the observer is said to have made a false-negative response; $P(N|s)$ is called the false-negative ratio.

It is desirable that the observer responds N when no signal is present in the image. The specificity of the diagnostic test is the fraction of images that is correctly identified as having no signal, or $P(N|n)$.

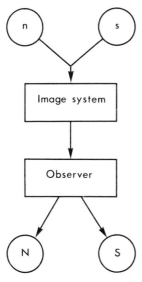

Figure 1
The simple diagnostic imaging system. The input to the system is either noise n or signal plus noise s. On seeing the image, the observer decides whether an image is present S or absent N.

Finally, the observer may erroneously think that a signal is present when the image does not in fact contain a signal. This is called a false-positive response, and the probability of its occurrence $P(S|n)$ is called the false-positive ratio.

An ideal imaging system has unity sensitivity and specificity, and zero false-positive and false-negative ratios. A good diagnostic test has high sensitivity and specificity, and low false-positive and false-negative ratios.

3. Receiver-Operating-Characteristic Curve

Once an image has been displayed, the observer must respond either S or N. This 2×2 matrix therefore has only two degrees of freedom since

$$P(S|s) + P(N|s) = 1$$

and

$$P(S|n) + P(N|n) = 1$$

The information in the matrix cells can therefore be represented by a point A on a two-dimensional graph. The coordinates of A are conventionally given by $P(S|s)$ and $P(S|n)$, the true-positive and false-positive ratios (Fig. 2).

The point A on the graph shows how the observer can be expected to respond should he be shown a series of diagnostic images with the experimental conditions unchanged. What would happen if the observer were told that a clinical management decision rested upon his findings? For example, the observer might be told that the patient would be discharged from hospital if the observer does not find any abnormality in the diagnostic image, i.e., if the observer responds N. The observer can be expected to modify his decisions accordingly and to adopt a stricter criterion for responding N—he is conscious of the need to make doubly sure that no false negatives occur. Because of this he will make more false-positive responses as he tends to respond S to those images where he is not sure whether or not a signal is present.

The result of performing the experiment again without changing the physical parameters of the stimulus and noise is to derive new conditional probabilities which can be represented on the graph of $P(S|s)$ against $P(S|n)$ by a new point B (Fig. 2). Because of the stricter criteria for responding N, the observer responds S more often both when the input to the imaging system constitutes a signal and when it constitutes noise. The point representing the observer's responses is therefore displaced towards the top right-hand corner of the graph. By successively changing the observer's criteria a curve can be generated: the curve is called the receiver operating characteristic (ROC) curve. Any point on the curve describes the observer's criteria for distinguishing between what he perceives as signal and what he perceives as noise—the point is called the operating position on the ROC curve.

Note that the entire ROC curve is generated without changing any physical parameter of the stimulus situation. The signal level and the noise level are the same for all points on the curve. The curve is generated by changing the way the observer responds, in fact by changing a threshold value of the perceived signal. The a priori signal probability determines the position of the operating point on the ROC curve but does not affect the shape of the curve.

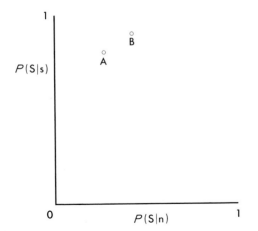

Figure 2
The information contained the stimulus–response matrix expressed in graphical form using $P(S|s)$ and $P(S|n)$ as the coordinates of the point A. When the observer's criteria are changed, a new point B is generated. Further changes in the observer's criteria result in the generation of the ROC curve

4. Generation of Practical ROC Curves

ROC curves were initially produced by a single-threshold yes/no simple detection experiment (Green and Swets 1966). An observer took part in a large number of trials (typically 600) in each of which he had to state the presence or absence of a signal. The results of these trials gave one point on the ROC curve. Further points on the ROC curve were obtained either (a) by altering the a priori probability of signal occurrence, or (b) by altering the value of the decision outcomes, e.g., by making it more important to respond N when noise alone is present. Such experiments were lengthy and tedious.

Fortunately it was shown that an equivalent ROC curve could be generated by a rating method. The observer is required to respond to each trial by indicating a measure of his certainty that a signal is present. The observer may be given a sliding scale with a pointer on which he is required to indicate the a posteriori probability of the signal being present. More commonly the observer is required to rate his certainty verbally so that his opinion varies from almost definitely present to almost definitely not present through probably present, unsure, and probably not present. The analysis of the results of the rating task is performed by using the division between two categories as a threshold thus producing a sequence of simple yes/no responses.

5. Quantification of Differences between ROC Curves

5.1 Area Under the Curve

At first glance the area under the ROC curve appears to be a measure of how detectable the signal is. If $P(S|s)$ is close to unity for most values of $P(S|n)$, the area under the curve approaches unity and the observer is nearly always correct. If the signal is not distinguishable from the noise, the observer will guess N or S at random and the ROC curve will be the positive diagonal of the graph. The area under the curve is then 0.5, the observer's probability of responding correctly.

In general, the total probability P_c of a correct response is

$$P_c = P(S|s) P(s) + P(N|n) P(n)$$

This expression does not appear to be useful in relation to the ROC curve. It can be shown, however, that the area under the ROC curve is equal to the probability of a correct decision in a two-alternative forced-choice experiment (Green and Swets 1966). In this type of experiment the observer is required to choose in which of two presentations the signal resides knowing that the other presentation contains noise alone. These experimental conditions are not appropriate for the examination of medical images. The area under the ROC curve is not in general the probability of the correct decision in a simple yes/no detection experiment.

When two ROC curves intersect, the curve having the smaller area may offer a higher probability of true positives $P(S|s)$ for a given $P(S|n)$ and so the area under the ROC curve is not a reliable measure of signal detectability.

5.2 Image Information Content from ROC Curve Data

Suppose the imaging system is a noisy communication channel so that the input is a set of messages $\{y_i\}$ and the output is a set of responses $\{Y_j\}$. When a ROC is generated $\{y_i\}$ is either s or n, and $\{Y_j\}$ is either S or N.

Using Shannon and Weaver's (1949) information-theory approach, the average information content I of one image may be defined as the reduction in uncertainty about the input that is provided by the output, and is equal to the average uncertainty about $\{y_i\}$ before seeing the image minus the average uncertainty about $\{y_i\}$ given the output Y_j.

Note that if the communication channel, i.e., the image, is so good that the uncertainty about the message transmitted is zero, the information content of the image

is equal to the a priori uncertainty, whereas if the uncertainty about the input has not changed after observing the image, then the information content of the image is zero.

It can be shown that the information content can be written as a function of $P(S|s)$, $P(S|n)$ and $P(s)$:

$$I = f[P(S|s), P(S|n), P(s)]$$

The average information content of an image using the imaging system at the operating point on the ROC curve given by $P(S|s)$ and $P(S|n)$ can therefore be calculated for each a priori probability of a signal being present $P(s)$.

For each probability $P(s)$, isoinformation curves can be drawn using the same axes as the ROC curve. The amount of information available from viewing the image will vary along the ROC curve with the operating point and, using the isoinformation curves I_{max} the maximum information content for a given a priori probability $P(s)$, can be derived. I_{max} can therefore be used as a measure of image system quality for the experimental conditions used to generate the ROC curve.

Furthermore the amount of information transmitted by the image at two operating points on one ROC curve, or at two points on different ROC curves, can be compared quantitatively. Although this measure depends on the object configuration and on the a priori probability of the signal being present, these are realistic constraints. Some imaging systems may be better than others at imaging certain structures and it was shown that when ROC curves intersect some systems are better than others only if there is a high probability of a signal being present.

6. Selection of Operating Points on ROC Curves: Efficacy Analysis

The selection of an operating point on a ROC curve can allow for the possibly quite different consequences of the several decisions when expressed as risks or costs. A false-negative test where the patient may suffer seriously from his disease remaining undetected and therefore untreated must be assessed in relation to a false-positive result where a patient may be subjected to further costly and risky procedures before the absence of the disease is finally established.

Following Metz et al. (1975, 1976) the expected value $\bar{C}$ of a decision can be expressed in terms of the following costs:

$$\bar{C} = \bar{C}_o + \sum_i \sum_j C_{ij} P(D_i|d_j) P(d_j)$$

where $\bar{C}_o$ is the average overheads cost of performing the test, $P(d_j)$ is the probability of the patient having the disease d_j, $P(D_i|d_j)$ is the conditional probability of the decision D_i resulting from the diagnostic test on a patient having the disease d_j, and C_{ij} is the cost of the decision D_i being made on a patient with disease d_j. Cost is a term embracing financial, morbidity and mortality costs; their inclusion in the one equation implies a common set of units. Morbidity and mortality must therefore be expressed in financial units, an explicit decision contrasting with the everyday implicit decisions of clinical practice.

By expressing the decision D_i as S or N, but now using s and n to signify the presence or absence of a disease (i.e., there are two states of d_j), the equation may be rewritten thus:

$$\bar{C} = \bar{C}_o + C_{TP} P(S|s) P(s) + C_{FN} P(N|s) P(s)$$
$$+ C_{FP} P(S|n) P(n) + C_{TN} P(N|n) P(n)$$
$$= \bar{C}_o + C_{FN} P(s) + C_{TN} P(n)$$
$$+ (C_{TP} - C_{FN}) P(S|s) P(s)$$
$$+ (C_{FP} - C_{TN}) P(S|n) P(n)$$

where C_{TP}, C_{FN}, C_{TN} and C_{FP} are the cost of making a true-positive decision, a false-negative decision, a true-negative decision and a false-positive decision, respectively.

The average net benefit may be defined as the reduction in the average costs made by performing the diagnostic test. It is clearly the difference between the average cost of performing the test and the average cost of not performing the test.

If the test is not performed, $\bar{C}_o = 0$ and all patients are called "disease absent," hence in this instance $\bar{C}$ (no test) $= C_{FN} P(s) + C_{TN} P(n)$. The average net benefit $\overline{NB}$ is therefore:

$$\overline{NB} = \bar{C} \text{ (no test)} - \bar{C}$$
$$= (C_{FN} - C_{TP}) P(S|s) P(s)$$
$$- (C_{FP} - C_{TN}) P(S|n) P(n) - \bar{C}_o$$

Provided the costs associated with each decision and the overheads cost are known, this equation can be used to calculate the net benefit associated with any point on a ROC curve.

$\overline{NB}$ can be maximized by solving $\partial\overline{NB}/\partial P(S|n) = 0$. A maximum $\overline{NB}$ is attained when the slope of the ROC curve is given by

$$\frac{\partial P(S|s)}{\partial P(S|n)} = \frac{P(n)}{P(s)} \frac{(C_{FP} - C_{TN})}{(C_{FN} - C_{TP})}$$

The maximum average net benefit is a measure of the usefulness of the diagnostic test. Not only is it a function of the relative costs of the decisions made but it also depends on the incidence of the disease in those patients subjected to the diagnostic test. When the disease is rare (e.g., in screening programs) $P(n)/P(s)$ is great, and the maximum occurs when the slope of the ROC curve is great, i.e., towards the lower left-hand side of the graph.

When the results of treating the disease are of marginal value and the health costs of treating a patient without the disease are high, then $(C_{FP} - C_{TN})$ is high and the maximum net benefit again occurs when the slope is steep. Thus the theory offers a rational basis for the guidance given by intuition that when the cost of treating a false-positive is high, the observer favors a low false-positive rate. ROC curves are usually monotonically increasing functions of $P(S|n)$ and therefore $\partial P(S|s)/\partial P(S|n)$ is positive. If the cost of a true-negative decision is taken to be zero (i.e., nothing further is done) the average net benefit can only be maximized if $(C_{FN} - C_{TP})$ is positive, that is, the cost of a false-negative decision must exceed the cost of a true-positive decision.

7. Concluding Remarks

ROC curve analysis enables the diagnostic imaging process to be assessed as a complete system. The concept of accuracy as the fraction of correct diagnoses made gives a limited description of the usefulness of a test procedure because it gives no indication of whether the errors are false positives or false negatives. A single definition of a false-positive fraction and a false-negative fraction is also unsatisfactory for it implies a single arbitrary decision threshold. The ROC curve method relates the number of errors the observer makes to the rate of correct detection of the presence of a signal and provides a complete description of the effects of changing the decision threshold. It has been shown that one observer stays on one ROC curve in repeated experiments but different observers have different ROC curves so that the observer has become part of the "system" under examination. Construction of a practical ROC test necessitates the absolute knowledge by the experimenter of the presence or absence of a signal in the input to the imaging system. Such methods are therefore convenient for phantom generated or simulated lesion images but this constraint becomes more demanding when real clinical material is used.

Despite the difficulty of quantifying the information on an ROC curve these analyses have become very useful as databases for clinical decision-making studies.

See also: Image Analysis: Extraction of Quantitative Diagnostic Information; Quality Assurance in Diagnostic Radiology

Bibliography

Green D M, Swets J A 1966 *Signal Detection Theory and Psychophysics*. Wiley, New York

Metz C E 1978 Basic principles of ROC analysis. *Semin. Nucl. Med.* 8: 283–98

Metz C E, Starr S J, Lusted L B 1976 Quantitative evaluation of visual detection performance in medicine: ROC analysis and determination of diagnostic benefit. In: Gray H A (ed.) 1976 *Medical Images: Formation, Perception and Measurement*. Institute of Physics, Bristol, pp. 220–41

Metz C E, Starr S J, Lusted L B, Rossmann K 1975 In: Raynaud C, Todd-Pokropek A 1975 *Proc. 4th Int. Conf.*

Information Processing in Scintigraphy. Commissariat à l'Energie Atomique, Orsay, France, pp. 420–36

Shannon C E, Weaver W 1949 *The Mathematical Theory of Communication*. University of Illinois Press, Urbana, Illinois

A. L. Evans

Image Analysis: Transfer Functions

A powerful technique for defining the system response (e.g., the resolution) of a linear stationary system having an output which is the convolution of the input and the impulse response function is to determine its transfer function. A Fourier or Laplace transform will convert the convolution into a multiplication. The transfer function is the transform of the system's impulse response function. Much image processing then becomes simple in the transform (or spatial frequency) domain. One aim of image processing is to find an estimate of the true image subject for some, for example least squares, constraint about noise. Many linear stationary filters may be treated in terms of the transfer function of some processing stage generating the "best" estimate of the true image, or of some feature of it. Such filters have been used in the following ways: to correct for instrumentation degradation; to alter the signal-to-noise ratio; and for feature extraction, segmentation, etc. Many types of nonlinear image processing which may not be analyzed in this manner (e.g., homomorphic filtering) are also of considerable interest and are briefly presented here. Many of these basic tools form part of more complex general procedures used to analyze medical images.

1. Transfer Functions

Resolution is an important attribute of a system. However, it may not be trivial to measure. One reason for this is that the definition of resolution is dependent on the manner in which it is measured. The classical definition often used is that of the separation of two "points" (i.e., delta functions) such that the system can just establish the existence of two such separate objects, or some equivalent measure dependent on a knowledge of the point spread function (PSF) such as full-width at half-maximum (FWHM). This type of definition may prove inadequate, particularly in a photon-limited situation, as is often the case with systems used in medicine. (It should be noted that the type of resolution being considered here is that of spatial or temporal resolution, rather than contrast resolution as described by Rose (1973).

Rather than attempting to define resolution by a single number, a standard and more powerful technique is to define the transfer function of the system, often called the modulation transfer function, but more accurately known as the optical or system transfer function. This function defines the ability of a system to transfer information with respect to a set of orthogonal functions, normally sine and cosine functions at various frequencies, between its input and its output. The system must be linear and stationary for the use of a transfer function to be meaningful.

Let $f(x)$ be the input to a system where x may represent time or space. Let $g(x)$ be the output from a system. Then

$$g(x) = \int_{-\infty}^{+\infty} f(x')\,h(x; x')\,\mathrm{d}x'$$

where $h(x; x')$ is the general transfer function relating the output at x to an input at all values of x'. As expressed here, such a system is linear. A further constraint which may be added is that the system is stationary, that is, h is not a function of both x and x'. In this case the equation can be reexpressed as the convolution of $f(x)$ and $h(x)$:

$$g(x) = \int_{-\infty}^{+\infty} f(x')\,h(x - x')\,\mathrm{d}x'$$

As is well known such a convolution can be transformed by use of the Fourier (or Laplace) transform such that, if $G(u)$ is the transform of $g(x)$, $F(u)$ of $f(x)$ and $H(u)$ of $H(x)$, then

$$G(u) = F(u)\,H(u)$$

that is, a simple pointwise multiplication. $H(u)$ is now the normally used form of the transfer function of the system in question. $H(u)$ is just the Fourier transform of $h(x)$, the impulse response function. Note also that if $r(x)$ is the autocorrelation function of $f(x)$, then the power spectrum of $f(x)$ is the Fourier transform $R(u)$ of $r(x)$ and is equal to $|F(u)|^2$.

In two dimensions, the same relationships hold and $f(x, y)$ and $F(u, v)$ replace $f(x)$ and $F(u)$, and correspondingly two-dimensional forms are substituted for g, G, h and H.

An alternative approach is to consider the same problem in matrix notation. Let f, g be the input and output functions which, if two dimensional, must be expressed as column vectors of the form

$$\tilde{g} = [g(0, 0)\, g(1, 0) \ldots g(N - 1, 0)\, g(1, 0)$$
$$\times g(1, 1) \ldots g(N - 1, N - 1)]$$

and let H be a matrix of size $N^2 \times N^2$ composed of N^2 submatrices of form

$$H = \begin{bmatrix} [H_0] & [H_{N-1}] & \ldots & [H_1] \\ [H_1] & [H_0] & \ldots & [H_2] \\ \ldots & \ldots & \ldots & \ldots \\ [H_{N-1}] & [H_{N-2}] & \ldots & [H_0] \end{bmatrix}$$

where

$$[H_j] = \begin{bmatrix} h(j,0) & h(j,N-1) & \ldots & h(j,1) \\ h(j,1) & h(j,0) & \ldots & h(j,2) \\ \ldots & \ldots & \ldots & \ldots \\ h(j,N-1) & h(j,N-2) & \ldots & h(j,0) \end{bmatrix}$$

then

$$\tilde{g} = H\tilde{f}$$

H is block Toeplitz and block circulant.

A well-known property of the Fourier transform is to diagonalize circulant matrices (Andrews and Hunt 1977, Pratt 1978). Thus, after a Fourier transform, the matrix H is diagonal, and as before, a simple pointwise multiplication suffices to determine G from F, when H is known. This illustrates, in a more general manner, that a convenient description of the transfer function in frequency terms can be found provided the system is linear and stationary.

As illustrated in Fig. 1, the transfer function can also be thought of as the function, in terms of spatial frequency, which shows the transfer of contrast between input and output.

A (linear) filtering operation is defined as the transformation $G'(u) = G(u)\,\mathrm{FILT}(u)$ where the filter $\mathrm{FILT}(u)$ is defined as a weighting function in the frequency domain. This is obviously equivalent to a convolution by $\mathrm{filt}(x)$ where $\mathrm{filt}(x)$ is the inverse Fourier transform of $\mathrm{FILT}(u)$.

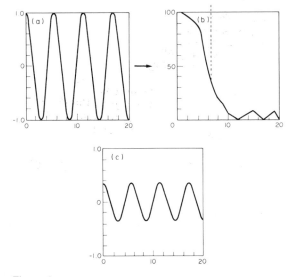

Figure 1
A sinusoidal input function (a) to a system described by the transfer function (b) results in an output function as shown (c). The dotted line marks the value of the spatial frequency of the input and output functions

2. Linear Image Processing

Much image processing then becomes readily comprehensible in the spatial frequency domain. For example, using the above terminology, in a noise-free situation, an unknown $F(u)$ can be determined immediately from an observed $G(u)$ by dividing by $H(u)$. The equivalent filter $1/H(u)$ is known as the inverse (resolution recovery) filter. However, it is of little practical value as such for the following reasons. In most systems, the signal power spectrum will tend to look like $|H(u)|^2$. However, the noise power spectrum is likely to be quite different. For example, in many systems the noise is white, that is, the expected noise power at any frequency is constant. In some systems, such as in computerized tomography, the noise power tends to increase with frequency. The inverse filter will therefore in such cases severely decrease the signal-to-noise ratio. For image processing techniques to be of practical value, a constraint must be added so as to preserve, or if possible to improve, the signal-to-noise ratio. Thus one aim in image processing is to find an estimate $\hat{f}$ of f, subject to an appropriate constraint, for example $|\hat{f} - f|^2 < \epsilon$.

There are some other important limitations. When such techniques are developed, apart from the assumption that the system is linear and stationary, there is usually also an assumption that the noise is additive, and of known power spectrum. In addition, the exact form of any constraint used, such as that of least squares difference, makes a severe assumption about the form of the signal. An usual feature of medical images is that the form of a signal which it is desired to enhance may be largely unknown.

One general form for a linear stationary filter used in image processing may therefore be treated as the transfer function of some processing stage used to generate $\hat{f}$ from g. The general shape of such a class of useful filters is shown in Fig. 2. In a noise-free situation, the ideal enhancing filter is the inverse filter. As the signal-to-noise ratio becomes poorer, filters with greater noise-

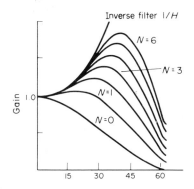

Figure 2
A family of bandpass "Metz" filters ranging from the inverse filter ($1/H$) for $N = \infty$ to a pure smoothing filter ($N = 0$), which is in fact the transfer function (H) of the system

reducing properties are desired. Such families of filters can be generated from the solution of the equation

$$g = Hf + n$$

which may be demonstrated (Pratt 1978) to be given by

$$\hat{f} = (H^{*T}H + \gamma Q^T Q)^{-1} H^{*T} g$$

where n is an additive noise vector, γ is a constant and Q is a linear operator minimizing $\| Q\hat{f} \|^2$ subject to $\| g - H\hat{f} \|^2 = \| n \|^2$. The symbols $*$ and T represent the complex conjugate and transpose, respectively, and $\| \ \|^2$ is the squared norm. Various values for γ and the operator Q may be chosen to generate various classes of linear enhancing filters. It should be noted that here the aim is to obtain an estimate of f, the unknown input, as accurately as possible (in the least squares sense) in the presence of noise.

However, many other goals are possible. It might be desired to produce a function $p(x)$ which has value at x dependent on the likelihood of some signal $s(x')$ being present in the observed $g(x)$ at the point x. The matched filter does this, and may be shown to be given by $s*(x)\exp(-iux)$, in otherwords by a convolution with the desired signal after a complex shift.

An area of particular interest is that of computerized tomography and emission computed tomography reconstruction. Here, the most common technique employed is that of filtered back-projection, where a ramp filter is employed to correct for the blurring resulting from the

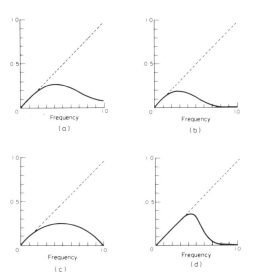

Figure 3
A series of filters used for tomographic reconstruction. Each is composed of a ramp filter (given by the dotted line) multiplied by a window function. These are: (a) a Hamming window, (b) a Parzen window and (c) a Bartlett window, each with a cutoff frequency of 1.0, the Nyquist frequency, and (d) a Butterworth window with cutoff frequency of 0.5 and order 5. The x axis is the ratio of the frequency to the Nyquist frequency

back-projection. This is a normal linear filtering operation. However, when used as such, the noise properties of the reconstruction are very poor. Therefore, a noise-reducing window function is normally employed. A variety of window functions have been used, as illustrated in Fig. 3. Typical of such window functions is the Butterworth window,

$$\text{FILT}(u) = \sqrt{\{1/[1 + (u/u_c)^{2N}]\}}$$

where u_c is a parameter, the cutoff frequency, and N is a constant, the order of the filter. The use of such window functions is widespread.

3. Nonlinear Image Processing

Many types of nonlinear image processing are of considerable interest, but cannot be treated at any length here. They cannot directly be analyzed in terms of their transfer functions. Homomorphic filtering, for example, is the name given to the technique where a (normally linear) filter is applied to the log of the data to be processed, after which the exponential of the result is taken. Note also that, in many cases, Poissonian noise can be made approximately additive by means of the Anscombe transform (Anscombe 1948), that is, by performing the filtering operation after taking the square root of the data, after which the variance of any Poisson distributed variable is constant to a good approximation.

Another widely used nonlinear filtering process is that of median filtering. Here, for all possible positions within a window, the central value is replaced by the median of all the values within that window. This type of filter will tend not to blur edges to the same extent as a linear smoothing filter. Such filters can be tailored to respect known properties of the image. If only for these reasons, such nonlinear filtering operations are frequently employed.

Not only is the transfer function approach of considering the filter in the frequency domain meaningless for such nonlinear filters, the calculation of signal-to-noise ratios before and after nonlinear processing can be misleading. It is difficult to compare linear and nonlinear filters directly. For example, the calculation of the noise properties of a median filter cannot in general be estimated (Pratt 1978).

Detailed consideration of other nonlinear techniques will not be given here. It should be noted, however, that many nonstationary techniques, where an operator is used which varies as a function of position, can be considered simply as conventional linear operators acting within a local region, and their effect interpreted in terms of "local" transfer functions.

4. Image Analysis

Many of the basic tools described above form part of more complex general procedures used to analyze medical images. Filters have been used for many purposes: to correct for some instrumentation degradation, to alter signal-to-noise ratios, to enhance some aspect of the data, and so on. For example, part of any attempt to quantify an image involves some kind of "segmentation," that is, an operator which divides the image into different regions, several of which may correspond to anatomical regions of interest, and some to "background." The most common operation used in segmentation is that of edge detection. Differentiation of the image by taking the gradient or some higher derivative serves to enhance edges, and may be performed by a simple linear filter. This is illustrated in Fig. 4 for a

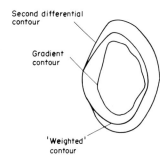

Figure 4
A radionuclide cardiac study, with (superimposed) the left ventricular contour determined from a gradient filter (inner curve), a second differential Laplacian filter (outer curve), and an angularly weighted interpolation between the two

radionuclide cardiac image. However, signal-to-noise ratios normally encountered suggest the use of more sophisticated techniques, and in particular, nonstationary methods where the filter adapts itself to local conditions. It should be noted that edge detection is normally subdivided into four sections: processing, edge enhancement (such as described), thresholding (which is typically a nonstationary operation), and edge following.

A second example relates to the generation of functional images, where as is common in many medical imaging procedures, a set of two-dimensional images, which have been collected over a period of time (or with respect to some other parameter), are to be reduced to a single two-dimensional image by fitting some function for every position in the image. Such transformations tend to be very noisy, and considerable care must be taken over error propagation. In other words, as part of the procedure to generate such images, noise reduction, either in terms of simple linear filters or in more general terms of regularizing the mathematical formulation (Tikhonov and Arsenin 1977), needs to be employed.

A third important example is that of deconvolution. For example, from a knowledge of the input and output functions to some system, it is possible to derive a transfer function. Such techniques have been widely used for kidneys, brains (e.g., to evaluate cerebral blood flow), etc. An important assumption in such a method is that the system can be modelled and is linear and nonstationary. The use of regularization (Tikhonov and Arsenin 1977) and the establishment of suitable mathematical models (Rescigno et al. 1983) is important. While deconvolution techniques have often been performed for a complete organ system, they may also be employed on a local basis. The idea of analyzing a three-dimensional set of data on a very local basis, for example pixel by pixel, has led to the use of a number of other techniques such as factor or principal component analysis (Di Paola et al. 1982). Here, a small set of orthogonal functions have been determined (so as to minimize their covariance), and the set of images decomposed into the weights associated with each of these functions. Oblique transformations may be required to modify such functions so that they have some reasonable physiological meaning.

An important question is how to evaluate such methods. One such method of evaluation is that of constant stimulus, where a given signal at various amplitudes is presented and the outputs compared. Alternative approaches such as the use of receiver operating characteristic (ROC) curves based on assessing a large number of processed images can be employed (see *Image Analysis: Receiver Operating Characteristic Curves*).

See also: Image Analysis: Extraction of Quantitative Diagnostic Information

Bibliography

Andrews H C, Hunt B R 1977 *Digital Image Restoration.* Prentice-Hall, Englewood Cliffs, New Jersey
Anscombe F J 1948 The transformation of Poisson, binomial and negative-binomial data. *Biometrika* 35: 246–54
Di Paola R, Bazin J P, Aubry F, Aurengo A, Cavailloles F, Herry J Y, Kahn E 1982 Handling of dynamic sequences in nuclear medicine. *IEEE Trans. Nucl. Sci.* 29: 1310–21
Pratt W K 1978 *Digital Image Processing.* Wiley, New York
Rescigno A, Lambrecht R M, Duncan C C 1983 Mathematical methods in the formulation of pharmacokinetic models. In: Lambrecht R M, Rescigno A (eds.) 1983 *Tracer Kinetics and Physiologic Modeling,* Lecture Notes in Biomathematics Vol. 48. Springer, Berlin, pp. 59–119
Rose A 1973 *Vision: Human and Electronic.* Plenum, New York
Tikhonov A N, Arsenin V Y 1977 *Solutions of Ill-Posed Problems.* Winston, Washington, DC

A. Todd-Pokropek

Implanted Prostheses: Tissue Response

Implanted prostheses are used for a variety of purposes in numerous clinical situations. Certain basic conditions may be identified as potentially warranting treatment by implant surgery:

(a) gross congenital defects which have functional consequences, such as hydrocephalus;

(b) developmental defects, such as scoliosis, which also have implications of impaired body function;

(c) organic diseases, including osteoarthritis and atherosclerosis;

(d) tumors requiring resection;

(e) gross tissue atrophy, such as alveolar ridge resorption;

(f) trauma, after which tissues (such as bone) require temporary support while healing takes place; and

(g) psychological problems, where reconstruction of the body leads to alleviation of the condition, such as congenital hypoplasia of the breast or parts of the face, treated by augmentation.

The function required of these prostheses is naturally different in each case, but will involve one or more of the following features:

(a) load transmission (e.g., dental implants, bone fracture plates);

(b) a bearing (joint replacement);

(c) control of fluid flow (heart valves, ureteral prostheses, hydrocephalus shunts);

(d) passive space filling (augmentation mammaplasty);

(e) generation and application of electrical stimuli (cardiac pacemakers);

(f) transmission of electrical signals from external sources to internal organs (neurological electrodes); and

(g) sound or light transmission (intra-auricular and intraocular prostheses).

Some of these functions require the properties of strength and rigidity that can only be supplied by metals, whereas others require the flexibility of elastomers, the optical transparency of glasses or amorphous polymers, the electrical resistivity of ceramics or plastics, and so on. Generally, however, these functional requirements are not particularly onerous and can be satisfied by a relatively large number of engineering materials. Intrinsic biofunctionality is, therefore, not a serious challenge.

Of far greater significance are the interactions between the implanted material and the tissue, constituting the biocompatibility characteristics of the material (Table 1). Many materials have been investigated for implant use, but very few have proved suitable. The success or failure of a device is determined by many factors, including design parameters and surgical skill, but is often largely determined by the biocompatibility of the material; indeed, biocompatibility can, and often does, influence long-term biofunctionality. The subject of biocompatibility is discussed here in relation to the various phenomena that are involved, and summarized in terms of the specific materials currently used.

The term biocompatibility is used to describe the ability of a material to exist within the physiological environment without either the material adversely and significantly affecting the body, or the environment of the body adversely and significantly affecting the material. Such a definition is open to wide interpretation and there is no single parameter by which biocompatibility can be quantified. It is generally considered prudent to view this in its broadest sense and to assume that every interaction, be it a chemical, biochemical, electrochemical, biomechanical or any other reaction, is a potentially important factor in biocompatibility.

A few important points must be stressed. First, biocompatibility involves effects on both the material and the tissue and we cannot divorce these two aspects. Frequently a bulk material may have little effect on the tissues, but as the physiological environment causes degradation of that material, the release of degradation products initiates an adverse response from the tissue. Both aspects, therefore, must be considered. Secondly, and in connection with this, the tissue response is a time-dependent phenomenon, and both acute and chronic phases may be involved. Thirdly, although the basic interactions will take place at or near the interface between tissue and material, the effects of the interaction may not be localized but may, either in addition or instead, be of systemic character. This is very important since it is often very difficult to attribute systemic conditions or symptoms localized in a different part of the body to an implant located elsewhere.

1. Metallic Corrosion

The physiological environment, being basically an isotonic saline solution, is very hostile to metallic materials, the corrosiveness of salt being well known to metallurgists. This hostility is so great, and the requirement of negligible corrosion in implanted devices is so stringent, that only the most corrosion-resistant alloys can be used in this situation. At present this restriction means that only austenitic stainless steel, some dilute titanium alloys and various cobalt-based alloys can be used. This excellence in corrosion resistance is essential, not so much because of the effects of corrosion on the implant itself, but more because of the effects of corrosion products on tissue. Very slow rates of corrosion may be undetectable in the majority of engineering situations but the accumulation of even small amounts of these products can significantly influence the tissue response. It is now established that of the three alloys mentioned above, stainless steel is the most susceptible to corrosion—as many as 50% of all stainless steel orthopedic implants corroding to some extent (Fig. 1). Crevice corrosion is the most common mechanism but fretting, galvanic and intergranular corrosion may also be observed. The problem with stainless steel is that the oxide film breakdown potential is so close to the resting potential that there is

Table 1
Implant materials

Material	Uses	Comments on biocompatibility
Stainless steel	joint prostheses, orthopedic-fracture plates, heart valves	generally good but susceptible to corrosion
Titanium	joint prostheses, orthopedic-fracture plates dental implants, heart valves pacemaker encapsulation	physiologically indifferent, excellent biocompatibility
Cobalt–chromium alloys	joint prostheses, orthopedic-fracture plates, dental implants, heart valves	generally good but hypersensitivity may be observed if large amounts of wear products released
Carbon	dental implants, heart valves, transcutaneous implants	excellent
Alumina	joint prostheses, dental implants, heart valves	excellent
Apatite-like ceramics	dental implants, bone reconstruction	variable, depending on Ca–P ratio and rate of resorption
Silicone rubber	joint prostheses, heart valves, catheters and other tubes, intraocular implants, augmentation prostheses	generally excellent but occasional problems of lipid absorption, cracking or excessive fibrosis
Ultra-high-molecular-weight polyethylene	joint prostheses	excellent
Poly(methyl methacrylate)	joint prostheses, dental implants, intraocular implants	good
Hydrogels	heart valves, vascular prostheses	depends on degree of hydration and counter-ions present
Thermosetting resins	pacemaker encapsulation	variable
Polyurethanes	vascular prostheses	variable, but can be excellent
Polyesters	heart valves, vascular prostheses	variable, but very good with some fibers (e.g., Dacron)

always a risk of film breakdown. Stainless steel is passivated only marginally by virtue of its chromium content. To attain a more suitable state of passivation the chromium has to be alloyed with a less susceptible element than iron; this is achieved with the cobalt–chromium alloys. Alternatively, the chromium can be discarded altogether in favor of another passive element; this is the basis of the titanium alloys. Corrosion following film breakdown is rarely seen in cobalt–chromium alloys and never in the titanium alloys.

Film breakdown is not the only mechanism by which metal can be released into the tissue for there is a finite rate of ion diffusion through an intact oxide layer. Thus, whatever alloy is placed in the tissue some metal ions will be released with a potential for accumulation. This occurs even with titanium, and pigmentation of the surrounding tissue occurs as a consequence. Fortunately no adverse effects appear to arise from the accumulation of titanium in the tissue. These titanium-containing deposits are not usually associated with any inflammatory response and there is no obvious correlation between the titanium content and the presence of leukocytes which normally deal with harmful foreign bodies. In the case of stainless steel, slight amounts of interfacial corrosion do not appear to lead to clinically significant effects, but when large amounts are involved a pattern of tissue change may be observed when phagocytosis of the particles is attempted leading to fibrosis and tissue necrosis, promoting inflammation in the surrounding tissue with exudation, edema and pain.

2. Polymer Degradation

The environmental conditions under which polymers normally degrade include physical agencies such as heat, light and ionizing radiations which are not operative within the human body. Polymers are, therefore, generally stable after implantation. If degradation does occur it is usually by hydrolysis, so that polymeric materials which are hydrophilic and which contain hydrolyzable bonds are most likely to suffer. Hydrophobic polymers

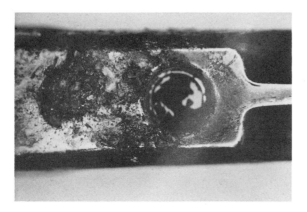

Figure 1
Extensive corrosion on a stainless-steel orthopedic implant

or hydrophilic polymers without hydrolyzable bonds, including poly(tetrafluoroethylene) and poly(methyl methacrylate) are far less susceptible to degradation *in vivo* and elicit very little in the way of an adverse tissue response. Polyesters such as poly(ethylene terephthalate) contain labile ester linkages but are relatively stable because of their hydrophobicity. Polyamides and polyalkyl acids are hydrophilic and hydrolyzable and therefore suffer some degradation which impairs tissue compatibility.

There is some indication that the above generalization is an over-simplification for some hydrophobic polymers which do not hydrolyze but do degrade slowly. Evidence is accumulating that enzymes may be responsible for such degradation.

3. Ceramic Degradation

Ceramics with a wide range of activities are used for implantation and they may be conveniently classified according to their reactivity in the physiological environment. The more inert ceramics are typified by alumina and undergo little or no degradation *in vivo*, being used for permanent implants. At the other extreme are the absorbable or degradable ceramics, such as some calcium phosphates which dissolve in the body over a period of months or years. In between these extremes are the ceramics of controlled surface activity where there is a self-limiting reaction between the ceramic surface and the tissue which can lead to adhesion at the interface.

4. Soft-Tissue Response

The response to the intramuscular or subcutaneous implantation of engineering materials ranges from the simple formation of an acellular fibrous capsule around

the implant to the acute inflammatory response that leads to tissue necrosis. The extent of the response is a reflection of the degree of irritancy of the implant, which will depend on the chemical composition of the material and the surface morphology, on any corrosion or degradation that takes place, and on many other factors. The nature and extent of the interaction can be judged by the following parameters:

(a) the morphology and extent of the cellular layer interposed between the fibrous capsule and the implant,

(b) the cellularity and density of collagen within the fibrous capsule,

(c) the degree of capillarity surrounding and invading the capsule, and

(d) the presence of the inflammatory response.

The minimal response, that of encapsulation with a minimum of fibrous tissue, is often referred to as the ideal response. There is a range of materials that can be considered minimally irritant in this context and are, therefore, used clinically without inducing the formation of anything less benign than this thin fibrous capsule. Titanium, alumina, glassy carbon, polyethylene and silicone rubber are good examples. In most cases this response is the most appropriate but it must be remembered that encapsulation in this way is a manifestation of a rejection phenomenon—the body recognizing the implant material as foreign and trying to wall it off from the normal tissue. True incorporation cannot be achieved in this way although it must be said that the parameters of the tissue reaction that favor incorporation are poorly understood.

More extensive responses are seen either when the nominally inert materials described above degrade in some way (either chemically or by abrasion), when the materials are used in some morphological form which is associated with a different response (e.g., a powder), or when more reactive substances are used.

A material yielding particulate degradation products provides a good example of a more extensive response. While an uncorroding stainless-steel implant elicits the minimal response described above, corrosion of this alloy leads to an excessive cellular response. Similarly, fragments of wear particles from a polymer such as polyethylene or polytetrafluoroethylene provoke a greater cellular reaction (Fig. 2). When the degradation results in soluble products, the reaction will vary with the cytotoxicity and immunological status of the substance. As noted above, although titanium is corrosion resistant it slowly releases titanium ions into the tissue, but these are so physiologically indifferent that they do not appear to influence the cellular response. However, the presence of heavier metal ions, such as cobalt and nickel, can have a significant local effect arising from their cytotoxicity. It is also possible for such ions to induce a hypersensitivity response, for although the ions

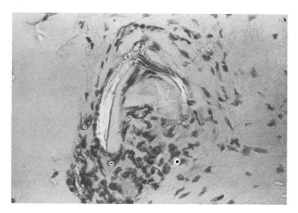

Figure 2
Cellular response to particulate ultra-high-
molecular-weight polyethylene released from a hip
prosthesis

themselves cannot be antigenic, they may bind to proteins *in vivo*, constituting haptens which are capable of inducing such a response. This may be of extreme clinical significance if large amounts of certain metals (especially cobalt) are released into the tissue.

The more chemically reactive the substance, the more irritant it tends to be. Experience with the homologous series of cyanoacrylate adhesives clearly shows that the lower molecular weight, more reactive derivatives provoke the greatest response. The stimulus for an inflammatory response may also be physical or mechanical; the thermal damage associated with an exothermic reaction, and the mechanical irritation associated with an unstable implant are well known.

An inflammatory response tends to have the same general features irrespective of the nature of the stimulus, and so the appearance of the acute inflammation associated with a reactive implant may not be too different to that seen with, for example, an infection. Polymorphonuclear leukocytes are often dominant in the acute cellular infiltrate and, in a reaction involving a marked chronic response, macrophages are frequently observed, possibly leading to foreign-body giant cells. Many other cell types can be present, however, including mast cells, plasma cells and lymphocytes.

Although implant-induced carcinogenicity is a well established phenomenon in experimental animals, such a possibility in humans has yet to be observed to any significant extent.

5. Hard-Tissue Response

Bones and teeth also come into contact with implanted materials although these two tissues present quite different problems. The hard tissues of teeth are enamel, dentine and cementum (see *Dental Enamel: Crystallography*). These are more highly mineralized than bone,

are avascular and, apart from the odontoblastic processes in dentinal tubules, are acellular. There is, therefore, little in the way of a reaction to synthetic materials—adverse responses of teeth being associated with the soft-tissue components comprising the pulp and periodontal membrane. With bone a number of different factors control the overall response. First there may be a chemical effect directly analogous to that seen in soft connective tissue but usually taking place at a slower rate. Secondly there may be a change in the blood supply to bone after implantation. For example, the efferent blood vessels of the cortex of long bones are blocked when a fracture plate is secured to the bone, suppressing the centrifugal flow of blood in the cortex and causing effective devascularization. However, it does not normally take very long for the medullary supply to be re-established. On the other hand, tight-fitting medullary nails which contact virtually all of the endosteal surface, devascularize and necrotize the full thickness of the medial, anterior and lateral cortices after mid-shaft fracture, leading to extensive periosteal callus. Thirdly, there is the question of biomechanical compatibility. It is well known that the ultrastructure and mechanical integrity of bone is dependent on the transmission of stress through the bone and that the balance between osteoblastic and osteoclastic activity is upset if the stress is altered significantly. The main point of concern has been the influence of a rigid implant in supporting bone, which, by virtue of the large dissimilarity between the elastic moduli of bone and metal, can result in a considerable reduction in the stress transmission in the bone, and a degree of so-called disuse atrophy.

There has recently been much discussion concerning the role of implanted materials in osteogenesis, when these materials have been used for bone reconstruction. It is unlikely that any synthetic material will be truly osteogenic but several may be osteoconductive. When such a material is placed in a bony cavity it will facilitate the filling of that cavity by new bone derived from the walls of the cavity. When synthetic hydroxyapatite is used for this purpose, the bone appears to grow right up to its surface and produce a strong bond. Materials which interact with bone in this way are described as bioactive.

6. Blood Compatibility

Blood compatibility refers to a variety of phenomena which may be observed when a foreign material comes into contact with blood, the most important feature of which is thrombogenicity, or the ability of the material to activate the clotting of the blood. Also included are interactions with plasma proteins, red cells, enzymes and the electrolyte.

When a material is placed in the blood stream, two related sequences of events may be initiated. Firstly, a layer of proteins is adsorbed onto the surface. This occurs instantaneously, the nature of the adsorption depending on the type of surface and the rheological characteristics. Although several different plasma proteins may be involved, the adsorption appears to be dominated by fibrinogen. The change in conformation of these proteins may then activate the coagulation cascade, leading to the formation of a blood clot. At the same time platelets start to adhere to the proteinaceous layer on the surface, the morphological changes which then take place favoring an acceleration of the platelet aggregation and further activation of the clotting sequence.

Clearly, several different features of the material surface can influence the thrombogenicity, including electrical surface charge and interfacial free energy. As a result of fundamental work on these interactions and on empirical data, several relatively nonthrombogenic materials have been prepared. At least four different approaches have been made, involving:

(a) the use of chemically inert materials such as carbon, which may be able to maintain a structurally unaltered layer of protein on their surface;
(b) the use of hydrogels which contain large amounts of water;
(c) the use of negatively charged surfaces which irritate the electrical characteristics of the natural intima; and
(d) the use of anticoagulants such as heparin, bonded to the surface.

See also: Artificial Joints, Implanted; Heart Valves; Cardiac Pacemakers, Implantable; Dental Materials

Bibliography

Albrektsson T 1985 The response of bone to titanium implants. *CRC Crit. Rev. Biocompat.* 1: 53–84
Baier R E 1975 Blood compatibility of synthetic polymers: Perspectives and problems. In: Kronenthal R L, Oser Z, Martin E (eds.) 1975 *Polymers in Medicine and Surgery.* Plenum, New York, pp. 139–59
Hench L L 1980 The interfacial behaviour of biomaterials. *J. Biomed. Mater. Res.* 14: 803–11
McNamara A, Williams D F 1981 The responses to the intramuscular implantation of pure metals. *Biomaterials* 2: 37–40
McNamara A, Williams D F 1984 Enzyme histochemistry of the tissue response to pure metals. *J. Biomed. Mater. Res.* 18: 185–206
Vernon-Roberts B, Freeman M A R 1977 The tissue response to total joint replacement prostheses. In: Swanson S A V, Freeman M A R (eds.) 1977 *The Scientific Basis of Joint Replacement.* Pitman, London, pp. 86–129
Williams D F 1976 Corrosion of implant materials. *Annu. Rev. Mater. Sci.* 6: 237–66
Williams D F 1981 *Biocompatibility of Clinical Implant Material*, Vols. 1 and 2. CRC Press, Boca Raton, Florida
Williams D F (ed.) 1981 *Fundamental Aspects of Biocompatibility*, Vols. 1 and 2. CRC Press, Boca Raton, Florida
Williams D F (ed.) 1981 *Systemic Aspects of Biocompatibility*, Vols. 1 and 2. CRC Press, Boca Raton, Florida
Williams D F 1982 Degradation of surgical polymers. *J. Mater. Sci.* 17: 1233–46
Williams D F, Roaf R 1973 *Implants in Surgery.* Saunders, London

D. F. Williams

Intra-Aortic Balloon Pumps

An intra-aortic balloon pump is a device used for the mechanical support of the failing heart. If it is employed early in the failure process, further cardiac deterioration can be slowed down or arrested, giving valuable time for medical and surgical intervention.

1. The Concept

Inflation of a gas-filled balloon in the descending aorta will displace blood both distally towards the abdomen and proximally towards the heart. Conversely, deflation of a balloon will draw blood from both these regions (Fig. 1). The failing heart produces a low cardiac output and will maintain only a low aortic blood pressure. The coronary arteries which are the main channel for oxygen and metabolic fuel for cardiac work fill principally during diastole, the phase of the cardiac cycle during which the heart is relaxed and awaiting filling for the next work stroke. A low filling pressure gives poor coronary blood flow, and hence a sparse supply of metabolic fuel which further reduces the capacity for cardiac work and causes a progressive and usually fatal deterioration. Synchronized inflation of a balloon in the descending aorta during the phase of cardiac diastole displaces blood towards the heart, augmenting the pressure in the aorta at the level at which the coronaries join it, increasing coronary blood flow. If the balloon is now deflated to synchronize with systole (the work stroke), help will be given to empty the cardiac chamber, both increasing the ejection fraction of blood and reducing the metabolic requirements of the heart by off-loading about 25% of its work. Generally, cardiac output will increase with continued support until normal physiological pressures are maintained. In addition, drugs or surgery (or both) will be necessary, depending on the cause of the cardiac failure.

2. Balloon-Pump Systems

Balloon-pump systems are electronically controlled and pneumatically driven. Usually, the patient's electrocardiogram (ECG) is used as a trigger to synchronize the pump with the action of the heart, although it is possible

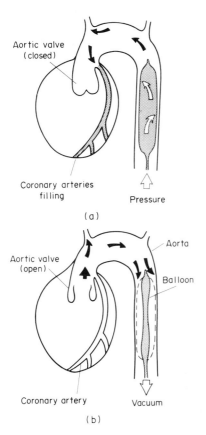

Aortic valve
(closed)

Coronary arteries
filling

Pressure

(a)

Aortic valve
(open)

Aorta

Balloon

Coronary artery

Vacuum

(b)

Figure 1
The action of an intra-aortic balloon upon the heart:
(a) diastole: heart relaxed, balloon inflating, and (b)
systole: heart ejecting, balloon deflating

to use other inputs such as aortic blood pressure sensed by a blood-pressure transducer. The control of balloon inflation and deflation is achieved using electrically actuated valves switched by a logic system linked with override and alarm circuitry which limits the balloon function should certain alarm conditions develop. If an alarm is activated, due for example to the loss of ECG synchronization, the intra-aortic balloon will deflate to present the minimum obstruction to blood flow.

The timing of the balloon's inflation–deflation cycle is related to the cardiac cycle times. Deflation is triggered by the ECG QRS complex, and a delay may be added to compensate for delayed contraction so that the balloon remains deflated until a predetermined period corresponding to the contraction time. Balloon filling is timed to commence with aortic valve closure and continues during early diastole until the balloon is inflated to the correct volume and pressure. Care is taken to avoid overinflation which would damage the aorta. Balloon volume is maintained constant until the next ECG QRS complex is received or a present time limit is exceeded. There are different logic modes for the oper-

ation of balloon pumps, but the one described is that generally preferred since it automatically deflates the balloon just before the heart contracts. This affords two great advantages over other methods; namely, that it is almost impossible for the balloon to obstruct ventricular outflow thus increasing the work load of the heart; and it is easy to optimize the assist phase of the cycle, when deflation of the balloon helps the heart to eject.

The patient balloon is constructed at the end of a semirigid plastic catheter and takes the form of a cylinder of highly flexible but indistensible transparent polymer, usually a type of polyurethane. The catheter where it passes through the balloon is perforated to admit driving gas into the balloon. Some types of balloon are equipped with a second balloon section of smaller capacity which, mounted downstream from the main balloon, serves to restrict blood flow past the device while it is inflating. This restrictor balloon is provided with a greater number of inflation ports per unit volume than the pump balloon to ensure early inflation. The advantages of such dual-chambered balloons over their single-chambered counterparts are difficult to demonstrate in clinical practice.

The gas used for inflating the balloon must be soluble in blood at body temperature in case of any small leak. Helium and carbon dioxide are both widely used. Although helium has a lower viscosity than carbon dioxide, the latter is cheaper and more readily available, and has proved to be of sufficiently low viscosity for this purpose. The patient balloon is isolated from the driving circuit by a safety chamber balloon. This balloon, mounted in a rigid chamber which is itself connected to the pressure–vacuum system, confines the driving gas between itself and the patient balloon. External pressure applied to the safety chamber transfers gas into the patient balloon. Conversely, the application of a vacuum has the effect of sucking the patient balloon flat, reinflating the safety chamber balloon.

When pumping is correctly instituted, a slight drop in presystolic pressure should be noted, followed by a marginally lower systolic pressure. This is a good indication that the work stroke of the heart is being assisted but the effect may not be as apparent on a peripheral-artery blood-pressure trace as on a tracing made from the aortic arch pressure. Diastolic pressure should be elevated considerably and is often significantly higher than systolic, which is a reversal of the normal situation.

3. Clinical Use

The balloon catheter is inserted through a small incision into the femoral artery and advanced until the tip is just below the level of the aortic arch in the descending aorta. Anticoagulants are usually administered to prevent blood clotting on the balloon and being thrown off as systemic emboli. After preliminary adjustments have been made and the balloon primed with driving gas, pumping can be started and fine adjustments made to obtain optimum assistance to the heart and effective ventricular unloading. It is necessary from time to time to empty and refill the balloon with a fresh charge, as gas is lost by diffusion into the driving system and to a smaller extent into the patient.

It is now generally agreed that intra-aortic balloon pumping is of value in selected cases of cardiac failure but the rationale for selection differs widely. The most enthusiastic proponents advocate the use of balloon pumping early in cases of cardiogenic shock, concurrent with drug therapy. Others wait until there is no alternative treatment. When it proves difficult to take a patient off cardiopulmonary bypass (see *Heart–Lung Machines*), a prompt decision to balloon pump can bring about a rapid and dramatic improvement in cardiac function, facilitating removal from the extracorporeal circuit usually without prolonged balloon support. In view of the poor survival statistics for cardiogenic shock (many sources report only a 20% survival), it is surprising that balloon pumping is not more widely used. Use of this support technique increases survival to over 40% and if it were not for complications due to the invasive-balloon positioning, these figures could be higher.

The most common complications are associated with the impairment of peripheral circulation caused directly by balloon positioning. If the balloon catheter is inserted into the common femoral artery there is a risk of leg ischemia on that side and this has resulted in amputation of the limb. Reports have been published of ischemic damage to the spine and abdominal organs. The incidence of damage to the aorta is very low, and complications of removal of the catheter from the femoral arteriotomy are also few. The use of a Dacron vascular graft at the arterial incision greatly simplifies catheter removal, as it is only necessary to oversew the graft to form a stump and the artery is closed. A similar graft technique can be used to insert the catheter through the ascending aorta at the time of thoracotomy, the balloon is thus reversed and pushed around the aortic arch into the descending aorta. This allows fast counterpulsatory support to be given during surgery without the need for a femoral cutdown. Removal of the balloon can be performed later at another thoracotomy or the Dacron graft can be lead out to the skin where removal can be performed under local anesthesia. The graft is then oversewn and buried.

See also: Cardiac Function: Noninvasive Assessment; Cardiac Catheterization

Bibliography

Bonchek L I, Olinger G N 1979 Intra-aortic balloon counterpulsation for cardiac support during noncardiac operations. *J. Thorac. Cardiovasc. Surg.* 78: 147–49.
Braunwald E (ed.) 1974 *The Myocardium: Failure and Infarction.* H P Publishing, New York
Bregman D 1974 Dual-chambered intra-aortic balloon counterpulsation. In: Ionescu M I, Wooler G H (eds.) 1974 *Current Techniques in Extracorporeal Circulation.* Butterworth, London, pp. 407–516
Hines G L, Delaney T B 1979 Intra-aortic balloon pumping. *J. Thorac. Cardiovasc. Surg.* 78: 140–46
Swank M, Singh H M, Flemma R J, Mullen D C, Lepley D 1978 Effect of intra-aortic balloon pumping on nutrient coronary flow in normal and ischemic myocardium. *J. Thorac. Cardiovasc. Surg.* 76: 538–44

K. J. Maxted

Intracerebral Electrodes

In experimental animals and in certain circumstances in man, the introduction of electrodes into the brain provides information about the electrical activity of the central nervous system that cannot be obtained using scalp electrodes (see *Electroencephalography*). The choice of materials for such electrodes is determined primarily by the toxicity of the metal. Electrical polarization of the electrode surface may distort the electroencephalogram (EEG) and this should be taken into account when choosing a metal. Chronic indwelling microelectrodes may be used but only in very special circumstances. The electrodes described here are large compared with the cortical neurons; they record the mass action of conglomerates of cells.

Many materials are toxic to brain tissue; their implantation causes inflammatory reactions and abnormal electrical activity. Such reactions can be caused by the insulating material used and/or by the metal of the electrode. Since most intracerebral electrodes are made from thin insulated wire (so that their insertion causes minimum damage), enamel insulation is preferred. Very thin, extruded Teflon insulation is now also available which is very effective but expensive. Most modern plastics used for coating fine wire are more-or-less inert but often toxicity data are not available from the manufacturers. Toxicity measurement is a laborious process, involving histological examination of brain tissue over various periods of time after the implantation of the materials.

Chlorided silver, the most satisfactory substance for electrodes in general, is very toxic and should therefore not be used. The inflammatory reaction induced by chlorided silver occurs within a few days of implantation and may give rise to large-amplitude, low-frequency electrical activity in the brain which subsides slowly over a few months (Cooper and Crow 1966).

Copper, tungsten and platinum all evoke reactions when left in brain tissue for weeks or months. Reactions induced by tungsten and platinum are relatively mild compared with the violent reaction induced by silver and copper. Gold and stainless steel are inert and are thus the materials of choice (Fischer et al. 1961, Robinson and Johnson 1961).

The electrical double layer surrounding certain electrode surfaces (including gold and stainless steel) acts as a barrier impeding the flow of current, and polarizing the electrodes. Such electrodes have a high (dc) resistance and a relatively low (ac) impedance. When used in amplifier input circuits they act like capacitors, differentiating the electrical activity (Cooper et al. 1980). Accurate low-frequency ac and dc signal recording is possible only if the input impedance of the amplifiers is much higher ($> 50\,\text{M}\Omega$) than the electrode resistance. The impedance of intracerebral electrodes should be measured with alternating current at an appropriate frequency (say 40 Hz for the electroencephalograph). Electrode impedance values range from $20\,\text{k}\Omega$ to $0.5\,\text{M}\Omega$ or more, depending on the electrode metal and the size of the electrode. Electrically, an electrode in solution can be represented by a capacitor and a resistor in parallel. As in a parallel-plate capacitor, the capacitance ($0.1\,\mu\text{F}$ or more for intracerebral electrodes) relates to the surface area of the electrode. The resistance is high for polarized electrodes ($\text{M}\Omega$) and low ($\text{k}\Omega$) for non-polarized (reversible) electrodes. Stainless steel is "stainless" because of a thin surface film of chromic oxide that acts like a capacitor in series with the electrical double layer, resulting in a smaller net value of capacitance. Typical values of capacitance and resistance for 38-gauge stainless steel wire ($150\,\mu\text{m}$ diameter, 2 mm bared) are $0.05\,\mu\text{F}$ and $10\,\text{M}\Omega$, respectively. A method of measuring electrode characteristics may be found in Cooper et al. (1980).

Intracerebral electrodes are made usually in sheaves of several insulated wires twisted together, with the ends cut at various distances from the tip of the sheaf. The insulation is stripped for a fixed length at the tip of each wire. The diameter, length and spacing of the electrodes are determined by the particular application envisaged, but are usually less than $150\,\mu\text{m}$ in diameter, 1–2 mm long and with the tips separated by a few millimeters, respectively. A much smaller electrode can be made by simply cutting the wire and not removing any of the insulation. (Gold is very soft and difficult to handle if its diameter is less than $75\,\mu\text{m}$.)

The wires are usually soldered into a subminiature plug that is mounted on the skull (in animals). Ray (1966) has described a rigid probe, affixed to the skull, that has been used in man. Otherwise flexible wires exit through the scalp and terminate on a socket linked to the skull. If the scalp entry point is kept clean the risk of infection is very small. Low-temperature solder should be used when using gold wire, since gold alloys with lead at normal soldering temperatures. Thin stainless steel can be difficult to solder because of the chromic oxide layer. Details of electrode construction are given in Myers (1971).

Intracerebral electrodes are often used for the electrical stimulation of brain tissue. Since the capacitative characteristic of polarized electrodes distort square waveform pulses (Weinman and Mahler 1964), the effects of stimulation can depend, to a certain extent, on the metal used.

Intracerebral electrodes are sometimes used for making electrolytic lesions in animals and in man. The size of the lesion formed, for a given quantity of electricity, depends upon the electrode metal; it is large if gold is used. If coagulating currents are passed through stainless steel electrodes, the metal erodes and is absorbed by the tissue.

See also: Evoked Potentials

Bibliography

Cooper R, Crow H J 1966 Toxic effects of intra-cerebral electrodes. *Med. Biol. Eng.* 4: 575–81.
Cooper R, Osselton J W, Shaw J C 1980 *EEG Technology*, 3rd edn. Butterworth, London
Fischer G, Sayre G P, Bickford R G 1961 Histologic changes in the cat's brain after introduction of metallic and plastic coated wire used in electroencephalography. In: Sheer D E (ed.) 1961 *Electrical Stimulation of the Brain: An Interdisciplinary Survey of Neurobehavioral Integrative Systems*. University of Texas Press, Austin
Myers R D (ed.) 1971 *Methods in Psychobiology*. Academic Press, London
Ray C D 1966 A new multipurpose human brain depth probe. *J. Neurosurg.* 24: 911–21
Robinson F R, Johnson H T 1961 Histopathological studies of tissue reactions to various metals implanted in cats' brains, Aeronautical Systems Division Technical Report No. 61/397. US Department of Commerce, Washington, DC
Weinman J, Mahler J 1964 An analysis of electrical properties of metal electrodes. *Med. Electron. Biol. Eng.* 2: 299–310

R. Cooper

Intracranial Pressure Measurement

The brain and spinal cord are bounded by membranous coverings and a layer of thick bone which help to isolate the central nervous system (CNS) and protect it from external injury. The cerebrospinal fluid bathes the brain and spinal cord and circulates freely within the CNS. The pressure of this fluid is above atmospheric pressure and is transmitted equally in all directions within the intracranial space. The intracranial pressure (ICP) is considered to be the pressure of the cerebrospinal fluid (CSF) and an understanding of the ICP requires a thorough understanding of the CSF system, the mechanism of CSF formation, storage and absorption, and the hydraulic interaction between the fluid and other tissue compartments. This article focuses on the practical aspects of ICP measurements, with a brief overview of CSF physiology.

1. Central Nervous System

The brain and spinal cord, which together form the central nervous system, are soft semigelatinous organs encased in the bony skull and vertabrae. They are protected by three membranous coverings or meninges: the dura mater, arachnoid mater and pia mater (Fig. 1).

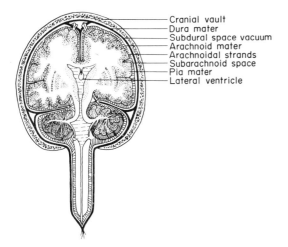

Figure 1
Anterior–posterior projection of the brain, showing the meninges

The dura is the outermost covering and is a tough, fibrous membrane which adheres to the wall of the cranium. In the spine, the dura is separated from the vertabrae by a layer of fat. The next inner layer is the arachnoid—a delicate covering of connective tissue which bridges the inner folds (sulci) of the brain, thus forming compartments of fluid (cisterns). The pia follows the contour of the brain and spinal cord accurately and is an extremely thin layer of tissue. The space between the arachnoid and the pia is designated the subarachnoid space and includes the major cisterns. Between the dura and arachnoid is the subdural space which is moistened by fluid but has no direct communication with the subarachnoid space.

In the adult human, the intracranial volume is approximately 1900 ml and can be subdivided into three main tissue compartments; brain, blood and cerebrospinal fluid. The brain occupies 80% of the total volume, while blood and cerebrospinal fluid compartments each occupy 10%. Under normal conditions these tissue compartments are in a state of pressure and volume equilibrium such that an increase in size of one compartment must be accompanied by a decrease in one or more of the remaining compartments so that the total intracranial volume remains fixed. (The tissues capable of rapid volume interchange include the blood and the cerebrospinal fluid.)

2. Cerebrospinal Fluid

The cerebrospinal fluid (CSF), a clear, colorless liquid of specific gravity close to water, occupies two major compartments of the central nervous system: the ventricles of the brain and the subarachnoid spaces (Fig. 2).

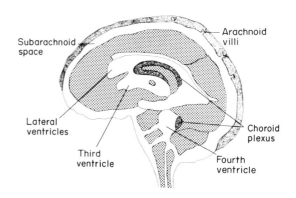

Figure 2
Longitudinal section of the brain, showing the major fluid-containing compartments

The ventricles consist of four interconnecting cavities located within the brain and which communicate with the subarachnoid space via three exit points. In man the total CSF volume is approximately 140 ml. The subarachnoid space occupies 120 ml of the total volume whereas only 20 ml is contained within the ventricular system.

It has been widely accepted that the greater part of the CSF is elaborated by the choroid plexus (fringe-shaped tufts of small capillary vessels which are covered by very fine layers of ependymal cells). A choroid plexus is found in each of the four ventricles. The CSF is not considered to be a simple filtrate of plasma—in general it has higher sodium, chlorine ion and magnesium concentrations and lower potassium, calcium, urea and glucose content than would be expected in a simple plasma dialysate. Hence the CSF resembles a modified ultrafiltrate in which both secretory and diffusion processes are involved.

The rate of CSF formation varies considerably in different species—from 2.2 μl min^{-1} in rabbits to 300 μl min^{-1} in man. An infant will produce approximately 200–300 ml per day so that the entire CSF volume is replaced within a 24 h period. This clearly emphasizes the delicate balance that must be maintained by the absorption mechanisms.

3. Circulation and Absorption of CSF

The CSF is generally considered to circulate from the ventricles to the subarachnoid space and about the hemispheres to the arachnoid villi, which are the sites of fluid absorption. The arachnoid villi are small, finger-like bodies which protrude into the lumina of the large venous drainage conduits, the dural sinuses. Histological studies (Welch and Friedman 1960) conducted in monkeys showed that the villi are best described as a labyrinth of coapted tubes, some 4–12 μm in diameter, connecting with each other and providing communication between the subarachnoid space and the venous channels of the dura. The villi act as one-way valves, opening at approximately 20 mm H_2O of CSF pressure. When opened, the valve resistance offered to the passage of fluid appears constant and independent of pressure.

4. Intracranial Pressure

The newly formed CSF circulates within the ventricles and subarachnoid space and eventually reaches the arachnoid villi where it is absorbed. The formation and absorption mechanisms maintain a constant CSF volume. The intracranial pressure (ICP) can be defined as the hydrostatic pressure of the CSF measured in relation to atmosphere. By international convention, the pressure is expressed in mm Hg. This definition is applicable when the CSF is freely circulating and communication between all regions of CSF storage is maintained. The CSF pressure under these conditions acts equally in all directions and represents a true measure of ICP.

In pathological circumstances obstruction of the CSF pathway may occur, which subdivides the CSF space into two or more compartments. With continued displacement of tissue, transient or sustained gradients may develop and, as a result, a single measure of ICP cannot adequately describe the biomechanics of the intracranial space. This is a complex problem and would require mapping of the instantaneous local tissue pressure of each compartment to define regional tissue stress accurately.

5. Components of ICP

With free communication of fluid throughout the CSF space, the ICP can be measured at any point within the ventricular or subarachnoid regions. The ICP is above atmospheric pressure and consists of a pulsatile wave superimposed upon a baseline level. The baseline or diastolic level is commonly referred to as the ICP, whereas the rhythmic components are associated with cardiac and respiratory activity (Fig. 3). At present, there is no standardization of nomenclature. In order to describe the ICP waveform, the measurement site (i.e., ventricular, lumbar, cisterna magna), the magnitude of the baseline or steady-state level, and the amplitude of the pulsatile components should be specified, as well as the gauge reference level. More often than not, the pulsatile levels and gauge reference are excluded, and only the baseline is reported. Intracranial pressure can

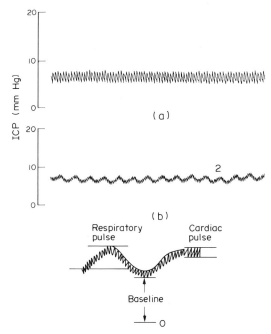

Figure 3
ICP recorded during steady-state conditions (ventricular catheter, gauge reference at right heart level, human patient in recumbent position). The peak/base value equals 7.9/5.9 mm Hg. The average ICP, calculated as base plus 1/3 pulse height, equals 5.9 + 1/3(2) or 6.6 mm Hg (b) ICP recording from cisterna magna of adult cat to illustrate both respiratory and cardiac pulsation superimposed upon steady-state level. The animal was anesthetized and artificially ventilated (gauge reference at right heart level, respiratory pulse of 1 mm Hg, ICP baseline at 6.0 mm Hg)

be recorded in the same manner as blood pressure (i.e., peak pulse pressure over minimum pulse pressure which is analogous to a systolic over diastolic value).

6. Normal Level of ICP

The closed calvarium is partially exposed to the atmosphere via the venous system. Thus the intracranial cavity may be regarded as a partially closed box. As a result, the static level of ICP is variable depending upon the height of the fluid column formed by the geometry of the craniospinal axis. With the patient lying in a horizontal plane the fluid pressure measured in the ventricle and spinal subarachnoid space will be equal and considered normal up to an average level of 15 mm Hg. In the erect position, the ventricular pressure will be negative with reference to the pressure of the fluid at the height level of the right heart while pressure in the lumbar region will be positive and variable with patient height. Intracranial pressure levels are generally reported with the patient in the recumbent position and the gauge set at the level of the heart or the intracranial line.

The ICP of infants is lower than values reported for older children or adults. Salmon et al. (1977) using fontanel transducers reported infant pressure values of 8 mm Hg. Welch (1980), using visual estimates of fontanel tension, reported normal ICP of approximately 3 mm Hg. Taken in concert with the lower pressure found during infancy, these values are probably reasonable approximations of the true steady-state ICP of younger children.

7. ICP Instrumentation and Recording Methods

Intracranial pressure measurement techniques may be classified into two major groups: direct and indirect. In the direct method, a pressure gauge is hydraulically coupled directly to the CSF using plastic tubing filled with sterile saline or artificial CSF solution. Intraventricular or lumbar recordings of ICP fall into this category. With the indirect method, a transducer is mechanically positioned to sense the pressure transmitted by a diaphragm in contact with the dura or arachnoid. The indirect method includes both epidural and subdural pressure measuring devices.

8. ICP Measurement: Direct Methods

The physician gains access to the CSF by needle puncture of the spinal subarachnoid space or cannulation of left or right lateral ventricles. These are the most common routes of entry. Measurement of ICP by the spinal route was routinely performed by investigators in the early 1900s. They found the procedure to be a useful diagnostic tool and a convenient means of draining CSF to decrease pressure. Other workers were concerned that lumbar puncture could be dangerous if there was increased pressure above a point of blockage between cranial and spinal compartments. They argued that lowering of pressure in the lumbar region would increase the differential pressure and promote brain herniation toward the lower pressure region.

Lundberg (1960) and Janny (1950) proposed that measurement of ICP using a catheter placed in the ventricle would yield a more direct indication of the ICP. Their classic works describe the technique of measurement and categorization of pressure waves which set the stage for most of the current progress in the field of ICP monitoring.

Lundberg developed a method for continuous recording of ICP using a cannula inserted into the anterior horn of one of the lateral ventricles through a burr hole in the skull. The burr hole was plugged with a rubber stopper to prevent leakage and reduce the risk of infection. There is general agreement that ventricular cannulation, the direct method, is probably the most

reliable technique for continuous monitoring of ICP and is the reference measure by which all other methods are compared. The main advantages of the direct method are its simplicity and ease in absolute calibration. Another major advantage is that the fluid pathway established for pressure measurement provides a means of CSF drainage for relief of pressure in patients with raised ICP.

There are some disadvantages. It is difficult in certain types of brain injury to place the catheter in a ventricle which is displaced or collapsed. The problem of catheter placement, the need for penetration of brain tissue during cannulation and the increased risk of infection associated with the perforation of the meninges are the primary drawbacks of the direct recording method. To circumvent this problem and reduce the incidence of infection, Fleischer et al. (1975) used an implanted reservoir attached to an intraventricular catheter with exchangable external connections for monitoring ICP in head injured patients. After positioning the reservoir in a burr opening, the 8 cm incision was closed and a needle inserted percutaneously into the reservoir and connected to a pressure transducer. With this technique fluids could be removed quickly if necessary. One disadvantage of the method is that the insertion and removal requires minor surgical procedures.

Vries et al. (1973) developed a device which provides continuity with the CSF without ventricular cannulation. A bolt was designed for mounting through a trephine opening in the skull. The hole is threaded and the bolt inserted to provide a watertight seal. Tubing attached to the head of the bolt is connected to a conventional pressure transducer and polygraph. Before mounting the bolt, the dura is punctured with a special instrument to provide communication with the subarachnoid space. The device functions well as long as tight fluid coupling between the transducer and brain surface is maintained. Other devices based on a similar principal have been used by other workers with varying degrees of success. The subarachnoid bolt originally described by Vries et al. (1973) and modified by James et al. (1975) is particularly useful for children with small ventricles. The main disadvantage of the bolt system is that fluids cannot easily be removed for therapeutic purposes.

9. ICP Instrumentation: Direct Methods

The fluid-filled tubing communicating with the CSF is usually routed to a pressure-measuring device consisting of a simple U-tube manometer or the more conventional electromanometer systems. The standard U-tube manometer remains in use; however, considerable error is introduced because of the relatively large volumes of fluid required to fill the manometer tube. The measurement error introduced by the displaced CSF volume is greater at elevated intracranial pressure because of the exponential pressure–volume relationship of the fluid space (Marmarou et al. 1978).

A low-volume displacement instrument is required for accurate measurement of the ICP. The most popular device in both clinical and research use today is the strain-gauge pressure transducer which displaces less than 10^{-7} ml of fluid per mm Hg. There are a variety of commercial strain-gauge transducers currently available which meet the requirements necessary for ICP measurement, that is, low electrical drift, low compliance, high sensitivity and mechanical stability. Devices which are gaining more popularity are those which include disposable diaphragms. The use of a disposable diaphragm attached to a conventional strain-gauge housing reduces the risk of infection.

The main requirement of the hydraulic interface between site of measurement and gauge is that the system be of low compliance and free of air bubbles. This can be accomplished by using rigid plastic tubing and exercising care during the connection procedure.

The deflection of the diaphragm of the pressure gauge produces a voltage in the microvolt range. This output is routed to a transducer coupler which amplifies the gauge signal to a level which is suitable for polygraph recording. The strain-gauge amplifier or transducer coupler is an important link of the measurement system and must be stabilized for low electrical drift and have sufficient gain and bandwidth (0–100 Hz) to process the low-level pressure signal. The coupler must also provide adjustments for balancing the pressure gauge at zero pressure.

The polygraph selected for recording of the ICP must be drift-free to ensure the accuracy of the pressure baseline. Frequency response of the driver amplifier and pen system must be uniform from 0 to 30 Hz to record properly fundamental frequencies of the ICP pulsatile components. A variable chart speed is desirable for recording of both instantaneous ICP waveform (25 mm s^{-1}) and compressing long-term ICP recordings (0.5 mm s^{-1}).

The pressure response of the system can be tested during calibration and adjustment of transducer amplifier and polygraph gain. A valve can be inserted in the patient monitor line to momentarily disconnect the patient from the system and expose the gauge to a fixed level of pressure. Care must be taken to maintain sterility during this calibration procedure. Full-scale polygraph deflection is adjusted for 50 mm Hg and the gauge system is exposed to 50 mm Hg pressure from a sphygmomanometer connected to a mercury-tube manometer. This is carried out in stepwise fashion by turning the valve rapidly to check the response of the pressure monitoring system. A sluggish response indicates that air is trapped in the connecting line on the gauge and must be flushed. The gauge is cycled from atmosphere to the preset pressure level to adjust the zero and full-scale deflection of the monitoring system.

Most patient monitors have the capability of electrical zero and calibration. These adjustments are useful in checking the electrical portion of the system but should not replace a mechanical calibration, that is, actual pressurization of the transducer followed by adjustment of pen deflection or recording device gain.

10. ICP Measurement: Indirect Methods

The low risk of infection and minimal trauma associated with indirect methods of ICP monitoring have resulted in the development of a large number of implantable transducers (Shulman et al. 1980). These devices have been used with varying degrees of success. Ideally, the requirements for a transducer of the indirect type should include: extradural placement, no risk of infection, totally implantable, small size and mass, leakage free, low drift, high sensitivity, and provision for external calibration. Comparative studies of extradural and intraventricular pressure measurements have shown good correlation of epidural and ventricular pressure in the 0–20 mm Hg range (Zierski 1980). At pressures greater than 20 mm Hg, gradients appear and the epidural pressure tends to be higher. In general, further development and experience with newer models is required to validate the clinical usefulness of epidural monitoring systems.

A major problem in the application of these devices is that the sensing diaphragm must be coplanar with the membrane enclosing the pressure cavity. When coplanarity between the sensor and duramater exists, effects of changes in the tension of the membrane that accompany changes in intracavitary pressure are eliminated. The sensor, under this condition, is a reliable measure of the intracavitary pressure. Loss of coplanarity, due to mechanical shift of transducer or brain tissue, will invalidate pressure readings and gradients of epidural pressure and ventricular pressure will develop.

Another major drawback of epidural pressure measurement devices is the inability of absolute calibration. This, of course, must be accepted as a design compromise in a totally implantable system. The inability to calibrate under conditions of zero pressure requires that the electromechanical features of the device be ultrastable and virtually drift-free.

One proposed solution of the calibration problem was to incorporate an internal stop within the sensor to enable confirmation of the zero point *in vivo* (Zervas et al. 1980). The two-diaphragm epidural sensor developed by Zervas is placed in the skull so that the lower diaphragm senses the ICP while the upper diaphragm contacts the scalp. A tuning piston couples the two membranes and ICP is sensed by a change in the resonant frequency of the circuit in the sensor. A stop of the tuning piston relative to the circuit at zero pressure position enables an *in vivo* check of zero pressure. The sensor calibration at pressure levels other than zero can be checked by pressing on the scalp with a known pressure using an inflatable cuff. The system is, of course, subject to the conditions of coplanarity; however, the results of preliminary tests conducted so far are encouraging.

The attempt to measure intracranial pressure via the open fontanel can also be classified as an indirect technique and one which is least invasive. In the neonatal patient with a thin skull and open cranial sutures it is not practical to measure ICP with a subarachnoid bolt. Ventricular puncture in infants with small ventricles is hazardous. An adaption of a displacement tonometer for measurement of fontanel pressure is in use. The device consists of a solid-state force transducer which is encased within a coplanar ring. Pressure measurement is obtained by depressing the fontanel to a point where the output voltage of the device levels out: this represents the point of coplanarity. The transducer is calibrated externally under similar conditions with known pressures. The device has not been fully tested in children.

11. Summary

Intracranial pressure monitoring is gradually being accepted as a standard tool in the routine care of patients with head injuries. Although the history of ICP dates back many years, it is a relatively new field of investigation which owes much of the recent advancement to the international conferences on intracranial pressure started in 1972 (Brock and Dietz 1972). Many problems remain unsolved: the origin of the pressure, the mechanisms for sustained increases of ICP and the most appropriate means of treatment. Hopefully, with the proper measurement of ICP and the refinement of analytic techniques, we can begin to solve some of these problems.

See also: Neurosurgery: Physiological Monitoring; Radionuclide Cisternography; Cerebral Blood Flow: Regional Measurement

Bibliography

Brock M, Dietz H (eds.) 1972 *Intracranial Pressure: Experimental and Clinical Aspects.* Springer, Berlin
Edwards J 1974 An intracranial pressure tonometer for use on neonates: Preliminary report. *Dev. Med. Child Neurol.* 16(6) Suppl. 32: 38–39
Fleischer A S, Patton J M, Tindall G T 1975 Monitoring intraventricular pressure using an implanted reservoir in head injured patients. *Surg. Neurol.* 3: 309–11
James H E, Bruno L A, Schut L, Shalna E 1975 Intracranial subarachnoid pressure monitoring in children. *Surg. Neurol.* 3: 313–15
Janny P 1950 La pression intracranielle chez l'homme. Thesis Auborne, Cleremont, France
Lundberg N 1960 Continuous recording and control of ventricular fluid pressure in neurosurgical practice. *Acta Psychiatr. Neurol. Scand.* 36 (Suppl. 149)
Marmarou A, Shulman K, Rosende R M 1978 A nonlinear analysis of the cerebrospinal fluid system and intracranial pressure dynamics. *J. Neurosurg.* 48: 332–44
Salmon J H, Hajjar W, Bada H S 1977 The Fontogram: A noninvasive intracranial pressure monitor. *Pediatrics* 60: 721–25
Shulman K, Marmarou A, Miller J D, Becker D P, Hochwald G M, Brock M (eds.) 1980 *Intracranial Pressure IV.* Springer, Berlin
Vries J K, Becker D P, Young H F 1973 A subarachnoid screw for monitoring intracranial pressure. *J. Neurosurg.* 39: 416–19
Welch K 1980 The intracranial pressure in infants. *J. Neurosurg.* 52: 693–99
Welch K, Friedman V 1960 The cerebrospinal fluid valves. *Brain* 83: 454–69
Zervas N T, Chapman P H, Cosman E R 1980 Clinical application of telemetric intracranial pressure monitoring. In: Shulman et al. 1980, pp. 436–37
Zierski J 1980 Extradural ventricular and subdural pressure recording comparative clinical study. In: Shulman et al. 1980, pp. 371–73

A. Marmarou

Intraocular Fluid Dynamics

The movement of fluid in the eye is important because of its relevance to sight-threatening pathological conditions such as glaucoma, a condition in which the intraocular pressure is abnormally high, and retinal detachment, in which the retina is displaced from its normal apposition against the underlying pigment epithelium. The aqueous humor fills the anterior (0.25 ml) and posterior (0.06 ml) chambers of the eye while the vitreous humor occupies the vitreous cavity (3.9 ml) (Fig. 1). These fluids, which together account for some 60% of the total volume of the eye, fulfil several roles: they are carriers of nutrients and waste, they maintain

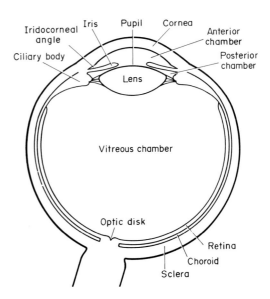

Figure 1
Section of the eye

the normal distension of the eye, and they are part of the optical system of the eye.

1. Aqueous Humor

The aqueous humor in man (pH = 7.3; viscosity = 1.0 relative to water; osmolality = 306 mmol per kg H_2O; refractive index = 1.3336) is normally a clear, colorless liquid separated from its source (plasma) by a blood–aqueous barrier. However, a breakdown in this barrier (e.g., after injury) results in the fluid of the anterior chamber becoming plasmoid, with an increased protein content.

1.1 Formation

Aqueous humor is formed by secretion and ultrafiltration in the ciliary processes, which are highly vascularized ridges projecting from the ciliary body towards the posterior chamber. Secretion is an active process in which molecular carriers effect the transport of substances across a membrane which is otherwise impermeable to these substances. Ultrafiltration is a passive mechanism, that is, it does not require energy from the cell, and is defined as hydrostatic pressure combined with dialysis (the net transfer of water and solvents across a membrane permeable to salt and water but not to protein) (see *Artificial Membranes*).

It has been demonstrated that secretion contributes to the presence of ascorbic acid in aqueous humor. Ascorbic acid concentration in aqueous humor rises as its concentration in serum rises until a point is reached where further increase in serum concentration produces no change in the level of ascorbic acid in the humor. The saturation of carrier molecules is evidenced by this phenomenon from which it follows that secretion plays a role in aqueous-humor formation.

The relative contribution of secretion and ultrafiltration to the total production of aqueous humor is not known exactly. Secretory components varying between 35 and 100% have been reported, 70% being the most widely accepted figure. This broad range of values may be explained from evidence that, with slowly developing glaucoma, compensatory factors work to keep the rate of aqueous inflow approximately constant. Thus, the relative importance of ultrafiltration and secretion may depend on the pathological state of the eye.

Aqueous humor enters the eye in the posterior chamber and passes between the lens and the iris before entering the anterior chamber through the pupil. The normal bulk flow rate in man is from 2 to 3 µl min^{-1}. Constituents of the posterior-chamber fluid exchange with the vitreous humor and lens while those of the anterior chamber exchange with the cornea and iris. As a result, some constituents are present in higher concentrations in the posterior chamber (e.g., ascorbate), while others such as phosphate occur in higher concentrations in the anterior chamber.

1.2 Dynamics

Assuming that a substance enters the anterior chamber both by flow from the posterior chamber and by diffusion through the anterior surface of the iris, the rate of change of the concentration in the anterior chamber c_{Aa} is given by

$$dc_{Aa}/dt = k_f c_{Ap} + k_{da}(c_p - c_{Aa}) - k_f c_{Aa} \qquad (1)$$

where c_{Ap} is the concentration in the aqueous humor in the posterior chamber, c_p is the concentration in plasma, k_f is the coefficient of transfer by flow into and out of the anterior chamber, and k_{da} is the coefficient of transfer by diffusion between blood and anterior chamber.

If the rate of entry of a substance into the posterior chamber is independent of the concentration in the chamber, and if it leaves partly by flow into the anterior chamber and partly by diffusion into the surrounding tissue (environment), the rate of change of the concentration in the posterior chamber is given by

$$dc_{Ap}/dt = k_{sp} c_p - K_f c_{Ap} - k_{dpe}(c_{Ap} - c_e) \qquad (2)$$

where k_{sp} is the coefficient of transfer by secretion from blood to posterior chamber, K_f is the coefficient of transfer from posterior chamber by flow, k_{dpe} is the coefficient of transfer by diffusion from posterior chamber into environment, and c_e is the concentration in environment.

Alternatively, if a substance enters the posterior chamber by diffusion from the blood and leaves as described above, its rate of change of concentration in the posterior chamber is given by

$$dc_{Ap}/dt = k_{dp}(c_p - c_{Ap}) - K_f c_{Ap} - k_{dpe}(c_{Ap} - c_e) \qquad (3)$$

where k_{dp} is the coefficient of transfer by diffusion between blood and posterior chamber.

Efforts have been made (Kinsey and Palm 1955) to determine the nature of the transport mechanisms involved in intraocular flows by measuring the rates of accumulation of two test substances, sodium (a positively charged ion) and thiocyanate (a negatively charged ion) in the anterior and posterior chambers of the rabbit eye. Thiocyanate was given as a 1.2% solution of sodium thiocyanate and the sodium (radioactive ^{24}Na) was injected in the form of isotonic sodium bicarbonate. The fractional rate of flow out of the anterior chamber was in the range 0.017–0.020 min^{-1}. The results supported the idea that both the negatively charged ion (thiocyanate) and the positively charged ion (sodium) enter the anterior chamber almost equally by flow from the posterior chamber and by diffusion from capillary blood in the iris (Eqn. (1)). As to the mode of entry into the posterior chamber, the data for sodium fitted a curve derived from a secretory (from blood) hypothesis (Eqn. (2)), whereas the data for thiocyanate fitted a curve derived from the diffusion hypothesis (Eqn. (3)). Although the findings do not demonstrate the exact nature of the transport mechanisms involved, they justify the conclusion that the mechanisms are different for the two ions.

1.3 Outflow Facility

Facility of inflow in the eye, C_{ps} (commonly called the pseudofacility), is defined by the equation

$$C_{ps} = \frac{F_{in} - S}{P_c - P_e} \qquad (4)$$

where F_{in} is the rate of formation of aqueous humor, P_c is the blood pressure in the ciliary vessels, P_e is the intraocular pressure, and S is the rate of secretory inflow. A similar equation defines the outflow facility, C_{out}:

$$C_{out} = \frac{F_{out} - U}{P_e - P_v} \qquad (5)$$

where F_{out} is the rate of outflow of aqueous humor, P_v is the episcleral venous pressure, and U is the rate of uveoscleral outflow (see Sect. 1.4).

The outflow facility is important clinically because it is a measure of the ease with which aqueous humor leaves the eye through the normal drainage apparatus. If the outflow F_{out1} is associated with an intraocular pressure of P_{e1} and F_{out2} with P_{e2}, then

$$C_{out} = \frac{F_{out1} - F_{out2}}{P_{e1} - P_{e2}} \qquad (6)$$

To measure facility in experimental animals, the anterior chamber is cannulated and connected to an external reservoir at a fixed height above the eye. The ratio of the change in fluid flow to the change in pressure, on raising the reservoir, is a measure of the total facility C_{tot} related to the outflow facility

$$C_{tot} = C_{ps} + C_{out} \qquad (7)$$

since raising the intraocular pressure not only affects the outflow but also reduces the rate of aqueous-humor formation. It is for this reason that the facility of inflow is usually termed pseudofacility. Clinical measurement is accomplished by the technique of tonometry in which a small weight indents or flattens the cornea. The indentation volume V_{i1} raises the intraocular pressure from P_{e0} to P_{e1}. After time t, fluid is expelled from the eye, the pressure drops to P_{e2} and the indentation volume is V_{i2}. Facility is derived from the expression:

$$C_{tot} = \frac{(V_{i2} - V_{i1}) + (1/E)(\log P_{e1} - \log P_{e2})}{t(\overline{P_e} - P_{e0} - \Delta P_v)} \qquad (8)$$

where E is the ocular rigidity, $\overline{P_e}$ is the average raised intraocular pressure, and ΔP_v is the increase in episcleral venous pressure. This equation includes components for distensibility of the eye and compression of the episcleral veins. Pseudofacility is about 20% of the total facility in human eyes.

1.4 Drainage

The exit route for more than 80% of aqueous humor in the human eye is through the trabecular meshwork into Schlemm's canal, which is located in the vicinity of the junction between the iris and the cornea, usually called the iridocorneal angle (Fig. 1). This is usually referred to as the conventional drainage system. The aqueous humor is then led into the venous system, either in the aqueous veins or the episcleral veins. Open-angle glaucoma, in which there is no blockage in the anterior chamber angle, is thought usually to be a result of increased resistance in the conventional drainage system.

Another exit pathway for aqueous humor is uveoscleral drainage, whereby aqueous humor enters the iris root and passes between the muscle bundles in the ciliary body to the choroid and out through the episcleral tissues. Although this pathway contributes little to outflow in rabbits, it accounts for 35% of aqueous-humor removal in cynomolgus monkeys at normal intraocular pressure. Unlike the conventional route, this system is relatively pressure insensitive.

The canal of Schlemm is lined with endothelial cells and several routes have been proposed for the drainage of fluid through the endothelium into the lumen of the canal, for example Sondermann's canals, intercellular spaces and micropinosomes.

It is now generally accepted that the main pathway for the passage of fluid into Schlemm's canal is by way of large swellings (giant vacuoles) within the endothelial cells. Most vacuoles are blind invaginations but some have pore openings into the lumen of the canal and thus constitute transcellular channels. To establish whether these pores are sufficient to account for aqueous-humor drainage, it is necessary to calculate their resistance to fluid flow.

1.5 Pore-Invagination Resistance

Quantitative scanning-electron-microscopy (SEM) studies on the lining endothelium of Schlemm's canal in human eyes at normal or near-normal intraocular pressure have produced a range of values between 350 and 1800 pores mm^{-2} with pore widths up to 5 μm.

An early mathematical model of the transcellular channel treated the pore and invagination as small tubes and applied Poiseuille's law:

$$Q = p\pi r^4 / 8l\eta \qquad (9)$$

where Q is the flow, p is the pressure drop, r is the radius of tube, l is the length of tube, and η is viscosity. Application to SEM data showed that the resistance through the transcellular flow channels was less than 10% of the total human outflow resistance. More complicated shapes of the invagination were subsequently modelled mathematically and experimental studies were performed on brass or plexiglas models of giant vacuoles at a magnification of 2000 using aqueous humor simulated by glycerol with a viscosity 2000 times that of water. Since resistance values were about half those of the earlier model, it was concluded that the upper limit for pore-invagination resistance was 5% in humans. This implied acceptance of the SEM data and some of the assumptions used in the earlier simplified model.

In a more detailed treatment of the pore opening, a venturi tube was proposed as a particularly appropriate model. A venturi tube is a confocal hyperboloid of revolution with flow given by

$$Q = \left[\frac{Pr^3(1 + 2\cos\alpha)(1 - \cos\alpha)^2}{3\eta} \right] \qquad (10)$$

where α is the angle of convergence of the venturi tube. This model was applied to one of the sets of SEM data and it was shown that the total pore-invagination resistance in the rhesus monkey eye was 3.30 mm Hg min μl^{-1} at an intraocular pressure of 9 mm Hg, and 0.26 mm Hg min μl^{-1} at an intraocular pressure of 15 mm Hg. The resistance evidently decreased substantially with increasing pressure, a fact which must contribute to the stability of the intraocular pressure. In human eyes, the pore-invagination resistance in a control group was 0.72 mm Hg min μl^{-1}, whereas in a group receiving pilocarpine, a drug used in the treatment of glaucoma, it was 0.14 mm Hg min μl^{-1}. The effect on pore resistance must, therefore, be at least a factor in the effectiveness of pilocarpine in the treatment of glaucoma. The contribution of the pore-invagination to the total normal human outflow resistance was 24%. This is higher than obtained with earlier methods and is a result of the application of different mathematical models and the adoption of different SEM data. However, the various models agree that the pore-invagination system is sufficient for the passage of aqueous humor into the canal of Schlemm.

2. Vitreous Humor

Vitreous humor in man (pH = 7.5; viscosity relative to water = 1.5 centrally and 2.1 peripherally; osmolality = 290 mmol per kg H_2O; refractive index = 1.3347) is a clear, colorless gel with structure and viscosity derived from a network of collagen and hyaluronic acid. The varying concentrations of these substances within the vitreous humor is reflected in viscosity changes whereby the central portion is more liquid than at the periphery. A membranous layer of dense collagen fibrils forms the outermost part of the cortical vitreous humor, which is attached to neighboring tissue behind the lens, at the peripheral retina and around the optic disk.

2.1 Dynamics

Many similarities exist in the composition of the vitreous and aqueous humors. However, some substances, such as lactate and glucose, are found in significantly different concentrations in the two fluids. Vitreous humor not only exchanges with the aqueous humor but also with the lens and the retina. Studies of the penetration into the vitreous humor and removal therefrom of a variety of substances are summarized in Fig. 2. These results are from many sources, some of which are conflicting and few are definitive.

Penetration into the vitreous humor depends on lipoid solubility and on electrical charge. The tracers, iodo-

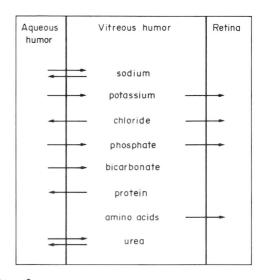

Figure 2
Exchange of substances in the vitreous humor with the aqueous humor and retina

pyracet and fluorescein, do not normally enter the vitreous humor from the blood but are rapidly removed from the vitreous across the retina. This movement may be obstructed by the use of probenecid, which is known to block active transfer in the kidney. This suggests that a pump, probably located at the retinal vessels or pigment epithelium, is responsible for the removal of organic anions from the vitreous humor. An active transport mechanism of iodite has also been proposed.

A study of the movement of a dye in rabbit eye showed that a posterior flow may exist across the retina. Colloidal iron, on the other hand, moved towards the optic disk and this led to the hypothesis that there was no flow across the retina but that, instead, there was a posterior flow in the direction of the optic disk. These findings may be explained on the basis of diffusion combined with posterior flow, due consideration being given to molecular size and possible carrier mechanisms.

2.2 Water Transfer

With the existence of active transport mechanisms, extrapolation of the above tracer studies to the movement of water in the vitreous humor could be misleading. In 1942, Kinsey et al. published results showing the rate of appearance of deuterium-labelled water in the vitreous humor after systemic administration of the tracer. It was concluded that half of the vitreous water was replaced every 10–15 min and that there was, therefore, a rapid turnover of water in the vitreous.

A more detailed mathematical analysis of water movement in the vitreous has recently been undertaken (Moseley 1981). To determine the transfer mechanism, a mathematical model of diffusion was developed. In the eye model, the tracer was assumed to diffuse through the vitreous, retina, choroid and sclera with removal occurring in the vascular layer of the choroid. A numerical solution was adopted and the results of the model concurred with experiments in which ^{133}Xe was injected into the vitreous humor in rabbit eye.

Similar experiments have shown that $100 \pm 16\%$ of intravitreal tritiated water was removed from the midvitreous of rabbit eye by the choroidal circulation with a mean transit time of 30 ± 2 min. Since tritiated water was believed to behave in a similar fashion to ordinary water, it was concluded that there was a rapid removal of ordinary water molecules from the vitreous by the choroidal blood. Again, there was good agreement between experiment and model, suggesting that diffusion was the limiting process in the normal eye. This does not imply that diffusion was the only mechanism involved since, for example, there is evidence for the existence of a sodium pump in the retinal pigment epithelium. Interference with this normal outward passage of water may result in the accumulation of subretinal fluid, the etiology of which is not understood.

See also: Vision

Bibliography

Cant J S (ed.) 1977 The intraocular fluids. *Trans. Ophthal. Soc. UK* 97: 535–780

Cole D F 1966 Aqueous humour formation. *Doc. Ophthalmol.* 21: 116–238

Davson H, Graham L T (eds.) 1974 *The Eye*, Vol 5, *Comparative Physiology*. Academic Press, New York

Jarus G, Blumenkranz M, Hernandez E, Sossi N 1985 Clearance of intravitreal flourouracil. *Ophthalmol.* 92: 91–96

Kinsey V E, Grant M, Cogan D G 1942 Water movement and the eye. *Arch. Ophthalmol.* 27: 242–52

Kinsey V E, Palm E 1955 Posterior and anterior chamber aqueous humor formation. *Arch. Ophthalmol.* 53: 330–44

Moseley H 1981 Mathematical model of diffusion in the vitreous humor of the eye. *Clin. Phys. Physiol. Meas.* 2: 175–81

Moses R A (ed.) 1970 *Adler's Physiology of the Eye: Clinical Application*, 5th edn. Mosby, St Louis, Missouri

H. Moseley

Ionizing Radiation: Absorption in Body Tissues

When penetrating electromagnetic radiation is absorbed in body tissues, any ensuing biological effects are essentially a consequence of the release of secondary charged particles. The nature of the processes giving rise to these particles depends upon the energy of the radiation and the elemental composition of the tissue concerned. When the ionizing radiation is itself particulate, the important interactions are between the primary particles and the atoms making up the tissue. In this case, the exact mode of absorption depends upon the mass of the primary particle, its energy, and whether the particle carries an electrical charge.

In this article, the fundamental interaction processes for both electromagnetic and particulate radiation are considered, followed by an outline of the basic quantities and units used to characterize radiation absorption in body tissues. Finally, some practical implications are described.

1. The Interaction of Radiation with Body Tissues

When x rays or γ rays with energies in the range 10 keV–100 MeV pass into an organ or tissue, interactions occur by an energy-dependent combination of photoelectric absorption, coherent and incoherent scattering, and pair production. In soft tissue, at 10 keV, photoelectric absorption predominates, with the scattering processes accounting for only 7% of the total effect. The coherent process reaches its peak (12%) at ~ 30 keV, with the photoelectric and incoherent effects making up 39 and 49%, respectively. At 1 MeV, incoherent scattering predominates ($>99\%$) in soft tissue, while at 100 MeV it has fallen to 16% and pair production (84%) is the most important effect. Except for incoherent scattering which is independent of the atomic number Z of the tissue, all the other processes are Z dependent.

Charged particles (α particles, β particles, electrons, pions and protons) lose energy when passing through body tissues by collision and radiative (bremsstrahlung) processes. For electrons with energies of 10–500 keV, electronic collisons are the major cause of loss of energy, while above 500 keV, radiative processes become important. For heavy charged particles such as protons, electronic collisions predominate over a wide range of energies. As charged particles slow down near the ends of their tracks, the amount of ionization per millimeter of tissue traversed increases sharply; in graphical form this is referrred to as the Bragg peak. Negatively charged pions exhibit an interesting property as they are stopped, when the tissue nuclei disintegrate (star formation) emitting protons, neutrons, deuterons and other heavy particles.

Neutrons in the energy range 100 eV–30 MeV lose their energy while traversing tissue by a series of elastic and inelastic effects, capture processes and fragmentation of nuclei, termed spallation. Certain elements have resonance regions within this energy interval, including oxygen (0.3–9 MeV) and carbon (2–9 MeV). Up to 14 MeV, elastic scattering with hydrogen is the most important effect, whereas the capture processes ^{1}H(n, γ)^{2}H and ^{14}N(n, p)^{14}C are significant at low and thermal neutron energies.

For all types of radiation the elemental composition of the tissue being traversed is an important factor governing the magnitude of the interaction process. For

Table 1
Elemental composition of some body tissues

Body tissue	Elemental composition (wt%)												Mass density (kg m^{-3})
	H	C	N	O	Na	Mg	P	S	Cl	K	Ca	Fe	
Adipose tissue	12.00	64.00	0.80	22.90	0.05		0.02	0.07	0.12	0.03			920
Average breast	11.70	38.04		50.26									960
Brain	10.70	15.33	1.29	71.40	0.18		0.34	0.17	0.23	0.30			1030
Cortical bone	3.39	15.50	3.97	44.10	0.06	0.21	10.20	0.31			22.20		1850
Lung	9.90	12.30	2.80	74.00	0.18	0.01	0.08	0.22	0.26	0.19	0.01	0.04	260–1050
Muscle	10.20	12.30	3.50	72.89	0.08	0.02	0.20	0.50	0.08	0.30	0.01		1000–1040
Skin	10.00	22.70	4.62	61.50	0.18	0.01	0.03	0.16	0.27	0.08	0.02		1100

x rays and γ rays the strong Z dependence of the photoelectric effect produces enhanced absorption in tissues such as cortical bone which contains appreciable concentrations of high-Z elements. Collision interactions by charged particles have an inverse dependence on Z, resulting in tissues with large quantities of hydrogen, giving increased absorption. For neutrons, the interactions are also dependent upon hydrogen content, with nitrogen concentration being important at certain low energies. For soft tissues containing ~ 10 wt% hydrogen, a small change of only 1% in this concentration can result in $\sim 10\%$ change in the magnitude of the neutron interactions taking place. An illustration of the elemental compositions and mass densities of a few important body tissues is given in Table 1.

A useful listing of the approximate, average ranges of secondary ionizing particles in soft tissue for the frequently used radiations is summarized in Table 2. Only an approximate estimate of the average range can be given since the ionizing particles usually have complex energy spectra. The extent of the range as compared with the size of the biological entity in question is an important consideration when energy absorption in irradiated tissues is being evaluated.

2. Radiation Quantities

A number of internationally accepted quantities have been introduced during the past few decades and are proving to be invaluable in the specification of the complete absorption process from incident radiation beam to final energy deposition in the tissue (see *Radiation Quantities and Units*).

The initial radiation may be described by the fluence (Φ) which relates to the number of particles or photons entering a sphere of given cross-sectional area. Energy fluence (Ψ) is similarly defined but is based upon the particles' or photons' total energy, exclusive of rest energies.

A quantity which expresses the probability that indirectly ionizing radiation (x rays or γ rays and neutrons) will interact with tissue is the mass attentuation coefficient (μ/ρ), relating to the fraction of photons or particles experiencing interactions while traversing a tissue of given thickness and mass density ρ. The quantity mass energy absorption coefficient (μ_{en}/ρ) takes into account all of the modes of energy transfer during these interactions. For directly ionizing radiation (e.g., electrons and protons), the total mass stopping power (S/ρ)

Table 2
Average ranges of ionizing particles

Primary radiation	Energy (MeV)	Ionizing particle in tissue	Average range of ionizing particles in soft tissue (cm)
Beta particles	0.015–5	electron	10^{-4}–1.0
Electron beams	2–20	electron	1–10
Gamma rays	0.05–2.9	electron	5×10^{-4}–0.6
X rays	0.01–0.4	electron	10^{-4}–5×10^{-3}
X rays	1–10	electron	5×10^{-2}–1.2
X rays	10–30	electron	1.2–3.5
Fast neutrons	0.1–10	proton	10^{-4}–6×10^{-2}
Slow neutrons	10^{-7}	0.6 MeV proton	10^{-3} (protons)
		(+ 2.2 MeV γ ray)	0.5 (electrons)
Proton beams	5–400	proton	3×10^{-2}–10^2
Alpha particles	5–10	alpha particle	3×10^{-3}–10^{-2}

Source: Spiers (1956)

considers the average energy lost by a charged particle traversing a tissue of given thickness and mass density.

Energy absorption is fundamental to the interpretation of the biological consequences of irradiating body tissues. The absorbed dose D of any ionizing radiation is defined in terms of the mean energy imparted to a given mass of tissue. The gray (Gy) is the unit of absorbed dose and is equal to $1 \, J \, kg^{-1}$. Another related quantity frequently used in neutron studies is kerma K, based upon the initial kinetic energies of all the charged particles liberated by indirectly ionizing particles in a given mass of tissue (see *Neutron Kerma Values*). The unit of kerma is, once again, the gray.

Although it is evident that ionizing radiation can bring about changes in tissue only by virtue of the energy absorbed, the biological effect also depends on the spatial distribution of the energy released along the tracks of ionizing particles. The linear spacing of ions is usually discussed in terms of the ion density or the number of ion pairs formed per unit length of track. In tissue, the spatial distribution of events along the tracks is often described by the linear energy transfer (LET) expressed in keV μm^{-1}.

In considering the absorption of energy by the body as a whole, the concept of energy imparted (integral dose) is valuable. It refers to the absorbed dose in tissue integrated over the mass irradiated. It is expressed in joules.

3. Practical Implications

3.1 X Rays and γ Rays

The implications of the enhanced energy absorption in bone compared with soft tissue at low photon energies have been studied in attempts to derive the dosage to the living soft-tissue structures contained within the canals which traverse the bone matrix. The number of ion pairs released in a small soft tissue inclusion within a bone cavity is derived from the electron emissions of bone and soft tissues at various radiation energies in relation to the size of cavity considered. The analysis indicates that only for photons of 0.5 MeV and above does the ionization in the living tissue structures within bone approximate to that in other soft tissues. At lower radiation energies the absorption within the bone cavities may be several times that in soft tissues alone. It is thus clear that in the radiotherapeutic treatment of a tumor in or near bone, higher-energy radiation (~ 1 MeV) should be selected to avoid considerable nonuniformity of dosage and its possible harmful consequences.

3.2 Particulate Radiations

Since the biological effect in tissue of x rays and γ rays is due to the secondary electrons produced, it would not be expected that the use of electrons directly will produce a different response. The difference is that their penetration is limited to the maximum range for the energy concerned. Beta-particle emitters are used in radiotherapy for irradiating tissue to a depth of a few millimeters, while avoiding damage to deeper tissues (see *Radiotherapy: Beta Particles*). High-energy electron beams are also used clinically; they have the advantage that the beam terminates abruptly in the body so that the energy imparted is less than it would be for the same tumor dose delivered by x rays or γ rays.

The radiation fields can be made very uniform and the energy imparted is readily assessed; doses in grays to bone, muscle and adipose tissue are nearly the same at the energies used.

Examples of heavy charged particles which have been used in biological experiments and patient treatments include protons, deuterons, α particles and pions (see *Radiotherapy: Heavy Ions, Mesons, Neutrons and Protons*). Heavy particles often have greater biological effectiveness than light ones because of the considerable difference in ion density. Thus β particles from a radium salt incorporated in tissue would give an average of 7 ions μm^{-1}, whereas the accompanying α particles would give rise to some 500 times this ion density; if the biological results of equal doses (i.e., equal total ionization) are compared, the α particles are found to be several times more effective than the β particles. The absorption of proton beams in tissue has been studied both theoretically and experimentally. The information obtained is also germane to considerations of the neutron irradiation of tissue.

When fast neutrons irradiate soft tissue, most of the energy absorbed is by elastic scattering with the hydrogen nuclei present and recoil protons are produced (see *Fast Neutron Therapy*). Other interactions give rise to recoil oxygen and carbon nuclei. The ion density for the secondary heavy particles is far higher than for electrons of the same energy because of the much lower velocities of the former; hence neutrons have the greater biological effectiveness. At rather higher energies, inelastic scattering and capture are also possible both accompanied by γ-ray emission but this does not appreciably affect the dose distribution because of the low cross section for such reactions. Those neutrons which are reduced to thermal energies, however, will produce protons from (n, p) reactions with nitrogen and γ rays from (n, γ) reactions with hydrogen. γ-Ray emission then tends to spread the total dosage through a greater volume.

3.3 Radiations from Radionuclides

The absorption of radiations from sources of radioactivity within the body forms the basis of the therapeutic use of administered radionuclides (see *Radionuclides: Clinical Uses*) and also the consideration of the permissible levels of ingested or inhaled radioactivity (see *Radiation Protection: Internal Exposure*). The dose to tissue from an incorporated radionuclide depends on the type and energy of radiation, on the rate of radioactive decay and on the rates of biological uptake, and on its concentration in and elimination from the tissue.

Consideration of α emitters applies to cases of accidental ingestion, for example, radium salts followed by α irradiation of bone, within which radium is accumulated. Theoretical estimates of the dosage resulting from such deposition have been supported by studies employing autoradiography and track-counting techniques.

With β emitters the dose is delivered within the tissue in which the radionuclide is incorporated and its value may be estimated from the average effective life (including biological elimination as well as physical decay) of the radionuclide and the mean energy released per disintegration.

With γ emitters, the dosage is spread throughout a larger volume and attempts have been made to calculate the doses to organs concerned for various models simulating actual distributions in the body.

For radionuclides emitting both β and γ radiation, the dosage from the former will usually predominate because of the short range and complete absorption of the β particles. The γ ray contribution must, however, be considered when the energy imparted to the whole body is required.

The establishment for different radionuclides of the maximum burdens which the body may safely tolerate has been based on calculations of the type indicated, together with knowledge of the metabolism and toxicity of the substances concerned.

See also: Radiobiology: Charged and Uncharged Particles; Radiotherapy: Radiation Dose, Time and Fraction Number Formulae; Radiobiology: Kinetic Basis of Normal Tissue Response to Radiation; Dosimetry of Internally Administered Radioactive Substances

Bibliography

Attix F H, Roesch W C, Tochilin E 1968 *Radiation Dosimetry*, 2nd edn. Academic Press, New York

Fowler P H 1965 π mesons versus cancer (1964 Rutherford memorial lecture). *Proc. Phys. Soc.* 85: 1051–66

International Commission on Radiation Units and Measurements 1976 *Determination of Absorbed Dose in a Patient Irradiated by Beams of X or Gamma Rays in Radiotherapy Procedures*, ICRU Report No. 24. ICRU, Washington, DC

International Commission on Radiation Units and Measurements 1977 *Neutron Dosimetry for Biology and Medicine*, ICRU Report No. 26. ICRU, Washington, DC

International Commission on Radiation Units and Measurements 1979 *Methods of Assessment of Absorbed Dose in Clinical Use of Radionuclides*, ICRU Report No. 32. ICRU, Washington, DC

International Commission on Radiation Units and Measurements 1979 *Quantitative Concepts and Dosimetry in Radiobiology*, ICRU Report No. 30. ICRU, Washington, DC

International Commission on Radiation Units and Measurements 1984 *Radiation Dosimetry: Electron Beams with Energies Between 1 and 50 MeV*, ICRU Report No. 35. ICRU, Washington, DC

International Commission on Radiological Protection 1975 *Reference Man: Anatomical, Physiological and Metabolic Characteristics*, ICRP Publication No. 23. Pergamon, Oxford

Spiers F W 1956 Radiation units and theory of ionization dosimetry. In: Hine G J, Brownell G L (eds.) 1956 *Radiation Dosimetry*. Academic Press, New York

White D R, Fitzgerald M 1977 Calculated attenuation and energy absorption coefficients for ICRP reference man (1975) organs and tissues. *Health Phys.* 33: 73–81

D. R. White

L

Laser Physics

A laser (an acronym for light amplification by stimulated emission of radiation) is a device for producing a beam of coherent monochromatic radiation in the ultraviolet, visible or infrared regions of the electromagnetic spectrum. Since the beam is coherent (i.e., the waves are in phase), it has very little divergence and as a result has an almost constant power density in space. Laser beams also have the advantage of a high concentration of energy per unit area. This article discusses the theory of laser action and the most common types of lasers in use.

1. Theory of Laser Action

Atoms exist in discrete energy states. When an atom in its ground (lowest) state absorbs energy it is raised to an excited energy state, subsequently to return to the ground state either directly or via intermediate energy states, with the emission of photons of radiation of energy (and wavelength) characteristic of the energy differences between the states. The atoms of many elements have metastable energy states. Photon emission from atoms in metastable energy states is significantly delayed leading to the emission of fluorescent or phosphorescent radiation. Atoms suitable for laser action have at least one such metastable state. When a photon emitted from an atom in a metastable state passes close to an atom in the same state, it can stimulate that atom to emit a photon of radiation which has the same energy (and wavelength), direction, polarization and phase as itself. Each such stimulated photon can itself stimulate the emission of yet another similar photon. This process, which is the basis of laser action, is cumulative and can be encouraged by providing the correct conditions.

The provision of a sufficient number of atoms in the correct metastable energy state is a fundamental requirement for laser action. Given a population of atoms in a state of thermal equilibrium at a particular temperature, the statistical distribution of the atoms over the allowable energy states is known. More atoms exist in the ground state than in any other state, and the higher the energy state the fewer the number of atoms in it.

Laser action depends on the creation of a population inversion, in which there are many more atoms in the metastable state than in the ground state. Energy has to be "pumped" into the population to produce the required inversion—the metastable state then being reached directly or by decay from a higher state.

2. Types of Laser

2.1 Ruby Lasers

The first laser was produced in 1960. The lasing medium used was a single crystal of synthetic ruby, 3–10 cm in length and approximately 0.5 cm in diameter. In the synthetic ruby (a transparent crystal of corundum (Al_2O_3) doped with approximately 0.05% of trivalent chromium ions in the form of Cr_2O_3), the atoms of aluminum and oxygen are inert, the stimulated emission taking place in the chromium atoms which are termed the active ingredients. Chromium ions have two metastable states, the red light of wavelength 694.3 nm emitted by the ruby laser being derived from one substantially favored state. The required population inversion is achieved by pumping energy into the ruby from a white-light flash tube wrapped around the crystal and flashed at a repetition rate of 1 MHz. The energy to operate the flash is derived from the repetitive charging and discharging of a bank of capacitors. In the original laser the parallel ends of the ruby crystal were ground optically flat. One end was silvered so that it was essentially totally reflecting while the other end was half silvered. In this configuration the rod between the mirrors acts as a "cavity resonator," the parallel mirrors preferentially selecting axially directed photons and repeatedly reflecting the resultant beam which is built up as a consequence of successive (axially directed) stimulated emission. After several thousand passages of the rod, the intensity of the beam is sufficiently high to be transmitted through the partially-silvered mirror. The emitted beam is well collimated and highly temporally coherent (monochromatic) and also highly spatially coherent, i.e., there is a constant phase difference between relatively widely separated points along the beam.

Figure 1 shows a currently available commercial ruby laser system. The resonance cavity consists of a ruby rod (ends not silvered) with two externally mounted mirrors. The optical path incorporates a Pockels cell Q-switch and a polarizer. The Pockels cell is an electrooptic crystal

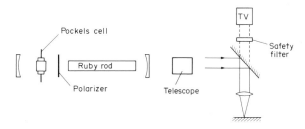

Figure 1
Commercial ruby laser system

of, for example, potassium dihydrogen phosphate (KDP). A voltage is applied across the KDP crystal so that plane polarized light incident upon it is converted into circularly polarized light. The polarizer produces the initial plane polarized light which strikes the Pockels cell (see Fig. 1) and is converted into right-hand (say) circularly polarized light. This light so polarized is then reflected at the rear mirror and converted into left-hand circularly polarized light which passes through the KDP crystal to be converted to plane polarized light, polarized at 90° to its original plane of polarization. So polarized, it is not transmitted by the polarizer. The voltage applied to the KDP crystal controls the output of light: when the voltage is on, there is no light output, and vice versa, the crystal having no effect on the beam when the voltage is switched off. This process is referred to as Q-switching. With a high-quality KDP crystal a voltage of some 5 kV is required. Q-switching essentially controls the resonance of the ruby light between the two mirrors of the cavity, the switching rate being synchronized to the flash rate of the white-light source used for pumping. This is possible in that the voltage across the crystal can be reduced to zero in less than 10 ns. Q-switching maximizes the storage of energy in the lasing medium, there being no output while the ruby rod is being energized. When the system is made resonant by removing the Q-switch voltage, the stored energy is emitted suddenly at high intensity. The ruby-laser power output per pulse can be increased from around 10 kW to 1 GW by Q-switching, the duration of the pulse decreasing from milliseconds to nanoseconds.

The output beam from the resonant cavity is first expanded using a beam-expanding telescope (see Fig. 1). It is then directed to the point of impact by a 45° mirror and a focusing lens system. The mirror is opaque to the ruby light, being coated with a wavelength specific dielectric material. The point of impact can therefore be safely viewed by eye from the position of the TV monitor depicted in Fig. 1. As an extra precaution a Corning-4 safety filter (optical density 6 at 694.3 nm) is placed in the optical path.

The efficiency of the ruby laser is very low, being of the order of 0.1%.

2.2 Helium–Neon Lasers

The next development, chronologically, in laser technology was the production of the helium–neon laser, the lasing medium being a mixture of helium and neon contained in a glass tube. The He–Ne laser emits visible red light of wavelength 632.8 nm, corresponding to an energy transition between two of the upper quantum levels of neon. A dc power supply connected to electrodes positioned within the glass tube is used to excite the helium atoms to a metastable energy state. The energy is transferred therefrom by a knock-on mechanism to the upper energy states of neon, thereby creating the necessary population inversion in the neon. Since the population inversion can be maintained continuously, the He–Ne laser is a continuous-wave (cw)

laser of low power (less than some 2 mW). A He–Ne laser beam has a diameter of typically 2 mm. Thus the energy per unit area is quite high—certainly high enough to cause severe eye damage were the beam to be viewed either directly or after reflection. The laser's resonant cavity is formed by plane parallel mirrors (one fully silvered and one partially silvered) mounted on either end of the tube. However, other arrangements (see Fig. 2) are now more commonly used to produce resonance, a divergent beam being emitted.

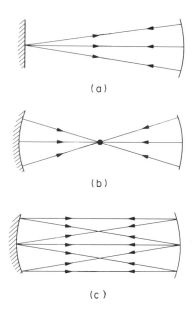

Figure 2
Alternative resonant arrangements for lasers:
(a) hemispherical, (b) spherical, and (c) confocal.
In each case the fully silvered mirror is shaded and the partially silvered mirror is not

With the end plates of a laser positioned normally to its axis, there is a reflection loss detrimental to coherence of approximately 4% at each of the interfaces. However, if tilted to the polarizing angle (57° for glass of refraction index 1.50), these end plates (called Brewster windows) have a 100% transmission for light waves with an electric vector parallel to the plane of incidence. Repeated reflection thus produces a plane-polarized output beam. To achieve the same effect in a solid laser, the ends are bevelled at the polarizing angle. The two cases are shown in Fig. 3.

The low power output of the helium–neon laser restricts its application in medicine. However, the He–Ne laser is very useful for aligning the optics of laser beams in the invisible infrared or ultraviolet regions of the electromagnetic spectrum. The He–Ne laser is also useful for Doppler measurements. By reflecting the laser beam from a moving object and measuring the subsequent change of wavelength, the velocity of the moving

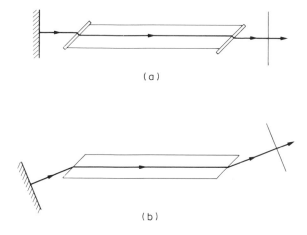

(a)

(b)

Figure 3
(a) Gas laser with Brewster windows, and (b) solid laser with ends bevelled at Brewster angle

Using this technique a 100 W cw laser can produce 100 kW pulses approximately 150 ns long at a pulse repetition rate of 400 pulses per second.

The CO_2 laser exemplifies a spectrum-rich type of energy source, a very large number of laser transitions being possible. Lasers with such a characteristic are said to be tunable. Modern CO_2 lasers are tunable over some 85 different wavelengths. Tuning may be accomplished by the well-established technique of the Littrow spectrograph in which either a prism or a diffraction grating is used for dispersion. An example of this is shown in Fig. 4. The fully silvered prism is rotatable; placed at one

Figure 4
Tunable carbon dioxide laser

object can be accurately assessed. This is of some limited importance in medicine.

2.3 Carbon Dioxide Lasers

The carbon dioxide (CO_2) laser is an example of a high-powered molecular gas laser. The output beam when focused can cut through diamond and thick steel plates in a matter of seconds.

The energy level diagram of a molecular gas is considerably more complicated than that of an atom. The energy states previously discussed are referred to as electronic energy levels, transitions between which lead to the emission of visible light. Each electronic level in a molecular gas has, in addition, sublevels corresponding to allowed vibrations of the molecule, and each of these vibrational levels has further sublevels corresponding to allowed rotations of the molecule. Laser action is possible by transitions between different vibrational–rotational levels, the output radiation being in the infrared (low-energy photons). The CO_2 laser exploits this type of transition to give, for example, an output beam of wavelength 10.6 μm in a cw operation. The design of the CO_2 laser is not unlike that of the He–Ne laser except that the gas mixture (9% carbon dioxide, 15% nitrogen, 76% helium) is passed continuously through the tube. As in the He–Ne case, pumping is by dc excitation. The lasing tube must be cooled, usually by passing water continuously through a water jacket placed round the tube.

The lasing action in a CO_2 laser is due to the transfer of energy from excited nitrogen atoms to adjacent energy levels of the CO_2 molecules. The high power efficiency ($\sim 15\%$) of the CO_2 laser is attributable to the low-lying vibrational and rotational energy states of the carbon dioxide which require little energy for excitation.

By incorporating a Q-switch, the CO_2 laser can be converted from continuous-wave to pulsed operation.

end of the laser it disperses the light so that only one spectrum line is collinear with the laser axis and consequently amplified, standing waves being set up.

The CO_2 laser is widely used in medical applications. It is very versatile in that it can be operated as either a CO or N_2O laser by changing the nature of the gas flow.

2.4 Argon and Krypton Lasers

The argon laser is also widely used in medicine. The energy levels of the excited argon ion are used to produce a blue–green output with the strongest emission at prism-selectable wavelengths of 457.9 nm and 514.5 nm. Argon lasers need to be cooled. In most of them the hot gas (plasma) is contained in a graphite tube, which requires large amounts of water for cooling. A more expensive version uses beryllium oxide plasma tubes which can be air-cooled. The argon laser can have a large power output, unlike its close relative the krypton laser which is very similar to it in design and construction. The krypton laser, because of its low power output, has a somewhat restricted use in medicine. It produces a variety of wavelengths: visible red light at 676.4 nm is generally selected.

2.5 Other Types of Lasers

The neodymium–YAG laser is also much used in medicine. Like the ruby laser its output is pulsed and it is similarly pumped with a flash tube. The lasing medium is yttrium–aluminum–garnet (YAG) doped with neodymium. A wavelength of 1.06 μm (in the near infrared) is selected. The efficiency (1%) of the neodymium–YAG laser is much higher than that of the ruby laser.

Mention must be made of dye lasers. These use a liquid containing dye molecules. Such molecules contain broad energy absorption bands in the visible region of the spectrum. They can be pumped in many ways. It is indeed quite common now to pump them using another laser, e.g., an argon laser. Tuning is possible over a wide range of wavelengths, by varying the concentration of the dye. Dye lasers are being increasingly used as research tools, particularly in the field of ophthalmology.

See also: Lasers in Medicine; Lasers: Safety Aspects

Bibliography

Bridges W B 1964 Laser oscillation in singly ionised argon in the visible spectrum. *Appl. Phys. Lett.* 4: 128–30
Jawan A, Bennett W R, Herriott D R 1961 Population inversion and continuous optical maser oscillation in a gas discharge containing a helium–neon mixture. *Phys. Rev. Lett.* 6: 106–10
Lengyel B A 1966 *Introduction to Laser Physics.* Wiley, New York
Liberman I 1969 High power Nd: YAG continuous laser. *IEEE J. Quantum Electron.* QE-5: 345
Maiman T H 1960 Stimulated optical radiation in ruby masers. *Nature (London)* 187: 493–94
Schawlow A L 1968 Laser light. *Sci. Am.* 219: 120–40
Schawlow A L, Townes C J 1958 Infrared and optical masers. *Phys. Rev.* 112: 1940–49
Sobolev N N, Sokovikov V V 1967 CO lasers. *Sov. Phys. Usp.* 10: 153–70

W. G. Ferrier

Lasers in Medicine

Numerous medical uses for lasers have been found in recent years, aided by the availability of several types of lasers with differing operating parameters and a reduction in their cost. The important parameters of lasers used in medicine are listed in Table 1. The physics of lasers is discussed elsewhere (see *Laser Physics*).

1. Ophthalmology

Argon-ion lasers are widely used in ophthalmology outpatient clinics. The ophthalmologist views the patient's retina through a slit lamp, a device with binocular viewing optics and a light source. The slit lamp can be adjusted independently in height and horizontal displacement, as well as rotationally. The laser beam is directed at the retina via the slit lamp (Fig. 1). Thus

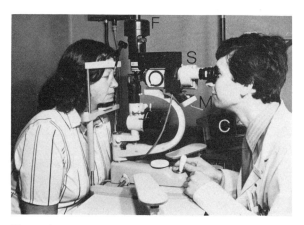

Figure 1
Argon laser connected to slit lamp for ophthalmic surgery: F is the fiberoptic cable connection, S the slit lamp, M the micromanipulator, and C the control module

throughout the treatment, the ophthalmologist can observe the area of the retina that is being treated. To facilitate the use of the slit lamp, the refractive power of the cornea is negated by placing a contact lens on the patient's eye. The target spot on the retina is selected using a low-power aiming beam, the blue–green argon light being readily visible. With early argon lasers, the laser output was coupled to an attachment on the slit lamp by means of a delivery system that incorporated mirrors mounted on articulated arms. Flexible fiber-optic light guides are now used for this purpose.

Table 1
Relevant parameters of various types of lasers used in medicine

Type of laser	Wavelength	Typical power	Typical spot diameter	Principal areas of application
CO_2	10.6 µm	25 W	0.75 mm	gynecology
Argon	488 nm, 514.5 nm	3 W[a], 10 W[b]	100 µm	ophthalmology gastroenterology
Krypton	647.1 nm	1 W	100 µm	ophthalmology
Nd–YAG	1064 nm	70 W	300 µm	gastroenterology
Ruby	694.3 nm	3 J in 30 ns[c] or 200 µs[d]	250 µm	dermatology ophthalmology
He–Ne	632.8 nm	1 mW	1 mm	laser alignment

a Power level in ophthalmic surgery b Power level in gastric surgery c Q-switched
d Non-Q-switched

1.1 Retinal Detachment

A ruby laser was first used, experimentally, in 1963 for the treatment of detached retina, the retina being "spot-welded" to the underlying choroid. Today, it is customary to use argon-ion lasers for the treatment of retinal detachment. "Welding" is an over-simplified description of the process whereby the two contiguous surfaces are united: a complex inflammatory reaction is involved.

1.2 Fundus Photocoagulation

The need for photocoagulation of the fundus (inner back surface) of the retina is indicated, for example, when a growth of new blood vessels spreads over the surface of the retina and invades the vitreous humor. If unchecked, these new blood vessels would obstruct the passage of light to the retinal rods and cones (photoreceptors) and thus severely reduce visual acuity. This proliferation of new blood cells is triggered by a stimulus generated when the blood supply to the retina is inadequate. If parts of the retina can be destroyed—by photocoagulation—then the retinal demand for blood decreases and the stimulus subsides.

In fundus coagulation, an intense short-exposure beam of light is focused on successive small areas of the retina. The ocular media transmit visible and near infrared light with very little loss, absorption occurring mainly in the melanin of the retinal pigment, epithelium and choroid. Lasers have supplanted the xenon arc lamps that were formerly used for fundus photocoagulation. The main disadvantage of the xenon arc lamp was that it emitted light over a very wide band of wavelengths, which is less efficiently absorbed than the blue–green light of the argon-ion laser—particularly by the hemoglobin that gives the red blood cells their color.

The argon-ion laser is used for photocoagulation to treat diabetic retinopathy, a condition in which the unwanted growth of new blood vessels (described above) occurs. Fundus coagulation is an effective—indeed the only effective—form of therapy for this condition. The treatment consists of some 2000 panretinal exposures, given over several sessions in the clinic. Vision is not appreciably diminished by the laser burns, provided the macular area of the retina (the yellow spot and the region of central vision) is avoided. The therapeutic effect of the treatment derives from the destruction of poorly transfused tissue and not to any direct sealing off of individual blood vessels.

The type of argon-ion laser in common use in ophthalmology emits blue light of wavelength 488 nm and green light of wavelength 514.5 nm. It cannot be used to treat the macular area of the retina, since the blue component is appreciably absorbed by macular xanthophyll. As there is xanthophyll in nerve fibers, these fibers would be at risk. For this reason, an argon laser has now been introduced which incorporates a filter that passes only the 514.5 nm spectral line. It is too soon to comment on its effectiveness; the green light is more efficiently transmitted through the ocular media than the blue light,

particularly through the crystalline lenses of older people.

The krypton-ion laser has recently been introduced in ophthalmic surgery. Krypton lasers are usually part of a two-tube (argon/krypton) system. Either the argon laser or the krypton laser is chosen—at the press of a button. The red krypton beam is less absorbed in its passage through the retina than the argon beam, particularly if red blood cells are encountered. Thus the krypton laser is particularly well suited to the treatment of the pigment epithelium and the choroid. It is also of value in the treatment of the macular area of the retina because its light is less well absorbed in comparison with conventional argon light by xanthophyll. The risk of burning nerve fibers is therefore reduced. In practice, the krypton laser is usually reserved for the relatively few cases where it is necessary to treat the area near the macula—xanthophyll being present only in the macular area of the retina.

1.3 Glaucoma

Lasers are also used for the treatment of glaucoma, a condition in which the intraocular pressure is raised (see *Intraocular Fluid Dynamics*). In closed-angle glaucoma, the normal flow of aqueous humor is prevented (e.g., by the adhesion of the lens to the iris), and this causes a build-up in pressure in the posterior chamber. The laser is used to make a hole in the iris (iridectomy) and so form an artificial pupil. A success rate of 95% has been claimed for producing iridectomies using argon-ion lasers. However, one observer (Pollock 1981) has found that 34% of the cases required retreatment but in no instance did the hole close after remaining open for six weeks.

A laser may be used to make holes in the trabecular meshwork tissue (see *Intraocular Fluid Dynamics*) if a blockage at this site is preventing the drainage of aqueous humor (chronic simple glaucoma). Results, using argon lasers, have been disappointing. However, recent experimental work (Frankhauser et al. 1982) with Nd–YAG lasers has produced encouraging results. The use of dye lasers for iridectomies and trabeculectomies is being investigated.

1.4 Other Uses

Laser beams are being used in some centers to divide fibrous strands in the vitreous humor. The Nd–YAG laser would seem to be a more promising tool than the CO_2 laser for this work. A more controversial role of lasers is their use in ophthalmic surgery for the removal of malignant melanomas in the choroid (Foulds 1983).

2. Gynecology

2.1 Carbon Dioxide Lasers

The CO_2 laser is a useful surgical tool. Because there is no suitable optical fiber that will transmit its 10.6 µm wavelength (infrared) beam, its use is limited to treatment sites that are accessible to laser light which is emitted externally to the body. However, by the same token, it finds useful application in gynecological surgery. Almost all the energy of a CO_2 laser beam is absorbed by a tissue thickness of some 100 μm. (Deeper penetration is, of course, achieved by successive superimposed applications of the laser beam.) The beam is absorbed mainly by body water which makes up some 85% of body weight. Cells are ruptured—and thus destroyed—at a temperature of 100 °C due to the sudden vaporization of cellular water. Damage to extracellular tissue is slight although there is some burning of cell residue. When diseased tissue is destroyed, the surrounding and underlying healthy tissue is relatively unaffected. Very little bleeding occurs in CO_2 laser surgery, as the beam coagulates blood vessels up to 0.5 mm in diameter. Discomfort is less in comparison with conventional surgery; healing times are shorter; no general anesthesia is required; little or no scar tissue forms; postoperative edema is slight; and trauma to remaining tissue is reduced due to it being manipulated less. Laser surgery is intrinsically aseptic because it is a noncontact procedure and because bacteria and viruses are destroyed by the high-intensity irradiation.

In general use, the CO_2 laser is attached to an operating microscope. Its optics are so designed that the beam is focused in the plane that is in sharp focus in the field of view. The laser beam is directed to the operating microscope by mirrors mounted on each point of rotation of an articulated arm. Very little power is lost if the mirror surfaces are kept clean. A collinear helium–neon (He–Ne) laser beam is used to align the invisible CO_2 laser light.

2.2 Cervical Intraepithelial Neoplasia

For gynecological surgery, the CO_2 laser is attached to a colposcope (see Fig. 2). Before deciding on laser surgery (a treatment which is primarily aimed at precancerous lesions and which requires a depth of destruction of some 7 mm), the tissue is carefully examined by an experienced colposcopist, and a biopsy performed. If malignant cells are present, conventional surgery is preferable to laser surgery because of the likelihood that malignant cells have invaded the tissue surrounding the lesion. Occasionally, however, the laser is used for the palliation of malignant tumors. Good results have been obtained in the treatment of intraepithelial neoplasia in cases in which the epithelial membrane is intact (Fig. 3). Success rates of 96% have been reported (Jordan and Mylotte 1982) for the ablation of cervical intraepithelial neoplasia. These results include patients who required a second or third treatment. Similar results have been realized in treating neoplastic tissue of the vagina and vulva. The CO_2 laser has also been used successfully for the treatment of condylomata acuminata (dry, pointed warts) on the vulva, the vestibule of the vagina and the perianal region, and for the treatment of herpes simplex

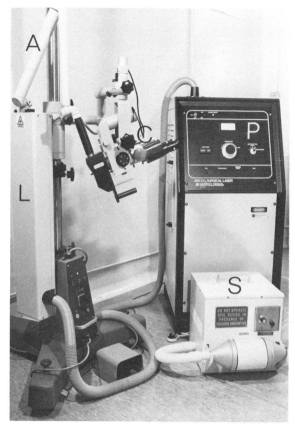

Figure 2
CO_2 laser used for gynecological surgery: L is the housing containing CO_2 and He–Ne laser tubes, C the colposcope, A the articulated arm, P the control panel, S the suction device

lesions, as well as capillary hemangioma (malformation of blood vessels) of the cervix.

3. Gastroenterology

Lasers may be used to stop bleeding in the upper gastrointestinal tract. Two types of lasers have been used, the argon-ion laser and the Nd–YAG laser. Peptic ulcers and esophageal varices have been treated, the laser beam being transmitted along an optical fiber (usually quartz) positioned in the biopsy channel of an endoscope. Thus no surgical dissection is necessary. Since the laser beam diverges from the end of the fiber light guide, the spot size and intensity of the impinging beam is dependent on the distance between the target tissue and the end of the guide. A coaxial flow of CO_2 through the endoscope removes blood and debris from the operating site.

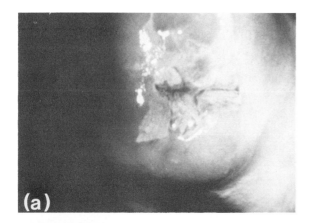

Figure 3
Cervical intraepithelial neoplasia (a) before treatment,
and (b) after ablation with a CO_2 laser

The results obtained in treating bleeding ulcers with
Nd–YAG lasers have been encouraging. In a trial at one
center (MacLeod et al. 1983), patients were carefully
selected to include only those with major hemorrhage
that was unlikely to abate naturally. While the control
group all required conventional surgery, it was required
for only 20% of the cases treated by laser. Almost all the
laser failures were due to limitations in the endoscopes
available. Another center reported a 93% success rate
(Kiefnaber et al. 1977).

High success rates have also been reported for the
treatment of bleeding ulcers using an argon laser. Comparison between the effectiveness of argon and Nd–YAG
lasers is difficult because of possibly differing criteria
applied in the selection of patients. The action mechanism of the two types of lasers is different: while the
argon light is absorbed in the red blood cell, the
Nd–YAG beam is believed to be absorbed in the blood
vessel wall, the temperature of the wall rising to $100\,^{\circ}C$
or more and the wall shrinking in consequence. The
penetration of the Nd–YAG beam through tissue and
blood is five times greater than that of the argon beam.

4. Otolaryngology

An attractive feature of CO_2 laser therapy in oto-
laryngology is the minimal damage to tissue adjacent to
the treatment site, arising from surgical trauma. An
articulated arm links the laser to an operating microscope; a He–Ne beam is used as a means of visualizing
the target site.

In surgery of the larynx, regions to be treated are in
direct line of sight, but mirrors have been used to reflect
the beam onto the dorsum of the soft palate or up on
the roof of the nasopharynx. For treating lesions of the
trachea and bronchi, the laser is coupled to a bronchoscope.

The CO_2 laser has been used to advantage for treating
papillomas in the oral cavity, oropharynx, larynx, trachea and bronchial areas, and in the nasal cavities. It has
been particularly successful in ablating papillomas in the
respiratory tract. Laryngeal polyps and nodules on the
vocal cords have been removed, as have granulomas.
Laryngeal stenoses have also been effectively treated
with CO_2 lasers. Premalignant carcinomas have been
removed from the larynx, tongue, floor of mouth, upper
nasal cavity, soft palate, and retromolar and para-
pharyngeal regions.

Leukoplakia (white thickened patches on mucous
membranes) have been similarly ablated as have sub-
glottis hemangiomas. Another related application of
lasers is in middle-ear surgery. Argon lasers have been
used for stapedectomies (removal of the median bone),
for ossicular reconstruction, and for the "spot welding"
of grafts to repair perforated ear drums.

A cautionary note must be added. There is a particular
hazard associated with laser surgery of the larynx. The
endotracheal tube introduced by the anesthetist is close
to the treatment site, and there is a real danger of it
catching fire. Precautions must be taken to avoid such an
occurrence; for example, by wrapping the endotracheal
tube in aluminum foil.

5. Dermatology

An unwanted tattoo can be a source of embarrassment
and social discomfort—even a cause of severe psychological stress. Treatment by CO_2 laser is effective if it is
only a particular tattooed word or name that needs to
be removed. In general, the removal of tattoos with a
CO_2 laser is not appropriate as the laser vaporizes the
tissue and there is no regrowth of normal skin. Some
success has been reported, however (Goldman 1983), in
the eradication of tattoos using a pulsed ruby laser. The
radiation is absorbed selectively by pigment-containing
areas and damage to the surrounding skin is thus
minimized.

Hemangioma (port-wine stain or birthmark) is another cause of stress. Early attmepts to remove such
marks using ruby lasers were not successful. While some
early results with CO_2 lasers were acceptable, the situation has improved with the introduction of argon lasers

and dye lasers. Results of an extensive study (Apfelberg et al. 1979) on argon-laser irradiation of superficial vascular lesions were very encouraging, lightening or disappearance of the blemish occurring in most cases. Follow-up studies confirmed that minimal damage was caused to normal dermal tissue. The argon beam is absorbed in the red blood cells, heat is generated and blood vessels are occluded. As a result, the blood vessels in the lesion are generally narrower than before the treatment. Of the conditions treated, which also included superficial varicose veins, capillary and cavernous hemangiomas, and superficial telangiectasis (dilatation of superficial capillaries), the port-wine stains responded best.

Unfortunately, the argon beam is not very effective in removing port-wine stains from children's skin—the stain is paler in juveniles than in adults.

6. Laser Angioplasty

Several groups have begun investigations of the use of lasers in ischemic heart disease, which is a major cause of death in the Western world. Ischemic heart disease results from narrowing of the coronary arteries, which supply the myocardium, by the deposition of atheroma (a complex mixture of lipids, fibrous tissue and sometimes calcium) on the walls of coronary arteries. The symptoms can be alleviated by drug therapy; in selected patients, stenoses can be treated surgically by "bypassing" the obstruction with section of vein taken from the patient's leg.

A technique known as balloon angioplasty has been introduced in the past few years (Gruntzig et al. 1979). A special catheter is passed across the stenosis and a balloon tip is inflated to crush aside the atheroma and widen the vessel. The laser angioplasty experiments investigate whether the atheromatous plaque can be destroyed by laser radiation passed down a fiberoptic light guide contained in a catheter. For this application, CO_2 lasers cannot be used since there is no available fiberoptic delivery system for *in vivo* use. Most investigations have employed argon lasers, typically at a power of 3–10 W (Lee et al. 1983, Choy et al. 1982), although Nd–YAG lasers are a possibility (Abela et al. 1982). Nd–YAG laser emission has an enhanced absorption in blood, while intermediary agents may further enhance surface absorption (Abela et al. 1982)—with the possibility of sparing surrounding tissue.

Successful experiments to recanalize obstructed arteries have been carried out on excised specimens (Lee et al. 1981, 1983) and animals (Choy et al. 1982): the first *in vivo* human experiments have been carried out on the exposed heart at the time of bypass surgery (Choy et al. 1983). Percutaneous treatment (via a catheter inserted into the femoral artery), perhaps carried out at the time of investigative coronary catheterization, could become available within 2–3 years. One factor which will facilitate this is the development of small-diameter, coherent fiber-optic bundles for direct visualization of the treatment area (*Wall Street Journal Europe* 1983).

7. Urology

The Nd–YAG laser has been used in urological surgery, for the destruction of tumors of the bladder, the penis and the urethra. No general anesthetic is needed. Bladder tumors up to some 15 mm in diameter have been removed without causing perforation of the bladder wall. While the destruction of carcinomas and malignant melanomas have been reported, it is necessary to await the long-term results before considering the technique to be an acceptable form of treatment. (One fear is the possible dissemination of viable malignant cells.)

8. Neurology

While a variety of tumors have been removed from the brain, spinal cord and peripheral nerves, using the CO_2 laser, the use of lasers in neurosurgery is in its early stages. The CO_2 laser has also been used for median commissurotomy (division of fibrous band) of spinal cord in the treatment of intractable pain. Small tumors have been excised from within the neuroaxis of the cord. Lobectomies and neurectomies have been performed.

9. General Surgery

In general surgery, the CO_2 laser has been used to ablate benign tumors, such as fibromas. In surgery on the colon and rectum, benign polyps and villous adenomas have been removed, as well as carcinomas. The potential of the CO_2 laser for debridement (removal of foreign matter and injured or infected tissue) has been explored—for example, in the treatment of gangrenous lesions. In some centers it is used as a cutting tool, and there are a wide variety of cardiac and thoracic procedures in which it functions as a "bloodless" scalpel.

10. Diagnostic and Other Applications of Lasers

Low-power He–Ne lasers are used for the optical alignment of radiotherapy treatment units and computerized axial tomography scanners. They are used, too, for blood-flow measurement using Doppler shift principles. Laser nephelometers are used for the quantitative *in vitro* determination of proteins. Systems using He–Ne laser light are being developed for the assessment of visual acuity, and also for the assessment of retinal function, even with opacities present in the lens or vitreous humor of the eye.

See also: Lasers: Safety Aspects; Nonionizing Electromagnetic Radiation: Potential Hazards

Bibliography

Abela, G S, Normann S, Cohen D, Feldman R L, Geiser E A, Conti C R 1982 Effects of carbon dioxide, Nd–YAG and argon laser radiation on cornary atheromatous plaques. *Am. J. Cardiol.* 50: 1199–205.

Apfelberg D, Maser M, Lash H 1979 Extended clinical use of the argon laser for cutaneous lesions. *Arch. Dermatol. Res.* 155: 719–21

Beesley M J 1978 *Lasers and their Applications*, 2nd edn. Taylor and Francis, London

Choy D S Stertzer S H, Rotterdam H Z, Sharrock N, Kaminow I P 1982 Transluminal laser catheter angioplasty. *Am. J. Cardiol.* 50: 1206–08

Choy D S, Stertzer S H, Rotterdam H Z, Sharrock N, Bruno M, Kaminow I 1983 Laser angioplasty: Experience in in vitro human and in vivo animal arteries. In: *Fifth Int. Congr. of Laser Medicine and Surgery, Detroit*

Department of Health and Social Security 1984 *Guidance on the Safe Use of Lasers in Medical Practice*. HMSO, London

Foulds W S 1983 Current options in the management of choroidal melanoma. *Trans. Ophthalmol. Soc. U.K.* 103: 28–30

Frankhauser F, Lortscher H, Van der Zypen E 1982 Clinical studies on high and low power laser radiation upon some structures of the anterior and posterior segments of the eye. *Int. Ophthalmol.* 5: 15–32

Goldman L 1983 Differential effects of different lasers on the skin. In: Arndt K A, Noe J M, Rosen S (eds.) 1983 *Cutaneous Laser Therapy: Principles and Methods*. Wiley, Chichester, pp. 65–68

Gruntzig R A, Senning A, Siegenthaler W E 1979 Non-operative dilatation of coronary artery stenosis: percutaneous transluminal coronary angioplasty. *N. Engl. J. Med.* 301: 61–68.

Jordan J A, Mylotte J 1982 Treatment of CIN by destruction: Laser. In: Jordan J A, Sharp F, Singer A (eds.) 1982 *Pre-Clinical Neoplasia of the Cervix*. Royal College of Obstetricians and Gynaecologists, London, pp. 205–11

Kiefnaber P, Nath G, Moritz K 1977 Endoscopical control of massive gastrointestinal haemorrhage by irradiation with a high power AG Neodymium-YAG laser. *Prog. Surg.* 15: 140–55

Lee G, Ikeda R M, Kozina J, Mason D T 1981 Laser dissolution of coronary atherosclerotic obstruction *Am. Heart. J.* 102: 1074–75

Lee G, Ikeda R, Herman I, Dwyer R M, Bass M, Hussein H, Kozina J, Mason D T 1983 The qualitative effects of laser radiation on human arteriosclerotic disease. *Am. Heart. J.* 105: 885–89

Lerman S 1980 Radiant energy and the eye. Macmillan, New York

MacLeod I A, Mills P R, MacKenzie J F, Joffe S N, Russell R I, Carter D C 1983 Neodymium yttrium aluminium garnet laser photocoagulation for major haemorrhage from peptic ulcers and single vessels: A single blind controlled study. *Br. Med. J.* 286: 345–48

McKenzie A L, Carruth J A S 1984 Lasers in surgery and medicine. *Phys. Med. Biol.* 29: 619–41

Pollock I P 1981 Use of argon laser energy to produce iridotomies. *Ophthalmic Surg.* 11: 506–15

Swain C P, Bown S G, Storey D W, Northfield T C, Kirkham J S, Salmon P R 1981 Controlled trial of argon laser photocoagulation in bleeding peptic ulcers. *Lancet* ii: 1313–16

Venkatesh S, Guthrie S, Foulds W S, Lee W R, Cruikshank F R, Bailey R T 1985 In vitro studies with a pulsed neodynium YAG laser. *Br. J. Ophthalmol.* 69: 86–91

Wall Street Journal Europe (3 April 1983) FDA approves catheter to be sold by Trimedyne, p. 4

World Health Organisation 19 *WHO Environmental Health Criteria Document: Laser and Optical Radiation*. WHO, Geneva

H. Moseley

Lasers: Safety Aspects

Early theories assumed that injury caused by laser radiation was due solely to the heating effect of the radiation. It is now considered proven that the photochemical and thermoacoustic effects of the radiation are also important factors. Damage may result, too, from electrical breakdown of tissue. The effect which predominates in a particular circumstance depends on a number of biological and physical parameters; for example, the biological tissues involved, the wavelength of the radiation, the energy absorbed by the tissue and the duration of the exposure, and the effective area of the laser beam where it interacts with the tissue.

Safety standards governing the use of lasers take all these factors into account. The maximum permissible exposures (MPEs) which have been derived are based on experimental data relating to threshold injuries, with theoretically obtained values interpolated between the experimental data points. The experimental data relate to acute effects which persist for 24–48 h at least, the observation of the lesions being delayed to await possible enhancement of the initial injury.

1. Laser-Induced Injuries

1.1 Photochemical-Effect Injury

A photochemical reaction involves electronic excitation within an atom or molecule as a result of the absorption of a photon of appropriate and sufficient energy. Over a wide range of irradiances ($W m^{-2}$) and durations of exposure, a given biological effect results from a given aggregate radiant exposure ($J m^{-2}$) to a particular wavelength of radiation. Depending on wavelength, reciprocity between irradiance and exposure duration may fail for short exposures if the irradiance is sufficiently high to cause either thermal injury or thermal enhancement of the photochemical injury. Reciprocity can fail also for low irradiances if the exposure duration is sufficiently long for repair mechanisms to operate effectively.

Photochemical effects do not occur if the photon energy of the radiation is less than the threshold value needed to initiate photoelectric absorption of the radiation by the target molecule. The action spectra of the target tissues display resonances over broad bands of frequencies. The shorter wavelengths are generally more heavily attenuated by any intervening tissues (e.g., the stratum corneum of the skin). For example, photochemical injury to the retina of the human eye seems limited, in practice, to exposures to wavelengths of 400–500 nm of durations longer than 1–10 s. Radiation of wavelengths <400 nm is too heavily attenuated by the intervening ocular media. At wavelengths >500 nm, retinal injury is observed only if the irradiance is sufficiently intense to cause thermal injury (Lawwill et al. 1977, Ham et al. 1980).

1.2 Thermal Injury

Threshold thermal injury occurs when a temperature increase raises the kinetic energy in a complex biological molecule sufficiently to disrupt its weakest bonds. These bonds dictate the shape of the molecule and therefore its function. Protein denaturation and enzyme inactivation are the results of the damage.

For irradiation of the skin, maximum permissible exposures are based on the premise that only relatively small areas of the body are involved, with overheating confined to the area irradiated. For irradiation of the eyes, the target tissue being the retina, two categories of exposure are considered by safety standards (British Standards Institution 1983, American National Standards Institute 1980):

(a) when the image on the retina is small (<100 μm in diameter) with rapid cooling of tissue effected by three-dimensional thermal cooling; and

(b) when the image on the retina is larger, with cooling effected by essentially one-dimensional thermal diffusion.

Much higher irradiances are required to produce a given temperature rise with decreasing size of the retinal image.

1.3 Thermoacoustic Injury

Thermoacoustic damage occurs when rates of thermal expansion are sufficiently high to generate acoustic shock waves. Very high irradiances are needed for this phenomenon to occur. Its occurrence is likely only in the retina of the eye, with damage possibly extending to other tissues of the eye. This thermoacoustic mechanism underlies threshold tissue damage for the range of exposure durations 10^{-9}–10^{-5} s.

1.4 Electrical-Breakdown Injury

Data for exposures of duration less than 10^{-9} s are scarce, but those which do exist (e.g., Goldman et al. 1975, 1977) suggest a hitherto unsuspected mechanism—electrical breakdown—as the limiting factor in relation to safe exposure levels. The radiant exposures for threshold retinal lesions are lower for electrical-breakdown injury than for thermoacoustic injury although the irradiances required are higher. A plausible explanation is that it is the electric field strengths associated with these high irradiances that cause the electrical breakdown of

the tissue. As with thermoacoustic injury the phenomenon depends on the high optical gain of the eye. For these exposures of less than 10^{-9} s duration, maximum permissible exposures are based on constant irradiance values (W m^{-2}), not on constant radiant exposure values (J m^{-2}).

2. Maximum Permissible Exposures

Maximum permissible exposures (MPEs) are determined in relation to minimal lesions of the skin and eye because lesions at these locations can be reliably observed. Because there is no satisfactory animal surrogate, the data on skin injury used as a basis for MPEs have been obtained using volunteer subjects. (Much of the data have been obtained using noncoherent (therefore nonlaser) uv radiation sources.) Some photokeratitic thresholds have been determined with volunteers. Because eye injuries are generally much more disabling (injuries to the lens and retina usually being more likely than not irreversible), retinal intrabeam MPEs are based entirely on work with experimental animals. Rhesus monkeys have been used in later studies in preference to rabbits whose optics and physiology make them unsuitable for retinal studies. Although a considerable amount of data has resulted from experience with laser treatment of diseased eyes (retinal coagulation and retinal ablation), the physiology of the diseased retina (including the diabetic retina) is so different from the normal retina that such data are not suitable for deriving safe exposure levels.

2.1 Ultraviolet Radiation (200–315 nm)

The limiting values for continuous exposure to uv radiation of wavelength 200–315 nm are based on the acute photochemical effects of photokeratitis and erythema. The MPEs recommended are the same for eyes and skin. Threshold data for these effects are shown in Figs. 1 and 2, together with the BS 4803 maximum permissible exposure curve (British Standards Institution 1983) for continuous exposure to uv laser radiation. The envelope curve formulated by Sliney (1972) for noncoherent uv radiation and subsequently adopted by the National Institute for Occupational Safety and Health (NIOSH) and the American Conference of Governmental Industrial Hygienists (ACGIH) is also shown. This curve now constitutes the threshold limit value (TLV) curve for occupational exposure to uv radiation promulgated by the ACGIH (ACGIH 1980) and recommended as a voluntary standard in the UK by the Health and Safety Executive and the National Radiological Protection Board.

2.2 Ultraviolet Radiation (315–400 nm)

For uv radiation in the wavelength range 315–400 nm (UVA), the limits set on exposure are based on corneal and lenticular damage thresholds. The relatively high transmission of UVA by the cornea and its absorption

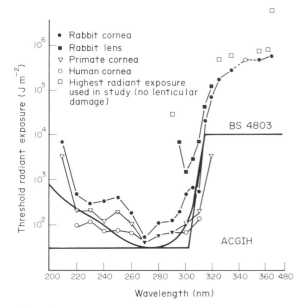

Figure 1
Comparison of threshold data for corneal and
lenticular injury with ACGIH and BS4803 maximum
permissible exposures

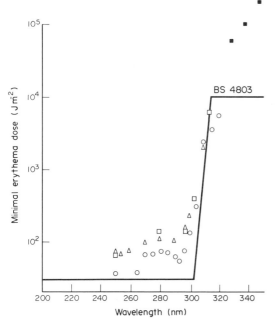

Figure 2
Minimal erythema dose for humans as a function of
wavelength together with BS4803 maximum
permissible exposures (sources: △ Sayre et al. (1966);
□ Berger et al. (1968); ■ Magnus (1964);
○ Cripps and Ramsey (1970))

by the lens, particularly at wavelengths between 320 and
370 nm, suggests that the lens could be the critical target
for this radiation (Zuclich 1980a). However, UVA is not
completely absorbed by the lens and retinal damage
from 325 nm He–Cd laser radiation has been demonstrated in the eyes of rhesus monkeys (Zuclich and
Taboada 1978, Zuclich 1980b). Figure 3 shows the

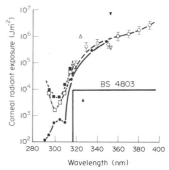

Figure 3
Action spectra for corneal and lenticular injury by uv
radiation: ●——● photokeratitis (Pitts et al. 1977);
■----■ permanent lenticular opacities; □–·–·–□
temporary lenticular opacities; ○ corneal threshold
injury (Zuclich and Kurtin 1977). Additional points:
△ corneal cataract He–Cd laser (Ebbers and Sears
1975); ▽ corneal lesion Kr laser (Zuclich 1980a);
▲ corneal exposure to produce retinal lesion in
monkeys with a He–Cd pulsed laser (Zuclich and
Taboada 1978); ▼ Kr laser, retinal lesion
(Zuclich 1980b)

action spectra for lenticular, corneal and retinal lesions
from UVA together with recommended MPEs (British
Standards Institution 1983).

2.3 Visible and Infrared Radiation (400–1400 nm)

Visible and infrared radiations in the wavelength range
400–1400 nm are refracted and focused onto the retina.
Because of the optical gain of the eye, the MPEs for
ocular exposure to these radiations are low. Most focusing takes place at the air–cornea interface because of the
large difference between the two refractive indices—1.00
and 1.38. The radiation then passes through the cornea,
the pupil and the aqueous humor to the lens, where it
is further refracted and focused through the vitreous
humor onto the retina.

The size of the retinal image, the wavelength of the
radiation and the duration of the radiation pulse, are the
most important factors for determining whether a given
energy absorption by the retina will be followed by
injury and for interpreting experimental threshold lesion
studies and their applicability in formulating maximum
permissible exposure levels. Thermal damage to the
retinal cells depends both on the temperature to which
the cells are heated and the time for which they are
maintained at that temperature.

For long exposures (of duration comparable with
cooling times) the threshold values of irradiance that

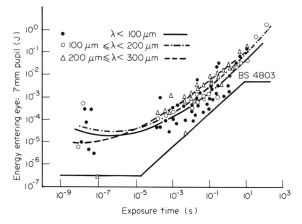

Figure 4
Ocular intrabeam viewing. Comparison of threshold data and regression line plot (Van Pelt et al. 1973) with BS4803 (British Standards Institution 1983)

cause thermal injury to the retina are markedly dependent on the size of the retinal image. This can be readily explained on the basis of the volume diffusion of heat from the target cells for small images compared with the linear heat loss from large planar images. For a given retinal irradiance the smaller the retinal image the smaller the temperature rise and the faster the establishment of thermal equilibrium.

Safety standards (e.g., British Standards Institution 1983, American National Standards Institute 1980) give maximum permissible exposure levels both for the viewing of small laser sources (intrabeam viewing MPEs, see Figs. 4, 5). Extended-source viewing MPEs are based mostly on sources of some 800–1000 μm in diameter. If

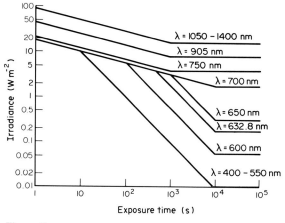

Figure 5
Ocular maximum permissible exposure for intrabeam viewing as a function of exposure time and wavelength

the output beam of a laser is effectively conjugate to a point, then it is this point which is to be regarded as the source rather than, for example, the exit aperture of the laser. In this latter instance, intrabeam viewing conditions apply. Whether a particular viewing condition conforms to intrabeam viewing or to extended-source viewing depends on the magnitude of the angle subtended by the source at the eye of the observer—the angular subtense. The MPEs for intrabeam viewing and extended-source viewing are identical when the angular subtense equals the limiting value of angular subtense for a particular exposure time (Fig. 6).

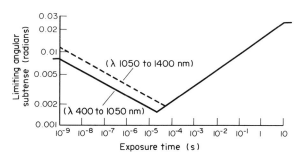

Figure 6
Limiting angular subtense as a function of exposure time

2.4 Infrared Radiations (1.4 μm–1.0 mm)

The same maximum permissible exposures apply for the eyes as for the skin for wavelengths in the range 1.4 μm–1.0 mm. In this range possible disabling injury to the cornea of the eye is of particular concern. The cornea is sensibly opaque to this range of wavelength except in the region 1.6–1.8 μm where it transmits up to 70% of the incident radiation. The effective depth of penetration into the cornea decreases rapidly for radiations of wavelengths > 2.6 μm, being some tens of micrometers, for example, in the case of CO_2 laser radiation (wavelength 10.6 μm). For 10.6 μm radiation the MPEs (in relation to the cornea) are approximately a factor of ten lower than the threshold exposure for corneal injury for exposures between 10^{-3} s and 10^3 s duration.

3. Standards of Laser Classification

Standards are intended to be practical guides to the safe use of lasers. They require manufacturers of lasers and laser systems to design their equipment for safe use including the use of enclosures, warning devices and interlocks. On the basis of the maximum permissible exposures they are required to classify their equipment and label it accordingly.

Lasers are classified into four classes according to their potential for causing injury. Class 1 lasers are those that are inherently safe either because of their low power or because their engineering design is such that in any event the maximum permissible exposure level cannot be exceeded.

Class 2 lasers are low-power lasers which emit visible radiation (400–700 nm) and no other radiation than that which is allowed for a class 1 laser. Class 2 lasers are not inherently safe but are safe for accidental exposures where the exposure time is limited by the aversion responses, such as the blink response. For continuous-wave lasers (emission duration >0.25 s) the output power of class 2 is limited to 1 mW averaged over 0.25 s.

Class 3 is split into two categories, class 3A and class 3B. Class 3A lasers constitute a relaxation of the criteria applying to class 2 lasers. They may emit up to 5 mW of visible radiation provided that the beam irradiance does not exceed 25 W m^{-2}. Protection for the unaided eye is afforded by the aversion responses but direct viewing with optical aids may be hazardous.

Class 3B lasers should not emit more than 0.5 W for continuous-wave lasers and 10^5 J m^{-2} for pulsed lasers. Directly viewing a class 3B laser beam may be hazardous but under certain conditions viewing is safe via a diffuse reflector. These conditions are: (a) a minimum viewing distance of 50 mm between the diffuse reflector and the cornea, (b) a maximum viewing time of 10 s, and (c) a minimum diffuse image diameter of 5.5 mm.

Class 4 lasers are high-output devices. They are capable of producing hazardous diffuse reflections. They may cause skin injuries and can also constitute a fire hazard. Their use requires extreme caution.

In general, laser users must adopt control measures appropriate to the hazard potential of the class to which the laser belongs. The parallel requirements of classification and labelling by the manufacturer and implementation of appropriate control measures by the user obviate the need for radiometric measurements on the part of most users of laser systems.

See also: Laser Physics; Lasers in Medicine; Nonionizing Electromagnetic Radiation: Potential Hazards; Ultraviolet Radiation: Potential Hazards

Bibliography

American Conference of Governmental Industrial Hygienists 1980 *TLVs: Threshold Limit Values for Chemical Substances and Physical Agents in the Workroom Environment with Intended Changes for 1980.* American Conference of Governmental Industrial Hygienists, Cincinnati, Ohio

American National Standards Institute 1980 *American National Standard for the Safe Use of Lasers*, ANSI Z136.1. American National Standards Institute, New York

Berger D, Urbach F, Davies R E 1968 The action spectrum of erythema induced by ultraviolet radiation, preliminary report. *Dermatologica* 2: 1112–17

British Standards Institution 1983 *Radiation Safety of Laser Products Systems.* British Standards Institution, London

Cripps D J, Ramsey C A 1970 Ultraviolet action spectrum with a prism–grating monochromator. *Br. J. Derm.* 82: 584–93

Ebbers R W, Sears D 1975 Ocular effects of a 325 nm ultraviolet laser. *Am. J. Optom. Physiol. Opt.* 52: 216–23

Friedman E, Kuwabara T 1968 The retinal pigment epithelium. IV. The damaging effects of radiant energy. *Arch. Ophthalmol.* 80: 265–80

Goldman A I, Ham W T, Mueller H A 1975 Mechanisms of retinal damage resulting from the exposure of rhesus monkeys to ultrashort laser pulses. *Exp. Eye Res.* 21: 457–69

Goldman A I, Ham W T, Mueller H A 1977 Ocular damage thresholds and mechanisms for ultrashort pulses of both visible and infrared laser radiation in the rhesus monkey. *Exp. Eye Res.* 24: 45–65

Ham W T, Mueller H A, Ruffolo J J 1980 Retinal effects of blue light exposure. *Proc. Soc. Photo-Opt. Instrum. Eng.* 229: 46–50

Lawwill T, Crocket S, Currier G 1977 Functional and histological measures of retinal damage in chronic light exposure. *Doc. Ophthalmol. Proc. Ser.* 15: 285–95

Magnus I A 1964 Studies with a monochromator in the common idiopathic photodermatoses. *Br. J. Derm.* 76: 245–65

National Institute for Occupational Safety and Health 1972 *Criteria for a Recommended Standard . . . Occupational Exposure to Ultra Violet Radiation.* National Institute for Occupational Safety and Health, Cincinnati, Ohio

Pitts D G, Cullen A P, Hacker P D, Parr W H 1977 *Ocular Ultra Violet Effects for 295 to 400 nm in the Rabbit Eye*, Contract CDC-99-74-12. National Institute for Occupational Safety and Health, Cincinnati, Ohio

Sayre R M, Olson R L, Everett M A 1966 Quantitative studies on erythema. *J. Invest. Dermatol.* 46: 240–44

Sliney D H 1972 The merits of an envelope action spectrum for ultraviolet radiation exposure criteria. *Am. Ind. Hyg. Assoc. J.* 33: 644–53

Van Pelt W F, Payne W R, Peterson R W 1973 *A Review of Selected Bioeffects Thresholds for Various Spectral Ranges of Light*, DHEW (FDA) 74-8010. US Government Printing Office, Washington, DC

Zuclich J A 1980a Cumulative effects of near-uv induced corneal damage. *Health Phys.* 38: 833–88

Zuclich J A 1980b Hazards to the eye from uv. *Proc. Soc. Photo-Opt. Instrum. Eng.* 229: 11–20

Zuclich J A, Kurtin W E 1977 Oxygen dependence of near-ultraviolet induced corneal damage. *Photochem. Photobiol.* 25: 133–35

Zuclich J A, Taboada J 1978 Ocular hazard from uv laser exhibiting self-mode-locking. *Appl. Opt.* 17: 1482–84

A. F. McKinlay

M

Mammography

Of all the malignant diseases that affect women in the Western world, breast cancer is the most common. In Western Europe and North America it accounts for about one fifth of cancer deaths and for about one twenty-fifth of all deaths. It is the leading single cause of death for women in the 35–54 year age group. In the US, one in twelve women suffer from it during a lifetime; one in seventeen in the UK. The earlier the disease is detected and treated, the better the prognosis. These facts underline the importance of mammography (breast radiography), which is the most reliable of the physical methods used to detect breast cancer.

During puberty, the connective tissue of the pre-pubescent breast, under a subcutaneous layer of adipose tissue, begins to be replaced by glandular and fibrous tissue. Once maturity is reached the fibroglandular tissue is replaced by adipose tissue until, at the post-menopausal stage, the breast tissue is almost entirely adipose. Mammography is thus a particularly exacting imaging technique because soft tissues have similar radiation absorption characteristics, and also because there is an important need to detect small image details such as microcalcifications of dimension 150–350 μm.

1. Radiographic Techniques

In current practice, mainly craniocaudal, mediolateral and oblique radiographs of the breast are taken.

The technique, which was first used in 1913 but did not become commonplace until the 1950s, has changed considerably over the last decade, the known hazardous effects of ionizing radiation having stimulated a great deal of research into ways of enhancing the efficiency of the procedure in terms of the radiation dose administered to the patient.

The original use of tungsten-target x-ray tubes with no added filtration, in order to provide a soft, more easily absorbed x-ray beam, in addition to the use of slow x-ray film, to provide an image with fine detail visible, meant inevitably that the radiation dose absorbed by the patient was high. Later, equipment specially designed for mammography was introduced: molybdenum-target x-ray tubes with breast compression devices (1967); much faster film–(fluorescent) screen image receptors (from 1970 onwards); xeroradiographic image receptors (1972); automatic exposure control (AEC) (1973); electron radiography (or ionography) and microfocal tubes for magnification (1977); and scatter-reducing grids (1979).

Some 80% of the approximately 200 departments in the UK that offer mammography use specialized equipment which generally consists of a molybdenum-target x-ray tube centered on a small table supporting the breast close to the image receptor. A cone or adjustable diaphragm is used to limit the x-ray field size. Most sets feature AEC and some have adjustable filtration and a grid. In the UK, 90% of centers use film–screen receptors, the remainder using xeroradiography (Fitzgerald and White 1981). In the USA, a much higher percentage of centers use xeroradiography. Xonics Inc. of the USA has made commercially available an ionography system which uses a high-pressure chamber, and has also developed an experimental ionography system that uses an atmospheric pressure chamber (see Sect. 3.3).

A device for compressing the breast is provided in some 95% of UK centers. The compressed average breast is generally assumed to be about 5 cm thick and approximately 100 cm^2 in surface area. Compression is vital because it produces a more uniform breast thickness and therefore a reduced radiographic density gradient. It reduces, too, the amount of (tissue-volume related) scattered radiation which reaches the image receptor, thereby improving contrast. Patient movement and geometric blurring are also reduced.

2. Physical Factors

The most important physical factors that determine the quality of the radiographic image are the primary x-ray spectrum, the size of the focal spot, the imaging geometry, the type of image receptor used and the proportion of scattered radiation reaching it.

2.1 X-Ray Spectrum

The primary spectrum is determined by the target material, the tube potential and filtration, and it is important in film mammography to optimize these to achieve high subject contrast. The most common choice of tube is one with a molybdenum target and a 0.03 mm molybdenum filter, operated at a potential of 25–35 kV. The spectrum is dominated by the characteristic radiation of molybdenum at 17.9 keV and 19.5 keV: the molybdenum filter preferentially absorbs photons of energy >20 keV because of the molybdenum K-edge x-ray absorption at this energy. High subject contrast is produced in small and average-sized breasts. However, because the breast itself hardens the x-ray beam considerably, contrast is reduced in the thicker, dense breast. Beaman et al. (1983) have shown that a tungsten-target tube operated at about 30 kV with a 0.04 mm palladium filter provides improved contrast with thicker breasts (7–8 cm), and comparable contrast to the molybdenum target/molybdenum filter combination with average (~5 cm) breasts. A clinical assessment confirmed this; dose reductions of 30–50% are thereby possible. If xeroradiographic image receptors are used, then because of their inherent enhancement of edges in the image, subject contrast is less important. Higher exposures are required, however, so higher energy spectra are used to minimize the surface dose. These are obtained by operating molybdenum- or tungsten-target tubes at 35–50 kV, with at least 2.0 mm aluminum filtration, to minimize the low-energy component of the spectrum ($\leqslant 15$ keV) which otherwise contributes an unnecessary skin dose and gives no image information.

2.2 Focal Spot and Geometry

Image details have fuzzy edges and tend to blur into each other because of geometric unsharpness which is influenced by the size, shape and intensity distribution of the focal spot and also by the focus-to-object and object-to-image receptor distances. Specialized mammography tubes usually have a nominal focal spot size of about 0.6 mm² but measured sizes are generally larger, often as much as 1 mm in one direction. Haus (1982) suggests that if a 1 mm focal spot is used to image a 5 cm thick breast placed directly on the image receptor, then a focus–skin distance of 66 cm is necessary to reduce geometric blurring so that it is not the limiting factor in the resolution of very small calcifications. Specialized units have focus-to-skin distances of 35–60 cm. In magnification mammography, focal spot sizes of approximately 0.1 mm are necessary to minimize geometric blurring.

2.3 Scattered Radiation

Scattered radiation carries no image information and if it reaches the receptor it can significantly reduce contrast. Scatter is increased by lowering the tube kilovoltage, increasing the breast thickness and increasing the field size. The ratio of scattered radiation to primary radiation can vary from 0.35 to 1.0, hence contrast improvements of 35–100% are possible if scatter can be eliminated (Barnes 1979).

Mammography grids are available to reduce scatter, but exposures have to be increased, if these are used. However, the extra dose to the breast can be offset partly by increasing the kilovoltage or the filtration and by using a more sensitive receptor.

3. Image Receptors

3.1 Film–Screen Combinations

As previously stated, slow direct-exposure films were used in the early days of film mammography; now, specialized film–screen combinations for mammography have gained wide acceptance from radiologists. They are about 20 times faster than direct-exposure film

which provides for lower doses, shorter exposure times and automatic film processing. However, image quality is impaired to some extent. A single emulsion film is exposed in contact with a single high-definition intensifying screen within a specially designed plastic cassette or vacuum envelope. The intensifying screen is made of calcium tungstate or a rare earth oxysulfide and it is largely the (fluorescent) light emitted by the screen that blackens the film. The absorption of x rays by the screen is high. Some 10^3 light photons are generated for every x-ray photon absorbed.

Blurring and quantum mottle determine the visibility of small details. Unless the film and screen are in close contact, light spreads before it reaches the film and produces a blurred image. Equally important are screen cleanliness and a film processor which does not produce artefacts in the film. Regular quality control tests are essential. Quantum mottle or noise is the fluctuation in film density caused by statistical variation in the number of x-ray photons absorbed by the screen. It is reduced by using slower screens because slower screens require higher x-ray exposures. However, patient dose is then increased. The current generation of film–screen receptors represents a compromise between the conflicting requirements of information content of the mammogram and breast dose.

3.2 Xeroradiography

In xeroradiography, a charged photoconductive selenium plate is the image receptor. The x rays emerging from the breast selectively dissipate the electric charge on the plate by ionization to form a position and x-ray intensity related charge distribution of the plate. This latent image is made visible by means of blue thermoplastic powder which adheres in proportion to the charge distribution and which is made permanent by thermal transference to white paper. The powder development is controlled by a back-bias voltage applied to the rear of the plate which can produce both positive and negative xeroradiographs. The negative mode corresponds in phase, but not color, to a conventional radiograph. It requires about 30% less exposure than the positive mode and is increasing in popularity.

The characteristic edge enhancement of xeroradiographs is the pronounced increase in image contrast which occurs in response to any sharp change in subject contrast; it is caused by the distribution of lines of electric force created by the charge. This emphasis of borders makes xeroradiography superior to current film–screen systems for imaging microcalcifications (150–250 μm detectability compared with 250–350 μm for film–screen systems). However, in xeroradiographs, large area contrast is poorer. One problem that occurs is powder deficiency spots—areas of the plate that will not accept an electric charge. These can mask or simulate clinical details. They can be revealed by dark dusting, that is, processing the plate without exposure to x rays.

3.3 Ionography

Ionography, sometimes called electron radiography, is an electrostatic imaging method in which a layer of a high atomic number gas, usually at high pressure, is contained in a thin disk-shaped chamber and is used to absorb the x rays emerging from the breast. The x-ray image is converted by ionization to a charge distribution which is collected by an insulating foil stretched between and parallel to the electrodes of the chamber. The image is rendered visible by powder-cloud or liquid toner development and in this respect the method is similar to xeroradiography. Breast doses using ionography are comparable to or less than the doses obtained with film–screen mammography and xeroradiography, but the visibility of detail is currently not as good as that achieved with xeroradiography.

The advantage of the atmospheric pressure chamber experimental system, produced by Xonics Inc., over their commercial high-pressure chamber system is that the imaging foil can be easily removed (Moores et al. 1980).

4. Measurement and Specification of Dose

There are two general methods for deducing the radiation dose to the patient's breast.

(*a*) *Direct measurements using thermoluminescence dosimetry (TLD) on the patient.* One problem is that the attenuation characteristics of the most popular LiF and LiB_4O_7 thermoluminescent dosimeters are sufficiently different at low energies to make them visible on the x-ray images and may therefore mask clinical detail. Suitably calibrated square TL ribbons should be used because their shape is readily distinguishable; these are best placed on the superior medial surface of the breast for craniocaudal and oblique radiographic views, and on the inferior medial surface for mediolateral views. This avoids superimposition on the images of structures within the superior lateral quadrant of the breast, where at least 50% of malignancies occur. If AEC is used, the milliamps × seconds (mA s) exposure can be recorded (a meter may have to be fitted) and the dose deduced from indirect measurements.

(*b*) *Indirect measurements.* X-ray output (air kerma) is measured using an ionization chamber placed at the position occupied by the breast surface, that is, with the measuring point of the chamber about 5 cm above the image receptor surface. The chamber must have a suitably flat response for x rays generated at tube potentials over the range 20–50 kV, it must not be too large (<10 cm^3) and it must have a calibration traceable to a primary standard ionization chamber. Surface dose is obtained by applying a backscatter factor relating dose in tissue to dose in air (Dubuque et al. 1977), and an air kerma to dose in tissue factor (previously the röntgen-to-rad or f-factor) (Hammerstein et al. 1979). To deter-

mine the latter and also the chamber calibration factor, it is necessary to know the beam quality, that is, its half-value thickness (HVT). (The quality of an x-ray beam is conventionally and very suitably determined in terms of the thickness of material (aluminum at diagnostic x-ray kilovoltages) which, when placed in the path of the beam, reduces its intensity by one-half: hence the expression half-value thickness.)

Alternatively, a phantom can be used to represent the breast. A chamber is positioned "half-immersed" in the surface to record full radiation backscatter or just above the surface in which case the scatter contribution is less. TLDs can also be used. This method is suitable for both manual and automatic exposure control. When AEC is used, it is important that the cone, compression device and receptor are in place so that the examination is replicated exactly, otherwise considerable error can be introduced. If a transversely sectioned phantom is available (Fitzgerald et al. 1981), percentage depth-doses can be measured directly using TLD. The choice of phantom material is also critical: doses may be over-estimated by as much as 40% if water or Perspex is used rather than an "average" composition. Guidance on the use of TLDs can be found in McKinlay (1981).

The radiation dose to the breast has been specified in a number of ways by different authors. This can be misleading when results are compared unless what exactly is meant by dose is stated. The surface dose is the easiest quantity to deduce but it is not the most appropriate indicator of risk because the dose decreases markedly with depth in the breast, its magnitude being dependent upon the beam quality (HVT), the distance from the x-ray tube focus and the composition of the tissue. Exit dose can vary from about 1% to 15% of the surface dose depending on the HVT of the radiation. Air kerma-to-dose in tissue factors can vary from about 0.6 to 1.0, depending upon breast composition (Hammerstein et al. 1979). Other quantities used include the mid-breast dose, the mean dose to ductal parenchyma, fibroglandular tissue or whole breast, and the integral dose. This variety reflects the uncertainty surrounding the exact tissues at risk. The difficulty is avoided if the mean dose to the whole breast is used (see, for example, Dance 1980).

The assumed composition of equal parts by weight of fat and water for the "average breast" which stems from White et al. (1977) is one commonly used. This is equivalent to the composition used by Hammerstein et al. (1979). Mid-breast dose is recommended by NCRP (1980) although when measured in the "average" breast, the mean dose to the glandular tissue may be under-estimated by 15–30%.

A summary of data is given in Table 1 which enables the calculation of relevant doses from direct and indirect measurements. Currently used fast film–screen systems specifically designed for mammography and used with a molybdenum target tube (HVT ≈ 0.4 mm Al) should require an air kerma of about 2.5–5 mGy measured at the surface of a 5 cm thick breast of

Table 1
Summary of data for dose calculations

	Mo target tube and fast film–screen receptor	W target tube and xerox receptor (negative mode)
Typical tube settings		
kilovoltage (kV)	25–30	45–50
total filtration	Be window + 0.03 mm Mo or glass window + 0.5 mm Al	2.5 mm Al
first half-value layer (mm Al)	0.4	2.5
focus–receptor distance (cm)	45–60	60
Dose conversion factors[a]		
exposure to air kerma factor (mGy R^{-1})	8.7	8.7
backscatter factor	1.08	1.3
air kerma to dose in tissue factor	0.76	0.76
air kerma at surface (without backscatter)		
to mean breast dose factor	0.21	0.57
Typical depth dose data[a]		
depth at which surface dose is		
reduced to 50% (cm)	~1.0	~3.0
mid-plane (2.5 cm) depth dose (%)	~15	~55
exit-plane (5.0 cm) depth dose (%)	~3	~25
Typical breast dose ranges (mGy)[a]		
air kerma at surface (without backscatter)	3.0–6.0	4.0–7.0
air kerma at surface (with backscatter)	3.2–6.5	5.2–9.1
dose at surface	2.4–4.9	3.9–6.9
mid-plane (2.5 cm) dose	0.5–1.0	2.1–3.8
exit-plane (5.0 cm) dose	0.1–0.2	1.0–1.7
mean breast dose	0.6–1.2	2.2–4.0

a Assumed breast characteristics: compressed breast thickness = 5 cm; surface area = 100 cm^2; composition = 50% fat, 50% water by weight

"average" composition. A Xerox 125 xeroradiographic system (negative mode) used with a tungsten-target tube and 2.5 mm Al total filtration (HVT $\approx$ 1.5 mm Al) should require an air kerma of about 7.5–10 mGy under the same conditions. Corresponding mean doses are shown in Table 1.

5. Carcinogenic Risk in Mammography

The possibility that mammography may induce cancers as well as detect them, especially if the patients are a population of women who are well, has attracted much study and conjecture in recent years (Bailar 1976).

That x rays can cause breast cancer was first demonstrated in 1965 when it was noticed that a number of women who had contracted breast cancer had previously undergone multiple chest x-ray fluoroscopies during treatment for tuberculosis. Since then at least eight irradiated groups of women have been identified that might lend data on breast cancer induction. These include the survivors of the A-bomb, luminous-dial painters who ingested radium, patients who received radiotherapy for non-neoplastic breast disease or postpartum mastitis, and tuberculosis patients.

The evidence from three major groups has been analyzed extensively, notably by Boice (1982). The shortcomings include the fact that none of the groups

relates directly to a breast screening population. Moreover, all the groups were exposed to much higher doses than accrue from present-day mammography. There is much debate on the shape of the curves relating risk to dose level and the problem of extrapolation from high doses to low doses. A linear dose response could be applied to each group and although other relations could not be ruled out their application made only a small difference to the risk estimate. The main findings were that: (a) there is a linear dose response even below 0.5 Gy, (b) dose fractionation does not diminish risk, (c) there is a latent period of at least ten years, (d) risk continues throughout life or at least for 40 years, and (e) age at exposure influences the subsequent risk most.

Boice et al. (1979) have derived the absolute risk as 6.6 cancers per 10^4 women per year per gray for women aged 20 or more at irradiation and 10.4 per 10^4 women per year per gray for age 10 to 19 at irradiation. This assumes a linear model and a ten year latency. Some evidence suggests for the first time that the prepubescent breast is also at risk from radiation although not, of course, from mammography. From the foregoing, it follows that if 10^6 women aged 35 are x rayed annually for 30 years, then assuming for simplicity a mean breast dose of 1.2×10^{-3} Gy per examination (two views using a film–screen system; see Table 1) and that all the radiation is received at the outset, about 480 cases of breast cancer will be induced in the 30 year period. If the naturally occurring breast cancers are assumed to be about 800 per 10^6 women per year this amounts to 24 000 naturally occurring breast cancers in the same period. The radiation-induced cancers, therefore, represent about 2% of the natural incidence, which seems to be an acceptably small risk. It must be emphasized that this low risk results partly from the good mammographic techniques that can now be practised with very much smaller doses to patients than were prevalent when the breast screening debate first began, and partly due to the acceptance of mean breast dose, which may be an order of magnitude less than the surface dose, as the measure of risk. A perspective on radiation risks is often provided by comparing them with the risks incurred in everyday activities or pursuits in which we indulge without thought. It can be shown, by adapting some figures of Pochin (1978) and using the concept of dose equivalent, that two films involving a mean breast dose of 1.2×10^{-3} Gy are equivalent to a 320 mile air flight, a 50 mile car journey, smoking 2/3 of a cigarette, climbing a mountain for 72 s, or being a man aged 60 for 16 min.

The risk to organs of the body other than the breast should be negligible provided that the breast table absorbs the primary x-ray beam, and that the x-ray field is well collimated and carefully directed.

See also: Radiography and Fluoroscopy in Medicine

Bibliography

Bailar J C 1976 Mammography: A contrary view. *Ann. Intern. Med.* 84: 77–84

Barnes G T 1979 Characteristics of scatter. In: Logan and Muntz 1979, pp. 223–42

Beaman S, Lillicrap S C, Price J L 1983 Tungsten anode tubes with K-edge filters for mammography. *Br. J. Radiol.* 56: 721–27

Boice J D 1982 Risk estimates for breast cancer. In: *Critical Issues in Setting Radiation Dose Limits*, NCRP Proceedings Series No. 3. National Council on Radiation Protection and Measurements, Bethesda, pp. 164–81

Boice J D, Land C E, Shore R E, Norman J E, Tokunaga M 1979 Risk of breast cancer following low-dose radiation exposure. *Radiology* 131: 589–97

Bureau of Radiological Health 1978 *Breast Exposure: Nationwide Trends. A Mammographic Quality Assurance Program.* Bureau of Radiological Health, Rockville, Maryland

Dance D R 1980 The Monte Carlo calculation of integral radiation dose in xeromammography. *Phys. Med. Biol.* 25: 25–37

Dance D R, Davis R 1983 Physical aspects of mammography. In: Parsons C A (ed.) 1983 *Diagnosis of Breast Disease.* Chapman and Hall, London

Dubuque G L, Cacak R K, Hendee W R 1977 Backscatter factors in the mammographic energy range. *Med. Phys.* 4: 397–99

Fitzgerald M, White D R, 1981 Physical aspects of mammography in the UK. In: Drexler G, Eriskat M, Schibilla H (eds.) 1981 *Patient Exposure to Radiation in Medical X-Ray Diagnosis: Possibilities for Dose Reduction.* Commission of European Communities, Brussels

Fitzgerald M, White D R, White E, Young J 1981 Mammographic practice and dosimetry in Britain. *Br. J. Radiol.* 54: 212–20

Hammerstein G R, Miller D W, White D R, Masterson M E, Woodard H Q, Laughlin J S 1979 Absorbed radiation dose in mammography. *Radiology* 130: 485–91

Haus A G 1982 Physical principles and radiation dose in mammography. *Med. Radiogr. Photogr.* 58: 70–80

Jones C H 1982 Methods of breast imaging. *Phys. Med. Biol.* 27: 463–99

Logan W W (ed.) 1977 *Breast Carcinoma: The Radiologist's Expanded Role.* Wiley, New York

Logan W W, Muntz E P (eds.) 1979 *Reduced Dose Mammography.* Masson, New York

McKinlay A F 1981 *Thermoluminescence Dosimetry.* Hilger, Bristol

Moores B M, Ramsden J A, Asbury D L 1980 An atmospheric pressure ionography system suitable for mammography. *Phys. Med. Biol.* 25: 893–902

National Council on Radiation Protection and Measurements 1980 *Mammography*, NCRP Report No. 66. National Council on Radiation Protection and Measurements, Washington, DC

Pochin E E 1978 Estimates of industrial and other risks *J. R. Coll. Physicians London* 12: 210–18

Symposium on mammography 1983 *Radiol. Clin. North Am.* 21: 1–194

White D R, Martin R J, Darlison R 1977 Epoxy resin based tissue substitutes. *Br. J. Radiol.* 50: 814–21

M. Fitzgerald

Mathematical Modelling in Biology

The term modelling, used in connection with applications of mathematics to biology, reflects a clear awareness that the mathematical systems involved are extreme simplifications of the complex biological processes to which they refer. Unlike physics or engineering, it is unusual in biology to have a well-tested set of mathematical equations which are known to describe phenomena accurately and can be used as a starting-point for modelling. The difference, however, is one of degree: even the most widely used physical laws are simplified models too, and are in no sense exact descriptions of nature, but they are often many orders of magnitude closer to exactness than is common in biological modelling. Workers in those fields are thus allowed to talk of their models as if they were exact.

The purpose of a model is to provide a simplified description of some essential part of the phenomenon under study, which may be manipulated mathematically to derive logical consequences. These are in turn interpreted as implications for experimental observation of the real-world system. In principle, every part of the mathematical model should be treated as being quite distinct from the corresponding real phenomenon; in practice it is common to abuse language by identifying the two, and this is harmless provided it is not done in such a way as to cause confusion. For example, even in simple physical laws such as Boyle's law ($PV = RT$ for a perfect gas), the symbols P, V, R and T in principle denote mathematical quantities thought of as corresponding to pressure, volume, the gas constant and temperature; they do not denote the actual physical quantities (which, in fact, are not related exactly by such an equation at all, but only approximately). It may seem unnecessary pedantry to insist on the distinction in such cases; but, for example, the fact that the physical V must always be positive does not carry mathematical consequences for the model V. Mathematical manipulations involving negative values of V, for example, might be highly desirable as part of some investigation of some physical system for reasons of mathematical technique. This in no way invalidates the conclusions of such reasoning, even though some of the mathematical steps will have no physical interpretation.

Biology does not yet possess a repertoire of models to the extent that physics does; nor indeed would physics if it were not able to restrict itself to artificially simple systems (such as isolated electrons, single atoms and bars of uniform material). However, a wide range of models have been used, with varying effectiveness, and these have illuminated several biological problems.

1. Propagation of Nerve Impulses

One of the earliest successes of mathematical biology was the description of the propagation of nerve impulses. The subject was first treated by Rashevsky (1933). Later, Hodgkin and Huxley (1952) performed a classic series of experiments on the giant axon of the squid *Loligo*. From their data they derived a system of empirical equations, describing the way that local membrane currents propagated along a (uniform) axon. The equations take the following general form:

$$C_m \frac{\partial v}{\partial t} = \frac{d}{4R_i} \frac{\partial^2 v}{\partial x^2} - \bar{g}_{Na} m^3 h(v - \bar{v}_{Na})$$
$$- \bar{g}_K n^4(v - \bar{v}_K) - \bar{g}_L(v - \bar{v}_L)$$
$$\partial m/\partial t = (m_\infty(v) - m)/\tau_m(v)$$
$$\partial n/\partial t = (n_\infty(v) - n)/\tau_n(v)$$
$$\partial h/\partial t = (h_\infty(v) - h)/\tau_h(v)$$

where x represents distance along the axon; t is time; $v(x, t)$ is the deviation from resting value of the membrane potential at position x and time t; $\bar{g}_{Na}$, $\bar{g}_K$, and $\bar{g}_L$ are sodium, potassium, and leakage conductances, respectively, and the $\bar{v}$'s are corresponding Nernst potentials; m, n, and h are variables introduced to model the local membrane activity, being referred to as sodium activation, potassium activation and sodium inactivation. The auxiliary functions m_∞, n_∞, h_∞, τ_m, τ_n and τ_h are defined so as to make the equations fit experimental data.

The form of the model is a system of partial differential equations. This is an extremely common way

to model dynamic phenomena, in which quantities well approximated by a continuum vary with time; it gains respectability through the widespread occurrence of such equations in the physical sciences. The equations are deliberately chosen to give a good fit to data from voltage clamp experiments. However, mathematical manipulation of them under assumptions corresponding to unclamped voltage lead to the prediction of a travelling wave solution in the form of a propagating spike, whose form and speed are in good agreement with those observed in the squid axon.

This result is not so much strong evidence for the truth of the equations as a vindication of their utility: they provide insight into the mechanism by which impulses may propagate. The problem is that many similar equations also possess travelling wave solutions. In fact, if taken purely at face value, the Hodgkin–Huxley equations are now known to be wrong: they predict incorrect behavior in prepulsing experiments. Nonetheless they provide useful insights and are widely used in theoretical investigations of neural impulses. There is an extensive survey in Rinzel (1978).

At a slightly more complex level of structure, systems of differential or integro-differential equations have been used to model the behavior of neural networks. Again, good phenomenological insights are sought, rather than quantitative agreement to large numbers of decimal places. For example, it can be seen that a network consisting only of cells of one excitatory type can act as a bistable switching device, but cannot be locally addressed. Addition of inhibitory cells permits local addressing and oscillatory behavior.

2. Reaction–Diffusion Equations

Another widely encountered biological model concerns reaction–diffusion equations, of a type used by Rashevsky (1940) and Turing (1952), as a model for pattern formation. Under suitable conditions a spatially homogeneous solution to the model equations can become unstable. It is then replaced by a solution exhibiting periodicity, say, in one dimension (stripes on a zebra?), two dimensions (spots on a leopard?), or in time (chemical waves as in the Zhabotinskii–Belousov reaction). Mathematically, such phenomena come under the general headings of bifurcation theory (at the quantitative level) and catastrophe theory (the qualitative level). The precise equations should not be taken too seriously; but the effects they predict are important for their suggestive value. Nonlinearity in the equations, in particular, leads to interesting new phenomena. As a general rule in biology, linear equations should be treated with skepticism: they are likely to arise only as simplified approximations, and linearity is in many ways too simple to be of much value. Unfortunately, it is linear equations that are most readily soluble, so this approximation is often made purely for that reason.

3. Population Dynamics

An area that has received extensive mathematical treatment is that of population dynamics. While continuum models are often used, only discrete cases are considered here. At the simplest level, a population $p(t)$ of creatures is considered as evolving, from generation to generation, by a law of the form

$$p(t + 1) = F[p(t)]$$

where F is some suitably chosen function. For example, if we allow unlimited growth by a constant multiple A:

$$p(t + 1) = Ap(t)$$

then we find that

$$p(t) = p(0) \cdot A^t$$

which gives the well-known case of exponential growth. If $A > 1$, the population increases exponentially towards infinity; if $A < 1$, the population dies out. From even as simple a model as this, it may be concluded that a population must have a sufficiently high reproductive rate to survive—a conclusion which, though trite, does show that the mathematical model is not too far adrift in some ways.

Nonlinear terms produce far more varied and interesting effects. One widely discussed example is the equation

$$p(t + 1) = k \cdot p(t)[1 - p(t)]$$

in which p is scaled to the interval $[0, 1]$ and $0 \leqslant k \leqslant 4$. The final term $1 - p(t)$ simulates a cutoff in population growth as the size increases; this may be considered as modelling under-supply of food or habitable space. The mathematics of this model are discussed in Guckenheimer et al. (1977). If $k < 1$ the population dies out. If $1 < k < 3$ it settles to a steady value, oscillating about this if $k > 2$. For k a little larger than 3 the population oscillates between two values. As k increases, this changes to oscillations through 4, 8, 16, . . . , 2^n values; and at a particular value around 3.7 apparently random variations set in. This kind of deterministic randomness is fashionably called chaos and is acquiring considerable prominence in many fields of science: it poses deep questions about determinism and predictability, which no longer seem to correspond simply to known equations.

Random effects are often incorporated into models, not as consequences of chaotic dynamics, but as model input by means of stochastic terms. For example, the motion of a bacterium through a fluid may be modelled by assuming certain random motions due to the movement of the fluid being transmitted to the bacterium.

4. Conclusion

Mathematical models have been devised in biology at all levels from the molecular to the whole organism, or even

to populations of organisms. Particular hierarchical similarities may be observed; for instance, an organism is a structured population of cells, a cell is a structured population of molecules, and a society is a structured population of organisms. These similarities are sometimes reflected in the types of models encountered.

A great variety of models may be found in Levin (1978). In its current state, mathematical biology is more a collection of specific models than an organized discipline. A certain amount of coherence is beginning to emerge, but the majority of biological works remains experimental, and this is unlikely to change without the infusion of new mathematical ideas. Biology is a challenging field to the mathematical modeller; but it is also a highly daunting one.

See also: Biostatistics

Bibliography

Guckenheimer J, Oster G, Ipaktchi A 1977 Dynamics of density-dependent population models. *Theor. Popul. Biol.* 4: 101–47
Hodgkin A L, Huxley A F 1952 A quantitative description of membrane current and its application to conduction and excitation in nerve. *J. Physiol.* (*London*) 117: 500–44
Levin S A (ed.) 1978 *Studies in Mathematical Biology, Studies in Mathematics*, Vols. 15, 16. Mathematical Association of America, Oberlin, Ohio
Rashevsky N 1933 Some physico-mathematical aspects of nerve conduction. *Physics* 4: 301–49
Rashevsky N 1940 Further contributions to the theory of cell polarity and self-regulation. *Bull. Math. Biophys.* 2: 65–67
Rinzel J 1978 Integration and propagation of neuroelectric signals. In: Levin 1978, pp. 1–66
Turing A M 1952 The chemical basis of morphogenesis. *Philos. Trans. R. Soc. London B* 237: 37–72

I. Stewart

Mechanical Devices in Medicine and Rehabilitation

The range of mechanical devices in medicine, surgery and rehabilitation is very wide. It includes surgical instruments, implants, orthotic devices, external prostheses, diagnostic and research instruments, life-support systems, and also aids to mobility, physiotherapy and the handling and rehabilitation of the physically handicapped.

1. Surgical Instruments

A large proportion of surgical instruments are of traditional form, developed over many centuries. More recent changes have arisen either from the use of newer materials or from extension of their use to new fields such as microsurgery and prosthetics. A comprehensive catalog of one manufacturer of surgical instruments and appliances edited just before the First World War contains over 10 000 entries of items, nearly all of which are mechanical. A sample of instruments listed in the catalog consisting of one rigid element only is given in Table 1. A similar sample of two-element instruments could start with scissors, forceps, dilators, shears, etc.

Table 1
A selection of surgical instruments consisting of one rigid element only

awls	files	pus seekers
bone hooks	gouges	rasps
bone levers	hammers	rasparatories
broaches	knives	retractors
burrs	lenticulars	rongeurs
catheters	nails pins	saws
chisels	needles	scalpels
curettes	nerve protectors	scoops
directors	ossicle hooks	scrapers
drainage tubes	osteotribes	speculae
elevators	perforators	vaccinators
enucleators	probes	

The development of new materials originally for use in fields other than medicine and surgery has enabled construction of new instruments and equipment. Through their application to mechanical implants, these materials created the need for new devices and at the same time made the answers to those needs more feasible. Biocompatibility, coupled with such mechanical features as strength, surface hardness, resistance to wear and corrosion, gave these materials an important place in general surgery and in orthopedic and implantation surgery in particular.

Microsurgical instruments and equipment have gained widespread use, in neurosurgery, brain and vascular surgery, and particularly in ophthalmic surgery. A partial list of newer devices in eye surgery (Table 2) illustrates these developments. Similar lists of devices can be quoted for a whole range of surgical procedures, such as those associated with the brain, larynx, thorax,

Table 2
A partial list of devices used in eye surgery

localizers for retinal detachment	corneal trephines
microsurgical needle holders	devices for obtaining
microsurgical operating tables	donor corneal disks
microsurgical reciprocating saws	intraocular forceps
pediatric intraocular knives	iris retractors
spreading forceps for drainage	lacrimal dilators
of subretinal fluid	scleral rings
vitreoretinal dissection	vitrectomy instruments
instruments	

heart, blood vessels, abdomen, bones and joints, skin, viscera, urogenital organs and with virtually any other organ and system, or with any distinct speciality such as oncology. Surgical procedures carried out on experimental animals (e.g., guinea pigs, rats, dogs) require their own sets of special instruments.

As distinct from surgical instruments many mechanical devices form a major part of surgical equipment, such as operating tables both for universal and for specific tasks, such as knee or transurethral surgery. During the pioneering of certain important surgical operations, new life-support systems have been developed. In heart surgery a heart–lung machine with its oxygenerators, peristaltic pumps and other components has been a major development (see *Heart–Lung Machines*). Similarly, the development of renal dialysis machines, which have some parts in common with heart–lung machines, has contributed greatly to therapeutic advance (see *Renal Dialysis*). Indeed, the medical profession has been greatly aided in its endeavors in recent years by physical scientists and engineers. The burgeoning use of instrumentation in medical practice has made high demands on scientific skills and expertise. The mechanical engineer has also made significant contributions, since almost all instrumentation has an element—if not a major and primarily essential part—which is mechanical.

2. Prostheses

In joint replacement, cardiovascular prosthetics and plastic surgery, some polymers such as polytetrafluoroethylene, high-density polyethylene, poly(methyl methacrylate), the biopolymers, and silicone rubber have provided mechanical solutions to many problems. Developments in hip-joint replacement, both partial and total, spearheaded an increase in prosthetic procedures in other joints such as the knee, elbow, finger, shoulder and wrist (see *Artificial Joints, Implanted*). Replacement in more complex areas such as the spine, foot and hand are nearing a stage where they could become standard procedures. Orthotics of the congenital deformities of the spine has required many new instruments and appliances to be used both internally and externally.

A more detailed discussion of prostheses may be found elsewhere (see *Implanted Prostheses: Tissue Response*; *Prostheses, Myoelectrically Controlled*).

3. Mechanical Devices in Rehabilitation

Mechanical devices in rehabilitation include devices intended to improve the mobility and manipulative power of the disabled person, and also devices which assist with his or her management. Simpler devices such as the walking stick, the crutch and some external prostheses have their origin beyond explicit historical record. Two World Wars and rapid technological development have contributed to more and improved appliances and equipment. Wealthy and social-welfare-orientated countries (e.g., Sweden) have devoted significant resources to the development of aids for the disabled. In the USA, one of the outcomes of World War II and the Korean and Vietnam wars was to initiate more provision of such devices. In the UK, the National Health Service Act of 1946, the National Assistance Act of 1948 and the Chronically Sick and Disabled Persons Act of 1970 intensified similar work.

The simpler devices are more numerous, possibly because they are more acceptable and useful. Thus sticks, crutches, walking frames, wheelchairs, commodes, nursing baths, rising seats within geriatric chairs or wheelchairs, and posture beds compete with many small and varied gadgets for domestic use, or for feeding and personal hygiene. Mobile lifts, stair-climbing machines, overhead cranes, adaptations to road vehicles, exoskeletal systems, folding-up baths and various mechanical devices for exercise, communication, leisure, teaching, entertaining and special one-off solutions which often draw a great deal of attention are very rewarding for the inventor (although not necessarily financially), and stimulate interest in the field.

In recent years, efforts to develop the walking stick have resulted in such modification as better gripping contact with the floor by using a high-friction rubber for the end tip, while spring loading is also a helpful idea. Multilegged versions for helping with maintaining balance or for stair climbing and descending, with a facility for altering the length of the stick and adaptations of the handle for specific needs arising either from hand deformity, muscular inaction, or of preferred posture of the body or direction of support have been another important development. With reference to crutches, recent research on adverse effects on the blood circulation in the axillary region has resulted in modifications involving redistribution of pressure between the axilla and the hand, or using the elbow for the entire support.

Walking frames are of at least two kinds: those with four legs rigidly connected and those where the front and rear leg on each side are rigidly connected, but can be advanced forward while the legs on the opposite side are in contact with the ground. Thus a walking motion is possible which does not require all four legs to leave the ground at the same time.

Wheelchairs and pushchairs vary in many respects. Pushchairs, as the name suggests, are pushed by a helper. Some pushchairs have small wheels, but the majority have larger wheels at the rear and small castor-type wheels at the front. Wheelchairs are either self-propelled or electrically driven. In the former type a hand-rim fixed to each large propelling wheel facilitates propulsion. For persons who have the use of only one hand, two hand-rims are fitted on the side of the good arm, a larger one connected to the wheel in the usual manner, and a smaller hand-rim to the opposite wheel by an axle.

When they are both grasped together, the chair can be propelled in a straight line. To turn the chair the hand-rims are moved independently. Most wheelchairs have removable arm rests so that the user can be transferred to another support, such as a chair, a bed or a toilet, by sliding sideways. Here, simple sliding boards are of use. Some wheelchairs permit the user to slide through the back rest, which can be open, in the front-to-rear direction—a feature which can be particularly helpful in the use of the toilet.

There are many different motorized wheelchairs powered by rechargeable batteries. In simpler versions, propulsion and steering are accomplished by a single front wheel. More often, the two rear wheels, which are of larger diameter, are propelled independently with provision for forward, reverse, or circular motion. The radius of this latter circular motion can be minimized by imparting opposite motions to the wheels. Joystick-operated potentiometers provide all possible configurations of progression. An additional control regulates the speeds. Motorized wheelchairs often include facilities for different body postures according to the need of the user.

Many difficulties concerning wheelchairs still exist, particularly with regard to their outdoor use. Rough or soft ground presents one type of difficulty and a sloping path another. The latter makes special demands on propelling power and on the efficacy of the brakes. To improve their power-to-weight ratio, some chairs are made of light materials or designed with a minimum of heavy material. However, such lightening necessitates giving up some other advantageous features partly or totally, such as low cost or comfort, or accepting a reduced level of safety.

Rising seats assist the chair user in rising from a seated to a standing posture. The majority of these devices are fitted to stationary chairs, some of which are known as geriatric chairs. Again some rising seats derive their power from an arrangement of springs which store the occupier's energy during the phase of descending into a seated posture, while others use an external power source. There are also a few wheelchairs which successfully incorporate rising seats. In one of these a combination of springs and a small, battery-driven motor makes use of the stored energy and at the same time compensates for the friction losses and provides a positive control of posture (i.e., an ability to be positively arrested at any intermediate phase of rising).

A significant contribution towards the increase of the disabled person's manipulative power is made by mobile arm supports. They are normally attached to the wheelchair and, although the weight of the arm is fully supported, sufficient freedom of motion is left to the arm in order to perform reasonably successfully important functions such as feeding. The efficacy of the device, coupled with simplicity, helps to preserve the personal dignity of the disabled person by increasing his or her independence without overburdening the user with intricacy.

Mobile hoists are indispensable for the transporting of adult patients by a single helper. A large range of hoists are available. Most types have three, small, castor-type wheels arranged so that the patient can be brought above a bed, a seat, a bath or a toilet pedestal. The power for lifting and lowering the patient is supplied usually by the helper, and the lifting mechanism is either hydraulic or simply mechanical in the form of a jack. The patient either sits on a rigid stool which may be raised or lowered, or is supported by a harness which is suspended within the hoist. Certain types of harness can be put on a patient for his subsequent lifting and transport while he is virtually undisturbed from a sitting posture.

Stationary patient-handling devices include swinging booms which are particularly useful for transfer of a patient from a wheelchair into a fixed location such as a bath or a bed, and overhead rails whose function is to secure transport within a rather limited range between two or more locations, such as bed, bath, toilet and possibly some physiotherapeutic arrangement such as parallel bars. Such rails incorporate trolleys which ride on the rails.

Transfer of a patient from a wheelchair or a mobile hoist into a road vehicle—a private car in particular—has created a need for a special type of a lift which usually consists of a boom which can be raised and lowered, and one end of which is hinged either on the roof of the vehicle or inside the vehicle just under the roof. The mobile end of the boom is equipped with a hook or equivalent and is raised again either hydraulically, mechanically, or by means of a small electric motor powered by the car batteries. In swimming baths, water energy is used in at least one design for maneuvers such as lifting to avoid any danger which could arise from the use of electrical power.

Some important disabilities, such as hemiplegia or traumatically originated tetraplegia, arise suddenly and create a need for a means of transportation for the disabled person from the ground floor to a first-floor bedroom or toilet. This need can be obviated by moving into a ground-floor bungalow or a single-floor flat, but otherwise either a lift or a stair-climbing machine is needed. One of the simpler solutions to this problem is a counter-balanced lift. It occupies very little space, consisting of a platform just large enough to take a wheelchair with its occupier and a suitable skeleton cage suspended between two tubular stanchions which house the counter-balance weights. The latter are made equal to the combined weight of the disabled person and his or her wheelchair. Friction (and the slight momentum) losses are made up either by the helper, or where possible by the passenger himself. To install this type of lift, the most major alteration to the house is provision of a rectangular opening in the bedroom floor or landing. A rectangular shallow well for the platform, or a ramp, is necessary at the ground floor. No alteration is needed as regards the bedroom or landing ceiling, because in a normal house there is sufficient room for equipment such

as pulleys above the lift when it is in its uppermost position.

Most other solutions depend on devices utilizing the existing stairs. Single-run stairs normally provide least problems. Where they consist of two or three sections, various arrangements are used. One of these is in the form of a chair moving in a helical track within a spiral staircase while it is cantilevered from the central column. Another solution consists of a small, caterpillar conveyor, which can ride on normal stairs, and a seat suspended so that it remains horizontal regardless of the inclination of the conveyor.

A large amount of effort and high-quality engineering was devoted in the 1960s and 1970s to various attempts at compensating for congenital deformities arising mostly from spina bifida, or the effects of thalidomide. Two remarkable results have been the swivel walker and the power arm. The former is an unenergized pair of mechanically connected, small horizontal platforms which are caused alternately to move forward by a transverse swaying motion of a legless ambulant. The latter consists of mechanically elaborate artificial arms incorporating several links, bearings, cylinders and pistons, valves, pulley, cables, small motors and levers, all accommodated inside the articulating space which would normally be occupied by the healthy arm, to provide motion of several degrees of freedom, closely approximating those of the normal limb. The function of the hand is replaced by that of a split hook or, in some designs, by a considerably more complex cosmetic hand with all its five digits. Cosmesis brings with it at least two problems. One of these is an occasionally unavoidable need for compromise between the cosmesis and the effectiveness. The other problem, which is more subtle and not of an engineering nature, is that some physically impaired persons prefer their impairment to be recognized at an early stage of new social encounter whereas others prefer discretion.

See also: Artificial Limbs and Locomotor Aids: Evaluation; Communication Aids for the Physically Handicapped

Bibliography

Black M M 1972 *Developments in Biomedical Engineering.* Sussex University Press, London

Boswell D M, Wingrove J M (eds.) 1974 *The Handicapped Person in the Community: A Reader and Sourcebook.* Tavistock Publications, London

Clynes M, Milsum J H 1970 *Biomedical Engineering Systems.* McGraw-Hill, New York

Copeland K (ed.) 1974 *Aids for the Severely Handicapped.* Sector, London

Goble R E A, Nichols P J R 1971 *Rehabilitation of the Severely Disabled,* Vol. 1: *Evaluation of a Disabled Living Unit.* Butterworth, London.

Kenedi R M (ed.) 1965 *Biomechanics and Related Bio-Engineering Topics.* Pergamon, Oxford

Kenedi R M, Paul J P, Hughes J 1979 *Disability.* Macmillan, London

National Fund for Research into Crippling Diseases 1966 *Equipment for the Disabled: An Index of Equipment, Aids and Ideas for the Disabled*, 2nd edn. National Fund for Research into Crippling Diseases, London
Nichols P J R 1971 *Rehabilitation of the Severely Disabled*, Vol. 2: *Management*. Butterworth, London
Oxfordshire Area Health Authority 1975–80 *Equipment for the Disabled*, 4th edn. Oxfordshire Area Health Authority, Oxford
Williams D F, Roaf R 1973 *Implants in Surgery*. Saunders, London

J. Skorecki

Medical Electrical Equipment: Safety Aspects

The increasing use of medical electrical equipment in the clinical environment, while contributing much to improved diagnosis and therapy, has exposed the patient —and to a lesser extent the operator—to new hazards.

Medical diagnostic equipment is very varied in application, ranging in function from the passive procurement of biovoltages (e.g., the electrocardiograph) to the measurement of other parameters (such as blood pressure) and to instruments such as the impedance cardiograph, which pass small electric currents through the patient in order to perform their function.

Therapy, on the other hand, usually involves the deposition of significant amounts of energy in the patient from a variety of sources such as high-frequency current generators (short-wave therapy), high-frequency acoustic sources (ultrasound therapy) and sources of electromagnetic radiation ranging from the infrared to γ rays. Alternatively, the support of a particular patient function is involved, as with renal dialysis equipment or lung ventilators.

Medical electrical equipment may introduce hazards in a variety of ways, either due to the energies delivered in normal working or when a fault occurs or as a result of functional failure of life-support equipment. The hazard can be imposed on the patient, operator or surrounding personnel but the patient is often particularly vulnerable.

Although the passage of unwanted electric currents through the body constitutes the principal hazard, a variety of other hazards can also arise such as excessive mechanical forces on the patient due to improper use of equipment, excessive temperature on accessible surfaces causing burns or startle reactions, human error resulting in a faulty sequence of operation, component breakdown in life support equipment and explosive hazards due to the use of flammable anesthetics.

This article discusses how the international community has attempted to minimize these and other hazards.

1. Safety Standards

Since the early 1960s many countries, mindful of the safety problems created by the increasing use of medical technology, have developed Standards or Safety Codes to guide equipment manufacturers in the production of safe equipment. For example, in 1963, the British Ministry of Health published a *Safety Code for Electromedical Apparatus*, Hospital Technical Memorandum No. 8, which was of some value but often lacked precision of statement.

The multiplicity of National Standards on the safety of electromedical equipment created many problems for manufacturers and, in 1969, Technical Committee TC62 of the International Electrotechnical Commission (IEC) commenced work on an International Standard on the Safety of Medical Electrical Equipment leading to the publication of IEC 601-1, *Safety of Medical Electrical Equipment*, Part 1: *General Requirements*. Several countries have now adopted the IEC Standard as a National Standard but in the UK adoption has been through the publication of BS 5724 Part 1 (1979), which is almost identical to the IEC document.

IEC 601 has been followed by Part 2: *Particular Standards*, which specifies the special safety requirements for particular types of equipment. This will eventually be followed by Part 3 *Standards* which specifies appropriate performance requirements.

2. Physiological Effects of Electric Current on the Human Body

The accumulation of data from a variety of sources has established the effect of alternating electrical currents (frequency 50–60 Hz) of various magnitudes on the average human body. The findings are summarized in Table 1.

Table 1

Effects of 60 Hz current on an average human body (one-second contact)

Current intensity (mA)	Effect
1	Threshold of perception
5	Accepted as maximum harmless current intensity
10–20	"Let-go" current before sustained muscular contraction
50	Pain. Possible fainting, exhaustion, mechanical injury; heart and respiratory functions continue
100–300	Ventricular fibrillation will start but respiratory center remains intact
6000	Sustained myocardial contraction followed by normal heart rhythm. Temporary respiratory paralysis. Burns if current density is high

Source: Bruner (1967)

Above 5 mA, many sensory nerves are stimulated and the sensation becomes increasingly painful, often forcing the subject away from the source of stimulation. At 20–100 mA the subject may be unable to release his grip on the electrical conductor because stimulation of motor nerves causes the associated muscles to contract. Although physical injury may result from the powerful contraction of skeletal muscles, the heart and respiratory function usually continue because the current is uniformly spread throughout the trunk and therefore the current density at the heart is relatively low.

Above 100 mA, life-threatening phenomena appear, the most common of which is ventricular fibrillation. In this condition the muscle fibers of the heart cease to act in a coordinated fashion, inhibiting the heart's ability to pump blood and having a fatal outcome unless corrected within minutes. Work by Raftery et al. (1975) and Watson et al. (1973) on human subjects has established a figure of 60 μA as the minimum 50 Hz current likely to cause a disturbance of heart rhythm, when passed directly through the right ventricle.

Stimulation of motor nerves by electric current steadily decreases for frequencies above 50 Hz. At 10 kHz a current of 10 mA can be passed through the body safely. On the other hand, the passage of direct currents in the microamp range through the skin can lead to tissue destruction, leaving an open sore. The effect depends on current density and duration of application and can arise in equipment such as eneuresis alarms.

3. Electric-Shock Hazards

3.1 Macroshock

With the exception of the hospital environment, most accidental contacts with the public electricity supply occur through an intact skin surface. Skin, with its relatively high electrical resistance, normally affords a measure of protection to susceptible internal organs.

Measurements of the electrical resistance of the human body made between the hands produce values varying from about 1 kΩ when the skin is damp to over 1 MΩ for dry skin. These resistance values would result in current flows of 240 mA and 240 μA, respectively, in contact with a supply voltage of 240 V ac; the former current could be fatal but the latter is entirely safe.

Current inevitably flows through a body in contact with both poles (live and neutral) of the electricity supply but a hazard can still arise even if only the live pole is contacted because most supply systems are "earth" referenced. A typical substation used to supply mains voltage electricity to a hospital in the UK is shown in Fig. 1. The primary side of the transformer is fed with three-phase current at 11 kV and this is transformed down to 415/240 V in the secondary side. It is normal practice in the UK to bond permanently the star-point or center of the three-phase secondary winding (the neutral) to earth. In these circumstances a hazardous current will flow through a subject, connected to earth by a low resistance, if contact is made with one of the "lines" (Fig. 2). The degree of hazard will depend on the voltage present, the electrical resistance of the body and on the current pathway through the body.

Although in domestic and industrial circumstances as many as 40 million people use electricity daily in the UK, the number of deaths due to electrocution has never exceeded 200 per annum. Many people survive contact with the mains because involuntary muscle contraction causes contact with the supply to be broken.

3.2 Microshock

In the hospital environment the patient is particularly at risk because of: (a) the absence of normal reactions in unconscious, anesthetized or immobilized patients, (b) the absence of the relatively high impedance of normal skin which may have been penetrated or treated to obtain low skin resistance, (c) direct application of electric current either to the skin or by insertion of probes into internal organs such as the heart, and (d)

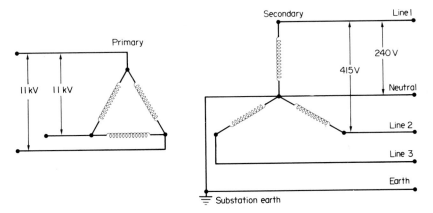

Figure 1
A typical hospital substation in the UK

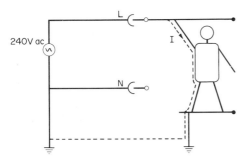

Figure 2
Path of current if contact with one of the lines is
made

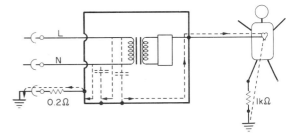

Figure 3
Leakage current paths

simultaneous connection of more than one item of mains
powered equipment.

In a hospital two types of electrical hazard can arise.
The first of these is obvious and no different to the
domestic situation and is caused by faults, such as frayed
power cords, broken plugs, and wrongly wired sockets,
which allow personal contact with the live wire. Such
faults are equally dangerous to staff and patients.

The second type of hazard, often termed microshock,
is more subtle but just as dangerous to a catheterized
patient as contact with a live wire. These hazards are
caused by "leakage currents," which are nonfunctional
currents flowing to earth through or across insulators.
The word leakage is perhaps unfortunate since it implies
the presence of a fault, but in fact leakage currents exist,
to a greater or lesser extent, in all electrically powered
equipment. In general, leakage currents in mains-
powered equipment contain two components, one in-
phase and the other out-of-phase with the supply voltage. The in-phase or resistive component arises from
current flow through and across imperfect insulation but
is normally insignificant in modern equipment. The
dominant out-of-phase or capacitative current flows in
the capacitance which exists between any two conductors in close proximity. Leakage capacitance to earth
arises principally between the primary winding and the
core of the mains transformer, between the line and
earth conductors of the mains supply cord and in any rf
filter which is present. In a typical instrument with an
earthed conductive enclosure (electrical safety Class I,
see Sect. 5) the leakage currents—called earth leakage
currents—flow to earth through the protective earth
conductor in the mains supply cord.

Consider a medical instrument of Class I construction
connected to electrodes inserted in a patient's heart, one
of which is connected to the ground (Fig. 3). Assume
that the patient presents an alternative path to ground
with resistance $1 k\Omega$ and that the resistance of the
protective earth conductor in the mains supply cord is
0.2Ω. If the earth leakage current of the instrument is
$200 \mu A$ then only $0.04 \mu A$ flows through the patient's
heart, as long as the integrity of the protective earth
system is maintained. If, however, the protective earth is

broken, then all the leakage current flows to the patient
and could cause ventricular fibrillation.

Figure 4 illustrates how a leakage current of $500 \mu A$
can cause a hazard to a patient when another person is
involved. The patient has a transvenous pacing catheter,
connected to a battery-operated external pacemaker,
inserted in his heart and is also connected to earth by
way of the right leg lead of an ECG monitor. The
bedside lamp—earth leakage current $500 \mu A$—has a
disconnected protective earth conductor. If a nurse
makes simultaneous contact with one of the pacemaker
connections and with the metal body of the lamp, the
leakage current from the lamp will be diverted through
the patient's heart. The current will be undiminished in
strength provided that the resistance of the path through
the nurse, catheter and right-leg lead is negligible compared with the $0.5 M\Omega$ leakage impedance of the lamp.
The nurse will be unaware of the hazard since a current
of $500 \mu A$ cannot be sensed through intact skin.

Many other hazardous situations can be identified.
For example, when several instruments are attached to
a patient, the breakdown of insulation between the
mains and the patient circuit of one instrument can cause
large currents to flow through the patient to the protective earth systems of the others. Alternatively, small but
hazardous potential differences can develop between the
protective earths of two patient-connected instruments
which are supplied from different mains outlets when a
large fault current flows in the hospital protective earth
system.

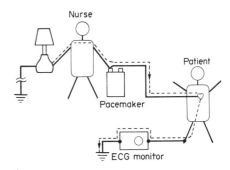

Figure 4
Hazardous leakage current path through a patient's
heart

4. Improved Safety and the New IEC Standard

The concept of safety is a relative one; absolute safety, however assiduously pursued, can never be achieved. The philosophy of the IEC on the safe use of medical equipment in medical practice is discussed in Publication 513 (1976). Several categories of safety are recognized and can be provided by the means listed below in descending order of desirability:

(a) by appropriate equipment design (unconditional safety),

(b) as additional external protective measures (conditional safety), and

(c) by the definition of conditions under which safe operation is possible (descriptive safety).

The conscientious application of IEC 601-1 to the design and construction of medical electrical equipment guarantees that a high degree of safety will be incorporated. The basic philosophy of the Standard is that the failure of a single means of protection or the appearance of a single abnormal condition shall not result in a safety hazard to either patient or operator. Inherent in this approach is the requirement for a routine test and maintenance program throughout the life of the equipment, aimed at the detection and correction of single-fault conditions before the occurrence of a second fault. In addition to the recommendations on electrical safety, other sections of the Standard deal with protection against mechanical, radiation and explosion hazards, protection against excessive temperatures, and advice on constructional requirements.

Although many of the requirements can be found in earlier documents, such as Hospital Technical Memorandum No. 8, some of the material is new and each requirement is more precisely stated and accompanied by a test for compliance. These tests, while adding to the complexity of the Standard, are vital in providing clear recommendations not subject to interpretation in different countries. Electrical safety is achieved by reliance on protective earthing, insulation, and by the limitation of leakage currents; these parameters form the basis of most of the safety tests.

5. Equipment Classification

Part 1 of IEC 601 classifies equipment in several different ways; for example, according to the type and the degree of protection against electric shock, the extent of the protection against ingress of liquids and the degree of electrical connection with the patient. Part 2 Standards recommend which classification and methods of protection are appropriate for particular equipment.

The General Standard recognizes three classes of externally powered equipment in which different means of protection are employed to guard against the possibility of electric shock caused by the failure of basic insulation.

(a) *Class I construction.* In this class the additional protection is achieved by connecting accessible conductive parts to the protective earth conductor. Therefore such parts cannot become live in the event of a failure of basic insulation. Although it is usual for Class I equipment to have a totally conductive enclosure, partially conductive, or even totally insulating enclosures are possible.

(b) *Class II construction.* In Class II equipment the additional protection is provided ideally by a layer of insulation supplementary to, but independent from, the basic insulation or, if this is impracticable, by using a single insulation system with an equivalent degree of protection. These insulation systems are called double and reinforced insulation, respectively.

Class II equipment must not have a protective earth connection although a functional earth is permitted when necessary; for example, to limit electrical interference. Although Class II construction has been successfully applied to domestic appliances, its use is not always practicable in medical equipment because of the difficulty in obtaining low leakage currents in the normal condition.

(c) *Class III construction.* In this class, protection relies on the provision of an isolated supply not exceeding 24 V ac or 50 V dc (medical safety extra low voltage). Requirements for earthing and enclosures are similar to those for Class II. However, since the safety of Class III equipment depends on factors beyond the control of the user, no Part 2 Standard has yet allowed for the use of this class of construction and it may eventually be deleted from the General Standard.

The Standard recognizes equipment incorporating, for example, a rechargable battery as being internally powered only if either no external connection to the power supply is possible or, alternatively, if such a connection can be made only after the internal power source has been physically and electrically separated from the remainder of the equipment. If these conditions cannot be met then the equipment must be classified as Class I, Class II or Class III.

6. Leakage Current Definition

The Standard defines three types of leakage current:

(a) Earth leakage current which is, as we have already seen, the current flowing from the mains through and across insulation to the protective earth conductor. This is, of course, relevant only to Class I construction.

(b) Enclosure leakage current, which can flow from accessible parts of the enclosure through a fortuitous external conductive connection, such as the equipment user, to earth. In the case of Class I equipment, the enclosure leakage current is negligible in normal circumstances but becomes equal to the earth leakage current when the protective earth conductor is broken.

(c) Patient leakage current which flows from the mains through insulation, the patient leads (called applied parts in the Standard) and the patient to earth. Alternatively, in the case of equipment with applied parts isolated from earth, patient leakage current can also flow in the reverse direction from a patient, inadvertently in contact with another source of mains voltage, through the patient leads and through the isolation impedance to earth.

These leakage currents are illustrated diagrammatically in Figs. 5 and 6. It can be seen that, in the case of the Class I equipment depicted in Fig. 5, enclosure leakage current is negligible except if the protective earth conductor is accidentally disconnected.

7. Degrees of Protection Against Electric Shock

Three equipment types with different levels of protection against electric shock are defined. These are, in increasing order of safety:

(a) Type B equipment, which has an adequate degree of protection regarding allowable leakage currents and

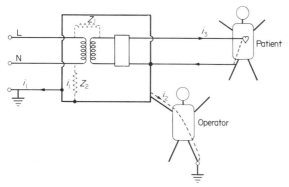

Figure 5
Leakage current paths, nonisolated applied parts. Z_1 is the leakage impedance between mains and patient, Z_2 is the earth leakage impedance, i_1 is the earth leakage current, i_2 is the enclosure leakage current, and i_3 is the patient leakage current

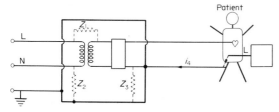

Figure 6
Alternative patient leakage current path, isolated applied parts. Z_3 is the patient isolation impedance and i_4 is the patient leakage current

reliability of any protective earth connection. This type is appropriate for all general medical applications excluding direct cardiac application.

(b) Type BF equipment which is similar to Type B but with the added safety factor of an isolated patient circuit, which limits patient leakage current, with mains voltage on the patient leads, to a maximum of 5 mA.

(c) Type CF equipment which has lower allowable leakage currents than Type B equipment and the isolation impedance of the patient circuit is one hundred times greater than that of Type BF equipment. It is intended for direct cardiac application.

IEC 601-1 tabulates the maximum permissible earth, enclosure and patient leakage currents under normal conditions and a number of specified single fault conditions such as disconnection of the protective earth conductor and the interruption of one supply conductor at a time, for the three types of equipment. The allowable values of enclosure and patient leakage currents in Types B and BF equipment are 0.1 mA in the normal condition, and 0.5 mA in the single-fault condition. For Type CF equipment the value, in the normal condition, is reduced to 0.01 mA for both enclosure and patient leakage currents and the value, in the single-fault condition, to 0.05 mA for patient leakage current and to 0.5 mA for enclosure leakage current.

8. Compliance Tests

Detailed recommendations are given in the Standard on the tests required to assess compliance with its provisions. For example, in the case of leakage current measurements, the Standard specifies the measuring supply circuit and its mode of connection to the equipment under test, the measuring arrangement, and the measuring device, the characteristics of which are designed so that its frequency response takes account of the reduced excitability of tissues at elevated frequencies.

Other compliance tests included measurement of the impedance of any protective earth conductor, using a current of not less than 10 A and not exceeding 25 A, and 10–16 dielectric strength tests of insulation, the higher number for equipment having applied parts. It must be emphasized that all these are type tests to be performed on a single representative sample of a particular instrument. Some of these tests, in particular the dielectric strength tests, are potentially destructive and must not be performed on equipment intended for subsequent use.

The Standard also gives recommendations on identification, markings and documents, and on the constructional requirements. Compliance with some of these recommendations can be checked by simple inspection.

9. Particular Standards

A series of Part 2 Standards of IEC 601 will eventually be produced to cover the additional safety requirements for individual equipment categories. A Particular Standard, the requirements of which take priority over the requirements of the General Standard, may modify or delete causes of the General Standard or insert new clauses.

One of the early Particular Standards issued by the IEC related to high-frequency surgical equipment, no doubt because of the special safety problems associated with this type of equipment, used for the cutting or coagulation of biological tissue during the performance of surgical operations (see *Electrosurgery*). Large high-frequency leakage currents flow readily from parts of the patient that make accidental contact with ground, or that have a high leakage capacitance to ground if the normal ground return path to the generator is broken. Since the current density at these sites is usually much higher than at the ground return electrode, with its large surface area, burning of the patient can occur.

Among the extra provisions of the Part 2 Standard (Particular requirements for the safety of high-frequency surgical equipment), are that the output circuit should be isolated at low frequency (Types BF or CF only) and optionally at high frequencies. A limit of 150 mA is specified for the high-frequency leakage current from a ground-referenced return electrode, or from both active and return electrodes if the output circuit is designed to be isolated at high frequency. Because of the environment in which the equipment is used the Standard requires that the generator is protected against ingress of liquids and that any foot switch is watertight.

10. Conclusion

Referring back to Fig. 3, IEC 601 requires that the equipment in the illustration, with its direct connection to the heart, should be type CF. Because of the isolation of the patient circuit from ground and the equipment enclosure, no enclosure leakage current will flow through the patient's heart on disconnection of the protective earth conductor, although a small increase in patient leakage current might occur. The situation shown in Fig. 4 could not arise with equipment designed to IEC 601—it would not be possible for the nurse to make direct electrical contact with the pacemaker circuit and neither would the patient's leg be grounded by the monitor.

To conclude, the continuing safety of medical electrical equipment throughout its working life relies on:

(a) a design which complies with IEC 601-1 and with the appropriate Particular Standard, if available;

(b) confirmation of compliance with IEC 601-1 by type testing, preferably by a recognized test house; and

(c) an adequate maintenance and safety checking program throughout the working life.

See also: Static Electricity in Hospitals; Defibrillators

Bibliography

Bruner J M R 1967 Hazards of electrical apparatus. *Anesthesiology* 28: 396–425

Department of Health and Social Security 1963 (revised 1969) *Safety Code for Electro-Medical Apparatus*, Hospital Technical Memorandum No. 8. Department of Health and Social Security, London

International Electrotechnical Commission 1976 *Basic Aspects of the Safety Philosophy of Electrical Equipment Used in Medical Practice*, Publication 513. International Electrotechnical Commission, Geneva

International Electrotechnical Commission 1979 *Safety of Medical Electrical Equipment*, Part 1: *General Requirements*, IEC 601-1. International Electrotechnical Commission, Geneva

International Electrotechnical Commission 1982 *Safety of Medical Electrical Equipment*, Part 2: *Particular Requirements for the Safety of High Frequency Surgical Equipment*, IEC 601-2. International Electrotechnical Commission, Geneva

Raftery E B, Green H L, Yacoub M H 1975 Disturbances of heart rhythm produced by 50 Hz leakage currents in human subjects. *Cardiovasc. Res.* 9: 263

Watson A B, Wright J S, Loughman J 1973 Electrical thresholds for ventricular fibrillation in man. *Med. J. Aust.* 1: 1179

Webster J G, Cook A M (eds.) 1979 *Clinical Engineering: Principles and Practice*. Prentice Hall, Englewood Cliffs

D. J. Mackinnon

Medical Gases: Measurement and Analysis

The choice of method for analyzing a particular gas is determined by the nature of any other gases which may be present. Nonspecific methods of analysis, which are based on the measurement of the velocity of light or sound in gases or the change in the thermal conductivity of a gas mixture, are restricted to the measurement of the partial pressure of a pure gas or the relative concentrations of the gases in a two-component mixture. For multicomponent gas mixtures a method of measurement specific to the gas of interest or a method capable of measuring each component individually must be chosen.

Having decided on possible methods, one must consider the sample volume of gas required. To a great extent, the volume required is determined by the sensitivity needed for the measurement, trace level measurements requiring large sample volumes. The sample volume in turn determines the method of sampling, and also to some extent the response time of the measurement if continuous measurement is undertaken.

The sample must be stored if immediate analysis is not possible. Both rigid and nonrigid gas sample storage containers are available. Leakage of gas from the storage container and the adsorption of components of the sample on the container surface are the two most important adverse factors affecting the accuracy of subsequent analysis.

The calibration of a gas-analysis system is achieved by using the system to measure gas mixtures of known concentrations. These mixtures may be stored in gas cylinders and used repeatedly over a period of time. The adsorption of mixture components of low concentration on the wall of the storage cylinder can be reduced by storing the calibration gas mixture at a low pressure. An alternative approach is to use a dynamic calibration system in which the mixture is prepared continuously from pure components. This latter method gives greater accuracy.

1. Refractometry (Interferometry)

The measurement of the refractive index of a two-component gas mixture, the components of which are known, is one of the simplest nonspecific methods of gas analysis. The technique depends upon the interference of two light beams from the same light source, one beam passing through a sample chamber and the other through a reference chamber. The two light beams are recombined optically and viewed through an eyepiece as a set of alternating light and dark bands. If the refractive index of the gas in the sample chamber alters (by alteration of the proportions in a gas mixture, for example), the effective optical path length for the light passing through the sample chamber alters and the light and dark bands viewed through the eyepiece are seen to move. A scale provided in the eyepiece allows the measurement of the change in refractive index and hence the concentration of the gas. Alternatively, the scale may be directly calibrated for a particular two-component gas mixture. Measurement accuracy is of the order of a few percent.

2. Thermal Conductivity Measurement

Hydrogen and helium have relatively high thermal conductivities. Thus their concentrations are readily obtained by measuring the temperature-dependent resistance of a heated wire filament enclosed with the gases in a sample chamber.

3. Paramagnetic Oxygen Analyzer

Oxygen is strongly paramagnetic (that is, it is attracted into a magnetic field) and has a magnetic susceptibility several hundred times that of other medical gases. This property is used in the paramagnetic oxygen analyzer.

In a typical paramagnetic oxygen analyzer a non-uniform magnetic field is created in a sample chamber by a C-shaped permanent magnet with wedge-shaped pole pieces (Fig. 1). Between the pole pieces a platinum–iridium suspension wire carries a dumbbell with a small mirror at its center. Oxygen in the sample cell is attracted into the magnetic field. The movement of the gas tends to force the dumbbell out of the field, twisting it on its suspension. The movement of the

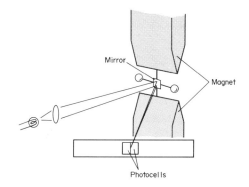

Figure 1
Paramagnetic oxygen analyzer. A nonuniform magnetic field is created by wedge-shaped pole pieces between which a dumbbell with mirror is suspended on a wire. The magnitude of a null-restoring current (see text) in a coil around the dumbbell is a measure of oxygen concentration

dumbbell is picked up by a light beam reflected from the mirror at its center, the light beam normally being centered on the division between two photocells. The changing light intensity on the photocells is used to generate a current which flows through a coil wound around the dumbbell and which holds the dumbbell in its normal position. The magnitude of the current required to do this is proportional to the oxygen concentration in the sample cell. An accuracy of 0.1% oxygen is obtainable with this null method of measurement.

An alternative form of paramagnetic analyzer has a sample chamber in which there is a heated platinum wire in a magnetic field. Oxygen present in the sample chamber is attracted into the magnetic field and is heated by the platinum wire. As paramagnetism decreases with increasing temperature, a flow of oxygen is generated over the platinum wire. The cooling of the platinum wire is a measure of the oxygen concentration.

The response time of paramagnetic analyzers is rather slow—usually a few seconds.

4. Zirconium Oxide Cell Oxygen Analyzer

The zirconium oxide cell oxygen analyzer contains a pencil-sized, closed-end, hollow tube exposed to the atmosphere or reference gas on one side and the sample gas on the other. The tube itself is a calcium-stabilized zirconium oxide cell with electrical connections on inside and outside surfaces. The operating temperature of the cell is about 850 °C.

The side of the cell exposed to the higher oxygen pressure becomes the anode at which oxygen molecules are ionized according to the reaction:

$$O_2 + 4e^- \rightarrow 2O^{2-}$$

and transported through the cell to the cathode where the reverse reaction:

$$2O^{2-} \rightarrow O_2 + 4e^-$$

takes place. The emf across the cell is proportional to the logarithm of the ratio of the partial pressure of oxygen on each side of the cell. A fast response time of a few milliseconds is possible, with an accuracy of about 1%.

At the temperature at which the cell operates, any inflammable gases will be oxidized resulting in an erroneously low measure of oxygen concentration.

5. Micro-Fuel Cell

The micro-fuel cell is also used for oxygen analysis. It has a gold-plated cathode and a lead anode in a potassium hydroxide electrolyte. The cathode is mounted behind a thin Teflon membrane through which oxygen in a gas sample diffuses to be reduced according to the reaction:

$$4e^- + O_2 + 2H_2O \rightarrow 4OH^-$$

The hydroxyl ions formed at the cathode diffuse to the lead anode which is oxidized according to the reaction:

$$Pb + 2OH^- \rightarrow PbO + H_2O + 2e^-$$

The current generated by the cell is proportional to the oxygen concentration. The cell has a positive temperature coefficient, the current increasing by 2.5% K^{-1}. Compensation is usually provided by a thermistor, giving an accuracy of about 5%. This figure can, of course, be improved with better methods of temperature compensation.

The cell has a limited lifetime (typically, 12 months) because of the gradual oxidation of the lead anode. The typical response time of the cell is 90% of oxygen concentration in about 30 s.

6. Polarographic Oxygen Analyzer

The polarographic cell consists usually of a gold cathode and a silver anode immersed in an electrolyte and protected by a membrane through which oxygen diffuses into the cell. A voltage of 0.6 V is applied across the cell. The oxygen present at the cathode is reduced according to the reaction:

$$O_2 + 2H_2O + 4e^- \rightarrow 4OH^-$$

At the anode, the reverse reaction occurs:

$$4OH^- \rightarrow O_2 + 2H_2O + 4e^-$$

The current flowing through the cell is proportional to the oxygen partial pressure. The polarographic cell is smaller than the fuel cell and has a faster response time (97% in 12 s).

7. Infrared Spectroscopy

As a general rule, gases with two or more different atoms per molecule (e.g., carbon dioxide, carbon monoxide), absorb infrared radiation at a number of different wavelengths characteristic of the particular gas (see *Spectroscopy*). Thus the infrared absorption spectrum of a gas mixture can be used for its analysis. Alternatively, by selecting a particular wavelength and measuring its intensity, the concentration of one component of a gas mixture may be analyzed continuously.

In a typical infrared analyzer (see Fig. 2), infrared radiation generated by a heated element passes through an interference filter, which selects a particular wavelength, and then through a sample chamber before impinging on a semiconductor detector. It is customary to make use of a reference beam of radiation to compensate for changes in the output of the infrared source and for differences in the transmission characteristics of the chamber windows at different wavelengths. Chamber windows and lenses are necessarily made of infrared-transparent materials (e.g., sodium chloride).

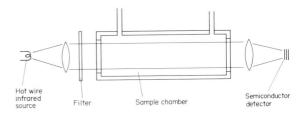

Figure 2
A single-beam infrared gas analyzer. Lenses and chamber windows are made of an infrared transparent material such as a sodium halide. A filter selects infrared radiation of the wavelength needed

Absorption wavelengths tend to be characteristic of particular molecular bonds and hence related compounds can exhibit similarities in their spectra. The wavelengths used for analysis must therefore be chosen carefully with regard to the components in the mixture being analyzed.

Response time of an infrared analyzer can be about 100 ms for concentrations of gases of a few percent and therefore such analyzers may be used for some respiratory measurements involving, for example, gases such as carbon dioxide, nitrous oxide and the anesthetic agent, halothane. Infrared analyzers are suitable also for trace level measurements but the larger sample chamber necessary for such analyses increases the response time.

See also: Anesthesia Physics; Respiratory Function: Physiology

Bibliography

Hill D W, Dolan A M 1982 *Intensive Care Instrumentation.* Academic Press, London

Parbrook G D, Davis P D, Parbrook E O 1982 *Basic Physics and Measurement in Anaesthesia.* Heinemann, London

Sykes M K, Vickers M D, Hull C J 1981 *Principles of Clinical Measurement.* Blackwell Scientific, Oxford

P. D. Davis

Medical Photography

Photography was first applied to medicine within months of the publication of the daguerreotype process in 1840, and when Albert Londe published *La Photographie Moderne* in 1888, a chapter was actually devoted to medical photography. W H Berend first used photography for serial medical records in 1852 when he took photographs of orthopedic patients before and after treatment, and in 1864 A B Squire published hand-colored photographs of dermatological conditions. From the 1870s onwards much progress was made in medical photography, and in 1893 the first textbook on the subject was published—*La Photographie Medicale,* by Londe.

Specialized techniques were devised, such as endoscopic photography—good photographs of the larynx, for example, were taken as early as 1862. However, at this stage, medical photography was still in the hands of the local photographer who was commissioned by the physician. Indeed, some doctors were keen amateur photographers and used their hobby to record their most interesting cases. Professional medical illustration first became available in the period immediately after World War II, when a number of hospitals tentatively pioneered the setting up of actual departments of medical photography.

Broadly speaking there are three main areas of work: clinical records, research, and teaching and publication. The first two areas are discussed here.

1. Clinical Photography

Photographs are taken of patients for various reasons, most commonly perhaps for making serial records of a patient's progress, e.g., pre- and postoperatively, or pre- and post-therapy. Many diseases have outward appearances which are best photographed rather than written about, e.g., the whole range of dermatological conditions. Such photographs, however, are only of use to doctors if they highlight the essential features of a case: for example, is a lesion raised or flat? Does it have a square or rolled edge? Is it erythematous? Is it hemorrhagic? Is it dry and scaly? When looking at photographs of a patient taken over a period of time the differences that are observable should result only from changes in the patient's condition. For this reason, general record photographs have to be standardized with reference to

the scale of reproduction, the angle of view, the lighting direction, balance and contrast, and the color or tone reproduction. For practical details of how this is achieved the reader is referred to texts by Hansell (1979) and Williams (1984).

Clinical photographs may be taken for diagnostic purposes. Infrared photography, which utilizes the ability of infrared radiation to penetrate the skin, is used to show the superficial vascular system (e.g., in the detection of breast tumors and in studies of thrombosis and venous stasis). Ultraviolet photographs may be taken (for inclusion in patient records) to delineate areas of depigmentation which are otherwise invisible—vitiligo, for example. Ultraviolet fluorescence photographs are taken of dye infiltration into the retinal vessels, using a retinal camera.

Photographic methods of measurement may be used in the clinic to assess a patient's progress. Special methods like double exposure are employed to measure the range of joint movements, e.g., the objective assessment of knee movement in progressive rheumatoid arthritis. Monophotogrammetry (measurement from accurately scaled, single photographs) is used to measure the degree of proptosis in exophthalmos, and stereophotogrammetry (measurement in three dimensions from stereo pairs) is used to measure facial growth and change during orthodontic treatment (see *Photogrammetry*). Special photographic apparatus may have to be developed; for example, to photograph the pressure areas of the foot in the standing and walking positions.

2. Research

Photography in medical research often involves the use of many specialized photographic techniques. On the other hand, the camera may be used merely as a recording instrument—to photograph dials, cathode-ray-tube waveforms, computer readouts, or fluoroscopic or thermographic images. Often a permanent camera with an instant-picture facility is fitted by the photographer to recording apparatus.

The following sections describe some of the more interesting specialized techniques.

2.1 Photomicrography

In photomicrography, a camera is attached to a compound microscope—practically any compound microscope can be used but usually purpose-built photomicrographic equipment is utilized. Rigidity, optical excellence and perfect lighting are all fundamental requirements for the equipment. In photomicrography, resolving power, determined by the numerical aperture (NA) of the objective lens, is a more important factor than magnification. For satisfactory results special attention must be given to a number of factors.

(a) *Optics.* Only the finest, most corrected, objectives should be used—planapochromatic objectives are recommended.

(b) *Illumination.* Good photomicrography is dependent upon good illumination of the subject that is being photographed. Thus every component of the system—from the light source to the film plane—must be correctly adjusted. Misalignment of, say, the condenser will seriously degrade the image, although this may not be apparent, visually. The Köhler system of illumination is most often employed. In this system the field condenser lens is used to focus an image of the lamp filament onto the substage condenser, which in turn focuses an image of the lamp condenser in the plane of the specimen. This gives even illumination and high resolution, even with nonuniform light sources.

(c) *Quality of the specimen.* Most histological materials examined are prepared by sectioning them into very thin slices, which are stained and mounted in a mounting media between a glass slide and a glass coverslip. The thickness of the section, the density of staining, and the thickness of the coverslip are important factors in relation to image quality. In general, a "good" section is one which is very thin (1 or 2 μm), well stained (or heavily stained for low-power work), and mounted in a clear mounting medium of high refractive index on a thin slide; a high quality, thin glass coverslide is used. A "bad" section is one which is thick (5 μm), understained for lower powers of magnification or overstained for high powers, its thickness or the staining is uneven, the microslides or cover glasses are thick (or worse still, made of plastic) and the section is mounted in a medium with a color cast.

There is a wide range of specialized photomicrographic techniques; for example, phase contrast, epillumination, dark-ground, interference, and polarization photomicrography. The reader is referred to texts by Lawson (1972) and Loveland (1970) for further reading.

2.2 Infrared Photography

The range of infrared radiation applicable to photography extends from a wavelength of 700 nm, the end of the red part of the visible spectrum, to a wavelength of 900 nm, this latter limit being set by the photographic emulsion sensitivity. Infrared images may also be visualized on cathode-ray tube screens (see *Thermography*).

Infrared radiation has two advantageous characteristics in respect of its use in medical photography. First, it can penetrate the superficial layers of the epidermis and reveal structures underneath. Secondly, it is possible to distinguish between veins and arteries because venous blood absorbs infrared radiation and well-oxygenated arterial blood reflects it. Thus vascular disorders, such as varicose veins or venous obstruction, may be revealed. Penetration below disturbances of the skin surface (below a scab, for example) makes it possible to assess the healing process. Whether pigmentation is superficial or deep can be determined. Infrared photography is used to examine the vascular patterns of the female breast, during pregnancy or when neoplastic conditions exist.

Infrared radiation can penetrate an opaque cornea which is obscuring the pupil of the eye. Thus, the shape and position of the pupil so obscured can be determined by means of infrared photography.

Infrared photography can be useful for recording morbid specimens. An example is the increased amount of detail shown by infrared photography in a specimen of a lung with heavy dust deposits. Injection techniques (mercuric sulfide into arteries and suspended carbon into veins) can be used to record arteries and veins as white (reflecting infrared) or black (absorbing infrared). Although tungsten lamps, such as photoflood and quartz–iodine (tungsten–halogen) lamps have a high infrared emission, the excessive heat they produce weighs against their use. Electronic flash tubes, however, emit infrared in sufficient quantity for clinical photographs to be taken with the camera lens stopped down to small apertures. These latter sources are therefore the usual choice for this type of photography. Note that variations in the absorption characteristics of tissues cause image density differences in infrared photography. Note also that shadows cast by lighting are a source of confusion. The subject must therefore be evenly illuminated, using several sources of radiation. Some workers recommend the use of a white room or tent to diffuse the radiation.

For monochrome infrared photography, the use of an infrared transmission filter that is opaque to visible light (Wratten 88A) is essential. It is placed in a light-tight mount over the camera lens. Black-and-white infrared-sensitive emulsions are available in sheet and miniature (35 mm) film sizes. Color infrared film requires a special filter (Wratten 12), and interprets infrared as pale green. The shelf life of infrared emulsions, both black-and-white and color, is limited, and care must be taken to ensure that only fresh stock is used. Standard photographic lenses are suitable for infrared photography, but the infrared image is displaced beyond the visible focus. Gibson (1978) gives a detailed description of the principles and practice of infrared photography.

2.3 Ultraviolet Photography

Ultraviolet radiation wavelengths extend from 14 nm to 400 nm. Using conventional glass lenses, photography is restricted to the 320–400 nm region. The use of fluorite or quartz lenses extends the range down to 150 nm. Below 250 nm, a special low-gelatin emulsion must be used because conventional emulsions absorb the shorter wavelengths.

There are two distinct forms of ultraviolet photography: reflected or direct ultraviolet photography, and ultraviolet fluorescence photography. In reflected ultraviolet photography, the exposure is made by uv radiation only. In uv fluorescence photography, the exposure is made by the visible radiation which is produced when the subject is irradiated with ultraviolet radiation, mainly in the 350–400 nm band of wavelengths.

Reflected ultraviolet photography may be usefully employed in certain applications for the following reasons:

(a) It shows slight changes in skin pigment more clearly than conventional methods. The extent of both hypo- and hyperpigmentary conditions (e.g., vitiligo and pigmented nevi) is clearly delineated.

(b) High-resolution photographs of skin and other surfaces are produced.

(c) Some tissues have dissimilar reflection and absorption characteristics for uv.

(d) Surface blood vessels on the sclera and conjunctiva of the eye and on some visceral surfaces of internal organs can be mapped with exceptional clarity.

(e) Chromatograms which contain ultraviolet-absorbing chemicals can be photographed.

(f) In photomicrography, resolving power is increased using transmitted ultraviolet radiation.

Ultraviolet fluorescence photography may be used for the following applications:

(a) The spores of Tinea capitis (*Microsporum canis* infection) and Erythrasma (*Microsporum minutissimum*), for example, can be made to fluoresce when irradiated by ultraviolet. Such disorders can thus be diagnosed photographically.

(b) Some chemicals in chromatograms may fluoresce, so demonstrating their presence and situation. In starch gels the chromatogram deposits may show up against the overall fluorescence of the starch base itself, a fluorescence which is stimulated by wavelengths below the 350–400 nm band.

(c) Injection of fluorescent dyes (fluorescein) can be used to delineate, for example, ischemic tissue, or blood vessels of the retina, or to outline parts of the body served by certain vascular pathways. Fluorescein drops may also be used to assess the fit of contact lenses.

(d) Fluorescence microscopy (a well-established technique).

(e) Many normal and pathological tissues may fluoresce and thus reveal their characteristics and contour.

Two types of light source can be used conveniently for ultraviolet fluorescence photography: continuously running sources such as low, medium or high-pressure mercury-vapor arc lamps, and discharge lamps, e.g., an electronic flash. Ultraviolet light, particularly the 280–310 nm wavelength band, can cause severe burning and acute conjunctivitis (see *Ultraviolet Radiation: Potential Hazards*). The use of short-duration electronic-flash, ultraviolet light sources is therefore highly recommended.

In reflected ultraviolet photography it is essential that only ultraviolet radiation reaches the film. The simplest way to ensure this is to fit an optically flat ultraviolet transmission filter (Chance OX1) over the lens in a light-tight mount. Any visible radiation emitted by the ultraviolet source can thus be ignored because it is absorbed by the filter. The photographer's routine task of focusing the subject prior to taking a photograph must be accomplished before the filter is attached. If a hand-held single-lens reflex camera is used, some form of flap filter mechanism must be devised so that the filter is rapidly positioned after focusing and prior to the photograph being taken.

Another method of filtration is the use of an ultraviolet transmission filter, perhaps made of rolled glass, that is fitted to the source. The photograph is taken in a darkened room so that the film is not affected by visible light.

Close-up photography (larger than half size) in ultraviolet needs an adjustment of the image conjugate to account for "focus shift." Focus shift is the difference between the visible focus and the ultraviolet focus. For some lenses the difference is such that realignment of the focal plane is required.

In ultraviolet fluorescence photography, the first step is to ensure that only ultraviolet radiation reaches the subject; thus an ultraviolet transmission filter (Chance OX1) is fitted over the light source. It should be noted at this point that the subject is (a) reflecting uv radiation, (b) reflecting any ambient or focusing lights and (c) emitting its own fluorescence. The next step, therefore, is to fit an ultraviolet absorbing filter (Wratten 2B) over the camera lens to eliminate the reflected uv radiation. The focusing light is then turned off prior to exposure, eliminating any ambient or focusing lights. The film emulsion can now only be affected by the fluorescence (c). In practice, some of the area surrounding the subject may show even though it does not fluoresce. Should the filtration be perfect, however, there may be difficulty in orientating some subjects in the absence of visible light. This of course can be overcome by providing a low level of overall illumination, but it must be carefully balanced in order to avoid swamping the fluorescent areas (Ruddick 1979, Tredinnick 1961).

2.4 Time-Lapse Photography

Events which happen too slowly to appreciate (e.g., the growth of bacterial cultures, the eruption of teeth, the growth of finger nails, and the movements of neonates or of adults during sleep) have all been studied by condensing the time scale of the events using time-lapse photography (Babcock 1954, Burke 1958, Hansell 1962, Tyson et al. 1981).

An essentially cinematographic technique is employed. Still frames are taken at relatively long intervals, then reprojected at the normal speed of 24 frames per second. Intervals may vary between two frames per second and one frame per week, depending upon the duration of the event. The framing rate is calculated by dividing the desired running time by the actual event time and multiplying by 24. Some cine cameras may be run continuously at speeds as low as 2 frames per second, and some provide single-shot facilities. Manual operation is practical for short periods but automatic control, using a separate intervalometer, is more usual. The intervalometer may control exposure duration, exposure frequency, and the switching of lights. Lighting must be consistent from frame to frame—small variations in illumination are most annoying—as are small movements of the subject. A rigid camera mounting is essential, and some means of ensuring that the subject remains stationary is highly desirable. An electronic flash is a very useful light source. It is sometimes helpful to present the finished result as a cine loop so that the event can be viewed repeatedly.

2.5 High-Speed Photography

High-speed photography is of great value in studying events which occur too rapidly for the eye to follow, e.g., the movement of muscles during running, the vibration of the vocal chords, the vibration of various types of probes (Ernford 1981, Fraser and Dombrowski 1962, Holwill 1967, Kodak 1981, Moore 1975). Many short exposures are made in rapid sequence but then reprojected at normal speed—or even hand analyzed—to measure subject movement.

Usually, exposure times are very short. Obtaining bright enough illumination can be a serious problem with biological subjects if a continuous record is required. Tungsten–halogen lights, particularly when focused by condenser lenses, can be used down to exposures of about 10 µs duration but here heat filters should be used to avoid burning the subject.

High-speed cameras come in a variety of types according to the framing rate required—the rotating prism camera being the commonest. It records 4000 to 16 000 frames at framing rates up to 10 000 frames per second. Analysis of high-speed film is accomplished using an analysis projector with a single-frame facility, or by using special film analyzers which can handle varying degrees of complexity, such as coordinate and angular measurement.

2.6 Equidensitometry

Equidensitometry is a technique in which a continuous tone image is converted into single isodensity lines that delineate points of equal density. The original method utilized the Sabattier effect—combining high contrast positive and negative images and printing the resulting sandwich onto high contrast bromide paper. The technique is particularly useful when it is necessary to map equal densities, e.g., for the assessment of cataracts *in vitro*. Reported applications of the technique include detailed investigation of malignant melanoma cell colonies, the assessment of ambient illumination, and the differentiation of chromosomes (Lau and Krug 1968, Nelson 1980).

2.7 Photoelastic Stress Analysis

The technique of photoelastic stress analysis uses polarized light to detect the stress distribution in objects subjected to complex loading. Plane polarized light is passed through a transparent model of the subject which is then photographed through a polariscope. Two types of lines may be observed: isochromatic lines, which are contours of stress and are colored when white light is used, and isoclinic lines, which are superimposed on the isochromatic lines, are always black, and show the directions of the principal stresses.

The fact that an isotropic transparent material becomes double refracting when stressed is the basis of the technique. The polariscope has the usual crossed polarizing filters, and, in addition, condenser lenses directing a parallel beam through the subject, and quarter-wave plates to remove the isoclinic lines.

Transparent models of subjects are often made from epoxy resins, the sensitivity of the technique increasing with model thickness. A reflected light technique can be used to study surface strain, the surface being coated with a birefringent material. The surfaces of teeth, bones and joints have all been successfully studied using this technique (Gurjian and Lissner 1961, Haboush 1952, Hollinger 1958, Window 1963).

2.8 Schlieren Photography

Schlieren photography detects small changes of refractive index in transparent subjects, presenting the changes as light and dark or colored bands.The name comes from the German word "Schliere," meaning streaks. A typical system employs a parallel beam of light formed by a small source placed at the focus of a lens or a spherical mirror. The beam passes through the transparent object and is brought to a focus again by a similar lens or mirror. Part of this focused image is obscured, or cut off, by a knife edge. The light which passes over the knife edge falls on a screen on which an auxilliary lens focuses the image of the transparent object. If any part of the light is deflected in a direction at right angles to the knife edge by refractive index gradients in the transparent object, the image of the light source at the knife edge is displaced in that direction. More or less light passes over the knife edge, and corresponding areas of the image lighten or darken in proportion to the displacement. By using a multicolored filter instead of a knife edge, the differences in refractive index are represented as different colors. The width of the schlieren field is limited by the diameters of the condenser lenses or mirrors used. Heat flow over the body and the invisible output from ultrasonic probes used in ENT surgery (Croot and Robbins 1967, Kodak 1977, Stephens and Start 1972) have been studied by this technique.

2.9 Interferometry

Interferometry is another technique for recording small changes in refractive index within transparent subjects. Interferometry can also be used to map the surface topography of very small objects (see *Photogrammetry*).

The technique depends on the fact that two coherent rays of light, polarized in the same plane, interfere when they are superimposed. The interference may be partial or complete, constructive or destructive, depending on the relative phase of the two beams at the point of combination. In practice, a collimated beam is divided into two beams using a partial reflector, one of the beams being passed through the test area and then recombined with the other which serves as a reference beam. The combination is focused into the recording plane by a condenser lens. The Mach–Zender double-beam interferometer is normally used (Arnold et al. 1971). The interferometer can be used to obtain quantitative values of gas density, from which pressure and velocity can be calculated—of anesthetic gases, for example.

2.10 Photometry

In photometry, the density of an image on a photographic film is a measure of the radiation intensity that produced it. Equal brightnesses in the subject are rendered as equal densities on the film provided, of course, that a film/developer combination with a linear response is chosen. Thus it is possible to measure the subject reflectance (or luminance in the case of a self-luminant subject) by measuring densities on the film and relating them to the densities produced by known standard or reference light sources placed within the scene. The strictest standardization of photographic materials and strict processing control are needed if the measurements taken are to be of use. The image points that are compared should be positioned similarly in relation to the lens axis to avoid the effects of uneven image illumination caused by the camera. The image areas must be of similar size to avoid complications due to development adjacency effects. These latter development effects occur at the edges of widely differing densities on the film and produce an incorrect record of the illumination distribution. Previous applications of photometry in medicine have been the quantification of ultraviolet fluorescence (in relation to intraoral fluorescence, and for the dansyl chloride test for skin "turnover time"), the quantification of skin surface roughness, and the assessment of pigmented lesions of the skin (Marshall 1980, 1982, Marshall and Marshall 1983).

See also: Photogrammetry

Bibliography

Arnold C R, Rolls P J, Stewart J C J 1971 *Applied Photography*. Focal Press, London

Babcock M 1954 Method for measuring fingernail growth rates in nutrional studies. *J. Nutr.* 55: 323–30

Burke P 1958 A photographic method of measuring eruption of human teeth. *Am. J. Orthod.* 44(8): 590–602

Croot C F J, Robbins R 1967 Schlieren photography of an ultrasonic beam. *Med. Biol. Illus.* 17(3): 202–7

Ernford L 1981 A fracture study of the diametral compression test by means of high-speed photography. *Acta Odontol. Scand.* 39(2): 71–77

Fraser R P, Dombrowski N 1962 The selection of a photographic technique for the study of movement. *J. Photogr. Sci.* 10: 155–69

Gibson H L 1978 *Photography by Infrared: Its Principles and Practice*. Wiley, New York

Gurjian E S, Lissner H R 1961 Photoelastic confirmation of the presence of shear strains at the craniospinal junction in closed head injury. *J. Neurosurg.* 18: 58–59

Haboush E J 1952 Photoelastic stress and strain analysis in cervical fractures of the femur. *Bull. Hosp. Jt. Dis.* 8: 252–57

Hansell P 1962 Growth of the human finger nail—a time-lapse study. *Res. Film* 4(3): 219–23

Hansell P (ed.) 1979 *A Guide to Medical Photography*. MTP, Lancaster

Hollinger H 1958 Photography in photoelastic stress analysis of restorations. *Dent. Radiogr. Photogr.* 31: 31–36

Holwill M E J 1967 High speed kinephotography of flagellated micro-organisms *J. Photogr. Sci.* 15: 299–302

Kodak 1977 *Schlieren Photography*, Publication P-11. Eastman Kodak, Rochester, New York

Kodak 1980 *Photography through the Microscope*, 7th edn., Publication P.2. Eastman Kodak, Rochester, New York

Kodak 1981 *High-Speed Photography*, Publication G-44. Eastman Kodak, Rochester, New York

Lau E, Krug W 1968 *Equidensitometry: Methods of Two-Dimensional Photometry Principles and Fields of Application*. Focal Press, London

Lawson D 1972 *Photomicrography*. Academic Press, London

Londe A 1893 *La Photographie Medicale: Application aux Sciences Médicales et Physiologiques*. Gauthier–Villars, Paris

Loveland R P 1970 *Photomicrography: A Comprehensive Treatise*, Vols. 1, 2. Wiley, New York

Marshall R J 1980 Evaluation of a diagnostic test based on photographic photometry of infrared and ultraviolet radiation reflected by pigmented lesions of the skin. *J. Audiov. Media Med.* 3(3): 94–98

Marshall R J 1982 Photographic photometry of ultraviolet fluorescence. *Br. J. Photogr.* 129: 958–60

Marshall R J, Marshall R W 1983 Quantification of skin surface texture by macrophotography and computer aided scanning densitometry. *J. Audiov. Media Med.* 6(3): 98–103

Moore G P 1975 Ultra high speed photography in laryngeal research. *Can. J. Otolaryngol.* 4: 793–99

Nelson M N 1980 Image enhancement using equidensitometry. *Br. J. Photogr.* 127: 868–71

Ruddick R F 1979 Ultraviolet fluorescence photography. *Photogr. J.* 119: 381–85

Stephens D B, Start I B 1972 Schlieren photography of the piglet's microenvironment. *Cornell Vet.* 62(1): 20–26

Tredinnick W D 1961 Further advances in fluorescence colour photography. *Med. Biol. Illus.* 11: 16–21

Tyson J E, Clarkson J E, Sinclair J C, Leitch R 1981

Analysis of newborn intensive care by time-lapse photography. *Crit. Care Med.* 9(11): 780–84

Williams A R 1984 *Medical Photography—A Guide to Study*. MTP, Lancaster

Window A L 1963 Photostress. *J. Photogr. Sci.* 11: 186–93

A. R. Williams

Microcomputers

Microcomputers are being used increasingly in medicine and biology as in other branches of science. In this article microcomputer technology and terminology are discussed. For an account of an application of microcomputers in medicine, the reader is referred to *Microcomputers: An Application in the Clinical Laboratory*.

A microcomputer is a modern type of electronic computer, built using up-to-date integrated-circuit technology. A computer (in this article the term computer is synonymous with digital electronic computer) is a device that can accept, store, process, and release information under the direction of an ordered list of instructions known as a program. Programs are readily swapped in and out of the computer and are readily modifiable by the computer user. Therefore, in theory at least, a given computer can be used, say, to play games one minute, and to control a power station the next. The effectiveness and power of a computer lie in its ability to execute its programmed instructions with a speed and accuracy that far exceeds human capability. It should be said, however, that no computer yet devised can approach human capability in respect of intelligence, imagination and innate adaptability.

Programs (and the data on which they operate) are stored in the memory. The control unit and the arithmetic and logic unit (h) ALU together make up the computer's central processing unit (CPU). The CPU can obtain access to and manipulate the information that is stored in the memory. Peripherals are devices that enable the CPU (and hence the programs that ultimately control the CPU) to communicate with the world outside it (e.g., visual display units (VDUs) and printers). An important form of peripheral, the magnetic storage device, is used to extend the computer's data storage capability. The most common examples of this type of device are the floppy disk drive and the tape recorder.

The term microcomputer is widely used for any computer in which a single-chip microprocessor serves as the CPU. ("Chip" is the colloquial term used to describe a large-scale integrated circuit.) This interpretation of the term microcomputer was unambiguous until about 1980, when manufacturers succeeded in encapsulating not only the CPU but also memory and peripheral interface functions in a single package—thus creating the single-chip computer. It is essentially their small size and low cost which distinguish microcomputers from their larger brethren. Appropriately, they are used wherever low-cost, responsive, flexible and reliable control of information is required. They are being increasingly used to control machines. In this mode of operation, the data which they generate are converted by means of an interface (see Sect. 4) to electrical signals which drive the machines.

1. Computer Architecture

Although computers are complex digital electronic devices, a knowledge of electronics is not a prerequisite for understanding how they handle data. Leaving aside the actual electronic aspects of a computer, their operation may be described using a logical representation. Such a representation is not merely a means of description. Logic is an essential branch of mathematics with respect to the design, programming and use of computers.

The main parts of a computer (Fig. 1) will now be considered in terms of their logical representation.

1.1 Memory and Registers

A computer can deal only with numbers, all information held in the memory being in number form. Thus before non-numerical data can be stored or processed by a computer, it must first be converted into number form. The conversion process is known as coding; decoding is the reverse process—numbers in memory are converted to a form of information capable of ready interpretation by the user and displayed, for example, on a screen.

Memory is organized as a set of slots, each slot being used to store one number. The size of number that may be stored in each slot is determined by the size of the slot. For the common 8-bit microcomputers the number stored can have any integer value between 0 and 255 (decimal). Larger computers generally have slots that can hold larger numbers. Memory may be represented by a "map" as shown in Fig. 2a. Microcomputers in common use today can handle 65 536 slots. This is more conveniently represented by redrawing Fig. 2a in the form shown in Fig. 2b. Computer designers and programmers make use of these maps to allocate program space in memory.

Unless it is actually executing a program, the computer treats the list of instructions that comprise the program in exactly the same way as it treats any other data, that is, the program is stored as a set of numbers. If it were possible to look into the computer memory and see the number representations, the numbers representing the program would look almost the same as those used to represent the data. The CPU is capable of

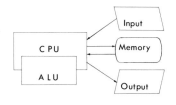

Figure 1
Block diagram of a computer

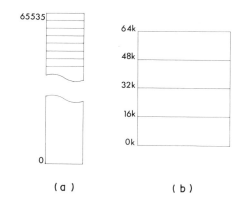

Figure 2
(a) A one-dimensional memory map, each slot
corresponds to one byte for an eight-bit computer,
and (b) a two-dimensional representation, where each
row is equivalent to many bytes (k = 1024)

understanding and dealing with numbers only if they
occur in particular sequences.

Registers are memory stores which have particular
uses. They are often integral with the CPU and are used
as temporary stores for numbers during calculations and
other special operations. They can usually hold the
contents of at most two "ordinary" memory locations.
Registers are used, too, as counters and pointers in
certain modes of CPU operation.

1.2 Central Processing Unit

The combination of control unit and ALU is indeed the
central part of any computer. The CPU fetches instructions from the program, decodes them, and converts each into a sequence of basic electronic steps that
it alone can perform. If the instruction says "add two
numbers" then the instruction decoder and the control
unit bring the two numbers into the CPU and feed them
into the ALU. The ALU contains circuits that are able
to carry out simple operations on numbers (e.g., add,
subtract, compare) in millionths of a second. The result
of any such operation is often calculated in a special
register called the accumulator and data can be sent back
from the accumulator to memory for use at a later stage.
The control unit is also responsible for keeping the entire
computer in step with the CPU. To follow a list of
instructions the CPU must be able to keep track of its
own progress as it works through the list. This task is
given to a special register in the control unit called the
program counter (PC). The PC is used to note the
position of the instruction that is currently being carried
out. When the instruction has been executed, the number
in the PC is increased by one to point to the very next
instruction in the list. The CPU can then carry out that
next instruction and this process continues until either
the list is finished or a special instruction (a jump) causes
the PC to point to any other particular instruction in the

list. This basic cycle of operations—fetching an instruction, carrying it out, and saving the result for later
use—is known as the fetch–process–store cycle.

1.3 Peripherals

Peripherals have two basic functions: as an interface
between the user (or machine in the case of an automatic
control system) and the computer system, and as a
means of storing masses of data to be assessed either
instantly or as occasion requires.

The interface between user and computer can be
further subdivided into (a) devices which are used
interactively (on-line), and (b) devices used for off-line
communication with the computer.

(*a*) *Visual display units.* VDUs are still the standard
computer terminal for interactive use. They can be
linked directly to the computer, or, alternatively, remotely via telephone lines or microwave links in conjunction with another device known as a MODEM
(*mo*dulator and *dem*odulator) used at each end of the
line. Modems modulate the carrier signal used for
sending data and demodulate it at the receiving end; they
operate from 300 baud (bits per second) to 19 200 baud,
depending on the sophistication of the modem. Increasingly, VDUs are being augmented by newer forms
of input or output. Touch screens, for example, allow the
user to point to an item on, say, a menu rather than
having to use the typewriter-like keyboard. Voice input
is also now available in certain situations.

(*b*) *Printers.* A vast range of printers is available. There
are two two main types: line printers, where all the
characters in a whole line are typed out simultaneously,
with writing speeds of up to 3000 lines per minute or
4000 characters per second (cps), and single-character
printers, where up to 200 cps can be printed. Printers
often have internal memory buffer stores to enable a
large amount of data to be transmitted at a high data
transfer rate, and then printed at the printer speed from
the buffer. Some printers print characters as a dot
matrix, enabling, when full software control is available,
good, accurate graphical output to be obtained, although the type quality of the characters is poorer. Other
printers use ink jets or heated pens on thermal paper,
giving much quieter operation suited perhaps for
patient-oriented microcomputer systems.

(*c*) *Magnetic tape.* Magnetic tape is the cheapest and
most reliable storage medium for large amounts of data
which do not require instant access. It is often used as
a back-up to hard-disk systems, with data not frequently
used being unloaded from disk to tape. The most
common magnetic-tape units use 0.5 in. or 1 in. wide
magnetic tape on 7 or 9 track, reel-to-reel drives. (Tracks
give either 6 or 8 bits plus a parity or check bit.) Several
recording densities are possible, 200, 800 or 1600 frames
per inch being common. Thus a standard 2400 ft, 9 track
tape at a density of 1600 frames per inch can store 46
megabytes of data. In practice, gaps have to be inserted

in the tape between records to permit access, thus reducing data storage capacity. Data transfer rate to the tape is high, but data retrieval can be slow since it may take several minutes to wind the tape to the required place.

(d) Cassette and cartridge tape units. Ordinary cassette tapes have been used for data storage. These are relatively cheap and easy to load, but are prone to error. They can store up to 1 megabyte of information at reasonable transfer rates. Access time to any piece of information is up to some 30 s (rewind time); bits are recorded serially on the tape. Thus cassettes are a useful cheap store for moderate amounts of information, and are often used with the many small, low-cost microcomputers available today.

Magnetic-tape cartridge units, more specifically designed for computer storage with 4 tracks recorded on typically 0.25 in. wide tape allow up to 10 megabytes to be stored with data transfer rates of up to 10 000 bytes per second. These have similar applications to cassette and magnetic tape units and are intermediate in price.

(e) Floppy-disk units. Floppy disks are made of thin plastic, coated with magnetic material as in magnetic tape. The disk is housed in a sleeve of low-friction material and rotates at 360 rpm inside the sleeve when inserted in the disk drive. Two sizes, 5.25 in. and 8 in. diameter, are most common, with variants allowing one or both sides of the disk and different storage densities to be used. Storage capacities range from 70 kilobytes to 1.5 megabytes; the information is in random access form, since the read/write head can move to any track of the disk and read any sector of that track within approximately 400 ms. This is considerably faster than a cartridge system, in some ways the main competitor of "floppies," and thus floppies are very commonly the main storage medium in modern small computer systems.

(f) Hard-disk units. Rigid disks coated with magnetic material and driven at speeds of 1500 rpm are also used for data storage. Read/write heads can access any track and sector within 10 ms, the heads floating on an air cushion and never actually making contact with the disk. Hard disks are often mounted in packs with several disks mounted together, or in drives with one nonremovable program disk and one removable data disk. To reduce access time even further, some drives have read/write heads over every track, which need only wait for a sector to rotate to them to retrieve or write a piece of information. Hard-disk storage ranges from approximately 5 megabytes of data for small systems to several hundred megabytes for large systems. Most large computer systems use hard disks as their main storage medium due to their large storage capacity, the high speed of data access and high data transfer rate. Magnetic tape is used with hard disk units as a back-up and for archival work.

(g) Analog-to-digital (ADC) converters and digital-to-analog (DAC) converters. In medical physics applications there are two very important uses of computers. One is the real-time automatic analysis of data from an analyzer or device (e.g., autoanalyzer, gamma camera, pressure transducer) which gives some sort of electrical signal output in analog form which must be converted to a digital binary form for computer analysis. ADCs perform this task. The second is in process control where data analyzed by computer is used to control or modify a system. To do this, information in digital form must be electrically converted to analog signals which can then be amplified suitably to control, say, temperature and humidity in, for example, a burns unit. DACs perform this operation.

The number of bits produced by the ADC depends on the accuracy required; for example, if 6 bits are used, the highest number available is 63 and thus accuracy of only 1 in 63 is obtained. Conversion times for ADCs are typically 1 µs, so up to perhaps 100 000 samples per second can be obtained if required.

DACs are designed to convert computer-output, digital signals to an analog voltage, a voltage which varies in discrete steps, determined by the number of bits used in the DAC. The use of ADCs and DACs in equipment interfacing with microcomputers for control or data logging is one of the most rapidly expanding areas of computing in health care.

2. A Typical Computer System

The operation of a typical personal microcomputer can be considered in terms of the above component parts. Switch on virtually any personal "micro" and it will immediately write a greeting on the screen as well as a "prompt" symbol to indicate that it is ready for commands to be input via its built-in, typewriter-like keyboard. Acceptable commands are quite "English-like;" it is rarely necessary to input the cryptic numbers that only the CPU can appreciate. How can the computer do all this at the throw of the mains power switch?

First, programs can remain stored in memory, even when the computer is switched off. The programs in question are held in a special form of memory called read-only memory (ROM)—of which more later. Secondly, CPUs are designed to carry out the fetch–process–store cycle as soon as the computer is switched on. Special start-up circuits ensure that when the CPU starts it begins at a fixed address in memory. If the ROM is placed at this address then, on start-up, the CPU must execute whatever program is in ROM.

Common ROM programs are the boot loader and the monitor. The monitor program, as the name implies, monitors the keyboard and converts the English-like commands into a set of instructions that can be executed by the CPU. The boot loader is a special program that is used only on start-up, to bring in an operating system program from, say, magnetic disk. An operating system is simply a more sophisticated form of monitor program. Typically it contains a number of smaller programs that can carry out the efficient transfer of programs or data from and to disk store. The task of the monitor or operating system is to make the CPU, memory and peripherals easy to use. If all programs had to operate in number form then computers would be by no means in such common use as they are today.

3. Interconnection of Functional Units

Large-scale integrated circuits that comprise hundreds or thousands of tiny transistors make up the CPU, memory, peripheral controllers, etc., that are the functional units of the microcomputers. Let us consider how these functional units are interconnected electrically.

Microcomputers in common use employ an 8-bit number (known as a byte) as the basic unit of number storage and transfer. Larger computers and some of the more recent microcomputers use 16 and 32 bit numbers, more often referred to as words. In an 8-bit computer, a set of 8 wires carries the actual data, while other sets of wires are used to control the transfer of data and direct the data to its correct destination.

Consider a well-understood information transfer system—the telephone network. Each subscriber has not only his or her own telephone but also a dedicated pair of wires which connect that telephone to the exchange. When the subscriber dials a number, it is transmitted along the pair of wires and picked up by the exchange. The exchange equipment decodes the number and automatically connects the wires of the dialler to those of the subscriber to be called. The same wires are then used to carry the ensuing conversation. This approach would be uneconomic if used to connect a CPU to 65 536 memory locations. With eight wires going to each location more than half a million wires would be needed.

The actual approach is analogous to a special form of telephone service—the "party line." Each memory location is connected to one set of signal lines and a separate system of control lines is used to decide which location actually uses the signal lines at any given time. Less than 30 wires divided into three sets can cope with the 65 536 memory locations. The sets of wires are known as buses, i.e., the data bus, the address bus, and the control bus. The signals (or number patterns) that appear on these wires determine what effect the bus has on any connected device.

The address bus consists of 16 lines and is used to select memory locations from the memory map. All the possible permutations of zeros and ones of 16 lines add up to 65 536 and hence this is the number of locations that can be uniquely identified by a 16-bit address bus. Each location responds to only one pattern on the address bus. When that pattern appears, the location becomes active and is ready to copy data to or from the data bus. Whether data goes from the memory to the CPU (a read operation) or from the CPU to memory (a write operation), it is controlled by the third bus, the control bus. It, too, is connected to all devices all of the time; however, only a memory location that has been correctly addressed will actually obey the control bus. All other locations ignore the control bus signals until addressed by the correct 16-bit pattern.

4. Interfacing

Before leaving the computer hardware it is worth considering how the computer sends information to the outside world. Not surprisingly perhaps, the buses come into play again. Just as some of the memory locations are used for ROM (and usually cannot be used for anything else), it is possible to use other memory locations as "windows" to the outside world. Instead of memory slots it is possible to put interface integrated circuits (ICs) at predetermined addresses in the memory map. These devices have the task of converting the 8-bit codes into a form suitable for transmission to, say, a VDU or a printer. The 8 wires of the data bus feed 8 inputs to the interface IC. The address bus is monitored by this interface device. When the correct address pattern is detected, the device copies the numbers from the data bus and sends them right out of the computer on yet another set of wires which are connected to, say, a peripheral. The peripheral itself responds to the data and, in the case of a VDU, converts the 8-bit number into a character on the display screen. A similar process occurs in reverse when a key is pressed on the VDU keyboard; the data travels down the wires to the interface IC in the computer, the IC copies the data onto the data bus and the CPU is able to read whatever character is represented.

Interfacing the computer to the outside world is no small task. Inside the computer (a relatively orderly environment) information is moved around as bytes or words. Outside the computer, information is handled in a large variety of ways. Interfacing into or out of a computer means changing the form of the information to achieve the required goal. For example, for a computer to control a steel rolling mill, the tiny electronic signals must be made powerful enough to drive large mechanical actuators—hence a power amplifier is required. For a computer to measure most things in the real world it must be able to handle analog data—in this case the data must be sampled and digitized before the computer can attempt any measurements. All interfacing requires information to be manipulated—signals made more powerful, slower or faster, understandable to humans, understandable to the computer, and so on.

5. Programming

Typical application programs (e.g., a stock-control program or a word processor program) may consist of 50–100 000 instruction codes that together make up the program suite. Such programs can manipulate masses of data many times their own memory size. For example, patient records on a regional computer "code" into thousands of millions of bytes or words of stored data. Writing a program by putting together 50 000 number codes would obviously be a difficult, error-prone task. Before going on to consider just how such programs are written, it is perhaps worth restating the function of the CPU.

A typical CPU can carry out some one hundred basic instructions. Sequences of these basic instructions enable the computer to carry out the required functions. The simple instructions that the computer "understands" are number codes. How does the CPU deal with these codes? Consider again the fetch–process–store cycle.

(*a*) *Fetch.* The PC keeps track of the current instruction to be executed. The 16-bit value in the PC is copied onto the address bus. The memory location corresponding to the PC value, that is, where the next instruction is stored, responds to the address and becomes active. The CPU sends out a read signal. The active memory location copies its own contents onto the data bus. The CPU then copies that value from the data bus into the instruction register.

(*b*) *Process.* The CPU decodes the instruction and sets about carrying out the sequence of steps that the instruction specifies. These steps are the computer's lowest-level operations. They are referred to as micro-operations and are controlled by what is termed a microprogram, which is part and parcel of the CPU.

(*c*) *Fetch again?* For some instructions more data may need to be brought in from memory. Thus one or more additional memory reads may take place before the current instruction is completely carried out. The address bus is supplied not necessarily with the PC value, but with the contents of some of the temporary registers which are used to hold the addresses of the required data.

(*d*) *Process continued.* The data, from whatever source, are manipulated in the ALU, the result probably finding its way to an accumulator register. Since an accumulator is used in nearly every processing operation, the result of the last operation cannot be left there as it would be overwritten when the next operation is carried out. Hence the need for a store cycle.

(*e*) *Store.* The address of the location where the data are to be stored is copied onto the address bus. The memory location so identified becomes active. The CPU sends a write signal on the control bus and the contents of the accumulator are copied onto the data bus and thence into memory where they can be retrieved at will.

6. A Simple Program

Figure 3a shows the decimal representation of a simple program that instructs the CPU to add two 8-bit num-

bers. The result, formed in the accumulator, is then stored in memory. To the uninitiated, the number codes in Fig. 3a convey little information on their own. The addition of a comment alongside each number shows that the numbers do have a hidden meaning. Even with the addition of such comments writing programs in decimal is awkward and tedious. A more acceptable approach is to write programs using mnemonics in place of the codes. Each mnemonic is chosen to convey the action taken by the CPU on encountering the corresponding machine code. The series of mnemonics is passed to a special assembler program, it carries out the task of replacing each mnemonic with the corresponding machine code equivalent. Figure 3b shows an assembly language representation of Fig. 3a. The mnemonics, although chosen to convey the meaning of the instruction, are still somewhat cryptic. Thus the use of program comments is crucial to remembering or understanding how a program works.

33	LHLD A	;Get number 'A' into temporary registers H and L
(0002)		;
17	LXI D,6	;Get the constant to be added
(0006)		;
25	DAD B	;Add the constant to 'A'
34	SHLD B	;Move result from H,L to memory 'B'
(0008)		;
201	RET	;and finish
(a)	(b)	

Figure 3
(a) A machine-code program written using decimal numbers to represent the machine codes, and (b) the same program as shown in (a) but written in assembly language. Note how the comments written in English are needed to fully describe the program's operation

A typical program development sequence is as follows:

(a) the assembly language program is created using an editor. (The editor is a program—usually supplied with the computer—designed to help the user by making it easy to enter and edit alphanumeric text);

(b) the edited text is passed to the assembler. The assembler converts each operational code into its binary equivalent, and builds up a sequence of bytes that constitute the final machine code program;

(c) the CPU's program counter is loaded with the start address of the newly created program and the program is executed and tested.

The obvious disadvantage of assembly language is that, by its very nature, it refers to the architecture of the CPU. Hence would-be assembly language programmers must have detailed knowledge of the CPU chip in order to write meaningful programs.

7. Higher-Level Languages

Higher-level computer languages are available which allow the computer user to communicate with the computer and program it without any need for the user to understand the internal workings of a CPU or even to appreciate how assembler programs work. In general, high-level languages require fewer mnemonics or lines of program than do their assembly language or machine-code equivalents. Whereas the CPU decodes each CPU instruction into a set of microoperations, and an assembler decodes each assembly language mnemonic into a CPU instruction, a compiler or an interpreter converts mnemonics that are more "English-like" than assembly language mnemonics into the equivalent of many CPU instructions.

One popular high-level language is called BASIC (beginners all-purpose symbolic instruction code). Designed as a language that is easy to learn, BASIC can be used (is available) on virtually every computer, particularly personal computers. In BASIC, a program equivalent to that of Fig. 3 could be written on one line, in the form

$$B = A + 6$$

Two main types of BASIC could be used to execute such a program—compiled BASIC, or interpreted BASIC. Compilers and interpreters are two high-level techniques for achieving the same end—execution of the BASIC program. The difference between the two is best explained by describing their use.

As with assembly language, the BASIC program may be created using an editor. The edited text is passed to the compiler which converts the program statements into machine code. To test the program, the CPU is directed to the beginning of the newly created machine-code program.

The interpreter is a program that controls the computer during the whole time that editing, conversion (coding) and execution are taking place. The interpreter's built-in editor allows the program to be entered by the user. Only when the user instructs the interpreter to "run" the program does any conversion to machine code take place. This is the major difference between the compiler and the interpreter. Whereas the compiler takes the entire source program and forms a corresponding machine-code program, the interpreter generates machine code only as each line of the BASIC program is executed. When the interpreter has converted a line of BASIC into machine code, only then does it let the CPU actually execute the newly generated code. Once the code equivalent to that line of BASIC has been executed, control of the CPU returns to the interpreter. The interpreter finds the next line in the BASIC program and the conversion process is repeated.

Whether compiler, interpreter or assembly language is used, the function of the resulting CPU instructions can be made the same. The differences arising from the three techniques are: the amount of knowledge and experience required to write the program in the first place, the ease of program creation and test, and the ease with which the program can be modified.

An interesting extension of higher level programming is the program generator. The idea is that, given a set of sentences written almost entirely in standard English, the generator program converts these into a program at high or assembly level. Such techniques are gaining favor and the aims of higher level interface are being successfully pursued.

8. Conclusion

The foregoing text is but the briefest introduction to several fields which are still expanding rapidly. The advent of low-cost 32-bit processors, higher level languages, semicustom integrated circuits (i.e., powerful "chips" made to order), artificial intelligence and expert systems software will have significant effects on the power, size and usability of future computers. Gone are the days when someone can buy a microcomputer, apply it to a real job of work and learn all about computing at the same time. One should pay as much as one can afford for a system that is powerful yet easy to use; in this way the application can be concentrated on, the computer being merely the tool, not the end.

Bibliography

Burns A 1981 *Computers and Their Applications*, Vol. 11, *The Microchip: Appropriate or Inappropriate Technology?* Halsted Press, New York

Chandor A, Graham J, Williamson R 1977 *The Penguin Dictionary of Computers*, 2nd edn. Penguin, Harmondsworth

Lewin D 1972 *Theory and Design of Digital Computers*. Nelson, London

G. J. Dunlop

Microcomputers: An Application in the Clinical Laboratory

An inexpensive antenatal monitoring service can be provided by measuring the relative amounts of estrogen and creatinine in the urine of expectant mothers over the last three months of pregnancy. Over this period the estrogen/creatinine ratio should remain within accepted limits: the ratio is thus a useful index of placental function and hence of fetal well-being (Rao 1977). This article describes the design and function of a microcomputer-based measurement system which was developed as a much-needed replacement for a semi-automated system handling 1000 patient samples per week.

A continuous-flow analyzer is used to measure the concentrations of estrogen and creatinine in a sample of urine from each patient. The sample is split into two portions and the addition of specific reagents to each portion produces color changes which are proportional to the concentrations of each analyte. These color changes are converted to electrical signals by the analyzer's colorimeters. The signals are amplified and displayed as a sequence of peaks and troughs on a chart recorder, the peak heights representing the concentration of the analytes in the samples. Actual values of concentration are obtained by passing samples of known concentration through the analyzer.

In the past, the laboratory chemist measured the heights of the peaks manually and then calculated the estrogen/creatinine ratios. These values were then added to the patient's cumulative record cards and separate reports were written for the obstetrician. These tasks are now performed by the microcomputer-based system.

The automated measuring system was developed to perform two main tasks: data acquisition and cumulative reporting. The system digitizes the voltage signals from the chart recorder, determines the peak values, and calculates the estrogen/creatinine ratios. The results are then stored on floppy disk where they are available for instant recall. The system also produces a printed graph which shows clearly the trend in placental function and hence fetal well-being.

1. System Design

The system is made up of two microcomputers connected via an RS232 serial data link. Each microcomputer may be designed and tested separately during the project's development phase; this modular approach simplifies system design. Because digitization of the chart-recorder signals must take place at regular intervals it is convenient to dedicate one microcomputer, the data acquisition system (DAS), to perform this task. Once started, this microcomputer needs no further operator intervention. It acquires data from both estrogen and creatinine chemistry channels, detects the peaks for each sample and stores their values in its memory prior to transmitting them to the second microcomputer.

The second microcomputer performs the cumulative reporting tasks and is controlled by the operator from the keyboard of a visual display unit (VDU). On receiving peak data from the DAS this microcomputer calculates the ratio values and stores them along with previous results in patient files on floppy disk. In addition to creating and updating patient files the operator may request a printed copy of an updated file: this is the report that is issued to the antenatal clinic.

2. The Data Acquisition System

The continuous-flow analyzer generates two asynchronous analog signals. Each signal is a regularly timed series of peaks that represent chemical concentrations. The DAS digitizes the signals to 8 bit accuracy every

2.6 s; at this digitization rate there are 32 points in a typical peak. The DAS also calculates the height of each peak and stores the values in its memory. The first two peaks in each channel are calibration standards, the DAS uses these known, good peaks to initiate timing. If subsequent peaks are not detected within the predicted time interval then their height values are marked as being doubtful. Peak height values that are off-scale are similarly marked.

The DAS can store 150 patients' results in its memory. This means that no data are lost if the second microcomputer is too busy to accept results from the DAS, for example, while the second microcomputer is accessing disks. The DAS can be switched to output peak results directly to a printer; this facility is useful for checking DAS operation or if the second microcomputer is not operational.

2.1 Data Acquisition Hardware

The DAS microcomputer is a S100 bus-based system made up of the following printed circuit boards (Fig. 1): Intel 8080 CPU; 16K RAM; 4K ROM; serial and parallel interfaces; ADC and analog multiplexer.

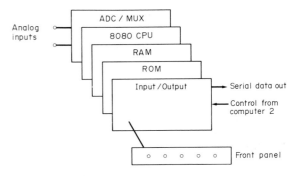

Figure 1
Schematic diagram of the S100 boards comprising the data acquisition system

When the unit is switched on or reset, test routines check the essential subunits: the RAM, the ROM and the input–output ports. Defects in any of these subunits will cause the test routines to illuminate indicator light-emitting diodes (LEDs) on the DAS front panel. Other LEDs indicate the changing status of the DAS while it is running normally. One flashes at each digitization; another indicates the detection of a peak; a third indicates when data is available for output. A final LED is normally held "off" by the program—it lights only in the event of a complete program "crash."

2.2 Data Acquisition Software

The DAS programs are written in 8080 Assembly language and are organized in two major segments—the main segment and the interrupt segment (Fig. 2).

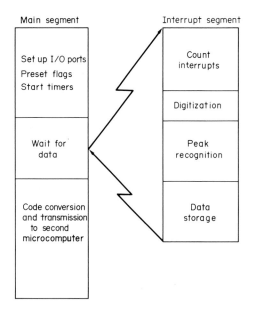

Figure 2
DAS program structure: timer generated interrupts are used for data acquisition, completed results are sent to the second microcomputer as soon as they are ready

The main segment starts by initializing the system: it sets up flags, address pointers and timers. It then waits in a program loop until data has been received from the interrupt segment. The task of the interrupt segment is to digitize the signals, detect the peaks, measure the peak heights and store them in memory. An output routine, called by the main segment, converts the peak values to decimal numbers suitable for transmission to the second microcomputer or for direct output to a printer. The output routine also generates warning characters for any value that has been marked as doubtful.

3. The Cumulative Reporting System

The second microcomputer enables the operator to manipulate patient details and files while the DAS processes signals from the continuous-flow analyzer. Patient details may be typed in from the VDU and then printed out as a "worksheet." The second microcomputer also performs the final calculations at the end of each run and the worksheet can then be printed out again with the patient results filled in. Disk storage for up to 2000 patients has been allocated on one of the microcomputer's two disk drives.

3.1 Cumulative Reporting Hardware

This microcomputer (Fig. 3) is also S100 bus-based, it is made up of the following boards: Z80 CPU; parallel and

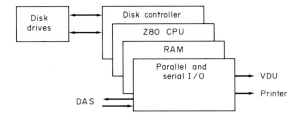

Figure 3
Schematic diagram of the S100 boards comprising the cumulative reporting microcomputer

serial interfaces; 64K RAM; disk controller. In addition to the DAS, the microcomputer is linked to a printer, a VDU and two 1 Mbyte, double-sided floppy disk drives.

3.2 Cumulative Reporting Software

The microcomputer runs under the standard CP/M operating system. The programs are written in BASIC and are compiled to increase their speed of operation.

The main BASIC program displays a menu of subprograms on the VDU. These subprograms are utilities which allow various data manipulations to be performed (e.g., "start new tray," "display and print worksheet," "enter patient records," "print graphs," etc.). Data continues to be transmitted to the second microcomputer while any of the subprograms are being run.

The DAS creates an interrupt when it wishes to send data to the cumulative reporting microcomputer. A burst of digitized data is transferred from the DAS to a protected buffer area of memory in the second microcomputer. The data are then transferred to peak value areas of memory in the second microcomputer leaving the buffer area free for the next line of data. When results are to be transferred to or from the floppy disks, the second microcomputer disables interrupts thereby preventing the DAS from interrupting the time-critical disk read/write processes.

One of the two disk drives is reserved for programs; the other drive is used for patient data. The main patient file contains 2000 records. Each patient record is 128 bytes long—holding up to 25 patient results as well as personal details (Fig. 4). Only the 25 most recent results are retained on file.

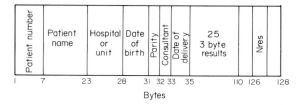

Figure 4
Patient record structure. Nres is the number of results in the patient's file

An index file capable of holding two thousand 6-byte patient numbers is also stored on the data disk. This allows the record for an individual patient to be found much more quickly than by searching through the whole disk. At the end of each day, graphs are printed of those patient records which have been updated during that day. These graphs form the reports that are sent back to the antenatal clinic (Fig. 5).

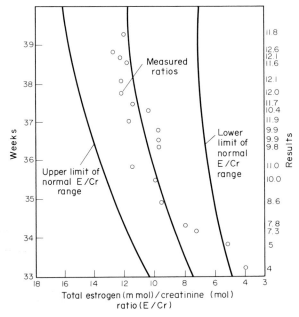

Figure 5
Example of output: patient estrogen to creatinine ratios are plotted as circles. The mean and normal range are also shown

4. Discussion

Before it was accepted for routine use the DAS was, for some time, run in parallel with the manual system of peak measurement. Replicate analyses showed that the two systems were of comparable precision. The DAS was thus considered to be performing adequately. Its advantages over the manual method are speed, the elimination of fatigue and also the elimination of the occasional human errors which occur in manually processed data.

The DAS has been in routine operation since 1980 and the cumulative reporting system since 1982. Both systems have proved highly reliable.

The equipment is used to provide a comprehensive reporting system for a busy laboratory and it demonstrates how microcomputers can be used for data acquisition, data processing, cumulative reporting and data presentation.

See also: Microcomputers; Computers in Clinical Biochemistry

Bibliography

Chandor A, Graham J, Williamson R 1977 *The Penguin Dictionary of Computers*, 2nd edn. Penguin, Harmondsworth

Furman W B, 1976 *Continuous Flow Analysis Theory and Practice*, Clinical and Biochemical Analysis Vol. 3. Dekker, New York

Keating D, Dunlop G J, Evans A L, Gowdie R A, Gregory N L, Lee T, Rao L G S 1983 A microcomputer data-acquisition and cumulative reporting system for oestrogen and creatinine continuous-flow analysis. *J. Autom. Chem.* 5: 14–17

Rao L G S 1977 Predicting fetal death by measuring oestrogen:creatinine ratios in early morning samples of urine. *Br. Med. J.* 2: 874–76

G. J. Dunlop and A. L. Evans

Microdosimetry

Microdosimetry is a subspecialty of radiological physics that is concerned with the pattern of energy deposition in microscopic domains within matter exposed to ionizing radiation.

1. Concepts and Quantities

When matter is exposed to ionizing radiation it is traversed by charged particles of various kinds (e.g., electrons, protons). These particles may originate from external or internal sources, but they may also be produced by various interactions between uncharged particles (e.g., photons, neutrons) and the material. The charged particles impart energy to matter by the processes of ionization and excitation; the energy loss per unit distance in the particle track is termed the linear energy transfer (LET) which is usually specified in units of keV μm^{-1}. The absorbed dose D is defined as

$$D = \mathrm{d}\bar{\mathscr{E}}/\mathrm{d}m \qquad (1)$$

where $\mathrm{d}\bar{\mathscr{E}}$ is the mean energy and $\mathrm{d}m$ is the mass to which it is imparted.

The SI unit of energy divided by mass is the joule per kilogram and for use in radiological physics it has been given the special name gray (Gy).

At absorbed doses of interest in radiobiology and medicine, the number of charged particles depositing energy in a tissue mass of 1 g or more is generally very large and fluctuations are negligible. However, the discreteness of energy deposition is a fundamental aspect of the action of radiations on individual cells. This is especially so for high LET where a single particle can injure or kill mammalian or other cells. The quantitative characterization of differences in energy deposition at the cellular level is an essential requirement for the understanding of the action of radiation on cells as well as on the tissues constituted by them.

The central quantity of microdosimetry is the specific energy z, which is defined as

$$z = \mathscr{E}/m \qquad (2)$$

where $\mathscr{E}$ is the energy imparted to the mass m that is contained in a volume of specified shape and dimensions. It is usually assumed that the mass m is in a sphere of unit-density material.

The stochastic quantity z may have a wide range of values in a uniformly irradiated medium. It is impossible to predict its value for any given volume, but it is possible to provide the probability distribution $f(z)$, which is the probability of occurrence of any value of z (per unit interval of z). Thus $f(z)$ is normalized:

$$\int_0^\infty f(z)\,\mathrm{d}z = 1 \qquad (3)$$

and its mean value $\bar{z}$ is the absorbed dose D

$$\bar{z} = \int_0^\infty zf(z)\,\mathrm{d}z = D \qquad (4)$$

The need to know $f(z)$ rather than only its mean value D is due to the fact that biological effects are usually not proportional to the energy deposited in a cell. In most of the cells of higher organisms the response seems to be, at least approximately, quadratic. For this reason the mean value of z^2 is an important quantity, given by

$$\overline{z^2} = \int_0^\infty z^2 f(z)\,\mathrm{d}z \qquad (5)$$

The function $f(z)$ is basically governed by three quantities. Two of these determine the mean number of events n (energy deposits by individual charged particles and/or their secondaries) in m. These are the absorbed dose D, and the mean diameter $\bar{d}$ of the volume containing m. The dependence of the event frequency on D is one of simple proportionality. When the charged particle range is small compared with $\bar{d}$ the event frequency depends on $\bar{d}^3$. In the more common case where $\bar{d}$ is much less than the mean particle range the dependence is on $\bar{d}^2$. Since events are statistically independent of each other they follow a Poissonian distribution with $P(v)$, the probability of v events, given by

$$P(v) = \frac{\mathrm{e}^{-n}n^v}{v!} \qquad (6)$$

The third factor influencing $f(z)$ is the magnitude of individual events, which depends on the radiation quality. This in turn can be roughly expressed in terms of the LET distribution of the charged particles. The characterization is only approximate because a particle of a given LET can deposit a range of energies in m. There are many reasons for this, including the varying length of the track segment intercepted in m, variations of the LET during traversal, finite range, track curvature, statistical fluctuations of energy loss and the pattern of δ rays (i.e., electrons ejected by the primary particle that have sufficient energy to ionize in turn).

These complex processes determine the probability distribution of specific energy produced by single events $f_1(z)$, and the determination of this function is the primary task of microdosimetry since it is relatively simple to derive $f(z)$ from it. Once $f_1(z)$ is known, the so-called frequency average $\bar{z}_F$ defined by

$$\bar{z}_F = \int_0^\infty zf_1(z)\,\mathrm{d}z \qquad (7)$$

may be determined. This is the mean specific energy produced in single events. Since the product of $\bar{z}_F$ and the mean number of events is equal to the mean specific energy and since the latter is equal to the absorbed dose, it follows that

$$n = D/\bar{z}_F \qquad (8)$$

$P(v)$, the probability of v events, is thus given by

$$P(v) = \frac{\mathrm{e}^{-(D/\bar{z}_F)}.(D/\bar{z}_F)^v}{v!} \qquad (9)$$

The probability that a specific energy z is produced by v events is denoted by $f_v(z)$. This function can be obtained by convolution:

$$f_v(z) = f_{v-1}(z) * f_1(z) = \int_0^z f_{v-1}(z')f_1(z-z')\,\mathrm{d}z' \qquad (10)$$

This relation permits the evaluation of $f_2(z)$ and higher terms if $f_1(z)$ is known. Summing over all values of v:

$$f(z) = \sum_{v=0}^\infty P(v)f_v(z) \qquad (11)$$

A variety of computer codes have been written for this procedure which yields $f(z)$ in terms of $f_1(z)$ for any value of D. In certain biological applications of microdosimetry another characteristic value of the $f_1(z)$ spectrum is needed. This is the dose average energy $\bar{z}_D$ defined by

$$\bar{z}_D = \int_0^\infty z^2 f_1(z)\,\mathrm{d}z/\bar{z}_F \qquad (12)$$

It can be shown that

$$\overline{z^2} = \bar{z}_D D + D^2 \qquad (13)$$

This relation is of basic importance in the theory of dual radiation action (see Sect. 3).

In the measurement of energy depositions by single events, a different microdosimetric quantity is usually employed. This is the lineal energy y, defined as the energy deposited in a volume divided by its mean diameter:

$$y = \mathscr{E}/\bar{d} \qquad (14)$$

The specific energy imparted by one event is related to the lineal energy by

$$z = ky/\bar{d}^2 \qquad (15)$$

In the case of a spherical volume in unit density material $k = 0.204$ when z is expressed in grays, y in keV μm^{-1} and $\bar{d}$ in μm. Equation (15) also relates $\bar{y}_F$ and $\bar{y}_D$, the frequency and the dose average of lineal energy, to $\bar{z}_F$ and $\bar{z}_D$ the corresponding single event averages of specific energy.

2. Measurement

Because of the complexities of energy deposition by charged particles, theoretical determinations of micro-dosimetric spectra have rarely been attempted, but calculations based on recently available track structure data give promise to such efforts.

By far the most important experimental method utilizes the proportional counter because of its unique sensitivity and adaptability. In this approach a volume (usually spherical) of unit-density tissue is represented by a gas-filled cavity having a diameter that is larger by a factor typically of 10^4. A wall of tissue-equivalent plastic surrounds the cavity and is the principal source of charged particles, which are produced by photons or neutrons and which appear in the cavity. The pattern of energy loss is substantially the same in a sphere of unit density and in the cavity which represents it. However, separate events in the former can occur simultaneously in the latter. This source of error is reduced in more advanced designs by establishing a collecting volume at some distance from the solid wall. This is the so-called wall-less counter shown in Fig. 1.

Its major components are a shell and a concentric high transmission grid that are made from tissue-equivalent plastic, and a collecting wire and a concentric helical grid made of stainless steel. The function of the helical electrode is to provide a more uniform electrical field in the immediate vicinity of the wire. Typically both the shell and the collecting wire are at ground potential. The spherical grid is at about -1000 V and the helical grid at about -800 V.

The counting gas is a tissue-equivalent mixture at a pressure of the order of 10 torr. Since the multiplication in the gas depends critically on its purity, continuous gas flow is frequently employed. The diameter of the unit-density sphere simulated is proportional to gas pressure. The minimum diameter that can be simulated with adequate performance of the system is about 0.25 μm.

The gas gain (i.e., the multiplication in the electron avalanche) is typically 1000. Employment of low-noise charge-sensitive preamplifiers and conventional linear amplifiers provides adequate sensitivity to detect a single ion pair with high probability. The pulses are sorted by a multichannel analyzer. Figure 2 shows examples of microdosimetric spectra obtained for 1 μm diameter tissue spheres for various radiations.

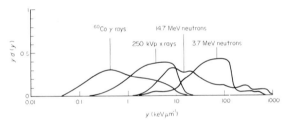

Figure 2
Lineal energy distributions in 1 μm spherical tissue regions for various radiations. The ordinate is $d(y)$ the fraction of the dose (rather than that of events) at y and is multiplied by y to provide a normalization per log interval

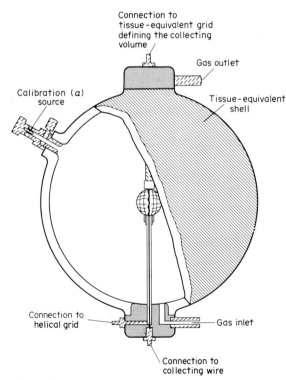

Figure 1
Simplified diagram of a wall-less counter employed in microdosimetry

3. Radiobiological Significance

Microdosimetry data can be applied to radiobiology in two ways. The first, and far simpler way, relates biological effects to the frequency of events and particularly to their presence or absence. Figure 3 is a plot of the event frequency $\Phi(0)$, which is the average number of events (of any size) per unit of absorbed dose, against the mean diameter $\bar{d}$ of a spherical volume of tissue. Applications of this kind of information have shown that certain cells in irradiated tissues are killed when no particle traverses them; that at least one type of radiation carcinogenesis must be related to injuries inflicted on more than one cell; and that cellular injury must in many cases be due to the interactions of affected subnuclear entities that are separated by average distances of the order of 1 μm.

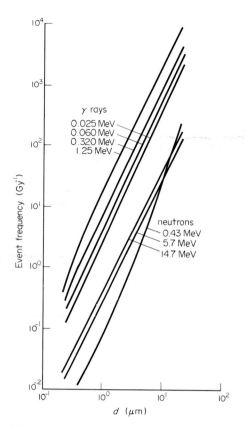

Figure 3
The event frequency per gray for a range of diameters of spherical regions in unit density tissue

Applications that are more difficult, but also more fundamental, are based on the magnitude rather than the frequency of events. The theory of dual radiation action is a comprehensive approach to radiobiology that is directly based on microdosimetry. It postulates that the cellular response to radiation depends on the square of the specific energy in sites within the nucleus. It has applications to such phenomena as the relative biological effectiveness of different radiations, the dependence of biological effects on absorbed dose rate and the shape of the dose–effect curve—especially at low doses.

See also: Radiation Dosimetry; Ionizing Radiation: Absorption in Body Tissues; Radiobiology: Kinetic Basis of Normal Tissue Response to Radiation

Bibliography

Booz J, Ebert H G (eds.) 1978 *Proc. 6th Symp. Microdosimetry*, CEC Report EUR-6064. Commission of European Communities, Luxembourg

Booz J, Ebert H G (eds.) 1983 *Proc. 8th Symp. Microdosimetry*, CEC Report EUR-8395. Commission of European Communities, Luxembourg

Booz J, Ebert H G, Eickel R, Wacker A (eds.) 1974 *Proc. 4th Symp. Microdosimetry*, CEC Report EUR-5122. Commission of European Communities, Luxembourg

Booz J, Ebert H G, Hartfiel H D (eds.) 1981 *Proc. 7th Symp. Microdosimetry*, CEC Report EUR-7147, Commission of European Communities, Luxembourg

Booz J, Ebert H G, Smith B G R (eds.) 1976 *Proc. 5th Symp. Microdosimetry*, CEC Report EUR-5452. Commission of European Communities, Luxembourg

Ebert H G (ed.) 1968 *Proc. 1st Symp. Microdosimetry*, CEC Report EUR-3747. Commission of European Communities, Luxembourg

Ebert H G (ed.) 1970 *Proc. 2nd Symp. Microdosimetry*, CEC Report EUR-4452. Commission of European Communities, Luxembourg

Ebert H G (ed.) 1972 *Proc. 3rd Symp. Microdosimetry*, CEC Report EUR-4810. Commission of European Communities, Luxembourg

International Commission on Radiation Units and Measurement 1984 *Microdosimetry*, ICRU Report No. 36. ICRU, Washington, DC

Kellerer A M, Chmelevsky D 1975 Concepts of microdosimetry I, II, III. *Radiat. Environ. Biophys.* 12: 61–69, 205–16, 321–35

Kellerer A M, Rossi H H 1978 A generalized formulation of dual radiation action. *Radiat. Res.* 75: 471–88

H. H. Rossi

Microphotometry

Microphotometry is the measurement of the intensity of light emanating from a region of a microscope specimen. Since the light may be transmitted, emitted or reflected, microphotometry may be used for estimating absorption (microdensitometry), fluorescence (microfluorimetry), or reflection (microreflectometry). Each of these is discussed below: but the first, which is at the same time the most widely used biomedically, and yet arguably the worst beset by nonobvious pitfalls, is discussed in the greatest detail.

Two remarks, however, apply throughout. The first is that since the great majority of biomedical photometry has the purpose of microchemical assay, principles will be illustrated here entirely from such examples. The second is that absolute measurement of light intensity is almost never attempted; the photometric parameter for any one specimen is instead the dimensionless ratio of its intensity to that of some standard (e.g., the transmitted intensity of a clear glass slide, or the reflected intensity of a fully-exposed film). Chemical quantification can be made absolute only if one reference specimen has been independently (not microphotometrically) quantified.

The section on equipment should enable the reader to set up and use the simplest instrumentation likely to be worthwhile. More sophisticated classes of instrument are treated only sufficiently to give initial guidance on their selection. This article thus seeks to counter-weight the research literature (such as that cited in the Bibliography of this article) which rather emphasizes sophistication.

1. Absorption

When light is absorbed by an object, the ratio of transmitted to incident intensity is termed the transmittance T. The key function in almost all absorptiometry (microscopical or otherwise) is what is variously termed the optical density, extinction or absorbance A, where

$$A = -\log_{10} T$$

It is this function which, when monochromatic light traverses a homogeneous layer of material, is proportional to the quantity of material (concentration multiplied by path length) traversed (Beer–Lambert law): hence the potential for chemical assay based on absorptiometry. To see why the expression is logarithmic, consider two identical slabs of material in series: the second reduces the intensity incident on it by the same proportion as the first. Differentiation of the Beer–Lambert relation shows that relative error dA/A is minimal at $A = 0.43$ and acceptably low for the range $0.1 < A < \sim 1.2$ (Zimmer 1973). These limits should be taken into account in experiment design.

In biological microphotometry, the absorbing substance (chromophore) may be natural or the product of a laboratory reaction, and the wavelength (with image converters for focusing where necessary) may be anything from fairly short uv to near ir. Examples are assays of nucleic acids by their own absorbance at 340 nm (originally performed by Caspersson), and of oxidative enzyme activities by the absorbances at 550–585 nm of diformazan granules deposited in a prior, histochemical reaction. The latter example immediately calls attention to two sources of error which can be present in macroscopic assay systems also: nonlinearity of the prior reaction, and nonlinearity of the light absorption. The first of these faults (product formation less than proportional to enzyme activity) is usually due to some form of biochemical inhibition; the second (absorbance less than proportional to quantity of product) is caused by molecular interactions in the product. Thus, both these nonlinearities can be kept within bounds by reduction of reaction time and/or product density. A preliminary trial can usually establish an acceptably linear range for a given investigation.

Many other errors occur for practical purposes only at the microscale. Refraction within the specimen is minimized by use of suitable (usually high-refractive-index) mounting media. Spread of stray light into the photometric field (glare) is minimized by illuminating a field not much larger than that to be metered (especially if the periphery contains bright/refractive features) and setting condenser aperture below that of objective (so that light does not strike the lens mount). However, the most substantial error, which is demon-

strable on the macroscale but a problem only on the microscale, is distribution error. This arises wherever the field sampled in a single reading is internally patterned into denser and less dense areas. Since light, not darkness, is what a photometer reads, light passing through the less-dense areas dominates the reading increasingly as the dense areas become denser.

The different materials shown in Fig. 1 have optical properties as follows: (a) $T = 0.80$, $A = 0.097$; (b) $T = 0.08$, $A = 1.097$; and (c) $T = 0.0008$, $A = 3.097$. Yet, if background transmittance is 1.0, the transmittances T' of the whole samples, each measured in a single gross observation ($T' = 0.2 \times T + 0.8 \times 1.0$), are 0.96, 0.816, 0.800, respectively; whence apparent absorbances ($A' = -\log_{10} T'$) are: (a) 0.018, (b) 0.088, and (c) 0.097. Thus such a measurement (Fig. 2a) barely detects the difference of density between the material in (c) and that in (b). If, however, absorbances of a large number of tiny points are measured separately, then 80% show zero absorbance and 20% absorbance A. The mean absorbances, weighted by area, which would be arrived at by summation of all such points ($\bar{A} = 0.2 \times A + 0.8 \times 0.0$) are: (a) 0.019, (b) 0.219, and (c) 0.619. So $\bar{A} \propto A$. It is $\bar{A}$ which is the reading provided by scanning densitometry (Figs. 2b, c) from a pattern which is fully in focus.

More generally, if p is the proportion of area with transmittance T_1, and $(1 - p)$ that with T_2, then

$$A' = -\log\left[pT_1 + (1 - p)T_2\right]$$

whereas

$$\bar{A} = -\left[p \log T_1 + (1 - p) \log T_2\right]$$

Three troublesome, and more or less expensive, categories of technique for overcoming distribution error exist; each works well in optimum circumstances, but not universally. Instrumentally easiest and cheapest, but most time-consuming, is to photograph the specimen. If this is done on the normal (log–linear) portion of the emulsion's exposure curve and, after development, the silver or dye is extracted from the image area of interest the chemical quantity of image material (itself assayable by one or another form of macrophotometry) relates linearly to the integrated absorbance of the original

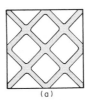

(a) (b) (c)

Figure 1
Distribution error illustrated by three nets having identical geometry, each occupying 20% total area of the region sampled

481

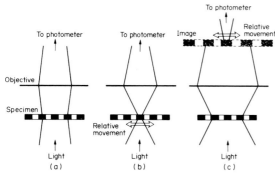

Figure 2
Methods of absorption microphotometry. (a) Gross
measurement, in which a beam of light, broad
compared with specimen details, is projected through
the specimen onto the photometer cathode. (Actually
a somewhat larger beam is projected; the area for
measurement is selected by a diaphragm in the image
plane.) The condenser aperture is usually kept low, so
lateral rays are shown only slightly oblique. In normal
absorptiometry—but not fluorescence or reflectance
measurement—such readings are subject to
distribution error. (b) Object scan. (c) Image
scan—see text

specimen area. An alternative is to expose at the toe of
the curve where there is an (inverse) linear relation
between image transmittance and original absorbance;
photometry performed on the negative is then free
from distribution error. These and other photographic
techniques may be found in Rost (1980).

A second group of techniques comprises the two-
wavelength methods (Mendelsohn 1966). In these direct,
photometric measurements of the gross sort (Fig. 2a) are
made from the specimen at two wavelengths, for which
the absorbance of the uniformly deposited chromophore
has been found to differ—originally by a factor of
exactly two but recently, with benefit (van Oostvelt and
Boecken 1977), by somewhat smaller ratios. The underlying model assumes an area of high, fairly uniform
density in a clear surround: cell nuclei, set in unstained
cytoplasm, are thus the most suitable and common
objects of study. Photometric stability is essential and
computerization highly desirable.

Finally, at further increased cost in capital but less in
operator time, distribution error may be avoided by
scanning. In an object scan (Fig. 2b) the object is
scanned with a spot of light smaller than the object
details, and the point readings are separately measured
and automatically summed. Either the object or the
beam may actually move, and the beam can be projected
either upward or downward. High apertures are required
to image small points—so lateral rays are oblique.
Alternatively, the image can be scanned, optically (as
Fig. 2c) or by TV raster. Provided the photometer
diaphragm or the TV picture point is smaller than image
details, effect on distribution error is equivalent to that

in the object scan. High aperture is again required,
this time to resolve fine object detail at the image.
These methods are applicable to chromophores of any
absorption spectrum and specimens of any geometry
in the plane of the scan. Their limitation (insufficiently
recognized) is that they apply only to specimens no
thicker than the objective's depth of field (Mendelsohn
1966, Spurway et al. 1985).

There remains the simplest, quickest and cheapest way
of reducing distribution error. It is applicable, without
any of the above exceptions, to all circumstances in
which the measured optical density can be controlled by
the experimenter. The density is simply kept low: for
example, a natural constituent is measured at a wavelength which is not that of its peak absorption; or a
prior, histochemical reaction is carried less far than it
would be for ordinary viewing or photography. (In the
latter case, the linear range of increase of measured
absorbance with reaction intensity, recommended earlier, avoids distribution error too.) The limitation in this
method is that the signal-to-noise ratio within the specimen is decreased. Nevertheless, the measurement may be
as accurate as that obtainable from an image with
greater contrast by a technique of greater sophistication.
Furthermore, in the histochemical case, the effects of
most forms of within-specimen variation can be eliminated if background readings are taken, not as a random
sample, which will be averaged to give a $T = 1.0$ value,
but individually, before the reaction is set going, from
the exact areas of subsequent study—whatever absorbing or refractile features these areas happen to contain.
When this can be done, the weak-reaction method
becomes, for most histochemical purposes, unequivocally the best way of circumventing distribution
error. Regrettably, if more than a very few photometric
fields per section are to be studied, expense and sophistication creep back in a nonphotometric form, for a
microscope reaction stage with microprocessor-
controlled positioning is required (Pette 1981).

This section has been concerned with absorption approximating, even in the denser areas, to that described
by the Beer–Lambert law. To conclude, however, it is
worthwhile to note the alternative case of "all-or-none"
absorption, where the absorbing material is opaque
($A = \infty$). Here, provided the specimen is thin enough
so that pieces of material do not obscure each other,
distribution does not cause error. This is because the
quantity of material (assuming uniform thickness) or
number of particles (of constant size) is estimated not
by a logarithmic function but by the obvious linear
function, $1 - T$ (termed absorption).

Comprehensive reviews of absorption microphotometry can be found in Pollister et al. (1969), Rost (1980)
and Zimmer (1973).

2. Fluorescence

When emitted light is seen against a dark ground, much
greater sensitivity is attainable than when absorption is

studied against a bright background. Examples of fluorescence specimens showing this feature, but otherwise analogous to those considered above, are catecholamines, which fluoresce specifically after fixation by formaldehyde vapor (Falck); and antibodies, made fluorescent by conjugation with fluorescein, which bind stoichiometrically to specific tissue antigens and so make possible their assay. In addition to sectioned material, however, fluorescence may be followed in the surface layers of opaque, even living, matter (e.g., the cyclically changing redox state of NAD in beating hearts (investigated by B Chance)). In every instance the procedure is to illuminate in a short waveband (usually uv) whose upper limit is set by exciter filter(s), and to view and/or measure in a longer band, lower-limited by barrier filtration.

Since it is light that is being collected, particles too small to be visible in transmitted light may be seen (exactly as are stars at night). Also, distribution error does not arise. Dense regions of fluorescent material (fluorochrome) do, however, present a group of problems with similar consequence: self-absorption and quenching of fluorescence by interactions within the specimen (Prenna et al. 1974). The worst form of self-absorption, the inner filter effect, occurs when absorption and emission spectra overlap. Denser regions of fluorochrome then have emission peaks of greater wavelength than less dense regions. The effect can be reduced by illumination from above, and completely eliminated (where sufficient intensity is available) by measuring in wavelengths longer than those of overlap.

The other major problem with biological fluorochromes is fading due to photodecomposition. This imposes rigorous time disciplines on the investigator, and makes highly desirable the provision of means for prior survey of the specimen (e.g., by phase contrast) at a nonexciting wavelength. It also means that fluorimetry reaches its optimum potential when inspection is not required and each specimen is measured once in a brief, standardized pulse—as when the DNA contents of a stream of cells are assayed, in flow cytofluorimetry.

Provided all the above errors are avoided, the total intensity of emission from a fluorochrome is proportional to the absorbed intensity, $1 - T$. Yet in the preceding section it was found that not this function but the logarithmic one, absorbance A, was proportional to fluorochrome concentration. Writing

$$1 - T = 1 - 10^{-A}$$

it follows that only when A is small enough for the right-hand side to be proportional to A will intensity of fluorescence vary linearly with concentration. This condition applies when $A < \sim 0.1$ (Prenna et al. 1974). Microfluorimetry is reviewed by Rost (1980).

3. Reflection

Reflectometry of natural surfaces in air presents several consistency problems, especially of angle (Piller 1977).

Most of them are encountered in fields like petrology and industrial product control, but rarely in biology. Several of the difficulties disappear if it becomes possible to take the separate, though related, steps of using an immersion objective and illuminating through it (see Sect. 4). In particular, the latter move ensures that illumination geometry is reproduced every time an image is focused. Such a system, for use in studies on dental amalgam, has been described by Darvell (1977). The same mode of illumination is also the best one for quantifying autoradiographs. This is the most common biomedical use of microreflectometry.

The autoradiographer's interest is in the exposure—usually just the relative exposure—of different areas of emulsion to β radiation. Typical autoradiographic exposure differs from that used in ordinary photography both in being weaker and in depending essentially on single-hit rather than multi-hit quantum events. Accordingly, it may be taken as linearly indicated by each of several criteria of silver density, but to different upper limits. The sequence, with figures for one common emulsion per 10 µm diameter circle (a representative cell size) quoted as a guide, is (Goldstein and Williams 1971):

(a) grain count—to ~ 30 grains per 10 µm circle,

(b) reflectance—to ~ 150 grains per 10 µm circle,

(c) absorbance ("blackening")—to > 300 grains per 10 µm circle ($A \sim 0.8$).

Absorbance measurements, however, are less accurate in weakly exposed regions, particularly if the stained biological section underlies the emulsion. Reflectance is thus generally accepted as the measure which provides the greatest range without interference, and is, incidentally, quicker and capitally cheaper than either scanning (the absorptiometric technique most appropriate here) or automated grain counting. Reflectance of fully fogged, developed emulsion is set as 1.0 and reflectance relative to this (R) is the parameter employed. In the above tabulation, the limit of linearity quoted corresponds to $R \sim 0.4$. Exposure E can even be quantified by reflectometry above the linear range, using the relation (Goldstein and Williams 1971):

$$R \propto 1 - 10^{-E}$$

(This follows from the expressions presented above, given that $R \propto$ absorption, provided the fully fogged emulsion is totally opaque and $E \propto A$.) Alternatively, the relation between E and R may be linearized yet further, and interference from larger reflecting particles (such as melanin) reduced at the same time, by viewing through crossed polars (Goldstein and Williams 1974).

Unlike absorbance and fluorescence, reflectance may usually be studied in white light. This requires the least lamp power or photometric sensitivity. However, some common histological stains fluoresce in these circumstances, notably eosin, which when excited by green emits red. Filters, deployed according to an inverse logic from that of the previous section, suppress such interference.

Reflectiometric assessment of autoradiograph exposure is reviewed thoroughly by Dormer (1972), though with some differences of approach to that followed here.

4. Equipment

In principle, any light-measuring device which can be attached to a microscope in fixed geometrical relationship, and responds to the instrument's normal levels of light intensity, will turn it into a microphotometer. In particular, the exposure-determining systems of most photomicroscopes serve well for gross measurements (Fig. 2a), except where extreme sensitivity is required. Modern, commercial instruments offer a "detail measurement" mode of operation, sampling an object circle say 10 µm diameter through a 100× objective. Failing this, a diaphragm of appropriate size to limit the measured field could be placed by the user in the intermediate image plane of the projective. With simple exposure meters, not coupled to the shutter, the procedure is then obvious. With the now common, automatic coupled systems, a "trial exposure" setting giving a suitably precise readout may be available; alternatively, film can be omitted from the camera, "exposures" made of the empty film-holder, and the duration of the shutter-opening electrical square wave measured. (The shutter is opened for a time that is inversely proportional to metered intensity.) A digital timer, giving a range of 0.01–9.99 s, serves generally. Using the film speed setting to control sensitivity, almost all histochemical absorptiometry and much reflectance work can be done with this device. Zs-Nagy (1975) has described such a system, but with a slower clock, as suitable for fluorescence studies. Only for low-intensity reflectance and fluorescence, or when fading is to be avoided, are the high-sensitivity photomultipliers of commercial microphotometers really necessary.

Ordinary microscope illumination is also acceptable for visible-spectrum work, particularly if the lamp is of quartz-halogen type (which gives greater intensity in the blue). However, a voltage-stabilized power supply, which may be inexpensively added, is a great asset; without it, repeat readings and very frequent checks of illuminating intensity (T or $R = 1.0$) have to be included in the protocol to guard against emission fluctuations. This is especially true at the blue end of the spectrum: a cooler lamp runs redder. To provide for uv absorption studies, and to excite most fluorescence, discharge lamps are required. Xenon is often preferable to mercury, if its extra cost can be accepted. The xenon discharge is less spatially unstable, and provides a wider, more continuous emission spectrum; mercury emits more short-wave energy, but almost all in narrow lines. Current-stabilized power supplies are essential for discharge lamps.

Spectral selection is of importance. The procedure involves the selection of those waveband(s) at which the required chromophore or fluorochrome is optimally discriminated, or the reflectance is least complicated by fluorescence. Dye filters can be used in preliminary experiments, but interference filters are almost always preferable. Monochromators are, however, hardly ever necessary except for microspectrophotometry (the critical examination of absorption or emission spectra). The continuous wedge type of interference filter, which allows selection of any peak transmission within a wide range (e.g., the visible spectrum), has a sufficiently narrow bandpass for almost all other purposes and yet usually gives brighter illumination than a monochromator, at fractional cost.

Absorbance measurements are affected by the size of field illuminated and the numerical aperture of the condenser. The diameter of the circle of specimen illuminated should normally be 2–3× the diameter of the circle to be metered: larger fields increase glare but smaller fields put too much of a premium on the accuracy with which the position of the photometer diaphragm is indicated to the microscopist. Thus, for measurements of 10 µm circles, a 25 µm illuminated field is appropriate. Few normal, iris-type field diaphragms achieve this, and even fewer do so reproducibly. The better provision is a set of interchangeable fixed-size apertures, one of which may be positively registered in the light path for measurements but swung out when the specimen is to be explored. Further improvement is achieved by a condenser objective; compared with an ordinary condenser, this gives greater reduction and sharper imaging of the field stop on the specimen. As to the aperture at which any condensing lens is operated, how far to close below that of the objective is best determined empirically: provided adequate light is available, one should simply use the setting giving greatest contrast within a typical specimen. (There is a further consideration that, with wide cones of light, the lateral rays, being oblique, have a greater path length through the specimen than axial rays. This only matters directly when absolute absorption coefficients are to be calculated; for relative absorbances it just constitutes one more reason why all measurements of a series must be made under identical optical conditions.) Unfortunately, not all manufacturers fit their condenser objectives with click-stop or clampable aperture irises, but calibrated control levers are usual.

For fluorescence, transparent specimens were formerly excited through dark-field condensers. With a proper array of modern barrier filters, the more intense and definable bright-field (dia-) excitation can safely be adopted, and is the best mode at low magnifications. However, objectives of greater power than perhaps ×10 can better excite the specimen from above: incident (epi-) illumination. One gain is that, during explorations, the transparent specimen can then be dia-illuminated by a nonexciting wavelength in phase contrast or dark field. With opaque specimens, only the incident excitation route is feasible. Older epi-illuminators projected a hollow cone of rays from an annulus around the outside of the objective front element. Recent technology allows the objective itself to direct excitation downward as a solid cone. There are three advantages in the latter: exact focus; exact definition of illuminated field—the area seen, or less with an appropriate field stop; and greatest image brightness at highest magnification. This last, invaluable feature arises because objectives of higher power have higher numerical apertures.

For all epi-illumination, the exciter beam enters the microscope axis between objective and viewing head, and there strikes a 45° reflector. With a through-lens system, this reflector is optimally a dichroic mirror. Such a device deflects most of the lamp's short wavelengths downwards, through the objective; yet it allows longer waves, emitted by the specimen, to pass predominantly upward, undeflected, for viewing. Thus it contributes a basic stage of both exciter and barrier filtration, though each should be supplemented.

Somewhat similar comments apply to reflectance measurement. Diffuse reflectance of grain layers, as in autoradiographs, can be demonstrated with a dark-field condenser. Intensity of illumination is, however, dependent on the transmittance of the underlying biological specimen, so that the chief advantage of reflectance over absorption measurements is sacrificed. Illumination for quantification should therefore be from above, even when the object of study is not totally opaque. Crude experiments, with long-working-distance objectives, can be based on two external lights, shining down obliquely from opposite sides. Integral epi-illuminators are nevertheless better for all quantitative purposes, and essential at high magnifications. Hollow-cone systems are often preferred with objectives up to $20 \times$; as in absorbance work, the measured field is then limited by the photometer diaphragm. Above this, through-lens illumination has the same superiorities as for fluorescence; the illuminated-field diaphragm should now be made limiting (Entingh 1974). For reflectance, however, the 45° beam deflector above the objective must be half-silvered, not dichroic.

To suppress rear-surface reflections under incident lighting, reflectances of autoradiographs and similar specimens should be measured with a light absorber, such as black glass, oiled to the back of the slide.

Irrespective of whether they also constitute the route of illumination, oil immersion objectives improve signal-to-noise ratios in all forms of microphotometry; many manufacturers supply them down to 10 or $20 \times$ magnification.

The component least likely to be worth constructing from separate components is the photometer head. Hughes et al. (1977), however, describe how to achieve adequate sensitivity for white-light reflectance with a photodiode; and Entingh's (1974) light pipe might assist such developments. Nevertheless, for most purposes, and particularly where narrow wavebands have to be selected, the sensitivity of at least a general purpose photomultiplier is required. The most crucial feature is

then the spectral response of the unit. Standard range is about 300–650 nm. Extension further into uv (say 185 nm) is a matter of window transmittance; to deep red and near ir (e.g., 930 nm) needs a special cathode—usually one containing several species of alkali cation. Note that a flat response curve is neither attainable nor necessary. Accurate knowledge of the curve is rarely required: only the range of adequate sensitivity needs to be known. The parameter recorded is almost always the ratio of two meter readings at the same wavelength.

The foregoing discussion refers to the use of single-beam instruments for gross measurements. Instruments with reference beams, enabling performance variations to be monitored, also exist (Piller 1977). Modern electronics stablize both photomultipliers and filament lamps so well that it is difficult to conceive a biomedical use for visible-light photometry which would justify the extra precision obtained. By contrast, all work utilizing discharge lamps gains considerably from such a provision (Rost 1980).

The other, more common direction in which to increase sophistication applies only to thin-specimen absorptiometry: it is to substitute scanning (Figs. 2b, c) for gross measurement. Altman (1975) appraised the majority of systems then available. He found that systems not scanning the image by TV fell into two groups: one comprised those using movable mirrors, whether to shift a narrowly converged probe beam across the object or to move an image of the field across a narrow photometer aperture; the other involved moving the specimen mechanically. Those in the first group were cheaper and quicker to use, but ultimately less versatile. The only class of instruments in which the available choice has extended substantially is that of TV scanners. These are expensive instruments, but provide for a gamut of image-analytic procedures, geometric and enumerative as well as photometric. They also have a facility not available in other systems: measurement of the whole of an object whose outline, however awkward, has been traced with a light pen. This has great utility, even with sections too thick for their distribution error to be overcome by the scanning. Regrettably, no traditional TV system appears to have an intensity response as linear as that obtainable from a photomultiplier. Nonlinearities are introduced, for each point-value of $A > 1$–1.4, by the characteristics of the camera tube (Jarvis 1981); and for very low integrated values of A by a 6 bit digitization of point brightnesses adopted by most contemporary software systems. Solid-state detectors are however starting to replace camera tubes; it seems likely that, by combining wider linear range with finer intensity resolution, they will effectively eliminate both problems.

See also: Fluorimetry

Bibliography

Altman F P 1975 Quantitation in histochemistry: A review of some commercially available microdensitometers. *Histochem. J.* 7: 375–95

Darvell B W 1977 Practical aspects of a quantitative microphotometer system. *Microsc. Acta* 79: 353–62

Dormer P 1972 Photometric methods in quantitative autoradiography. In: Lüttge U (ed.) 1972 *Micro-autoradiography and Electron Probe Analysis: Their Application to Plant Physiology*. Springer, New York, pp. 7–48

Entingh D 1974 Performance characteristics of a micro-reflectometer for measuring autoradiographic grain density. *J. Microsc.* 101: 9–19

Goldstein D J, Williams M A 1971 Quantitative auto-radiography: An evaluation of visual grain counting, reflectance microscopy, gross absorbance measurements and flying-spot microdensitometry. *J. Microsc.* 94: 215–39

Goldstein D J, Williams M A 1974 Quantitative assessment of autoradiographs by photometric reflectance micros-copy: An improved method using polarised light. *Histo-chem. J.* 6: 223–30

Hughes H C, Meyer P M, Meyer J W, Meyer D R, Bresnahan J C 1977 An inexpensive microphotometer system for measuring silver grain densities in auto-radiographs. *Stain Technol.* 52: 79–83

Jarvis L R 1981 Microdensitometry with image analyser video scanners *J. Microsc.* 121: 337–46

Mendelsohn M L 1966 Absorption cytophotometry: Com-parative methodology for heterogeneous objects, and the two-wavelength method. In: Wied G L (ed.) 1966 *Intro-duction to Quantitative Cytochemistry*. Academic Press, New York, pp. 201–14

Pette D 1981 Microphotometric measurement of initial maximum reaction rates in quantitative enzyme histo-chemistry *in situ. Histochem. J.* 13: 319–27

Piller H 1977 *Microscope Photometry*. Springer, New York

Pollister A W, Swift H, Rasch E 1969 Microphotometry with visible light. In: Pollister A W (ed.) 1969 *Physical Techniques in Biological Research*, 2nd edn., Vol. 3c. Academic Press, New York

Prenna G, Mazzini G, Cova S 1974 Methodological and instrumentational aspects of cytofluorometry. *Histochem. J.* 6: 259–78

Rost F W D 1980 Fluorescence microscopy; Quantitative histochemistry. In: Pearse A G E (ed.) 1980 *Histo-chemistry, Theoretical and Applied*, 4th edn., Vol. 1. Churchill Livingstone, Edinburgh

Spurway N C, Chapman J N, Steele J D 1985 The effect of section thickness on distribution error. *J. Microsc.*, in press

Van Oostvelt P, Boecken G 1977 Two-wavelength cytophotometry: The choice of the wavelengths from a practical point of view. *J. Histochem. Cytochem.* 25: 1337–44

Zimmer H G 1973 Microphotometry. In: Neuhoff V (ed.) 1973 *Micromethods in Molecular Biology*. Springer, New York, pp. 297–328

Zs-Nagy I 1975 Transformation of Leitz Orthomat W into cytophotometer. *Mikroskopie* 31: 241–45

N. C. Spurway

Monitoring Equipment in Coronary and Intensive Care

In this article the electronic patient-monitoring equip-ment commonly used in present-day coronary care units (CCU) and intensive care units (ICU) is described. The discussion includes the "classical" individual bedside units used to monitor the electrocardiogram (ECG), blood pressure, temperature and respiration, and also central station equipment that collects and displays information gathered from a number of bedside monitors.

1. General Aspects

Today, the management of the critically ill in coronary and intensive care units is heavily dependent upon the availability of relevant data and information that de-scribe the condition of individual patients. Over the past decade, the explosion in electronic technology—the development of large-scale integrated circuits, micro-processors and multitask computers—has had an enormous impact on the equipment used for patient monitoring.

2. The Coronary Care Unit

The CCU is primarily used for treatment of heart diseases such as myocardial infarction. The most im-portant physiological signal monitored is the ECG, although in particular instances blood pressure and temperature may also be monitored.

When patients improve they are often transferred to a separate area of the CCU equipped with ECG radio-telemetry systems: each patient in this area moves around freely and carriers a small transmitter that allows his ECG to be monitored at a receiving station.

The CCU typically has four to sixteen beds, each equipped with a bedside ECG monitor. A number of the beds may have telemetry installations. The patients' ECGs are routed to a central station and displayed there on oscilloscopes which are constantly observed by a coronary care nurse. If any of the physiological signals (e.g., the heart rate) exceed limits preset for each individ-ual patient, an alarm is activated and the patient's ECG is automatically recorded on paper.

In some CCUs the central station equipment includes a digital computer which performs an ongoing analysis of the patients' ECGs and automatically detects any abnormalities (arrhythmias) in the ECGs and records them.

3. The Intensive Care Unit

An ICU is usually organized to treat a wider spectrum of diseases than the CCU, but is mainly used for the treatment of circulatory and respiratory conditions. The ECG is monitored for all patients and, depending on the patient's condition, additional parameters such as blood pressure, temperature and respiration are also mon-itored. A patient's condition may change continuously and thus the need to monitor the various parameters changes likewise. An example of the average monitoring of the various parameters is shown in Fig. 1. The

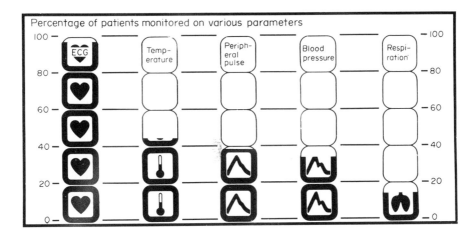

Figure 1
A comparison of the various parameters monitored in an intensive care unit

constant change of monitoring needs for the individual patient has led to the development of modular bedside monitoring equipment.

The number of beds in an ICU is normally between four and twelve. Some patients may be isolated to reduce the risk of infection. The ICU is not usually provided with a central station, but if it is, the station is often equipped with a digital computer capable of storing the data acquired over several days of monitoring. This data, relating to the patients' heart rates, systolic and diastolic pressures, temperatures, etc., can be displayed as trend curves. The computer can also handle fluid balance, medication, laboratory data and other relevant data that describe the patients' conditions and treatments during hospitalization.

4. ECG Monitoring

ECG monitoring is routinely performed in many departments of a modern hospital (see *Electrocardiography*). The ECG is the electric signal pattern generated by the heart. It is usually measured by means of three adhesive skin electrodes of the disposable silver/silver chloride type placed on the patient's chest. The quality of the electrode as well as careful preparation of the skin (to lower the skin/electrode impedance) is important for reliable ECG monitoring. In a modern hospital environment, there are many sources of electrical noise, such as ac interference from other equipment and installations, static discharges, and patient movement, that can degrade the ECG signal.

The ECG has typically a peak-to-peak amplitude of 1 mV, and an information content for monitoring purposes in the frequency range 0.1–30 Hz. A typical ECG signal is presented in Fig. 2. Picked up by the three electrodes, the signal is routed to a differential input

amplifier with an input impedance of several megaohms and a high common mode rejection ratio (minimum 60 dB).

The signal is preamplified, and coupled to the main amplifiers. There is thus, in effect, an isolation barrier ("floating input") created which protects the patient from any lethal voltage developed via the ECG monitor as a result of an electrical fault condition. The signal is amplified and processed before being displayed either on a cathode ray tube or on an ECG recorder.

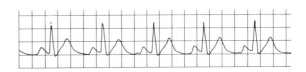

Figure 2
A typical ECG showing a normal sinus rhythm

5. Blood-Pressure Monitoring

Direct (invasive) blood pressure monitoring is performed, primarily in the ICU (and in the operating theater) by introducing either a catheter or a hypodermic needle into the patient's venous or arterial circulatory system. The needle or catheter is connected to a external pressure transducer. Catheters are available in diameters of 1–10 mm and in many different lengths to suit their location for particular measurements.

Physiological pressures do not exceed 300 mm Hg. The pressure measured is the pressure at a hole in the catheter close to its rounded tip. The catheter tip can be directed to almost any position in the circulatory system, including the heart. Entry into the body is usually at an arm or a leg where the arteries and veins are close enough to the body surface for easy introduction of a catheter or needle.

The transducers used for blood pressure measurement are normally based on semiconductor strain-gauge elements that form a Wheatstone bridge. The change of resistance in the pressure-sensitive parts of the bridge produces a change in output voltage of the bridge. The sensitivity of modern pressure transducers is some $30\,\mu V/mmHg$ per volt of bridge excitation voltage. The signal from the transducer is connected to the pressure monitor via a cable. In the monitor the signal is amplified and processed in a "floating input" isolation amplifier. Signal processing extracts characteristic features from the pressure waveform, such as maximum (systolic) pressure, average (mean) pressure and minimum (diastolic) pressure. As in the ECG monitor, an alarm is generated if manually preset limits are exceeded.

Indirect (noninvasive) blood pressure measurement is much more common. The well-known sphygmomanometer—incorporating a hand pump, an inflatable arm cuff, a pressure meter and a stethoscope—is used. An electronic version of this familiar device has gained increased popularity over recent years. The inflation and deflation of the cuffs, and the calculation of systolic, mean and diastolic pressures are done automatically at preset intervals. The instrument displays the most recent pressure values digitally. In some cases, the values and waveforms are recorded on paper.

6. Peripheral-Pulse Monitoring

Measurement of peripheral blood flow (usually in the fingers and toes) is based on the measurement of blood's absorption of light in the red and infrared region of the spectrum. The method gives a rough assessment of the local blood flow near the skin surface; it is not in widespread use because it is not very accurate and gives only relative values for local circulation.

7. Temperature Monitoring

Measurement of the patient's temperature is probably the oldest routine physiological measurement performed in clinical practice (see *Clinical Temperature Measurement*). The temperature in different parts of the patient's body gives valuable information on blood circulation, perfusion, and metabolism and infection.

Different types of temperature-sensing elements are employed. A thermistor is commonly used built into the tip of a probe that is shaped according to the site where it is used (i.e., oral, skin, rectal, esophageal). In some cases PTC resistive materials are used instead of thermistors. Thermocouples are employed for probes with very small dimensions (e.g., needle probes). The signal is carried from the probe to the monitor via a patient cable and amplified. The measured value is subsequently dis-

played. Again an alarm is generated if the temperature recorded exceeds the preset high or low limit.

8. Respiration Monitoring

Respiration monitoring is clinically important in several critical care areas of the hospital including the ICU and the neonatal intensive care unit. For ICU patients on a ventilator basic respiration monitoring includes the measurement of tidal volume, minute volume and respiration rate.

Respiration monitoring of patients, who are breathing without assistance, is usually by means of a thermistor probe placed in the patient's nostril. Alternatively the changing impedance between two ECG type electrodes placed on the patient's thorax is monitored. Both methods are used to calculate the number of respiration cycles per minute (i.e., respiration rate). The impedance method may give a relative value of air volume as the impedance change is approximately proportional to the volume of air inspired (see *Respiratory Function: Methods of Assessment*).

9. Transcutaneous Gas Monitoring

Measurement of the partial pressures of oxygen and carbon dioxide in a blood sample is a useful and important clinical procedure (see *Blood Gas Analysis*). For many years, such measurement of blood gases has been done in laboratories and wards using samples of blood withdrawn from the patient. The inconvenience and time delay involved in analyzing a blood sample has been partly overcome by a technique for measuring the gases that diffuse from the blood vessels through the skin. Oxygen and carbon dioxide skin electrodes have been developed over recent years, and the measurement of both gases has gained popularity in certain areas of the hospital, primarily the neonatal intensive care unit. Although for adult patients there seems to be a poorer correlation between the transcutaneous (skin) measurements and the corresponding blood sample values, the new method does give very valuable clinical information.

10. Future Trends in Patient Monitoring

Patient monitoring equipment is closely related to electronics technology. The rapid development in this latter field will obviously have an enormous impact on future management of physiological signals and patient data.

The availability of microcomputers has already brought more intelligence to both the bedside and central station equipment. The result will be more signal processing giving higher quality (artefact-free) signals; combination of more criteria to produce qualified alarms of high reliability; increased data storage and data management, primarily at the bedside; and more diagnostic programs that automatically check the equipment itself, improving the in-serviceability and operating reliability. Monitoring of respiratory data, fluid balance and other therapeutic data together with on-line access to laboratory data will eventually automate the patient's record and facilitate the decision processes that relate to management.

See also: Neonatal Intensive Care Equipment; Physiological Measurement

Bibliography

Hayes B, Healy T E J 1973 Equipment for intensive care. *Br. J. Hosp. Med.* (Equipment Suppl.) 9: 4–32
Hill D W, Dolan A M 1982 *Intensive Care Instrumentation*, 2nd edn. Academic Press, London
Rithalia S 1978 Equipment in the ITU. *Nurs. Times* 74: 713–16
Rolfe P 1976 Monitoring equipment for the neonate. *Br. J. Clin. Equip.* 1: 189–205
Sheppard L C 1979 The computer in the care of critically ill patients. *Proc. IEEE* 67:1300–6

B. Holte

Muscle

In complex animals, muscles are the sources of all but a tiny fraction of total motive power. This is true of internal motions, such as that of blood, or ingested food, as well as external motions, such as that of limbs. Different classes and subclasses of muscle are specialized for these various functions. Those of mammals (including man), with which this article will be concerned, are shown in Fig. 1. Only minor additions would be needed to extend the account to other vertebrates, but among invertebrates much greater diversity occurs.

Appearance	striated		smooth	
Location	skeletal	heart	visceral	vascular
Control	voluntary	involuntary		
Cell size	immense	large	medium	small
Response	fast	relatively slow	slow	very slow	maintained 'tone'

Figure 1
The major classes of mammalian muscle. The three classical classes are skeletal, heart and smooth muscle; only major subclasses are indicated. The last line indicates typical values, relative to that of fast, skeletal muscle, of unloaded shortening velocity (V_0 in Fig. 5)

Each form of muscle may properly be thought of as a linear motor, with parameters adapted, over a wide range, to particular requirements. It is important to appreciate, however, that in many cases tension with little or no motion is what is produced when a muscle is activated: the muscles controlling blood-vessel diameter, or maintaining posture, may remain sensibly static for minutes or hours. Even a high-speed gait is not quite what it appears, for recent studies show that a large part of the energy required for each stride comes from the elastic recoil of tendons which have been more importantly stressed by muscles than moved by them. Muscles also have rather limited ranges of possible motion. Skeletal muscles characteristically change their length by only about 20% in a maximum movement; and the length-flexibility of smooth muscles in organs like the bladder is only an order of magnitude greater, yet is attained at considerable cost in speed and control. Thus it is perhaps most fruitful to regard muscles essentially as force-generators, though possessing some tolerance of movements in the structure to which the force is applied.

The long-established word, contraction, for the events which follow muscle activation in this sense embodies an outdated view of what is essential; in any case, it has always been applied to static force generation, and even to the condition of a muscle opposing a superior extending force, as well as to active shortening.

1. Structure

The unit of growth in muscle, as in all biological tissues, is the cell. Skeletal muscle cells, or fibers, are cylinders of cytoplasm which may be many centimeters long, each containing many hundred nuclei; fiber diameters typically range from 15 μm in an external eye muscle to 150 μm in a large leg muscle. Cardiac and smooth muscles each have much smaller cells, governed by one nucleus each.

1.1 Contractile System

The major feature of the cytoplasm in all classes of muscle is an array of protein filaments, parallel to the fiber axis. Two kinds of filament are of interest: actin, a protein widely found in motile cells; and myosin, found in substantial concentrations only in muscle. The actin filaments are thinner (about 5 nm everywhere) while those of myosin range upwards from 10 nm. In skeletal and cardiac muscles, ranks of one kinds of filament are arrayed transversely across the whole cell widths, alternating and interdigitating with equivalent ranks of the other kind of filament (Fig. 2). The resultant alternations of refractive index and staining characteristic, being of periodicity about 2.5 μm, are resolvable with the light microscope and give rise to the designation "(cross)-striated" for these two classes of muscle. By contrast, in smooth muscle, actin and myosin filaments appear randomly overlapped throughout the cytoplasm.

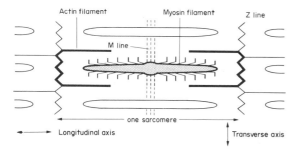

Actin filament   Myosin filament   Z line

M line

one sarcomere

Longitudinal axis   Transverse axis

Figure 2
Force-generating unit of striated muscle. The element shown bold normally occupies about $2\frac{1}{2}$ μm longitudinally, repeating $\sim 10\,000$ times in all three axes in a single skeletal-muscle fiber. Transverse dimensions are exaggerated relative to longitudinal; only about 1/3 of the head-groups (cross-bridges) on either side of the myosin filament, in the plane of the paper, are represented. Head-group spacing in this plane is 42.9 nm; other head-groups project 30° either side of the normal to the paper plane in a 6/2 helical array, longitudinal spacing 14.3 nm. Transverse support components (Z and M lines), not named in text, are also shown

Interactions between actin and myosin are the means by which muscle generates force (see Sect. 2), and the sliding of one kind of filament past the other is the mechanism of length change. (This sliding filament account of muscle action, substantiated particularly by H E Huxley in the 1950s, represented one of the first triumphs of the electron microscope in biology; previously the assumption had been that the individual protein filaments changed their length, probably in concertina fashion.) Chemically reactive knobs, called head-groups, protruding at regular intervals from the myosin filaments of all classes of muscle, are now regarded as the sites of force-generating interaction with actin. In striated muscles, each myosin filament is bare of head-groups at its center; on respective sides of this bare region the groups are oppositely oriented, so that from either end their action is to pull the overlapped actin filaments towards the middle.

1.2 X-Ray Diffraction

This crystallographic technique has contributed as much in recent years to the understanding of muscle as has electron microscopy, principally because it provides information (though not always unambiguous information) about the movements of the head-groups during force generation. Results from active muscle are often compared, not only with the relaxed state, but with the other extreme, rigor; this is a state of inextensibility, representing incipient death, in which the maximum possible number of head-groups (over 50% of the total) are considered to be attached to actin.

The basic repeat periodicity along the muscle axis (meridional) reflections is 14.3 nm. This represents the longitudinal separation of adjacent head-groups, and is a value which changes only about 1% between one state and another (changing actually in the direction which suggests slight extension of the myosin on activation). Transverse dimensions also stay constant unless muscle length has been allowed to change; in which case they vary with the square root of the length indicating that the lattice volume, like that of the whole fiber, remains constant. However, the transverse (equatorial) reflections indicate that mass moves from the thick toward the thin filaments upon activation, and even more so in rigor, suggesting movement of head-groups away from the myosin cores to make contact with adjacent actin filaments. (When a head-group is thus bound to actin, what is properly called a cross-bridge (XB) exists. However, for want of a separate term, a head-group plus the flexible arm which links it to the myosin filament is called an XB, whether attached to actin or not.) Other changes, particularly of the oblique (off-meridian) reflections, indicate that the head-groups swivel in every possible sense such that their outer ends adopt, in rigor, the periodicity of the actin filaments (which differs from that of the myosin) while in activity they oscillate asynchronously between this and the resting position.

These x-ray observations have contributed powerfully to present views about the mechanism of force generation (Sect. 2). The essential events are assumed to be similar in smooth muscle, which lends itself less favorably to crystallographic investigation. Before the question of force generation is taken further, three kinds of intermediate-scale structure must be described. All are concerned with activation.

1.3 Membranous Systems

In every class of muscle a portion of the intracellular volume is closed off from the rest of the cytoplasm (and thus doubly isolated from the extracellular space) by an internal membrane system called the sarcoplasmic reticulum (SR). In biochemical isolates, SR membranes display a highly developed ability to take up calcium ions (Ca^{2+}). In intact muscle, at rest, the space within the SR membrane has long been regarded as sequestering Ca^{2+} and this has been confirmed by Somlyo et al. (1981) using x-ray emission microanalysis. Activation of the contractile process occurs when Ca^{2+} is released to diffuse among the protein filaments. The SR is so arranged that the diffusion distance never exceeds a few micrometers: in cells of greater radius the filament system is divided longitudinally into fibrils, themselves of 0.5–1.5 μm radius, which run the whole length of the cell and are enveloped throughout that length by a collar of SR (Fig. 3). Thus the activation (and relaxation) processes are not seriously retarded by diffusion delays.

Wherever this fibrillar structure is encountered (e.g., in all mammalian skeletal muscle fibers and in the main, "ventricular" cells of almost all mammalian hearts) an additional system of membranes, the transverse tubular system (T system), interrupts the SR at regular intervals,

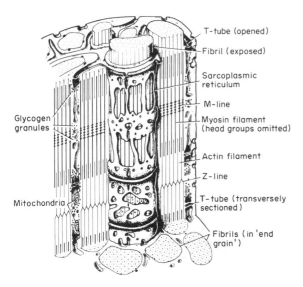

Glycogen granules

Mitochondria

T-tube (opened)

Fibril (exposed)

Sarcoplasmic reticulum

M-line

Myosin filament (head groups omitted)

Actin filament

Z-line

T-tube (transversely sectioned)

Fibrils (in 'end grain')

Figure 3
Ultrastructure of skeletal muscle. Part of the edge of one fiber is depicted. At left, two metabolic features which may be seen in electron micrographs are shown: glycogen is the stored form of glucose, mitochondria are the sites of aerobic ATP synthesis. This is a fast fiber. Slow skeletal muscle, and even the majority of cardiac muscle fibers, would differ only in details, but smooth muscle is much more loosely organized

once or twice per sarcomere of the enveloped fibril, and the two membranes make contact at ultrastructurally specialized junctions. The T tubes are invaginations of the surface membrane, containing extracellular fluid within their lumens; they act to convey electrical excitation from the surface to the depth of the fiber.

1.4 Muscle Cell–Muscle Cell Junctions

Cardiac and many smooth muscles have each cell linked to its neighbors by regions of closely apposed surface membranes. The best-recognized form of apposition is the gap junction (formerly known as the nexus). Here the two membranes are only about 1.5 nm apart and are cross-linked by what appear to be hollow protein plugs, through which the cells can communicate both electrically and chemically. Recently, intermediate junctions have also received attention. These are regions of moderately close apposition ($\sim$20 nm) but much greater membrane area; they are found where cells are electrically (not chemically) coupled, yet gap junctions are absent. Neither of the junctions found in smooth muscle has mechanical strength—smooth muscle cells are tied together by connective tissue fibers. However, cardiac gap junctions are always closely associated with regions of cell–cell adhesion; the total zone of close contact therefore affords both electrical and mechanical coupling. Such a zone, appearing like a butt joint between

two cells in a light microscope preparation, is termed an intercalated disk.

No such cell–cell junctions occur in mature skeletal muscle, but there the multinucleate fiber itself represents the total fusion of perhaps several thousand precursor cells.

1.5 Nerve Cell–Muscle Cell Junctions

The neuromuscular junction is a form of synapse (see *Bioelectricity*). At its largest, where a motor-nerve branch terminates against a skeletal muscle fiber, it is specifically termed a motor end plate. As well as in its exceptional size, this formation is an unusual synapse in that the subsynaptic (muscle) membrane is deeply infolded. In the large, strong muscles of limbs and trunk, one motor axon may branch to supply several hundred muscle fibers. Nevertheless, after infancy, mammals hardly ever have more than one end plate per muscle fiber, each supplied by a single nerve branch. The motor axon and the family of muscle fibers under its control are together termed a motor unit.

In cardiac and smooth muscles, which possess means to spread excitation from muscle cell to muscle cell, nerves directly affect only a small sample of the total population of cells: their influence is then relayed through the mass. There are few well-formed synapses, and no infoldings of the underlying membrane, in these tissues.

2. Force Generation

The electron-microscope and x-ray diffraction data described above (Sects. 1.1, 1.2) are widely interpreted as indicating that force is generated when myosin head-groups reach out, bond to actin, and then swing backward. Thus they create a force tending to pull the temporarily attached actin filament towards a position of greater overlap with the myosin. This is analogous to an oar-stroke, though in this case the "stroke" is only some 10–12 nm. Then, the view is, the head-group detaches, and swings forward again (just as the oarsman comes forward on his seat) before reaching out, re-attaching to another site on the actin filament, and repeating the cyclic process. The questions which arise as to the nature of the force causing the XB to change its conformation (the "pull stroke"), and the overall source of energy for the cycle, are as much biochemical as they are biophysical.

It is certain that the ultimate energy source, as in most other endergonic cellular processes, is the free energy released when adenosine triphosphate (ATP) is hydrolyzed to the diphosphate (ADP). It might have been assumed that the hydrolysis would occur during the pull and that automatic detachment would occur at the end. In fact, results of recent research have suggested a mechanism as shown in Fig. 4. It has become clear that:

(a) Only the final stages of a complex hydrolytic reaction sequence take place during the pull stroke,

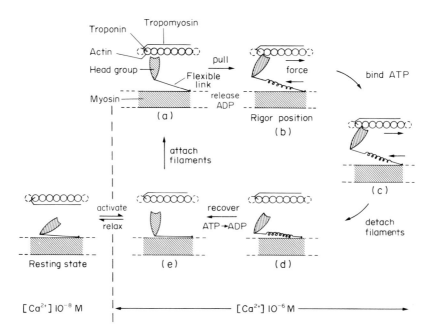

Figure 4
The cross-bridge cycle in striated muscle. Diagrams indicate main geometrical changes attributed (principally on the basis of x-ray diffraction data) to the myosin head-group and its flexible link (the cross-bridge, XB) and to the troponin–tropomyosin complex. Parts (a) to (e) in the cycle refer to text paragraphs in Sect. 2

though these are indeed the stages which release the greatest energy.

(b) Far from detaching automatically, head-groups remain tightly bound to actin when left to themselves at the end of the pull stroke; indeed this is their position in rigor.

(c) The only molecule for which myosin has a higher affinity than it has for actin is ATP. It is the availability of this substance in live muscle which allows the bridge to be broken and the head-group to swing forward again, in preparation for the possibility of another stroke; and it is the unavailability of further ATP that leads to rigor when metabolism has ceased.

(d) The initial hydrolysis of the ATP molecule takes place when it is bound to myosin alone—probably during the head-group's forward swing. The main energy release, however, is deferred until the products of hydrolysis detach (the next repeat of process (a)).

(e) Provided activation persists the head-group now reattaches to the actin filament, and the cycle repeats.

In Fig. 4, steps (b) and (c), no filament movement is shown to have taken place in response to the force

generated; that is, the diagrams refer to an ideal isometric contraction. The representation of XB compliance as residing in the flexible arm (biochemically known as the S_2 subunit) is a commonly adopted diagrammatic convenience; bending of the head-group itself (S_1 subunit) would be equally compatible with the evidence.

The treatment above is simplified. For example, strong arguments may be offered for the occurrence of not less than two physically and chemically identifiable intermediates on the pull stroke, and for at least one chemically identifiable intermediate on the recovery sequence. Moreover, although thermodynamic analyses (developed to a particularly sophisticated level by Eisenberg and Hill 1978) show that the cycle will be driven by the empirically observed energetics of the various chemical steps, the detailed molecular mechanisms of these steps and of the associated conformation changes in the myosin are not understood.

There are even some who doubt whether force is created by angular head-group movement at all. Other theories range from shortening S_2 subunits to electrostatic attraction between myosin and actin. It must certainly be admitted that spectroscopic probes have failed to provide consistent evidence of appropriate XB movement. For more information on this subject, the reader should compare Huxley (1980), for the orthodox view, with papers and discussions concerning difficulties in Pollack and Suai (1984) and Morel (1985).

3. Activation and Relaxation

The processes described in Sect. 2 take place only when Ca^{2+} ion concentration within the myofibrils is of order 10^{-6} M or more. In striated muscles, Ca^{2+} is required to bring about an adjustment on the thin filament, without which head-groups are unable to attach to actin. The adjustment is in one of two minor constituent proteins of the thin filament, troponin—a molecule which readily takes up Ca^{2+} and changes shape in doing so. Each troponin molecule partners one molecule of the other minor constituent, the rod-like tropomyosin, which is thought to be held by the troponin just above the surface of a length of the actin filament (Fig. 4). At 10^{-6} M Ca^{2+} the tropomyosin is held clear of the sites at which myosin can attach. At 10^{-8} M, however, the shape of the troponin is such that tropomyosin rods completely block access of myosin to the actin.

Calcium ions are released into the myofibrils of skeletal muscle from the SR, which responds in this way to depolarization of the adjacent T-tube—such as that caused by an action potential. In smooth, and to a lesser extent cardiac, muscle, substantial additional Ca^{2+} entry occurs through the surface membrane of an active cell from the extracellular fluids. When excitation ceases in any form of muscle, resorption of Ca^{2+} into the SR, or its reextrusion through the surface membrane, being processes directed against concentration gradients, require energy and so themselves involve (directly in the case of the SR, partly indirectly at the surface membrane) the consumption of further ATP.

Once the troponin/tropomyosin block has been reestablished, the muscle, though it never actively lengthens, will be extensible again by the least external force. However, studies of, for example, a slow, late phase of relaxation, suggest that the total transition between rest and activity may also involve changes in the myosin, as indicated diagrammatically in the bottom left of Fig. 4. In smooth muscle, it is probable that troponin and tropomyosin play no part and myosin changes are the principal changes involved in activation.

4. Excitation

All mammalian muscle cells can be electrically stimulated and most normally are. (The major exceptions are the smooth muscle cells of arterial walls.) The ionic mechanisms of their respective action potentials (APs) are described elsewhere in this Encyclopedia (see *Bioelectricity*); so are the mass external effects of these APs, which give rise to the possibilities of electromyography and electrocardiography.

4.1 Electrical Excitation

Skeletal muscles, except in disease or injury, are excited only electrically, and only by their incoming (motor) nerves. Since motor-nerve activity originates in the spinal cord, and cannot (except, perhaps, when branches fatigue) avoid invading all branches of the nerve at the periphery, no smaller group of muscle fibers than those comprising one whole motor unit can be activated. Thus the motor unit is the unit of function; small units (having a few, small muscle fibers) are used for fine movements, while both synchronization and economy are assisted by the availability of large units (many, large fibers) which are recruited for large forces. (Infoldings of the subsynaptic membrane are most pronounced in these large fibers. Presumably this is because they have the lowest input resistance, and so require an increased area of current sink to succeed in triggering an AP.)

Heart and electrically active smooth muscles are "autorhythmic," driven by action potentials which arise spontaneously in their own pacemaker regions and spread throughout the tissue by muscle cell/muscle cell junctions. Consequently, the action of nerves in these tissues is not to initiate activity, but only to modulate it. The frequency of activation can be modulated only at the pacemaker(s); the mechanical strength of the response is influenced elsewhere (e.g., by the innervation of the ventricles of the heart).

4.2 Neurotransmitters

The transmitter substance at the skeletal neuromuscular junction is acetylcholine (ACh); its action here can be blocked by curare (the drug of Amazonian-Indian poisoned arrows) with consequent paralysis. In the heart, ACh has an almost opposite effect—it slows the beat, by increasing potassium conductance during the interval between APs, and so is the transmitter substance of the inhibitory (parasympathetic) nerves. The excitatory (sympathetic) nerves employ chiefly noradrenaline (NA); this accelerates the frequency (a pacemaker effect) and enhances the strength of the beat (ventricular effect). Both these NA effects appear to arise from increased inflows of Ca^{2+} during the AP. (These mechanisms have been elucidated chiefly in the laboratories of Reuter and Scholz (1977), and Noble (1979).)

Of smooth muscles, it is necessary to consider at least two groups: those which are electrically activated and those which are not. The largest group of electrically activated smooth muscles are those forming the walls of the intestines and stomach. These have been studied particularly by E Bulbring and her colleagues. They are autorhythmic, and their contraction waves are both weakened and slowed by NA (sympathetic), but enhanced in both respects by ACh (parasympathetic). (Parasympathetic effects of ACh are everywhere blocked by atropine, not curare.) The principal electrically quiescent tissues are the majority of large blood vessels. These are only sympathetically (NA) innervated and the effect of the transmitter is to enhance their steady tension ("tone"), and so constrict the vessels. There is some evidence that this effect is achieved entirely chemically, without even a dc change of membrane potential.

4.3 Hormones and Metabolites

Hormones are information-carrying substances which are carried in the blood, rather than released from nerve terminals near the site of action. Many hormones affect the metabolism of skeletal muscle but none have strong influences on its contraction. The case for heart is different only in respect of adrenaline, the "emergency hormone," which is chemically similar to NA and has similar effects. Smooth muscles, however, whether electrically active or quiescent, are all profoundly sensitive to a wide range of hormones; those inducing labor in the uterus are an important example in which—this being an autorhythmic tissue—the effects are electrically as well as mechanically striking. Finally, many smooth muscles are highly sensitive to the chemical consequences of metabolism; for example, blood vessels dilate when the tissue which they supply runs short of oxygen or accumulates carbon dioxide.

5. Mechanics

Muscle contractions are usually studied under one of two conditions: *isometric* (tension-changes recorded at constant length) or *isotonic* (length-changes at constant tension). In skeletal muscle work there are also two extreme modes of activation: the single twitch (which is produced by one AP) and the tetanus. This latter is a sustained, smooth contraction, produced by repetitive electrical stimulation at sufficient frequency that myofibrillar Ca^{2+} levels do not fall below 10^{-6} M after one AP before being raised again by the next. Because twitches do not allow time for internal compliances (series elasticity) to be taken up, tetanic behavior is easier to analyze and will be implied in the next three subsections. In cardiac and smooth muscles, one or other of the above modes of activation is often not available, but reference in what follows is confined to the longest achievable activation.

Evidence that the slow achievement of full tension is due to internal elasticity, and not to slow activation, was provided in the 1940s by A V Hill. If a muscle is loaded a few milliseconds after activation to the tension it is capable of holding during a steady tetanus, it will—within a few percent tension and a few milliseconds resolution—hold that tension till activation ceases. The interpretation is that, the series elasticities having been stretched by the load, the muscle has no internal work to do; therefore it immediately shows its full internal tension to the external transducer.

5.1 Length–Tension Plots

Every muscle, even when unstimulated, resists stretch when it has been extended (or the hollow organ it bounds has been distended) beyond a certain length. For skeletal and many smooth muscles this length is close to the mean of its anatomical working range, though it is above that range for a healthy heart. The resistance is not due to the muscle fibers themselves, but to connective tissue (collagen) strands which enwrap them. The resulting stress–strain curve is termed the passive length–tension plot, and is very non-Hookean (Fig. 5, relaxed curve).

The equilibrium length for activated muscles is well to the left of the previous curve, but represents a degree of shortening never attained by any muscle *in vivo* and never approached by intact skeletal muscle. Whether the overall active length–tension plot (Fig. 5, active curve) has an inflexion, shoulder or clear-cut maximum, before approximating to the resting curve at high lengths, it varies greatly from muscle to muscle, according to the exact length at which the passive component becomes substantial.

The two curves so far discussed are those which must be considered along with the continuously varying contributions of skeleton geometry, in all attempts to relate body statics to muscle performance. For the understanding of muscle itself, however, the difference curve in Fig. 5, representing the tension contributed by the muscle's activity, is far more important. The absolute linearity of the long declining part AB, as actin filaments are drawn out of overlap with myosin head-groups, is exactly in accord with the concept of completed XBs as

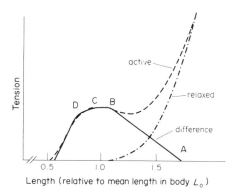

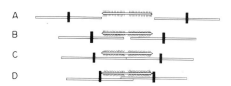

Figure 5

Length–tension plots for skeletal muscle: the active and difference plots are for isometric stimulations at a series of preset lengths. The difference curve was obtained from the central sarcomeres of a single fiber (Gordon et al. 1966). For these sarcomeres, the degrees of filament overlap at points A–D are shown beneath. With nonskeletal muscles, mean in-body length is not precisely definable, but L_0 is usually taken as that at which the difference curve is at its maximum

force generators; even more strikingly so is the short preceding plateau, arising where actins overlap the parts of the myosin filaments bereft of head-groups. As the fibers shorten below this plateau length, actins interfere, first with each other, then with myosin heads oriented to work in the opposite sense. Below the plateau, moreover, activation itself falls off. This is particularly important in heart muscle, as the healthy heart works on this part of the curve. Thereby it automatically has more force at its disposal when filled further—an inbuilt regulation, first recognized by E H Starling in 1914.

Smooth muscle, with its less precise juxtaposition of actin and myosin filaments, has a plot of length against actively produced tension which can be up to twice as wide as the difference curve in Fig. 5, and usually shows the ascending part less steep than the descending.

The maximum force producible by a muscle's activity is proportional to its cross section. A widely quoted upper limit is $400 \, \text{kN m}^{-2}$, but values only a little more than half of this are usually obtained experimentally. Interestingly, there is no convincingly established difference between the types of muscle in this respect.

5.2 Force–Velocity Plots

For the shortening of all types of muscle, except possibly heart, force-velocity plots (Fig. 6) are hyperbolic, conforming to Hill's formula

$$V = b(F_0 - F)/(F + a)$$

where F_0 is the isometric force, F the force at shortening velocity V, and a and b are constants. Experimental difficulties are encountered in analyzing smooth muscle, although similar curves to those in Fig. 6 (well fitted by rectangular hyperbolae for $F \not> F_0$) have been obtained. For heart muscle, however, many preparations seem to

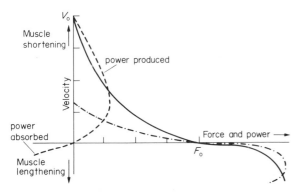

Figure 6

Force–velocity and power–velocity curves for two forms of skeletal muscle: —, force–velocity curve for fast muscle; –·–, force–velocity curve for slow muscle. The curves apply to both active shortening and forced extension. Power curve (– – –), plotted for fast skeletal muscle but with arbitrary abscissal units, indicates form for any tissue having hyperbolic F–V relation

show peak V values when $F > 0$. Unlike F_0 per unit cross section, V_0 (V for $F = 0$) varies greatly between specimens. Compared with a fast skeletal muscle, a slow skeletal muscle from the same species may be able to move 1/3 as fast, heart muscle 1/10, and smooth muscles from 1/30 to 1/100. However, for all muscles, it follows from the hyperbolic form of the relation that maximum mechanical power ($\sim 0.1 \, F_0 V_0$) is obtained when F and V are both about 30% of peak values. Although this purely mechanical maximum is to be distinguished from chemomechanical conversion efficiency, studies of the few forms of whole-body exercise which can be accurately investigated, notably cycling, suggest that proficient performers, even in events lasting tens of minutes, work remarkably near to peak power rates.

Shortening velocities are informative in relation to the concept of XBs as independent force generators. Firstly, V_0 proves independent of the degree of filament overlap. This is to be expected if movement is generated by individual XBs acting in parallel. Secondly, the form of the force–velocity curve has been accounted for by conjunction of two features—that when sliding occurs a number of XBs become overbent, resisting shortening until they detach, and that the total number of links formed decreases with shortening speed. Supporting this latter, sudden stretches applied while a muscle is shortening rapidly find it more extensible than when at rest.

Conversely, XBs which form and then are bent backwards by the muscle being actively stretched also require force to break them—more force, this time, than they would exert during an isometric pull stroke, such as is illustrated in Fig. 3. Hence the forces $> F_0$, with which a muscle resists forcible extension.

5.3 Tension Transients

Detailed information about the kinetics of XB movement is being obtained (Huxley 1980) by imposing step length changes (positive and negative) on active muscle fibers, and studying the time course of tension restoration with sub-millisecond resolution. Many of the conclusions from such work have been implied in the earlier parts of this section, and in that on force generation.

6. Metabolism

6.1 Biochemical Basics

Muscles make their own immediately utilizable fuel, adenosine triphosphate (ATP). In this they are unlike most man-made chemical machines, but are somewhat analogous to diesel-electric engines. Alternatively, and much more closely—since the diesel engine is a heat engine, which no part of the biological system is—they may be likened to electric motors driven by fuel cells.

Smooth muscles can make ATP as fast as they use it. This is not true of striated muscles, all of which—especially the faster skeletal muscles—make use of a "buffer" system (in the electrical analogy, an accumulator) to extend by an order of magnitude the time for which they can work at maximal rate. The energy currency is the third phosphate group of ATP: the "high-energy phosphate" ($\sim$P). The buffer molecule is creatine, which can reversibly exchange $\sim$P with ATP: creatine is present to only modest extents in smooth muscles but in striated muscles it is found at much greater concentration than ATP and its precursors. Thus the concentration of ATP itself does not fall detectably in striated muscles until absolute exhaustion is imminent—a fact which formerly caused confusion. Several advantages of the buffering can be proposed. One stems from the fact that the activities of many metabolic enzymes are strongly controlled by the levels of ATP and its dephosphorylated relatives; a cell storing large quantities of ATP would therefore have had to evolve a set of enzymes radically modified from those of other cells, to do the same jobs.

ATP can be made in two ways: aerobically (using oxygen) and anaerobically (without oxygen). The former is, in its use of primary fuels, both more versatile and more economical: glucose (principally from carbohydrate foods) and triglycerides (from fat) can both be utilized, and both are oxidized fully to carbon dioxide and water—which is as far as biologically available exergonic processes can go. Yet to make this possible in all the muscle fibers ever likely to work simultaneously would require an immense cardiac and respiratory apparatus—so big as to be an evolutionary disadvantage for the majority of the time. So muscle fibers which are rarely used (a group which includes the large motor units, recruited only for extreme effort, in most muscles) are run on the alternative, anaerobic system. This utilizes no substrate but glucose, and so is termed glycolysis. It yields in the short term one thirteenth as much ATP per mole of glucose used (and in the long term still less), and leads to the rapid accumulation of lactic acid—a substance of which only limited quantities can be tolerated. Nevertheless, bursts of force or power output can be obtained by recruiting anaerobic ATP supply which are 2–3 times greater than those achievable by remaining in aerobic balance. Glycolysis is the mechanism used to maintain $\sim$P levels by 100 m sprinters or weight-lifters, who may not breathe at all during a race or single lift. Lactate is, however, subsequently reoxidized, mainly in the liver. The oxygen necessary to do this after an anaerobic event, and to rephosphorylate muscle creatine after any event, constitutes the "oxygen debt" repaid by the performer during the period of vigorous breathing which follows exercise.

Standard biochemical techniques, which have provided most of the data underlying the above outline, require destruction of the specimen studied. Two sophisticated biophysical methods both avoid this: they are fluorescence spectrophotometry and nuclear magnetic resonance (NMR). The former, perfected by B Chance, has given valuable information about the redox states of critical metabolic intermediates, and has time-resolution

sufficient to follow, for example, the variations during a heart-beat. Phosphorus-31 NMR, as performed by D R Wilkie and colleagues, is significantly advancing knowledge of $\sim$P metabolism in the muscles of intact human limbs.

6.2 Heat

It is common knowledge that muscle activity generates heat. Measurements of muscle heat output, especially the highly sensitive ones performed in A V Hill's laboratory, were a prime component of biophysical research one or two generations ago. Interpretation, however, has never been easy—all chemical and many physical processes occurring contemporaneously within the muscle are pooled; so are both entropic and energetic changes. To minimize problems of time resolution, many studies have employed muscles of cold-blooded animals, which can operate at 0 °C, but extrapolation to mammalian preparations and higher temperatures has proved hazardous. Finally, apart from the incidental interest of calculating efficiencies (Sect. 7), the point of thermal measurements is to provide a check on the completeness of the balance sheet of processes associated with contraction; this requires that thermochemical parameters be accurately estimated for the conditions prevailing in cytoplasm, not test-tubes. Despite all these difficulties, the challenge of thermal measurements has attracted many able investigators.

Hill divided the heat output into initial heat, which is metabolism-independent and directly associated with the contraction itself, and recovery heat, which happens to be of similar magnitude in some preparations under aerobic, isometric conditions, and signifies the metabolic processes which follow. (The two classes of process overlap in mammalian skeletal muscle and virtually coincide in the rapidly metabolizing mammalian heart.) Initial heat was itself subdivided into the heats of activation of any contraction, of maintenance of any tetanus, of shortening if allowed to occur, and of relaxation. It is a matter for research how far these phenomenological labels represent separate intracellular processes. Activation heat now seems principally ascribable to the binding of Ca^{2+}, released from the SR, onto troponin and other proteins. This recently recognized source of heat would, of course, be present in all forms of muscle activity, and has very lately come to be seen as explaining many previously unexplained shortfalls of thermal accounting. Maintenance heat is probably due largely to ATP hydrolysis—partly on the SR as Ca^{2+} is recycled, and partly at the XBs as they maintain tension. Shortening heat (that released, in excess of the maintenance level, when muscle shortens) shows many forms of variation not originally envisaged. Also Homsher et al. (1981) have demonstrated that it can involve little or no hydrolysis of $\sim$P. Current thinking is that, during rapid shortening, XBs lose phosphate (which dissolves and gives heat) but not ADP; slowly, thereafter, the ADP is displaced by fresh ATP, and only then can a new cycle commence. Isotonic relaxation heats include a large contribution from work done on the muscle by the load. Isometric relaxation heats are attributed mainly to reversal of the activation and force-generating processes.

One other classical result, first appreciated by Fenn in the 1920s, retains heuristic value in that, since heat is evolved rather than absorbed during shortening against load, muscle is a machine which mobilizes greater energy when doing external work. This basic result is normally considered to hold for both cardiac and smooth muscles. In smooth muscle, however, energy output at maximum work rates is only about twice that at rest; for striated muscles, the figure ranges from 10- to 100-fold.

7. The Muscle as a Machine

7.1 Efficiency and Economy

Within a muscle cell there are no significant temperature gradients; as stressed above it is a fuel cell not a heat engine. Therefore, the net heat output of a contraction–relaxation–recovery sequence constitutes waste, from the standpoint of force production. (From the standpoint of body thermoregulation it is beneficial in cold environments but problematic in hot ones.)

Consider the processes reflected in heat outputs: during slow shortening, individual XBs probably use $\sim$P with something like 90% efficiency, but every other hydrolysis of $\sim$P—notably that by SR—is parasitic. Also only about 50% of the chemical free energy released by degrading glucose and fat is converted to $\sim$P in the metabolic (recovery) processes: aerobic metabolism differs little from anaerobic in this respect. Overall, skeletal muscles can attain peak efficiencies of some 20%; the optimum conditions are shortening, during brief tetanus, at about $0.2V_0$ (i.e., against roughly $0.5F_0$). However, $\sim$P consumption being reasonably commensurate with need between perhaps 0.25 and $0.7F_0$, a range of working speeds, including that of peak power (see Sect. 5) are only a little less efficient. What is less efficient is a single twitch, where much of the work done is internal.

Cardiac muscle appears able to attain percentage efficiencies in the high 20s; it releases less Ca^{2+}, allows it to act longer before resorption, and perhaps also metabolizes rather more efficiently than skeletal muscle. Smooth muscle, on the other hand, with its high fraction of metabolism going on even at rest, and so contributing nothing to the doing of work, probably rarely attains 10% efficiency.

What matters more for most smooth muscles and many skeletal ones (e.g., for posture and resilient gaits), is the economy with which steady tensions are maintained. (As no work is done, efficiency then is zero.) Tonic forms of smooth muscle may, by this criterion, be at least 100 times better than fast skeletal muscle: generally, as between muscle types, there is rough reciprocity between steady-tension economies and the values of V_0, listed earlier.

7.2 Whole-Body Power

Fit young adults, of normal (70 kg) body size, can work at perhaps 4 kW during a single lift (when they are principally utilizing $\sim$P stores). Over 10 s they can maintain $\sim$1 kW (supplementing the stores almost entirely by anaerobic processes). For exercise lasting 10–60 min, the rate might be 0.3 kW. Guideline figures for untrained but healthy middle-aged subjects would be about half the above, in each type of performance. Since, for periods longer than 10 min, the energy supply is almost totally aerobic, the power sustainable over these longer terms by any individual can be calculated fairly satisfactorily from the much more readily measured parameter, oxygen consumption. Each liter extracted from respired air supports a mechanical power output of about 70 W in all persons with normal metabolic processes, irrespective of age, size or sex.

7.3 Negative Work

This term refers to the forced extension of active muscle by a load which exceeds F_0. Consumption of $\sim$P per newton of tension may then be one quarter of the isometric rate, and a lower fraction still of that during shortening. We have already seen (force–velocity plot, Fig. 6) that $F > F_0$. Economy results from the fact that the bent-back XBs never reach the position at which they dissociate phosphate, let alone ADP. This is no academic property; without it there would be no significant respect in which locomotion on legs, as distinct from wheels, would be any easier downhill than up!

8. Types of Striated Muscle Fiber

8.1 Working Fibers

Aerobic metabolism occurs in organelles within the fiber called mitochondria, which show prominently in electron micrographs. Under the light microscope—and to the naked eye, when seen en masse—fibers well equipped for oxidative metabolism show red, because they contain an oxygen-transporting and storing pigment, myoglobin. It used to be thought that all red fibers were slow-contracting. This is incorrect: virtually all slow striated fibers are red (cardiac muscle may be thought of as the extreme form of both slowness and redness), all white (anaerobic) fibers are fast, but a trained performer has a very large number of fast red fibers too. Almost all human skeletal muscles contain a mixture of fiber types. However, since the properties are determined by the pattern of use, fibers within any one motor unit are identical.

There is a regular sequence of recruitment. Slow, red units are used for posture and low-force movements. Fast units are called upon when more power is required, but if used often will still be aerobic. Those motor units recruited only in extremes are, however, not equipped for aerobic metabolism—when they are required, there will never be spare oxygen available. This recruitment-pattern explains why it is that one produces not just *less* lactic acid when standing still rather than sprinting, but in the former case none at all.

8.2 Muscle Spindles

As well as the main, working fibers of skeletal muscles and the large, or "α," motor nerves by which they are innervated, skeletal muscles also contain small clusters of fibers, called muscle spindles (so-named because of their shape). Spindle properties have been extensively characterized by Boyd (1984) and Matthews. Sensory nerves run to the spinal cord from around the middle of each muscle fiber within a spindle; spindles thereby serve as detectors, reporting muscle extension to the central nervous system (CNS). Small, or γ, motor nerves also innervate the spindle muscle fibers; by altering the lengths of these muscle fibers, the γ nerves adjust the sensitivity with which lengths and length changes of the whole muscle are detected. As the sensory nerves in turn are arranged to relay excitatorily in the cord onto the α motor neurons of the same muscle, a feedback loop is established. Merton, in the 1950s, pointed out that, on occasions when γ activation alone is initiated by the CNS, the spindle–loop–muscle system will act as a follow-up length servo, and adjust the working contraction automatically to variations of external resistance. The extent to which this happens physiologically is, however, still a subject of debate.

See also: Biomechanics; Electromyography; Biophysics

Bibliography

Åstrand P-O, Rodahl K 1977 *Textbook of Work Physiology.* McGraw-Hill, New York

Bagshaw C R 1982 *Muscle Contraction.* Chapman and Hall, London

Bohr D F, Somlyo A P, Sparks H V (eds.) 1980 *Handbook of Physiology*, Sect. 2, Vol. 2, *Vascular Smooth Muscle.* American Physiological Society, Bethesda, Maryland

Boyd I A 1984 Muscle spindles and stretch reflexes. In: Swash M, Kennard C (eds.) 1984 *Scientific Basis of Clinical Neurology.* Churchill-Livingstone, Edinburgh

Bulbring E, Bolton T B (eds.) 1979 Smooth muscle. *Br. Med. Bull.* 35 (3): 292

Bulbring E, Brading A F, Jones A W, Tomita T (eds.) 1981 *Smooth Muscle: An Assessment of Current Knowledge.* Arnold, London

Curtin N A, Woledge R C 1978 Energy changes and muscular contraction. *Physiol. Rev.* 58: 690–761

Droogmans G, Raeymaekers L, Casteels R 1977 Electro-mechanical and pharmacomechanical coupling in the smooth muscle cells of the rabbit ear artery. *J. Gen. Physiol.* 70: 129–48

Edwards R H T, Dawson M J, Wilkie D R, Gordon R E, Shaw D 1982 Clinical use of NMR in the investigation of myopathy. *Lancet* i: 725–31

Eisenberg E, Hill T L 1978 A cross-bridge model of muscle contraction. *Prog. Biophys. Mol. Biol.* 33: 55–82

Gibbs C L 1978 Cardiac energetics. *Physiol. Rev.* 58: 174–254

Goldspink G 1977 Mechanics and energetics of muscle in animals of different sizes, with particular reference to the muscle fibre composition of vertebrate muscle. In: Pedley T J (ed.) 1977 *Scale Effects in Animal Locomotion.* Academic Press, London

Gordon A M, Huxley A F, Julian F J 1966 The variation in isometric tension with sarcomere length in vertebrate skeletal muscle fibres. *J. Physiol.* 184: 170–92

Homsher E, Irving M, Wallner A 1981 High-energy phosphate metabolism and energy liberation associated with rapid shortening in frog skeletal muscle. *J. Physiol.* 321: 423–36

Huxley A F 1980 *Reflections on Muscle.* Liverpool University Press, Liverpool

Lymn R W 1979 Kinetic analysis of myosin and actomyosin ATPase. *Annu. Rev. Biophys. Bioeng.* 8: 145–63

McMahon T A 1984 *Muscles, Reflexes and Locomotion.* Princeton University Press, Princeton, New Jersey

Morel J E 1985 Models of muscle contraction and cell motility: A comparative study of the usual concepts and the swelling theories. *Proc. Biophys. Mol. Biol.* 46: 97–126

Noble D 1979 *The Initiation of the Heartbeat*, 2nd edn. Oxford University Press, Oxford

Peachey L E, Adrian R H (eds.) 1983 *Handbook of Physiology*, Sect. 10, *Skeletal Muscle.* American Physiological Society, Bethesda, Maryland

Pollack G H, Suai H 1984 *Contractile Mechanisms in Muscle.* Plenum, New York

Reuter H, Scholz H 1977 The regulation of the slow inward conductance of cardiac muscle by adrenaline. *J. Physiol.* 264: 49–62

Somlyo A V, Gonzales-Serratos H, Shuman H, McClellan G, Somlyo A P 1981 Calcium release and ionic changes in the sarcoplasmic reticulum of tetanized muscle: An electron-probe study. *J. Cell Biol.* 90: 577–94

Squire J M 1981 *The Structural Basis of Muscular Contraction.* Plenum, New York

Wilkie D R 1976 *Muscle*, 2nd edn. Arnold, London

N. C. Spurway